# Cometary Science

Nicolas Thomas • Björn Davidsson • Laurent Jorda •
Ekkehard Kührt • Raphael Marschall •
Colin Snodgrass • Rafael Rodrigo
Editors

# Cometary Science

## Insights from 67P/Churyumov-Gerasimenko

Previously published in *Space Science Reviews* in the Topical Collection "Comets: Post 67P / Churyumov-Gerasimenko Perspectives"

Springer

*Editors*
Nicolas Thomas
Space Research and Planetology Division
University of Bern
Bern, Switzerland

Björn Davidsson
NASA Jet Propulsion Laboratory
Pasadena, CA, USA

Laurent Jorda
Laboratoire d'Astrophysique
Aix-Marseille University
Marseille, France

Ekkehard Kührt
Institute of Optical Sensor Systems
German Aerospace Center (DLR)
Berlin, Germany

Raphael Marschall
Southwest Research Institute
Boulder, CO, USA

Colin Snodgrass
Institute for Astronomy
University of Edinburgh
Edinburgh, UK

Rafael Rodrigo
INTA-CSIC, ESA, ESAC
Villanueva de la Cañada, Spain

ISSN 1385-7525 Space Sciences Series of ISSI
ISBN 978-94-024-2089-0



*Front cover image*: Comet 67P/Churyumov-Gerasimenko, as seen by the OSIRIS narrow angle camera onboard ESA's Rosetta mission on August 20, 2015 at 00:14 UTC. To show both the nucleus surface and the much fainter dust coma simultaneously, the image has been unevenly stretched. On nucleus pixels are scaled to the reflectance range [0.00, 0.10], while off nucleus pixels are scaled to the range [0.00, 0.01]. Image credit: ESA/Rosetta/MPS for OSIRIS Team MPS/UPD/LAM/IAA/SSO/INTA/UPM/DASP/IDA.

Printed on acid-free paper

This Springer imprint is published by the registered company Springer Nature B.V.
The registered company address is: Van Godewijckstraat 30, 3311 GX Dordrecht, The Netherlands

# Contents

# Preface to the Topical Collection: Comets: Post 67P / Churyumov-Gerasimenko Perspectives

**Nicolas Thomas[1] · Björn Davidsson[2] · Laurent Jorda[3] · Ekkehard Kührt[4] · Raphael Marschall[5] · Colin Snodgrass[6] · Rafael Rodrigo[7]**



The Rosetta mission to comet 67P/Churyumov-Gerasimenko was one of the major highlights in space science over the past decade. Orbiting a cometary nucleus for 2 years and monitoring its activity was a remarkable achievement after more than 20 years of work. The experiment suite onboard Rosetta and its lander element, Philae, was designed to make a detailed analysis of the nucleus and the innermost coma. Given that our knowledge of comets at the time of selection and implementation was restricted to data acquired during fast flybys, it was inevitable that the mission would not only produce remarkable new scientific results but also would provide data analysis challenges and reveal inadequacies in our initial concept for the mission.

While many of the early publications from Rosetta focused on the results from individual instruments, it was clear that the Rosetta/Philae suite could provide a great deal of new science by exploiting synergies between the data products of different instruments. With this in mind, a proposal was made to the EU Horizon 2020 programme called MiARD (Multi-instrument Analysis of Rosetta Data), which aimed at pulling several data sets together to constrain the physical processes evident on the nucleus and in the coma. For programmatic reasons, it was necessary to start the MiARD programme earlier than was perhaps ideal with the instrument teams still actively calibrating and publishing. Nonetheless, the MiARD project produced a series of high quality papers that exceeded its target.

[1] Space Research and Planetology Division, Physikalisches Inst., University of Bern, Bern, Switzerland

[2] Jet Propulsion Laboratory, Pasadena, CA, USA

[3] Laboratoire d'Astrophysique de Marseille, 38 rue Frédéric Joliot-Curie, Marseille, France

[4] Deutsches Zentrum für Luft- und Raumfahrt (DLR), Institut für optische Sensorsysteme, Rutherfordstraße 2, Berlin, Germany

[5] Department of Space Studies, Southwest Research Institute, Boulder, CO, USA

[6] Institute for Astronomy, University of Edinburgh, Royal Observatory, Edinburgh, UK

[7] International Space Sciences Institute, Hallerstrasse 5, Bern, Switzerland

**Fig. 1** The ISSI workshop participants

The final element of MiARD was the preparation and execution of a workshop that was held at the International Space Sciences Institute (ISSI) in Bern from 15-19 January, 2018 (see Fig. 1). As is customary for ISSI, it extended participation to include a limited number of scientists (up to the limit that could be supported by ISSI) from outside the MiARD project group. This collection of papers is the result of that workshop. It should be noted that the limitation on the size of the workshop made it difficult to include all aspects of Rosetta science and hence the papers presented here are focussed on elements that were also addressed by the MiARD project.

The aim was to look at several aspects of the nucleus and innermost coma in a multi-disciplinary way but also including links to previous observations of comets and taking a look into the implications for future missions. These were itemized as

1. To review the progress made on multi-instrument data analysis. This included
    a. Current status of the development of an integrated 3D shape model and its use for mapping properties and scientific measurements to its facets
    b. Current status and results from gas dynamics modelling including activity distributions
    c. Current status of dust emission and brightness modelling
    d. Current status of analysis of surface physical and morphological structure and thermal balance
    e. Current understanding of the coma and surface chemical composition
2. To assess and possibly revise current models of nucleus activity (including potential evidence for different mechanisms) and evolution.
3. To assess our current understanding of risk from comets (through perturbation of nuclei by non-gravitational forces and particle impact on interplanetary spacecraft).

4. To draw conclusions about the origin and evolution of the planetary system. To re-assess the (extensive) study work performed on future cometary missions, especially for comet nucleus sample return concepts and provide guidelines as to how this work should be updated in the light of Rosetta's results.

It should be noted that the plasma measurements (which are very interesting as illustrated by many papers in recent literature) and studies of coma chemistry (i.e. reaction chemistry) were not addressed.

Many of the more animated discussions surrounded interpretation of Rosetta data when looking at different instrument results. The papers by El-Maarry et al., Filacchione et al., Vincent et al., and Groussin et al. address the nucleus and its properties in a direct and thorough manner. The relationship of the coma to the activity of the nucleus has been discussed in a paper led by one of several young scientists who attended the workshop (see Marschall et al.) while the work on the refractory to ice mass ratio, which produced the most controversy, is presented in a balanced way by Choukroun et al. Some of the controversies remain but significant steps have been taken to resolve them. The discussion of the gas composition by Rubin et al. complements these two works. The influence of activity on orbit perturbations has been presented by Mottola et al. and we note that this effort clearly revealed that there is work still to do in determining the exact orbit of the nucleus from Rosetta data. Weissman et al. have presented the implications for studies of the origins of cometary nuclei – a field that has been blown open by the Rosetta observations of the two lobes and the results from New Horizons. The two papers by Keller and Kührt and Thomas et al. cover the degree to which Rosetta achieved its goals and what could be achieved by future missions. The former is a critical and possibly provocative assessment while the latter shows how the outstanding questions have arisen and what types of mission could be thought of to address these issues. The work presented, both here and elsewhere, has also shown that we can still extract significant science from the existing data sets and that these results may yet influence our understanding of 67P and comets in general.

Comets remain vitally important objects. It is still probable that they represent the only pristine physico-chemical link to the formation of our Solar System. While the Rosetta mission has provided a huge leap in our knowledge of these remarkable objects, uncertainties remain. It is to be hoped that the next generation of space scientists will provide further insights in the coming decades.

**Acknowledgements** The editors would like to thank the European Union for providing important parts of funding for our MiARD project and the staff at ISSI for their friendly and generous support (in particular Silvia Wenger, Alexandra Lehmann, Andrea Fischer and Saliba F. Saliba). The support of Springer and the referees for the articles are also gratefully acknowledged.

N. Thomas, B. Davidsson, L. Jorda, E. Kührt
R. Marschall, C. Snodgrass, R. Rodrigo

Space Sci Rev (2020) 216:14
https://doi.org/10.1007/s11214-020-0634-6

# Cometary Nuclei—From Giotto to Rosetta

**Horst Uwe Keller[1,2] · Ekkehard Kührt[2]**

Received: 30 May 2019 / Accepted: 6 January 2020 / Published online: 29 January 2020


**Abstract** We will briefly recapitulate the beginning of modern cometary physic. Then we will assess the results of the cometary flyby missions previous to ESA's Rosetta rendezvous with comet 67P/Churyumov–Gerasimenko. Emphasis is given to the physical properties of cometary nuclei. We will relate the results of the Rosetta mission to those of the flybys. A major conclusion is that the visited cometary nuclei seem to be alike but represent different stages of evolution. Coma composition and appearance are not only controlled by the composition of the nucleus but also strongly influenced by the shape and rotation axis orientation of the nucleus and resulting seasons that generate varying surface coverage by back fall material. Rosetta showed that the coma composition is not only varying spatially but also strongly with time during the perihelion passage. Hence past interpretations of cometary coma observations have to be re-considered. Finally, we will try to assess the impact of the cornerstone mission leading to a critical evaluation of the mission results. Lessons learned from Rosetta are discussed; major progress and open points in cometary research are reviewed.

**Keywords** Comets · 67P/Churyumov–Gerasimenko · Cometary nuclei · Rosetta mission

## 1 History of Cometary Physics

Comets display a unique spectacle on the night sky. Usually they appear like a faint star during their discovery and later move fast across the night sky becoming brighter when

Comets: Post 67P / Churyumov-Gerasimenko Perspectives
Edited by Nicolas Thomas, Björn Davidsson, Laurent Jorda, Ekkehard Kührt, Raphael Marschall, Colin Snodgrass and Rafael Rodrigo

✉ H.U. Keller
keller@linmpi.mpg.de

1 Institut für Geophysik und extraterrestrische Physik (IGeP), Technische Universität Braunschweig, Mendelssohnstr. 3, 38106 Braunschweig, Germany

2 Deutsches Zentrum für Luft- und Raumfahrt (DLR), Institut für Planetenforschung, Asteroiden und Kometen, Rutherfordstraße 2, 12489 Berlin, Germany

they are best seen in the early morning or late evening. They can become spectacular as diffuse objects sometimes covering a large part of the visible sky. But most of all their unpredictable appearance, their apparent size, and their similarly fast disappearing make these unique. They are intruders into the eternity and regularity of space. No wonder that they were considered harbingers of imminent disasters over centuries.

What sets comets apart from the fixed stars and the planets, slowly moving on regular orbits, is the diffuse appearance of their coma and frequently their long tails. They are so different that it was only consequent for Aristotle (384–322 BC) in his Cosmology to consider them not being members of the sky but as temporary exhalation of the earth's atmosphere (Heidarzadeh 2008). It took more than 1800 years before the parallax measurements by Tycho Brahe (1546–1601) positioned the Great Comet of 1577 beyond the orbit of the Moon. Comets were understood as members of our planetary system. It took another 100 years when Edmund Halley (1656–1742) predicted the return of a bright comet orbiting the Sun on a very elongated ellipse with a period of 75 years. This validated the law of gravity far beyond the realm of our planets suggesting its universal ruling.

But what is the physical nature of these strange wanderers in our planetary system? There are comprehensive reviews of the history of comets studies (Festou et al. 1993a, 2004) and therefore we mention only some highlights. G. Schiaparelli (1866) connected comets with meteor streams showing that their diffuse appearance is due to small dust particles within the coma. F.A. Bredikhin and F.W. Bessel developed their mechanical model of cometary tails based on a repulsive force from the sun that was later identified by S. Arrhenius (1900) as radiation pressure. In the second half of the 19th century the first spectroscopic observations were performed and revealed that carbon was an important constituent of comets. Schwarzschild and Kron (1911) explained the brightness distribution of the tail of comet 1P/Halley by fluorescence (absorption and reemission) of solar radiation. Only in 1941 P. Swings (Swings 1941) explained why the bright CN emission of the cometary coma differed from laboratory observations. A concentration of solar absorption lines influences the molecular bands according to the Doppler shift of the cometary velocity. Wurm (1943) developed his concept of "parent" molecules because the observed radicals and ions are not stable enough to survive storage in the nucleus.

In the fifties of the last century the foundations of our present day understanding of the nature of comets were created. Whipple (1950) modelled the cometary nucleus as a solid conglomerate of ices and imbedded dust (non-volatile compounds). About 40 years later the images of the nucleus of comet 1P/Halley strongly suggested that the matrix of the nucleus material was dominated by refractories, an icy dirtball rather than Whipple's dirty snowball Keller (1989).

Oort (1950) postulated a cometary reservoir at the fringes of the solar gravitational influence, and Biermann (1951) predicted the solar wind that is accelerating cometary ions into tails of small aberration angles, a ingenious interpretation that was confirmed by measurements of a lunar space probe in 1959.

New technologies in spectroscopy covering the extended wavelengths from the UV to IR, radio dishes and space borne observations provided the means for systematic and quantitative comet observations. A striking observational fact is that, at first sight, all comets look alike (Cochran et al. 1989; Festou et al. 1993b). There are, however, a few noticeable exceptions such as comet 29P/Schwassmann–Wachmann 1 (mainly $CO^+$) that display a completely different composition of the coma. In addition, no systematic changes of comets with evolution, not even a clear difference between new Oort cloud comets and Jupiter family comets could be established; see for instance the more recent discussion of the abundance of the highly volatile CO and $CO_2$ molecules by A'Hearn et al. (2012) and about the

D/H ratio (Altwegg and the ROSINA Team 2017). The situation was already reflected by Delsemme (1982) in the Comets book who concluded: Cometary material originating from different depth of the nuclei shows (a) the same dust to gas pattern, (b) the same spectral composition of the volatile fraction, (c) the same structural strength against fragmentation, and (d) the same vaporization pattern after fragmentation. Consequently cometary nuclei are homogeneous and show little variation—notwithstanding that there could be large surface differentiation effects during their aging. This was proven again in 2006 by observations of the split comet 73P/Schwassmann–Wachmann 3 (Dello Russo et al. 2007).

At that time the first missions to encounter comet 1P/Halley in 1986 were prepared. Of particular interest for the design of the mission instruments were the estimated sizes of cometary nuclei and their albedos. The results of Delsemme and Rud (1973) for the long period comets C/1969 T1 (Tago–Sato–Kosaka) albedo, $A = 0.6$ and radius, $R = 2.2$ km and C/1969 Y1 (Bennett) $A = 0.7$ and $R = 3.8$ km served as guideline for what was to be expected for 1P/Halley.

## 2 The Giotto and VEGA Missions (1P/Halley)

The upcoming reappearance of comet 1P/Halley excited public interest in reminiscence of its last spectacular passage near earth in 1910. This trigger and the generally strong interest in comets as witnesses of the physical and chemical conditions of the early solar system resulted in the proposal of the grandiose and technically advanced Halley Flyby/Tempel 2 Rendezvous mission, a collaboration of NASA and ESA, in 1979. Unfortunately, NASA had to withdraw from the endeavor due to budget cuts, but ESA designed its own dedicated mission, based on a GEOS spacecraft (earth orbiting). Cooperation with the Soviet Union missions VEGA 1 and 2 facilitated the aiming for a close flyby of the nucleus of comet 1P/Halley. The Giotto mission became ESA's first interplanetary mission and by its success opened the door to its further planetary research program. The worldwide enthusiasm helped to overcome the risks inherent in the super high flyby speed of 70 km/s and the very narrow launch window of less than 3 weeks.

Both three axis stabilized VEGA space probes passed the comet nucleus in several thousand kilometers distance after their successful flybys of planet Venus. Their camera systems were troubled by some technical problems. The VEGA 1 narrow angle camera was defocused. The dynamic range of the images of both cameras were severely limited caused by electronic problems (Merényi et al. 1990; Stooke and Abergel 1991). Hence it was up to the Giotto mission to reveal details of the nucleus of comet 1P/Halley providing images with a resolution down to 45 m per pixel. A comprehensive overview of the results of the nucleus of 1P/Halley claimed by the VEGA and Giotto probes is given by Keller (1990). The observational results of these missions confronted the model concepts developed over the previous decades with the reality and set the next step to our understanding of comets (see Newburn et al. 1991).

In particular

- Cometary nuclei are very dark (albedo 0.04) containing refractories, organic material, and ices (Keller et al. 1986; Kissel et al. 1986)
- The rough surface reaches nearly black body temperature ($>$360 K) what rules out large amounts of subliming ice on the surface (Emerich et al. 1987)
- Cometary nuclei are bigger than required to produce the observed water production rates (limited areas of sublimation or quenched activity)

- Cometary nuclei are porous and of low density and tensile strength
- Abundant super volatiles (CO, $CO_2$) require nucleus formation temperatures below 30 K
- Comets provide access to the most primitive (pristine) material out of which our solar system formed

## 3 The NASA Comet Flybys

After the success of the Halley campaign and during the preparation of Rosetta (see next chapter) NASA sent 3 spacecraft to 5 cometary flybys within a decade. DEEP SPACE 1 flew by comet 19P/Borrelly in 2001. It was followed by STARDUST to 81P/Wild 2 in 2004 and DEEP IMPACT to 9P/Tempel 1 in 2005. EPOXI (the renamed DEEP IMPACT) passed comet 103P/Hartley 2 in 2010 and STARDUST NEXT re-visited 9P/Tempel 1 in 2011.

**DEEP SPACE 1 (19P/Borrelly)** The DEEP SPACE 1 images of 19P/Borrelly confirmed the impression that cometary nuclei do not dominantly consist of ice (dirty snowball) but that their topography and morphology are rough down to the scale of the resolution and probably beyond. On the well illuminated surface of the elongated nucleus geologic features such as mesas, pits, mottled terrain, ridges, and fractures could be discerned within varying units. No ice was observed on the surface. So the question how cometary activity works from beneath an inert layer of refractories became more mysterious but also more crucial for the understanding of cometary physics. Britt et al. (2004) suggested that the major but still relatively weak jet-like activity originated from collapsing walls of the prominent mesa-like formation near the rotation pole. The walls should not be covered by residual or back fall refractory material. Overall the nuclei of comets 1P and 19P seemed rather similar (e.g. the elongation and overall roughness). The major difference between both comets is their level of activity.

**STARDUST (81P/Wild 2)** STARDUST was the first mission to bring cometary material back into Earth's laboratories. It flew through the coma of 81P/Wild 2 with a minimum distance of 236 km and collected particles in aerogel at a relative speed of 6 km/s (Brownlee et al. 2004). The minute amounts of cometary refractory material trapped in the aerogel foil triggered a burst of analytic papers that investigated the mineralogy and petrology. Overall the composition of the particles was found to be similar to that of CI chondrites with some excess of more volatile atoms. Highly processed material such as CA inclusions show that the material in the outer protoplanetary regime where comets formed was strongly mixed providing a large mineralogical diversity (Flynn 2008). Furthermore, Elsila et al. (2009) identified extraterrestrial glycine in the sample what was the first detection of a cometary amino acid.

Contrary to both earlier comets the shape of 81P is more spherical, its total volume is comparable to that of 19P/Borrelly. The topographic relief on 81P/Wild 2, however, is substantially rougher (>200 m) than that on Borrelly (about 100 m). Slopes on Wild 2 are steep, even overhangs are visible. The nucleus of 81P is evidently not a rubble pile but rather a solid body. The surface layers seem strong and cohesive relative to the gravitational force. This and the multitude of deep crater-like features suggest that 19P is more evolved than 81P that arrived in the inner solar system only 30 years before. The crater density lies beyond the saturation curve (Basilevsky and Keller 2006). The genesis of the craters is unclear but could possibly be attributed to impacts during the formation and early lifetime of the nucleus.

**DEEP IMPACT (9P/Tempel 1)** Only one year later the DEEP IMPACT experiment and flyby of comet 9P/Tempel 1 took place. The prime goal of the mission was to observe possible compositional variations of cometary volatiles and to determine the structural properties and strength of the surface layers by excavating a crater on the cometary surface. This was done with a 372 kg impactor mainly made of copper that hit the surface with 10.2 km/s. Except for a slight increase in the $CO_2/H_2O$ ratio no deviation from the ambient coma was found. Local gravitational field and average nucleus density ($\sim 600\ \mathrm{kg/m^3}$) were estimated from ejecta fallback of particles in the range of 1 to 100 μm of negligible tensile strength (<65 Pa) (A'Hearn et al. 2005). This was later questioned by Holsapple and Housen (2006) who found that any strength from 0 to 12 kPa could furnish the ejecta data. The size of the crater, a measure of the physical properties of the ground, could not be derived after the impact since it was hidden by the ejecta.

The analysis of the ejecta spectra showed that water ice (and other species) were uniformly distributed below a transition layer of <1 m down to 10 to 20 m (Sunshine et al. 2007).

Only very thin patches of water frost could be detected on the surface (Sunshine et al. 2006). Relatively detailed temperature maps of the illuminated surface were achieved by scans with the IR spectrometer. Groussin et al. (2007) modelled the temperature maps and derived a very low value for the thermal inertia ($< 50\ \mathrm{W\,K^{-1}\,m^{-2}\,s^{1/2}}$). By using re-calibrated data (Davidsson et al. 2013) found terrains with thermal inertias up to $200\ \mathrm{W\,K^{-1}\,m^{-2}\,s^{1/2}}$. Asymmetries of the coma were observed in the spatial distribution of $CO_2$ and $H_2O$ indicating chemical heterogeneities within the nucleus (Feaga et al. 2007).

**STARDUST NEXT (9P/Tempel 1)** After delivering its sample package the Stardust spacecraft was re-directed to perform a second flyby of 9P/Tempel 1 as STARDUST NEXT in 2011. This improved the determination of its nucleus shape, density, and geology in more detail (Thomas et al. 2013a). For the first time indications of interplay of erosion and depositions were suggested. No explanation for the strong erosional features could be given. Furthermore, the DEEP IMPACT crater, produced a cometary orbit ago, could now be estimated to be $49 \pm 12$ m in diameter (Richardson and Melosh 2013). It was surrounded by an area of slightly brightened material of 85–120 m in diameter yielding an effective target strength of 1–10 kPa. The Tempel 1 observational results, most model dependent, confirmed that cometary nuclei are of low density (see also Rickman et al. 1987), low tensile strength, low thermal inertia and are dominated by refractory material.

**EPOXI (103P/Hartley 2)** The Deep Impact spacecraft was re-directed after its visit of Tempel 1 and the mission was re-named EPOXI. The images of comet 103P/Hartley 2 taken during the flyby of EPOXI in 2010 (A'Hearn et al. 2011) revealed its bilobate, elongated, nearly axially symmetric shape, 2.33 km in length. A smooth but variegated region forms a "waist" between the two lobes (Thomas et al. 2013b). It is covered by back fall of icy grains that do not contain any $CO_2$ anymore because this volatile ice can sublime even from the center of the grains. Bigger centimeter and decimeter size chunks, some with speeds above escape could be followed optically on their trajectories (Kelley et al. 2013). Most of the activity of the comet is driven by $CO_2$ sublimation which explains the hyperactivity by extended sources. More water is found in the coma than can be produced by water ice sublimation on the surface. Above the illuminated lobe icy grains driven by $CO_2$ sublimation strongly contribute to the coma signal. Above the waist, however, mainly water vapor and no $CO_2$ was observed.

In summary, the nuclei of the 5 comets (4 JFC comets) look quite diverse. Most of them suggest a bilobate origin. The morphology of the surface (where resolved) is quite different.

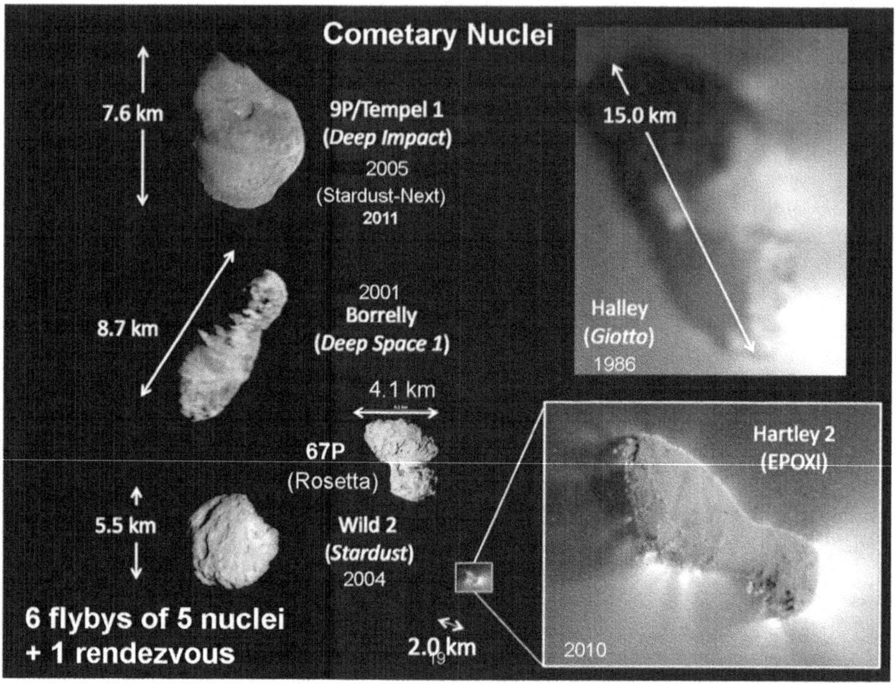


**Fig. 1** Cometary nuclei imaged during flybys and by Rosetta

The strongly pitted (cratered) surface of 81P looks much different to the smooth terrains observed on 9P (Fig. 1). Their activity levels are very diverse, encompassing the absolutely high activity of 1P, the relatively high activity of the 103P to typically (for JFC comets) low activity of 9P. Are these nuclei intrinsically diverse or are their appearances an expression of their evolution? The Rosetta mission will provide the answer.

A summary of the missions, listing target, date of encounter, and distance of closest approach is given by Vincent et al. (2019) (Table 2) in this issue.

## 4 The Rosetta Mission

### 4.1 Background of the ESA Cornerstone Mission

The flybys of comet 1P/Halley had not even taken place when NASA and ESA appointed a science definition team to investigate a comet nucleus sample return mission (CNSR). Originally NASA did not consider comet flybys as scientifically rewarding. This was the major reason why NASA did not participate in the Giotto mission but was interested in a very advanced and complex sample return mission. In Europe the Solar System Working Group of ESA recommended a comet nucleus sample return (CNSR) mission in 1985, the year of the Giotto launch, as a follow up to the NASA CRAF (Comet Rendezvous Asteroid Flyby) mission based on the American Mariner Mark II spacecraft. When NASA cancelled their CRAF mission to comet 22P/Kopff because of budgetary issues, it became evident that the ambitious sample return mission was too expensive for ESA alone. Instead, ESA followed a similar concept and redesigned its Cornerstone mission as an asteroid flyby followed by a comet rendezvous with in-situ examination, including two landing modules. But again NASA dropped out from the US/French lander project Champollion (Kerridge et al. 1997; Neugebauer and Bibring 1998) and finally one European lander (Philae) remained (Ulamec et al. 1997).

The new Rosetta mission was to replace the laboratory investigations of the returned sample by in situ analyses near the nucleus surface for an extended time (rendezvous) and direct contact using surface science packages (probes). The main science goals remained unchanged:

1. Determination of the global (mass, shape, volume) and dynamic properties (rotation state) of the nucleus
2. Determination of the nucleus physical properties and mineralogy
3. Investigation of the composition and origin of comets
4. Understanding and characterization of cometary activity and evolution

ESA's cornerstone mission was finally approved in November 1993, originally targeting comet 46P/Wirtanen, based on the Red Report SCI(93)7. The launch was scheduled for 2003 but could not be met because of technical problems with the launcher. Comet 67P/Churyumov–Gerasimenko, slightly larger than the original target, was chosen for a launch in 2004. The resources for the remaining surface science package (Philae) were increased by a factor two following the strong recommendation of the Rosetta Science Team. The finally proposed complex lander (with 10 instruments) with a mass of 100 kg comprised 40% of the science payload. For details of the mission and payload see Glassmeier et al. (2007).

## 4.2 The Journey

After the launch in 2004, two flybys at asteroids Šteins and Lutetia and in total four gravity assist maneuvers at Earth and Mars were conducted. The target comet 67P/Churyumov-Gerasimenko was reached in August 2014 (Accomazzo et al. 2016). The rendezvous lasted for more than 2 years and about 300 GB of data were collected by the eleven instruments and the navigation system of the spacecraft. The landing of Philae was scheduled for November 12, 2014 at a heliocentric distance 3 au. Due to a failure of the anchoring systems Philae bounced over the surface and came to rest in an initially unknown and unfavorable position (Biele et al. 2015). It was detected by the OSIRIS science camera only shortly before the end of the mission.

The observations of the nucleus of 67P by Rosetta started at a heliocentric distance beyond 3.5 au when activity was still mostly dormant. Rosetta followed the development of the activity through perihelion and terminated when the cometary activity had almost died again. The concept of the Rosetta rendezvous mission was best and uniquely suited to investigate the physical processes of activity.

## 4.3 What Have We Learned from Rosetta?

The objectives of points 2 and 3 (Sect. 4.1) would have been achieved by the original Rosetta CNSR mission by returning nucleus material into the laboratory. Now the composition of the nucleus had to be inferred from the analysis of the compounds found in the coma during the rendezvous. The success of this step strongly depends on the knowledge of the physical conditions and properties of the nucleus matrix material. The only way to derive these properties from the outside is the analysis of the physical processes on and near the surface that

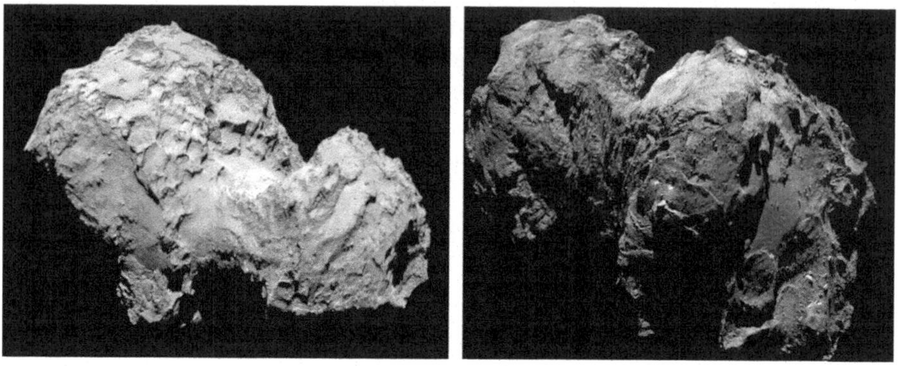

**Fig. 2** The nucleus of comet 67P/Churyumov–Gerasimenko. Left: the north hemisphere is dust covered. Image taken pre-perihelion early in the mission. Right: the south hemisphere is very rough. Image taken after perihelion (images ESA/Rosetta/MPS for OSIRIS Team MPS/UPD/LAM/IAA/SSO/INTA/UPM/DASP/IDA)

**Table 1** Global and dynamic properties of 67P

| Property | Value | | | Remarks | Reference |
|---|---|---|---|---|---|
| Mass [kg] | $(9.982 \pm 3) \times 10^{12}$ | | | before perihelion | Pätzold et al. (2019) |
| Mass loss [kg] | $(10.5 \pm 3.4) \times 10^{9}$ | | | during orbit | Pätzold et al. (2019) |
| Volume [$km^3$] | $18.56\ km^3 \pm 0.02$ | | | before perihelion | Preusker et al. (2017) |
| Density [$kg\,m^{-3}$] | $537.8 \pm 0.6$ | | | Ratio mass/vol. | see above values |
| Rotation period [h] | $12.4043 \pm 0.0007$ | | | before perihelion | Mottola et al. (2014) |
| | $12.05506 \pm 0.00002$ | | | after perihelion | Godard et al. (2017) |
| Tensor of inertia [$kg\,m^2$] | $9.3408457 \times 10^{18}$ | $5.6695663 \times 10^{16}$ | 0 | see discussion in the reference | Kramer et al. (2019) |
| | $5.6695663 \times 10^{16}$ | $1.6562414 \times 10^{19}$ | 0 | | |
| | 0 | 0 | $1.8192083 \times 10^{19}$ | | |

lead to activity and erosion. Here the surface science probe was most important (Goesmann et al. 2015; Wright et al. 2015). It is time to ask what have we learned from Rosetta with respect to the original goals of the mission.

### *4.3.1 Determination of the Global and Dynamic Properties of the Nucleus*

The shape of 67P is unexpectedly complex. The nucleus seems to be composed from 2 rather evolved sub nuclei covered with pits, faults, cliffs, and mountains (see Fig. 2). The bigger of which looks like a strongly oblate sphere. The mass, volume, density, rotation axis orientation, rotation period, and moments of inertia of 67P could be determined to high accuracy by the extended rendezvous mission (Table 1).

The limited radar observations with a resolution in the m-range by CONSERT (Kofman et al. 2015) did not find any substantial voids, neither did the radio science instrument RSI (Pätzold et al. 2016). Considering the extremely low gravity on the nucleus surface a large scale tensile strength of the material of much less than 100 Pa (Groussin et al. 2019) is sufficient to support high cliffs (Hathor) and overhangs (Groussin et al. 2015; Attree et al. 2018). It turns out that the material strength to gravity ratio on the comet is quite similar to the equivalent on earth. It is therefore not surprising that the eroded nucleus surface on many places looks earth-like.

The geology of the surface is quite diverse, about 60 different regions could be discerned (Thomas et al. 2018) with rocky consolidated nucleus matrix material, very rough, boulder covered, with steep cliffs and huge cliff walls on one hemisphere and dust covered with pits, terraces, spires, and flats that seemed to be covered by thick layers of back fall material on the other hemisphere. All types of surface morphologies observed on the nuclei previously passed by flyby missions can be found on comet 67P (El-Maarry et al. 2019). This strongly suggests that at least the normal JFC are alike and their surfaces reflect different states of evolution with morphologies that depend on orbit, shape, and rotation of the individual comet nucleus.

The strong dichotomy between the northern regolith covered surface and the consolidated "rocky" southern hemisphere (Fig. 2) demonstrates that it is almost impossible to draw conclusions about the composition and physical processes of individual nuclei from remote sensing observations, from Earth orbit, or even from flybys. Many observations and conclusions of the past have to be reassessed.

Less suggestive is the indication of onion shell-like structures of both lobes separately (Massironi et al. 2015). If true, it would confirm that the bilobate shape was formed by the merger of two bodies rather than by erosion (Jutzi and Asphaug 2015; Vavilov et al. 2019). No differences of the physical and chemical properties of the bodies were found.

The surface color of 67P is rather uniformly red. VIRTIS detected water ice mainly as frost right after local sunrise (De Sanctis et al. 2015). While this frost is only short lived more and more icy patches were found when the comet approached perihelion. Based on mutual observations less red ("blue") bright patches were observed by OSIRIS and interpreted as water ice. The lifetime of some lasted weeks and months. Overall the surface color changed from red towards blue while the comet became more and more active (Fornasier et al. 2016). The trend reversed after perihelion; a strong indication that the surface properties systematically changed. How this influences the activity remains an open question.

Changes of the spin state were measured with high accuracy and interpreted in terms of non-gravitational forces caused by cometary activity (Keller et al. 2015b; Kramer et al. 2019).

### *4.3.2 Determination of the Nucleus Physical Properties and Mineralogy*

VIRTIS measurements revealed that the surface of the comet is characterized by a dark refractory polyaromatic carbonaceous component mixed with opaque minerals (Quirico et al. 2016). VIRTIS data do not provide direct insights into the nature of these opaque minerals but they may consist of Fe-Ni alloys and FeS sulfides. A semi-volatile component, consisting of a complex mix of low weight molecular species is likely a major carrier of the 3.2 μm band (Quirico et al. 2016). Water ice significantly contributes to this feature in the neck region but not in other regions of the comet. COOH in carboxylic acids is the only chemical group that encompasses the broad width of this feature.

Both lobes are similar and in fact the whole nucleus is homogeneous at a meter-scale (Herique et al. 2016). The density of the nucleus of 538 $kg/m^3$ confirmed previous strongly model dependent estimates based on the effect of non-gravitational forces (NGF) on cometary orbits (e.g. Rickman et al. 1987; Davidsson and Gutiérrez 2005). This leads to an overall porosity between 50 and 80%.

VIRTIS found the thermal inertia to be $< 170\ \mathrm{J\,K^{-1}\,m^{-2}\,s^{-0.5}}$ (Marshall et al. 2018), or even $< 50\ \mathrm{J\,K^{-1}\,m^{-2}\,s^{-0.5}}$ in most regions (Tosi et al. 2019). This is consistent with a highly porous structure.

The unclear environmental circumstances of Philae during its short observation campaign question the reliability of the measurements (Böhnhardt et al. 2017). The few high resolution images (Bibring et al. 2015; Mottola et al. 2015; Schröder et al. 2017), the SESAME experiment (Möhlmann et al. 2018; Knapmeyer et al. 2018) and the MUPUS penetrator (Spohn et al. 2015) provided limited information on the mechanical properties of the surface. Material parameters (tensile and compressive strength) were as well derived from modelling the impacts during the touch downs (Biele et al. 2015; Heinisch et al. 2019). The resulting material properties are quite controversial and difficult to reconcile with the very low values derived from the observations of cliff collapses and morphological structures such as overhangs.

The dust to ice ratio in cometary nuclei has been of particular interest since it turned out that the nucleus matrix is dominated by refractories and not by ice (Keller 1989). Ground based observations of comets reveal dusty and gaseous comets indicating a wide range of this ratio. However, the determination of the dust amounts is rather uncertain because large particles (carrying most of the mass) cannot be observed in the visible wavelength range. Even after the rendezvous the intrinsic refractory to ice ratio of 67P remains puzzling since a direct measurement is missing and derived values are strongly model dependent (see Choukroun et al. 2020, this issue).

The potential for erosion by water ice sublimation is 4 times higher for the southern than for the northern hemisphere (Keller et al. 2017). Therefore, if more than 25% of the dust liberated on the south falls back on the north it will accumulate there from orbit to orbit. Seasonal variations of the cover by back fall were observed at several places particularly at northern mid latitudes. The thickness of the cover was estimated to 0.5 to 1 m (Hu et al. 2017). A more precise measure of 1.4 m could be derived for the variation of the ground level at Hapi using the height of outcrops (Cambianica et al. 2020). How these relatively large erosion values from the area with the least overall solar insolation can be conciliated with a total loss of less than 0.5 m averaged over the whole surface (Keller et al. 2017; Pätzold et al. 2019) remains a puzzle.

About half of the total mass loss of the nucleus of 67P determined by RSI (Pätzold et al. 2019) is attributed to the integrated gas production (Biver et al. 2019; Läuter et al. 2019). This results in a dust to gas ratio in the coma of about 0.6 to 1.5 (Choukroun et al. 2020, this issue). How this does correlate with the dust to ice ration in the nucleus is under debate. A large amount of refractories were ejected in the centimeter and decimeter size range (Fulle et al. 2016; Ott et al. 2017). A sizable percentage fell back on the then inactive northern hemisphere and possibly accumulated there from orbit to orbit. The dust chunks are "wet", contain water ice (but no super volatiles) that then sublimes from the northern hemisphere. The water driven activity may not be strong enough to remove all the dust. Activity on the northern hemisphere could still be maintained from orbit to orbit by a cover of fresh material transferred during perihelion from the south. In this scenario the dust to gas mass ratio derived from the loss of material does not reflect the dust to ice ratio of the original cometary material. The erosion is controlled by super volatiles on the southern hemisphere and not by water ice sublimation. The overall erosion of the cometary surface (mainly on the south) would be much stronger than estimated from the total mass loss. A varying amount

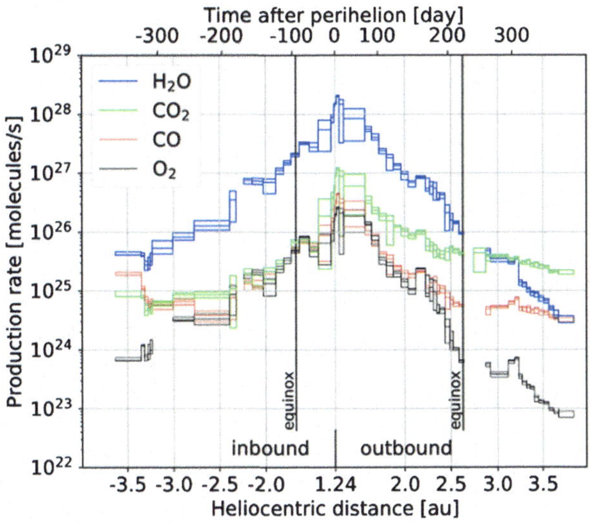


**Fig. 3** The production rates of $H_2O$, $CO_2$, CO, and $O_2$ of 67P as function of heliocentric distance (Läuter et al. 2019)

of back fall material controls the varying observed dust to gas ratios in the coma and tail of comets that could still have similar intrinsic dust/ice ratios.

### *4.3.3 Investigation of Composition and Origin of Comets*

The long term observations during the 2 years of rendezvous produced an abundance of compositional data. However, the seasonal mass transfer not only influences the observed dust to ice ratio and morphology of the surface but it also causes a seasonal and spatial variation of gas compounds in the near nucleus coma what makes an interpretation difficult. This is particularly obvious for the $CO_2$ to $H_2O$ ratio (Läuter et al. 2019) if one compares the pre-perihelion to the post-perihelion ratios. The ratio changes from about 5% before perihelion to more than 100% later during the rendezvous (Fig. 3). CO behaves similarly. Stochastic, time limited, observations of the ratios of super volatiles to water are not conclusive for the intrinsic ratios. This also explains why no correlation of these ratios with cometary type (JFC, Oort cloud) and possibly with evolutionary age or place of origin was found (A'Hearn et al. 2012). One of the key results of the rendezvous mission Rosetta is that time limited incidental observations of the coma composition do not necessarily reflect the intrinsic composition of the nucleus, neither in the content of volatiles nor in the refractory to volatile ratio. The summary given by Delsemme (1982) in the Comets book that cometary nuclei are alike and homogeneous still holds.

ROSINA took advantage of the long measuring cycles and approximately doubled the number of detected compounds found in the coma. These include aromatic ring compounds as well as long chain hydrocarbons which can be linked to gas phase and grain surface chemistry in dark molecular clouds (Hasegawa et al. 1992). Larger molecules beyond mass 60 are inferred by the IR VIRTIS observations (Capaccioni et al. 2015) but do not sublime easily. Correlating the cometary abundance ratios with those observed in the ISM suggest formation under similar circumstances. Comparable results were first derived for comet C/1995 O1 (Hale–Bopp) (Bockelée-Morvan et al. 2000). The large number of sulfur species ($H_2S$, $S_2$ etc.) provide a further hint of molecular cloud chemistry on dust grains (Calmonte et al. 2016).

ROSINA confirmed that the various highly volatile compounds ($N_2$, CO, Ar, Xe, etc.) require the formation of nuclei at a temperature below 25 K and that they were never warmed up, i.e. as part of a large body by radiogenic heating (Rubin et al. 2015). This was already concluded after the comet Halley flybys (Altwegg et al. 1999). Surprising was the finding of the highly volatile and reactive molecule $O_2$ in the percentage range well correlated with $H_2O$, not with $CO_2$ like most other highly volatile species. Its formation process is unclear but points to an interstellar origin.

The extended observations of ROSINA yielded a large number of isotope ratios. The observed D/H ratio of $1.5 \times 10^{-5}$ is much higher than found in any other comet. The variability of this ratio from comet to comet indicates that the various comet classes (JFC, Oort cloud) originate from the same extended reservoir but undergo different dynamical evolutions (Altwegg et al. 2015). This confirms the earlier conclusions derived from varying $CO_2$ and CO ratios relative to water for a variety of comets (A'Hearn et al. 2012).

Their homogeneity, their low density and tensile strength may suggest that comets are formed by accretion of a collapsing cloud (streaming instability) of pebbles (Blum et al. 2014; Davidsson et al. 2016). However, using the same Rosetta data basis different authors come to different conclusions about the importance and effects of impacts in the early history of comets (comp. Rickman et al. 2015; Davidsson et al. 2016).

The confirmation of the amino acid glycine (Altwegg et al. 2016) could be a hint that comets may have brought ingredients of life to Earth in an early phase of the Late Heavy Bombardment (LHB) but this remains speculative and is not at all proved.

Results from ROSINA considerably strengthen the non (pre) solar origin of some cometary compounds. Isotope ratios of Xe, Si, S, and H as well as the presence of highly volatile compounds such as $S_2$ and the correlation of $O_2$ with $H_2O$ require pre solar ice. Comets are formed at low temperature and contain original interstellar material and compounds processed in the inner solar system (see also the Deep Space results A'Hearn 2006) but mixing of proto solar/proto planetary nebula material was inefficient.

The very high D/H ratio compared to that of Earth rules out that comets of 67P type played an important role in delivering water to Earth. The abundance ratios of various deuterated water molecule suggested that water was inherited from the pre solar cloud (Furuya et al. 2016). Overall the observed isotope ratios are not consistent, some ratios are enhanced relative to the solar value, some are depleted (Hässig et al. 2015; Calmonte et al. 2016; Rubin et al. 2017). Of particular interest are the Xe isotopes that seemed to have been trapped in the cometary ice before the solar system formed. Their ratios match that of the unknown source of U-Xenon in the Earth's atmosphere (Marty et al. 2017).

### *4.3.4 Understanding and Characterizing Cometary Activity and Evolution*

The determination of the dust to ice ratio from coma and tail observations is difficult (see Sect. 4.3.2) and the dust to ice mass ratio was long underestimated and considered to be less than one. The predominance of the refractory material was inferred from results of the Giotto mission and confirmed by observations of cometary trails that consist mainly out of centimeter to decimeter particles (Sykes and Walker 1992). Whipple (1950) assumed that large dust particles (meteoritic material) form a mantle on the surface if they cannot be removed by the subliming ice. The particles do not interact and are retained by the gravity of the nucleus. Kührt and Keller (1994) showed that mantles of loose grains are stabilized by van der Waals forces that are high enough to withstand the gas pressure. This would lead to a growing permanent dust mantle and was the starting point of the "activity paradox" in the literature (Blum et al. 2014; Vincent et al. 2019): how can a comet stay active when such an insulating mantle is formed on its surface? Rosetta was supposed to disentangle the physical processes of activity.

The rendezvous mission did bring the complexity of activity and erosion processes of cometary nuclei to our attention. It revealed that mass transfer by back fall material is important for our understanding and interpretation of cometary appearances and as a consequence will lead to re-assessment of many earlier papers and conclusions. This could well be one of the most significant results of the mission. This complexity was mainly triggered by the strong obliquity of the rotation axis of 52° that causes pronounced seasons on the nucleus. Their impact was enhanced by the near coincidence of the southern solstice with the perihelion passage resulting in a short ($< 1$ y) but intense southern summer while the northern summer lasts for the rest of the orbit. Rosetta approached the comet during northern summer when the southern hemisphere was not illuminated. This imbalance of the seasons produces a remarkable dichotomy in the morphologies of the hemispheres. The northern hemisphere is in large parts covered by grained material that fell back when the southern hemisphere was strongly active. The south shows mostly the consolidated nucleus matrix and is very rough.

The dark surface produces dust activity whenever and wherever the sun shines. The strength of the gas activity is variable within an order of magnitude (Kramer et al. 2019). A correlation with surface morphology could not be established. Some dust features can be traced to specific areas on the nucleus, and result conceivably from locally enhanced outgassing and/or dust emission (Shi et al. 2018). Linking observed coma morphology to the distribution of activity on the nucleus remains difficult.

It has not been possible to discern active from inactive surfaces, neither by remote sensing nor by the in situ coma measurements. This is not astonishing because from a distance of 10 nucleus radii ($\sim$20 km) the whole surface contributes to the signals observed by instruments measuring coma compounds, the nucleus behaves like a point source (Rezac et al. 2019). Unfortunately the dust near the nucleus could only be observed on the limb or against shadowed areas but not directly at the source. Only little amounts of $CO_2$ originate from the northern dust covered surface. The regolith is depleted in $CO_2$ relative to water.

## 4.4 What Do We Miss?

Rosetta provided an unprecedented wealth and detail of data (about 300 GB) of the cometary nucleus of 67P. We followed the comet evolution along its orbit through perihelion. We determined the global properties of the body, observed activity and erosion, detected many more compounds in the coma. But some basic questions remain unanswered.

We did not properly characterize the physical properties of the matrix material, mainly because the lander Philae failed to provide reliable data. From ROLIS image we can conclude that the pores at the surface are smaller than the resolution limit of 0.8 mm (Schröder et al. 2017). Recent models cover the range from submicron, e.g. (Keller et al. 2015a; Skorov et al. 2017) to cm (Gundlach et al. 2015) pores. The strength at small scales and the distribution of ice and dust below the surface were not measured. The derived thermal inertia of the surface from VIRTIS and MIRO data is not detailed enough to calculate the heat transport through the upper layers. For this, one needs the depth- and temperature-dependent heat conduction coefficient that could not be measured by MUPUS since the thermal sensors were not hammered into the ground as planned (Spohn et al. 2015). Consequently, modeling work of the temperature profile and the activity behavior is limited since its outcome strongly depends on these unconstrained parameters (Prialnik et al. 2008).

Little was found beyond the results from the 1P/Halley flybys about the semi- or nonvolatile hydrocarbon components that constitute probably the majority of the refractories

(Wright et al. 2015). The mineral components are compatible with a mixture of silicates and fine-grained opaque compounds, including Fe-sulfides like troilite and pyrrhotite. The organics contain COOH and OH-groups and a refractory macromolecular material bearing aliphatic and polycyclic aromatic hydrocarbons (Filacchione et al. 2019). A more detailed composition of the refractories and organics remains puzzling (Quirico et al. 2016).

Icy patches and frost were observed but no criteria could be established that would allow to characterize the strength of activity, or even any inactivity of surface regions. Systematic measurement of local surface temperature variations as function of insolation and illumination geometry in good resolution as an indicator of sublimation processes are missing.

Unfortunately, the orbit parameters of 67P were not determined by ESA. Just the trajectory of the spacecraft was evaluated based on piece wise approximations. As a consequence the acceleration of the nucleus caused by its activity cannot be derived. This forfeits the unique chance to model the non-gravitational forces (NGF) for a nucleus with well-known mass, shape, and rotation parameters. The attempt by Kramer and Läuter (2019) to overcome this shortcoming by approximation of the perturbed cometary orbit does not constrain parameters well enough.

Although new details of cometary activity are identified such as active pits with lateral sublimation from their walls, cliff collapses, and night/day terminator region activity (see Vincent et al. 2015; Pajola et al. 2017), the more descriptive than analytic observations and measurements of Rosetta do not provide the information required to understand cometary activity in more detail.

## 4.5 Why Did We Not Make a Real Breakthrough in Understanding Activity?

A breakthrough could not be achieved in respect to the important goal of understanding cometary inactivity in detail. We still do not know how the interplay of dust and gas near the surface takes place, which role water and super volatiles play for the activity, or how outbursts are triggered. The lack of knowledge of the microphysical parameters and processes prevent realistic modelling of the dust ejection.

Why did that happen?

The spacecraft was not near the nucleus when activity dominated. We did not see where and how frequently large particles in the centimeter and decimeter or even meter size range were lifted although the limiting resolution of the camera was easily sufficient. In the active phase of the comet the spacecraft was hundreds of kilometers away. The relatively long legs of the hyperbolic spacecraft trajectories did not provide enough control for systematic local observations. As a consequence the lift-off mechanisms of the dust of different sizes remain unresolved. Furthermore, this is one of the reasons that the essential refractory to volatile (ice) ratio of the nucleus matrix is not well constrained.

Rosetta's many close flybys were still only flybys and made it difficult to impossible to achieve systematic observations of specific points of interest and of particular activities. Just to give an example. Very early in the mission a long and wide fracture along the Aswan cliff was observed. Talus on its bottom showed that break offs or collapses had taken place in the past. So it was anticipated that here a collapse was imminent and could take place during this perihelion passage. The collapse took place resulting in a dust cloud activity that was serendipitously observed by the navigation camera (Pajola et al. 2017). It took 5 days to confirm the collapse but more than three months to look again at the cliff with somewhat better resolution and improved illumination conditions. It took more than a year before high resolution images of the cliff wall were taken. Of course we missed all the information that could have been derived from the fresh surface revealing the nucleus interior immediately after the collapse.

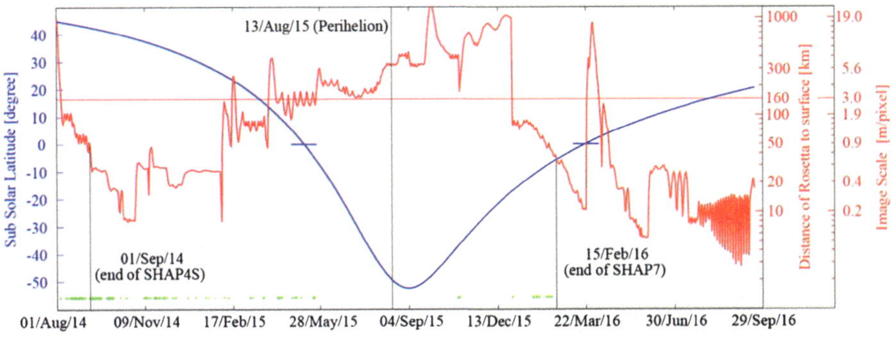


**Fig. 4** Distance of the Rosetta spacecraft relative to the nucleus of 67P and image scale of the NAC OSIRIS camera on the right ordinates (red curve) and the sub solar latitude on the left ordinate (blue curve), green dots indicate images used for shape modelling (Preusker et al. 2017). Abscissa shows the date. Time of perihelion is indicated

The limitations of the spacecraft star trackers to operate in the dust environment near the nucleus strongly hampered the analysis of the activity processes. The design concept of the spacecraft safety rules (adapted from missions around planets) turned out at least as disadvantageous. They required Rosetta to stay on hyperbolic orbits or on stable closed orbits. This made systematic and repeated observations of points of interest difficult or impossible. In addition, much time was wasted (for most instruments) far away from the comet nucleus at the return points of the hyperbolic arcs. The very low escape velocity (1 m/s) from the nucleus of 67P would allow the spacecraft to hover above it and to move towards the Sun in case of emergency (safe mode). This concept was unfortunately rejected during the design phase of the Rosetta spacecraft. Operational turnaround times at the beginning of the mission in the order of months, later still one or two weeks prevented any fast reaction to activity events.

A further big shortcoming was that the beta angle (angle between the direction towards the sun and the orbital plane of the spacecraft) had to be > 60°. This induced navigation requirement led to solar incident angles of more than 60°, not at all optimal for mapping and surface observations such as temperature determinations. Even much less favorable were the terminator orbits of the spacecraft. Interpretation of images for phase angles close to or beyond 90° is difficult and quantitatively ambiguous, never mind that a major part of the images is taken up by black shadows. Of course here systematic repetitive observations are possible but they take place above the least illuminated and hence least active surface. In addition, the active hemisphere on one side and the dark inactive hemisphere on the other side make modeling the local coma difficult and uncertain.

Figure 4 shows that the spacecraft spent less than 20% of the rendezvous interval at a distance below 10 nucleus radii (< 20 km) and this only during times of low activity levels of the comet at large heliocentric distances. This is too far off the surface for the instruments to map the chemical homogeneity of the morphological regions by in situ measurements of compounds found in the coma. Beyond 20 km the nucleus behaves essentially like a point source. This holds for the in situ density measurements of ROSINA as well as for the remote sensing instruments such as VIRTIS and MIRO (Rezac et al. 2019). Originally it was envisaged to reach distances as close as one nucleus radius above the surface (< 2 km) and map most of the surface with a resolution < 10 cm/px (Schwehm and Schulz 1999).

The high phase angles of the surface observations complicated measurements of the temperature but also of other radiative properties of the surface. As discussed earlier the derived (not directly measured by VIRTIS) surface temperatures were not accurate enough to constrain the thermal properties of the surface that would be important for activity. The interpretation of the MIRO temperature data depends on the little constrained material properties and their homogeneity.

The top level answer to why we missed a breakthrough is that the spacecraft performance and its operation did not allow the instruments to be used to their capabilities and in a systematic way. One could characterize the Rosetta mission as a series of flybys interrupted by bound orbits in the least optimal configuration along the terminator.

In summary, a breakthrough in our understanding of comets requires the understanding of cometary activity processes. Only then meaningful models can convert the observations of the coma composition and phenomena into knowledge about the nucleus composition and its structure.

## 5 Lessons to Be Learned for Future Cometary Missions

ESA did not discuss the important lessons learned report within the science community after the end of the Rosetta mission. We therefore summarize our points of view.

As discussed before, major progress could be achieved in the detection and measurement of chemical and isotopic compounds of the comet, in plasma physics (Goetz et al. 2016; Mandt et al. 2016) but not in understanding how activity works. Basic deficiencies of the mission stem from its early stages when the science objectives were discussed and the spacecraft and mission were designed. What are the shortcomings and what can be done better in the future?

1. The science goals were not precisely enough defined that they could be used as compelling design and performance criteria and finally as success criteria. In fact success criteria were not defined. Debilitating was also the heritage from the originally conceived CNSR. The Red Report (SCI(93)7) intended to demonstrate that most of the original science goals of the CNSR mission could still be achieved by the rendezvous concept, rather than concentrating on the physical processes of activity that could uniquely be investigated by the rendezvous.
2. The in situ chemical analysis of coma compounds was presented as a surrogate for the laboratory investigations of a nucleus sample. It was not realized that a comprehensive coma analysis still requires the knowledge of how to draw conclusions on the composition and physics of the nucleus itself. This also holds true for deriving the refractory/ice ratio in the nucleus and of outgassing rates from coma measurements.
3. A major deficiency of the science payload on Rosetta was the absence of a thermal IR instrument that could provide the surface temperatures with adequate spatial resolution and accuracy over a temperature range from $< 100$ to $> 350$ K. Temperature variations are most analytic for subliming ice, even under the surface and even when intimate dust/ice mixtures mask the ice bands in spectra (Yoldi et al. 2017). Omitting the proposed thermal mapper ("Thema") was probably also a consequence of the imbalance of the science goals.
4. The over-complexity of the Philae lander loaded with sophisticated in situ analysis instruments investigating chemical aspects is also a consequence of the desire to achieve laboratory like investigations at the comet. Early in the mission design two surface packages were discussed, adequate for a completely unknown environment with all its risks (Kührt et al. 1997). Simple clam shell like probes were proposed that operated on

batteries and concentrated on the physical properties like tensile strength and porosity applying high resolution imaging. Two packages were suggested to mitigate the risk and to probe the diversity of the surface (Kerridge et al. 1997; Ulamec et al. 1997). Leaving the design concept and science criteria in the hands of a PI and his coinvestigators (Wittmann et al. 1999) resulted in a too ambitious concept of the lander without adequate funding and spacecraft resources. A major part of the science payload (40% by mass) was not under the control of ESA. The advertised capabilities made this instrument so important to the mission that the PI lander became mission critical and significantly influenced the operation of Rosetta over many months, not only before the release of Philae but also afterwards during the desperate attempts to re-establish contact and even by the search for the dead lander (Ashman et al. 2016).

5. An important design limitation for instruments is the maximum data rate and total data budget. It is not helpful to created data that cannot be transmitted to earth. The envisaged budget for Rosetta during the design phase was more than an order of magnitude smaller than what was finally achieved. For instance a camera with longer focal length that could have achieved images of similar quality as the NAC from larger distances would have been very useful and reasonable. This often practiced conservatism in the announced resources, particularly in the present extreme, leads to sub optimal instrument design and performance.
6. Two significant deficiencies of the spacecraft design had the strongest impact on the success of the Rosetta mission. The dust environment in the near nucleus coma was not a surprise. Actually, the dust production during the Rosetta apparition of 67P closely followed the amounts observed during earlier orbits (Snodgrass et al. 2017). The impact of the dust particles on the performance of the star trackers was not realized before arrival at the comet and little was done to mitigate the problem (Accomazzo et al. 2017). This led to distances from the comet during the peak of activity about a factor 100 beyond what was anticipated during the design phase with all its consequences for the resolution of the instruments.
7. An even worse influence on the success of the mission resulted from the fact that the spacecraft could not hover near the nucleus but had to move on hyperbolic or bound orbits to avoid the risk of crashing into the nucleus in case of malfunctioning. This made mapping difficult and systematic observations almost impossible. One can characterize the Rosetta mission as a series of flybys interrupted by bound orbits mostly in the least optimal configuration along the terminator. This complex scheme posed a major challenge for the operational team requiring a huge effort that was skillfully mastered, but it was definitely not optimal or even adequate for science.
8. The situation was aggravated by the requirement that the minimum beta angle had to be $> 60°$, required by navigation constraints. This restricted measurements of the optical properties and mapping of the surface. In addition, eclipse or near eclipse observations would have been very helpful to derive dust grain properties and size distribution.
9. The spacecraft and instruments operation software, specifically developed for Rosetta, was late, too late to be finished in time, and had to be replaced by an adaptation of existing software developed for multiple missions (MAPPS). The foreseen time during hibernation (31 months) was not sufficient for the development and implementation. Early warnings based on experience with the Cassini–Huygens mission were disregarded.
10. The operation of the science instruments had turnaround times of about one month at the beginning of the mission. Again, the spacecraft operation was very conservative and science requirements were not given proper attention during the operational phase. At the end of the mission celebrating a PR event for the final crash of the spacecraft on the

surface was more important than to take risks to finally meet the originally promised near surface observations at good lighting conditions.

11. A continuous effective leadership to coordinate and optimize the science operations and data transfer rates for all instruments according to the scientific goals of the mission is the prerequisite to achieve maximum scientific return from a complex planetary mission like Rosetta. ESA decided to replace the Rosetta Project Scientist, a comet observer, by a plasma physicist after the spacecraft came out of hibernation in 2013, about a year before the comet rendezvous started. In addition, the ESA mission manager, a cometary scientist already involved in the Giotto mission, was also replaced. The deprivation of comet science expertise and the lack of a clear strategy how to meet the scientific requirements (see #10) led to suboptimal mission planning at the target where safety aspects always dominated but coordination and promotion of cometary science goals did not receive the required attention. In addition, the organization and leadership of the Rosetta Science Working Team was weak and hence not focused.

It is clear that no mission design and performance is perfect, but Rosetta shows more than the usual deficiencies many of which could have been avoided by better planning this cornerstone mission. The long travel time of 10 years was not adequately used.

## 6 Summary and Outlook

The Rosetta mission, staying over two years with comet 67P, has dramatically kindled the interest in cometary physics in the science community but also in the public. The harvest of the mission results is not over yet and further progress will be achieved. It can be expected that many of the previous comet observations and models will be reassessed in the light of the very detailed and copious measurements of comet 67P by Rosetta.

### 6.1 Flybys Versus Rendezvous

A comparison between the programs of both agencies, ESA and NASA, suggests itself. The total cost of the three American small missions with five flybys and the Rosetta cornerstone mission seem to be comparable. Rosetta could follow the evolution of cometary activity from its onset through perihelion and revealed the importance of seasons for the nucleus evolution. On one hand the interpretation of the extensive Rosetta campaign with its long term observations of the complete surface is more involved and more significant for the understanding of comets as members of the solar system and their early formation. On the other hand many inferred results achieved in the brink of the short flybys can now be considered consolidated. The detailed observations during the Rosetta rendezvous suggest that the apparent diversity of the admittedly still small number of "typical" JFC comets can be understood as the expression of cometary nucleus evolution in the inner solar system. Their diverse morphology is caused by their specific orbit parameters, rotation, shape, and seasonal effects of the more complex than previously modeled cometary activity. The question which of the programs—several flybys or the single extended rendezvous—provided more and/or better results is certainly legitimate from the agency and science political points of view but is difficult to assess objectively. It is, however, obvious that the diverse flybys are a very good and important complement to the detailed results of the Rosetta mission for one single comet. But it is also true that Rosetta yielded results that cannot be achieved by a flyby mission.

## 6.2 Major Scientific Conclusions from Rosetta

1. Time limited incidental observations of the coma composition do not necessarily reflect the intrinsic composition of the nucleus, neither in the content of volatiles nor in the refractory to volatile ratio. Mass transfer by back fall material is important for our understanding and interpretation of cometary appearances. This insight will lead to a reassessment of many papers that discuss cometary composition and its temporal and/or spatial variation or compare the global composition of specific comets. It is obviously difficult to prove that comets are intrinsically different.
2. The detailed observations of the surface revealed that most morphologies have already been found on nuclei passed by flybys. This strongly suggests that all (so far visited) comets are alike. The diverse morphologies are hence an expression of evolution.
3. Hydrocarbons, semi volatile or refractory compounds, contribute a major, if not the major, fraction of the mass of the nucleus. Rosetta observations confirmed their importance.
4. Findings of the mass spectrometry strongly suggest that the formation of the comet took place at very low temperatures ($< 25$ K) and a considerable part of its molecules is of interstellar origin.
5. Isotopic investigation based on ROSINA data suggest that comets significantly contributed to the inert gases in the Earth atmosphere but not to the terrestrial water.
6. A new paradigm for cometary activity seems to emerge. The activity and erosion is controlled by super volatiles ($CO_2$, CO), more than by water ice sublimation. The overall erosion of the cometary surface (mainly on the south in the case of 67P) is stronger than estimated from the total observed mass loss. Varying amounts of back fall material control the observed dust to gas ratios (smaller than the refractory to ice ratio) in the coma and tail of comets. The background (continuous) dust activity is driven mainly by $H_2O$ that does not remove all dust on the northern hemisphere during northern summer—in fact may not be able to maintain cometary activity. Desiccated dust accumulates on the north but notwithstanding the overall activity stays constant because the north is reactivated every orbit by wet back fall.

## 6.3 Major Open Points Remain

1. The composition measurements continue to indicate that comets carry material of interstellar/pre solar origin as well as solar processed compounds. Many more details were revealed but no clear concept has surfaced yet.
2. The existence and importance of amorphous water ice in the nucleus is still an open question.
3. Rosetta data are not conclusive concerning the collisional environment during the formation of the comet and right afterwards. This point is important to constrain evolution models of our planetary system such as the Nice model.
4. The refractory to ice ratio in cometary nuclei has been of particular interest since it constrains comet formation models in the solar nebula. However, it may differ from the dust/gas ratio in the coma and Rosetta did not provide reliable values for both.
5. Physical parameters such as the microstructure and small-scale strength of the surface layer are essential for the modeling and understanding of the physics of activity processes. Here Rosetta could not reach a breakthrough caused by the overall concept (one lander only) and the failure of Philae at a couple of essential measurements.
6. Very limited progress was made in revealing the physical processes of activity, both of the "normal" background activity and of the frequent outbursts. We did not see where

and how the frequent large particles in the centimeter and decimeter or even meter size range were lifted off.

7. Modelling the effects of the non-gravitational forces on the orbit provides additional constraint for the physical parameters of the nucleus. These model calculations based on the well-known properties of a nucleus would be highly desirable to calibrate the estimates for a large number of observed comets. Unfortunately, the orbit parameters of 67P have not been determined to the required accuracy.

Without any doubt Rosetta was a milestone in cometary research. It provided 300 GB of data that have only partly been analyzed up to date. Many new findings have been published in recent years. However, key questions have remained unanswered. Future rendezvous missions should optimize the operational scheme and bring original samples back to Earth (Thomas et al. 2019). Mission proposals such as CAESAR (NASA) that intends to visit 67P again and bring back samples (not selected), Comet Interceptor (ESA) to a dynamically new comet (selected www.cometinterceptor.space), the Japanese Destiny+ mission passing 3200 Phaeton (Kawakatsu and Iwata 2013), and a Chinese Asteroid Sample Return concept with an additional rendezvous of a main belt comet (see http://www.cnsa.gov.cn/english/n6465652/n6465653/c6805893/content.html?from=singlemessage&isappinstalled=0) are already in a selection process and could close the gap of open questions.

**Acknowledgements** Open Access funding provided by Projekt DEAL. We thank Dr. Imre Toth and an unknown reviewer. Their knowledgeable comments and suggestions clarified and substantiated some sections. This project has received funding from the European Union's Horizon 2020 research and innovation programme under grant agreement No 686709 (MiARD). This research has made use of NASA's Astrophysics Data System Bibliographic Services.



## References

A. Accomazzo, P. Ferri, S. Lodiot, J.-L. Pellon-Bailon, A. Hubault, R. Porta, J. Urbanek, R. Kay, M. Eiblmaier, T. Francisco, Acta Astronaut. **126**, 190 (2016)

A. Accomazzo, P. Ferri, S. Lodiot, J.-L. Pellon-Bailon, A. Hubault, J. Urbanek, R. Kay, M. Eiblmaier, T. Francisco, Acta Astronaut. **136**, 354 (2017)

M.F. A'Hearn, Science **314**, 1708 (2006)

M.F. A'Hearn, M.J.S. Belton, W.A. Delamere, J. Kissel, K.P. Klaasen, L.A. McFadden, K.J. Meech, H.J. Melosh, P.H. Schultz, J.M. Sunshine, P.C. Thomas, J. Veverka, D.K. Yeomans, M.W. Baca, I. Busko, C.J. Crockett, S.M. Collins, M. Desnoyer, C.A. Eberhardy, C.M. Ernst, T.L. Farnham, L. Feaga, O. Groussin, D. Hampton, S.I. Ipatov, J.-Y. Li, D. Lindler, C.M. Lisse, N. Mastrodemos, W.M. Owen, J.E. Richardson, D.D. Wellnitz, R.L. White, Science **310**, 258 (2005)

M.F. A'Hearn, M.J.S. Belton, W.A. Delamere, L.M. Feaga, D. Hampton, J. Kissel, K.P. Klaasen, L.A. McFadden, K.J. Meech, H.J. Melosh, P.H. Schultz, J.M. Sunshine, P.C. Thomas, J. Veverka, D.D. Wellnitz, D.K. Yeomans, S. Besse, D. Bodewits, T.J. Bowling, B.T. Carcich, S.M. Collins, T.L. Farnham, O. Groussin, B. Hermalyn, M.S. Kelley, M.S. Kelley, J.-Y. Li, D.J. Lindler, C.M. Lisse, S.A. McLaughlin, F. Merlin, S. Protopapa, J.E. Richardson, J.L. Williams, Science **332**, 1396 (2011)

M.F. A'Hearn, L.M. Feaga, H.U. Keller, H. Kawakita, D.L. Hampton, J. Kissel, K.P. Klaasen, L.A. McFadden, K.J. Meech, P.H. Schultz, J.M. Sunshine, P.C. Thomas, J. Veverka, D.K. Yeomans, S. Besse, D. Bodewits, T.L. Farnham, O. Groussin, M.S. Kelley, C.M. Lisse, F. Merlin, S. Protopapa, D.D. Wellnitz, Astrophys. J. **758**, 29 (2012)

K. Altwegg, (the ROSINA Team), *Proceedings of the International Astronomical Union*, vol. 13 (2017), p. 153

K. Altwegg, H. Balsiger, J. Geiss, in *Composition and Origin of Cometary Materials*, ed. by K. Altwegg, P. Ehrenfreund, J. Geiss, W.F. Huebner (Springer, Dordrecht, 1999), pp. 3–18

K. Altwegg, H. Balsiger, A. Bar-Nun, J.J. Berthelier, A. Bieler, P. Bochsler, C. Briois, U. Calmonte, M. Combi, J. De Keyser, P. Eberhardt, B. Fiethe, S. Fuselier, S. Gasc, T.I. Gombosi, K.C. Hansen, M. Hässig, A. Jäckel, E. Kopp, A. Korth, L. LeRoy, U. Mall, B. Marty, O. Mousis, E. Neefs, T. Owen, H. Rème, M. Rubin, T. Sémon, C.-Y. Tzou, H. Waite, P. Wurz, Science **347**, A387 (2015)

K. Altwegg, H. Balsiger, A. Bar-Nun, J.-J. Berthelier, A. Bieler, P. Bochsler, C. Briois, U. Calmonte, M.R. Combi, H. Cottin, J. De Keyser, F. Dhooghe, B. Fiethe, S.A. Fuselier, S. Gasc, T.I. Gombosi, K.C. Hansen, M. Haessig, A. Jäckel, E. Kopp, A. Korth, L. Le Roy, U. Mall, B. Marty, O. Mousis, T. Owen, H. Rème, M. Rubin, T. Sémon, C.-Y. Tzou, J.H. Waite, P. Wurz, Sci. Adv. **2**, e1600285 (2016)

M. Ashman, M. Barthélémy, L. O'Rourke, M. Almeida, N. Altobelli, M. Costa Sitjà, J.J. García Beteta, B. Geiger, B. Grieger, D. Heather, R. Hoofs, M. Küppers, P. Martin, R. Moissl, C. Múñoz Crego, M. Pérez-Ayúcar, E. Sanchez Suarez, M. Taylor, C. Vallat, Acta Astronaut. **125**, 41 (2016)

N. Attree, O. Groussin, L. Jorda, D. Nebouy, N. Thomas, Y. Brouet, E. Kuehrt, F. Preusker, F. Scholten, J. Knollenberg, P. Hartogh, H. Sierks, C. Barbieri, P. Lamy, R. Rodrigo, D. Koschny, H. Rickman, H.U. Keller, M.F. A'Hearn, A.-T. Auger, M.A. Barucci, J.-L. Bertaux, I. Bertini, D. Bodewits, S. Boudreault, G. Cremonese, V. Da Deppo, B. Davidsson, S. Debei, M. De Cecco, J. Deller, M.R. El-Maarry, S. Fornasier, M. Fulle, P.J. Gutiérrez, C. Güttler, S. Hviid, W.-H. Ip, G. Kovacs, J.R. Kramm, M. Küppers, L.M. Lara, M. Lazzarin, J.J. Lopez Moreno, S. Lowry, S. Marchi, F. Marzari, S. Mottola, G. Naletto, N. Oklay, M. Pajola, I. Toth, C. Tubiana, J.-B. Vincent, X. Shi, Astron. Astrophys. **611**, A33 (2018)

A.T. Basilevsky, H.U. Keller, Planet. Space Sci. **54**, 808 (2006)

J.-P. Bibring, Y. Langevin, J. Carter, P. Eng, B. Gondet, L. Jorda, S. Le Mouélic, S. Mottola, C. Pilorget, F. Poulet, M. Vincendon, Science **349**, 493 (2015)

J. Biele, S. Ulamec, M. Maibaum, R. Roll, L. Witte, E. Jurado, P. Muñoz, W. Arnold, H.-U. Auster, C. Casas, C. Faber, C. Fantinati, F. Finke, H.-H. Fischer, K. Geurts, C. Güttler, P. Heinisch, A. Herique, S. Hviid, G. Kargl, M. Knapmeyer, J. Knollenberg, W. Kofman, N. Kömle, E. Kührt, V. Lommatsch, S. Mottola, R. Pardo de Santayana, E. Remetean, F. Scholten, K.J. Seidensticker, H. Sierks, T. Spohn, Science **349**, 019816 (2015)

L. Biermann, Z. Astrophys. **29**, 274 (1951)

N. Biver, D. Bockelée-Morvan, M. Hofstadter, E. Lellouch, M. Choukroun, S. Gulkis, J. Crovisier, F.P. Schloerb, L. Rezac, P. von Allmen, S. Lee, C. Leyrat, W.H. Ip, P. Hartogh, P. Encrenaz, G. Beaudin, (the MIRO team), Astron. Astrophys. **630**, 58 (2019)

J. Blum, B. Gundlach, S. Mühle, J.M. Trigo-Rodriguez, Icarus **235**, 156 (2014)

D. Bockelée-Morvan, D.C. Lis, J.E. Wink, D. Despois, J. Crovisier, R. Bachiller, D.J. Benford, N. Biver, P. Colom, J.K. Davies, E. Gérard, B. Germain, M. Houde, D. Mehringer, R. Moreno, G. Paubert, T.G. Phillips, H. Rauer, Astron. Astrophys. **353**, 1101 (2000)

H. Böhnhardt, J.-P. Bibring, I. Apathy, H.U. Auster, A. Ercoli Finzi, F. Goesmann, G. Klingelhöfer, M. Knapmeyer, W. Kofman, H. Krüger, S. Mottola, W. Schmidt, K. Seidensticker, T. Spohn, I. Wright, Philos. Trans. R. Soc., Math. Phys. Eng. Sci. **375**, 20160248 (2017)

D.T. Britt, D.C. Boice, B.J. Buratti, H. Campins, R.M. Nelson, J. Oberst, B.R. Sandel, S.A. Stern, L.A. Soderblom, N. Thomas, Icarus **167**, 45 (2004)

D.E. Brownlee, F. Horz, R.L. Newburn, M. Zolensky, T.C. Duxbury, S. Sandford, Z. Sekanina, P. Tsou, M.S. Hanner, B.C. Clark, S.F. Green, J. Kissel, Science **304**, 1764 (2004)

U. Calmonte, K. Altwegg, H. Balsiger, J.J. Berthelier, A. Bieler, G. Cessateur, F. Dhooghe, E.F. van Dishoeck, B. Fiethe, S.A. Fuselier, S. Gasc, T.I. Gombosi, M. Hässig, L. Le Roy, M. Rubin, T. Sémon, C.-Y. Tzou, S.F. Wampfler, Mon. Not. R. Astron. Soc. **462**, S253 (2016)

P. Cambianica, M. Fulle, G. Cremonese, E. Simioni, G. Naletto, M. Massironi, Astron. Astrophys. (2020, submitted)

F. Capaccioni, A. Coradini, G. Filacchione, S. Erard, G. Arnold, P. Drossart, M.C. De Sanctis, D. Bockelee-Morvan, M.T. Capria, F. Tosi et al., Science **347**, aaa0628 (2015)

Choukroun et al., Space Sci. Rev. (2020, this issue)

A.L. Cochran, J.R. Green, E.S. Barker, Icarus **79**, 125 (1989)

B.J.R. Davidsson, P.J. Gutiérrez, Icarus **176**, 453 (2005)

B.J. Davidsson, P.J. Gutiérrez, O. Groussin, M.F. A'Hearn, T. Farnham, L.M. Feaga, M.S. Kelley, K.P. Klaasen, F. Merlin, S. Protopapa, Icarus **224**, 154 (2013)

B.J.R. Davidsson, H. Sierks, C. Güttler, F. Marzari, M. Pajola, H. Rickman, M.F. A'Hearn, A.-T. Auger, M.R. El-Maarry, S. Fornasier, P.J. Gutiérrez, H.U. Keller, M. Massironi, C. Snodgrass, J.-B. Vincent, C. Barbieri, P.L. Lamy, R. Rodrigo, D. Koschny, M.A. Barucci, J.-L. Bertaux, I. Bertini, G. Cremonese, V. Da Deppo, S. Debei, M. De Cecco, C. Feller, M. Fulle, O. Groussin, S.F. Hviid, S. Höfner, W.-H. Ip, L. Jorda, J. Knollenberg, G. Kovacs, J.-R. Kramm, E. Kührt, M. Küppers, F. La Forgia, L.M. Lara, M. Lazzarin, J.J. Lopez Moreno, R. Moissl-Fraund, S. Mottola, G. Naletto, N. Oklay, N. Thomas, C. Tubiana, Astron. Astrophys. **592**, A63 (2016)

M.C. De Sanctis, F. Capaccioni, M. Ciarniello, G. Filacchione, M. Formisano, S. Mottola, A. Raponi, F. Tosi, D. Bockelée-Morvan, S. Erard, C. Leyrat, B. Schmitt, E. Ammannito, G. Arnold, M.A. Barucci, M. Combi, M.T. Capria, P. Cerroni, W.-H. Ip, E. Kuehrt, T.B. McCord, E. Palomba, P. Beck, E. Quirico, V. Team, G. Piccioni, G. Bellucci, M. Fulchignoni, R. Jaumann, K. Stephan, A. Longobardo, V. Mennella, A. Migliorini, J. Benkhoff, J.P. Bibring, A. Blanco, M. Blecka, R. Carlson, U. Carsenty, L. Colangeli, M. Combes, J. Crovisier, P. Drossart, T. Encrenaz, C. Federico, U. Fink, S. Fonti, P. Irwin, Y. Langevin, G. Magni, L. Moroz, V. Orofino, U. Schade, F. Taylor, D. Tiphene, G.P. Tozzi, N. Biver, L. Bonal, J.-P. Combe, D. Despan, E. Flamini, S. Fornasier, A. Frigeri, D. Grassi, M.S. Gudipati, F. Mancarella, K. Markus, F. Merlin, R. Orosei, G. Rinaldi, M. Cartacci, A. Cicchetti, S. Giuppi, Y. Hello, F. Henry, S. Jacquinod, J.M. Rees, R. Noschese, R. Politi, G. Peter, Nature **525**, 500 (2015)

N. Dello Russo, R.J. Vervack, H.A. Weaver, N. Biver, D. Bockelée-Morvan, J. Crovisier, C.M. Lisse, Nature **448**, 172 (2007)

A.H. Delsemme, in *Comets*, ed. by L.L. Wilkening, M.S. Matthews (University of Arizona Press, Tucson, 1982), pp. 85–130

A.H. Delsemme, D.A. Rud, Astron. Astrophys. **28**, 1 (1973)

M.R. El-Maarry, O. Groussin, H.U. Keller, N. Thomas, J.-B. Vincent, S. Mottola, M. Pajola, K. Otto, C. Herny, S. Krasilnikov, Space Sci. Rev. **215**, 36 (2019)

J.E. Elsila, D.P. Glavin, J.P. Dworkin, Meteorit. Planet. Sci. **44**, 1323 (2009)

C. Emerich, J.M. Lamarre, V.I. Moroz, M. Combes, N.F. Sanko, Y.V. Nikolsky, F. Rocard, R. Gispert, N. Coron, J.P. Bibring, T. Encrenaz, J. Crovisier, Astron. Astrophys. **187**, 839 (1987)

L.M. Feaga, M.F. A'Hearn, J.M. Sunshine, O. Groussin, T.L. Farnham, Icarus **191**, 134 (2007)

M.C. Festou, H. Rickman, R.M. West, Astron. Astrophys. Rev. **5**, 37 (1993a)

M.C. Festou, H. Rickman, R.M. West, Astron. Astrophys. Rev. **4**, 363 (1993b)

M.C. Festou, H.U. Keller, H.A. Weaver, in *Comets II* (2004), pp. 3–16

G. Filacchione, O. Groussin, C. Herny, D. Kappel, S. Mottola, N. Oklay, A. Pommerol, I. Wright, Z. Yoldi, M. Ciarniello, L. Moroz, A. Raponi, Space Sci. Rev. **215**, 19 (2019)

G.J. Flynn, Earth Moon Planets **102**, 447 (2008)

S. Fornasier, S. Mottola, H.U. Keller, M.A. Barucci, B. Davidsson, C. Feller, J.D.P. Deshapriya, H. Sierks, C. Barbieri, P.L. Lamy, R. Rodrigo, D. Koschny, H. Rickman, M. A'Hearn, J. Agarwal, J.-L. Bertaux, I. Bertini, S. Besse, G. Cremonese, V. Da Deppo, S. Debei, M. De Cecco, J. Deller, M.R. El-Maarry, M. Fulle, O. Groussin, P.J. Gutierrez, C. Güttler, M. Hofmann, S.F. Hviid, W.-H. Ip, L. Jorda, J. Knollenberg, G. Kovacs, R. Kramm, E. Kührt, M. Küppers, M.L. Lara, M. Lazzarin, J.J.L. Moreno, F. Marzari, M. Massironi, G. Naletto, N. Oklay, M. Pajola, A. Pommerol, F. Preusker, F. Scholten, X. Shi, N. Thomas, I. Toth, C. Tubiana, J.-B. Vincent, Science **354**, 1566 (2016)

M. Fulle, F. Marzari, V. Della Corte, S. Fornasier, H. Sierks, A. Rotundi, C. Barbieri, P.L. Lamy, R. Rodrigo, D. Koschny, H. Rickman, H.U. Keller, J.J. López-Moreno, M. Accolla, J. Agarwal, M.F. A'Hearn, N. Altobelli, M.A. Barucci, J.-L. Bertaux, I. Bertini, D. Bodewits, E. Bussoletti, L. Colangeli, M. Cosi, G. Cremonese, J.-F. Crifo, V. Da Deppo, B. Davidsson, S. Debei, M. De Cecco, F. Esposito, M. Ferrari, F. Giovane, B. Gustafson, S.F. Green, O. Groussin, E. Grün, P. Gutierrez, C. Güttler, M.L. Herranz, S.F. Hviid, W. Ip, S.L. Ivanovski, J.M. Jerónimo, L. Jorda, J. Knollenberg, R. Kramm, E. Kührt, M. Küppers, L. Lara, M. Lazzarin, M.R. Leese, A.C. López-Jiménez, F. Lucarelli, E. Mazzotta Epifani, J.A.M. McDonnell, V. Mennella, A. Molina, R. Morales, F. Moreno, S. Mottola, G. Naletto, N. Oklay, J.L. Ortiz, E. Palomba, P. Palumbo, J.-M. Perrin, F.J.M. Rietmeijer, J. Rodríguez, R. Sordini, N. Thomas, C. Tubiana, J.-B. Vincent, P. Weissman, K.-P. Wenzel, V. Zakharov, J.C. Zarnecki, Astrophys. J. **821**, 19 (2016)

K. Furuya, E.F. van Dishoeck, Y. Aikawa, Astron. Astrophys. **586**, A127 (2016)

K.-H. Glassmeier, H. Boehnhardt, D. Koschny, E. Kührt, I. Richter, Space Sci. Rev. **128**, 1 (2007)

B. Godard, F. Budnik, G. Bellei, P.M. Oz, T. Morley, ISTS-2017-D-95/ISSFD-2017-095 16 (2017)

F. Goesmann, H. Rosenbauer, J.H. Bredehöft, M. Cabane, P. Ehrenfreund, T. Gautier, C. Giri, H. Krüger, L.L. Roy, A.J. MacDermott, S. McKenna-Lawlor, U.J. Meierhenrich, G.M.M. Caro, F. Raulin, R. Roll, A. Steele, H. Steininger, R. Sternberg, C. Szopa, W. Thiemann, S. Ulamec, Science **349**, aab0689 (2015)

C. Goetz, C. Koenders, I. Richter, K. Altwegg, J. Burch, C. Carr, E. Cupido, A. Eriksson, C. Güttler, P. Henri, P. Mokashi, Z. Nemeth, H. Nilsson, M. Rubin, H. Sierks, B. Tsurutani, C. Vallat, M. Volwerk, K.-H. Glassmeier, Astron. Astrophys. **588**, A24 (2016)

O. Groussin, M.F. A'Hearn, J.-Y. Li, P.C. Thomas, J.M. Sunshine, C.M. Lisse, K.J. Meech, T.L. Farnham, L.M. Feaga, W.A. Delamere, Icarus **191**, 63 (2007)
O. Groussin, L. Jorda, A.-T. Auger, E. Kührt, R. Gaskell, C. Capanna, F. Scholten, F. Preusker, P. Lamy, S. Hviid, J. Knollenberg, U. Keller, C. Huettig, H. Sierks, C. Barbieri, R. Rodrigo, D. Koschny, H. Rickman, M.F. A'Hearn, J. Agarwal, M.A. Barucci, J.-L. Bertaux, I. Bertini, S. Boudreault, G. Cremonese, V. Da Deppo, B. Davidsson, S. Debei, M. De Cecco, M.R. El-Maarry, S. Fornasier, M. Fulle, P.J. Gutiérrez, C. Güttler, W.-H. Ip, J.-R. Kramm, M. Küppers, M. Lazzarin, L.M. Lara, J.J. Lopez Moreno, S. Marchi, F. Marzari, M. Massironi, H. Michalik, G. Naletto, N. Oklay, A. Pommerol, M. Pajola, N. Thomas, I. Toth, C. Tubiana, J.-B. Vincent, Astron. Astrophys. **583**, A32 (2015)
O. Groussin, N. Attree, Y. Brouet, V. Ciarletti, B. Davidsson, G. Filacchione, H.-H. Fischer, B. Gundlach, M. Knapmeyer, J. Knollenberg, R. Kokotanekova, E. Kührt, C. Leyrat, D. Marshall, I. Pelivan, Y. Skorov, C. Snodgrass, T. Spohn, F. Tosi, Space Sci. Rev. **215**, 29 (2019)
B. Gundlach, J. Blum, H.U. Keller, Y.V. Skorov, Astron. Astrophys. **583**, A12 (2015)
T.I. Hasegawa, E. Herbst, C.M. Leung, Astrophys. J. Suppl. Ser. **82**, 167 (1992)
M. Hässig, K. Altwegg, H. Balsiger, A. Bar-Nun, J.J. Berthelier, A. Bieler, P. Bochsler, C. Briois, U. Calmonte, M. Combi, J. De Keyser, P. Eberhardt, B. Fiethe, S.A. Fuselier, M. Galand, S. Gasc, T.I. Gombosi, K.C. Hansen, A. Jäckel, H.U. Keller, E. Kopp, A. Korth, E. Kührt, L. Le Roy, U. Mall, B. Marty, O. Mousis, E. Neefs, T. Owen, H. Rème, M. Rubin, T. Sémon, C. Tornow, C.-Y. Tzou, J.H. Waite, P. Wurz, Science **347**, 276 (2015)
T. Heidarzadeh, *A History of Physical Theories of Comets, from Aristotle to Whipple* (Springer, Berlin, 2008)
P. Heinisch, H.-U. Auster, B. Gundlach, J. Blum, C. Güttler, C. Tubiana, H. Sierks, M. Hilchenbach, J. Biele, I. Richter, K.H. Glassmeier, Astron. Astrophys. **630**, A2 (2019)
A. Herique, W. Kofman, P. Beck, L. Bonal, I. Buttarazzi, E. Heggy, J. Lasue, A.C. Levasseur-Regourd, E. Quirico, S. Zine, Mon. Not. R. Astron. Soc. **462**, S516 (2016)
K.A. Holsapple, K.R. Housen, An interpretation of the deep impact experiment, in *Lunar Planet Sci. XXXVII Abstract*, vol. 1068 (2006)
X. Hu, X. Shi, H. Sierks, M. Fulle, J. Blum, H.U. Keller, M.L.E. Kühr, B. Davidsson, C. Güttler, B. Gundlach, M. Pajola, D. Bodewits, J.B. Vincent, N. Oklay, M. Massironi, S. Fornasier, C. Tubiana, O. Groussin, S. Boudreault, S. Höfner, S. Mottola, C. Barbieri, P.L. Lamy, R. Rodrigo, D. Koschny, H. Rickman, M. A'Hearn, J. Agarwal, M.A. Barucci, J.L. Bertaux, I. Bertini, G. Cremonese, V. Da Deppo, S. Debei, M. De Cecco, J. Deller, A. Gicquel, P. Gutierrez-Marques, P.J. Gutiérrez, M. Hofmann, S.F. Hviid, W.H. Ip, L. Jorda, J. Knollenberg, G. Kovacs, J.R. Kramm, M. Küppers, L.M. Lara, Astron. Astrophys. **604**, A114 (2017)
M. Jutzi, E. Asphaug, Science **348**, 1355 (2015)
Y. Kawakatsu, T. Iwata, Adv. Astronaut. Sci. **146**, 727–739 (2013)
H.U. Keller, in *Physics and Mechanics of Cometary Materials*, ed. by J.J. Hunt, T.D. Guyenne (1989), pp. 39–45
H.U. Keller, in *Physics and Chemistry of Comets*, ed. by W.F. Huebner (Springer, Berlin, 1990), pp. 13–68
H.U. Keller, C. Arpigny, C. Barbieri, R.M. Bonnet, S. Cazes, M. Coradini, C.B. Cosmovici, W.A. Delamere, W.F. Huebner, D.W. Hughes, C. Jamar, D. Malaise, H.J. Reitsema, H.U. Schmidt, W.K.H. Schmidt, P. Seige, F.L. Whipple, K. Wilhelm, Nature **321**, 320 (1986)
H.U. Keller, S. Mottola, B. Davidsson, S.E. Schröder, Y. Skorov, E. Kührt, O. Groussin, M. Pajola, S.F. Hviid, F. Preusker, F. Scholten, M.F. A'Hearn, H. Sierks, C. Barbieri, P. Lamy, R. Rodrigo, D. Koschny, H. Rickman, M.A. Barucci, J.-L. Bertaux, I. Bertini, G. Cremonese, V. Da Deppo, S. Debei, M. De Cecco, S. Fornasier, M. Fulle, P.J. Gutiérrez, W.-H. Ip, L. Jorda, J. Knollenberg, J.R. Kramm, M. Küppers, L.M. Lara, M. Lazzarin, J.J. Lopez Moreno, F. Marzari, H. Michalik, G. Naletto, L. Sabau, N. Thomas, J.-B. Vincent, K.-P. Wenzel, J. Agarwal, C. Güttler, N. Oklay, C. Tubiana, Astron. Astrophys. **583**, A34 (2015a)
H.U. Keller, S. Mottola, Y. Skorov, L. Jorda, Astron. Astrophys. **579**, L5 (2015b)
H.U. Keller, S. Mottola, S.F. Hviid, J. Agarwal, E. Kührt, Y. Skorov, K. Otto, J.-B. Vincent, N. Oklay, S.E. Schröder, B. Davidsson, M. Pajola, X. Shi, D. Bodewits, I. Toth, F. Preusker, F. Scholten, H. Sierks, C. Barbieri, P. Lamy, R. Rodrigo, D. Koschny, H. Rickman, M.F. A'Hearn, M.A. Barucci, J.-L. Bertaux, I. Bertini, G. Cremonese, V. Da Deppo, S. Debei, M. De Cecco, J. Deller, S. Fornasier, M. Fulle, O. Groussin, P.J. Gutiérrez, C. Güttler, M. Hofmann, W.-H. Ip, L. Jorda, J. Knollenberg, J.R. Kramm, M. Küppers, L.-M. Lara, M. Lazzarin, J.J. Lopez-Moreno, F. Marzari, G. Naletto, C. Tubiana, N. Thomas, Mon. Not. R. Astron. Soc. **469**, S357 (2017)
M.S. Kelley, D.J. Lindler, D. Bodewits, M.F. A'Hearn, C.M. Lisse, L. Kolokolova, J. Kissel, B. Hermalyn, Icarus **222**, 634 (2013)
S.J. Kerridge, B.K. Muirhead, M. Neugebauer, A. Mauritz, G. Tan-Wang, D. Sabahi, J.R. Green, J. Grimes, D.J.P. Moura, F. Bonneau, F.-X. Chaffaut, P. Rangeard, F. Rocard, J.-P. Bibring, Acta Astronaut. **40**, 585 (1997)

J. Kissel, R.Z. Sagdeev, J.L. Bertaux, V.N. Angarov, J. Audouze, J.E. Blamont, K. Buchler, E.N. Evlanov, H. Fechtig, M.N. Fomenkova, H. von Hoerner, N.A. Inogamov, V.N. Khromov, W. Knabe, F.R. Krueger, Y. Langevin, B. Leonasv, A.C. Levasseur-Regourd, G.G. Managadze, S.N. Podkolzin, V.D. Shapiro, S.R. Tabaldyev, B.V. Zubkov, Nature **321**, 280 (1986)

M. Knapmeyer, H.-H. Fischer, J. Knollenberg, K.J. Seidensticker, K. Thiel, W. Arnold, C. Faber, D. Möhlmann, Icarus **310**, 165 (2018)

W. Kofman, A. Herique, Y. Barbin, J.-P. Barriot, V. Ciarletti, S. Clifford, P. Edenhofer, C. Elachi, C. Eyraud, J.-P. Goutail, E. Heggy, L. Jorda, J. Lasue, A.-C. Levasseur-Regourd, E. Nielsen, P. Pasquero, F. Preusker, P. Puget, D. Plettemeier, Y. Rogez, H. Sierks, C. Statz, H. Svedhem, I. Williams, S. Zine, J. Van Zyl, Science **349**, 020639 (2015)

T. Kramer, M. Läuter, Astron. Astrophys. **630**, A4 (2019)

T. Kramer, M. Läuter, S. Hviid, L. Jorda, H.U. Keller, E. Kührt, Astron. Astrophys. **630**, A3 (2019)

E. Kührt, H.U. Keller, Icarus **109**, 121 (1994)

E. Kührt, J. Knollenberg, H.U. Keller, Planet. Space Sci. **45**, 665 (1997)

M. Läuter, T. Kramer, M. Rubin, K. Altwegg, Mon. Not. R. Astron. Soc. **483**, 852 (2019)

K.E. Mandt, A. Eriksson, N.J.T. Edberg, C. Koenders, T. Broiles, S.A. Fuselier, P. Henri, Z. Nemeth, M. Alho, N. Biver, A. Beth, J. Burch, C. Carr, K. Chae, A.J. Coates, E. Cupido, M. Galand, K.-H. Glassmeier, C. Goetz, R. Goldstein, K.C. Hansen, J. Haiducek, E. Kallio, J.-P. Lebreton, A. Luspay-Kuti, P. Mokashi, H. Nilsson, A. Opitz, I. Richter, M. Samara, K. Szego, C.-Y. Tzou, M. Volwerk, C.S. Wedlund, G. Stenberg Wieser, Mon. Not. R. Astron. Soc. **462**, S9 (2016)

D. Marshall, O. Groussin, J.-B. Vincent, Y. Brouet, D. Kappel, G. Arnold, M.T. Capria, G. Filacchione, P. Hartogh, M. Hofstadter, W.-H. Ip, L. Jorda, E. Kuehrt, E. Lellouch, S. Mottola, L. Rezac, R. Rodrigo, S. Rodionov, P. Schloerb, N. Thomas, Astron. Astrophys. **616**, A122 (2018)

B. Marty, K. Altwegg, H. Balsiger, A. Bar-Nun, D.V. Bekaert, J.-J. Berthelier, A. Bieler, C. Briois, U. Calmonte, M. Combi, J. De Keyser, B. Fiethe, S.A. Fuselier, S. Gasc, T.I. Gombosi, K.C. Hansen, M. Hässig, A. Jäckel, E. Kopp, A. Korth, L. Le Roy, U. Mall, O. Mousis, T. Owen, H. Rème, M. Rubin, T. Sémon, C.-Y. Tzou, J.H. Waite, P. Wurz, Science **356**, 1069 (2017)

M. Massironi, E. Simioni, F. Marzari, G. cremonese, L. Giacomini, M. Pajola, L. Jorda, G. Naletto, H. Sierks, C. Barbieri, P.L. Lamy, R. Rodrigo, D. koschny, H. Rickman, H.U. Keller, M.F. A'Hearn, Nature **526**, 402–405 (2015)

E. Merényi, L. Földy, K. Szegő, I. Tóth, A. Kondor, Icarus **86**, 9 (1990)

D. Möhlmann, K.J. Seidensticker, H.-H. Fischer, C. Faber, A. Flandes, M. Knapmeyer, H. Krüger, R. Roll, F. Scholten, K. Thiel, W. Arnold, Icarus **303**, 251 (2018)

S. Mottola, S. Lowry, C. Snodgrass, P.L. Lamy, I. Toth, A. Rożek, H. Sierks, M.F. A'Hearn, F. Angrilli, C. Barbieri, M.A. Barucci, J.-L. Bertaux, G. Cremonese, V. Da Deppo, B. Davidsson, M. De Cecco, S. Debei, S. Fornasier, M. Fulle, O. Groussin, P. Gutiérrez, S.F. Hviid, W. Ip, L. Jorda, H.U. Keller, J. Knollenberg, D. Koschny, R. Kramm, E. Kührt, M. Küppers, L. Lara, M. Lazzarin, J.J. Lopez Moreno, F. Marzari, H. Michalik, G. Naletto, H. Rickman, R. Rodrigo, L. Sabau, N. Thomas, K.-P. Wenzel, J. Agarwal, I. Bertini, F. Ferri, C. Güttler, S. Magrin, N. Oklay, C. Tubiana, J.-B. Vincent, Astron. Astrophys. **569**, L2 (2014)

S. Mottola, G. Arnold, H.-G. Grothues, R. Jaumann, H. Michaelis, G. Neukum, J.-P. Bibring, S.E. Schröder, M. Hamm, K.A. Otto, I. Pelivan, G. Proffe, F. Scholten, D. Tirsch, M. Kreslavsky, E. Remetean, F. Souvannavong, B. Dolives, Science **349**, aab0232 (2015)

M. Neugebauer, J.-P. Bibring, Adv. Space Res. **21**, 1567 (1998)

R.L. Newburn, M. Neugebauer, J. Rahe (eds.), *Comets in the Post-Halley Era* (1991)

J.H. Oort, Bull. Astron. Inst. Neth. **11**, 91 (1950)

T. Ott, E. Drolshagen, D. Koschny, C. Güttler, C. Tubiana, E. Frattin, J. Agarwal, H. Sierks, I. Bertini, C. Barbieri, P.I. Lamy, R. Rodrigo, H. Rickman, M.F. A'Hearn, M.A. Barucci, J.-L. Bertaux, S. Boudreault, G. Cremonese, V. Da Deppo, B. Davidsson, S. Debei, M. De Cecco, J. Deller, C. Feller, S. Fornasier, M. Fulle, B. Geiger, A. Gicquel, O. Groussin, P.J. Gutiérrez, M. Hofmann, S.F. Hviid, W.-H. Ip, L. Jorda, H.U. Keller, J. Knollenberg, G. Kovacs, J.R. Kramm, E. Kührt, M. Küppers, L.M. Lara, M. Lazzarin, Z.-Y. Lin, J.J. López-Moreno, F. Marzari, S. Mottola, G. Naletto, N. Oklay, M. Pajola, X. Shi, N. Thomas, J.-B. Vincent, B. Poppe, Mon. Not. R. Astron. Soc. **469**, S276 (2017)

M. Pajola, S. Höfner, J.B. Vincent, N. Oklay, F. Scholten, F. Preusker, S. Mottola, G. Naletto, S. Fornasier, S. Lowry, C. Feller, P.H. Hasselmann, C. Güttler, C. Tubiana, H. Sierks, C. Barbieri, P. Lamy, R. Rodrigo, D. Koschny, H. Rickman, H.U. Keller, J. Agarwal, M.F. A'Hearn, M.A. Barucci, J.-L. Bertaux, I. Bertini, S. Besse, S. Boudreault, G. Cremonese, V. da Deppo, B. Davidsson, S. Debei, M. de Cecco, J. Deller, J.D.P. Deshapriya, M.R. El-Maarry, S. Ferrari, F. Ferri, M. Fulle, O. Groussin, P. Gutierrez, M. Hofmann, S.F. Hviid, W.-H. Ip, L. Jorda, J. Knollenberg, G. Kovacs, J.R. Kramm, E. Kührt, M. Küppers, L.M. Lara, Z.-Y. Lin, M. Lazzarin, A. Lucchetti, J.J. Lopez Moreno, F. Marzari, M. Massironi,

H. Michalik, L. Penasa, A. Pommerol, E. Simioni, N. Thomas, I. Toth, E. Baratti, Nat. Astron. **1**, 0092 (2017)

M. Pätzold, T. Andert, M. Hahn, S.W. Asmar, J.-P. Barriot, M.K. Bird, B. Häusler, K. Peter, S. Tellmann, E. Grün, P.R. Weissman, H. Sierks, L. Jorda, R. Gaskell, F. Preusker, F. Scholten, Nature **530**, 63 (2016)

M. Pätzold, T.P. Andert, M. Hahn, J.-P. Barriot, S.W. Asmar, B. Häusler, M.K. Bird, S. Tellmann, J. Oschlisniok, K. Peter, Mon. Not. R. Astron. Soc. **483**, 2337 (2019)

F. Preusker, F. Scholten, K.-D. Matz, T. Roatsch, S.F. Hviid, S. Mottola, J. Knollenberg, E. Kührt, M. Pajola, N. Oklay, J.-B. Vincent, B. Davidsson, M.F. A'Hearn, J. Agarwal, C. Barbieri, M.A. Barucci, J.-L. Bertaux, I. Bertini, G. Cremonese, V.D. Deppo, S. Debei, M.D. Cecco, S. Fornasier, M. Fulle, O. Groussin, P.J. Gutiérrez, C. Güttler, W.-H. Ip, L. Jorda, H.U. Keller, D. Koschny, J.R. Kramm, M. Küppers, P. Lamy, L.M. Lara, M. Lazzarin, J.J.L. Moreno, F. Marzari, M. Massironi, G. Naletto, H. Rickman, R. Rodrigo, H. Sierks, N. Thomas, C. Tubiana, Astron. Astrophys. **607**, L1 (2017)

D. Prialnik, G. Sarid, E.D. Rosenberg, R. Merk, Space Sci. Rev. **138**, 147 (2008)

E. Quirico, L.V. Moroz, B. Schmitt, G. Arnold, M. Faure, P. Beck, L. Bonal, M. Ciarniello, F. Capaccioni, G. Filacchione, Icarus **272**, 32 (2016)

L. Rezac, Y. Zhao, P. Hartogh, J. Ji, D. Marshall, X. Shi, Astron. Astrophys. **630**, A34 (2019)

J.E. Richardson, H.J. Melosh, Icarus **222**, 492 (2013)

H. Rickman, L. Kamel, M.C. Festou, C. Froeschle, in *Diversity and Similarity of Comets*, ed. by E.J. Rolfe, B. Battrick, M. Ackerman, M. Scherer, R. Reinhard (1987)

H. Rickman, S. Marchi, M.F. A'Hearn, C. Barbieri, M.R. El-Maarry, C. Güttler, W.-H. Ip, H.U. Keller, P. Lamy, F. Marzari, M. Massironi, G. Naletto, M. Pajola, H. Sierks, D. Koschny, R. Rodrigo, M.A. Barucci, J.-L. Bertaux, I. Bertini, G. Cremonese, V. Da Deppo, S. Debei, M. De Cecco, S. Fornasier, M. Fulle, O. Groussin, P.J. Gutiérrez, S.F. Hviid, J. Jorda, L. andKnollenberg, J.-R. Kramm, E. Kührt, M. Küppers, L.M. Lara, M. Lazzarin, J.J. Lopez Moreno, H. Michalik, L. Sabau, N. Thomas, J.-B. Vincent, K.-P. Wenzel, Astron. Astrophys. **583**, A44 (2015)

M. Rubin, K. Altwegg, H. Balsiger, A. Bar-Nun, J.-J. Berthelier, A. Bieler, P. Bochsler, C. Briois, U. Calmonte, M. Combi, J.D. Keyser, F. Dhooghe, P. Eberhardt, B. Fiethe, S.A. Fuselier, S. Gasc, T.I. Gombosi, K.C. Hansen, M. Hässig, A. Jäckel, E. Kopp, A. Korth, L.L. Roy, U. Mall, B. Marty, O. Mousis, T. Owen, H. Rème, T. Sémon, C.-Y. Tzou, J.H. Waite, P. Wurz, Science **348**, 232 (2015)

M. Rubin, K. Altwegg, H. Balsiger, J.-J. Berthelier, A. Bieler, U. Calmonte, M. Combi, J. De Keyser, C. Engrand, B. Fiethe, S.A. Fuselier, S. Gasc, T.I. Gombosi, K.C. Hansen, M. Hässig, L. Le Roy, K. Mezger, C.-Y. Tzou, S.F. Wampfler, P. Wurz, Astron. Astrophys. **601**, A123 (2017)

S.E. Schröder, S. Mottola, G. Arnold, H.-G. Grothues, R. Jaumann, H.U. Keller, H. Michaelis, J.-P. Bibring, I. Pelivan, A. Koncz, K. Otto, E. Remetean, F. Souvannavong, B. Dolives, Icarus **285**, 263 (2017)

G. Schwehm, R. Schulz, in *Composition and Origin of Cometary Materials*, ed. by K. Altwegg, P. Ehrenfreund, J. Geiss, W.F. Huebner (Springer, Dordrecht, 1999), pp. 313–319

X. Shi, X. Hu, S. Mottola, H. Sierks, H.U. Keller, M. Rose, C. Güttler, M. Fulle, S. Fornasier, J. Agarwal, M. Pajola, C. Tubiana, D. Bodewits, C. Barbieri, P.L. Lamy, R. Rodrigo, D. Koschny, M.A. Barucci, J.-L. Bertaux, I. Bertini, S. Boudreault, G. Cremonese, V.D. Deppo, B. Davidsson, S. Debei, M.D. Cecco, J. Deller, O. Groussin, P.J. Gutiérrez, S.F. Hviid, W.-H. Ip, L. Jorda, J. Knollenberg, G. Kovacs, J.-R. Kramm, E. Kührt, M. Küppers, L.M. Lara, M. Lazzarin, J.J. Lopez-Moreno, F. Marzari, G. Naletto, N. Oklay, I. Toth, J.-B. Vincent, Nat. Astron. **2**, 562 (2018)

Yu.V. Skorov, L. Rezac, P. Hartogh, H.U. Keller, Astron. Astrophys. **600**, A142 (2017)

C. Snodgrass, M.F. A'Hearn, F. Aceituno, V. Afanasiev, S. Bagnulo, J. Bauer, G. Bergond, S. Besse, N. Biver, D. Bodewits, H. Boehnhardt, B.P. Bonev, G. Borisov, B. Carry, V. Casanova, A. Cochran, B.C. Conn, B. Davidsson, J.K. Davies, J. de León, E. de Mooij, M. de Val-Borro, M. Delacruz, M.A. DiSanti, J.E. Drew, R. Duffard, N.J.T. Edberg, S. Faggi, L. Feaga, A. Fitzsimmons, H. Fujiwara, E.L. Gibb, M. Gillon, S.F. Green, A. Guijarro, A. Guilbert-Lepoutre, P.J. Gutiérrez, E. Hadamcik, O. Hainaut, S. Haque, R. Hedrosa, D. Hines, U. Hopp, F. Hoyo, D. Hutsemékers, M. Hyland, O. Ivanova, E. Jehin, G.H. Jones, J.V. Keane, M.S.P. Kelley, N. Kiselev, J. Kleyna, M. Kluge, M.M. Knight, R. Kokotanekova, D. Koschny, E. Kramer, J.J. López-Moreno, P. Lacerda, L.M. Lara, J. Lasue, H.J. Lehto, A.C. Levasseur-Regourd, J. Licandro, Z.Y. Lin, T. Lister, S.C. Lowry, A. Mainzer, J. Manfroid, J. Marchant, A.J. McKay, A. McNeill, K.J. Meech, M. Micheli, I. Mohammed, M. Monguió, F. Moreno, O. Muñoz, M.J. Mumma, P. Nikolov, C. Opitom, J.L. Ortiz, L. Paganini, M. Pajuelo, F.J. Pozuelos, S. Protopapa, T. Pursimo, B. Rajkumar, Y. Ramanjooloo, E. Ramos, C. Ries, A. Riffeser, V. Rosenbush, P. Rousselot, E.L. Ryan, P. Santos-Sanz, D.G. Schleicher, M. Schmidt, R. Schulz, A.K. Sen, A. Somero, A. Sota, A. Stinson, J. Sunshine, A. Thompson, G.P. Tozzi, C. Tubiana, G.L. Villanueva, X. Wang, D.H. Wooden, M. Yagi, B. Yang, B. Zaprudin, T.J. Zegmott, Philos. Trans. R. Soc. A **375**, 20160249 (2017)

T. Spohn, J. Knollenberg, A.J. Ball, M. Banaszkiewicz, J. Benkhoff, M. Grott, J. Grygorczuk, C. Hüttig, A. Hagermann, G. Kargl, E. Kaufmann, N. Kömle, E. Kührt, K.J. Kossacki, W. Marczewski, I. Pelivan, R. Schrödter, K. Seiferlin, Science **349**, aab0464 (2015)

P.J. Stooke, A. Abergel, Astron. Astrophys. **248**, 656 (1991)
J.M. Sunshine, M.F. A'Hearn, O. Groussin, J.-Y. Li, M.J.S. Belton, W.A. Delamere, J. Kissel, K.P. Klaasen, L.A. McFadden, K.J. Meech, H.J. Melosh, P.H. Schultz, P.C. Thomas, J. Veverka, D.K. Yeomans, I.C. Busko, M. Desnoyer, T.L. Farnham, L.M. Feaga, D.L. Hampton, D.J. Lindler, C.M. Lisse, D.D. Wellnitz, Science **311**, 1453 (2006)
J.M. Sunshine, O. Groussin, P.H. Schultz, M.F. A'Hearn, L.M. Feaga, T.L. Farnham, K.P. Klaasen, Icarus **190**, 284 (2007)
P. Swings, Lick Obs. Bull. **19** (1941)
M.V. Sykes, R.G. Walker, Icarus **95**, 180 (1992)
P. Thomas, M. A'Hearn, M.J.S. Belton, D. Brownlee, B. Carcich, B. Hermalyn, K. Klaasen, S. Sackett, P.H. Schultz, J. Veverka, S. Bhaskaran, D. Bodewits, S. Chesley, B. Clark, T. Farnham, O. Groussin, A. Harris, J. Kissel, J.-Y. Li, K. Meech, J. Melosh, A. Quick, J. Richardson, J. Sunshine, D. Wellnitz, Icarus **222**, 453 (2013a)
P.C. Thomas, M.F. A'Hearn, J. Veverka, M.J.S. Belton, J. Kissel, K.P. Klaasen, L.A. McFadden, H.J. Melosh, P.H. Schultz, S. Besse, B.T. Carcich, T.L. Farnham, O. Groussin, B. Hermalyn, J.-Y. Li, D.J. Lindler, C.M. Lisse, K. Meech, J.E. Richardson, Icarus **222**, 550 (2013b)
N. Thomas, M.R. El Maarry, P. Theologou, F. Preusker, F. Scholten, L. Jorda, S.F. Hviid, R. Marschall, E. Kührt, G. Naletto, H. Sierks, P.L. Lamy, R. Rodrigo, D. Koschny, B. Davidsson, M.A. Barucci, J.L. Bertaux, I. Bertini, D. Bodewits, G. Cremonese, V. Da Deppo, S. Debei, M. De Cecco, S. Fornasier, M. Fulle, O. Groussin, P.J. Gutièrrez, C. Güttler, W.H. Ip, H.U. Keller, J. Knollenberg, L.M. Lara, M. Lazzarin, J.J. Lòpez-Moreno, F. Marzari, C. Tubiana, J.B. Vincent, Planet. Space Sci. **164**, 19 (2018)
N. Thomas, S. Ulamec, E. Kührt, V. Ciarletti, B. Gundlach, Z. Yoldi, G. Schwehm, C. Snodgrass, S.F. Green, Space Sci. Rev. **215**, 47 (2019)
F. Tosi, F. Capaccioni, M.T. Capria, S. Mottola, A. Zinzi, M. Ciarniello, G. Filacchione, M. Hofstadter, S. Fonti, M. Formisano, D. Kappel, E. Kührt, C. Leyrat, J.-B. Vincent, G. Arnold, M.C.D. Sanctis, A. Longobardo, E. Palomba, A. Raponi, B. Rousseau, B. Schmitt, M.A. Barucci, G. Bellucci, J. Benkhoff, D. Bockelée-Morvan, P. Cerroni, J.-P. Combe, D. Despan, S. Erard, F. Mancarella, T.B. McCord, A. Migliorini, V. Orofino, G. Piccioni, Nat. Astron. **3**, 649–658 (2019)
S. Ulamec, J. Block, M. Fenzi, B. Feuerbacher, G. Haerendel, P. Hemmerich, M. Maibaum, H. Rosenbauer, B. Schiewe, H.P. Schmidt, R. Schütze, K. Wittmann, Space Technol. **1**, 59 (1997)
D.E. Vavilov, S. Eggl, Y.D. Medvedev, P.B. Zatitskiy, Astron. Astrophys. **622**, L5 (2019)
J.-B. Vincent, D. Bodewits, S. Besse, H. Sierks, C. Barbieri, P. Lamy, R. Rodrigo, D. Koschny, H. Rickman, H.U. Keller, J. Agarwal, M.F. A'Hearn, A.-T. Auger, M.A. Barucci, J.-L. Bertaux, I. Bertini, C. Capanna, G. Cremonese, V. Da Deppo, B. Davidsson, S. Debei, M. De Cecco, M.R. El-Maarry, F. Ferri, S. Fornasier, M. Fulle, R. Gaskell, L. Giacomini, O. Groussin, A. Guilbert-Lepoutre, P. Gutierrez-Marques, P.J. Gutiérrez, C. Güttler, N. Hoekzema, S. Höfner, S.F. Hviid, W.-H. Ip, L. Jorda, J. Knollenberg, G. Kovacs, R. Kramm, E. Kührt, M. Küppers, F. La Forgia, L.M. Lara, M. Lazzarin, V. Lee, C. Leyrat, Z.-Y. Lin, J.J. Lopez Moreno, S. Lowry, S. Magrin, L. Maquet, S. Marchi, F. Marzari, M. Massironi, H. Michalik, R. Moissl, S. Mottola, G. Naletto, N. Oklay, M. Pajola, F. Preusker, F. Scholten, N. Thomas, I. Toth, C. Tubiana, Nature **523**, 63 (2015)
J.-B. Vincent, T. Farnham, E. Kührt, Y. Skorov, R. Marschall, N. Oklay, M.R. El-Maarry, H.U. Keller, Space Sci. Rev. **215**, 30 (2019)
F.L. Whipple, Astrophys. J. **111**, 375 (1950)
K. Wittmann, B. Feuerbacher, S. Ulamec, H. Rosenbauer, J.P. Bibring, D. Moura, R. Mugnuolo, S. diPippo, K. Szego, G. Haerendel, Acta Astronaut. **45**, 389 (1999)
I.P. Wright, S. Sheridan, S.J. Barber, G.H. Morgan, D.J. Andrews, A.D. Morse, Science **349**, aab0673 (2015)
Z. Yoldi, O. Poch, B. Jost, O. Brissaud, E. Quirico, P. Beck, B. Rousseau, A. Pommerol, N. Thomas, EPSC Abstr. **11**, 538 (2017)

Space Sci Rev (2019) 215:36
https://doi.org/10.1007/s11214-019-0602-1

# Surface Morphology of Comets and Associated Evolutionary Processes: A Review of Rosetta's Observations of 67P/Churyumov–Gerasimenko

**M.R. El-Maarry[1] · O. Groussin[2] · H.U. Keller[3] · N. Thomas[4] · J.-B. Vincent[5] · S. Mottola[5] · M. Pajola[6] · K. Otto[5] · C. Herny[4] · S. Krasilnikov[7,8]**

Received: 19 November 2018 / Accepted: 21 May 2019 / Published online: 12 June 2019


**Abstract** Comets can be regarded as active planetary bodies because they display evidence for nearly all fundamental geological processes, which include impact cratering, tectonism, and erosion. Comets also display sublimation-driven outgassing, which is comparable to volcanism on larger planetary bodies in that it provides a conduit for delivering materials from the interior to the surface. However, in the domain of active geological bodies, comets occupy a special niche since their geologic activity is almost exclusively driven by externally supplied energy (i.e. solar energy) as opposed to an internal heat source, which makes them "seasonally-active" geological bodies. During their active phase approaching the Sun, comets also develop a transient atmosphere that interacts with the surface and contributes to its evolution, particularly by transporting materials across the surface. Variations in solar energy input on diurnal and seasonal scale cause buildup of thermal stresses within consolidated materials that lead to weathering through fracturing, and eventually mass-wasting. The commonly irregular shapes of comets also play a major role in their evolution by leading to (1) non-uniform gravitational forces that affect material movement across the surface, and

Post 67P/Churyumov–Gerasimenko Perspectives
Edited by Nicolas Thomas, Björn Davidsson, Laurent Jorda, Ekkehard Kührt, Raphael Marschall, Colin Snodgrass and Rafael Rodrigo

✉ M.R. El-Maarry
m.elmaarry@bbk.ac.uk

1 Birkbeck College, University of London, WC1E 7HX, London, UK

2 Aix Marseille Univ., CNRS, CNES, LAM, Marseille, France

3 Institut für Geophysik und extraterrestrische Physik (IGEP), Technische Universität Braunschweig, 38106 Braunschweig, Germany

4 Physikalisches Institut, University of Bern, Sidlerstr. 5, 3012 Bern, Switzerland

5 Deutsches Zentrum für Luft- und Raumfahrt (DLR), Institut für Planetenforschung, 12489 Berlin, Germany

6 INAF Osservatorio Astronomico di Padova, Vicolo dell'Osservatorio 5, 35122 Padova, Italy

7 Vernadsky Institute of Geochemistry and Analytical Chemistry RAS, 119334 Moscow, Russia

8 Max-Planck-Institut für Sonnensystemforschung, 37077 Göttingen, Germany

(2) spatially heterogeneous outgassing patterns that affect the comet's orbital dynamics and lead to tidal stresses that can further fracture the nucleus. In this chapter, we review the surface morphology of comet 67P/Churyumov–Gerasimenko as well as its seasonal evolution as viewed by Rosetta from August 2014 to September 2016, their link to various processes, and the forces that drive surface evolution.

**Keywords** Comets · Geology · Rosetta · Surfaces · 67P, Churyumov–Gerasimenko · Geomorphology · Surface processes

## 1 Introduction

Previous flyby missions to comets had provided snapshots of morphologically diverse worlds that are home to unique geological processes. Close-up images of cometary nuclei are often challenging to interpret because comets offer an environment that is impossible to replicate in its entirety on any other planetary body. Comets combine a complex composition incorporating various forms of ice, dust, and organics, low density, high porosity, a microgravity environment, and transient activity to create a landscape that challenges comparative studies with other planetary bodies. Pre-Rosetta missions had flown by five different comets (Fig. 1) giving us remarkable insights into the diverse surface morphologies of cometary nuclei. These missions demonstrated that while all comets share common surface traits, they were also different from one another. While fly-by images from these diverse small bodies enabled theories to be developed about the processes controlling cometary activity and resurfacing, the Rosetta mission at Comet Churyumov–Gerasimenko was able to directly test these theories through longer-term reconnaissance of the surface.

The Giotto, Vega 1, and Vega 2 missions to comet Halley[1] were the first missions to acquire resolved images of a comet's nucleus (Keller et al. 1986; Sagdeev et al. 1986). Images from the Giotto mission in particular indicated that only minor parts of the surface were active (Keller et al. 1986) and suggested a morphologically diverse and rough surface that displayed putative ridges, mountains, and depressions. Higher resolution (10 s of meters/pixel) images of Borrelly's surface from the Deep Space 1 mission were the first to show unambiguously the morphological diversity of cometary nuclei. Borrelly displayed mottled (rough) and smooth terrains, and sub parallel sets of ridges and fractures, but lacked clear evidence of impact craters, which suggested a young active surface (Soderblom et al. 2002). In 2004, the Stardust mission flew by comet Wild 2 and returned images that showed a surface predominantly composed of irregular and circular depressions with variable sizes reaching up to 2 km in width, along with pinnacles and limited signs of mass-wasting (Brownlee et al. 2004).

In 2005, the Deep Impact mission flew by Tempel 1 and returned images showing a similarly diverse surface, yet distinct from that of Borrelly (A'Hearn et al. 2005). Tempel 1 displayed circular and irregular depressions, smooth units that suggested a flow behavior, and for the first time, evidence of layering (A'Hearn et al. 2005; Thomas et al. 2007). The mission also included an impact experiment to study the ejected materials (A'Hearn et al. 2005). The Deep Impact spacecraft followed its mission at Tempel 1 with a flyby of comet Hartley 2 in 2010. Hartley 2 displayed rough knobby terrains along the margins of the bilobed-shaped nucleus and smooth terrains in the waist/neck region (Fig. 1). The nucleus lacked circular

[1] For brevity, and unless we state otherwise, we will refer to comets in this section by their common names.

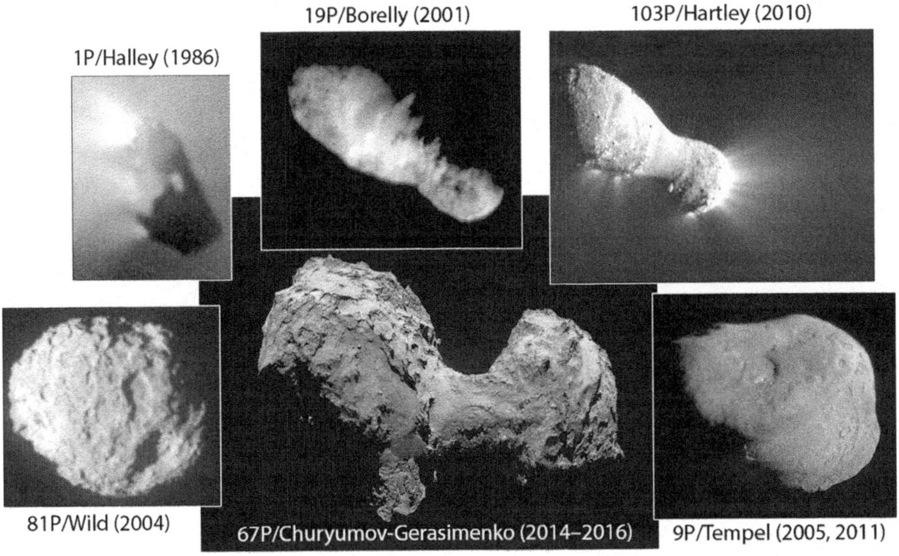


**Fig. 1** All comets that have been visited by spacecraft (not to scale). Dates in parenthesis indicate the time when the comets were visited. 9P/Tempel is the only comet to be visited by two different spacecraft at different times. 67P/Churyumov–Gerasimenko is the only comet to be investigated by an orbiting spacecraft continuously for more than 2 years

features that dominate Wild 2's landscape and are scattered around on Tempel 1. In addition, no evidence for layering or flowing material was observed (A'Hearn et al. 2011; Thomas et al. 2013b). However, correlations between jet activity and features such as pits, arcuate depressions, scarps, and rimless depressions suggested that mass wasting contributes to volatile sublimation at active comets (Bruck Syal et al. 2013). Distribution of high and low-albedo blocks (where high-albedo blocks were theorized to be enriched in surficial volatiles) near scarps and areas of enhanced dust jets suggested a formation mechanism related to jet activity and mass wasting. Finally, the Stardust spacecraft followed its earlier flyby of Wild 2 with a second visit of Tempel 1. The second flyby in 2011 allowed the imaging and mapping of previously unobserved areas of the comet in 2005. The expanded view demonstrated that the smooth units were preferentially present in gravitational lows (Thomas et al. 2013a; Veverka et al. 2013). In addition, the repeated viewing of Tempel 1 over two perihelia meant that for the first time a study of surface changes could be carried out. The most notable change that was observed was a ~50 m scarp retreat in one of the smooth units (Thomas et al. 2013a; Veverka et al. 2013).

The Rosetta mission to 67P/Churyumov–Gerasimenko (herein after referred to as 67P) expanded our view of cometary surfaces by offering us a near-continuous stream of data for more than two terrestrial years (from August 2014 to September 2016), which was highlighted by the first landing of a spacecraft (Philae) on the surface of a comet in November 2014, and perihelion passage in August 2015. During these two years, Rosetta spent a substantial amount of mission time orbiting 30 km away or less from 67P, which allowed the mission's main scientific camera, the Optical, Spectroscopic, and Infrared Remote Imaging

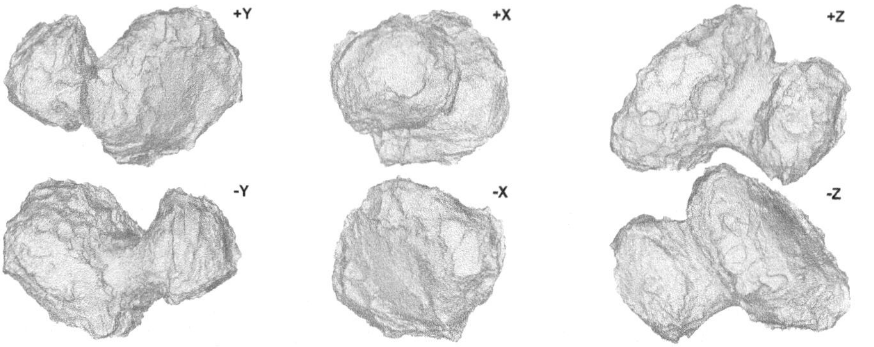


**Fig. 2** Different viewpoints for the most recent shape model of comet 67P prepared with stereo-photogrammetry techniques. North is up for views along the $X$ and $Y$ axes. Views along the $Z$-axis are polar views ($+Z$ for the northern polar view, $-Z$ for the southern one). Figure from Preusker et al. (2017)

System (OSIRIS) (Keller et al. 2007), to image the surface at resolutions of 50 cm/pixel or higher for most of the mission's lifetime.[2]

In the coming sections, we review the large body of work that was published using data from the Rosetta mission, particularly from OSIRIS, to showcase the various morphological elements that shape up the surface of 67P. Section 2 presents 67P's shape and large-scale morphology. Section 3 focuses on surface processes that affect the morphology on all scales. In Sect. 4, we summarise current hypotheses regarding the driving forces behind these active processes.

## 2 67P's Morphology

### 2.1 Overall Shape, Volume and Density

Comet 67P's nucleus has a bilobate shape (Fig. 2) with overall dimensions of approx. 4.3 × 2.6 × 2.1 km along its principle axes (Jorda et al. 2016). A larger lobe (approx. 4.1 × 3.5 × 1.6 km) is connected to a smaller lobe (approx. 2.5 × 2.1 × 1.6 km) by a narrow neck region that constitutes only 7% of the comet's total volume (Jorda et al. 2016). Early studies following Rosetta's arrival at comet 67P's vicinity (e.g., Preusker et al. 2015) reported only estimates of the comet's shape and volume since the southern hemisphere was not illuminated at the time (i.e. prior to May 2015). Following equinox in May 2015, these earlier estimates were refined to the values reported by Jorda et al. (2016) based on full shape models derived through stereo-photoclinometry methods. Preusker et al. (2017) used stereo-photogrammetric analysis to derive more accurate estimates of the total volume ($18.56\ \text{km}^3 \pm 0:02\ \text{km}^3$), as well as mean density ($537.8\ \text{kg/m}^3 \pm 0.7\ \text{kg/m}^3$). The density estimate is based on calculated mass of approx. $9.98 \times 10^{12}$ kg as measured by the Radio Science Investigation experiment (Pätzold et al. 2016). In 2014, the comet had a spin period of approx. 12.404 hours. The spin period increased to 12.430 hours in May 2015 before dropping rapidly (i.e. the comet's rotation speed increased) to 12.305 hours in Aug. 2015 (Jorda et al. 2016). The changes in the spin period resulting from changes in torques

[2]Images were acquired at lower resolutions when Rosetta was moved further away from the comet during perihelion to protect the spacecraft, or during far excursions to study outer coma structures.

caused by non-gravitational forces had an important role in driving some of the changes that affected the comet during the mission's lifetime (see Sect. 3).

## 2.2 Large-Scale Morphology

Images acquired from 67P reveal remarkably diverse surface morphologies and textures. This is possibly due to the higher spatial resolutions achieved by Rosetta's cameras than in most of these missions, and the longer duration and orbiting nature of the mission, which allowed for mapping of the entire nucleus. In terms of regional morphology, the principle terrains on 67P encompass exposed consolidated terrains, smooth plains, dust-covered units, and large-scale depressions (Thomas et al. 2015b, Fig. 3). The large morphological and structural boundaries between these terrain types delineate the identified geographic regions of the comet (Fig. 4, El-Maarry et al. 2015b, 2016; Thomas et al. 2015b, 2018).

Consolidated terrains are essentially the comet's exposed surface, which retain enough structural integrity to sustain fractures, cliffs, and overhangs (Fig. 3). The fragmentation of these consolidated materials results in mass wasting deposits and boulders of variable sizes. The surface of the comet displays regions of different strength where regions such as Seth and Anhur are relatively weakly consolidated as evident from the common occurrence of mass wasting deposits of various sorting (e.g., Giacomini et al. 2016; Lee et al. 2016; Fig. 3C), as opposed to more strongly consolidated regions that display a paucity of such deposits, and presence of large fracture systems (El-Maarry et al. 2016). In addition, consolidated terrains display evidence of layering and terracing on a whole-body scale (e.g., Massironi et al. 2015; Penasa et al. 2017).

The principle morphological trait of 67P is its hemispherical dichotomy, which is marked by a disparity of relief between the northern and southern hemispheres (see Sect. 2.1), as well as a morphological dichotomy highlighted by the uneven spatial distribution of non-consolidated materials (NCMs). The northern hemisphere appears to be predominantly covered by NCMs whose distribution and extent is most probably controlled by local gravitational slopes (Sect. 2.1). A thin layer (only a few metres thick at the most) of NCMs covers nearly 20% of the total comet's surface (Thomas et al. 2018), but is almost exclusively confined to the northern hemisphere. Smooth terrains represent locations where NCMs have accumulated into deposits that are thick enough to entirely cover the underlying topography and form quasi flat surfaces (e.g., Imhotep and Hapi regions, Fig. 3A). In contrast, the southern hemisphere appears to be devoid of large-scale deposits of NCMs (El-Maarry et al. 2016); this difference is consistent with modelled predictions of seasonal activity-driven redistributions of NCMs (Keller et al. 2015; Thomas et al. 2015a). NCMs generally appear smooth in meters/pixel images as the predominant grain sizes are below imaging resolution. Sub-meter resolution images from both the orbiter and lander can resolve NCM grains down to the $\sim$5 and 7 cm scale at the Agilkia (first landing site of Philae) (Pajola et al. 2016b) and Sais (final resting site of the Rosetta orbiter) (Pajola et al. 2017a), respectively.

The northern hemisphere and equatorial regions also include a number of large-scale depressions, which appear to be lacking in the southern hemisphere. Aten, one of the notable depressions in the large lobe is $\sim$0.12 km$^3$ in volume (Thomas et al. 2015b, Fig. 3B). Thomas et al. (2015a) calculated that the production of Aten by a sublimation mechanism alone would require 10–20 orbits in the current configuration. As a result, they concluded that large scale depressions were formed by major mass loss events. El-Maarry et al. (2015b) arrived at similar conclusions based on the morphology of the depression floors and the bounding scarps. Other less conspicuous, yet larger in size, regional depressions include

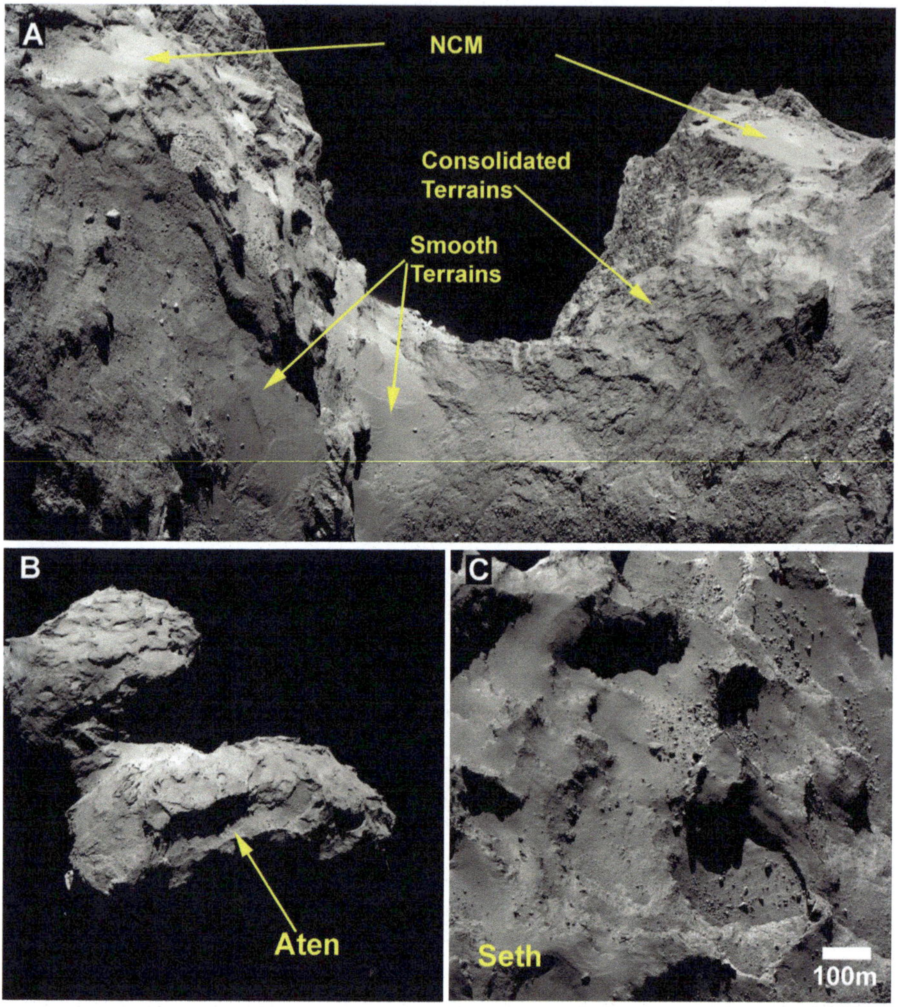


**Fig. 3** (**A**) Mosaic of two Narrow Angle Camera (NAC) images displaying various terrain types on the northern hemisphere of 67P. "NCM" is an acronym for Non-Consolidated Materials. (**B**) NAC image showing the Aten depression in the northern part of the large lobe. (**C**) NAC image of the Seth region in 67P's large lobe (in the northern hemisphere) showing weakly consolidated materials as evident from the numerous arcuate and circular features, as well as ubiquitous talus deposits. Parts of Seth are covered in NCM deposits

the equatorial regions Imhotep, which is partially filled with smooth deposits, and Khonsu (Fig. 5), which shows a unique pancake-shaped structure that appears to have been exhumed through intense erosion (El-Maarry et al. 2016).

Other notable geomorphological features of the surface include circular pits, albeit predominantly in the northern hemisphere (Vincent et al. 2015a, Sect. 3.4), large-scale (100 s of m) fracture systems, preferentially close to the neck region in both hemispheres (El-Maarry et al. 2015b, 2016), 10–100 s of meter-high cliffs (e.g., Hathor), and a dearth of unambiguous impact features. Comprehensive geological and/or geomorphological maps of the northern hemisphere, southern hemisphere and the nucleus as a whole have been published by Giacomini et al. (2016); Lee et al. (2016), and Birch et al. (2017), respectively.

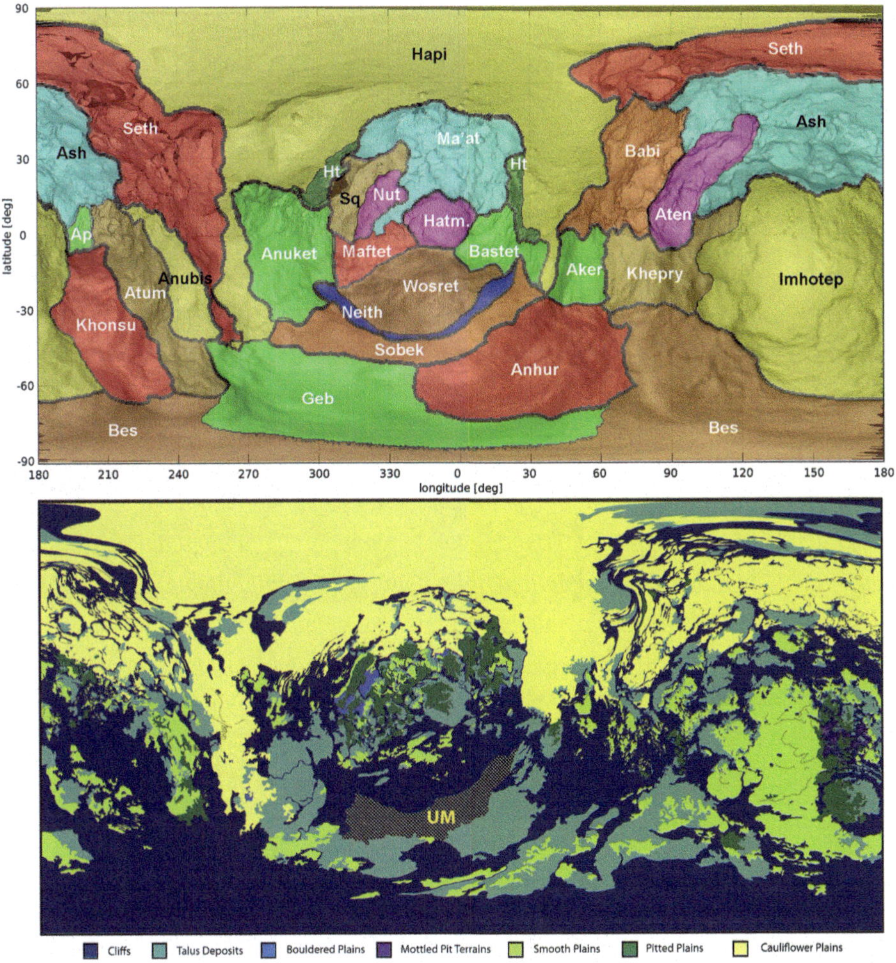


**Fig. 4** Map projections of defined regions (top) adapted from El-Maarry et al. (2016), and global geomorphological map (bottom) adapted from Birch et al. (2017). Acronyms are as follows: (top) Ap: Apis, Ht: Hathor, Sq: Serqet, Hatm.: Hatmehit. (bottom) UM: Unmapped region in Birch et al. due to shadows in images that were available for the study

Briefly, Giacomini et al. (2016) identified 4 main geological units that comprise the northern hemisphere: (1) fine particle deposits (corresponding to the term "NCM" used in this revision), (2) consolidated materials, (3) smooth terrains, and (4) mass wasting deposits. Giacomini et al. (2016) further noted that the mass -wasting deposits had differing origins and size-sorting and so where classified into different sub-classes. Lee et al. (2016) carried a similarly comprehensive geologic mapping of the southern hemisphere describing similar units but noted that the southern hemisphere landscape was dominated by outcropping consolidated terrains. Finally, Birch et al. (2017) carried out an independent geomorphological mapping of the entire nucleus (Fig. 4) in which they identified the following units: (1) cliffs, essentially most of the exposed consolidated surfaces (2) talus deposits, (3) boulder-filled plains, (4) mottled pit terrains, a specific site in Imhotep containing circular features, (5) smooth plains, (6) pitted plains, and (7) "cauliflower" plains, roughly

**Fig. 5** The Khonsu region in the equatorial region of 67P's large lobe, which displays elevated, and sometimes sharp, boundaries and houses a pancake-shaped feature that appears to have been exhumed by erosion. Adapted from El-Maarry et al. (2016)

corresponding to the distribution of NCMs. In the next section we briefly describe the processes that acted on these terrains, and outline the major morphological changes that were observed by Rosetta during its mission.

## 3 Surface Processes Affecting Morphology

Comets display evidence for all fundamental geological processes, which include impact cratering, tectonism, and erosion (e.g., Bennet et al. 2017). We use the term 'tectonism' here to refer to processes that result in internal stresses on a (near-) whole body scale, as well as surface features that result from deformation caused by these stresses. The fourth fundamental geological process is conventionally volcanism, which is understandably absent on comets due to their small size (i.e. lack of significant decaying radioactive elements to supply internal heat). However, comets display sublimation-driven outgassing (Vincent et al. this volume). This process is comparable to volcanism on larger planetary bodies in that it provides a conduit for delivering materials from the interior to the surface. However, in the domain of active geological bodies, comets occupy a special niche since their geologic activity is almost exclusively driven by externally supplied energy (i.e. solar energy) as opposed to an internal heat source, which makes them "seasonally-active" geological bodies.

In the following sections, we review a number of morphologic features and their evolution (if changes were observed during the mission's lifetime) that are particularly indicative of active processes. Multiple processes may affect certain features simultaneously or with time. Therefore, rather than investigate each process individually, we focus instead on particular geomorphologic units and review the processes that collectively affect them. In this way, we can present a concise picture of how cometary materials evolve on diurnal, seasonal, and possibly longer, scales. Section 3.1 discusses the processes that affect consolidated materials, which may also consequently create NCMs, as well as associated features indicative of these processes. Section 3.2 moves on to discuss how NCMs further evolve with time

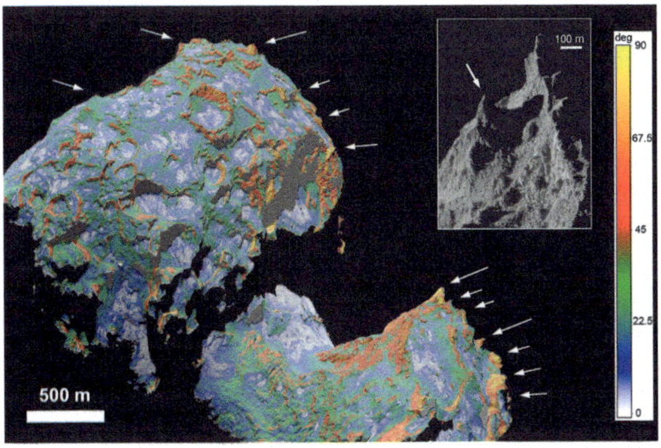


**Fig. 6** Potential pinnacles on comet 67P (arrows) with superposed color-coded values of gravitational slopes. Note that pinnacles appear to have slopes ranging from 40° to 90° (Basilevsky et al. 2017). The inset shows an example of a prominent pinnacle. Figure is adapted from Basilevsky et al. (2017)

and the features associated with the active processes acting on NCMs. Section 3.3 discusses impacts as a unique process since it is an exogenic process that should affect all terrain types.

## 3.1 Consolidated Terrains

While consolidated terrains appear to represent the bare surface of the comet, it is probable that physical and chemical processes are responsible for their formation in the first place. Earlier studies that used various modeling approaches and lab experiments to simulate the evolution of cometary nuclei suggest that cometary materials may become "consolidated" by sintering of ice or re-condensation of sublimated water vapor following sublimation. For instance, "Kometen-Simulation" (KOSI) experiments (e.g., Kochan et al. 1989; Sears et al. 1999), and follow-up experiments, demonstrated that such a process in the lab leads to (1) hardening of the near-surface ice layer (e.g., Kossacki et al. 1999; Kossacki et al. 1997; Seiferlin et al. 1995), (2) increase of thermal conductivity, especially in the presence of organics (e.g., Kömle et al. 1996), and (3) development of a millimeter-thick surface dust layer (e.g., Grün et al. 1993; Kossacki et al. 1997; Pommerol et al. 2015a). Indeed, measurements by other instruments on Rosetta such as MUPUS and SESAME (both on the Philae lander) are consistent with consolidated materials having properties resulting from sintering as described above (Lethuillier et al. 2016; Spohn et al. 2015). This is discussed further later in Sect. 4.3. Continuous imaging and follow-up morphological investigations have not yielded enough information about the activity and effectiveness of such a process with time. However, sintering appears to be currently the most reasonable hypothesis for the formation of the consolidated materials despite remaining poorly understood particularly beyond controlled laboratory settings (e.g., Keller and Markiewicz 1991).

Following their formation, and in the presence of active geologic processes, weathering is probably the first process that acts on 67P's consolidated materials. In the absence of liquid water, weathering would be constrained to mechanical breakdown of materials. Indeed, Rosetta has observed evidence for weathering as fracturing and fragmentation leading to eventual erosion through mass wasting. We discuss these features below.

### *3.1.1 Weathering and Pinnacles*

Pinnacles are local promontories of various shapes including spires with pointed tops (e.g., Migon 2004). They are likely erosional remnants created by a loss of material surrounding spots of the more-erosion-resistant material. Pinnacles (Fig. 6) were initially identified and described by Basilevsky et al. (2017) in 67P's northern hemisphere. Basilevsky et al. (2017) identified 49 pinnacles (Fig. 6) in 67P's northern hemisphere. They are generally asymmetric in shape with heights ranging from 10 m to 200 m (40 m as a mean value), and foot diameters ranging from 30 m to 300 m (60 m as a mean value) (Basilevsky et al. 2017). Pinnacles possibly represent an intermediate stage of weathering of strongly consolidated material whereas weaker consolidated units react to weathering more readily by fracturing and eventual fragmentation.

### *3.1.2 Fracturing and Polygonal Networks*

Fractures are ubiquitous on the exposed consolidated surfaces of 67P forming various patterns and topologies at different size-scales (El-Maarry et al. 2015a). Fracture networks on planar surfaces are the most common setting of fractures on 67P (Fig. 7). Sub-meter-wide fractures form 2-dimensional surface patterns with different topologies. They vary in length from 100 s of meters down to mm-scale as observed by the Philae lander (Bibring et al. 2015; Poulet et al. 2016).

Fractures often intersect to form polygonal networks, modes of intersection are variable but include orthogonal intersections, which are a common indicator of slowly building stress regimes in which fractures form slowly and consequent to one another instead of a more rapid and simultaneous manner (El-Maarry et al. 2015b). In a preliminary analysis of images from orbit, El-Maarry et al. (2015b) noted the presence of hummocky features embedded within fracture networks that could be indicative of high-centered thermal contraction-like polygons. A near-global mapping of more than 6000 of these features by Auger et al. (2018) indicated that they are ubiquitous at all latitudes and longitudes. The size of the polygons (90% of the mapped features) is predominantly 1–5 meters in width although few features exceed 13 m. The fractures intersect in both 3-ray (i.e., $\sim$120 degrees), and 4-ray (orthogonal or near orthogonal) manner, with 3-ray intersections being more common (Auger et al. 2018). El-Maarry et al. (2015b) favored an insolation weathering mechanism for the formation of fracture networks based on simple thermal models and the features' morphology and settings. Auger et al. (2018) utilised a more complex thermal model and concluded that polygonal fracture systems were consistent with a seasonal thermal contraction mechanism caused by variations in insolation on daily and seasonal scales, similar to what is observed in periglacial/glacial landscapes on Earth and Mars.

Fracture networks have also been observed at the mm scale by cameras onboard Philae at its final landing site. However, these smaller fractures form poorly-developed networks that lack well-defined polygonal shapes (Poulet et al. 2016). The longest fractures reach up to 10 s of cm in length with an average of $\sim$20 mm in rough units and $\sim$45 mm in relatively smoother units (Poulet et al. 2016). Given the wide-spread of fracturing in the surface of 67P, spanning many length scales, it is safe to assume that fracturing processes play a primary role in the mechanical weathering of consolidated surface materials. As fractures present preferential zones of weakness, they become the point of focus for continuing breakdown leading to eventual failure such as collapsing of cliffs and formation of poorly sorted non-consolidates debris (i.e. boulders).

### *3.1.3 Mass Wasting and Boulder Formation*

Fractures of a more irregular pattern are observed on the edges of scarps (Fig. 8) particularly in weakly consolidated units. The bases of fractured scarps display debris and talus deposits, indicating an ongoing process of mass wasting associated with weakly consolidated

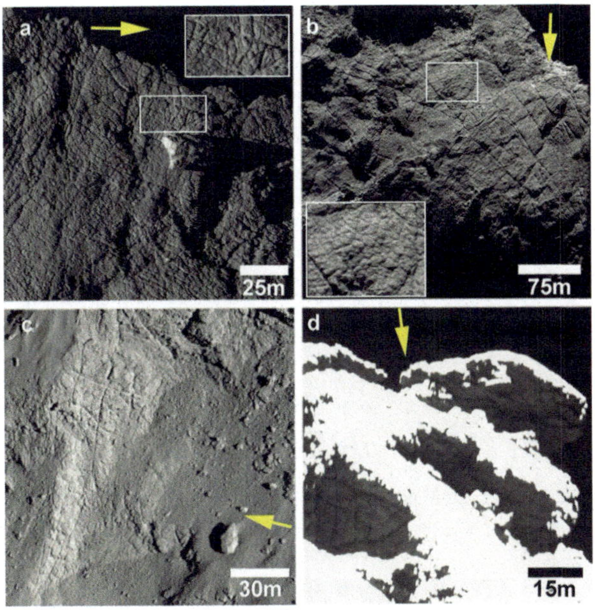


**Fig. 7** Polygonal fracture patterns in consolidated materials. Figure is adapted from El-Maarry et al. (2015b). Note the embedded features in panels **a** and **b**, which could be a result of thermal contraction processes

units (El-Maarry et al. 2015b). During the mission, two sites of fractured cliffs displayed mass wasting events marked by detachment of materials at the fractured site (Pajola et al. 2017b; El-Maarry et al. 2017, Fig. 9), highlighting the role of such fractures in mass wasting events. The approximate timing of theses collapses is consistent with periods of large diurnal temperature fluctuations, which are particularly pronounced close to perihelion when these cliffs likely collapsed (El-Maarry et al. 2017; Pajola et al. 2017b).[3] Such mass wasting events should be the primary process by which cometary materials are being fragmented. Indeed, one of the main morphological features characterising the nucleus of 67P is the widespread presence of boulders (Pajola et al. 2015).

Early in the mission, more than one hundred bright boulders were observed in the comet's northern hemisphere (Pommerol et al. 2015b). These boulders were up to ten times brighter than the average surface of the comet at visible wavelengths and their spectrophotometric properties were consistent with exposure of water ice (Pommerol et al. 2015b). These bright boulders were either observed as (1) clusters, particularly in association with talus, (2) isolated boulders, or (3) partially bright boulders that displayed only bright patches on the surface (Pommerol et al. 2015b). The clustered bright boulders in particular, offer direct evidence for common, and recent, mass wasting events.

Studies focused on boulder counting and characterisation resulted in global size-frequency distributions (SFD) of boulders $\geq 7$ m in size (Pajola et al. 2015, 2016b, Fig. 10A), as well as local SFD studies of sizes $\geq 2$ m (Pajola et al. 2016c, Fig. 10B) down to even the decimeter and centimeter scales (Pajola et al. 2016a, 2017a). The numerous studies focused on boulders suggest that their SFD can be indicative of their formation processes and how the surface of 67P evolves. Indeed, Pajola et al. (2015, 2016b) identified that whenever there is a collapse/pit formation and creation of depressions with subsequent escape

[3]Close to perihelion in August 2015, Rosetta was moved to higher orbits for its safety due to the intense activity of the comet. Unfortunately, this resulted in a decrease of image resolution and coverage. Therefore, the exact timing of certain surface changes is uncertain and can only be constrained to a certain time-frame.

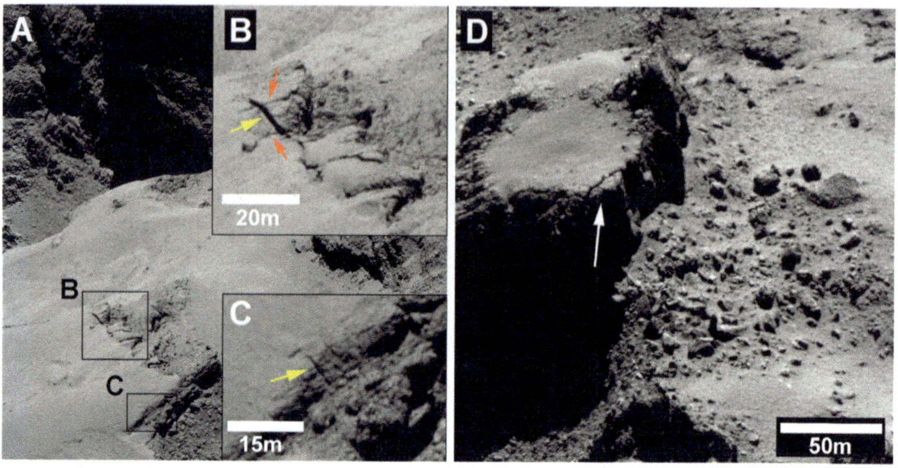


**Fig. 8** Fractures in proximity to cliffs or small scarps. Note the debris field in (**D**) at the base of a fractured cliff, which suggests a continuous process of mass wasting. Figures adapted from El-Maarry et al. (2015b)

**Fig. 9** The collapsing cliff in the area dubbed Aswan (Pajola et al. 2017b). The cliff collapse was triggered by the failure of a long arcuate fracture near the scarp. Figure adapted from El-Maarry et al. (2017)

of high-pressure volatiles and consequent high fracturing, the boulders SFD range between −5 and −6.5 (called type 1 boulders). On the other hand, gravitational events triggered by sublimation and/or thermal fracturing causing cliff collapses and regressive erosion result in power-law indices between −3.5 and −4.5 (called type 2 boulders). Further evolution of the original material formed during collapsing events results in indices of about −1 to −2 (Pajola et al. 2015). The global 67P boulder power-law index of $-3.6 + 0.2/-0.3$ suggests that type 2 boulders are dominant on the comet, and hence its surface may well be shaped by progressive erosion and receding of its cliffs (Pajola et al. 2015, 2017b).

### *3.1.4 Tectonism and Associated Features*

While most fractures observed in 67P's consolidated materials form polygonal networks suggestive of a global homogenous process, certain fracture systems show a unique setting that may be indicative of internal stresses acting on the whole body but leading to surface

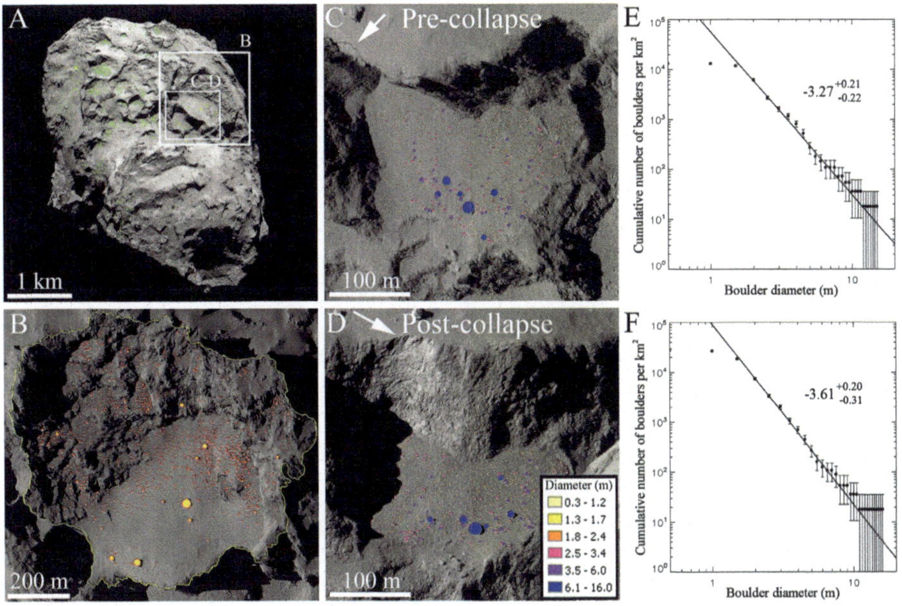


**Fig. 10** Fig. 1: (**A**) Boulders ≥7 m identified on specific areas on 67P (Pajola et al. 2015). Insets show location covered by panels (**B**)–(**D**). (**B**) Spatial distribution of the boulders ≥2 m identified in the Aswan site (Pajola et al. 2016c). (**C**) Pre-collapse spatial distribution of boulders ≥1.5 m identified below the Aswan cliff. Arrow points to the section that would later detach. (**D**) Post-collapse spatial distribution of the boulders ≥1.5 m located on the new talus formed below the Aswan cliff. Arrow points to location of detachment. (**E**) and (**F**) Cumulative number of boulders per $km^2$ identified in (**C**) and (**D**), respectively. The corresponding power-law indices of the fitting curves are also presented. Panels (**C**) and (**D**) adapted from Pajola et al. (2017b)

expressions at particular zones where stresses would be concentrated. The most prominent example of this process is the >500 m-long fracture system in the northern neck region (Thomas et al. 2015a). El-Maarry et al. (2017) reported that this fracture system had extended by ∼30 m in length at a time consistent with the period when the comet's orbital period decreased (i.e. its spin rate increased). Other fracture systems such as those observed in Hathor cliffs, the angular fracture observed in the Khepry region (El-Maarry et al. 2015a) and large fractures in the southern hemisphere close to the neck region and perpendicular to the comet's long axis (El-Maarry et al. 2016) could either be other manifestations of tectonic processes or the result of external collisions during 67P's past orbital evolution (El-Maarry et al. 2015a).

### 3.2 Non-consolidated Materials

As discussed earlier, non-consolidated materials (NCMs) are probably the heavier weathering products of consolidated materials that are initially ejected from the surface and then re-deposited predominantly in the northern hemisphere. They form two distinctive settings depending on topography, which controls the deposit thickness: thin (probably up to a few meters thick) deposits covering topographically high (with high gravitational potential) regions form the so-called dust layers (e.g., Thomas et al. 2015b). Areas with low gravitational potential allow NCMs to pond into thick deposits forming "smooth terrains" (e.g., Thomas et al. 2015b). While the NCMs forming both settings are expected to be of the same initial

**Fig. 11** Large-scale tectonic features on 67P. (**A**) Angular fracture system on the large lobe in the Aker region (El-Maarry et al. 2015a). (**B**) Long fracture system in the neck region, which was observed to lengthen during the course of the mission (El-Maarry et al. 2017). (**C**) Long hundreds-of-meters-long fractures in the small lobe's southern hemisphere in the Wosret region (El-Maarry et al. 2016)

composition (with respect to ice-to-dust ratio and concentration of organics), it is possible that their setting and thickness may affect the way they evolve and the degree to which they are affected by weathering and erosion. The NCMs display transient changes (potentially caused by weathering) that appear to be unique to comets. However, NCMs in particular locations also display evidence for erosion and re-mobilisation. In the case of NCMs, coma-surface interactions have arguably a bigger role in driving evolution than they do with consolidated materials.

### *3.2.1 Outgassing: Conventional Cometary Activity and Associated Features*

Historically, the term "activity" in cometary science was used exclusively to describe relatively bright (with respect to the nucleus and surrounding coma) sublimation-driven outgassing events that create and maintain a transient coma around the nucleus when the comet is sufficiently close to the Sun. Conventional cometary activity displays numerous morphologies ranging from highly collimated to diffuse, as well as a wide range of sizes depending partly on whether the features are resolved in close flyby/orbiting spacecraft images, or Earth-based observations (see Vincent et al., this volume). Vincent et al. (this volume) classify the classical form of cometary activity into perennial collimated streams, and transient plumes. The distinction is based essentially on their physical shape and temporal behavior irrespective of their size, source region, or sublimating species (e.g., $CO_2$ vs. $H_2O$). For this chapter, we are primarily interested in how such activity affects surface morphology and potentially drives its evolution. Cometary activity can act as both a depositional and erosional process. There is clear evidence for the activity's depositional after effects as evident

**Fig. 12** Outburst in the Imhotep region in July 2016 (adapted from Agarwal et al. 2017). (**A**) Bright outburst in the Imhotep region. (**B**) Close-up view of the event. (**C**) Area before the outburst in May 2016. Arrow points to the location of outburst that would occur nearly one month later. (**D**) View of the area a few weeks after the event. Note the bright patch of water–ice

from the ubiquitous distribution of NCM in the northern hemisphere and equatorial region. Materials that get entrained along with the energetic streams and plumes will either contribute to the comet's coma or in the case of heavier blocks, settle back on the surface to create the air-fall-like deposits (see Fig. 3). With respect to the activity's erosional effects, Rosetta's results have been more ambiguous. Two prominent examples that were reported are the dust plume in temporal association with the collapsing cliff of Aswan (Sect. 3.1.3, Pajola et al. 2017b), and the small outburst of activity reported by Agarwal et al. (2017) in the equatorial region of Imhotep (Fig. 12). While it is difficult to draw concrete conclusions from the cliff collapse since the available images do not possess adequate temporal resolution to verify whether the activity drove the cliff collapse or if the dust plume was a result of the cliff collapse, it is possible to infer from the outburst in Imhotep more information on how activity affects the surface. In this particular example, an outburst of activity in 3 July, 2016 lasted for 14–68 min in the Imhotep region, and left behind an ice patch of approx. 10 meters in size (Fig. 12, Agarwal et al. 2017). The resulting morphology would suggest that such outbursts result in the erosion of surface materials and exhumation of near-surface ice. Similarly, bright features were also apparent in another equatorial region, Khonsu, after a series of outbursts in the region (Deshapriya et al. 2016), which may suggest a similar mechanism despite lacking a similar temporally resolved dataset like that of Imhotep. In addition, the movement of a large 30 m-wide boulder in the Khonsu region may have been triggered by erosion from such events (El-Maarry et al. 2017). In short, conventional activity on comets appears to act as both an eroding and transporting agent removing materials from their source location and transporting them into the coma. Some materials would eventually resettle in areas of low gravitational potential. Other forms of erosion are discussed in a later section. As materials get ejected by streams and plumes and later are re-deposited, they undergo further evolution, which is discussed next.

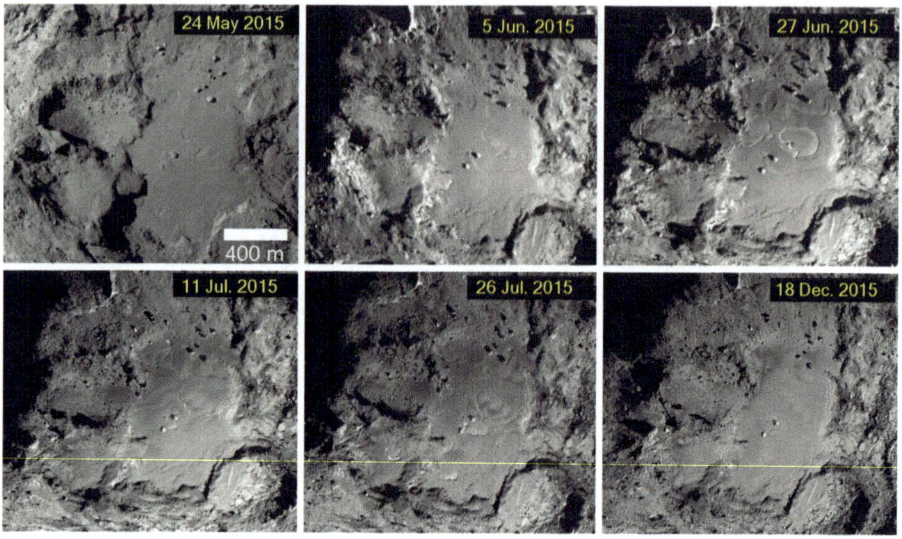


**Fig. 13** Transient surface changes in the smooth terrains of Imhotep first reported in Groussin et al. (2015). Figure adapted from El-Maarry et al. (2017)

### *3.2.2 Transient Surface Changes*

The NCM covers most of the northern hemisphere, and consists predominantly of re-deposited ejected materials. NCM evolved in various ways throughout the mission depending on their setting. Smooth terrains are located in the northern hemisphere, particularly within gravitational lows in the Imhotep, Anubis, and Hapi regions (Fig. 4; Sect. 2.2). An in-depth analysis of the smooth terrain at Imhotep indicates that the terrains have low gravitational slopes (<5 degrees) (Auger et al. 2015). The deposits are mainly composed of grains/blocks of variable sizes with larger grains reaching tens of cm in width (Auger et al. 2015). The smooth terrains of Imhotep displayed one of the earliest remarkable transient surface changes (Fig. 13, Groussin et al. 2015). In early June 2015, nearly two months before perihelion, roundish features appeared in the smooth terrain and expanded in nearly circular fashion in the following days. The features would eventually reach 100 s of meters in width and few meters in depth. Groussin et al. (2015) calculated expansion rates ranging from 4.3 to 7.6 m/day for two of such features, and noted that these rates were constant between observations. These rates are more than one order of magnitude faster than the erosion rate of pure $H_2O$ or $CO_2$ ice (Groussin et al. 2015), which suggests that erosion may not be occurring by water–ice sublimation. Within a few months, almost all the Imhotep smooth terrains (nearly 1 $km^2$, Auger et al. 2015) had been affected by these changes. If we consider an average height of mobilized materials of 5 m (Groussin et al. 2015), and an average density of 532 $kg/m^3$ (Jorda et al. 2016), this would correspond to roughly $2.7 \times 10^9$ kg of mobilized materials. This value is comparable to the total amount of water released by the comet in a single orbit (Bertaux 2015), which suggests that most of the materials mobilized during these transient changes did not escape the nucleus.

The transient changes in the Hapi region were morphologically similar to those observed at Imhotep, although they started 5 months before, at 2.6 a.u. from the Sun instead of 1.6 a.u. in Imhotep's case (El-Maarry et al. 2017). They also involve development of roughly circular depressions, expanding with time at a rate incompatible with the erosion rate of

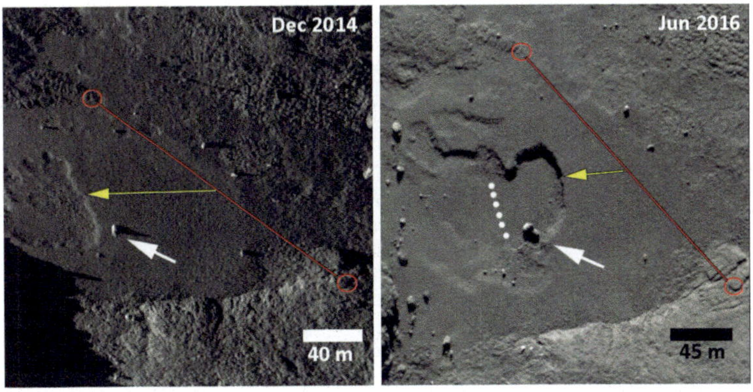


**Fig. 14** Erosion by scarp retreat in the Anubis region. The scarp shown has retreated by approx. 50 m in total (yellow arrow). A boulder is shown (white arrow), which may have been dislodged form its original location through the erosion. Figure is adapted from El-Maarry et al. (2017)

pure $H_2O$ or $CO_2$ ice at that heliocentric distance. One of the most prominent features that developed in the Hapi region was a circular feature that grew to nearly 100 m in width from April to July 2015. During its growth, its scarps displayed brightening suggestive of exposed water–ice (El-Maarry et al. 2017). However, when changes had stopped, the smooth terrains had nearly returned to its initial form prior to the changes. One possible reason for this would be deposition/ponding of new NCMs ejected from the southern hemisphere around perihelion (El-Maarry et al. 2017). El-Maarry et al. (2017) reported other changes in the smooth terrains of Anubis (Fig. 14). However, in the Anubis region, scarp retreats (also displaying brightening during their evolution) were observed in pre-existing shallow depressions, reaching 50 m (Fig. 14, El-Maarry et al. 2017), which were reminiscent of similar changes with respect to morphology, scale and setting, at comet Tempel 1 (e.g., Thomas et al. 2013a).

With regards to thinner deposits of NCM such as those in the Ma'at region, the nature of the transient changes in them was different from that in the smooth terrains. These changes occurred in the form of modification of the texture of the terrain, which became rougher during the period of changes (few months), forming "honeycomb" structures (Hu et al. 2017; Shi et al. 2016, Fig. 15). The eroded areas measure typically a few tens to a hundred meters and are all located in the northern hemisphere, between latitudes 20 °N and 40 °N. A digital terrain model for one of these patches indicated vertical erosion of 0.5 m on average (Hu et al. 2017). The dependency on solar insolation is supported by the fact that the latitude of these changes coincided temporally with sub-solar latitude over the surface (Hu et al. 2017). However, there is no sign of bluing (i.e. exposed ice) during the changes.

### *3.2.3 Aeolian-Like Processes and Associated Features*

The unambiguous observation of aeolian-like features at the surface of 67P was somewhat unexpected. The comet, and particularly its northern hemisphere, contains a substantial amount of NCMs, which in theory could be mobilized in a low gravity regime. However, comets display a very low gas density compared to planetary objects which possess such features, and most of the gases escape radially (Kuehrt and Keller 1994). Nevertheless, mechanisms of grain transport and aeolian erosion occurring at the surface of comets were

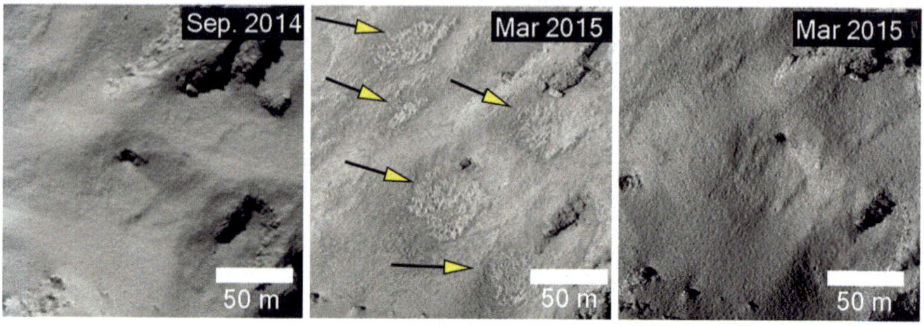


**Fig. 15** "Honey-comb" structures in the thin NCM coatings in the Ma'at region. Figure adapted from El-Maarry et al. (2017)

previously considered prior to the Rosetta mission, and evaluated by analogy with study on classical aeolian processes (Cheng et al. 2013).

The most striking examples of aeolian-like landforms on comet 67P are the ripples on the surface of the NCMs covering the neck region of Hapi, first identified in 2014 shortly following the arrival of Rosetta (Fig. 16, Thomas et al. 2015b). At that time, the length of the ripples field was ~100 m with a width of ~60 m (Thomas et al. 2015a). The bedforms had characteristic asymmetric profiles with small steep lee side and a longer and gentle slope on the stoss side (Jia et al. 2017; Thomas et al. 2015a). The longest axis of crests was roughly aligned E-W (Giacomini et al. 2016) with a mean wavelength of ~7–18 m for different bedforms (Thomas et al. 2015a, 2015b, Jia et al. 2017). The amplitude of the larger ripples was estimated to be $22 \pm 12$ cm (Thomas et al. 2015a). The aspect ratio, i.e. the ratio between the amplitude and the wavelength of the wave, lay between 0.02 and 0.04 (Thomas et al. 2015a) which is the order of magnitude of aspect ratios of ripples and dunes on other planets (e.g., Ellwood et al. 1975). As the comet approached perihelion these features disappeared (around April 2015) and circular pits developed instead, and extended (El-Maarry et al. 2017). After perihelion, the circular pits vanished and a new set of ripples developed at the same place of the previous ones (Fig. 16). The wavelength of the new ripples field is on average ~16–20 m (Jia et al. 2017). The formation of these ripples has been attributed so far to either a "splash" effect of dust grain due to dust air-fall (Thomas et al. 2015a), or saltation/reptation of grains due to lateral outgassing at the surface produced by pressure differences due to varying illumination conditions (Cheng et al. 2013; Thomas et al. 2015a; Jia et al. 2017). Other ripple formations have also been observed in other locations such as in Maftet and Ma'at (Jia et al. 2017).

Other candidate aeolian features were observed on the surface of the comet on both lobes (La Forgia et al. 2015; Giacomini et al. 2016; Hu et al. 2017; Jia et al. 2017). On the small lobe, dust deposits appear to have experienced dust mobilization/redistribution, especially in the regions of Ma'at, Nut, Maftet and Hatmehit (La Forgia et al. 2015; Giacomini et al. 2016; Keller et al. 2017; Thomas et al. 2015a). In these regions dune-like features formed on flat areas ($<10°$ of gravitational slope) covered by smooth material composed of centimeters scale grains (Jia et al. 2017; Mottola et al. 2015) that exhibit local heterogeneities (Thomas et al. 2015a). Aeolian-like bedforms are arranged in fields showing a preferential and consistent orientation of their longest axis through the whole area (La Forgia et al. 2015; Thomas et al. 2015a; Giacomini et al. 2016). They were interpreted to be analogous to longitudinal dunes (La Forgia et al. 2015; Thomas et al. 2015a; Giacomini et al. 2016).

Finally, more than 60 wind-tail like features and moats have been observed around large ($>5$ m) boulders located among the NCM deposits (Mottola et al. 2015; Thomas et al.

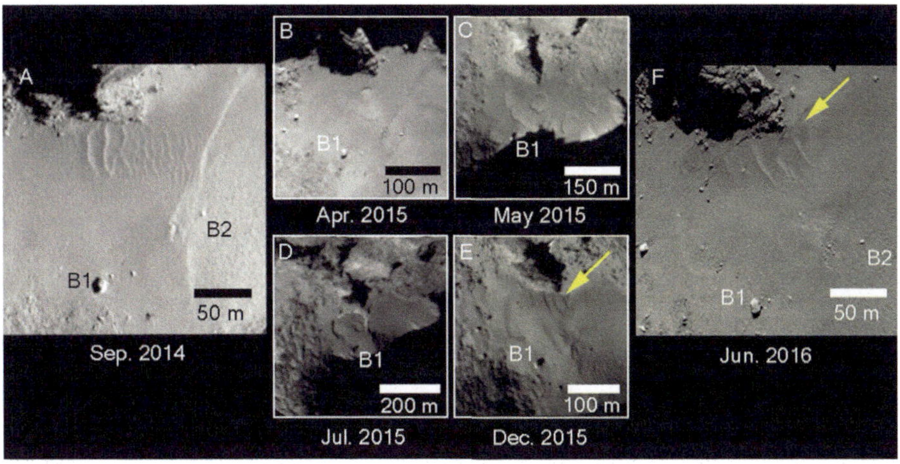


**Fig. 16** Evolution of the ripples (**A**) in the Hapi (northern neck) region of 67P. Surface modifications in the form of developing circular features (**B**)–(**D**) started in April 2015 at the earliest and continued through the perihelion passage (in Aug. 2015). New ripples developed in Dec. 2015 at the earliest (**E**), (**F**). Two prominent boulders (B1 and B2) are marked for orientation. Figure adapted from El-Maarry et al. (2017)

**Fig. 17** Wind-tail-like features on 67P. Figure adapted from Tirsch et al. (2017)

2015a,2015b; Tirsch et al. 2017, Fig. 17). These features appear as triangular shaped aprons of fine debris (Mottola et al. 2015). In the Ma'at region, the wind-tail lengths on Ma'at (near the initial Philae landing site) were reported to be 5–30 m in length, with a maximum height of approximately one third of the boulder height with the wind-tails generally pointing in the south-north direction (Mottola et al. 2015). Follow up analysis by Tirsch et al. (2017) of nearly 60 wind-tail features yields an average length of 13.7 m. Despite the apparent abundance of wind-tail-like features, many NCM-covered areas on 67P with an abundance of boulders do not show any aeolian-like features, suggesting that the processes that form these features are spatially constrained.

### *3.2.4 Localised Erosion and Exhumation*

Certain locations in the smooth terrains (i.e. accumulated NCMs) display evidence for higher than average vertical erosion that reaches a few meters. Examples of this localized erosion include the scarp retreats in the Anubis region (Sect. 3.2.2, Fig. 14) and the circular features that developed in the Hapi region that led to the destruction and reformation

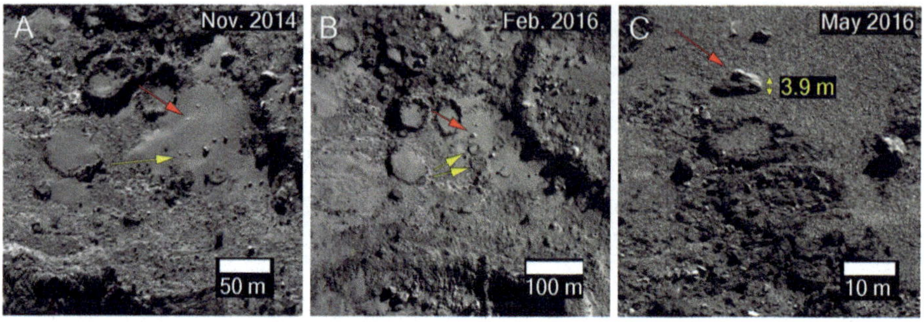


**Fig. 18** Erosion in the Imhotep region. A thin coating of NCM (**A**) was removed (mainly around perihelion) leading to the exhumation of previously covered circular features (yellow arrows), and partially buried boulders (**B**). One of such boulders (red arrow in all panels) was used to estimate the vertical extent of erosion in that particular location by measuring the height of its previously buried section (**C**). Figure adapted from El-Maarry et al. (2017)

of aeolian-like features (Sect. 3.2.3, Fig. 16). In Anubis, the vertical erosion resulted in the widening of a previously existing shallow irregular depression by scarp retreat (Fig. 14, El-Maarry et al. 2017). The depressions that have widened by ~50 meters have a depth of 4–5 meters (El-Maarry et al. 2017).

A similarly remarkable event occurred in the Imhotep region where a particular location underwent localised erosion reaching ~4 meters vertically (El-Maarry et al. 2017, Fig. 18). This intensive erosion resulted in the exhumation of previously covered circular structures and boulders (El-Maarry et al. 2017). The extent of vertical erosion in Imhotep appears to be consistent with what was observed in Anubis. In addition, the scarp retreats in Anubis and Hapi were observed to initiate with brightening of receding scarps, which would indicate the exposure of water ice (El-Maarry et al. 2017), which along with the consistent depth (4–5 m in Anubis and Imhotep) may indicate the presence a certain thermophysical boundary on the vertical extent of this type of erosion.

## 3.3 Circular Features and Possible Role of Impacts

Impact cratering is one of the fundamental geologic processes that act on planetary bodies. A discussion of the possible role of impacts/collisions in *forming* comet 67P (e.g., Jutzi and Asphaug 2015; Jutzi and Benz 2017; Rickman et al. 2015) is beyond the scope of this chapter as we merely discuss here surface features that might be impact-related. Circular depressions are a common morphological feature on solid surfaces. Comets are no exception and all nuclei visited by space missions have displayed crater-like features. On planets and asteroids, the large majority of circular depressions results from impacts (except for bodies that are large, and internally hot enough, to drive volcanism), and these craters reflect the collisional history of the body. On objects where active erosion takes place, some of these depressions can be associated to the formation of subterranean cavities and subsequent collapse.

Prior to Rosetta, Vincent et al. (2015b) reviewed the observations of crater-like features on cometary nuclei. They found that these morphologies appear as two main types: pit-halo (circular pit surrounded by a shallow depression) and deep pits (flat-floor features circled by steep walls, no rim). Both types display talus close to the pit's walls, indicating of post-formation wall collapse. The Rosetta mission returned similar observations and provided high resolution views of pits of ranging from a few meters to several hundred meters in

**Fig. 19** Circular features on 67P. (**A**) Cluster of circular features in the Imhotep region. Note the patch of NCM to the southeast shown previously in Fig. 18 where additional circular features were exhumed. The apparent clustering of these features argues against an impact origin. (**B**) One of the circular pits first reported by Vincent et al. (2015b). Vincent et al. (2015b) argued against an impact origin for these pits and favored a sinkhole collapse-like mechanism. (**C**) A candidate impact feature first reported by Thomas et al. (2015b) that either impacted the NCM or was covered by it. All images were acquired in Sep. 2014 and are adapted from El-Maarry et al. (2015a)

diameter. A particular set of deep circular pits was reported by Vincent et al. (2015a). These pits varied in size but exhibited clustering, and depth/diameter ratios of $\sim$0.78 and $\sim$0.26 for active and inactive pits, respectively (Vincent et al. 2015b).

While circular depressions on comets are well characterized, their formation mechanism remains challenging to establish. The Deep Impact mission (A'Hearn et al. 2005) and modeling experiments (e.g., Vincent et al. 2015b) have shown that similar morphologies can be created by impacts. Yet, the size and apparent clustering of circular features suggests that alternative mechanisms play a role. The current population of small impactors potentially encountered by a Jupiter Family Comet on its orbit does not match the size frequency distribution (SFD) of observed depressions (Vincent et al. 2015b). It has been argued that erosion, being much faster on comets than other bodies due to the sublimation of a major component of the object, is responsible for enlarging the pits and thus modifying the SFD. This is not confirmed by observations. Only a handful of mass wasting events over one orbit of comet 67P has been observed by Rosetta, and no preexisting deep pits were observed to grow.

In short, comet 67P is devoid of unambiguous impact features. Currently, the best candidate is a 35 m-wide bowl-shaped depression, which may have been covered by NCMs in the Ash region (Thomas et al. 2015b, Fig. 19). Any potentially new impacts would probably constitute micro-impacts primarily caused by the NCM back fall. However, these would be either too difficult to detect on consolidated materials, or would be promptly eroded away or covered by newly depositing NCMs.

## 4 Forces Driving the Active Processes

Having reviewed the setting and evolution of the numerous features on 67P, as well as the processes most likely to drive them, we briefly discuss here the main driving forces for the key processes that act on 67P specifically, and probably all active periodical comets in general.

## 4.1 Gravity and Shape Driven Internal Stresses

Gravity on 67P, while only $\sim 10^{-5}$ of Earth's, plays a main role in controlling the distribution of mass-wasted events, as well as the placement of NCMs that are deposited from the coma on the northern hemisphere. In addition, 67P's irregular bi-lobed shape complicates its gravitational field and creates zones where internal stresses may cause tectonic events. For instance, a long (>500 m-long) fracture system is located in the northern neck region (and may even extend further below the NCM cover in the Hapi region, Fig. 11B). This fracture was observed to lengthen by tens of metres, in addition to the possible development of new parallel fractures (El-Maarry et al. 2017). Numerical models reported in Hirabayashi et al. (2016) suggested that outgassing on 67P would generate torques that affect the rotational period of the comet, and that 67P's shape should lead to a concentration of these stresses in the neck region to levels that may lead to splitting and re-merging of the lobes. Indeed, the timing of the neck fracture growth on 67P was consistent with the time when 67P's rotation period changed (El-Maarry et al. 2017). A recent study by Matonti et al. (2019) suggests that bilobate comets such as 67P are subjected to shear deformation caused by shape-driven stresses that could lead to the formation of shear-fracture and fault networks both on the surface and in its interior. The study further suggests that certain fractures and ridges in 67P's southern hemisphere, particularly in the Wosret (large lobe, e.g., Fig. 11C) and Sobek (southern neck) regions are indicative of such stresses.

## 4.2 Seasonal Effects of Illumination and Orbital Configuration

67P orbits the sun on an ellipse with a perihelion distance of 1.2432 au and an aphelion of 5.6824 au with an inclination of 7.0402°, typical for Jupiter Family comets. The rotation period on the approach leg to the Sun was $\sim$12.4 h decreasing in the vicinity of perihelion due to non-gravitational forces (Keller et al. 2015, see Sect. 4.3). The large obliquity of 52° of the rotational axis of the nucleus combined with the close coincidence of the southern solstice with the perihelion passage leads to strongly varying seasonal insolation. Near perihelion, regions at high northern latitudes are not illuminated at all, while the total insolation of the comet reaches its maximum. This variation of the thermal input determines the seasonal activity pattern. The peculiar bi-lobate shape of 67P with a strong constriction between the lobes in the neck region (that forms a large concavity) strongly influences the diurnal insolation. Particularly near the north pole, the shadow of the steep and high rim wall of the small lobe (Fig. 20) is responsible for an additional interruption of the insolation during the day time of each rotation cycle. The polar day on southern latitudes of the nucleus near southern solstice and the extreme insolation changes caused by the wandering shadows provide an environment with strong thermal gradients and resulting stresses.

First estimates of the erosion by activity caused by water ice sublimation were calculated using a two-layer model by Keller et al. (2015) testing a range of different parameters. These models demonstrated that erosion averaged over the southern hemisphere was $\sim$15 m (Keller et al. 2015). These calculations highlighted the maximum capability of erosion by water ice sublimation. The models showed that the erosion in the southern hemisphere was about 4 times stronger than on the northern hemisphere. The erosion over one spin period (1 comet day) varied between a fraction of a millimeter on the northern hemisphere, when the comet was about 3 au away from the sun before perihelion, to several centimeters on the southern hemisphere around perihelion.

It should be noted that the models of Keller et al. (2015) assumed that the surface of the nucleus has homogeneous properties. The cometary activity creates torques that influence the rotation period of the nucleus. While it was possible to predict the variation of

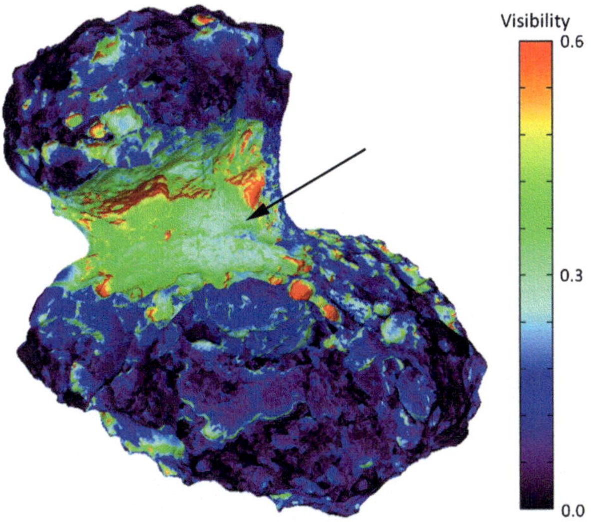


**Fig. 20** Viewing factors of insolation model for 67P by Keller et al. (2015). This model shows the degree to which each single facet of the shape model is exposed to irradiation from other facets (0 implying none, 1 implying that the entire field of view of a facet is filled by other facets. Note that the concavity of the northern neck (arrow) affects insolation patterns greatly and results in excessive self-heating in the northern neck

the rotation period over the perihelion passage (Keller et al. 2015) based on these simple models more thorough investigations showed that a fully homogeneous surface activity cannot explain all the details (Keller et al. 2017). The pronounced imbalance of the insolation, activity, and erosion between the south and the north hemispheres, and the observation that the northern hemisphere is widely covered by NCMs, suggests that mass is transferred from the south to the north. Observations with OSIRIS show that cm and dm sized particles are liberated from active surfaces on the south around perihelion (Fulle et al. 2016; Ott et al. 2017). The trajectories of some of these particles could be determined. At least 20% of them (and possibly more) return directly back to the nucleus surface (Agarwal et al. 2016; Keller et al. 2017). These chunks themselves show activity that leads to additional (non-gravitational) acceleration in anti-solar direction towards the surface. Modelling work (e.g., Keller et al. 2017) indicates that ejected dust particles in the cm to dm size range would be volatile rich; predominantly water ice-rich due to the higher volatility of other species such $CO_2$. The ability of the dust grains to retain water ice but not the more volatile CO or $CO_2$ is consistent with the fact that CO and $CO_2$ in 67P's coma were predominantly sourced from the southern hemisphere during the entire period of the Rosetta mission (Laeuter et al. 2018).

The northern neck region Hapi is in polar night during the perihelion passage. When the particles enter the seasonal shadow, they are "shock frozen" and maintain all their water ice content. This rapid freezing contributes to the exceptional activity level observed at Hapi when the comet returns from its aphelion, similar to what was observed early in the mission (e.g., Lara et al. 2015). If more than 25% of the lifted mass from the south falls back on the northern hemisphere, the back fall will accumulate orbit by orbit. In any case, the fall back in certain regions will accumulate because it cannot be removed entirely by the water sublimation-induced activity during northern summer.

In summary, the strong seasonal variation with an overwhelming activity level during southern summer near perihelion strongly influences the current evolution of the nucleus. Most of the erosion of the consolidated nucleus takes place on the south driven by water, $CO_2$, and CO sublimation, which is consistent with the observed hemispherical dichotomy in surface morphology.

### 4.3 Insolation-Driven Thermal Stresses

Cometary nuclei are subject to strong diurnal and seasonal surface and sub-surface temperature variations, which result from a combination of their elliptic orbit, fast rotation (i.e., few hours) (Lamy et al. 2004) and low thermal inertia (Groussin et al., this volume). In the case of 67P, these changes can be even more pronounced given its shape and orbit and how they affect the insolation patterns as discussed in Sect. 4.1. The distance of the comet to the Sun during its elliptical orbit also plays a major role. On a diurnal scale, temperature may vary by more than 200 K in a few hours during perihelion (El-Maarry et al. 2015a), and in regions with large projected shadows like Hapi, the temperature gradient can reach 30 K/min (Ali-Lagoa et al. 2015). These large and rapid temperature variations could cause severe thermal stresses on cometary materials. Indeed, the strongest case is offered by the detection of polygonal networks on the surface of 67P (see Sect. 3.1.2). By analogy with similar morphologies found on the Earth and Mars, Auger et al. (2018) interpreted them as thermal contraction crack polygons, resulting from the thermal stress of hard icy materials located in the sub-surface. Near subsurface water ice is expected to expand and contract due to temperature variations, which are large and fast enough to cause thermal stress that exceed the tensile stress of the surface (e.g., Lachenbruch 1962).

The presence of a hard (MPa) layer of water ice below the surface is supported by Philae observations, laboratory experiments and modeling. For instance, Spohn et al. (2015) reported on the first results from the "Multi Purpose Sensors for Surface and Subsurface Science" (MUPUS) instrument on Philae, and concluded that the uniaxial compressive strength of the sub-surface material is larger than 2 MPa. Knapmeyer et al. (2018) utilized the Cometary Acoustic Surface Sounding Experiment (SESAME/CASSE) observations to infer the presence of a hard layer (Young's modulus $E$ of 7.2 MPa $\leq E \leq$ 980 MPa) located 10–50 cm below surface. These findings are consistent with a processed water ice layer of a few centimeters to decimeters thick, which results from the sublimation/redeposition and/or sintering of water ice close to the surface. This hardening process has been studied and demonstrated for comets, both experimentally (Jessberger and Kotthaus 1989; Pommerol et al. 2015a) and theoretically (Kossacki et al. 2015), and leads to typical strengths of 1 MPa, much harder than the nucleus bulk tensile strength of $<100$ Pa (Groussin et al., this volume).

A complete thermo-mechanical model was developed by Attree et al. (2018) to study the stress induced by the diurnal temperature variations in such a hard water ice layer. The stress can reach tens of MPa in the first meter, exceeding the tensile strength of the processed icy layer at all latitudes, assuming a thermal inertia $>50\ \mathrm{J\,m^{-2}\,K^{-1}\,s^{0.5}}$ and an ice content $>45\%$. These results agree with an older analysis made by Kuehrt and Keller (1994) with a less elaborated thermal model. All the above arguments support the formation of the observed polygonal networks by thermal stress.

While polygons are easily recognizable as surface morphologies resulting from thermal stress, there are not unique examples. Many fractures, observed at all scales on the nucleus (Sect. 3.2), may also be caused by thermal stress. Their identification is however difficult because a mechanical stress can also be the origin of these fractures. It is nevertheless likely that the fractures observed on boulders, at the meter scale, are of thermal stress origin, as suggested for boulders on the Moon and airless bodies in general (e.g., Molaro et al. 2017). On 67P, because of the low thermal inertia, boulders present a strong temperature gradient between the side facing the Sun and the opposite one, making them a perfect case for a strong thermal stress leading to fracturing and breakdown. Numerous fractured boulders have been observed on the surface (Fig. 21).

**Fig. 21** A fractured boulder in the Hatmehit region. The weathering of this boulder has probably been caused by thermal stresses

It is also highly probable that opening new fractures leads to activity/outgassing since it exposes fresh, icy materials to sunlight. A study by Skorov et al. (2016) suggests that if the new fractures are shallow ($<1$ m) and expose water ice, the increase in activity can be negligible compared to the global activity. On the contrary, if the new fractures are deep ($>1$ m) and expose highly volatiles species like CO or $CO_2$ to sunlight, the increase in activity can be important, in the form of outbursts lasting for a few minutes to hours (Skorov et al. 2016).

## 5 Concluding Remarks

Historically, the term "activity" in cometary science has generally been associated with the outgassing events and the production of a coma (essentially what could be observable by Earth-based instruments). What space missions achieved, culminating with Rosetta, was to widen our perspective on the activity of comets. Rosetta in particular has demonstrated that when comets become active, they behave like a typical geologically active solar system body. Most of the processes and resulting morphological changes that have been discussed here occurred mainly when the comet was 2–3 a.u. away from the Sun, and preferentially in locations at or close to the sub-solar latitude (e.g., El-Maarry et al. 2017; Vincent et al. 2016). It is clear that solar insolation is the main energy input that feeds most of the activity. Its intensity directly controls the scale of erosion and outgassing activity that occurs (e.g., Keller et al. 2015, 2017), whereas its derivative/gradient (affected by distance from the sun and comet geometry) dictates the magnitude of thermal stresses that act to disintegrate consolidated materials through fracturing, followed by mass wasting (e.g., El-Maarry et al. 2015a; Pajola et al. 2017b).

With regards to cometary evolution, Belton (2010) had proposed an evolutionary sequence based on the diverse surface morphologies of visited comets where comets would evolve from a rougher Wild 2-type-like morphology to Tempel 1-like, predicting that Hartley 2 and 67P would exhibit surface morphologies similar to Tempel 1. The diversity in terrain types observed at Hartley 2 and 67P showed that comet nucleus evolution follows a more complicated path. Cheng et al. (2013) proposed an alternative evolutionary sequence where geologic activity with diverse outcomes would be responsible for the different morphologies depending on the intensity of the process (specifically, vapor outflow) driving the activity. The global spatial and temporal coverage of Rosetta allowed us to gain more understating of cometary activity that is consistent with the general ideas proposed by both Belton (2010) and Cheng et al. (2013). Comet 67P shows diverse morphologies ranging from rough terrains and high cliffs to smooth and flat terrains. It also displays evidence for many active geological processes. As such, comets may be evolving in the same way that any active planetary body would: Outgassing and tectonic events act to roughen the surface whereas active erosion (including weathering and transport) progressively smoothens it, and it's the intensity of these processes that controls surface diversity. The flattening out of 67P's southern hemisphere where higher rates of erosion are taking place (Keller et al. 2017) and statistical analyses of cliff heights for comet 67P (Vincent et al. 2017) are consistent with this evolutionary view for 67P, and probably all active comets in general.

**Acknowledgements** This project has received funding from the European Union's Horizon 2020 research and innovation programme under grant agreement No. 686709 (MiARD). This work was supported by the Swiss State Secretariat for Education, Research and Innovation (SERI) under contract number 16.0008-2. The opinions expressed and arguments employed herein do not necessarily reflect the official view of the Swiss Government.



## References

J. Agarwal, M.F. A'Hearn, J.-B. Vincent, C. Güttler, S. Höfner, H. Sierks, C. Tubiana, C. Barbieri, P.L. Lamy, R. Rodrigo, D. Koschny, H. Rickman, M.A. Barucci, J.-L. Bertaux, I. Bertini, S. Boudreault, G. Cremonese, V. Da Deppo, B. Davidsson, S. Debei, M. De Cecco, J. Deller, S. Fornasier, M. Fulle, A. Gicquel, O. Groussin, P.J. Gutiérrez, M. Hofmann, S.F. Hviid, W.-H. Ip, L. Jorda, H.U. Keller, J. Knollenberg, J.-R. Kramm, E. Kührt, M. Küppers, L.M. Lara, M. Lazzarin, J.J. Lopez Moreno, F. Marzari, G. Naletto, N. Oklay, X. Shi, N. Thomas, Mon. Not. R. Astron. Soc. **462**, S78 (2016)

J. Agarwal, V. Della Corte, P.D. Feldman, B. Geiger, S. Merouane, I. Bertini, D. Bodewits, S. Fornasier, E. Grün, P. Hasselmann, M. Hilchenbach, S. Höfner, S. Ivanovski, L. Kolokolova, M. Pajola, A. Rotundi, H. Sierks, A.J. Steffl, N. Thomas, M.F. A'Hearn, C. Barbieri, M.A. Barucci, J.L. Bertaux, S. Boudreault, G. Cremonese, V. Da Deppo, B. Davidsson, S. Debei, M. De Cecco, J.F. Deller, L.M. Feaga, H. Fischer, M. Fulle, A. Gicquel, O. Groussin, C. Güttler, P.J. Gutiérrez, M. Hofmann, K. Hornung, S.F. Hviid, W.H. Ip, L. Jorda, H.U. Keller, J. Kissel, J. Knollenberg, A. Koch, D. Koschny, J.R. Kramm, E. Kührt, M. Küppers, P.L. Lamy, Y. Langevin, L.M. Lara, M. Lazzarin, Z.Y. Lin, J.J. Lopez Moreno, S.C. Lowry, F. Marzari, S. Mottola, G. Naletto, N. Oklay, J.Wm. Parker, R. Rodrigo, J. Rynö, X. Shi, O. Stenzel, C. Tubiana, J.B. Vincent, H.A. Weaver, B. Zaprudin, Mon. Not. R. Astron. Soc. **469**, S606 (2017)

M.F. A'Hearn, M.J.S. Belton, W.A. Delamere, J. Kissel, K.P. Klaasen, L.A. McFadden, K.J. Meech, H.J. Melosh, P.H. Schultz, J.M. Sunshine, P.C. Thomas, J. Veverka, D.K. Yeomans, M.W. Baca, I. Busko, C.J. Crockett, S.M. Collins, M. Desnoyer, C.A. Eberhardy, C.M. Ernst, T.L. Farnham, L. Feaga, O. Groussin, D. Hampton, S.I. Ipatov, J-.Y. Li, D. Lindler, C.M. Lisse, N. Mastrodemos, W.M. Owen, J.E. Richardson, D.D. Wellnitz, R.L. White, Science **310**, 258 (2005)

M.F. A'Hearn, M.J.S. Belton, W.A. Delamere, L.M. Feaga, D. Hampton, J. Kissel, K.P. Klaasen, L.A. McFadden, K.J. Meech, H. Jay Melosh, P.H. Schultz, J.M. Sunshine, P.C. Thomas, J. Veverka, D.D. Wellnitz, D.K. Yeomans, S. Besse, D. Bodewits, T.J. Bowling, B.T. Carcich, S.M. Collins, T.L. Farnham, O. Groussin, B. Hermalyn, M.S. Kelley, J.Y. Li, D.J. Lindler, C.M. Lisse, S.A. McLaughlin, F. Merlin, S. Protopapa, J.E. Richardson, J.L. Williams, Science (80-) **332**, 1396 (2011)

V. Alí-Lagoa, M. Delbo', G. Libourel, Astrophys. J. Lett. **810**, L22 (2015)

N. Attree, O. Groussin, L. Jorda, Astron. Astrophys. **610**, 76 (2018)

A.-T. Auger, O. Groussin, L. Jorda, S. Bouley, R. Gaskell, P.L. Lamy, C. Capanna, N. Thomas, A. Pommerol, H. Sierks, C. Barbieri, R. Rodrigo, D. Koschny, H. Rickman, H.U. Keller, J. Agarwal, M.F. A'Hearn, M.A. Barucci, J.-L. Bertaux, I. Bertini, G. Cremonese, V. Da Deppo, B.J.R. Davidsson, S. Debei, M. De Cecco, M.R. El-Maarry, S. Fornasier, M. Fulle, P.J. Gutiérrez, C. Güttler, S.F. Hviid, W.-H. Ip, J. Knollenberg, J.-R. Kramm, E. Kührt, M. Küppers, F. La Forgia, L.M. Lara, M. Lazzarin, J.J. Lopez Moreno, S. Marchi, F. Marzari, M. Massironi, H. Michalik, G. Naletto, N. Oklay, M. Pajola, L. Sabau, C. Tubiana, J.-B. Vincent, K.-P. Wenzel, Astron. Astrophys. **583**, A35 (2015)

A.-T. Auger, O. Groussin, L. Jorda, M.R. El-Maarry, S. Bouley, A. Séjourné, R. Gaskell, C. Capanna, B. Davidsson, S. Marchi, S. Höfner, P.L. Lamy, H. Sierks, C. Barbieri, R. Rodrigo, D. Koschny, H. Rickman, H.U. Keller, J. Agarwal, M.F. A'Hearn, M.A. Barucci, J.-L. Bertaux, I. Bertini, G. Cremonese, V. Da Deppo, S. Debei, M. De Cecco, S. Fornasier, M. Fulle, P.J. Gutiérrez, C. Güttler, S. Hviid, W.-H. Ip, J. Knollenberg, J.-R. Kramm, E. Kührt, M. Küppers, L.M. Lara, M. Lazzarin, J.J. Lopez Moreno, F. Marzari, M. Massironi, H. Michalik, G. Naletto, N. Oklay, A. Pommerol, L. Sabau, N. Thomas, C. Tubiana, J.-B. Vincent, K.-P. Wenzel, Icarus **301**, 173 (2018)

A.T. Basilevsky, S.S. Krasilnikov, U. Mall, S.F.S. Hviid, Y.V. Skorov, H.U. Keller, Planet. Space Sci. **140**, 80 (2017)

M.J.S. Belton, Icarus **210**, 881 (2010)

J. Bennet, M. Donahue, N. Schneider, M. Voit, *The Cosmic Perspective*, 8th edn. (2017). Pearson

J. Bertaux, Astron. Astrophys. **38**, 1 (2015)

J.P. Bibring, Y. Langevin, J. Carter, P. Eng, B. Gondet, L. Jorda, S. Le Mouélic, S. Mottola, C. Pilorget, F. Poulet, M. Vincendon, Science (80-) **349**, 8 (2015)

S.P.D. Birch, Y. Tang, A.G. Hayes, R.L. Kirk, D. Bodewits, H. Campins, Y. Fernandez, R. de Freitas Bart, N.W. Kutsop, H. Sierks, J.M. Soderblom, S.W. Squyres, J.-B. Vincent, Mon. Not. R. Astron. Soc. **469**, S50 (2017)

D.E. Brownlee, F. Horz, R.L. Newburn, M. Zolensky, T.C. Duxbury, S. Sandford, Z. Sekanina, P. Tsou, M.S. Hanner, B.C. Clark, S.F. Green, J. Kissel, Science (80-) **304**, 1764 (2004)

M. Bruck Syal, P.H. Schultz, J.M. Sunshine, M.F. A'Hearn, T.L. Farnham, D.S.P. Dearborn, Icarus **222**, 610 (2013)

A.F. Cheng, C.M. Lisse, M. A'Hearn, Icarus **222**, 808 (2013)

J.D.P. Deshapriya, M.A. Barucci, S. Fornasier, C. Feller, P.H. Hasselmann, H. Sierks, M.R. El-Maarry, M. Pajola, C. Barbieri, P.L. Lamy, R. Rodrigo, D. Koschny, H. Rickman, J. Agarwal, M.F. A'Hearn, J.L. Bertaux, I. Bertini, S. Boudreault, G. Cremonese, V. Da Deppo, B.J.R. Davidsson, S. Debei, J. Deller, M. De Cecco, M. Fulle, A. Gicquel, O. Groussin, P.J. Gutierrez, C. Güttler, M. Hofmann, S.F. Hviid, W. Ip, L. Jorda, H.U. Keller, J. Knollenberg, R. Kramm, E. Kührt, M. Küppers, L. Lara, M. Lazzarin, J.J. Lopez Moreno, F. Marzari, S. Mottola, G. Naletto, N. Oklay, D. Perna, A. Pommerol, N. Thomas, C. Tubiana, J.B. Vincent, Mon. Not. R. Astron. Soc. **462**, S274 (2016)

M.R. El-Maarry, N. Thomas, A. Gracia-Berná, R. Marschall, A.-T. Auger, O. Groussin, S. Mottola, M. Pajola, M. Massironi, S. Marchi, S. Höfner, F. Preusker, F. Scholten, L. Jorda, E. Kührt, H.U. Keller, H. Sierks, M.F. A'Hearn, C. Barbieri, M.A. Barucci, J.-L. Bertaux, I. Bertini, G. Cremonese, V. Da Deppo, B. Davidsson, S. Debei, M. De Cecco, J. Deller, C. Güttler, S. Fornasier, M. Fulle, P.J. Gutierrez, M. Hofmann, S.F. Hviid, W.-H. Ip, J. Knollenberg, D. Koschny, G. Kovacs, J.-R. Kramm, M. Küppers, P.L. Lamy, L.M. Lara, M. Lazzarin, J.J. Lopez Moreno, F. Marzari, H. Michalik, G. Naletto, N. Oklay, A. Pommerol, H. Rickman, R. Rodrigo, C. Tubiana, J.-B. Vincent, Geophys. Res. Lett. **42**, 5170 (2015a)

M.R. El-Maarry, N. Thomas, L. Giacomini, M. Massironi, M. Pajola, R. Marschall, A. Gracia-Berná, H. Sierks, C. Barbieri, P.L. Lamy, R. Rodrigo, H. Rickman, D. Koschny, H.U. Keller, J. Agarwal, M.F. A'Hearn, A.T. Auger, M.A. Barucci, J.-L. Bertaux, I. Bertini, S. Besse, D. Bodewits, G. Cremonese, V. Da Deppo, B. Davidsson, M. De Cecco, S. Debei, C. Güttler, S. Fornasier, M. Fulle, O. Groussin, P.J. Gutiérrez, S.F. Hviid, W.-H. Ip, L. Jorda, J. Knollenberg, G. Kovacs, J.R. Kramm, E. Kuhrt, M. Küppers, F. La Forgia, L.M. Lara, M. Lazzarin, J.J. Lopez Moreno, S. Marchi, F. Marzari, H. Michalik, G. Naletto, N. Oklay, A. Pommerol, F. Preusker, F. Scholten, C. Tubiana, J.B. Vincent, Astron. Astrophys. **583**, A26 (2015b)

M.R. El-Maarry, N. Thomas, M. Pajola, M. Massironi, H. Sierks, C. Barbieri, P.L. Lamy, Astron. Astrophys. **593**, A110 (2016)

M.R. El-Maarry, O. Groussin, N. Thomas, M. Pajola, A.T. Auger, B. Davidsson, X. Hu, S.F. Hviid, J. Knollenberg, C. Güttler, C. Tubiana, S. Fornasier, C. Feller, P. Hasselmann, J.B. Vincent, H. Sierks, C. Barbieri, P. Lamy, R. Rodrigo, D. Koschny, H.U. Keller, H. Rickman, M.F. A'Hearn, M.A. Barucci, J.L. Bertaux, I. Bertini, S. Besse, D. Bodewits, G. Cremonese, V. Da Deppo, S. Debei, M. De Cecco, J. Deller, J.D.P. Deshapriya, M. Fulle, P.J. Gutierrez, M. Hofmann, W.H. Ip, L. Jorda, G. Kovacs, J.R. Kramm, E. Kührt, M. Küppers, L.M. Lara, M. Lazzarin, Z. Yi Lin, J.J. Lopez Moreno, S. Marchi, F. Marzari, S. Mottola, G. Naletto, N. Oklay, A. Pommerol, F. Preusker, F. Scholten, X. Shi, Science (80-) **355**, 1392 (2017)

J.M. Ellwood, P.D. Evans, I.G. Wilson, J. Sediment. Res. **45**, 554 (1975)

M. Fulle, F. Marzari, V. Della Corte, S. Fornasier, H. Sierks, A. Rotundi, C. Barbieri, P.L. Lamy, R. Rodrigo, D. Koschny, H. Rickman, H.U. Keller, J.J. López-Moreno, M. Accolla, J. Agarwal, M.F. A'Hearn, N. Altobelli, M.A. Barucci, J.-L. Bertaux, I. Bertini, D. Bodewits, E. Bussoletti, L. Colangeli, M. Cosi, G. Cremonese, J.-F. Crifo, V. Da Deppo, B. Davidsson, S. Debei, M. De Cecco, F. Esposito, M. Ferrari, F. Giovane, B. Gustafson, S.F. Green, O. Groussin, E. Grün, P. Gutierrez, C. Güttler, M.L. Herranz, S.F. Hviid, W. Ip, S.L. Ivanovski, J.M. Jerónimo, L. Jorda, J. Knollenberg, R. Kramm, E. Kührt, M. Küppers, L. Lara, M. Lazzarin, M.R. Leese, A.C. López-Jiménez, F. Lucarelli, E. Mazzotta Epifani, J.A.M. McDonnell, V. Mennella, A. Molina, R. Morales, F. Moreno, S. Mottola, G. Naletto, N. Oklay, J.L. Ortiz, E. Palomba, P. Palumbo, J.-M. Perrin, F.J.M. Rietmeijer, J. Rodríguez, R. Sordini, N. Thomas, C. Tubiana, J.-B. Vincent, P. Weissman, K.-P. Wenzel, V. Zakharov, J.C. Zarnecki, Astrophys. J. **821**, 19 (2016)

L. Giacomini, M. Massironi, M.R. El-Maarry, L. Penasa, M. Pajola, N. Thomas, S.C. Lowry, C. Barbieri, G. Cremonese, F. Ferri, G. Naletto, I. Bertini, F. La Forgia, M. Lazzarin, F. Marzari, H. Sierks, P.L. Lamy, R. Rodrigo, H. Rickman, D. Koschny, H.U. Keller, J. Agarwal, M.F. A'Hearn, A.T. Auger, M.A. Barucci, J.L. Bertaux, S. Besse, D. Bodewits, V. Da Deppo, B. Davidsson, M. De Cecco, S. Debei, S. Fornasier, M. Fulle, O. Groussin, P.J. Gutierrez, C. Güttler, S.F. Hviid, W.H. Ip, L. Jorda, J. Knollenberg, G. Kovacs, J.R. Kramm, E. Kührt, M. Küppers, L.M. Lara, J.J. Lopez Moreno, S. Magrin, H. Michalik, N. Oklay, A. Pommerol, F. Preusker, F. Scholten, C. Tubiana, J.B. Vincent, Mon. Not. R. Astron. Soc. **462**, S352 (2016)

O. Groussin, H. Sierks, C. Barbieri, P. Lamy, R. Rodrigo, D. Koschny, H. Rickman, H.U. Keller, M.F. A'Hearn, A.-T. Auger, M.A. Barucci, J.-L. Bertaux, I. Bertini, S. Besse, G. Cremonese, V. Da Deppo, B. Davidsson, S. Debei, M. De Cecco, M.R. El-Maarry, S. Fornasier, M. Fulle, P.J. Gutiérrez, C. Güttler, S. Hviid, W.-H. Ip, L. Jorda, J. Knollenberg, G. Kovacs, J.R. Kramm, E. Kührt, M. Küppers, L.M. Lara, M. Lazzarin, J.J. Lopez Moreno, S. Lowry, S. Marchi, F. Marzari, M. Massironi, S. Mottola, G. Naletto, N. Oklay, M. Pajola, A. Pommerol, N. Thomas, I. Toth, C. Tubiana, J.-B. Vincent, Astron. Astrophys. **583**, A36 (2015)

E. Grün, J. Gebhard, A. Bar-Nun, J. Benkhoff, H. Düren, G. Eich, R. Hische, W.F. Huebner, H.U. Keller, G. Klees, H. Kochan, G. Kölzer, H. Kroker, E. Kührt, P. Lämmerzahl, E. Lorenz, W.J. Markiewicz, D. Möhlmann, A. Oehler, J. Scholz, K.J. Seidensticker, K. Roessler, G. Schwehm, G. Steiner, K. Thiel, H. Thomas, J. Geophys. Res. **98**, 15091 (1993)

M. Hirabayashi, D.J. Scheeres, S.R. Chesley, S. Marchi, J.W. McMahon, J. Steckloff, S. Mottola, S.P. Naidu, T. Bowling, Nature **534**, 352 (2016)

X. Hu, X. Shi, H. Sierks, M. Fulle, J. Blum, H.U. Keller, E. Kührt, B. Davidsson, C. Güttler, B. Gundlach, M. Pajola, D. Bodewits, J.-B. Vincent, N. Oklay, M. Massironi, S. Fornasier, C. Tubiana, O. Groussin, S. Boudreault, S. Höfner, S. Mottola, C. Barbieri, P.L. Lamy, R. Rodrigo, D. Koschny, H. Rickman, M. A'Hearn, J. Agarwal, M.A. Barucci, J.-L. Bertaux, I. Bertini, G. Cremonese, V. Da Deppo, S. Debei, M. De Cecco, J. Deller, M.R. El-Maarry, A. Gicquel, P. Gutierrez-Marques, P.J. Gutiérrez, M. Hofmann, S.F. Hviid, W.-H. Ip, L. Jorda, J. Knollenberg, G. Kovacs, J.-R. Kramm, M. Küppers, L.M. Lara, M. Lazzarin, J.J. Lopez-Moreno, F. Marzari, G. Naletto, N. Thomas, Astron. Astrophys. **604**, A114 (2017)

H.L. Jessberger, M. Kotthaus, in *Phys. Mech. Cometary Mater*, ed. by J.J. Hunt, T.D. Guyenne (1989), pp. 141–146

P. Jia, B. Andreotti, P. Claudin, Proc. Natl. Acad. Sci. **114**, 2509 (2017)

L. Jorda, R. Gaskell, C. Capanna, S. Hviid, P. Lamy, J. Ďurech, G. Faury, O. Groussin, P. Gutiérrez, C. Jackman, S.J. Keihm, H.U. Keller, J. Knollenberg, E. Kührt, S. Marchi, S. Mottola, E. Palmer, F.P. Schloerb, H. Sierks, J.B. Vincent, M.F. A'Hearn, C. Barbieri, R. Rodrigo, D. Koschny, H. Rickman, M.A. Barucci, J.L. Bertaux, I. Bertini, G. Cremonese, V. Da Deppo, B. Davidsson, S. Debei, M. De Cecco, S. Fornasier, M. Fulle, C. Güttler, W.H. Ip, J.R. Kramm, M. Küppers, L.M. Lara, M. Lazzarin, J.J. Lopez Moreno, F. Marzari, G. Naletto, N. Oklay, N. Thomas, C. Tubiana, K.P. Wenzel, Icarus **277**, 257 (2016)

M. Jutzi, E. Asphaug, Science (80-) **348**, 1355 (2015)

M. Jutzi, W. Benz, Astron. Astrophys. **597**, A62 (2017)

H.U. Keller, W.J. Markiewicz, Geophys. Res. Lett. **18**, 249 (1991)

H.U. Keller, C. Arpigny, C. Barbieri, R.M. Bonnet, S. Cazes, M. Coradini, C.B. Cosmovici, W.A. Delamere, W.F. Huebner, D.W. Hughes, C. Jamar, D. Malaise, H.J. Reitsema, H.U. Schmidt, W.K.H. Schmidt, P. Seige, F.L. Whipple, K. Wilhelm, Nature **321**, 320 (1986)

H.U. Keller, C. Barbieri, P. Lamy, H. Rickman, R. Rodrigo, K.-P. Wenzel, H. Sierks, M.F. A'Hearn, F. Angrilli, M. Angulo, M.E. Bailey, P. Barthol, M.A. Barucci, J.-L. Bertaux, G. Bianchini, J.-L. Boit, V. Brown, J.A. Burns, I. Büttner, J.M. Castro, G. Cremonese, W. Curdt, V. Da Deppo, S. Debei, M. De Cecco, K. Dohlen, S. Fornasier, M. Fulle, D. Germerott, F. Gliem, G.P. Guizzo, S.F. Hviid, W.-H. Ip, L.

Jorda, D. Koschny, J.R. Kramm, E. Kührt, M. Küppers, L.M. Lara, A. Llebaria, A. López, A. López-Jimenez, J. López-Moreno, R. Meller, H. Michalik, M.D. Michelena, R. Müller, G. Naletto, A. Origné, G. Parzianello, M. Pertile, C. Quintana, R. Ragazzoni, P. Ramous, K.-U. Reiche, M. Reina, J. Rodríguez, G. Rousset, L. Sabau, A. Sanz, J.-P. Sivan, K. Stöckner, J. Tabero, U. Telljohann, N. Thomas, V. Timon, G. Tomasch, T. Wittrock, M. Zaccariotto, Space Sci. Rev. **128**, 433 (2007)

H.U. Keller, S. Mottola, B. Davidsson, S.E. Schröder, Y. Skorov, E. Kührt, O. Groussin, M. Pajola, S.F. Hviid, F. Preusker, F. Scholten, M.F. A'Hearn, H. Sierks, C. Barbieri, P. Lamy, R. Rodrigo, D. Koschny, H. Rickman, M.A. Barucci, J.-L. Bertaux, I. Bertini, G. Cremonese, V. Da Deppo, S. Debei, M. De Cecco, S. Fornasier, M. Fulle, P.J. Gutiérrez, W.-H. Ip, L. Jorda, J. Knollenberg, J.R. Kramm, M. Küppers, L.M. Lara, M. Lazzarin, J.J. Lopez Moreno, F. Marzari, H. Michalik, G. Naletto, L. Sabau, N. Thomas, J.-B. Vincent, K.-P. Wenzel, J. Agarwal, C. Güttler, N. Oklay, C. Tubiana, Astron. Astrophys. **583**, A34 (2015)

H.U. Keller, S. Mottola, S.F. Hviid, J. Agarwal, E. Kührt, Y. Skorov, K. Otto, J.B. Vincent, N. Oklay, S.E. Schröder, B. Davidsson, M. Pajola, X. Shi, D. Bodewits, I. Toth, F. Preusker, F. Scholten, H. Sierks, C. Barbieri, P. Lamy, R. Rodrigo, D. Koschny, H. Rickman, M.F. A'Hearn, M.A. Barucci, J.L. Bertaux, I. Bertini, G. Cremonese, V. Da Deppo, S. Debei, M. De Cecco, J. Deller, S. Fornasier, M. Fulle, O. Groussin, P.J. Gutiérrez, C. Güttler, M. Hofmann, W.H. Ip, L. Jorda, J. Knollenberg, J.R. Kramm, M. Küppers, L.M. Lara, M. Lazzarin, J.J. Lopez-Moreno, F. Marzari, G. Naletto, C. Tubiana, N. Thomas, Mon. Not. R. Astron. Soc. **469**, S357 (2017)

M. Knapmeyer, H.-H. Fischer, J. Knollenberg, K.J. Seidensticker, K. Thiel, W. Arnold, C. Faber, D. Möhlmann, Icarus **310**, 165–193 (2018)

H. Kochan, B. Feuerbacher, F. Joo, J. Klinger, W. Seboldt, A. Bischoff, H. Duren, D. Stoffler, T. Spohn, H. Fechtig, E. Grun, H. Kohl, D. Krankowsky, K. Roessler, K. Thiel, G. Schwehm, U. Weishaupt, Adv. Space Res. **9**, 113 (1989)

N.I. Kömle, G. Kargl, K. Thiel, K. Seiferlin, Planet. Space Sci. **44**, 675 (1996)

K.J. Kossacki, N.I. Kömle, J. Leliwa-Kopystyński, G. Kargl, Icarus **128**, 127 (1997)

K.J. Kossacki, S.Ł. Szutowicz, J. Leliwa-Kopystyński, Icarus **142**, 202–218 (1999)

K.J. Kossacki, T. Spohn, A. Hagermann, E. Kaufmann, E. Kührt, Icarus **260**, 464 (2015)

E. Kuehrt, H.U. Keller, Icarus **109**, 121 (1994)

F. La Forgia, L. Giacomini, M. Lazzarin, M. Massironi, N. Oklay, F. Scholten, M. Pajola, I. Bertini, G. Cremonese, C. Barbieri, G. Naletto, E. Simioni, F. Preusker, N. Thomas, H. Sierks, P. Lamy, R. Rodrigo, D. Koschny, H. Rickman, H.U. Keller, J. Agarwal, A.-T. Auger, M.F. A'Hearn, M.A. Barucci, J.-L. Bertaux, S. Besse, D. Bodewits, V. Da Deppo, B. Davidsson, S. Debei, M. De Cecco, M.R. El-Maarry, F. Ferri, S. Fornasier, M. Fulle, O. Groussin, P.J. Gutiérrez, C. Güttler, I. Hall, S.F. Hviid, W.-H. Ip, L. Jorda, J. Knollenberg, J.R. Kramm, E. Kührt, M. Küppers, L.-M. Lara, J.J. Lopez Moreno, S. Magrin, F. Marzari, H. Michalik, S. Mottola, A. Pommerol, C. Tubiana, J.-B. Vincent, Astron. Astrophys. **583**, A41 (2015)

A.H. Lachenbruch, in *Mech. Therm. Contract. Cracks Ice-Wedge Polyg. Permafr.*, ed. by A.H. Lachenbruch (Geol. Soc. Am., Boulder, 1962)

M. Laeuter, T. Kramer, M. Rubin, K. Altwegg, Mon. Not. R. Astron. Soc. **483**, S852 (2018)

P.L. Lamy, I. Toth, Y.R. Fernandez, H.A. Weaver, in *Comets II*, ed. by M.C. Festou, H.U. Keller, H.A. Weaver (Univ. Arizona Press, Tucson, 2004), pp. 223–264

L.M. Lara, S. Lowry, J.-B. Vincent, P.J. Gutiérrez, A. Rożek, F. La Forgia, N. Oklay, H. Sierks, C. Barbieri, P.L. Lamy, R. Rodrigo, D. Koschny, H. Rickman, H.U. Keller, J. Agarwal, A.-T. Auger, M.F. A'Hearn, M.A. Barucci, J.-L. Bertaux, I. Bertini, S. Besse, D. Bodewits, G. Cremonese, B. Davidsson, V. Da Deppo, S. Debei, M. De Cecco, M.R. El-Maarry, F. Ferri, S. Fornasier, M. Fulle, O. Groussin, P. Gutié rrez-Marques, C. Güttler, S.F. Hviid, W.-H. Ip, L. Jorda, J. Knollenberg, G. Kovacs, J.-R. Kramm, E. Kührt, M. Küppers, M. Lazzarin, Z.-Y. Lin, J.J. López-Moreno, S. Magrin, F. Marzari, H. Michalik, R. Moissl-Fraund, F. Moreno, S. Mottola, G. Naletto, M. Pajola, A. Pommerol, N. Thomas, M.D. Sabau, C. Tubiana, Astron. Astrophys. **583**, A9 (2015)

J.C. Lee, M. Massironi, W.H. Ip, L. Giacomini, S. Ferrari, L. Penasa, M.R. El-Maarry, M. Pajola, I.L. Lai, Z.Y. Lin, F. Ferri, H. Sierks, C. Barbieri, P. Lamy, R. Rodrigo, D. Koschny, H. Rickman, H.U. Keller, J. Agarwal, M.F. A'Hearn, M.A. Barucci, J.L. Bertaux, I. Bertini, G. Cremonese, V. Da Deppo, B. Davidsson, S. Debei, M. De Cecco, J. Deller, S. Fornasier, M. Fulle, O. Groussin, P.J. Gutiérrez, C. Güttler, M. Hofmann, S.F. Hviid, L. Jorda, J. Knollenberg, G. Kovacs, J.R. Kramm, E. Kührt, M. Küppers, L.M. Lara, M. Lazzarin, F. Marzari, J.J. Lopez Moreno, G. Naletto, N. Oklay, X. Shi, N. Thomas, C. Tubiana, J.B. Vincent, Mon. Not. R. Astron. Soc. **462**, S573 (2016)

A. Lethuillier, A. Le Gall, M. Hamelin, W. Schmidt, K.J. Seidensticker, R. Grard, V. Ciarletti, S. Caujolle-Bert, H.-H. Fischer, R. Trautner, Astron. Astrophys. **591**, A32 (2016)

M. Massironi, E. Simioni, F. Marzari, G. Cremonese, L. Giacomini, M. Pajola, L. Jorda, G. Naletto, S. Lowry, M.R. El-Maarry, F. Preusker, F. Scholten, H. Sierks, C. Barbieri, P. Lamy, R. Rodrigo, D. Koschny, H. Rickman, H.U. Keller, M.F. A'Hearn, J. Agarwal, A.-T. Auger, M.A. Barucci, J.-L. Bertaux, I. Bertini, S. Besse, D. Bodewits, C. Capanna, V. Da Deppo, B. Davidsson, S. Debei, M. De Cecco, F. Ferri, S. Fornasier, M. Fulle, R. Gaskell, O. Groussin, P.J. Gutiérrez, C. Güttler, S.F. Hviid, W.-H. Ip, J. Knollenberg, G. Kovacs, R. Kramm, E. Kührt, M. Küppers, F. La Forgia, L.M. Lara, M. Lazzarin, Z.-Y. Lin, J.J. Lopez Moreno, S. Magrin, H. Michalik, S. Mottola, N. Oklay, A. Pommerol, N. Thomas, C. Tubiana, J.-B. Vincent, Nature **526**, 402 (2015)

C. Matonti, N. Attree, O. Groussin, L. Jorda, S. Viseur, S.F. Hviid, S. Bouley, D. Nébouy, A.-T. Auger, P.L. Lamy, H. Sierks, G. Naletto, R. Rodrigo, D. Koschny, B. Davidsson, M.A. Barucci, J.-L. Bertaux, I. Bertini, D. Bodewits, G. Cremonese, V. Da Deppo, S. Debei, M. De Cecco, J. Deller, S. Fornasier, M. Fulle, P.J. Gutiérrez, C. Güttler, W.-H. Ip, H.U. Keller, L.M. Lara, F. La Forgia, M. Lazzarin, A. Lucchetti, J.J. López-Moreno, F. Marzari, M. Massironi, S. Mottola, N. Oklay, M. Pajola, L. Penasa, F. Preusker, H. Rickman, F. Scholten, X. Shi, I. Toth, C. Tubiana, J.-B. Vincent, Nat. Geosci. **12**, 157 (2019)

P. Migon, in *Encyclopedia of Geomorphology*, ed. by A.S. Goudie (Taylor & Francis, London, 2004)

J.L. Molaro, S. Byrne, J.-L. Le, Icarus **294**, 247 (2017)

S. Mottola, G. Arnold, H.-G. Grothues, R. Jaumann, H. Michaelis, G. Neukum, J.-P. Bibring, S.E. Schröder, M. Hamm, K.A. Otto, I. Pelivan, G. Proffe, F. Scholten, D. Tirsch, M. Kreslavsky, E. Remetean, F. Souvannavong, B. Dolives, Science (80-) **349**, 2.232 (2015)

T. Ott, E. Drolshagen, D. Koschny, C. Güttler, C. Tubiana, E. Frattin, J. Agarwal, H. Sierks, I. Bertini, C. Barbieri, P.I. Lamy, R. Rodrigo, H. Rickman, M.F. A'Hearn, M.A. Barucci, J.-L. Bertaux, S. Boudreault, G. Cremonese, V. Da Deppo, B. Davidsson, S. Debei, M. De Cecco, J. Deller, C. Feller, S. Fornasier, M. Fulle, B. Geiger, A. Gicquel, O. Groussin, P.J. Gutiérrez, M. Hofmann, S.F. Hviid, W.-H. Ip, L. Jorda, H.U. Keller, J. Knollenberg, G. Kovacs, J.R. Kramm, E. Kührt, M. Küppers, L.M. Lara, M. Lazzarin, Z.-Y. Lin, J.J. Ló pez-Moreno, F. Marzari, S. Mottola, G. Naletto, N. Oklay, M. Pajola, X. Shi, N. Thomas, J.-B. Vincent, B. Poppe, Mon. Not. R. Astron. Soc. **469**, S276 (2017)

M. Pajola, J.-B. Vincent, C. Güttler, J.-C. Lee, I. Bertini, M. Massironi, E. Simioni, F. Marzari, L. Giacomini, A. Lucchetti, C. Barbieri, G. Cremonese, G. Naletto, A. Pommerol, M.R. El-Maarry, S. Besse, M. Küppers, F. La Forgia, M. Lazzarin, N. Thomas, A.-T. Auger, H. Sierks, P. Lamy, R. Rodrigo, D. Koschny, H. Rickman, H.U. Keller, J. Agarwal, M.F. A'Hearn, M.A. Barucci, J.-L. Bertaux, V. Da Deppo, B. Davidsson, M. De Cecco, S. Debei, F. Ferri, S. Fornasier, M. Fulle, O. Groussin, P.J. Gutierrez, S.F. Hviid, W.-H. Ip, L. Jorda, J. Knollenberg, J.-R. Kramm, E. Kürt, L.M. Lara, Z.-Y. Lin, J.J. Lopez Moreno, S. Magrin, S. Marchi, H. Michalik, R. Moissl, S. Mottola, N. Oklay, F. Preusker, F. Scholten, C. Tubiana, Astron. Astrophys. **583**, A37 (2015)

M. Pajola, S. Mottola, M. Hamm, M. Fulle, B. Davidsson, C. Güttler, H. Sierks, G. Naletto, G. Arnold, H.-G. Grothues, R. Jaumann, H. Michaelis, J.P. Bibring, C. Barbieri, P.L. Lamy, R. Rodrigo, D. Koschny, H. Rickman, H.U. Keller, J. Agarwal, M.F. A'Hearn, M.A. Barucci, J.L. Bertaux, I. Bertini, S. Boudreault, G. Cremonese, V. Da Deppo, S. Debei, M. De Cecco, J. Deller, M.R. El Maarry, C. Feller, S. Fornasier, A. Gicquel, O. Groussin, P.J. Gutierrez, M. Hofmann, S.F. Hviid, W.H. Ip, L. Jorda, J. Knollenberg, J.R. Kramm, E. Kührt, M. Küppers, F. La Forgia, L.M. Lara, Z.Y. Lin, M. Lazzarin, J.J. Lopez Moreno, A. Lucchetti, F. Marzari, M. Massironi, H. Michalik, N. Oklay, A. Pommerol, F. Preusker, F. Scholten, N. Thomas, C. Tubiana, J.B. Vincent, Mon. Not. R. Astron. Soc. **462**, S242–S252 (2016a)

M. Pajola, A. Lucchetti, J.-B. Vincent, N. Oklay, M.R. El-Maarry, I. Bertini, G. Naletto, M. Lazzarin, M. Massironi, H. Sierks, C. Barbieri, P. Lamy, R. Rodrigo, D. Koschny, H. Rickman, H.U. Keller, J. Agarwal, M.F. A'Hearn, M.A. Barucci, J.-L. Bertaux, S. Boudreault, G. Cremonese, V. Da Deppo, B. Davidsson, S. Debei, M. De Cecco, J. Deller, S. Fornasier, M. Fulle, A. Gicquel, O. Groussin, P.J. Gutierrez, C. Güttler, M. Hofmann, S. Höfner, S.F. Hviid, W.-H. Ip, L. Jorda, J. Knollenberg, J.-R. Kramm, E. Kührt, M. Küppers, F. La Forgia, L.M. Lara, J.-C. Lee, Z.-Y. Lin, J.J. Lopez Moreno, F. Marzari, H. Michalik, S. Mottola, F. Preusker, F. Scholten, N. Thomas, I. Toth, C. Tubiana, Astron. Astrophys. **592**, L2 (2016b)

M. Pajola, N. Oklay, F. La Forgia, L. Giacomini, M. Massironi, I. Bertini, M.R. El-Maarry, F. Marzari, F. Preusker, F. Scholten, S. Höfner, J.-C. Lee, J.-B. Vincent, O. Groussin, G. Naletto, M. Lazzarin, C. Barbieri, H. Sierks, P. Lamy, R. Rodrigo, D. Koschny, H. Rickman, H.U. Keller, J. Agarwal, M.F. A'Hearn, M.A. Barucci, J.-L. Bertaux, G. Cremonese, V. Da Deppo, B. Davidsson, M. De Cecco, S. Debei, F. Ferri, S. Fornasier, M. Fulle, C. Güttler, P.J. Gutierrez, S.F. Hviid, W.-H. Ip, L. Jorda, J. Knollenberg, J.-R. Kramm, M. Küppers, E. Kürt, L.M. Lara, Z.-Y. Lin, J.J. Lopez Moreno, S. Magrin, H. Michalik, S. Mottola, N. Thomas, C. Tubiana, Astron. Astrophys. **592**, A69 (2016c)

M. Pajola, A. Lucchetti, M. Fulle, S. Mottola, M. Hamm, V. Da Deppo, L. Penasa, G. Kovacs, M. Massironi, X. Shi, C. Tubiana, C. Güttler, N. Oklay, J.B. Vincent, I. Toth, B. Davidsson, G. Naletto, H. Sierks, C. Barbieri, P.L. Lamy, R.R.D. Koschny, H. Rickman, H.U. Keller, J. Agarwal, M.F. A'Hearn, M.A.

Barucci, J.L. Bertaux, I. Bertini, G. Cremonese, S. Debei, M. De Cecco, J. Deller, M.R. El Maarry, S. Fornasier, E.F.A. Gicquel, O.G.P.J. Gutierrez, S. Höfner, M. Hofmann, S.F. Hviid, W.H. Ip, L. Jorda, J. Knollenberg, J.R. Kramm, E. Kührt, M. Küppers, L.M. Lara, M. Lazzarin, J.J. Lopez Moreno, F. Marzari, H. Michalik, F. Preusker, F. Scholten, N. Thomas, Mon. Not. R. Astron. Soc. **469**, S636 (2017a)

M. Pajola, S. Höfner, J.B. Vincent, N. Oklay, F. Scholten, F. Preusker, S. Mottola, G. Naletto, S. Fornasier, S. Lowry, C. Feller, P.H. Hasselmann, C. Güttler, C. Tubiana, H. Sierks, C. Barbieri, P. Lamy, R. Rodrigo, D. Koschny, H. Rickman, H.U. Keller, J. Agarwal, M.F. A'Hearn, M.A. Barucci, J.-L. Bertaux, I. Bertini, S. Besse, S. Boudreault, G. Cremonese, V. da Deppo, B. Davidsson, S. Debei, M. de Cecco, J. Deller, J.D.P. Deshapriya, M.R. El-Maarry, S. Ferrari, F. Ferri, M. Fulle, O. Groussin, P. Gutierrez, M. Hofmann, S.F. Hviid, W.-H. Ip, L. Jorda, J. Knollenberg, G. Kovacs, J.R. Kramm, E. Kührt, M. Küppers, L.M. Lara, Z.-Y. Lin, M. Lazzarin, A. Lucchetti, J.J. Lopez Moreno, F. Marzari, M. Massironi, H. Michalik, L. Penasa, A. Pommerol, E. Simioni, N. Thomas, I. Toth, E. Baratti, Nat. Astron. **1**, 92 (2017b)

M. Pätzold, T. Andert, M. Hahn, S.W. Asmar, J.P. Barriot, M.K. Bird, B. Häusler, K. Peter, S. Tellmann, E. Grün, P.R. Weissman, H. Sierks, L. Jorda, R. Gaskell, F. Preusker, F. Scholten, Nature **530**, 63 (2016)

L. Penasa, M. Massironi, G. Naletto, E. Simioni, S. Ferrari, M. Pajola, A. Lucchetti, F. Preusker, F. Scholten, L. Jorda, R. Gaskell, F. Ferri, F. Marzari, B. Davidsson, S. Mottola, H. Sierks, C. Barbieri, P.L. Lamy, R. Rodrigo, D. Koschny, H. Rickman, H.U. Keller, J. Agarwal, M.F. A'Hearn, M.A. Barucci, J.L. Bertaux, I. Bertini, G. Cremonese, V. Da Deppo, S. Debei, M. De Cecco, J. Deller, C. Feller, S. Fornasier, E. Frattin, M. Fulle, O. Groussin, P.J. Gutierrez, C. Güttler, M. Hofmann, S.F. Hviid, W.H. Ip, J. Knollenberg, J.R. Kramm, E. Kührt, M. Küppers, F. La Forgia, L.M. Lara, M. Lazzarin, J.C. Lee, J.J. Lopez Moreno, N. Oklay, X. Shi, N. Thomas, C. Tubiana, J.B. Vincent, Mon. Not. R. Astron. Soc. **469**, S741 (2017)

A. Pommerol, B. Jost, O. Poch, M.R. El-Maarry, B. Vuitel, N. Thomas, Planet. Space Sci. **109**, 106 (2015a)

A. Pommerol, N. Thomas, M.R. El-Maarry, M. Pajola, O. Groussin, A.-T. Auger, N. Oklay, S. Fornasier, C. Feller, B. Davidsson, A. Gracia-Berná, B. Jost, R. Marschall, O. Poch, M.A. Barucci, J.-L. Bertaux, F. La Forgia, H.U. Keller, E. Kührt, S.C. Lowry, S. Mottola, G. Naletto, H. Sierks, C. Barbieri, P.L. Lamy, R. Rodrigo, D. Koschny, H. Rickman, J. Agarwal, M.F. A'Hearn, I. Bertini, S. Boudreault, G. Cremonese, V. Da Deppo, M. De Cecco, S. Debei, C. Güttler, M. Fulle, P.J. Gutierrez, S.F. Hviid, W.-H. Ip, L. Jorda, J. Knollenberg, G. Kovacs, J.-R. Kramm, E. Küppers, L. Lara, M. Lazzarin, J.L. Lopez Moreno, F. Marzari, H. Michalik, F. Preusker, F. Scholten, C. Tubiana, J.-B. Vincent, Astron. Astrophys. **583**, A25 (2015b)

F. Poulet, A. Lucchetti, J.-P. Bibring, J. Carter, B. Gondet, L. Jorda, Y. Langevin, C. Pilorget, C. Capanna, G. Cremonese, Mon. Not. R. Astron. Soc. **462**, S23 (2016)

F. Preusker, F. Scholten, K.-D. Matz, T. Roatsch, K. Willner, S.F. Hviid, J. Knollenberg, L. Jorda, P.J. Gutiérrez, E. Kührt, S. Mottola, M.F. A'Hearn, N. Thomas, H. Sierks, C. Barbieri, P. Lamy, R. Rodrigo, D. Koschny, H. Rickman, H.U. Keller, J. Agarwal, M.A. Barucci, J.-L. Bertaux, I. Bertini, G. Cremonese, V. Da Deppo, B. Davidsson, S. Debei, M. De Cecco, S. Fornasier, M. Fulle, O. Groussin, C. Güttler, W.-H. Ip, J.R. Kramm, M. Küppers, L.M. Lara, M. Lazzarin, J.J. Lopez Moreno, F. Marzari, H. Michalik, G. Naletto, N. Oklay, C. Tubiana, J.-B. Vincent, Astron. Astrophys. **583**, A33 (2015)

F. Preusker, F. Scholten, K.-D. Matz, T. Roatsch, S.F. Hviid, S. Mottola, J. Knollenberg, E. Kührt, M. Pajola, N. Oklay, J.-B. Vincent, B. Davidsson, M.F. A'Hearn, J. Agarwal, C. Barbieri, M.A. Barucci, J.-L. Bertaux, I. Bertini, G. Cremonese, V. Da Deppo, S. Debei, M. De Cecco, S. Fornasier, M. Fulle, O. Groussin, P.J. Gutiérrez, C. Güttler, W.-H. Ip, L. Jorda, H.U. Keller, D. Koschny, J.R. Kramm, M. Küppers, P. Lamy, L.M. Lara, M. Lazzarin, J.J. Lopez Moreno, F. Marzari, M. Massironi, G. Naletto, H. Rickman, R. Rodrigo, H. Sierks, N. Thomas, C. Tubiana, Astron. Astrophys. **607**, 1 (2017)

H. Rickman, S. Marchi, M.F. A'Hearn, C. Barbieri, M.R. El-Maarry, C. Güttler, W.-H. Ip, H.U. Keller, P. Lamy, F. Marzari, M. Massironi, G. Naletto, M. Pajola, H. Sierks, D. Koschny, R. Rodrigo, M.A. Barucci, J.-L. Bertaux, I. Bertini, G. Cremonese, V. Da Deppo, S. Debei, M. De Cecco, S. Fornasier, M. Fulle, O. Groussin, P.J. Gutiérrez, S.F. Hviid, L. Jorda, J. Knollenberg, J.-R. Kramm, E. Kührt, M. Küppers, L.M. Lara, M. Lazzarin, J.J. Lopez Moreno, H. Michalik, L. Sabau, N. Thomas, J.-B. Vincent, K.-P. Wenzel, Astron. Astrophys. **583**, A44 (2015)

R.Z. Sagdeev, J. Blamont, A.A. Galeev, V.I. Moroz, V.D. Shapiro, V.I. Shevchenko, K. Szegó, Nature **321**, 259 (1986)

D.W.G. Sears, H.W. Kochan, W.F. Huebner, Meteorit. Planet. Sci. **34**, 497 (1999)

K. Seiferlin, T. Spobn, J. Benkbofp, Adv. Space Res. **15**, 35 (1995)

X. Shi, X. Hu, H. Sierks, C. Güttler, M. A'Hearn, J. Blum, M.R. El-Maarry, E. Kührt, S. Mottola, M. Pajola, N. Oklay, S. Fornasier, C. Tubiana, H.U. Keller, J.-B. Vincent, D. Bodewits, S. Höfner, Z.-Y. Lin, A. Gicquel, M. Hofmann, C. Barbieri, P.L. Lamy, R. Rodrigo, D. Koschny, H. Rickman, M.A. Barucci, J.-L. Bertaux, I. Bertini, G. Cremonese, V. Da Deppo, B. Davidsson, S. Debei, M. De Cecco, M. Fulle,

O. Groussin, P.J. Gutiérrez, S.F. Hviid, W.-H. Ip, L. Jorda, J. Knollenberg, G. Kovacs, J.-R. Kramm, M. Küppers, L.M. Lara, M. Lazzarin, J.J. Lopez-Moreno, F. Marzari, G. Naletto, N. Thomas, Astron. Astrophys. **586**, A7 (2016)

Y.V. Skorov, L. Rezac, P. Hartogh, A.T. Bazilevsky, H.U. Keller, Astron. Astrophys. **593**, A76 (2016)

A.L.A. Soderblom, T.L. Becker, G. Bennett, D.C. Boice, D.T. Britt, R.H. Brown, B.J. Buratti, C. Isbell, B. Giese, T. Hare, M.D. Hicks, R.L. Kirk, R.M. Nelson, J. Oberst, T.C. Owen, M.D. Rayman, B.R. Sandel, S.A. Stern, R.V. Yelle, Science (80-) **296**, 1087 (2002)

T. Spohn, J. Knollenberg, A.J. Ball, M. Banaszkiewicz, J. Benkhoff, M. Grott, J. Grygorczuk, C. Hüttig, A. Hagermann, G. Kargl, E. Kaufmann, N. Kömle, E. Kührt, K.J. Kossacki, W. Marczewski, I. Pelivan, R. Schrödter, K. Seiferlin, Science (80-) **349**, 2.464 (2015)

P.C. Thomas, J. Veverka, M.J.S. Belton, A. Hidy, M.F. A'Hearn, T.L. Farnham, O. Groussin, J.Y. Li, L.A. McFadden, J. Sunshine, D. Wellnitz, C. Lisse, P. Schultz, K.J. Meech, W.A. Delamere, Icarus **187**, 4 (2007)

P. Thomas, M. A'Hearn, M.J.S. Belton, D. Brownlee, B. Carcich, B. Hermalyn, K. Klaasen, S. Sackett, P.H. Schultz, J. Veverka, S. Bhaskaran, D. Bodewits, S. Chesley, B. Clark, T. Farnham, O. Groussin, A. Harris, J. Kissel, J.Y. Li, K. Meech, J. Melosh, A. Quick, J. Richardson, J. Sunshine, D. Wellnitz, Icarus **222**, 453 (2013a)

P.C. Thomas, M.F. A'Hearn, J. Veverka, M.J.S. Belton, J. Kissel, K.P. Klaasen, L.A. McFadden, H.J. Melosh, P.H. Schultz, S. Besse, B.T. Carcich, T.L. Farnham, O. Groussin, B. Hermalyn, J.Y. Li, D.J. Lindler, C.M. Lisse, K. Meech, J.E. Richardson, Icarus **222**, 550 (2013b)

N. Thomas, B. Davidsson, M.R. El-Maarry, S. Fornasier, L. Giacomini, A.G. Gracia-Berná, S.F. Hviid, W.-H. Ip, L. Jorda, H.U. Keller, J. Knollenberg, E. Kührt, F. La Forgia, I.L. Lai, Y. Liao, R. Marschall, M. Massironi, S. Mottola, M. Pajola, O. Poch, A. Pommerol, F. Preusker, F. Scholten, C.C. Su, J.S. Wu, J.-B. Vincent, H. Sierks, C. Barbieri, P.L. Lamy, R. Rodrigo, D. Koschny, H. Rickman, M.F. A'Hearn, M.A. Barucci, J.-L. Bertaux, I. Bertini, G. Cremonese, V. Da Deppo, S. Debei, M. de Cecco, M. Fulle, O. Groussin, P.J. Gutierrez, J.-R. Kramm, M. Küppers, L.M. Lara, M. Lazzarin, J.J. Lopez Moreno, F. Marzari, H. Michalik, G. Naletto, J. Agarwal, C. Güttler, N. Oklay, C. Tubiana, Astron. Astrophys. **583**, A17 (2015a)

N. Thomas, H. Sierks, C. Barbieri, P.L. Lamy, R. Rodrigo, H. Rickman, D. Koschny, H.U. Keller, J. Agarwal, M.F.A. Hearn, F. Angrilli, A. Auger, M.A. Barucci, J. Kramm, E. Kührt, M. Küppers, F. La Forgia, Science (80-) **347**, aaa0440 (2015b)

N. Thomas, M.R. El Maarry, P. Theologou, F. Preusker, F. Scholten, L. Jorda, S.F. Hviid, R. Marschall, E. Kührt, G. Naletto, H. Sierks, P.L. Lamy, R. Rodrigo, D. Koschny, B. Davidsson, M.A. Barucci, J.L. Bertaux, I. Bertini, D. Bodewits, G. Cremonese, V. Da Deppo, S. Debei, M. De Cecco, S. Fornasier, M. Fulle, O. Groussin, P.J. Gutièrrez, C. Güttler, W.H. Ip, H.U. Keller, J. Knollenberg, L.M. Lara, M. Lazzarin, J.J. Lòpez-Moreno, F. Marzari, C. Tubiana, J.B. Vincent, Planet. Space Sci. **164**, 19–36 (2018)

D. Tirsch, K. Otto, S. Mottola, S. Hviid, R. Jaumann, L. Jorda, E. Kührt, K.-D. Matz, F. Preusker, in *Eur. Planet. Sci. Congr.* (2017). EPSC2017-739

J. Veverka, K. Klaasen, M. A'Hearn, M. Belton, D. Brownlee, S. Chesley, B. Clark, T. Economou, R. Farquhar, S.F. Green, O. Groussin, A. Harris, J. Kissel, J.Y. Li, K. Meech, J. Melosh, J. Richardson, P. Schultz, J. Silen, J. Sunshine, P. Thomas, S. Bhaskaran, D. Bodewits, B. Carcich, A. Cheuvront, T. Farnham, S. Sackett, D. Wellnitz, A. Wolf, Icarus **222**, 424 (2013)

J.B. Vincent, D. Bodewits, S. Besse, H. Sierks, C. Barbieri, P. Lamy, R. Rodrigo, D. Koschny, H. Rickman, H.U. Keller, J. Agarwal, M.F. A'Hearn, A.T. Auger, M.A. Barucci, J.L. Bertaux, I. Bertini, C. Capanna, G. Cremonese, V. Da Deppo, B. Davidsson, S. Debei, M. De Cecco, M.R. El-Maarry, F. Ferri, S. Fornasier, M. Fulle, R. Gaskell, L. Giacomini, O. Groussin, A. Guilbert-Lepoutre, P. Gutierrez-Marques, P.J. Gutiérrez, C. Güttler, N. Hoekzema, S. Höfner, S.F. Hviid, W.H. Ip, L. Jorda, J. Knollenberg, G. Kovacs, R. Kramm, E. Kührt, M. Küppers, F. La Forgia, L.M. Lara, M. Lazzarin, V. Lee, C. Leyrat, Z.Y. Lin, J.J. Lopez Moreno, S. Lowry, S. Magrin, L. Maquet, S. Marchi, F. Marzari, M. Massironi, H. Michalik, R. Moissl, S. Mottola, G. Naletto, N. Oklay, M. Pajola, F. Preusker, F. Scholten, N. Thomas, I. Toth, C. Tubiana, Nature **523**, 63 (2015a)

J.B. Vincent, N. Oklay, S. Marchi, S. Höfner, H. Sierks, Planet. Space Sci. **107**, 53 (2015b)

J.-B. Vincent, M.F. A'Hearn, Z.-Y. Lin, M.R. El-Maarry, M. Pajola, H. Sierks, C. Barbieri, P.L. Lamy, R. Rodrigo, D. Koschny, H. Rickman, H.U. Keller, J. Agarwal, M.A. Barucci, J.-L. Bertaux, I. Bertini, S. Besse, D. Bodewits, G. Cremonese, V. Da Deppo, B. Davidsson, S. Debei, M. De Cecco, J. Deller, S. Fornasier, M. Fulle, A. Gicquel, O. Groussin, P.J. Gutiérrez, P. Gutiérrez-Marquez, C. Güttler, S. Höfner, M. Hofmann, S.F. Hviid, W.-H. Ip, L. Jorda, J. Knollenberg, G. Kovacs, J.-R. Kramm, E. Kührt, M. Küppers, L.M. Lara, M. Lazzarin, J.J. Lopez Moreno, F. Marzari, M. Massironi, S. Mottola, G. Naletto, N. Oklay, F. Preusker, F. Scholten, X. Shi, N. Thomas, I. Toth, C. Tubiana, Mon. Not. R. Astron. Soc. **462**, S184 (2016)

J.B. Vincent, S.F. Hviid, S. Mottola, E. Kuehrt, F. Preusker, F. Scholten, H.U. Keller, N. Oklay, D. de Niem, B. Davidsson, M. Fulle, M. Pajola, M. Hofmann, X. Hu, H. Rickman, Z.Y. Lin, C. Feller, A. Gicquel, S. Boudreault, H. Sierks, C. Barbieri, P.L. Lamy, R. Rodrigo, D. Koschny, M.F. A'Hearn, M.A. Barucci, J.L. Bertaux, I. Bertini, G. Cremonese, V. Da Deppo, S. Debei, M. De Cecco, J. Deller, S. Fornasier, O. Groussin, P.J. Gutiérrez, P. Gutiérrez-Marquez, C. Güttler, W.H. Ip, L. Jorda, J. Knollenberg, G. Kovacs, J.R. Kramm, M. Küppers, L.M. Lara, M. Lazzarin, J.J.L. Moreno, F. Marzari, G. Naletto, L. Penasa, X. Shi, N. Thomas, I. Toth, C. Tubiana, Mon. Not. R. Astron. Soc. **469**, S329 (2017)

Space Sci Rev (2019) 215:19
https://doi.org/10.1007/s11214-019-0580-3

# Comet 67P/CG Nucleus Composition and Comparison to Other Comets

**Gianrico Filacchione[1] · Olivier Groussin[2] · Clémence Herny[3] · David Kappel[4,5] · Stefano Mottola[5] · Nilda Oklay[5] · Antoine Pommerol[3] · Ian Wright[6] · Zurine Yoldi[3] · Mauro Ciarniello[1] · Lyuba Moroz[7,5] · Andrea Raponi[1]**

Received: 31 July 2018 / Accepted: 16 January 2019 / Published online: 14 February 2019


**Abstract** We review our current knowledge of comet 67P/Churyumov–Gerasimenko nucleus composition as inferred from measurements made by remote sensing and in-situ instruments aboard Rosetta orbiter and Philae lander. Spectrophotometric properties (albedos, color indexes and Hapke parameters) of 67P/CG derived by Rosetta are discussed in the context of other comets previously explored by space missions. Composed of an assemblage made of ices, organic materials and minerals, cometary nuclei exhibit very dark and red surfaces which can be described by means of spectrophotometric quantities and reproduced with laboratory measurements. The presence of surface water and carbon dioxide ices was found by Rosetta to occur at localized sites where the activity driven by solar input, gaseous condensation or exposure of pristine inner layers can maintain these species on the surface. Apart from these specific areas, 67P/CG's surface appears remarkably uniform in composition with a predominance of organic materials and minerals. The organic compounds contain abundant hydroxyl group and a refractory macromolecular material bearing aliphatic and aromatic hydrocarbons. The mineral components are compatible with a mixture of silicates and fine-grained opaques, including Fe-sulfides, like troilite and pyrrhotite, and ammoniated salts. In the vicinity of the perihelion several active phenomena, including

Comets: Post 67P / Churyumov–Gerasimenko Perspectives
Edited by Nicolas Thomas, Björn Davidsson, Laurent Jorda, Ekkehard Kührt, Raphael Marschall, Colin Snodgrass and Rafael Rodrigo

✉ G. Filacchione
gianrico.filacchione@inaf.it

1 Institute for Space Astrophysics and Planetology, INAF-IAPS, via del Fosso del Cavaliere, 100, 00133 Rome, Italy

2 Aix Marseille Univ., CNRS, CNES, LAM, Marseille, France

3 Physikalisches Institut, University of Bern, Bern, Switzerland

4 Institute of Physics and Astronomy, University of Potsdam, Potsdam, Germany

5 Institute of Planetary Research, German Aerospace Center (DLR), Berlin, Germany

6 Department of Physical Sciences, The Open University, Walton Hall, Milton Keynes, UK

7 Institute of Earth and Environmental Sciences, University of Potsdam, Potsdam, Germany

the erosion of surface layers, the localized activity in cliffs, fractures and pits, the collapse of overhangs and walls, the transfer and redeposition of dust, cause the evolution of the different regions of the nucleus by inducing color, composition and texture changes.

**Keywords** Comets · Composition · Ices · Organic matter · Minerals

## 1 Introduction

At the dawn of our solar system, $\approx$4.5 billion years ago, in the cold outer regions of the solar nebula, the first anhydrous dust grains started to aggregate with volatile ices and organics. Driven by accretion, these grains formed comets and planetesimals (Skorov and Blum 2012), the building blocks of the solid bodies of our solar system. Although processed by impacts, comets have not experienced a significant heating and alteration of their solid materials (de Niem et al. 2018). Differently from asteroids and larger bodies, on comets the imprint of pre-solar materials has been preserved (Mumma and Charnley 2011). The study of cometary nuclei is therefore the one of the best way to assess the synthesis of volatiles and organic materials. There is convincing evidence that comets played a key role in the evolution of the planets, because cometary impacts are known to have been much more common in the early solar system than today (Gomes et al. 2005). Comet flux is supposed to be one of the major sources of the water present in Earth's oceans and to have provided the organic material necessary for the life evolution on Earth (Delsemme 1984; Mayo Greenberg and Mendoza-Gómez 1992). So far, before Rosetta only few cometary missions were able to reach the proximity of comets and to infer locally their composition: Giotto at comet 1P/Halley (Reinhard 1982), Stardust at 81P/Wild 2 (Brownlee et al. 2003), Deep Space 1 at 19P/Borrelly (Nelson et al. 2004), Deep Impact at 9P/Tempel 1 and then EPOXI at 103P/Hartley 2 (A'Hearn et al. 2005a).

In this review we mainly focus on summarizing the current knowledge of the 67P/Churyumov–Gerasimenko (hereafter 67P/CG) nucleus composition which was one the major scientific objectives of ESA's Rosetta mission (Glassmeier et al. 2007) in the context of the results returned by previous space missions at others comets.

By orbiting around comet 67P/CG from August 2014 to September 2016, the Rosetta mission has provided unprecedented spatial, spectral and temporal coverage of the changes occurring on a cometary surface and surrounding inner coma before, during and after perihelion passage (Taylor et al. 2017). The remote sensing instruments aboard the orbiter have returned a comprehensive dataset about the nucleus composition while the in-situ payloads have analyzed the gas and dust environment in the surrounding coma. This wealth of information allows to constrain the nucleus composition from different prospectives: OSIRIS, the Optical, Spectroscopic, and Infrared Remote Imaging System (Keller et al. 2007) Wide and Narrow Angle Cameras have imaged the comet's nucleus through different UV, visible and near infrared color filters aiming to determine its volume, shape, bulk density, surface properties, morphology and activity. Three spectrometers operating at different wavelengths, ALICE, the Ultraviolet Imaging Spectrometer (Stern et al. 2007), VIRTIS, the Visible, InfraRed and Thermal Imaging Spectrometer (Coradini et al. 2007) and MIRO, the Microwave Instrument for the Rosetta Orbiter (Gulkis et al. 2007) have mapped the composition and temperature of the nucleus, analyzed the gases in the near-nucleus region, measured and monitored the comet's production rates of water, carbon dioxide and other gaseous species. Additional clues are given by the composition instruments aboard the Philae lander (Bibring et al. 2007) which were designed to infer elemental, isotopic, molecular and mineralogical composition. Unfortunately, due to the non-nominal landing of Philae, only gas chromatography and time-of-flight mass-spectrometry by COSAC (Goesmann et al. 2007) and PTOLEMY (Wright et al. 2007) have provided useful data on those fields. The coma environment, including dust particles and gaseous species, has been analyzed by the in-situ and remote sensing instruments aboard Rosetta. These species are naturally linked to the composition of the surface and subsurface layers from which they are released by active processes. A review of the compositional properties of cometary nuclei, and in particular for 67P/CG, as inferred from remote sensing and surface instruments, is given in Sect. 2. The surface composition appears dominated by three endmembers groups, including volatile species condensed as ices, organic material and minerals (Sect. 3). Moving along the orbit of the comet around the Sun, the surface composition of 67P/CG shows temporal changes caused by diurnal and seasonal effects which are discussed in Sect. 4. Conclusions are discussed in Sect. 5.

## 2 Comets Nuclei Properties

### 2.1 VIS-IR Albedo and Photometry

Cometary nuclei are characterized by very low geometric albedo ($A_{geo}$) at visible wavelengths: average value of $A_{geo} = 0.04$ was measured on 1P/Halley (Keller 1987), 81P/Wild 2 (Brownlee et al. 2003) and 9P/Tempel 1 (A'Hearn et al. 2005b). At $A_{geo} = 0.03$ (Soderblom et al. 2002; Buratti et al. 2004) the surface of 19P/Borrelly appears darker while 103P/Hartley 2 at $A_{geo} = 0.046$ (Li et al. 2013) is a little more brighter. On 67P/CG both OSIRIS (Fornasier et al. 2015) and VIRTIS (Ciarniello et al. 2015) have measured a geometric albedo of $A_{geo} = 0.06$ making this comet the brightest among the ones explored so far by spacecraft. Photometrically-corrected images and maps of the nucleus are shown in Fig. 1 where the illumination and viewing effects have been removed to better study the surface's physical properties and composition. VIRTIS measurements of the surface show a significant phase reddening (the surface becomes redder when the solar phase increases) at both VIS (0.44%/deg) and IR (0.73%/deg) wavelengths (Ciarniello et al. 2015). A similar reddening effect (0.95%/deg) has been reported by OSIRIS on visible images (Fornasier et al. 2015). Once the phase-dependence has been corrected, the surface spectral slope can be determined: VIRTIS has measured a spectral slope of 0.20/kÅ between 0.55–0.8 μm and 0.033/kÅ between 1–2 μm. The surface albedo increases in presence of water ice. In fact, the active areas in the Hapi and Seth regions appear brighter on both photometric-corrected OSIRIS images (Fornasier et al. 2015) and on VIRTIS albedo map (Filacchione et al. 2016b) shown in Fig. 1. These results were obtained during the inbound orbit at heliocentric distance of about 3.6 AU.

For this low-albedo surface where multiple scattering has a small contribution relative to single-scattering, the phase reddening is interpreted to be caused by a wavelength-dependent single-scattering phase function that might be related to the roughness of the regolith particle surfaces (Schröder et al. 2014; Ciarniello et al. 2015). A summary of the Hapke model (Hapke 1993) photometric parameters of 67P/CG compared with others comets is given in Table 1. Following Longobardo et al. (2017), the Phase Curve Slope (PCS) and the I/F at 0.75 μm retrieved at 30° phase angle (R30) parameters are given in Table 2 for 67P/CG and compared with values for other comets in Table 3.

Reddening in the visible is characterizing 67P/CG surface also at local scales and in many studies the color changes have been correlated with surface composition inhomogeneities and activity sources. Local water ice enrichments show higher albedo and a flat

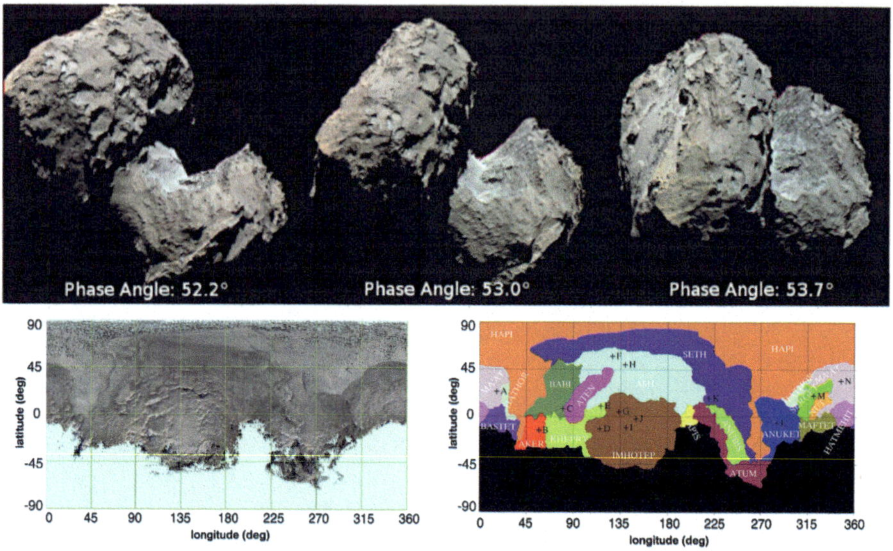


**Fig. 1** Photometrically corrected albedo images of 67P/CG's nucleus from August–September 2014 observations. Top panel: OSIRIS color images (filters B: 480 nm, G: 649 nm, R: 882 nm). Spatial resolution is about 2.5 m/px. From Fornasier et al. (2015). Bottom left panel: cylindrical map of single scattering albedo by VIRTIS (channels B: 440 nm, G: 550 nm, R: 700 nm). Spatial resolution is about 15 m/bin. From Filacchione et al. (2016c). Bottom right panel: boundaries of 67P/CG's nineteen geomorphological regions in cylindrical projection. From El-Maarry et al. (2015), Thomas et al. (2015). Missing data (light blue bottom left and black bottom right panels) represents the un-illuminated southern hemisphere at the given epochs

to blue spectral response in contrast to the more dark and red reflectance of the anhydrous terrains. Following these properties, water ice-rich areas have been recognized and studied on OSIRIS images by mapping the albedo and the bluening of the VIS slope (Oklay et al. 2016b) or by using multivariate analysis methods (Perna et al. 2017). By adopting this method the spatial inhomogeneities and temporal variations occurring on very different landforms of the nucleus have been traced, including: the grained deposits and outcrops of more consolidated materials on cuestas and steep walls in the Agilkia landing site (La Forgia et al. 2015); the water ice patches in the Khonsu (Deshapriya et al. 2016) and Anhur-Bes regions (Fornasier et al. 2016) in the south hemisphere; the fresh icy material visible on the exposed wall formed after the collapse of the Aswan cliff (Pajola et al. 2017), on Abydos (Lucchetti et al. 2016) and on the circular niches located in the Seth region (Lucchetti et al. 2017); the large water ice-rich clusters (BAP1 and BAP2 in Filacchione et al. 2016a) and isolated water ice features on Imhotep (Oklay et al. 2017), Hapi and its surrounding (Oklay et al. 2016b).

During few dedicated Rosetta's flybys designed to fly on the nucleus at low phase angles it has been possible to observe the opposition surge while swiping across the Imhotep region of 67P/CG (Hasselmann et al. 2017). A broad nonlinear phase curve has been measured at low phases resulting in the lack of a sharp opposition peak which is typically associated with the coherent-backscattering mechanism. A similar photometric behavior can be caused by the low albedo of the surface which is dominated by opaque, dehydrated and larger-than-wavelength irregular grains. The similarity between the 67P/CG surface phase function with carbon soot samples observed at extremely small angles has been recognized

**Table 1** Spectrophotometric properties and Hapke parameters of 67P/CG compared to other cometary nuclei

| Comet | B–V | V–R | R–I | SSA | $b$ | $\bar{\theta}$ | $A_{geo}$ | $\beta$ [mag/°] | $\alpha$(°) |
|---|---|---|---|---|---|---|---|---|---|
| 67P/CG[a] | $0.73 \pm 0.07$ | $0.57 \pm 0.03$ | $0.59 \pm 0.04$ | $0.052 \pm 0.013$ | $-0.42$ | $19°^{+4}_{-9}$ | $0.062 \pm 0.002$ | $0.077 \pm 0.002$[b] | 1.2°–111.5° |
| | – | – | – | - | – | – | – | $0.041 \pm 0.001$[c] | – |
| 67P/CG[d] | – | – | – | $0.037 \pm 0.002$ | $-0.42 \pm 0.03$ | 15° | $0.059 \pm 0.02$[e] | $0.047 \pm 0.002$ | 1.3°–53.9° |
| 67P/CG[f] | – | $0.52 \pm 0.05$ | – | – | – | – | – | – | 4.8° |
| 67P/CG[g] | $0.83 \pm 0.08$ | $0.54 \pm 0.05$ | $0.46 \pm 0.04$ | – | – | – | – | 0.061–0.076 | 0.5°–10.6° |
| 1P/Halley[h] | $0.72 \pm 0.04$ | $0.41 \pm 0.03$ | $0.39 \pm 0.06$ | – | – | – | – | – | 107° |
| 1P/Halley[i] | – | – | – | – | – | – | – | 0.04 | – |
| 1P/Halley[j] | – | – | – | – | – | – | – | $0.05^{+0.03}_{-0.01}$ | – |
| 1P/Halley[k] | – | – | – | – | – | – | – | $0.04^{+0.01}_{-0.02}$ | – |
| 19P/Borrelly[l] | – | – | – | $0.057 \pm 0.009$ | $-0.43 \pm 0.07$ | $22° \pm 5°$ | $0.072 \pm 0.020$ | 0.043 | 51°–75° |
| 19P/Borrelly[m] | – | – | – | – | – | – | – | $0.03 \pm 0.005$ | – |
| 19P/Borrelly[n] | – | – | – | – | – | – | – | $0.029 \pm 0.002$ | – |
| 19P/Borrelly[o] | – | $0.25 \pm 0.78$ | – | – | – | – | – | – | 1.98° |
| 103P/Hartley 2[p] | $0.75 \pm 0.05$ | $0.43 \pm 0.04$ | – | $0.036 \pm 0.006$ | $-0.46 \pm 0.06$ | $15° \pm 10°$ | $0.045 \pm 0.009$ | $0.046 \pm 0.002$ | 79°–93° |
| 9P/Tempel 1[q] | $0.84 \pm 0.01$ | $0.50 \pm 0.01$ | $0.49 \pm 0.02$ | $0.039 \pm 0.005$ | $-0.49 \pm 0.02$ | $16° \pm 8°$ | $0.051 \pm 0.009$ | $0.046 \pm 0.007$ | 63°–117° |
| 9P/Tempel 1[r] | – | – | – | – | – | – | – | 0.04 | – |
| 81P/Wild 2[s] | – | – | – | $0.038 \pm 0.04$ | $-0.52 \pm 0.04$ | $27 \pm 5°$ | 0.059 | $0.0513 \pm 0.0002$ | 11°–100° |

Legend: [a]from Ciarniello et al. (2015); [b]for phase angle $\alpha < 15°$; [c]for phase angle $\alpha > 15°$; [d]values for $w$, $b$ (0.535 μm), $\bar{\theta}$ and $\beta$ from Fornasier et al. (2015); [e]from Sierks et al. (2015); [f]Lamy et al. (2006); [g]from Tubiana et al. (2011). B–V has been computed from B–R and V–R index reported in Tubiana et al. (2011); [h]from Thomas and Keller (1989); [i]from Keller (1987); [j]from Hughes (1985); [k]from Sagdeev et al. (1986); [l]photometric parameters derived in R band from Li et al. (2007b), ground-based observations with solar phase angle down to 13° have been included to compute the slope parameter; [m]from Soderblom et al. (2002); [n]from Buratti et al. (2004); [o]from Lowry et al. (2003); [p]from Li et al. (2013), Gemini and HST observations at low phase angle have been included to constrain Hapke modeling and derive the slope parameter; [q]photometric parameters derived in V band from Li et al. (2007a), ground-based and HST observations with solar phase angle down to 4° have been included to compute the slope parameter; [r]from A'Hearn et al. (2005b); [s]photometric parameters derived in R band by Li et al. (2009)

**Table 2** Spectrophotometric parameters of 67P/CG at visible and near infrared wavelengths. The Phase Curve Slope (PCS) is defined as 1-R60/R20, where R60 and R20 are the reflectances (I/F) at 60° and 20° phase angles, respectively. The R30 (I/F at phase 30°) error is 0.001 up to 1.2 μm and 0.002 from 1.7 μm. An absolute error of 5% is affecting the PCS data. From Longobardo et al. (2017)

| λ (μm) | 0.55 | 0.75 | 1.2 | 1.7 | 2.0 | 2.8 | 4.0 |
|---|---|---|---|---|---|---|---|
| R30 | 0.013 | 0.017 | 0.031 | 0.035 | 0.036 | 0.041 | 0.041 |
| PCS | 0.61 | 0.60 | 0.61 | 0.60 | 0.60 | 0.57 | 0.55 |

**Table 3** Spectrophotometric parameters R30 and PCS for 67P/CG compared to other comets

| Comet | λ (μm) | R30 | PCS | Reference |
|---|---|---|---|---|
| 19P/Borrelly | 0.66 | 0.018 | 0.65 | Li et al. (2007b) |
| 103P/Hartley 2 | 0.62 | 0.013 | 0.67 | Li et al. (2013) |
| 9P/Tempel 1 | 0.67 | 0.015 | 0.71 | Li et al. (2007a) |
| 81P/Wild 2 | 0.7 | 0.014 | 0.78 | Li et al. (2009) |
| 67P/CG | 0.75 | 0.017 | 0.60 | Longobardo et al. (2017) |

by Masoumzadeh et al. (2017) who reported that the areas placed on the boundary of the Imhotep-Ash regions are made by a very dark and fluffy layer.

## 2.2 VIS-IR Spectroscopy from Remote Sensing

During Rosetta's initial observations taken during the pre-perihelion period, 67P/CG surface composition appeared quite homogeneous across the different regions illuminated by the Sun. A notable feature of the majority of the spectra is the absence of water–ice absorption bands (at 1.5, 2.0 μm). This indicates that water ice-rich patches are absent across the nucleus surface, at a scale larger than $\approx$10 m with an upper limit on the water–ice abundance of $\approx$1% (Capaccioni et al. 2015). Moreover, reflectance spectra acquired on different regions of the surface (Fig. 2, top panel) show very low visible albedo, equal to 0.06 at 0.55 μm (Fornasier et al. 2015; Ciarniello et al. 2015), a steep red color at visible wavelengths with a slope change around 1.0 μm and continuing more gently up to 2.7 μm Filacchione et al. (2016c), a broad absorption band in the range between 2.9 to 3.6 μm encompassing smaller structures compatible with aromatic and aliphatic signatures (Fig. 2, bottom panel), a wide thermal emission starting from 3.5 to 5.0 μm. After modelling the thermal emission on the radiance data (Groussin et al. 2019, this issue), a surface temperature in the range between 180 and 230 K during daytime (August–September 2014, heliocentric distance 3.6 AU, inbound orbit) is derived (Capaccioni et al. 2015).

After having derived and applied a photometric correction (Ciarniello et al. 2015), surface properties have been studied by means of Cometary Spectral Indicators (CSI) suitable to trace color and compositional units across the nucleus (Filacchione et al. 2016c). The more active areas in Hapi, appear brighter and bluer on maps of both 0.5–0.8 μm visible and 1.0–2.5 μm infrared spectral slopes than the rest of the surface (Fig. 3, top row panels).

The wide 3.2 μm absorption band observed by VIRTIS is compatible with a broad class of organic materials, detailed later in Sect. 3. In presence of water ice, the short-wavelength-wing of the band is affected by the strong fundamental symmetric and anti-symmetric stretching modes which causes a broadening of the band and a spectral shift towards shorter

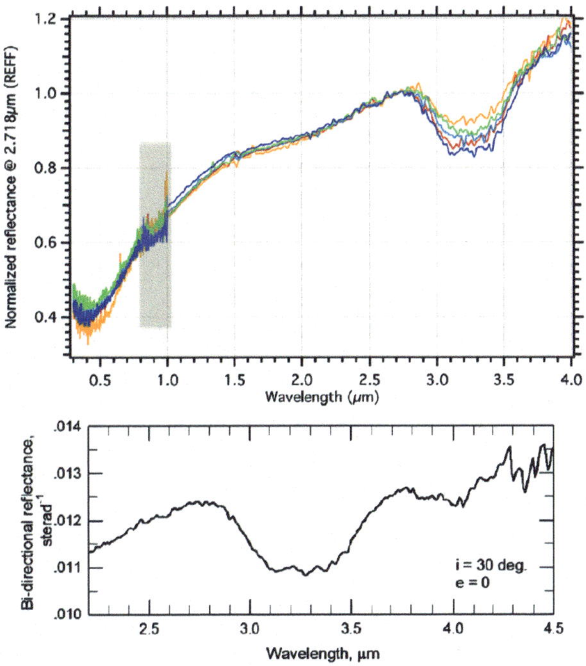


**Fig. 2** Top panel: Average visible and infrared reflectance spectra of 67P/CG from VIRTIS taken from different macroregions. Gray box indicates calibration residuals. Bottom panel: Bidirectional reflectance of the average dark terrain in the 2.2–4.5 μm spectral range. Within the wide 3.2 μm band are visible fainter bands associated to aromatic and aliphatic organic materials discussed in Sect. 3. Features at 4.4 μm are calibration artifacts. Thermal emission is removed from reflectance data. From Ciarniello et al. (2015), Quirico et al. (2016)

wavelengths. Thermal emission influences the long-wavelength-wing of the band with a contribution increasing with the surface temperature and emissivity. The 3.2 μm absorption band appears ubiquitous across the entire 67P/CG surface (Capaccioni et al. 2015) and has been extensively mapped during the pre-perihelion phases (Filacchione et al. 2016c; Ciarniello et al. 2016). As shown in Fig. 3-bottom left panel, the band is deeper in the Hapi region and some parts of Imhotep (see Fig. 1 for the locations of these named regions). These areas correspond to surface regions that have been shown to exhibit water ice, i.e. in Imhotep the bright area patches (Filacchione et al. 2016a) and roundish features (Auger et al. 2015), or that have been inferred from the deeper band to contain water ice (de Sanctis et al. 2015). The presence of water ice on these areas is confirmed by the corresponding shift of the band center towards shorter wavelengths (Fig. 3-bottom right map).

## 2.3 Surface Composition from Philae

One of the major achievements of the Rosetta mission has been the first landing on a cometary nucleus performed by the Philae lander. The landing occurred on 12th November 2014 when 67P/CG was at heliocentric distance of 2.99 AU. While the descent of the lander was nominal allowing to reach the planned touchdown site (called Agilkia), the successive part of the mission has not followed the planned schedule (Biele et al. 2015). This was caused by the failure of the cold gas thruster and harpoon systems necessary to anchor the lander on the nucleus' surface. Images returned by the lander's camera ROLIS have shown that the Agilkia site is characterized by grains and blocks with size distribution ranging between centimeters to 5 meters (Mottola et al. 2015). The presence of aeolian-like features, similar to wind-tails, suggests the mobilization of the regolith grains and erosion of the surface. The analysis of Agilkia data indicates a soft granular surface, with a compressive strength estimated of the order of 1 kPa, with a thickness of the order of 20 cm displaced

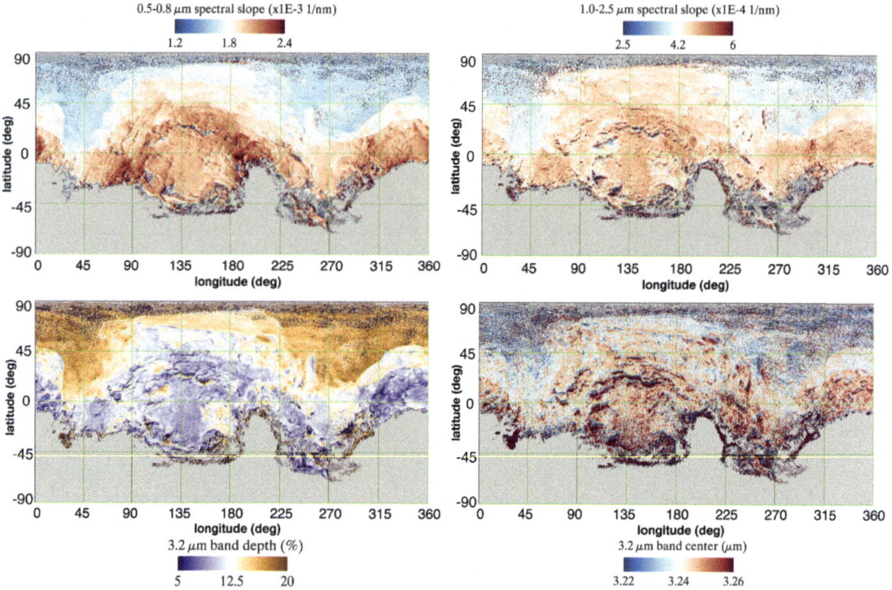


**Fig. 3** Cometary Spectral Indicators for the north and equatorial regions of 67P/CG surface measured by VIRTIS in August–September 2014. Top left panel: 0.5–0.8 μm visible spectral slope. Top right: 1.0–2.5 μm infrared spectral slope. Bottom left: 3.2 μm band depth. Bottom right: 3.2 μm band center. Adapted from Filacchione et al. (2016c)

above a more rigid layer (Biele et al. 2015). The failure of Philae's landing subsystem has caused three successive rebounds of the lander which finally stopped on the rim of the large Hatmehit depression. The final landing site, called Abydos, shows an exposed and hard surface (Biele et al. 2015) with the presence of fractured boulders, elongated cliffs and terrain composed by a texture of small agglomerates of dark granular material and bright (up to 10% reflectance) pebbles of larger dimensions (Bibring et al. 2015). A review of the major results based on measurements recorded by Philae is given in Boehnhardt et al. (2017). Here we focus on the measurements performed by the two gas analyzers on board the lander. COSAC, the Cometary Sampling and Composition experiment (Goesmann et al. 2007) and PTOLEMY, the gas chromatograph and mass spectrometer (Wright et al. 2007), have both operated during the landing phases according to the nominal schedule and therefore their results must be interpreted considering the real lander's trajectory and timing. COSAC has detected the presence of 16 organic compounds, including nitrogen-bearing species but no sulfur-bearing molecules. Among the organic materials, a total of four compounds (methyl isocyanate, acetone, propionaldehyde, and acetamide) never reported on other comets, have been identified (Goesmann et al. 2015). However, isocyanomethane, 2-hydroxyethanal, and propanal could not be confirmed on ROSINA data (Altwegg et al. 2017). A list of compounds observed by COSAC and their possible formation pathways are shown in Fig. 4.

PTOLEMY has reported a wealth of additional information about the nucleus composition (Wright et al. 2015), including: the presence of compounds with $-CH_2-$ and $-O-$ groups having a mass/charge ratios equal to 14 and 16, respectively; the possible presence of polyoxymethylene chains, a radiation-formed polymer, which, however, has not been confirmed by ROSINA (Altwegg et al. 2017); the depletion of aromatic compounds (like benzene) and of sulfur-bearing species; the aromatic molecule toluene found in both PTOLEMY

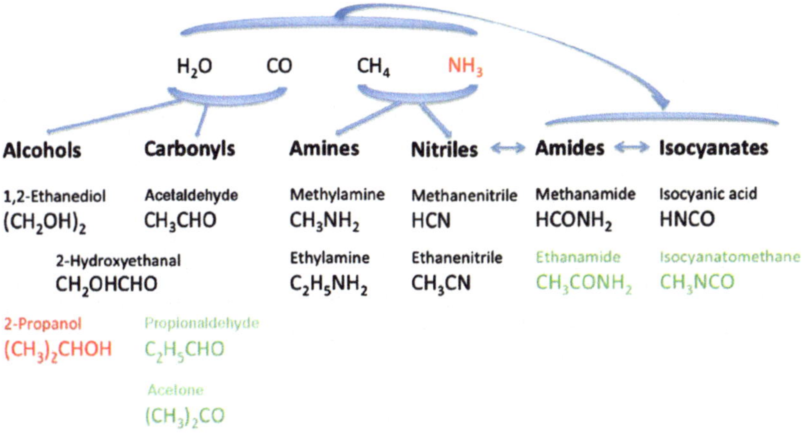


**Fig. 4** Organic compounds detected by COSAC and possible formation pathways. Species marked in red are not confidently identified; species in green were observed for the first time in comets by COSAC. Modified from Goesmann et al. (2015)

and ROSINA-DFMS spectra; very low concentrations of nitrogenous material. Philae lander was successful in detecting volatile species, including water, carbon monoxide and carbon dioxide measured directly at the surface of the comet (Wright et al. 2017). Thanks to the un-nominal Philae's operation it was possible to measure the ratios of the volatiles species sniffed by PTOLEMY on the two landing sites (Agilkia and Abydos). From this analysis results that the two sites have very different ratios which seems to indicate heterogeneity in nucleus composition (albeit measurements could be biased by the different surface morphology of the landing sites and different local times and illumination conditions occurring during Philae operations). Temporal evolution of the mass spectra have been analyzed by Krüger et al. (2017) who has reported that COSAC and PTOLEMY counts decay over time. The first spectra with the highest intensities were recorded by both instruments about 20–30 min after Philae's first touchdown at Agilkia. On the six consecutive measurements taken at Abydos the counts decay. Due to the instruments' designs and vent tubes configurations, the two experiments were preferentially sniffing in two different directions with COSAC collecting regolith samples from the surface while PTOLEMY measured gas species from the ambient coma. As a consequence of this configuration, the diagnostic m/z peak at 44 measured by both instruments has a different origin: for COSAC it is the signature of organic material embedded in surface regolith grains while for PTOLEMY it corresponds to gaseous $CO_2$ in the coma.

## 2.4 Coma as a Pathway to Constrain Surface Composition

As they have spent long time far from the Sun, comets are expected to have preserved their initial composition and be one of the most primitive object of the Solar System (Mumma and Charnley 2011). Therefore determining the composition of the nucleus provides precious information on the prevalent thermo-chemical and physical properties of the nebula in which comets have formed. One possibility to determine the composition of the nucleus is the direct sampling of cometary material. For the first time, the Rosetta mission has succeeded to reach the surface of a comet by landing Philae at the surface of 67P/CG. Initial plan was to drill the upper centimeters of the comet and to analyze the samples with the mass spectrometer PTOLEMY but unfortunately these steps could not be performed (Wright et al. 2015).

Another way to get information on the composition of the volatile species in the nucleus is to study the gaseous species and their outgassing evolution in the coma. Indeed, as the nucleus orbits around the Sun it experiences sublimation and gas release. Thus by measuring the composition in the coma we indirectly sample the nucleus composition. However, relationships between the coma and nucleus composition are complex (de Almeida et al. 1996; Huebner and Benkhoff 1999) because volatiles can be present as condensed and/or gaseous phase in the nucleus. Among volatiles, pure ices have their own thermodynamical properties and will sublimate at different temperature, pressure and depth (Huebner et al. 2006; Fray and Schmitt 2009; Marboeuf and Schmitt 2014). In addition, the structure of ices, i.e. amorphous, crystalline and clathrate hydrate, influence the outgassing pattern according to the way volatiles are trapped (Espinasse et al. 1991; de Sanctis et al. 2005; Huebner et al. 2006; Marboeuf and Schmitt 2014). Also the presence of a dust mantle at the surface of the nucleus will affect the intensity of the outgassing (Marboeuf and Schmitt 2014). Finally, gas release is strongly dependent on orbital parameters such as heliocentric distance, obliquity of the object, period of rotation and shape of the nucleus (Huebner and Benkhoff 1999; Fulle et al. 2016b). As a result of the combined effects of all these parameters, the mixing ratio measured in the coma often do not reflect directly the abundances of volatiles in the nucleus (Espinasse et al. 1991; Benkhoff 1995; de Almeida et al. 1996; Huebner 2008; Marboeuf and Schmitt 2014), and interpretation of the outgassing data should be done very carefully. For example, the fact that coma composition presents heterogeneities does not necessary imply similar compositional heterogeneities in the nucleus (Huebner and Benkhoff 1999; Hässig et al. 2015). Studies based on coma composition either interpreted it as a result of a homogeneous or either heterogeneous nucleus composition. Analyses of a fragment of the comet 73P/Schwassmann–Wachmann 3 revealed that the composition was rather homogeneous (Dello Russo et al. 2007). In contrast, observations of freshly exposed material of comet 9P/Tempel 1 during the Deep Impact mission rather suggest a heterogeneous composition (Feaga et al. 2007). Comet 103P/Hartley 2, which presents a bi-lobate shape comparable to the one of 67P/CG, exhibited a heterogeneous coma which has been interpreted to reflect variations of composition within the nucleus (A'Hearn et al. 2011). However, so far most of the observations of explored comets were sparse and cannot provide an insightful comprehension of the nuclei compositions and properties.

For the first time the Rosetta mission has extensively investigated the coma evolution of a comet during a long period spanning around the perihelion passage with a comprehensive set of instruments. Abundances of molecules in the coma of 67P/CG have been measured in detail by the ROSINA instrument (Balsiger et al. 2007) revealing a CO- and $CO_2$-rich nucleus (compared to other Jupiter Family Comets), the presence of highly volatile molecules and organic molecules (Le Roy et al. 2015). The coma monitoring at different heliocentric distances and above different latitudes on the surface of the nucleus (see Fig. 5) has revealed large fluctuations in the outgassing pattern (Hässig et al. 2015; Luspay-Kuti et al. 2015). The major species detected, $H_2O$, CO and $CO_2$ do not always evolve with the same periodicity and intensity. Conversely, minor species seem to follow either the $H_2O$ or the $CO_2$ emissions (Luspay-Kuti et al. 2015). Apart from in situ measurements, the spatial and temporal distributions of gaseous species in the 67P/CG coma have been measured also by remote sensing payloads, e.g. in the infrared by VIRTIS (Coradini et al. 2007) and at microwaves by MIRO (Gulkis et al. 2007). The fluorescences of water vapor (at 2.67 μm) and $CO_2$ (at 4.27 μm) measured by VIRTIS between November 2014 and January 2015 when the heliocentric distance was decreasing from 2.91 to 2.47 AU were analyzed by Bockelée-Morvan

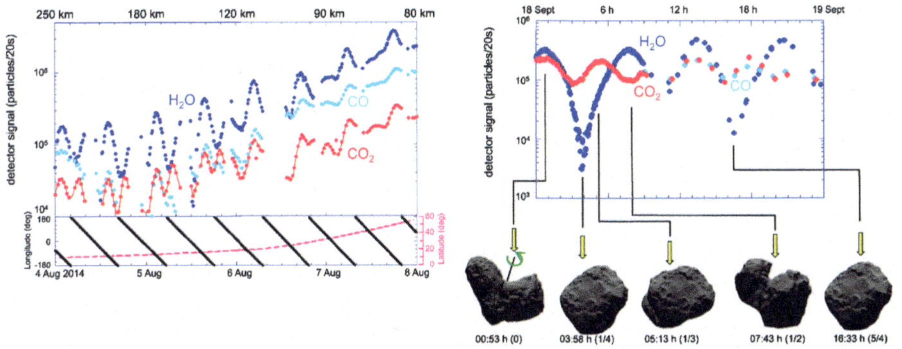


**Fig. 5** Left panel: $H_2O$, CO and $CO_2$ molecular densities from in-situ measurements by ROSINA/DFM within the coma of 67P/CG (upper panel) on August 4th–8th 2014. The spacecraft explored different longitude and latitude ranges during that time (lower panel). Right panel: ROSINA/DFMS molecules density measurements of $H_2O$, CO and $CO_2$ in-situ in the coma of 67P/CG for September 18th 2014 correlated to the spacecraft view of the nucleus. From Hässig et al. (2015)

et al. (2015). This study has shown a constant water production rate during this period of time, with higher column abundance detected above the active areas in Hapi. A similar distribution has been observed in the sub-mm domain by MIRO (Biver et al. 2015; Lee et al. 2015). While the water emission is correlated with solar illumination, $CO_2$ appears not to be influenced by this as it was observed on both illuminated and non-illuminated areas, including on the south hemisphere (Fink et al. 2016). The derived $CO_2/H_2O$ column density ratio varies from 2% (on illuminated areas) to 60%. The study of the fluorescence emissions of $H_2O$, $CO_2$, $CH_4$ and OCS species in the coma around the perihelion passage indicates an abrupt increase of the water production about six days after perihelion while the abundances of $CO_2$, $CH_4$ and OCS relative to $H_2O$ increase by a factor 2 (Bockelée-Morvan et al. 2016). This behavior can be explained by the erosion of the dust layer on the surface occurring at perihelion which exposes subsurface volatile-rich layers.

The heterogeneity of the gaseous species observed in the 67P/CG coma has been interpreted in terms of:

1. Homogeneous nucleus composition. The fluctuation of gas density in the coma shows strong correlation to the rotation period and the latitude of observation which suggests that the outgassing is mainly solar insolation-driven (Hässig et al. 2015). Variations of composition in the coma are enhanced by seasonal and diurnal effects due to the combination of the large spin obliquity and the complex shape of the nucleus (Preusker et al. 2015). Therefore, if 67P/CG nucleus is homogeneous in composition, the heterogeneities observed in the coma must be controlled by the illumination (Hässig et al. 2015; Bieler et al. 2015; Fougere et al. 2016; Fulle et al. 2016b; Hoang et al. 2017).
2. Heterogeneous nucleus composition. The lack of correlation suggests that CO and $CO_2$ are not embedded by $H_2O$ (Hässig et al. 2015). Therefore, $H_2O$ can be present at the surface while CO and $CO_2$, which are more volatile, will sublimate from deeper layers. Variation of composition will be generated in the sub-surface of the comet as the nucleus orbits around the Sun and receives different amount of solar energy. The complex shape of the nucleus will strongly enhance the variation of illumination and therefore sub-surface heterogeneities from one area to another (Hässig et al. 2015; Luspay-Kuti et al. 2015; Fink et al. 2016; Hoang et al. 2017). Some asymmetries detected in the $H_2O$ and $CO_2$ outgassing between the northern and southern regions could be attributed to dif-

ferent characteristics of the nucleus (Hässig et al. 2015; Le Roy et al. 2015; Luspay-Kuti et al. 2015). They can result from heterogeneities in the nucleus with somehow more $CO_2$ compared to $H_2O$ in the southern plains (Luspay-Kuti et al. 2015). An alternative hypothesis is that these variations are induced by local heterogeneities of the surface properties such as ice sintering or dust coverage due to local morphology (Keller et al. 2015; Luspay-Kuti et al. 2015; Hoang et al. 2017). The presence of a dust mantle in the northern plains of 67P/CG was observed by Thomas et al. (2015). Potential presence of condensed $H_2O$ in this layer (de Almeida et al. 1996; Rubin et al. 2014) can enhance the $H_2O$ outgassing while it inhibits the $CO_2$ release (Hoang et al. 2017). The presence of a dust mantle could influence the outgassing pattern by shifting the maximum of gas release in dependence on local time related to the mantle thickness (Marboeuf and Schmitt 2014). Another evidence of heterogeneities in the coma is the singular behavior of $CH_4$ which shows a poor correlation with other species (Luspay-Kuti et al. 2015). Numerical simulations suggest that $CH_4$, and possibly $C_2H_6$, can potentially be trapped in clathrates at shallow depths (Mousis et al. 2015; Luspay-Kuti et al. 2016). However, observations and numerical simulations seem to preclude the presence of amorphous phase (Luspay-Kuti et al. 2016) as it has been proposed by Vincent et al. (2015) and Mousis et al. (2015).

A consensus seems to emerge from different studies that the 67P/CG outgassing pattern is driven mainly by illumination. Heterogeneities detected in the coma are, at global scale, induced by the combination of the tilt of the rotation axis and the complex shape of the nucleus. The strong variation of illumination between different areas of the nucleus's surface is thought to produce heterogeneities in the sub-surface composition. However, insolation variations do not succeed to explain completely the behavior of the species' outgassing. Heterogeneities of the surface-subsurface properties of the nucleus such as the presence of a dust mantle or of clathrates could explain some of the gas behavior.

While gaseous emissions in the coma trace the volatile/ice species on the nucleus from which they have been released, dust grains maintain the minerals and organic materials present on the surface. Dust grains were measured by a dedicated suite of in-situ instruments onboard Rosetta: GIADA, the Grain Impact Analyser and Dust Accumulator (Colangeli et al. 2007), has measured the dust flux from grains larger than 100 μm and has inferred a dust-to-water ice mass ratio of six from a heliocentric distance of 3.6 AU to perihelion (Rotundi et al. 2015; Fulle et al. 2016a). The same authors state that the mass of dust particles is mainly contained within compact grains having a density of about 2000 $kg\,m^{-3}$ and sizes larger than 1 mm. As reported by Della Corte et al. (2015), coma dust grains appear in two distinct families: compact particles and fluffy porous aggregates having masses from $10^{-10}$ to $10^{-7}$ kg. COSIMA, the Cometary Secondary Ion Mass Analyser (Kissel et al. 2007) is a time-of-flight secondary-ion mass spectrometer (TOF-SIMS) instrument able to resolve particles with sizes from tens of microns to millimeters and to capture microscopic images of the grains after their impact on exposed targets (Langevin et al. 2016). By means of secondary ion mass spectroscopy, Bardyn et al. (2017) have modeled grains composition with organic matter for 45% of the mass and anhydrous mineral phases for the remaining 55% (see Fig. 6). The loss of volatiles in the dust is a consequence of the heating of the grains in the coma. The average atomic composition of the grains is modeled with H (30%), C (30%), O (30%) with Si, Fe, Mg and other species accounting for the remaining 10%. All these results were derived with certain assumptions on the H/C and O/C ratios which cannot be measured by the instrument. Noteworthily, within the dust size range (from 50 to 1000 μm) investigated by COSIMA, any particle that would consist of a single mineral or pure organic has never been seen. This means that all the single grains contained both

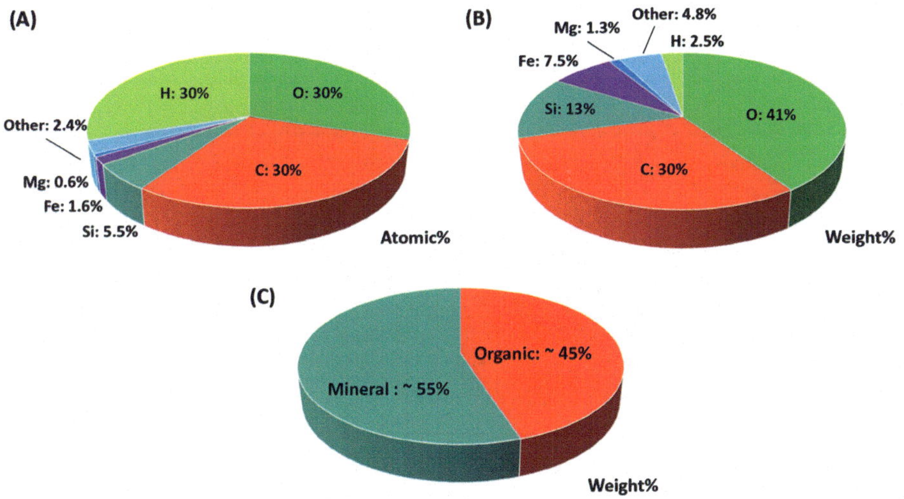


**Fig. 6** Averaged composition of 67P/CG's dust particles as deduced from COSIMA measurements: panel (**A**) by atomic fraction; panel (**B**) by atomic mass fraction. Panel (**C**) is the mineral and organic content estimated in mass fraction. Figure from Bardyn et al. (2017)

organics and minerals, which means that the non-icy components are very well-mixed at the intimate level up to a relatively small scale.

MIDAS, the Micro-Imaging Dust Analysis System (Riedler et al. 2007), used atomic-force microscopy techniques to characterize grains from nanometers to micron sizes which appear as hierarchical agglomerates of fractal and compact particles (Mannel et al. 2016; Bentley et al. 2016).

## 3 Compositional Endmembers

Following the 'dirty snowball' model of Whipple (1950) in which the cometary nuclei are described as a mixture of ices, CHON (carbon, hydrogen, oxygen, and nitrogen) organic material and rocky particles, in the following sections we review the current knowledge of these endmembers in the context of the aforementioned Rosetta's results.

### 3.1 Volatiles/Ices

So far, only two ice species have been identified on the surfaces of cometari nuclei by means of infrared spectroscopy: *crystalline water ice* and *carbon dioxide ice*. Water and carbon dioxide ices have been detected by Rosetta/VIRTIS-M on the surface of 67P/CG during the pre-perihelion time. Deep Impact/HRI has observed water ice deposits on 9P/Tempel 1. A summary of the properties of the ices is given in Table 4.

During the pre-perihelion phase, *crystalline water ice* has been spectroscopically identified at near infrared wavelengths in three distinct modalities on 67P/CG nucleus:

1. On the active areas of Hapi region where water ice changes its abundance with local time and illumination conditions, condensing during the night hours and sublimating during daytime (de Sanctis et al. 2015). During each rotation, when the surface is illuminated by

**Table 4** Infrared spectroscopic identification of ice species, abundances and distributions on the surfaces of cometary nuclei

| Comet | Region | Dimensions | Ice | Abundance | Grain size (Mixing[a]) | Reference |
|---|---|---|---|---|---|---|
| 67P/CG | Hapi | $\leq$100 m[b] | $H_2O$ (cryst.) | 1–14% | 1–3 µm (I) | de Sanctis et al. (2015) |
| 67P/CG | Imhotep[c] | 100 × 100 m | $H_2O$ (cryst.) | 6% | 33–72 µm (I), 1.4–2.6 mm (A) | Filacchione et al. (2016a) |
| 67P/CG | Various[d] | 1–60 m | $H_2O$ (cryst.) | 0.3–4% | 10 µm (I), 6.5 mm (A) | Barucci et al. (2016) |
| 67P/CG | Anhur[e] | 60 × 80 m | $CO_2$ | 0.1% | 50 µm (A) | Filacchione et al. (2016b) |
| 9P/Tempel 1 | Various | 0.55 km$^2$ | $H_2O$ | 3–6% | 30 ± 20 µm (I) | Sunshine et al. (2006) |

Note: [a]A = Areal mixing, I = Intimate mixing; [b]from shadow line. Labels [c], [d], [e] indicate similar areas in Table 5 for comparison

the Sun, the water ice sublimates mainly from the inner layers reached by the propagation of the thermal wave. As soon as the same surface goes into shadow (or moves into the night side) surface temperature rapidly drops. Since Hapi is a region with a gravitational minimum, the surface here is preferentially blanketed by a many meters-thick layer of dust (Keller et al. 2017). In these conditions a temperature inversion occurs between the colder surface layer and the interior which remains at higher temperature for a longer time. The entire process is driven by the illumination conditions by means of the duration of the shadow/night and daytime periods, and by the surface properties, e.g. surface thermal inertia and thermal skin depth. Because the shadow is cast onto Hapi alternating by 67P/CG's head and body at the time of the observation, and due to the short rotation period, equal to 12.4 hours (Mottola et al. 2014; Jorda et al. 2016), the sublimation and condensation process occurred two times at each rotation. As a consequence of this, very small water ice grains in intimate mixture with the dark terrain condense on the surface. The water ice particles have diameters of 1–3 µm and abundances up to 14% for the areas close to the shadow line where the measured temperature is 165–170 K (see Fig. 7). The condensation–sublimation process seems to be common among comets. It has been also proposed to explain the surge of outbursts observed on comet 9P/Tempel 1 from areas near sunrise (Prialnik et al. 2008).

2. On recent debris fields collapsed from two elevated structures in the Imhotep region where more fresh and pristine material is exposed (Filacchione et al. 2016a). These areas are named BAPs, i.e. Bright Area Patches (Fig. 8). The gravitational displacement of the surface caused by overhang falls and the disruption of elevated structures (Groussin et al. 2015a) allows the exposure of more pristine material from the interior. The best-modeling of the VIRTIS spectra of the BAPs (panel g in Fig. 8) corresponds with a solution in which 1.2% of the pixel area is occupied by patches of water ice particles with a diameter around 2 mm while the remaining 98.8% of the pixel area consists of an intimate mixture of water ice (3.4%, 56 µm grain diameter) and average dark terrain (95.4%). Therefore the average total amount of surface water ice detected on the BAPs corresponds to about 4.6% for a 2.5 × 2.5 m pixel footprint. The measured BAPs' surface temperature ($T < 160$–180 K) is about 20–25 K colder than in the surrounding areas but still warm enough to allow the presence of crystalline ice.

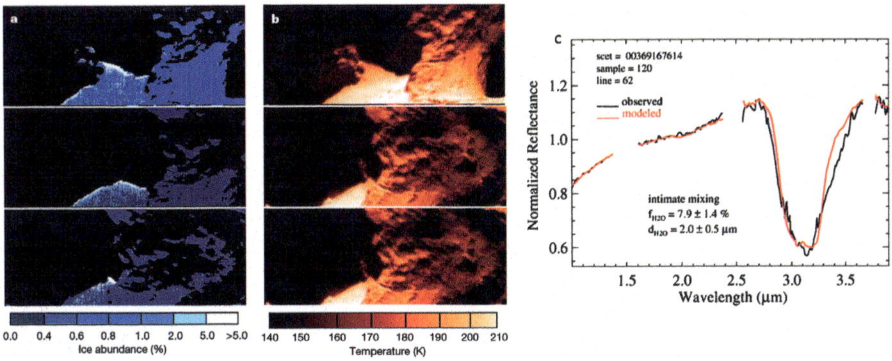


**Fig. 7** Panel (**a**): Ice abundance (by volume) maps on the Hapi active area; (**b**): Temperature images; (**c**): Best spectral fit to VIRTIS data corresponding to a mixture of average dark terrain and water ice grains. From de Sanctis et al. (2015)

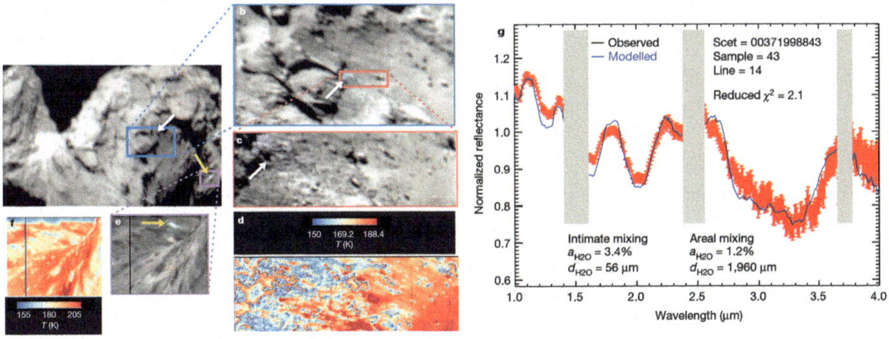


**Fig. 8** Panel (**a**): color image showing BAP1 and BAP2 (white and yellow arrows, respectively). (**b**): Water–ice-rich waste terrain on BAP1 is visible (blue) on the right side of the circular elevated structure. (**c**): Close-up image acquisition of BAP1 reveals the bluish color of the water–ice-rich unit. (**d**): Temperature map showing the water–ice-rich terrain at $T < 160$ K. (**e**): Hillside image showing BAP2. (**f**): Temperature map showing the water–ice-rich terrain at $T < 180$ K. All color images are obtained from the $B = 1.3$ µm, $G = 2.0$ µm and $R = 2.9$ µm channels. (**g**): BAP1 water–ice-rich VIRTIS spectrum (black curve with error bars in red) and best-fit synthetic model (blue curve). $a_{H_2O}$ is the percentage of crystalline water ice and $d_{H_2O}$ is the water–ice grain size. 'Scet' is the spacecraft time used to identify the observation; 'sample' and 'line' correspond to the position of the pixel on the image. The grayed-out spectral ranges are affected by instrumental order sorting filters. From Filacchione et al. (2016a)

3. On eight bright spots located in the Khonsu, Imhotep, Anhur, Atum and Khepry regions where single or multiple grouped icy patches are observed (Barucci et al. 2016). Some of those areas were first detected on high resolution OSIRIS images (Pommerol et al. 2015) and form a plethora of small bright spots, or much larger individual or twin bright patches spanning from few meters to about 60 m in size. The spots are characterized by high albedo and blue visible colors in striking contrast with the dark-red color of the average terrain. OSIRIS images and visible spectra are shown in Fig. 9 for one of the eight spots discussed by Barucci et al. (2016). The spots are preferentially located in various morphological terrains within the equatorial regions, including freshly exposed cliffs, alcoves, flat terraces, layers, accumulation deposits, ancient outgassing conduits and rocky terrains. In general these spots appear on consolidated and dust-free areas

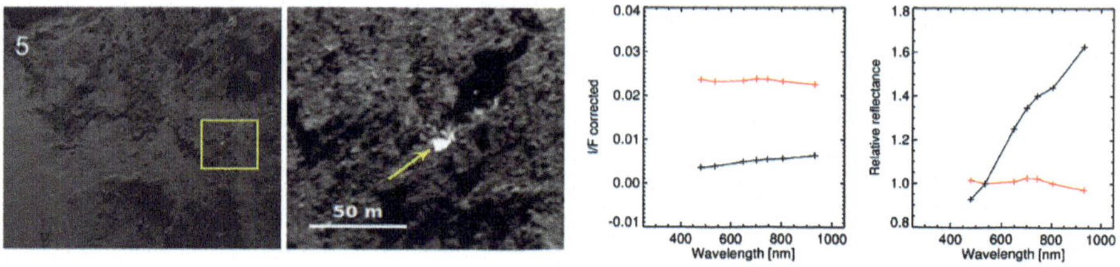


**Fig. 9** Optical identification of a water ice bright spot in the Imhotep region by OSIRIS with F22 filter (at 649.2 nm). First column: context image. Second column: zoomed image. Third column: I/F of the bright spots (red plot) shows higher reflectance than the surrounding non-icy area (black). Fourth column: relative reflectance (normalized to F23 filter at 535 nm) of the bright spot (red) is spectrally flat while the surrounding area (black) is very red. From Barucci et al. (2016), spot#5

and in some cases correspond to the sources of active processes (outbursts). On many of them the surface water ice continued to be visible for weeks and months, while others are more variable and correlated with the diurnal variation (Raponi et al. 2016). Co-located infrared spectroscopy by VIRTIS shows very variable water ice grain properties from spot to spot but within the limits given by the two aforementioned cases: water ice areal abundances are ranging between 0.3 to 4% with the presence of both small 10 μm grains in intimate mixing with the dark terrain or very large mm-sized grains in areal mixing. Additionally, another debris field (CF3) enriched in water ice has been detected in the OSIRIS images in Imhotep region and confirmed by the presence of the 2 μm absorption band on infrared spectra (Oklay et al. 2017).

A summary of the infrared spectroscopic identification of water ice, abundances and distributions on the surface of 67P/CG is given in Table 4). Using spectroscopy, visible photometry, and imagery many of those water ice-rich areas are observed for prolonged period of time during the entire duration of the Rosetta mission on OSIRIS images. The large clusters (BAP 1, 2 and CF1, 2, 3) close to the cliffs survived on well-illuminated surface locations for about 10 months, and several boulders enriched in water ice were still existing after 1.5 years of the first detection (Oklay et al. 2017). The isolated water ice-rich area coded IF2 (Oklay et al. 2017), which survived the entire mission, was also spectrally confirmed using early IR spectra.

Despite *carbon dioxide* being the second most abundant gaseous species detected on cometary comae after water vapor (Bockelée-Morvan and Biver 2017), the presence in the form of ice on nuclei surfaces has been always elusive due to its high volatility. Before Rosetta, in fact, it has never been observed on cometary nuclei. So far, carbon dioxide ice has been detected by VIRTIS only in a 60 × 80 m area in the Anhur region while it was exiting from a four year-long winter-night season (Filacchione et al. 2016b). The spectroscopic identification has been possible by means of the diagnostic 2.0 μm triplet and the 2.6–2.75 μm bands (Fig. 10). This ice deposit underwent rapid sublimation, disappearing within about one month after its initial detection in March 21 2015 (Fig. 10). A heat balance calculation for this specific area shows that an erosion of a 9 cm thick layer of carbon dioxide ice, for a total of 56 kg, has occurred. The ice patch was presumably formed after the comet's previous perihelion passage, in 2009, when the carbon dioxide outgassing from the nucleus' deeper layers condensed on the cold surface. The Anhur region is in fact located in the southern hemisphere which experienced a four year-long winter-night season during which the surface temperatures have dropped to sufficiently low values to allow the carbon dioxide to condense as ice. Moving towards perihelion, with the increase of the solar illumination, the carbon dioxide ice has rapidly disappeared.

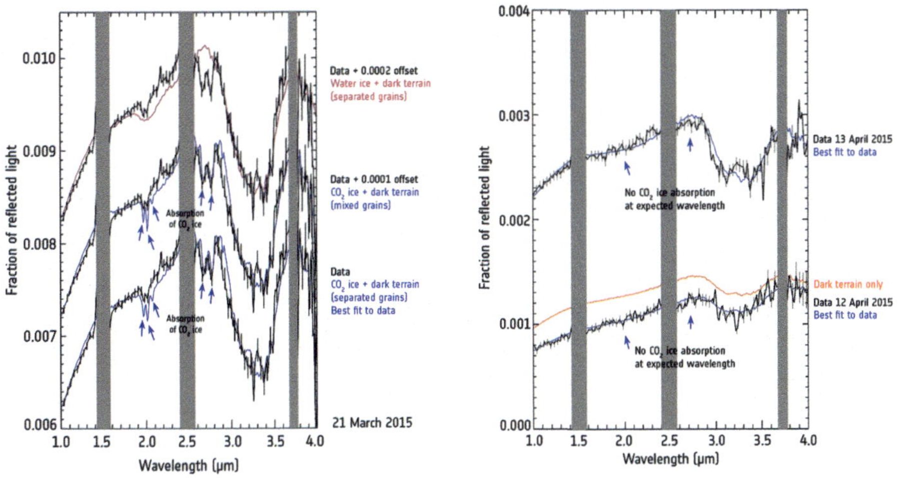


**Fig. 10** Spectroscopic identification of transient $CO_2$ ice in the Anhur region. Left panel: March 21 2015 spectra showing the detection of $CO_2$ ice and spectral modeling results. Note that the modeling with water ice (top plot) fails to match the observed data. Right panel: the $CO_2$ ice features are no more visible on April 12–13 2015 observations taken on the same area. The grayed-out spectral ranges are affected by instrumental order sorting filters. From Filacchione et al. (2016b)

The OSIRIS camera has observed two unusually large patches of water ice in the southern regions of Anhur and Bes (Fig. 11) resulting in 20–30% of water ice mixed with the dark material in a layer up to 30 cm thick (Fornasier et al. 2016). One of these patches has been located in the same area where the carbon dioxide ice sheet has been revealed by VIRTIS about one month before. Also the water ice patches have completely disappeared within about ten days after their initial detection. This indicates the presence of a dual layer, with superficial carbon dioxide ice overlayed above water ice, as result of the different condensation temperatures of the two species ($CO_2$ needs lower temperature than $H_2O$ to condense as ice) occurring during the winter-night season at aphelion. The double detection by VIRTIS and OSIRIS reveals the seasonal cycles of carbon dioxide and water ices, which unfolds during the 6.5-year orbit of 67P/CG, as opposed to the daily cycle of water ice occurring on Hapi (de Sanctis et al. 2015).

One of the major advancements of the Rosetta observations is the retrieval of the distribution of the superficial crystalline water ice grains which appear distributed in three populations: (1) small micron-sized grains are frequent on the active areas where the rapidly changing diurnal variations (de Sanctis et al. 2015) do not allow the formation of larger grains. These grains can be constrained only by means of the $H_2O$ stretch band superimposed on the deep 3.2 μm organic band and by the absence of the other diagnostic bands at 1.05, 1.25, 1.5, 1.65, 2.05 μm (see Fig. 7, panel c) which characterize larger grains; micron-sized grains are also candidates to constrain the average 1% amount of water ice visible on the dark dehydrated terrains (Capaccioni et al. 2015). (2) intermediate grains, e.g. tens of microns, appear preferentially in intimate mixing with the dark material in the water ice-rich patches (see Fig. 8, panel g). (3) large mm-size grains are built from the growth of secondary ice crystals from vapour diffusion in ice-rich colder layers and/or by ice grain sintering caused by thermal waves. Due to their dimensions, these grains are dispersed in areal mixing with the average dark terrain.

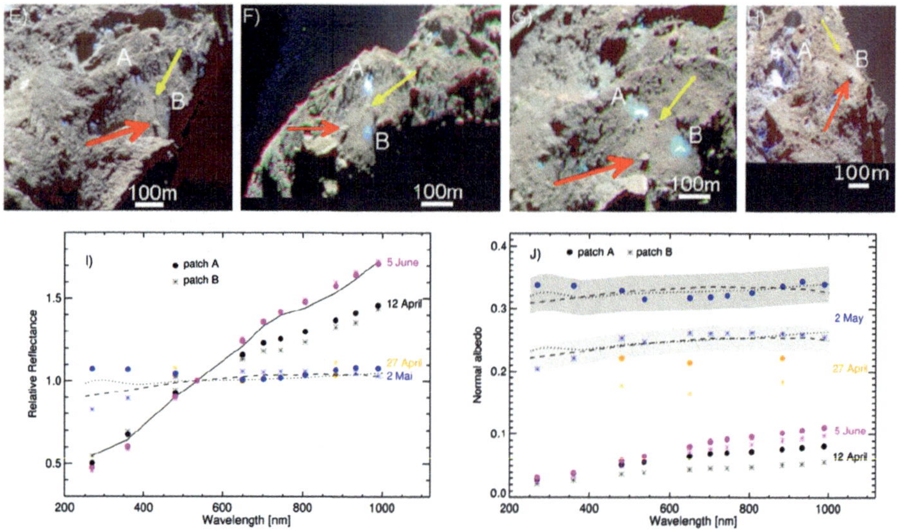


**Fig. 11** Composite OSIRIS images (882 nm, 649 nm and 480 nm filters) showing the appearance of the bright patches in the Anhur/Bes regions (panels **E**, **F**, **G**, **H**). Red and Yellow arrows indicate two common boulders. The reflectance relative to 535 nm and the normal albedo are represented in the (**I**) and (**J**) panels: black line is the average spectrum of the nucleus from a region outside the bright patches; dashed and dotted lines show the best fit spectral models for the patches (in gray the associated uncertainty) as given by a linear mixture of the cometary dark terrain enriched with 21 ± 3% of water ice (dashed line), or 23 ± 3% of water ice (dotted line) for patch B, and with 29 ± 3% of water ice (dashed line), or 32 ± 3% of water frost (dotted line) for patch A. Simulations are performed by assuming a water ice grain size of 30 µm. From Fornasier et al. (2016)

Spectral modeling allows to quantitatively correlate the amount of water ice on the surface with VIRTIS comet spectral indicators (CSI) as introduced by Filacchione et al. (2016c). The CSI method is capable to infer the abundance of the water ice from the measured VIS-IR spectra as well as to reduce the data set dimension. The standard CSI are:

- $S_{VIS}$ is the spectral slope in the VIS range, calculated by best fitting the reflectance in the 0.55–0.8 µm range by a linear function after normalization to 1 at 0.55 µm;
- $S_{IR}$ is the spectral slope in the IR range, calculated by best fitting the reflectance in the 1.0–2.35 µm range by a linear function after normalization to 1 at 1.0 µm;
- BA is the band area of the absorption at 3.2 µm computed after removal of the 2.7–3.6 µm continuum;
- BC is the band centre of the absorption at 3.2 µm computed after continuum removal as the position of the minimum of a third-degree polynomial fit in the 2.7–3.6 µm range.

The variability of these indicators as a function of the water ice abundance (up to 5%) is shown in the left panel of Fig. 12. Concurrent spectral modeling in the visible and infrared ranges allows to study the visible spectral slope value as a function of the water ice abundance determined by infrared absorption bands once the mixing modality (areal or intimate) and dark terrain properties have been defined (Raponi et al. 2016). In Fig. 12 are shown the water ice superficial abundances for different mixing modalities. The models allow a quantitative derivation of the amount of surface water ice by the measurement of the visible color: a completely anhydrous surface has an average spectral slope of 2.2 $\mu m^{-1}$. Slopes lower than 1.2 $\mu m^{-1}$ correspond to water ice abundance higher than 5% in intimate mix-

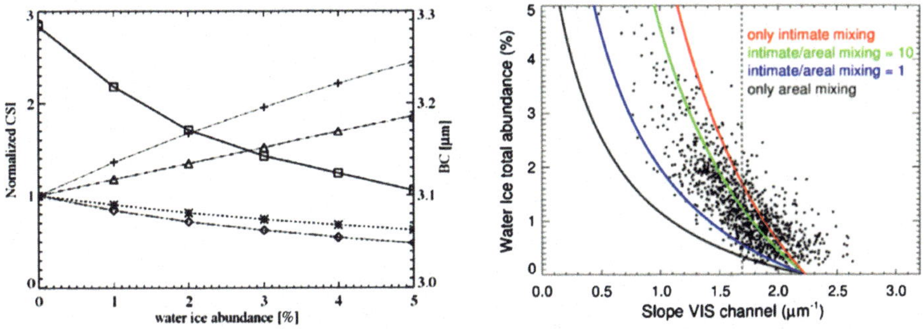


**Fig. 12** Left panel: Simulated VIRTIS comet spectral indicators (CSI) normalized at their value computed on 67P/CG average dark terrain (shown in Fig. 2) as a function of water ice abundance: $S_{VIS}$ (diamonds), $S_{IR}$ (asterisks), 3.2 μm BA (crosses) and single scattering albedo (triangles). The 3.2 μm BC (squares) is not normalized, and its scale is reported on the right axis. From Ciarniello et al. (2016). Right panel: theoretical abundance of water ice as a function of the slope in the VIS channel, as calculated by means of the Hapke (1993) radiative transfer model. The red and black lines correspond to water ice entirely in intimate or in areal mixture with the dark terrain, respectively. The blue and green lines correspond to the ratio between the amount of water ice in intimate and areal mixture equal to 1 and 10, respectively. Dots: water ice total abundance as a function of the measured slope in the VIRTIS VIS channel (for single acquisitions). Dashed grey line: average slope of the dots. The width of the dashed line is equal to the error on the average slope. From Raponi et al. (2016)

ing or 1% in areal mixing at 95° phase angle. To scale the result for different phase angles one should account for the phase function which produces slope variations of the order of $+0.007\ \mu m^{-1}\ deg^{-1}$, according to Ciarniello et al. (2015).

Operating at visible wavelengths, OSIRIS is not capable like VIRTIS to detect the diagnostic absorption features of the water ice and therefore cannot derive its physical properties (abundance, grain size, temperature). However, regions enriched in water ice are easily detectable due to their blue color on RGB composites generated by using the images taken at NIR, visible, blue wavelengths. Conversely, OSIRIS images allow to observe the nucleus at much higher spatial resolution than VIRTIS and to better correlate the color variations with local morphology. On OSIRIS decorrelated-stretch color images is possible to derive the visible spectral slope which can be used as a proxy for the presence of water ice (Oklay et al. 2016a):

$$slope\left(\frac{\%}{100\ \text{nm}}\right) = \frac{(R_{NIR} - R_{green}) \cdot 10.000}{(\lambda_{NIR} - \lambda_{green}) \cdot R_{green}} \tag{1}$$

where $R(\lambda)$ are the radiance factors at a given filter wavelength $\lambda$. Oklay et al. (2016a) have identified three surface classes on 67P/CG nucleus by means of OSIRIS spectral slopes values:

1. 8 to 13%/100 nm corresponds to the surfaces where activity is observed: Hapi, active pits, alcoves, and bright water ice-rich features in Imhotep;
2. 13 to 15%/100 nm characterizes either surfaces partially covered by material ejected from active spots or inactive areas peppered with small bright features;
3. >15%/100 nm is typical red color of the inactive surfaces where the dehydrated dust dominates.

In Fig. 13 is shown a sequence of color images and corresponding spectral slopes showing long-lasting (several months up to the entire mission) clusters and isolated icy features in

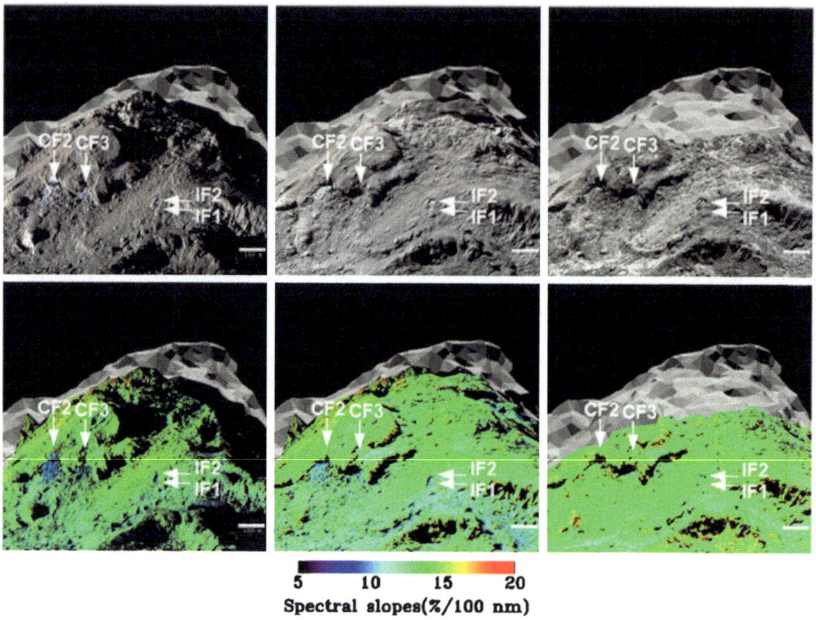


**Fig. 13** Top panel: sequence of images showing several cluster features (CF) and isolated features (IF) on three epochs of 2014, 2015 and 2016 (from left to right column). A 100 m scale-bar is shown in the right corner of each image. All images are projected onto a 3D shape model and displayed in a common frame to show the features in the same orientation. Images taken at 882.1 and 649.2 nm are used in the R and G channels, while 360.0 or 480.7 nm images are used for the B channel. Bottom panel: spectral slope images, showing the reduction of the ice on the surface. Retrieved water ice abundances for CF-IF features are given in Table 5. From Oklay et al. (2017)

Imhotep. All these regions enriched in water ice were existing on the surface when Rosetta reached the comet. The initial bluish color observed in 2014 on these features became much less evident in 2015 and finally disappeared in 2016 (top panel). However, the large isolated water icy feature (labeled IF) was existing on the surface since the first imaging to the last imaging of the area (Oklay et al. 2017). At the same time the spectral slopes from neutral become more and more red, similar to the non icy terrain, indicating that the ice fraction on the surface has sublimated (bottom panel). A further discussion about compositional changes caused by diurnal and seasonal effects is given in Sect. 4.

The properties of the water ice rich areas on several locations as derived from OSIRIS images are reported in Table 5. The spectral modeling in the visible spectral range performed by Fornasier et al. (2016) and Oklay et al. (2017) is based on the same assumption of areal mixing between 30 μm water ice grains and dark terrain. The thermal modeling performed in both studies gave more information about the water ice content, including the ice thickness estimations from several tens of centimeters up to several meters according to the lifetime of different water ice rich features. The locations for which retrievals from infrared observations are available are marked for comparison with Table 4. On average, water ice abundances derived from OSIRIS images are up to a factor 4 higher than those derived from VIRTIS data. This difference can be ascribed to three concurrent effects that need to be considered when comparing the results: (1) the spatial resolution between the two instruments is very different with OSIRIS NAC resolution about 7 times better than VIRTIS. As a consequence of this, OSIRIS can detect more pure areas within one VIRTIS pixel resulting in

**Table 5** Visible filters identification of water ice species, abundances and distributions on the surface of 67P/CG. Water ice abundance is computed assuming areal mixing and 30 μm water ice grain size

| Comet | Region | Area size | Ice | Abundance | Reference |
|---|---|---|---|---|---|
| 67P/CG | Imhotep CF1[c] | 8373.8 $m^2$ | $H_2O$ | 6 to 25% | Oklay et al. (2017) |
| 67P/CG | Imhotep CF2[c] | 14652.8 $m^2$ | $H_2O$ | 6 to 20% | Oklay et al. (2017) |
| 67P/CG | Imhotep CF3 | 4652.2 $m^2$ | $H_2O$ | 15 to 10% | Oklay et al. (2017) |
| 67P/CG | Imhotep IF1[d] | 126.5 $m^2$ | $H_2O$ | 6% | Oklay et al. (2017) |
| 67P/CG | Imhotep IF2 | 74.5 $m^2$ | $H_2O$ | 48% | Oklay et al. (2017) |
| 67P/CG | Anhur-Bes patch A[e] | $\approx$1500 $m^2$ | $H_2O$ | 21% (ice) or 23% (frost) | Fornasier et al. (2016) |
| 67P/CG | Anhur-Bes patch B[e] | $\approx$1500 $m^2$ | $H_2O$ | 29% (ice) or 32% (frost) | Fornasier et al. (2016) |

Labels [c], [d], [e] indicate similar areas in Table 4 for comparison

locally higher abundances; (2) the a-priori choice to use only areal mixing and fixed grain size to model OSIRIS data; (3) the different timing of OSIRIS and VIRTIS observations.

Apart from 67P/CG, water ice deposits have been observed on the surface of the 9P/Tempel 1 nucleus by Deep Impact (Sunshine et al. 2006) and identified by means of their blue colors at visible wavelengths and by absorption bands at 1.5, 2.0, 3.0 μm (Fig. 14). Spectral modeling indicates that the deposits are 3–6% enriched in water ice with grain sizes of $30 \pm 20$ μm. Nucleus morphology and illumination conditions limit the surface temperature between 285 and 295 K on the deposits, much colder than the maximum 330 K observed on the more illuminated areas. Moreover, the rapid appearance of large amounts of volatiles in the earliest ejecta after the impact of the projectile probe implies that water ice is quite near the surface (A'Hearn et al. 2005b). While detected on expelled chunks in the coma, water ice has not been directly detected by the EPOXI mission on the surface of 103P/Hartley 2 (A'Hearn et al. 2011). Also Deep Space 1 has not found spectral evidence of absorptions due to water ice in the 1.3–2.6 μm spectral interval when observing the 9P/Borrelly nucleus with a spatial resolution of 160 m (Soderblom et al. 2002).

## 3.2 Organic Materials

While cometary ice species are easily identifiable on VIS-IR reflectance data by means of their high albedo and numerous and well-known absorption bands, a much more challenging effort is to determine the composition of complex organic materials which can be grouped in the 'CHON' (Carbon, Hydrogen, Oxygen, Nitrogen) families (Whipple 1950; Clark et al. 1987). The emission features caused by the stretching and bending modes of organic (aliphatic and aromatic) compounds observed in the circumstellar envelope of evolved stars and in the interstellar medium (Kwok 2007) clearly indicate that plentiful C is available in the Universe (Pendleton and Allamandola 2002) and in the early solar system (Kerridge 1999). A large fraction of the carbon in the Universe is assembled in Polycyclic Aromatics (PAHs) resulting in the third most abundant molecular species after $H_2$ and CO (Wooden et al. 2004). PAH were identified by means of their UV-visible (Moreels et al. 1994) and 3.2–3.6 μm infrared emissions (Bockelée-Morvan et al. 1995) for many cometary comae. In Pizzarello et al. (2006) is given a thorough review of the chemical properties of the organic materials found in meteorites and interplanetary dust particles (IDPs).

Apart from their intrinsic low albedo, which implies lower signal, and red slopes at visible wavelengths, the majority of the organic materials in the 'CHON' inventory are char-

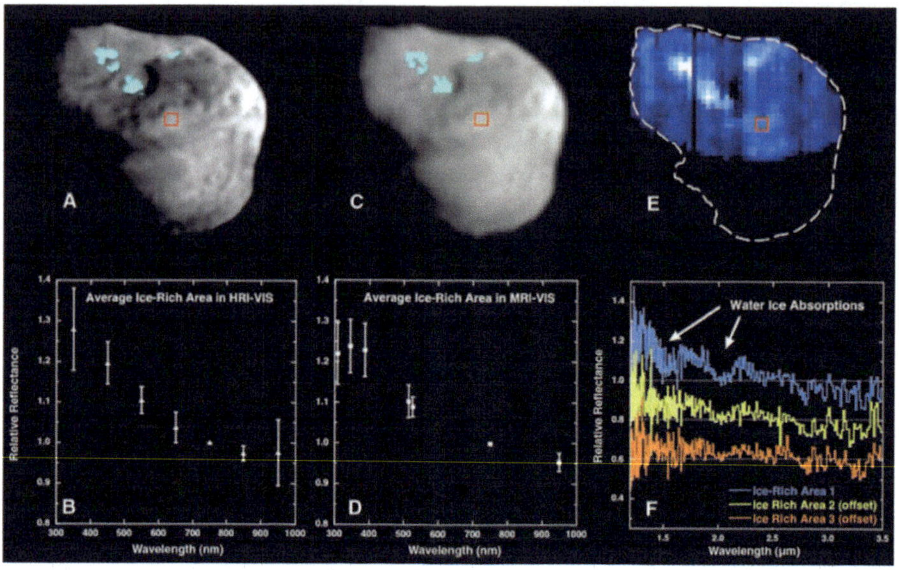


**Fig. 14** Optical and spectral identification of ice-rich areas relative to non-ice regions on the 9P/Tempel 1 nucleus by Deep Impact. Panels A–B: HRI visible data and filters spectra at 16 m/pixel. Panels C–D: MRI visible data and filters spectra at 82 m/pixel. Panels E–F: strength of the 2.0 μm water ice absorption and relative reflectance at 120 m/pixel. Note that the IR scan at the highest resolution only covers the upper half of the nucleus, as shown. The ice-rich areas are mapped in the visible images as combinations of high 450 nm/750 nm relative reflectance (387 nm/750 nm for MRI) and low 950 nm/750 nm relative reflectance. From Sunshine et al. (2006)

acterized by vibrational modes carried by hydrocarbons, oxygen/nitrogen-bonds and acids chemical groups within the 3.0–3.6 μm spectral range (see Table 6).

The wide 3.2 μm band visible on the average spectral reflectance of 67P/CG dark terrain (Fig. 2) has been attributed to a complex organic material with a variety of contributing species (Capaccioni et al. 2015). The broadness of the band may be explained by the contribution from carboxylic O–H with plausible contributions from aliphatic and aromatic C–H, as well as alcoholic/phenolic O–H (Capaccioni et al. 2015; Quirico et al. 2016). The 3.2 μm band appears distorted by two concurring effects, namely the presence of OH stretch which causes the broadening of absorption in the 2.7–3.0 μm range, and the surface thermal emission which influences the infrared spectral range starting from 3.5 μm range (Filacchione et al. 2016c). For this reason the best VIRTIS data to study the 3.2 μm band are those collected during Rosetta's prelanding phase when the surface was still relatively dehydrated and cold ($T < 230$ K). Approaching perihelion, the superficial water ice fraction increases (see discussion in Sect. 4) and the thermal emission becomes dominant with respect to the absorption bands exhibited by the reflectance spectra. In the following paragraphs we give a summary of the different organic materials which best-match the observed infrared spectral features.

The C–H bonds in methylene ($CH_2$), methyl ($CH_3$) aliphatic groups and in aromatic hydrocarbons can be identified by means of their symmetric and antisymmetric stretch vibration modes (Table 6). Some of these features are compatible with the secondary absorptions shown in Fig. 2—bottom panel (Raponi et al. 2018). Similar aliphatic features have been identified on other solar system objects, including Ceres (De Sanctis et al. 2017), Saturn's rings (Filacchione et al. 2014) and Iapetus (Cruikshank et al. 2014).

**Table 6** Diagnostic absorption bands of functional groups containing 'CHON' in the near infrared spectral range (Socrates 2001)

| Chemical group | Functional group | Vibration type | Absorption wavelength (μm) | Intensity |
|---|---|---|---|---|
| Alcohol | O–H | stretch H-bonded | 2.78–3.12 | strong-broad |
| Alcohol | O–H | stretch-free | 2-70–2.86 | strong-sharp |
| Alkane[a] | C–H | stretch | 3.33–3.51 | strong |
| Alkene[a] | =C–H | stretch | 3.23–3.32 | medium |
| Alkyne[a] | C–H | stretch | 3.03 | strong-sharp |
| Alkyne[a] | −C≡C− | stretch | 4.42–4.76 | variable |
| Aliphatic hydrocarbon | $CH_2$ | stretch | 3.42 (asym), 3.51 (sym.) | sharp |
| Aliphatic hydrocarbon | $CH_3$ | stretch | 3.38 (asym), 3.48 (sym.) | sharp |
| Aromatic hydrocarbon | C–H | stretch | 3.22–3.33 | medium |
| Amine | N–H | stretch | 2.86–3.03 | medium |
| Nitrile | C≡N | stretch | 4.35–4.55 | medium |
| IsoNitrile | C≡N | stretch | 4.59–4.67 | medium |
| Carboxylic acid | −COOH | stretch | 3.03–4.00 | strong-very broad |
| Aldehyde | =C–H | stretch | 3.63–3.68 | medium |
| Aldehyde | =C–H | stretch | 3.51–3.55 | medium |
| Amide | N–H | stretch | 2.85–3.22 | medium |

Note: [a]Aliphatic compounds can be saturated, joined by single bonds (alkanes), or unsaturated, with double bonds (alkenes) or triple bonds (alkynes)

Within the instrument limitations (calibration residual at 4.6 μm and removal of the thermal emission), VIRTIS data seems to indicate little to null contribution of the amine and nitrile groups (Capaccioni et al. 2015) confirming that 67P/CG is depleted in N with respect to other comets. A similar indication is in agreement with ROSINA which has measured an average ratio of $N_2/CO = (5.70 \pm 0.66) \cdot 10^{-3}$ corresponding to a depletion by a factor of $\approx 25.4 \pm 8.9$ as compared to the protosolar value (Rubin et al. 2015). These data indicate that 67P/CG must have formed at very low temperatures (<30 K) within the protosolar nebula resulting in a common origin with other Kuiper Belt objects.

Among the vast inventory of organic materials, *coals, bitumens* (Moroz et al. 1998; Quirico et al. 2016) and *aliphatic-rich tholins* (Quirico et al. 2008) share similar spectral features which make them a first class of analogs candidates for the identification of the hydrocarbon part of cometary organics. These compounds are characterized by a low reflectance and a neutral to red color at visible and near infrared wavelengths. Solid oil bitumens are complex macromolecular materials mostly consisting of polyaromatic and aliphatic hydrocarbons, that depending on their compositions and chemical structures are classified as asphaltites, kerites and anthraxolites (Moroz et al. 1998 and references therein). The fraction of kerogen relative to the fraction of solvent-soluble bitumen, carbon aromaticity, and C/H ratio increase in this series from low-temperature asphaltites through kerites to the most thermally evolved high antraxolites (Moroz et al. 1998 and references therein).

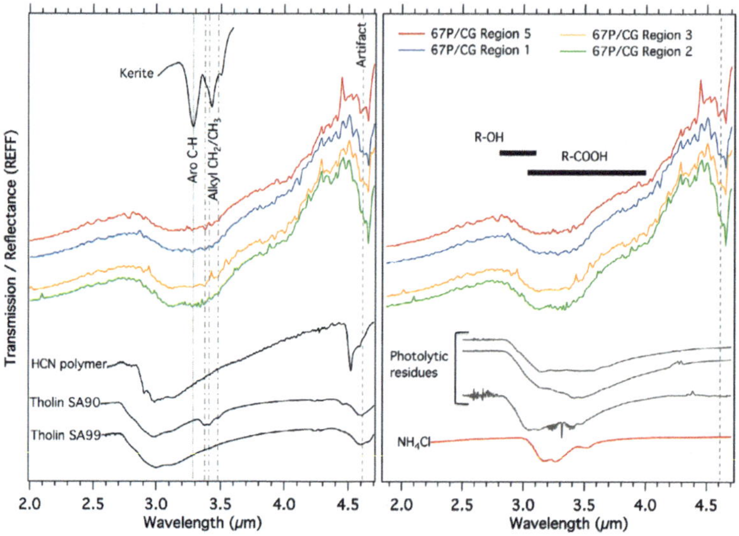


**Fig. 15** Left panel: representative reflectance spectra of 67P/CG regions compared with reflectance spectra of low-anthraxolite (Moroz et al. 1998), transmission spectra of tholins (Quirico et al. 2008) and HCN polymer (Bonnet et al. 2015) (data with offset for clarity). The spectral shift in the position of the aliphatic modes in the low anthraxolite with respect to those of the tholin SA90 (vertical dash-dotted lines) is due to the nitrogen content. The dashed line indicates the position of the 4.6 μm artifact on VIRTIS data which also fits the position of the OCN-anion in laboratory experiments. Right panel: further comparisons with transmission spectra of ammonium salt (from NIST database) and of photolytic residues produced under varying conditions (Muñoz Caro and Schutte 2003). From Quirico et al. (2016)

These changes are accompanied by a reduction of the spectral slope in the VNIR spectral range. High-kerites and low-anthraxolites show compositional and structural similarities to organic matter from CM2 and CI1 chondrites, while high-anthraxolites (shungites) resemble more carbonized organics in CV3 and CO3 chondrites (Moroz et al. 1998; de Bergh et al. 2008).

Tholins are synthesized by irradiation of simple C- and N-bearing compounds with solar UV radiation or energetic cosmic rays (Sagan and Khare 1979). The resulting compounds include amines, nitriles, and alkyne groups resulting in a prominent band around 3.2 μm and various NIR overtone and combination bands. The VIS-NIR spectra of these polymer-like materials are characterized by a steep red slopes controlled by their carbon structure. Anhydrous tholins can be distinguished from ice tholins by the presence of the C≡N stretches at 4.4–4.65 μm (Khare et al. 1994). In presence of methylene ($CH_2$)- and methyl ($CH_3$)-bonds, the aliphatic tholins are formed (Quirico et al. 2008). These are recognizable by means of their C–H stretches between 3.38–3.51 μm (see Fig. 15, left panel). The results of PTOLEMY and COSAC onboard Philae lander are compatible with the presence of polyoxymethylene (POM) chains in semi-volatile compounds (Wright et al. 2015; Goesmann et al. 2015). The possible POM (long debated as a possible parent molecule of formaldehyde in comets) detection by PTOLEMY and COSAC is, however, a controversial issue. While the teams have interpreted their data as fragments of POM, later on Altwegg et al. (2017) concluded that a similar interpretation of the results is not unique because other CHO-bearing molecules can be compatible with the measurements.

*Phenolic and alcoholic OH groups* (Pizzarello et al. 2006; Kolb and Liesch 2008) can contribute to the VIRTIS spectra of the 67P surface in the 2.7–3.1 μm spectral range (see Fig. 15, right panel).

*Carboxylic Acids* (R-COOH) are the main component in the soluble organic fraction in primitive meteorites, with several tens of carboxylic, dicarboxylic and hydroxylated acids identified so far (Sephton et al. 2002). Infrared spectra of these acids show a wide absorption in the 3.0–4.0 μm spectral range (see Fig. 15, right panel). Phenols and carboxylic acids can coexist in the same macromolecular structure, e.g., in solid bitumens and coals, but their contents are too low to show the intense absorption bands typical of the VIRTIS spectra. Quirico et al. (2016) suggested O–H in carboxylic acids as a highly plausible carrier of the broad absorption band in the VIRTIS spectra of the 67P/CG nucleus surface.

Among amines-bearing compounds, *HCN polymers* (Bonnet et al. 2015) have some similarities with 67P/CG spectra even if the high-abundance N species seems to be incompatible with the Rosina results (Altwegg et al. 2017). HCN polymers are characterized by the presence of an amine absorption band which extends from 2.7 to 3.7 μm and is much broader than the 3.2 μm band observed by VIRTIS towards shorter wavelengths (Fig. 15, left panel). Moreover, the strong nitrile absorption at 4.5 μm shown in the polymer spectrum cannot be safely identified in VIRTIS data.

Other viable candidates are *ammonium and cyanide salts*. Ammonium salts contain the ammonium ion $NH_4^+$ which absorbs between 2.94 and 3.34 μm. In the residues formed by irradiation or heating of ice mixtures, the ammonium ion is often associated with the $OCH^-$ group (Quirico et al. 2016). The cyanide salts form by annealing $NH_3 + HCN$ ices (Gerakines et al. 2004). Prolonged heating or irradiation of the ice mixtures by UV photons or energetic particles form refractory organic residues which include a variety of carboxylic acids and ammoniated salts, esters and amides (Muñoz Caro and Schutte 2003). The ammonium ion spectrum shown in Fig. 15, right panel, appears as a viable candidate, while photolytic residues may display a broad feature that encompasses the 3.2 μm band.

Apart form 67P/CG, organic materials have been detected in infrared spectra of other comets. For 19P/Borrelly Soderblom et al. (2002) reported about the presence of a spectral band at 2.39 μm about 0.02 μm wide. The origin of the absorption feature is unknown but it could be compatible with various hydrocarbons such as polyoxymethylene. Infrared emission features at 3.3–3.4 μm of organics have been detected in the coma of 103P/Hartley 2 (A'Hearn et al. 2011) where are mainly associated with $CO_2$-rich jets.

## 3.3 Inorganic Refractories

From the compositional analysis of 67P/CG dust grains (Bardyn et al. 2017) and from the inventory of mineral species detected on Stardust grains (Keller et al. 2006; Dobrică and Brearley 2011), interplanetary dust particles or IDPs (Engrand et al. 2016) and micrometeorites (Dobrică et al. 2011, 2012), we know that cometary inorganic refractories are dominated by a mixture of minerals including anhydrous Mg-rich silicates, Ni-bearing iron, Fe sulfides and glassy phases with embedded metals and sulphides or GEMS (Leroux et al. 2015). Other plausible refractory endmembers are amorphous and graphitic carbon present in the interstellar medium dust particles (Papoular 2001).

The presence of fine-grained opaques (including Fe-sulphides) is responsible for the reduction of the visible-near infrared albedo and of the reduced spectral contrast within the 3.2 μm absorption band (Moroz et al. 2017). A thorough spectral investigation of cometary refractories analogs in the laboratory within the 0.45–2.7 μm interval has shown that 67P/CG spectral properties in this range can be reproduced by a mixture of coal, pyrrhotite and dunite

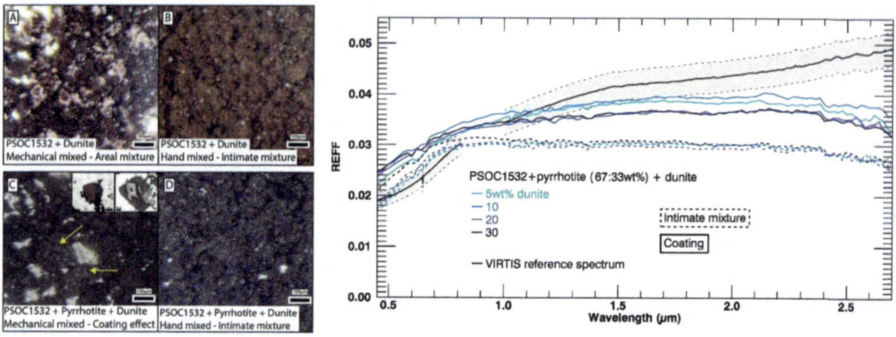


**Fig. 16** Left: Optical binocular images of different mixtures sieved to grains <25 μm after a colloidal grinding with a planetary grinder. (**A**) Areal mixture of coal PSOC 1532 and dunite mixed with grinder MM200 without milling balls. (**B**) Intimate mixture of PSOC 1532 and dunite hand-mixed into a mortar with a pestle. (**C**) PSOC 1532, dunite and pyrrhotite are mixed with the grinder MM200 as sample A. Pyrrhotite coats unbroken aggregates of dunite and organics (yellow arrows). Boxes in the upper right corner show broken aggregates of PSOC 1532 and dunite. (**D**) Intimate mixture of PSOC 1532, dunite and pyrrhotite, hand mixed into a mortar. Right: Reflectance spectra of mechanically mixed ternary mixtures of coal PSOC 1532, pyrrhotite and dunite. This protocol preserves clumps of organics and silicate which are coated by pyrrhotite grains (solid lines). The resulting spectra are similar to the ones of the intimate mixtures (dashed lines) in the visible while they are redder and brighter in the infrared range. The fraction of dunite varies from 5 wt% to 30 wt% but does not lead to strong variations of the spectral shape. From Rousseau et al. (2018)

grains (Rousseau et al. 2018). Noteworthy, the mixing state among these endmembers can drastically change the spectral properties of the sample (Fig. 16, left panel). This effect is mainly due to the coating of fine pyrrhotite particles occurring on the surfaces of larger clumps of coal and silicate grains. To date, the best spectral match to VIRTIS data in this range (Fig. 16, right panel) is achieved with a mixture of coal and pyrrhotite (in proportion 67:33 wt%) with a variable amount of dunite (between 5 to 30%). The sample fails to completely reproduce correctly the visible-near infrared spectral slope but this could be compensated by adding few percent of H-rich organic material which typically introduce a reddening at visible-near infrared wavelengths.

The spectral properties of mixtures of organic material with opaque phases have been investigated by Moroz et al. (2017). Pyrrhotites and troilites are sulphides characterized by a decrease in reflectance with decreasing grain size, which is typical for opaque phases (Hunt et al. 1971). Mixtures containing sub-micron sulfides grains have the capability to drastically reduce the visible and near infrared albedo. Coarse size sulfides grains are much less effective in reducing the albedo than fine-grained ones. This could help in explaining the low albedo of 67P/CG's nucleus, as discussed in Sect. 2.1. The presence of a small amount of fine-grained opaque phase can effectively reduce the overall albedo of a bright material. In a similar way, the spectral reflectance of the ubiquitous dark terrain observed by VIRTIS on the surface of 67P/CG could be explained by the presence of fine-grained opaques and organic refractories, while the presence of silicate is much less significant (Moroz et al. 2017; Rousseau et al. 2018). The analysis of reflectance spectra of kerite-troilite and kerite-pyrrhotite mixtures in different percentages made by Moroz et al. (2017) shows that Fe-sulfides intimately mixed with organic material significantly influence shapes and relative contrasts of absorption bands within the wide 3.2 μm band. The authors showed that the positions of individual aromatic C–H and aliphatic $CH_2$ and $CH_3$ absorption bands coincide with the fine structure observed within the average spectrum of the 67P/CG's surface shown in Fig. 2—bottom panel.

### 3.4 Spectrophotometry of Ice-Containing Cometary Analogs

It is commonly thought that while cometary nuclei are ice-rich in bulk, the low albedo of their surfaces is due to the presence of an optically thick desiccated layer, either a consequence of the sublimation of ice from the surface or rediposition of dry dust emitted by the cometary activity. Various laboratory experiments have reproduced the darkening of analogs for cometary matter, as the ice initially mixed with dark refractory dust sublimates, resulting in the formation and thickening of a porous sublimation lag (Oehler and Neukum 1991). The red spectral slope of these analogs also increases as the albedo decreases, resulting in the negative correlation between these parameters systematically observed at the surface of the nucleus of 67P (Pommerol et al. 2015; Poch et al. 2016a,b; Jost et al. 2017a).

However, many experiments show that it is actually possible to hide large amounts of water ice when intimately mixed with fine-grained dust which acts like a coating on larger grains. In these cases fine absorbing grains mixed with water ice have a darkening effect disproportional to their abundance. Measurements of the bidirectional visible reflectance of binary mixtures of calibrated water ice particles and reference lunar regolith simulant JSC1-AF (Fig. 17) have shown that it is possible to mix up to 75% of ice by weight within the dust without increasing its reflectance notably, therefore making this ice hardly detectable for visible imagers (Yoldi et al. 2015). Similar measurements with dust analogs more relevant to cometary matter show even more pronounced effects, with as little as 2 wt. % of fine-grained charcoal required to decrease the albedo of 70 μm diameter ice particles to the value measured at the surface the nucleus of comet 67P (Jost et al. 2017b).

The first consequence of these results is that the bright patches and regions observed at the surface of the nucleus must be very ice-rich if the ice is mixed intimately with the refractory dust. Depending on the size of the ice particles and both the size and composition of the dust particles, the measured reflectance factors of the bright patches that are several times higher than the average of the nucleus surface could indicate nearly pure ice. Alternative scenarios that would lead to lower inferred ice contents are areal mixtures at scales smaller than the resolution of OSIRIS NAC, i.e. centimeters to tens of centimeters, or the presence of frost at the surface of the dust grains. Both scenarios are plausible with ample evidence for frost deposition and sublimation at diurnal timescales (de Sanctis et al. 2015; Fornasier et al. 2016) and evidence for the formation of large particles of ice by re-condensation or sintering (Filacchione et al. 2016a). The second important implication of these experimental result is that the low albedo of the vast majority of the surface of the nucleus alone doesn't preclude the presence of large amounts of ice at the surface if it is intimately mixed with the dust, as expected for primordial ices that have condensed as layers around the surfaces of the dust grains. Therefore, the nucleus might be constituted of an ice-rich interior coated by a desiccated layer but punctuated by bright icy patches that form by re-condensation of sublimated water vapour and/or sintering of the ice and segregation from the dust.

The infrared spectral range contains more sensitive and diagnostic features for the presence and properties of water ice. Recent experiments by Yoldi et al. (2015) have been conducted to verify if the masking of the ice by dark dust seen in the visible spectral range is also efficient in the infrared. Binary mixtures between water ice and lunar regolith simulant (Fig. 17) show that the spectral signatures of water ice can be identified for lower ice contents, of about 1/3 by weight and possibly less. These infrared spectral signatures are thus significantly more sensitive to the presence of water ice than the visible albedo in the case of intimate mixtures. It still seems possible, however, to resolve at least 10 or 20 wt. % of ice in the regolith depending on the properties of the ice and dust components. Unfortunately,

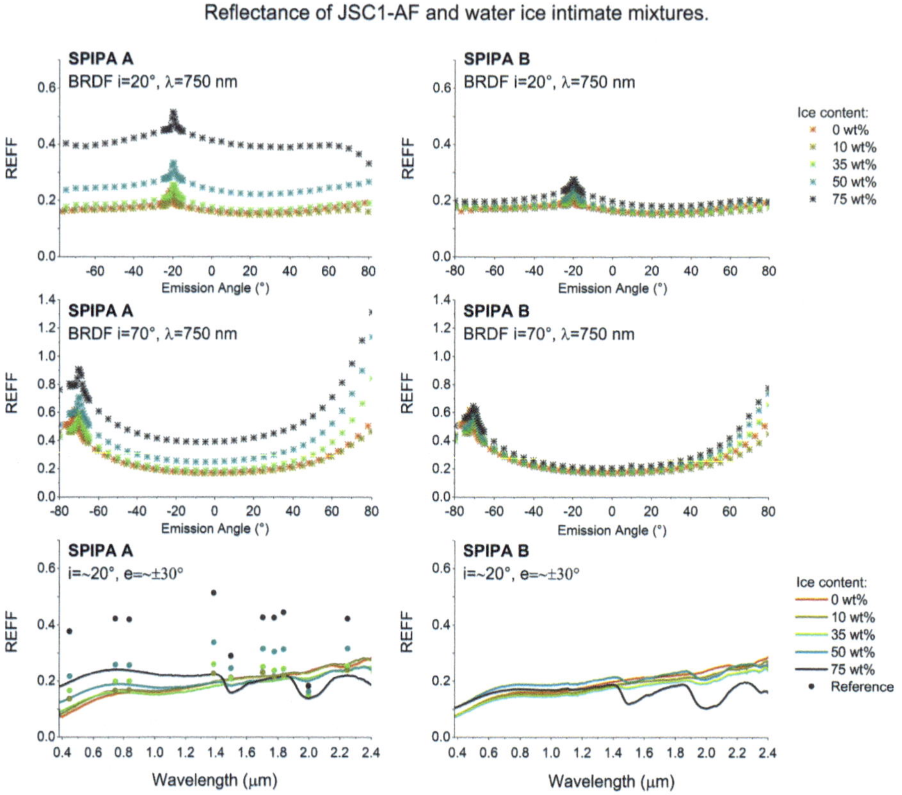


**Fig. 17** Measured VIS (750 nm) phase curves (top: incidence = 20° and middle: incidence = 70°) and VIS-NIR (0.4–2.4 μm) reflectance spectra (bottom) of binary mixtures of the JSC1-AF simulant for fine lunar dust (McKay et al. 1993) with fine-grained ice (SPIPA-A, left) and coarser-grained ice (SPIPA-B, right) in variable concentrations. The fine-grained ice (SPIPA-A) has a mean particle diameter of about 5 μm whereas the coarser-grained ice (SPIPA-B) has a mean particle diameter of about 70 μm. These well-calibrated ice samples have been produced using dedicated Setups for the Production of Icy Planetary Analogs (SPIPA) developed at the University of Bern. From Yoldi et al. (2015)

the 3 μm region, the most sensitive to the presence of water, was not investigated by Yoldi et al. (2015).

The intensity of the infrared signatures has been used to estimate both the ice content and the size of the ice particles by inverting the Hapke (1993) spectro-photometric model. However, this leads to contradicting implications between relatively low amounts of ice (a few wt. %) inferred from the infrared spectral features and much higher water ice contents required to explain the bright features in the visible range in the hypothesized case of purely intimate mixtures. This possibly means that the mode of association between ice and dust is generally not dominated by intimate mixtures at the grain scale and that re-condensation and sintering leading to segregation of the ice and the dust are active in the surface layers of the nucleus probed by optical remote-sensing. New controlled experiments using more realistic low-albedo cometary dust analogs and reflectance measurements spanning both the visible and infrared spectral ranges up to 4 μm need to be performed to better understand the amount and state of water ice compatible with Rosetta observations of 67P/CG's nucleus surface.

# 4 Compositional Changes Driven by Seasonal Cycle and Activity

By orbiting in the vicinity of 67P/CG for more than two years around the perihelion passage, from July 2014 to September 2016, Rosetta's remote sensing instruments had the opportunity to observe the nucleus with unprecedented spatial and temporal resolutions allowing to trace morphologic and compositional changes occurring on the surface at different heliocentric distances and local times. Thanks to these characteristics it was possible to follow the compositional variations caused by activity and seasonal cycles for different regions of the nucleus. 67P/CG surface composition changes are driven by the following mechanisms:

1. *Erosion of the surface* due to the sublimation of water ice and other volatile species in the subsurface layers. This is the case for the roundish features formed in the Imhotep flat area in June–July 2015 (Groussin et al. 2015b) as a consequence of the deflating of the surface. The largest of these features developed with a lateral velocity of the order of $5.6 \cdot 10^{-5}$–$8.1 \cdot 10^{-5}$ m/s reaching diameters of $\approx$140–220 m with rims of about 5 m height (Fig. 18). The features' rim contours appear distinctively neutral to blue colored compared to the rest of surface where the visible spectral slope is 16% per 100 nm. However, no unusual activity was observed during the formation or evolution of these large round features.
2. *Localized activity* in cliffs, fractures, pits, bright features areas appears correlated with the sources of dust jets (Vincent et al. 2016a). The jet activity was stable between successive nucleus rotations and is tuned by the local illumination conditions (Vincent et al. 2016b). As discussed in Sect. 3.1, the areas of local activity are enriched in water ice and are recognizable by means of their higher albedo and more neutral (8–13%/100 nm) visible spectral slope (Oklay et al. 2016a). This class encompasses the diurnal condensation and sublimation cycle of the water ice in Hapi discussed in Sect. 3.1.
3. *Collapse of elevated structures, overhangs and cliffs* resulting in the mechanical disruption of old terrains (Groussin et al. 2015a; Pommerol et al. 2015; Pajola et al. 2017) with consequent appearance of bright water ice-rich material in the exposed walls and nearby debris fields (Filacchione et al. 2016a; Vincent et al. 2016a; Oklay et al. 2016b, 2017).
4. *Transfer and redeposition of the dust* driven by the activity and local topography (Thomas et al. 2015; Keller et al. 2017; Hu et al. 2017). The material eroded and ejected from the south hemisphere is partially transferred to the northern regions during southern summer season (at perihelion). According to Lai et al. (2016), Hapi gains up to 0.4 m while Wosret loses up to 1.8 m of dust mantle per orbit. Some models indicate that about 20% of the eroded material in the south is deposited on the north hemisphere and that this airfall material is wet (Keller et al. 2017). This water ice rich airfall material was reported by Oklay et al. (2016b).

All these phenomena, occurring at different spatial and temporal scales, are affecting the surface composition: while the surface's disruption mechanisms give access to subsurface volatiles, and nocturnal condensation are able to enrich the outer layers with icy species, the transfer and consequent deposition of dust is responsible for the observed spatial homogeneity of the ubiquitous dark terrain composed of organic material and mineral dust grains (Capaccioni et al. 2015; Filacchione et al. 2016c; Ciarniello et al. 2016). However, the amount of water ice bound to dust grains is variable, depending on seasonal/diurnal cycle and local morphology. The major spectral changes were observed when 67P/CG was near perihelion. At this time VIRTIS has measured an increase of the average albedo (up to 30% on average), the progressive bluening and flattening of the visible slope, the increase of the 3.2 μm absorption band area caused by the OH absorption (Fig. 19) with the resulting shift

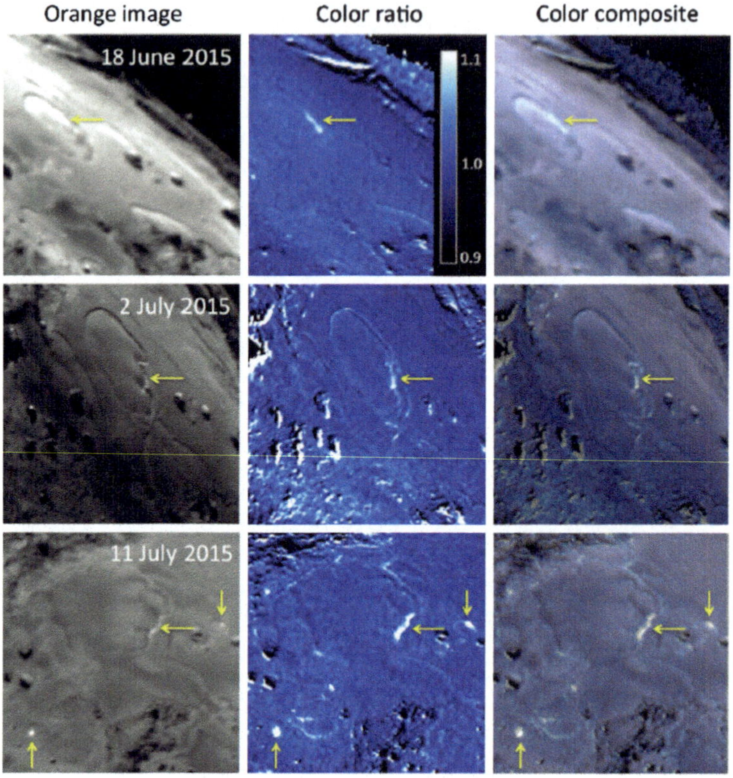


**Fig. 18** The evolution of the roundish features developed in a flat area of the Imhotep. Blue (481 nm) over orange (649 nm) or red (701 nm) color ratios of the evolving region on 18 June 2015 (upper row), 2 July 2015 (middle row), and 11 July 2015 (lower row). The first column shows a context image of the region, the second column shows the color ratio (blue-to-orange for 18 June and 2 July, blue-to-red for 11 July), and the third column shows a color-composite image made of the first two columns. The yellow arrows indicate where erosion features started with a material bluer than average (value above 1 in the color ratio). The images were acquired by OSIRIS/NAC. From Groussin et al. (2015b)

of the band center towards shorter wavelengths (Filacchione et al. 2016c; Ciarniello et al. 2016). In the visible spectral range, also OSIRIS observed a decrease of the spectral slopes around the perihelion passage (Fornasier et al. 2016). All these effects are the consequence of the exposure of more icy and pristine terrains caused by the removal of the surface dust.

Spectral analysis conducted by Longobardo et al. (2017) shows that the 3.2 µm band depth is independent from phase angle but the band centre is slightly shifting towards longer wavelengths when the incidence angle increases. This effect is probably due to the exposure of water ice fraction on the surface by weak cometary activity at increasing temperatures. The fact that the 3.2 µm band depth is uncorrelated to phase angle marks a substantial difference with the usual photometric responses of common materials which instead show an increase of the band depth at increasing phase angle. A similar effect is the consequence of the predominance of the multiple scattering on the band wings with respect to the single scattering on the band center occurring at high phases (Hapke 1993). This spectral behavior is interpreted by Longobardo et al. (2017) as the result of high abundance of fine-grained opaque species able to suppress not only the band itself, but also its dependence from the

**Fig. 19** Temporal evolution of 67P/CG composition maps by VIRTIS. Left column: visible spectral slope. Center: 3.2 μm band area. Right: 3.2 μm band center. From top to bottom are shown maps built from data acquired during MTP006 period (2014 Aug. 8–2014 Sept. 2, $r_h = 3.62$–$3.44$ AU), MTP007–MTP008 (2014 Sept. 2–2014 Oct. 24, $r_h = 3.44$–$3.12$ AU), MTP009–MTP010 (2014 Oct. 24–2014 Dec. 19, $r_h = 3.11$–$2.75$ AU), MTP011–MTP012 (2014 Dec. 19–2015 Feb. 10, $r_h = 2.73$–$2.35$ AU) and MTP014–MTP015 (2015 Mar. 10–2015 May 5, $r_h = 2.13$–$1.72$ AU). For each MTP period the surface coverage changes according to VIRTIS observation schedule along the Rosetta's orbital phases. All data are photometrically corrected and mapped with a resolution of $0.5 \times 0.5°$. Figure from Ciarniello et al. (2016)

viewing geometry or by a different phase reddening able to change the band depth properties with phase.

The temporal evolution of three large water ice rich areas, BAP1-2 (Filacchione et al. 2016a) and SPOT6 (Barucci et al. 2016), is analyzed by Raponi et al. (2016) on IR spectra along a period of time of about 180 days during which the heliocentric distance was decreasing from 3.6 to 2.2 AU. The three regions show similar life cycles: the patches increase their dimensions and surface ice abundance, reach a maximum and then rapidly decrease until, eventually, the ice completely disappears from the surface. Since these three areas are located very close to the equator they share similar thermal history and evolution. However, the local topography drives the timing of the maximum dimension and water ice abundance. A similar trend is interpreted by means of the exposure of deeper water ice-rich layers which is consistent with the local morphology of these areas where overhang falls or landslides have caused the exposure of fresh material. After the initial exposure of the ice, the local activity increases removing more and more surface dust thank to the ice sublimation giving a positive feedback on the exposure itself. The process continues while the nucleus approaches the Sun and it reaches a turning point when the sublimation rate prevails on the exposure rate, until the complete sublimation of the patch occurs (Raponi et al. 2016).

## 5 Conclusions

We have reviewed the principal results achieved by in situ and remote sensing payloads aboard the Rosetta orbiter and Philae lander aiming to determine the composition of comet 67P/CG and, where possible, to compare these results with data of other known comets. The determination of surface composition is a difficult task given the cometary active environment and the rough surface morphology. Moreover, the measurements were limited by the constraints given by Rosetta's orbit and Philae's not-nominal landing. In general the assessment of the volatile species, e.g. water and carbon dioxide, is more obvious than for the other materials: in gaseous phases they have been measured by mass spectrometers (Hässig et al. 2015), infrared spectrometer (Bockelée-Morvan et al. 2015; Fink et al. 2016) and microwave instruments (Biver et al. 2015; Lee et al. 2015); in solid phase at infrared wavelengths (de Sanctis et al. 2015; Filacchione et al. 2016a,b) with further evidence from the visible (Fornasier et al. 2016; Barucci et al. 2016; Oklay et al. 2017). The non-volatile fraction, which includes both mineral and organic materials, is much more difficult to characterize. While an infrared spectroscopic identification of the organic material (Capaccioni et al. 2015) present on the 67P/CG surface is still far from being fully achieved for the reasons previously described in Sect. 3, we can certainly assess that among the materials we have reviewed the most probable endmembers are included.

The nucleus particles consist of an assemblage of various organic components, minerals and water ice. The suggested organic compounds contain COOH and OH-groups and a refractory macromolecular material bearing aliphatic ($CH_2$ and $CH_3$) and polycyclic aromatic hydrocarbons which are the most plausible organic materials able to contribute to the reflectance spectra of 67P/CG's nucleus. The organic material appears mixed with minerals, including silicates, Fe-sulfides (pyrrhotite and/or troilite) and possibly ammoniated salt.

In the field of visible and infrared spectroscopy, the availability of optical constants measured on laboratory analogs is of the utmost importance. Spectral modeling efforts are in fact currently limited by the availability of these data which are known for only a relatively small number of the compounds that are likely to be present on the 67P/CG's nucleus surface. We must be aware that apart of composition also grain size distribution, mixing modalities among different endmembers and thermal evolution play fundamental roles in the interpretation of remotely-sensed data. Due to unexpected landing conditions of Philae, in-situ compositional studies could not be completed as planned. Therefore, a future sample return mission is strongly envisaged for the characterization of cometary nucleus composition.

**Acknowledgements** This project has received funding from the European Union's Horizon 2020 research and innovation programme under grant agreement no. 686709. This work was supported by the Swiss State Secretariat for Education, Research and Innovation (SERI) under contract number 16.0008-2. The opinions expressed and arguments employed herein do not necessarily reflect the official view of the Swiss Government. We gratefully thank the MIARD project team at ISSI, Bern, Switzerland led by Professor Nicolas Thomas from the University of Bern's Space Research and Planetology division. G.F., M.C. and A.R. acknowledge financial support from ASI, Italian Space Agency. L.M. acknowledges the DFG (Deutsche Forschungsgemeinschaft) grant MO 3007/1-1. D.K. acknowledges DFG-grant KA 3757/2-1.

## References


M.F. A'Hearn, M.J.S. Belton, A. Delamere, W.H. Blume, Deep impact: a large-scale active experiment on a cometary nucleus. Space Sci. Rev. **117**, 1–21 (2005a). https://doi.org/10.1007/s11214-005-3387-3

M.F. A'Hearn, M.J.S. Belton, W.A. Delamere, J. Kissel, K.P. Klaasen, L.A. McFadden, K.J. Meech, H.J. Melosh, P.H. Schultz, J.M. Sunshine, P.C. Thomas, J. Veverka, D.K. Yeomans, M.W. Baca, I. Busko, C.J. Crockett, S.M. Collins, M. Desnoyer, C.A. Eberhardy, C.M. Ernst, T.L. Farnham, L. Feaga, O. Groussin, D. Hampton, S.I. Ipatov, J.-Y. Li, D. Lindler, C.M. Lisse, N. Mastrodemos, W.M. Owen, J.E. Richardson, D.D. Wellnitz, R.L. White, Deep impact: excavating comet Tempel 1. Science **310**, 258–264 (2005b). https://doi.org/10.1126/science.1118923

M.F. A'Hearn, M.J.S. Belton, W.A. Delamere, L.M. Feaga, D. Hampton, J. Kissel, K.P. Klaasen, L.A. McFadden, K.J. Meech, H.J. Melosh, P.H. Schultz, J.M. Sunshine, P.C. Thomas, J. Veverka, D.D. Wellnitz, D.K. Yeomans, S. Besse, D. Bodewits, T.J. Bowling, B.T. Carcich, S.M. Collins, T.L. Farnham, O. Groussin, B. Hermalyn, M.S. Kelley, M.S. Kelley, J.-Y. Li, D.J. Lindler, C.M. Lisse, S.A. McLaughlin, F. Merlin, S. Protopapa, J.E. Richardson, J.L. Williams, EPOXI at comet Hartley 2. Science **332**, 1396 (2011). https://doi.org/10.1126/science.1204054

K. Altwegg, H. Balsiger, J.J. Berthelier, A. Bieler, U. Calmonte, S.A. Fuselier, F. Goesmann, S. Gasc, T.I. Gombosi, L. Le Roy, J. de Keyser, A. Morse, M. Rubin, M. Schuhmann, M.G.G.T. Taylor, C.-Y. Tzou, I. Wright, Organics in comet 67P—a first comparative analysis of mass spectra from ROSINA-DFMS, COSAC and Ptolemy. Mon. Not. R. Astron. Soc. **469**, 130–141 (2017). https://doi.org/10.1093/mnras/stx1415

A.-T. Auger, O. Groussin, L. Jorda, S. Bouley, R. Gaskell, P.L. Lamy, C. Capanna, N. Thomas, A. Pommerol, H. Sierks, C. Barbieri, R. Rodrigo, D. Koschny, H. Rickman, H.U. Keller, J. Agarwal, M.F. A'Hearn, M.A. Barucci, J.-L. Bertaux, I. Bertini, G. Cremonese, V. Da Deppo, B. Davidsson, S. Debei, M. De Cecco, M.R. El-Maarry, S. Fornasier, M. Fulle, P.J. Gutiérrez, C. Güttler, S. Hviid, W.-H. Ip, J. Knollenberg, J.-R. Kramm, E. Kührt, M. Küppers, F. La Forgia, L.M. Lara, M. Lazzarin, J.J. Lopez Moreno, S. Marchi, F. Marzari, M. Massironi, H. Michalik, G. Naletto, N. Oklay, M. Pajola, L. Sabau, C. Tubiana, J.-B. Vincent, K.-P. Wenzel, Geomorphology of the Imhotep region on comet 67P/Churyumov–Gerasimenko from OSIRIS observations. Astron. Astrophys. **583**, 35 (2015). https://doi.org/10.1051/0004-6361/201525947

H. Balsiger, K. Altwegg, P. Bochsler, P. Eberhardt, J. Fischer, S. Graf, A. Jäckel, E. Kopp, U. Langer, M. Mildner, J. Müller, T. Riesen, M. Rubin, S. Scherer, P. Wurz, S. Wüthrich, E. Arijs, S. Delanoye, J. de Keyser, E. Neefs, D. Nevejans, H. Rème, C. Aoustin, C. Mazelle, J.-L. Médale, J.A. Sauvaud, J.-J. Berthelier, J.-L. Bertaux, L. Duvet, J.-M. Illiano, S.A. Fuselier, A.G. Ghielmetti, T. Magoncelli, E.G. Shelley, A. Korth, K. Heerlein, H. Lauche, S. Livi, A. Loose, U. Mall, B. Wilken, F. Gliem, B. Fiethe, T.I. Gombosi, B. Block, G.R. Carignan, L.A. Fisk, J.H. Waite, D.T. Young, H. Wollnik, Rosina Rosetta Orbiter Spectrometer for Ion and Neutral Analysis. Space Sci. Rev. **128**, 745–801 (2007). https://doi.org/10.1007/s11214-006-8335-3

A. Bardyn, D. Baklouti, H. Cottin, N. Fray, C. Briois, J. Paquette, O. Stenzel, C. Engrand, H. Fischer, K. Hornung, R. Isnard, Y. Langevin, H. Lehto, L. Le Roy, N. Ligier, S. Merouane, P. Modica, F.-R. Orthous-Daunay, J. Rynö, R. Schulz, J. Silén, L. Thirkell, K. Varmuza, B. Zaprudin, J. Kissel, M. Hilchenbach, Carbon-rich dust in comet 67P/Churyumov–Gerasimenko measured by COSIMA/Rosetta. Mon. Not. R. Astron. Soc. **469**, 712–722 (2017). https://doi.org/10.1093/mnras/stx2640

M.A. Barucci, G. Filacchione, S. Fornasier, A. Raponi, J.D.P. Deshapriya, F. Tosi, C. Feller, M. Ciarniello, H. Sierks, F. Capaccioni, A. Pommerol, M. Massironi, N. Oklay, F. Merlin, J.-B. Vincent, M. Fulchignoni, A. Guilbert-Lepoutre, D. Perna, M.T. Capria, P.H. Hasselmann, B. Rousseau, C. Barbieri, D. Bockelée-Morvan, P.L. Lamy, C. De Sanctis, R. Rodrigo, S. Erard, D. Koschny, C. Leyrat, H. Rickman, P. Drossart, H.U. Keller, M.F. A'Hearn, G. Arnold, J.-L. Bertaux, I. Bertini, P. Cerroni, G. Cremonese, V. Da Deppo, B.J.R. Davidsson, M.R. El-Maarry, S. Fonti, M. Fulle, O. Groussin, C. Güttler, S.F. Hviid, W. Ip, L. Jorda, D. Kappel, J. Knollenberg, J.-R. Kramm, E. Kührt, M. Küppers, L. Lara, M. Lazzarin, J.J. Lopez Moreno, F. Mancarella, F. Marzari, S. Mottola, G. Naletto, M. Pajola, E. Palomba, E. Quirico, B. Schmitt, N. Thomas, C. Tubiana, Detection of exposed $H_2O$ ice on the nucleus of comet 67P/Churyumov–Gerasimenko. as observed by Rosetta OSIRIS and VIRTIS instruments. Astron. Astrophys. **595**, 102 (2016). https://doi.org/10.1051/0004-6361/201628764

J. Benkhoff, Numerical simulations of the gas flux at the surface of a 2060 Chiron, in *American Astronomical Society Meeting Abstracts*. Bulletin of the American Astronomical Society, vol. 27 (1995), p. 1338

M.S. Bentley, R. Schmied, T. Mannel, K. Torkar, H. Jeszenszky, J. Romstedt, A.-C. Levasseur-Regourd, I. Weber, E.K. Jessberger, P. Ehrenfreund, C. Koeberl, O. Havnes, Aggregate dust particles at comet 67P/Churyumov–Gerasimenko. Nature **537**, 73–75 (2016). https://doi.org/10.1038/nature19091

J.-P. Bibring, H. Rosenbauer, H. Boehnhardt, S. Ulamec, J. Biele, S. Espinasse, B. Feuerbacher, P. Gaudon, P. Hemmerich, P. Kletzkine, D. Moura, R. Mugnuolo, G. Nietner, B. Pätz, R. Roll, H. Scheuerle, K.

Szegö, K. Wittmann, The Rosetta Lander (Philae) investigations. Space Sci. Rev. **128**, 205–220 (2007). https://doi.org/10.1007/s11214-006-9138-2

J.-P. Bibring, Y. Langevin, J. Carter, P. Eng, B. Gondet, L. Jorda, S. Le Mouélic, S. Mottola, C. Pilorget, F. Poulet, M. Vincendon, 67P/Churyumov–Gerasimenko surface properties as derived from CIVA panoramic images. Science **349**(2) (2015). https://doi.org/10.1126/science.aab0671

J. Biele, S. Ulamec, M. Maibaum, R. Roll, L. Witte, E. Jurado, P. Muñoz, W. Arnold, H.-U. Auster, C. Casas, C. Faber, C. Fantinati, F. Finke, H.-H. Fischer, K. Geurts, C. Güttler, P. Heinisch, A. Herique, S. Hviid, G. Kargl, M. Knapmeyer, J. Knollenberg, W. Kofman, N. Kömle, E. Kührt, V. Lommatsch, S. Mottola, R. Pardo de Santayana, E. Remetean, F. Scholten, K.J. Seidensticker, H. Sierks, T. Spohn, The landing(s) of Philae and inferences about comet surface mechanical properties. Science **349**(1) (2015). https://doi.org/10.1126/science.aaa9816

A. Bieler, K. Altwegg, H. Balsiger, J.-J. Berthelier, U. Calmonte, M. Combi, J. De Keyser, B. Fiethe, N. Fougere, S. Fuselier, S. Gasc, T. Gombosi, K. Hansen, M. Hässig, Z. Huang, A. Jäckel, X. Jia, L. Le Roy, U.A. Mall, H. Rème, M. Rubin, V. Tenishev, G. Tóth, C.-Y. Tzou, P. Wurz, Comparison of 3D kinetic and hydrodynamic models to ROSINA-COPS measurements of the neutral coma of 67P/Churyumov–Gerasimenko. Astron. Astrophys. **583**, 7 (2015). https://doi.org/10.1051/0004-6361/201526178

N. Biver, M. Hofstadter, S. Gulkis, D. Bockelée-Morvan, M. Choukroun, E. Lellouch, F.P. Schloerb, L. Rezac, W.H. Ip, C. Jarchow, P. Hartogh, S. Lee, P. von Allmen, J. Crovisier, C. Leyrat, P. Encrenaz, Distribution of water around the nucleus of comet 67P/Churyumov–Gerasimenko at 3.4 AU from the Sun as seen by the MIRO instrument on Rosetta. Astron. Astrophys. **583**, 3 (2015). https://doi.org/10.1051/0004-6361/201526094

D. Bockelée-Morvan, N. Biver, The composition of cometary ices. Philos. Trans. R. Soc. Lond. Ser. A **375**, 20160252 (2017). https://doi.org/10.1098/rsta.2016.0252

D. Bockelée-Morvan, T.Y. Brooke, J. Crovisier, On the origin of the 3.2 to 3.6-micron emission features in comets. Icarus **116**, 18–39 (1995). https://doi.org/10.1006/icar.1995.1111

D. Bockelée-Morvan, V. Debout, S. Erard, C. Leyrat, F. Capaccioni, G. Filacchione, N. Fougere, P. Drossart, G. Arnold, M. Combi, B. Schmitt, J. Crovisier, M.-C. de Sanctis, T. Encrenaz, E. Kührt, E. Palomba, F.W. Taylor, F. Tosi, G. Piccioni, U. Fink, G. Tozzi, A. Barucci, N. Biver, M.-T. Capria, M. Combes, W. Ip, M. Blecka, F. Henry, S. Jacquinod, J.-M. Reess, A. Semery, D. Tiphene, First observations of $H_2O$ and $CO_2$ vapor in comet 67P/Churyumov–Gerasimenko made by VIRTIS onboard Rosetta. Astron. Astrophys. **583**, 6 (2015). https://doi.org/10.1051/0004-6361/201526303

D. Bockelée-Morvan, J. Crovisier, S. Erard, F. Capaccioni, C. Leyrat, G. Filacchione, P. Drossart, T. Encrenaz, N. Biver, M.-C. de Sanctis, B. Schmitt, E. Kührt, M.-T. Capria, M. Combes, M. Combi, N. Fougere, G. Arnold, U. Fink, W. Ip, A. Migliorini, G. Piccioni, G. Tozzi, Evolution of $CO_2$, $CH_4$, and OCS abundances relative to $H_2O$ in the coma of comet 67P around perihelion from Rosetta/VIRTIS-H observations. Mon. Not. R. Astron. Soc. **462**, 170–183 (2016). https://doi.org/10.1093/mnras/stw2428

H. Boehnhardt, J.-P. Bibring, I. Apathy, H.U. Auster, A. Ercoli Finzi, F. Goesmann, G. Klingelhöfer, M. Knapmeyer, W. Kofman, H. Krüger, S. Mottola, W. Schmidt, K. Seidensticker, T. Spohn, I. Wright, The Philae lander mission and science overview. Philos. Trans. R. Soc. Lond. Ser. A **375**, 20160248 (2017). https://doi.org/10.1098/rsta.2016.0248

J.-Y. Bonnet, E. Quirico, A. Buch, R. Thissen, C. Szopa, N. Carrasco, G. Cernogora, N. Fray, H. Cottin, L. Le Roy, G. Montagnac, E. Dartois, R. Brunetto, C. Engrand, J. Duprat, Formation of analogs of cometary nitrogen-rich refractory organics from thermal degradation of tholin and HCN polymer. Icarus **250**, 53–63 (2015). https://doi.org/10.1016/j.icarus.2014.11.006

D.E. Brownlee, P. Tsou, J.D. Anderson, M.S. Hanner, R.L. Newburn, Z. Sekanina, B.C. Clark, F. Hörz, M.E. Zolensky, J. Kissel, J.A.M. McDonnell, S.A. Sandford, A.J. Tuzzolino, Stardust: comet and interstellar dust sample return mission. J. Geophys. Res., Planets **108**, 8111 (2003). https://doi.org/10.1029/2003JE002087

B.J. Buratti, M.D. Hicks, L.A. Soderblom, D. Britt, J. Oberst, J.K. Hillier, Deep Space 1 photometry of the nucleus of Comet 19P/Borrelly. Icarus **167**, 16–29 (2004). https://doi.org/10.1016/j.icarus.2003.05.002

F. Capaccioni, A. Coradini, G. Filacchione, S. Erard, G. Arnold, P. Drossart, M.C. De Sanctis, D. Bockelee-Morvan, M.T. Capria, F. Tosi, C. Leyrat, B. Schmitt, E. Quirico, P. Cerroni, V. Mennella, A. Raponi, M. Ciarniello, T. McCord, L. Moroz, E. Palomba, E. Ammannito, M.A. Barucci, G. Bellucci, J. Benkhoff, J.P. Bibring, A. Blanco, M. Blecka, R. Carlson, U. Carsenty, L. Colangeli, M. Combes, M. Combi, J. Crovisier, T. Encrenaz, C. Federico, U. Fink, S. Fonti, W.H. Ip, P. Irwin, R. Jaumann, E. Kuehrt, Y. Langevin, G. Magni, S. Mottola, V. Orofino, P. Palumbo, G. Piccioni, U. Schade, F. Taylor, D. Tiphene, G.P. Tozzi, P. Beck, N. Biver, L. Bonal, J.-P. Combe, D. Despan, E. Flamini, S. Fornasier, A. Frigeri, D. Grassi, M. Gudipati, A. Longobardo, K. Markus, F. Merlin, R. Orosei, G. Rinaldi, K. Stephan, M. Cartacci, A. Cicchetti, S. Giuppi, Y. Hello, F. Henry, S. Jacquinod, R. Noschese, G. Peter, R. Politi, J.M. Reess, A. Semery, The organic-rich surface of comet 67P/Churyumov–Gerasimenko as seen by VIRTIS/Rosetta. Science **347**(1), 0628 (2015). https://doi.org/10.1126/science.aaa0628

M. Ciarniello, F. Capaccioni, G. Filacchione, A. Raponi, F. Tosi, M.C. De Sanctis, M.T. Capria, S. Erard, D. Bockelee-Morvan, C. Leyrat, G. Arnold, A. Barucci, P. Beck, G. Bellucci, S. Fornasier, A. Longobardo, S. Mottola, E. Palomba, E. Quirico, B. Schmitt, Photometric properties of comet 67P/Churyumov–Gerasimenko from VIRTIS-M onboard Rosetta. Astron. Astrophys. **583**, 31 (2015). https://doi.org/10.1051/0004-6361/201526307

M. Ciarniello, A. Raponi, F. Capaccioni, G. Filacchione, F. Tosi, M.C. De Sanctis, D. Kappel, B. Rousseau, G. Arnold, M.T. Capria, M.A. Barucci, E. Quirico, A. Longobardo, E. Kuehrt, S. Mottola, S. Erard, D. Bockelée-Morvan, C. Leyrat, A. Migliorini, A. Zinzi, E. Palomba, B. Schmitt, G. Piccioni, P. Cerroni, W.-H. Ip, G. Rinaldi, M. Salatti, The global surface composition of 67P/Churyumov–Gerasimenko nucleus by Rosetta/VIRTIS. II) Diurnal and seasonal variability. Mon. Not. R. Astron. Soc. **462**, 443–458 (2016). https://doi.org/10.1093/mnras/stw3177

B.C. Clark, L.W. Mason, J. Kissel, Systematics of the CHON and other light element particle populations in Comet p/ Halley. Astron. Astrophys. **187**, 779 (1987)

L. Colangeli, J.J. Lopez-Moreno, P. Palumbo, J. Rodriguez, M. Cosi, V. Della Corte, F. Esposito, M. Fulle, M. Herranz, J.M. Jeronimo, A. Lopez-Jimenez, E.M. Epifani, R. Morales, F. Moreno, E. Palomba, A. Rotundi, The Grain Impact Analyser and Dust Accumulator (GIADA) experiment for the Rosetta mission: design, performances and first results. Space Sci. Rev. **128**, 803–821 (2007). https://doi.org/10.1007/s11214-006-9038-5

A. Coradini, F. Capaccioni, P. Drossart, G. Arnold, E. Ammannito, F. Angrilli, A. Barucci, G. Bellucci, J. Benkhoff, G. Bianchini, J.P. Bibring, M. Blecka, D. Bockelee-Morvan, M.T. Capria, R. Carlson, U. Carsenty, P. Cerroni, L. Colangeli, M. Combes, M. Combi, J. Crovisier, M.C. De Sanctis, E.T. Encrenaz, S. Erard, C. Federico, G. Filacchione, U. Fink, S. Fonti, V. Formisano, W.H. Ip, R. Jaumann, E. Kuehrt, Y. Langevin, G. Magni, T. McCord, V. Mennella, S. Mottola, G. Neukum, P. Palumbo, G. Piccioni, H. Rauer, B. Saggin, B. Schmitt, D. Tiphene, G. Tozzi, Virtis: an imaging spectrometer for the Rosetta mission. Space Sci. Rev. **128**, 529–559 (2007). https://doi.org/10.1007/s11214-006-9127-5

D.P. Cruikshank, C.M. Dalle Ore, R.N. Clark, Y.J. Pendleton, Aromatic and aliphatic organic materials on Iapetus: analysis of Cassini VIMS data. Icarus **233**, 306–315 (2014). https://doi.org/10.1016/j.icarus.2014.02.011

A.A. de Almeida, W.F. Huebner, J. Benkhoff, D.C. Boice, P.D. Singh, The three sources of gas in the comae of comets, in *Revista Mexicana de Astronomia y Astrofisica Conference Series*, ed. by E. Falco, J.A. Fernandez, R.F. Ferrero, vol. 4 (1996), p. 110

C. de Bergh, B. Schmitt, L.V. Moroz, E. Quirico, D.P. Cruikshank, in *Laboratory Data on Ices, Refractory Carbonaceous Materials, and Minerals Relevant to Transneptunian Objects and Centaurs*, ed. by M.A. Barucci, H. Boehnhardt, D.P. Cruikshank, A. Morbidelli, R. Dotson (2008), pp. 483–506

D. de Niem, E. Kührt, S. Hviid, B. Davidsson, Low velocity collisions of porous planetesimals in the early solar system. Icarus **301**, 196–218 (2018). https://doi.org/10.1016/j.icarus.2017.09.024

M.C. de Sanctis, M.T. Capria, A. Coradini, Thermal evolution model of 67P/Churyumov–Gerasimenko, the new Rosetta target. Astron. Astrophys. **444**, 605–614 (2005). https://doi.org/10.1051/0004-6361:20053036

M.C. de Sanctis, F. Capaccioni, M. Ciarniello, G. Filacchione, M. Formisano, S. Mottola, A. Raponi, F. Tosi, D. Bockelée-Morvan, S. Erard, C. Leyrat, B. Schmitt, E. Ammannito, G. Arnold, M.A. Barucci, M. Combi, M.T. Capria, P. Cerroni, W.-H. Ip, E. Kuehrt, T.B. McCord, E. Palomba, P. Beck, E. Quirico, VIRTIS Team, G. Piccioni, G. Bellucci, M. Fulchignoni, R. Jaumann, K. Stephan, A. Longobardo, V. Mennella, A. Migliorini, J. Benkhoff, J.P. Bibring, A. Blanco, M. Blecka, R. Carlson, U. Carsenty, L. Colangeli, M. Combes, J. Crovisier, P. Drossart, T. Encrenaz, C. Federico, U. Fink, S. Fonti, P. Irwin, Y. Langevin, G. Magni, L. Moroz, V. Orofino, U. Schade, F. Taylor, D. Tiphene, G.P. Tozzi, N. Biver, L. Bonal, J.-P. Combe, D. Despan, E. Flamini, S. Fornasier, A. Frigeri, D. Grassi, M.S. Gudipati, F. Mancarella, K. Markus, F. Merlin, R. Orosei, G. Rinaldi, M. Cartacci, A. Cicchetti, S. Giuppi, Y. Hello, F. Henry, S. Jacquinod, J.M. Rees, R. Noschese, R. Politi, G. Peter, The diurnal cycle of water ice on comet 67P/Churyumov–Gerasimenko. Nature **525**, 500–503 (2015). https://doi.org/10.1038/nature14869

M.C. De Sanctis, E. Ammannito, H.Y. McSween, A. Raponi, S. Marchi, F. Capaccioni, M.T. Capria, F.G. Carrozzo, M. Ciarniello, S. Fonte, M. Formisano, A. Frigeri, M. Giardino, A. Longobardo, G. Magni, L.A. McFadden, E. Palomba, C.M. Pieters, F. Tosi, F. Zambon, C.A. Raymond, C.T. Russell, Localized aliphatic organic material on the surface of Ceres. Science **355**, 719–722 (2017). https://doi.org/10.1126/science.aaj2305

V. Della Corte, A. Rotundi, M. Fulle, E. Gruen, P. Weissman, R. Sordini, M. Ferrari, S. Ivanovski, F. Lucarelli, M. Accolla, V. Zakharov, E. Mazzotta Epifani, J.J. Lopez-Moreno, J. Rodriguez, L. Colangeli, P. Palumbo, E. Bussoletti, J.F. Crifo, F. Esposito, S.F. Green, P.L. Lamy, J.A.M. McDonnell, V. Mennella, A. Molina, R. Morales, F. Moreno, J.L. Ortiz, E. Palomba, J.M. Perrin, F.J.M. Rietmeijer, R. Rodrigo, J.C. Zarnecki, M. Cosi, F. Giovane, B. Gustafson, M.L. Herranz, J.M. Jeronimo, M.R. Leese, A.C. Lopez-Jimenez, N. Altobelli, GIADA: shining a light on the monitoring of the comet dust production from the nucleus of 67P/Churyumov–Gerasimenko. Astron. Astrophys. **583**, 13 (2015). https://doi.org/10.1051/0004-6361/201526208

N. Dello Russo, R.J. Vervack, H.A. Weaver, N. Biver, D. Bockelée-Morvan, J. Crovisier, C.M. Lisse, Compositional homogeneity in the fragmented comet 73P/Schwassmann–Wachmann 3. Nature **448**, 172–175 (2007). https://doi.org/10.1038/nature05908

A.H. Delsemme, The cometary connection with prebiotic chemistry. Orig. Life **14**, 51–60 (1984). https://doi.org/10.1007/BF00933639

J.D.P. Deshapriya, M.A. Barucci, S. Fornasier, C. Feller, P.H. Hasselmann, H. Sierks, M.R. El-Maarry, M. Pajola, C. Barbieri, P.L. Lamy, R. Rodrigo, D. Koschny, H. Rickman, J. Agarwal, M.F. A'Hearn, J.-L. Bertaux, I. Bertini, S. Boudreault, G. Cremonese, V. Da Deppo, B.J.R. Davidsson, S. Debei, J. Deller, M. De Cecco, M. Fulle, A. Gicquel, O. Groussin, P.J. Gutierrez, C. Güttler, M. Hofmann, S.F. Hviid, W. Ip, L. Jorda, H.U. Keller, J. Knollenberg, R. Kramm, E. Kührt, M. Küppers, L. Lara, M. Lazzarin, J.J. Lopez Moreno, F. Marzari, S. Mottola, G. Naletto, N. Oklay, D. Perna, A. Pommerol, N. Thomas, C. Tubiana, J.-B. Vincent, Spectrophotometry of the Khonsu region on the comet 67P/Churyumov–Gerasimenko using OSIRIS instrument images. Mon. Not. R. Astron. Soc. **462**, 274–286 (2016). https://doi.org/10.1093/mnras/stw2530

E. Dobrică, A.J. Brearley, Crystalline silicates in Comet 81P/Wild 2 from the Stardust Track 81. Meteorit. Planet. Sci. **74**, 5413 (2011)

E. Dobrică, C. Engrand, E. Quirico, G. Montagnac, J. Duprat, Raman characterization of carbonaceous matter in CONCORDIA Antarctic micrometeorites. Meteorit. Planet. Sci. **46**, 1363–1375 (2011). https://doi.org/10.1111/j.1945-5100.2011.01235.x

E. Dobrică, C. Engrand, H. Leroux, J.-N. Rouzaud, J. Duprat, Transmission electron microscopy of CONCORDIA UltraCarbonaceous Antarctic MicroMeteorites (UCAMMs): mineralogical properties. Geochim. Cosmochim. Acta **76**, 68–82 (2012). https://doi.org/10.1016/j.gca.2011.10.025

M.R. El-Maarry, N. Thomas, L. Giacomini, M. Massironi, M. Pajola, R. Marschall, A. Gracia-Berná, H. Sierks, C. Barbieri, P.L. Lamy, R. Rodrigo, H. Rickman, D. Koschny, H.U. Keller, J. Agarwal, M.F. A'Hearn, A.-T. Auger, M.A. Barucci, J.-L. Bertaux, I. Bertini, S. Besse, D. Bodewits, G. Cremonese, V. Da Deppo, B. Davidsson, M. De Cecco, S. Debei, C. Güttler, S. Fornasier, M. Fulle, O. Groussin, P.J. Gutierrez, S.F. Hviid, W.-H. Ip, L. Jorda, J. Knollenberg, G. Kovacs, J.-R. Kramm, E. Kührt, M. Küppers, F. La Forgia, L.M. Lara, M. Lazzarin, J.J. Lopez Moreno, S. Marchi, F. Marzari, H. Michalik, G. Naletto, N. Oklay, A. Pommerol, F. Preusker, F. Scholten, C. Tubiana, J.-B. Vincent, Regional surface morphology of comet 67P/Churyumov–Gerasimenko from Rosetta/OSIRIS images. Astron. Astrophys. **583**, 26 (2015). https://doi.org/10.1051/0004-6361/201525723

C. Engrand, J. Duprat, E. Dartois, K. Benzerara, H. Leroux, D. Baklouti, A. Bardyn, C. Briois, H. Cottin, H. Fischer, N. Fray, M. Godard, M. Hilchenbach, Y. Langevin, J. Paquette, J. Rynö, R. Schulz, J. Silén, O. Stenzel, L. Thirkell, Cosima Team, Variations in cometary dust composition from Giotto to Rosetta, clues to their formation mechanisms. Mon. Not. R. Astron. Soc. **462**, 323–330 (2016). https://doi.org/10.1093/mnras/stw2844

S. Espinasse, J. Klinger, C. Ritz, B. Schmitt, Modeling of the thermal behavior and of the chemical differentiation of cometary nuclei. Icarus **92**, 350–365 (1991). https://doi.org/10.1016/0019-1035(91)90058-2

L.M. Feaga, M.F. A'Hearn, J.M. Sunshine, O. Groussin, T.L. Farnham, Asymmetries in the distribution of $H_2O$ and $CO_2$ in the inner coma of Comet 9P/Tempel 1 as observed by Deep Impact. Icarus **191**, 134–145 (2007). https://doi.org/10.1016/j.icarus.2007.04.038

G. Filacchione, M. Ciarniello, F. Capaccioni, R.N. Clark, P.D. Nicholson, M.M. Hedman, J.N. Cuzzi, D.P. Cruikshank, C.M. Dalle Ore, R.H. Brown, P. Cerroni, N. Altobelli, L.J. Spilker, Cassini-VIMS observations of Saturn's main rings: I. Spectral properties and temperature radial profiles variability with phase angle and elevation. Icarus **241**, 45–65 (2014). https://doi.org/10.1016/j.icarus.2014.06.001

G. Filacchione, M.C. de Sanctis, F. Capaccioni, A. Raponi, F. Tosi, M. Ciarniello, P. Cerroni, G. Piccioni, M.T. Capria, E. Palomba, G. Bellucci, S. Erard, D. Bockelee-Morvan, C. Leyrat, G. Arnold, M.A. Barucci, M. Fulchignoni, B. Schmitt, E. Quirico, R. Jaumann, K. Stephan, A. Longobardo, V. Mennella, A. Migliorini, E. Ammannito, J. Benkhoff, J.P. Bibring, A. Blanco, M.I. Blecka, R. Carlson, U. Carsenty, L. Colangeli, M. Combes, M. Combi, J. Crovisier, P. Drossart, T. Encrenaz, C. Federico, U. Fink, S. Fonti, W.H. Ip, P. Irwin, E. Kuehrt, Y. Langevin, G. Magni, T. McCord, L. Moroz, S. Mottola, V. Orofino, U. Schade, F. Taylor, D. Tiphene, G.P. Tozzi, P. Beck, N. Biver, L. Bonal, J.-P. Combe, D. Despan, E. Flamini, M. Formisano, S. Fornasier, A. Frigeri, D. Grassi, M.S. Gudipati, D. Kappel, F. Mancarella, K. Markus, F. Merlin, R. Orosei, G. Rinaldi, M. Cartacci, A. Cicchetti, S. Giuppi, Y. Hello, F. Henry, S. Jacquinod, J.M. Reess, R. Noschese, R. Politi, G. Peter, Exposed water ice on the nucleus of comet 67P/Churyumov–Gerasimenko. Nature **529**, 368–372 (2016a). https://doi.org/10.1038/nature16190

G. Filacchione, A. Raponi, F. Capaccioni, M. Ciarniello, F. Tosi, M.T. Capria, M.C. De Sanctis, A. Migliorini, G. Piccioni, P. Cerroni, M.A. Barucci, S. Fornasier, B. Schmitt, E. Quirico, S. Erard, D. Bockelee-Morvan, C. Leyrat, G. Arnold, V. Mennella, E. Ammannito, G. Bellucci, J. Benkhoff, J.P. Bibring,

A. Blanco, M.I. Blecka, R. Carlson, U. Carsenty, L. Colangeli, M. Combes, M. Combi, J. Crovisier, P. Drossart, T. Encrenaz, C. Federico, U. Fink, S. Fonti, M. Fulchignoni, W.-H. Ip, P. Irwin, R. Jaumann, E. Kuehrt, Y. Langevin, G. Magni, T. McCord, L. Moroz, S. Mottola, E. Palomba, U. Schade, K. Stephan, F. Taylor, D. Tiphene, G.P. Tozzi, P. Beck, N. Biver, L. Bonal, J.-P. Combe, D. Despan, E. Flamini, M. Formisano, A. Frigeri, D. Grassi, M.S. Gudipati, D. Kappel, A. Longobardo, F. Mancarella, K. Markus, F. Merlin, R. Orosei, G. Rinaldi, M. Cartacci, A. Cicchetti, Y. Hello, F. Henry, S. Jacquinod, J.M. Reess, R. Noschese, R. Politi, G. Peter, Seasonal exposure of carbon dioxide ice on the nucleus of comet 67P/Churyumov–Gerasimenko. Science **354**, 1563–1566 (2016b). https://doi.org/10.1126/science.aag3161

G. Filacchione, F. Capaccioni, M. Ciarniello, A. Raponi, F. Tosi, M.C. De Sanctis, S. Erard, D.B. Morvan, C. Leyrat, G. Arnold, B. Schmitt, E. Quirico, G. Piccioni, A. Migliorini, M.T. Capria, E. Palomba, P. Cerroni, A. Longobardo, A. Barucci, S. Fornasier, R.W. Carlson, R. Jaumann, K. Stephan, L.V. Moroz, D. Kappel, B. Rousseau, S. Fonti, F. Mancarella, D. Despan, M. Faure, The global surface composition of 67P/CG nucleus by Rosetta/VIRTIS. (I) Prelanding mission phase. Icarus **274**, 334–349 (2016c). https://doi.org/10.1016/j.icarus.2016.02.055

U. Fink, L. Doose, G. Rinaldi, A. Bieler, F. Capaccioni, D. Bockelée-Morvan, G. Filacchione, S. Erard, C. Leyrat, M. Blecka, M.T. Capria, M. Combi, J. Crovisier, M.C. De Sanctis, N. Fougere, F. Taylor, A. Migliorini, G. Piccioni, Investigation into the disparate origin of $CO_2$ and $H_2O$ outgassing for Comet 67/P. Icarus **277**, 78–97 (2016). https://doi.org/10.1016/j.icarus.2016.04.040

S. Fornasier, P.H. Hasselmann, M.A. Barucci, C. Feller, S. Besse, C. Leyrat, L. Lara, P.J. Gutierrez, N. Oklay, C. Tubiana, F. Scholten, H. Sierks, C. Barbieri, P.L. Lamy, R. Rodrigo, D. Koschny, H. Rickman, H.U. Keller, J. Agarwal, M.F. A'Hearn, J.-L. Bertaux, I. Bertini, G. Cremonese, V. Da Deppo, B. Davidsson, S. Debei, M. De Cecco, M. Fulle, O. Groussin, C. Güttler, S.F. Hviid, W. Ip, L. Jorda, J. Knollenberg, G. Kovacs, R. Kramm, E. Kührt, M. Küppers, F. La Forgia, M. Lazzarin, J.J. Lopez Moreno, F. Marzari, K.-D. Matz, H. Michalik, F. Moreno, S. Mottola, G. Naletto, M. Pajola, A. Pommerol, F. Preusker, X. Shi, C. Snodgrass, N. Thomas, J.-B. Vincent, Spectrophotometric properties of the nucleus of comet 67P/Churyumov–Gerasimenko from the OSIRIS instrument onboard the ROSETTA spacecraft. Astron. Astrophys. **583**, 30 (2015). https://doi.org/10.1051/0004-6361/201525901

S. Fornasier, S. Mottola, H.U. Keller, M.A. Barucci, B. Davidsson, C. Feller, J.D.P. Deshapriya, H. Sierks, C. Barbieri, P.L. Lamy, R. Rodrigo, D. Koschny, H. Rickman, M. A'Hearn, J. Agarwal, J.-L. Bertaux, I. Bertini, S. Besse, G. Cremonese, V. Da Deppo, S. Debei, M. De Cecco, J. Deller, M.R. El-Maarry, M. Fulle, O. Groussin, P.J. Gutierrez, C. Güttler, M. Hofmann, S.F. Hviid, W.-H. Ip, L. Jorda, J. Knollenberg, G. Kovacs, R. Kramm, E. Kührt, M. Küppers, M.L. Lara, M. Lazzarin, J.J.L. Moreno, F. Marzari, M. Massironi, G. Naletto, N. Oklay, M. Pajola, A. Pommerol, F. Preusker, F. Scholten, X. Shi, N. Thomas, I. Toth, C. Tubiana, J.-B. Vincent, Rosetta's comet 67P/Churyumov–Gerasimenko sheds its dusty mantle to reveal its icy nature. Science **354**, 1566–1570 (2016). https://doi.org/10.1126/science.aag2671

N. Fougere, K. Altwegg, J.-J. Berthelier, A. Bieler, D. Bockelée-Morvan, U. Calmonte, F. Capaccioni, M.R. Combi, J. De Keyser, V. Debout, S. Erard, B. Fiethe, G. Filacchione, U. Fink, S.A. Fuselier, T.I. Gombosi, K.C. Hansen, M. Hässig, Z. Huang, L. Le Roy, C. Leyrat, A. Migliorini, G. Piccioni, G. Rinaldi, M. Rubin, Y. Shou, V. Tenishev, G. Toth, C.-Y. Tzou, Direct simulation Monte Carlo modelling of the major species in the coma of comet 67P/Churyumov–Gerasimenko. Mon. Not. R. Astron. Soc. **462**, 156–169 (2016). https://doi.org/10.1093/mnras/stw2388

N. Fray, B. Schmitt, Sublimation of ices of astrophysical interest: a bibliographic review. Planet. Space Sci. **57**, 2053–2080 (2009). https://doi.org/10.1016/j.pss.2009.09.011

M. Fulle, F. Marzari, V. Della Corte, S. Fornasier, H. Sierks, A. Rotundi, C. Barbieri, P.L. Lamy, R. Rodrigo, D. Koschny, H. Rickman, H.U. Keller, J.J. López-Moreno, M. Accolla, J. Agarwal, M.F. A'Hearn, N. Altobelli, M.A. Barucci, J.-L. Bertaux, I. Bertini, D. Bodewits, E. Bussoletti, L. Colangeli, M. Cosi, G. Cremonese, J.-F. Crifo, V. Da Deppo, B. Davidsson, S. Debei, M. De Cecco, F. Esposito, M. Ferrari, F. Giovane, B. Gustafson, S.F. Green, O. Groussin, E. Grün, P. Gutierrez, C. Güttler, M.L. Herranz, S.F. Hviid, W. Ip, S.L. Ivanovski, J.M. Jerónimo, L. Jorda, J. Knollenberg, R. Kramm, E. Kührt, M. Küppers, L. Lara, M. Lazzarin, M.R. Leese, A.C. López-Jiménez, F. Lucarelli, E. Mazzotta Epifani, J.A.M. McDonnell, V. Mennella, A. Molina, R. Morales, F. Moreno, S. Mottola, G. Naletto, N. Oklay, J.L. Ortiz, E. Palomba, P. Palumbo, J.-M. Perrin, F.J.M. Rietmeijer, J. Rodríguez, R. Sordini, N. Thomas, C. Tubiana, J.-B. Vincent, P. Weissman, K.-P. Wenzel, V. Zakharov, J.C. Zarnecki, Evolution of the dust size distribution of comet 67P/Churyumov–Gerasimenko from 2.2 au to Perihelion. Astrophys. J. **821**, 19 (2016a). https://doi.org/10.3847/0004-637X/821/1/19

M. Fulle, N. Altobelli, B. Buratti, M. Choukroun, M. Fulchignoni, E. Grün, M.G.G.T. Taylor, P. Weissman, Unexpected and significant findings in comet 67P/Churyumov–Gerasimenko: an interdisciplinary view. Mon. Not. R. Astron. Soc. **462**, 2–8 (2016b). https://doi.org/10.1093/mnras/stw1663

P.A. Gerakines, M.H. Moore, R.L. Hudson, Ultraviolet photolysis and proton irradiation of astrophysical ice analogs containing hydrogen cyanide. Icarus **170**, 202–213 (2004). https://doi.org/10.1016/j.icarus.2004.02.005

K.-H. Glassmeier, H. Boehnhardt, D. Koschny, E. Kührt, I. Richter, The Rosetta mission: flying towards the origin of the solar system. Space Sci. Rev. **128**, 1–21 (2007). https://doi.org/10.1007/s11214-006-9140-8

F. Goesmann, H. Rosenbauer, R. Roll, C. Szopa, F. Raulin, R. Sternberg, G. Israel, U. Meierhenrich, W. Thiemann, G. Munoz-Caro, COSAC, the cometary sampling and composition experiment on Philae. Space Sci. Rev. **128**, 257–280 (2007). https://doi.org/10.1007/s11214-006-9000-6

F. Goesmann, H. Rosenbauer, J.H. Bredehöft, M. Cabane, P. Ehrenfreund, T. Gautier, C. Giri, H. Krüger, L. Le Roy, A.J. MacDermott, S. McKenna-Lawlor, U.J. Meierhenrich, G.M.M. Caro, F. Raulin, R. Roll, A. Steele, H. Steininger, R. Sternberg, C. Szopa, W. Thiemann, S. Ulamec, Organic compounds on comet 67P/Churyumov–Gerasimenko revealed by COSAC mass spectrometry. Science **349**(2) (2015). https://doi.org/10.1126/science.aab0689

R. Gomes, H.F. Levison, K. Tsiganis, A. Morbidelli, Origin of the cataclysmic Late Heavy Bombardment period of the terrestrial planets. Nature **435**, 466–469 (2005). https://doi.org/10.1038/nature03676

O. Groussin, L. Jorda, A.-T. Auger, E. Kührt, R. Gaskell, C. Capanna, F. Scholten, F. Preusker, P. Lamy, S. Hviid, J. Knollenberg, U. Keller, C. Huettig, H. Sierks, C. Barbieri, R. Rodrigo, D. Koschny, H. Rickman, M.F. A'Hearn, J. Agarwal, M.A. Barucci, J.-L. Bertaux, I. Bertini, S. Boudreault, G. Cremonese, V. Da Deppo, B. Davidsson, S. Debei, M. De Cecco, M.R. El-Maarry, S. Fornasier, M. Fulle, P.J. Gutiérrez, C. Güttler, W.-H. Ip, J.-R. Kramm, M. Küppers, M. Lazzarin, L.M. Lara, J.J. Lopez Moreno, S. Marchi, F. Marzari, M. Massironi, H. Michalik, G. Naletto, N. Oklay, A. Pommerol, M. Pajola, N. Thomas, I. Toth, C. Tubiana, J.-B. Vincent, Gravitational slopes, geomorphology, and material strengths of the nucleus of comet 67P/Churyumov–Gerasimenko from OSIRIS observations. Astron. Astrophys. **583**, 32 (2015a). https://doi.org/10.1051/0004-6361/201526379

O. Groussin, H. Sierks, C. Barbieri, P. Lamy, R. Rodrigo, D. Koschny, H. Rickman, H.U. Keller, M.F. A'Hearn, A.-T. Auger, M.A. Barucci, J.-L. Bertaux, I. Bertini, S. Besse, G. Cremonese, V. Da Deppo, B. Davidsson, S. Debei, M. De Cecco, M.R. El-Maarry, S. Fornasier, M. Fulle, P.J. Gutiérrez, C. Güttler, S. Hviid, W.-H. Ip, L. Jorda, J. Knollenberg, G. Kovacs, J.R. Kramm, E. Kührt, M. Küppers, L.M. Lara, M. Lazzarin, J.J. Lopez Moreno, S. Lowry, S. Marchi, F. Marzari, M. Massironi, S. Mottola, G. Naletto, N. Oklay, M. Pajola, A. Pommerol, N. Thomas, I. Toth, C. Tubiana, J.-B. Vincent, Temporal morphological changes in the Imhotep region of comet 67P/Churyumov–Gerasimenko. Astron. Astrophys. **583**, 36 (2015b). https://doi.org/10.1051/0004-6361/201527020

O. Groussin, N. Attree, Y. Brouet, V. Ciarletti, B. Davidsson, G. Filacchione, H.-H. Fischer, B. Gundlach, M. Knapmeyer, J. Knollenberg, R. Kokotanekova, E. Kührt, C. Leyrat, D. Marshall, I. Pelivan, Y. Skorov, C. Snodgrass, T. Spohn, F. Tosi, The thermal, mechanical, structural, and dielectric properties of cometary nuclei after Rosetta. Space Sci. Rev. (2019, this issue)

S. Gulkis, M. Frerking, J. Crovisier, G. Beaudin, P. Hartogh, P. Encrenaz, T. Koch, C. Kahn, Y. Salinas, R. Nowicki, R. Irigoyen, M. Janssen, P. Stek, M. Hofstadter, M. Allen, C. Backus, L. Kamp, C. Jarchow, E. Steinmetz, A. Deschamps, J. Krieg, M. Gheudin, D. Bockelée-Morvan, N. Biver, T. Encrenaz, D. Despois, W. Ip, E. Lellouch, I. Mann, D. Muhleman, H. Rauer, P. Schloerb, T. Spilker, MIRO: microwave instrument for Rosetta orbiter. Space Sci. Rev. **128**, 561–597 (2007). https://doi.org/10.1007/s11214-006-9032-y

B. Hapke, *Theory of Reflectance and Emittance Spectroscopy* (1993)

P.H. Hasselmann, M.A. Barucci, S. Fornasier, C. Feller, J.D.P. Deshapriya, M. Fulchignoni, B. Jost, H. Sierks, C. Barbieri, P.L. Lamy, R. Rodrigo, D. Koschny, H. Rickman, M. A'Hearn, J.-L. Bertaux, I. Bertini, G. Cremonese, V. Da Deppo, B. Davidsson, S. Debei, M. De Cecco, J. Deller, M. Fulle, R.W. Gaskell, O. Groussin, P.J. Gutierrez, C. Güttler, M. Hofmann, S.F. Hviid, W.-H. Ip, L. Jorda, H.U. Keller, J. Knollenberg, G. Kovacs, R. Kramm, E. Kührt, M. Küppers, M.L. Lara, M. Lazzarin, J.J. Lopez-Moreno, F. Marzari, S. Mottola, G. Naletto, N. Oklay, A. Pommerol, N. Thomas, C. Tubiana, J.-B. Vincent, The opposition effect of 67P/Churyumov–Gerasimenko on post-perihelion Rosetta images. Mon. Not. R. Astron. Soc. **469**, 550–567 (2017). https://doi.org/10.1093/mnras/stx1834

M. Hässig, K. Altwegg, H. Balsiger, A. Bar-Nun, J.J. Berthelier, A. Bieler, P. Bochsler, C. Briois, U. Calmonte, M. Combi, J. De Keyser, P. Eberhardt, B. Fiethe, S.A. Fuselier, M. Galand, S. Gasc, T.I. Gombosi, K.C. Hansen, A. Jäckel, H.U. Keller, E. Kopp, A. Korth, E. Kührt, L. Le Roy, U. Mall, B. Marty, O. Mousis, E. Neefs, T. Owen, H. Rème, M. Rubin, T. Sémon, C. Tornow, C.-Y. Tzou, J.H. Waite, P. Wurz, Time variability and heterogeneity in the coma of 67P/Churyumov–Gerasimenko. Science **347**(1), 0276 (2015). https://doi.org/10.1126/science.aaa0276

M. Hoang, K. Altwegg, H. Balsiger, A. Beth, A. Bieler, U. Calmonte, M.R. Combi, J. De Keyser, B. Fiethe, N. Fougere, S.A. Fuselier, A. Galli, P. Garnier, S. Gasc, T. Gombosi, K.C. Hansen, A. Jäckel, A. Korth, J. Lasue, L. Le Roy, U. Mall, H. Rème, M. Rubin, T. Sémon, D. Toublanc, C.-Y. Tzou,

J.H. Waite, P. Wurz, The heterogeneous coma of comet 67P/Churyumov–Gerasimenko as seen by ROSINA: H2O, CO2, and CO from September 2014 to February 2016. Astron. Astrophys. **600**, 77 (2017). https://doi.org/10.1051/0004-6361/201629900

X. Hu, X. Shi, H. Sierks, M. Fulle, J. Blum, H.U. Keller, E. Kührt, B. Davidsson, C. Güttler, B. Gundlach, M. Pajola, D. Bodewits, J.-B. Vincent, N. Oklay, M. Massironi, S. Fornasier, C. Tubiana, O. Groussin, S. Boudreault, S. Höfner, S. Mottola, C. Barbieri, P.L. Lamy, R. Rodrigo, D. Koschny, H. Rickman, M. A'Hearn, J. Agarwal, M.A. Barucci, J.-L. Bertaux, I. Bertini, G. Cremonese, V. Da Deppo, S. Debei, M. De Cecco, J. Deller, M.R. El-Maarry, A. Gicquel, P. Gutierrez-Marques, P.J. Gutiérrez, M. Hofmann, S.F. Hviid, W.-H. Ip, L. Jorda, J. Knollenberg, G. Kovacs, J.-R. Kramm, M. Küppers, L.M. Lara, M. Lazzarin, J.J. Lopez-Moreno, F. Marzari, G. Naletto, N. Thomas, Seasonal erosion and restoration of the dust cover on comet 67P/Churyumov–Gerasimenko as observed by OSIRIS onboard Rosetta. Astron. Astrophys. **604**, 114 (2017). https://doi.org/10.1051/0004-6361/201629910

W.F. Huebner, Origins of cometary materials. Space Sci. Rev. **138**, 5–25 (2008). https://doi.org/10.1007/s11214-007-9299-7

W.F. Huebner, J. Benkhoff, From coma abundances to nucleus composition. Space Sci. Rev. **90**, 117–130 (1999). https://doi.org/10.1023/A:1005289829692

W.F. Huebner, J. Benkhoff, M.-T. Capria, A. Coradini, C. De Sanctis, R. Orosei, D. Prialnik (eds.), *Heat and Gas Diffusion in Comet Nuclei* (The International Space Science Institute/ESA Publications Division, Bern/Noordwijk, 2006). ISBN 1608-280X

D.W. Hughes, The size, mass, mass loss and age of Halley's comet. Mon. Not. R. Astron. Soc. **213**, 103–109 (1985). https://doi.org/10.1093/mnras/213.1.103

G.R. Hunt, J.W. Salisbury, C.J. Lenhoff, Visible and near-infrared spectra of minerals and rocks—IV. Sulphides and sulphates. Mod. Geol. **3**, 1–14 (1971)

L. Jorda, R. Gaskell, C. Capanna, S. Hviid, P. Lamy, J. Ďurech, G. Faury, O. Groussin, P. Gutiérrez, C. Jackman, S.J. Keihm, H.U. Keller, J. Knollenberg, E. Kührt, S. Marchi, S. Mottola, E. Palmer, F.P. Schloerb, H. Sierks, J.-B. Vincent, M.F. A'Hearn, C. Barbieri, R. Rodrigo, D. Koschny, H. Rickman, M.A. Barucci, J.L. Bertaux, I. Bertini, G. Cremonese, V. Da Deppo, B. Davidsson, S. Debei, M. De Cecco, S. Fornasier, M. Fulle, C. Güttler, W.-H. Ip, J.R. Kramm, M. Küppers, L.M. Lara, M. Lazzarin, J.J. Lopez Moreno, F. Marzari, G. Naletto, N. Oklay, N. Thomas, C. Tubiana, K.-P. Wenzel, The global shape, density and rotation of Comet 67P/Churyumov–Gerasimenko from preperihelion Rosetta/OSIRIS observations. Icarus **277**, 257–278 (2016). https://doi.org/10.1016/j.icarus.2016.05.002

B. Jost, A. Pommerol, O. Poch, Z. Yoldi, S. Fornasier, P.H. Hasselmann, C. Feller, N. Carrasco, C. Szopa, N. Thomas, Bidirectional reflectance and VIS-NIR spectroscopy of cometary analogues under simulated space conditions. Planet. Space Sci. **145**, 14–27 (2017a). https://doi.org/10.1016/j.pss.2017.07.009

B. Jost, A. Pommerol, O. Poch, Y. Brouet, S. Fornasier, N. Carrasco, C. Szopa, N. Thomas, Bidirectional reflectance of laboratory cometary analogues to interpret the spectrophotometric properties of the nucleus of comet 67P/Churyumov–Gerasimenko. Planet. Space Sci. **148**, 1–11 (2017b). https://doi.org/10.1016/j.pss.2017.09.009

H.U. Keller, The nucleus of Comet Halley, in *Diversity and Similarity of Comets*, ed. by E.J. Rolfe, B. Battrick. ESA Special Publication, vol. 278 (1987)

L.P. Keller, S. Bajt, G.A. Baratta, J. Borg, J.P. Bradley, D.E. Brownlee, H. Busemann, J.R. Brucato, M. Burchell, L. Colangeli, L. D'Hendecourt, Z. Djouadi, G. Ferrini, G. Flynn, I.A. Franchi, M. Fries, M.M. Grady, G.A. Graham, F. Grossemy, A. Kearsley, G. Matrajt, K. Nakamura-Messenger, V. Mennella, L. Nittler, M.E. Palumbo, F.J. Stadermann, P. Tsou, A. Rotundi, S.A. Sandford, C. Snead, A. Steele, D. Wooden, M. Zolensky, Infrared spectroscopy of Comet 81P/Wild 2 samples returned by Stardust. Science **314**, 1728 (2006). https://doi.org/10.1126/science.1135796

H.U. Keller, C. Barbieri, P. Lamy, H. Rickman, R. Rodrigo, K.-P. Wenzel, H. Sierks, M.F. A'Hearn, F. Angrilli, M. Angulo, M.E. Bailey, P. Barthol, M.A. Barucci, J.-L. Bertaux, G. Bianchini, J.-L. Boit, V. Brown, J.A. Burns, I. Büttner, J.M. Castro, G. Cremonese, W. Curdt, V. da Deppo, S. Debei, M. de Cecco, K. Dohlen, S. Fornasier, M. Fulle, D. Germerott, F. Gliem, G.P. Guizzo, S.F. Hviid, W.-H. Ip, L. Jorda, D. Koschny, J.R. Kramm, E. Kührt, M. Küppers, L.M. Lara, A. Llebaria, A. López, A. López-Jimenez, J. López-Moreno, R. Meller, H. Michalik, M.D. Michelena, R. Müller, G. Naletto, A. Origné, G. Parzianello, M. Pertile, C. Quintana, R. Ragazzoni, P. Ramous, K.-U. Reiche, M. Reina, J. Rodríguez, G. Rousset, L. Sabau, A. Sanz, J.-P. Sivan, K. Stöckner, J. Tabero, U. Telljohann, N. Thomas, V. Timon, G. Tomasch, T. Wittrock, M. Zaccariotto, OSIRIS the scientific camera system onboard Rosetta. Space Sci. Rev. **128**, 433–506 (2007). https://doi.org/10.1007/s11214-006-9128-4

H.U. Keller, S. Mottola, B. Davidsson, S.E. Schröder, Y. Skorov, E. Kührt, O. Groussin, M. Pajola, S.F. Hviid, F. Preusker, F. Scholten, M.F. A'Hearn, H. Sierks, C. Barbieri, P. Lamy, R. Rodrigo, D. Koschny, H. Rickman, M.A. Barucci, J.-L. Bertaux, I. Bertini, G. Cremonese, V. Da Deppo, S. Debei, M. De Cecco, S. Fornasier, M. Fulle, P.J. Gutiérrez, W.-H. Ip, L. Jorda, J. Knollenberg, J.R. Kramm, M. Küppers, L.M. Lara, M. Lazzarin, J.J. Lopez Moreno, F. Marzari, H. Michalik, G. Naletto, L. Sabau, N.

Thomas, J.-B. Vincent, K.-P. Wenzel, J. Agarwal, C. Güttler, N. Oklay, C. Tubiana, Insolation, erosion, and morphology of comet 67P/Churyumov–Gerasimenko. Astron. Astrophys. **583**, 34 (2015). https://doi.org/10.1051/0004-6361/201525964

H.U. Keller, S. Mottola, S.F. Hviid, J. Agarwal, E. Kührt, Y. Skorov, K. Otto, J.-B. Vincent, N. Oklay, S.E. Schröder, B. Davidsson, M. Pajola, X. Shi, D. Bodewits, I. Toth, F. Preusker, F. Scholten, H. Sierks, C. Barbieri, P. Lamy, R. Rodrigo, D. Koschny, H. Rickman, M.F. A'Hearn, M.A. Barucci, J.-L. Bertaux, I. Bertini, G. Cremonese, V. Da Deppo, S. Debei, M. De Cecco, J. Deller, S. Fornasier, M. Fulle, O. Groussin, P.J. Gutiérrez, C. Güttler, M. Hofmann, W.-H. Ip, L. Jorda, J. Knollenberg, J.R. Kramm, M. Küppers, L.-M. Lara, M. Lazzarin, J.J. Lopez-Moreno, F. Marzari, G. Naletto, C. Tubiana, N. Thomas, Seasonal mass transfer on the nucleus of comet 67P/Churyumov–Gerasimenko. Mon. Not. R. Astron. Soc. **469**, 357–371 (2017). https://doi.org/10.1093/mnras/stx1726

J.F. Kerridge, Formation and processing of organics in the early solar system. Space Sci. Rev. **90**, 275–288 (1999). https://doi.org/10.1023/A:1005222804192

B.N. Khare, C. Sagan, W.R. Thompson, E.T. Arakawa, C. Meisse, P.S. Tuminello, Optical properties of poly-HCN and their astronomical applications. Can. J. Chem. **72**, 678–694 (1994)

J. Kissel, K. Altwegg, B.C. Clark, L. Colangeli, H. Cottin, S. Czempiel, J. Eibl, C. Engrand, H.M. Fehringer, B. Feuerbacher, M. Fomenkova, A. Glasmachers, J.M. Greenberg, E. Grün, G. Haerendel, H. Henkel, M. Hilchenbach, H. von Hoerner, H. Höfner, K. Hornung, E.K. Jessberger, A. Koch, H. Krüger, Y. Langevin, P. Parigger, F. Raulin, F. Rüdenauer, J. Rynö, E.R. Schmid, R. Schulz, J. Silén, W. Steiger, T. Stephan, L. Thirkell, R. Thomas, K. Torkar, N.G. Utterback, K. Varmuza, K.P. Wanczek, W. Werther, H. Zscheeg, Cosima high resolution time-of-flight secondary ion mass spectrometer for the analysis of cometary dust particles onboard Rosetta. Space Sci. Rev. **128**, 823–867 (2007). https://doi.org/10.1007/s11214-006-9083-0

V.M. Kolb, P.J. Liesch, Astrobiological relevance of phenols and their silicates, in *Proceedings of the SPIE, Instruments, Methods, and Missions for Astrobiology XI*, vol. 7097 (2008), p. 709706. https://doi.org/10.1117/12.792665

H. Krüger, F. Goesmann, C. Giri, I. Wright, A. Morse, J.H. Bredehöft, S. Ulamec, B. Cozzoni, P. Ehrenfreund, T. Gautier, S. McKenna-Lawlor, F. Raulin, H. Steininger, C. Szopa, Decay of COSAC and Ptolemy mass spectra at comet 67P/Churyumov–Gerasimenko. Astron. Astrophys. **600**, 56 (2017). https://doi.org/10.1051/0004-6361/201630286

S. Kwok, Enrichment of the solar system by organic compounds delivered from evolved stars. Adv. Space Res. **40**, 1613–1619 (2007). https://doi.org/10.1016/j.asr.2007.05.016

F. La Forgia, L. Giacomini, M. Lazzarin, M. Massironi, N. Oklay, F. Scholten, M. Pajola, I. Bertini, G. Cremonese, C. Barbieri, G. Naletto, E. Simioni, F. Preusker, N. Thomas, H. Sierks, P. Lamy, R. Rodrigo, D. Koschny, H. Rickman, H.U. Keller, J. Agarwal, A.-T. Auger, M.F. A'Hearn, M.A. Barucci, J.-L. Bertaux, S. Besse, D. Bodewits, V. Da Deppo, B. Davidsson, S. Debei, M. De Cecco, M.R. El-Maarry, F. Ferri, S. Fornasier, M. Fulle, O. Groussin, P.J. Gutiérrez, C. Güttler, I. Hall, S.F. Hviid, W.-H. Ip, L. Jorda, J. Knollenberg, J.R. Kramm, E. Kührt, M. Küppers, L.-M. Lara, J.J. Lopez Moreno, S. Magrin, F. Marzari, H. Michalik, S. Mottola, A. Pommerol, C. Tubiana, J.-B. Vincent, Geomorphology and spectrophotometry of Philae's landing site on comet 67P/Churyumov–Gerasimenko. Astron. Astrophys. **583**, 41 (2015). https://doi.org/10.1051/0004-6361/201525983

I.-L. Lai, W.-H. Ip, C.-C. Su, J.-S. Wu, J.-C. Lee, Z.-Y. Lin, Y. Liao, N. Thomas, H. Sierks, C. Barbieri, P. Lamy, R. Rodrigo, D. Koschny, H. Rickman, H.U. Keller, J. Agarwal, M.F. A'Hearn, M.A. Barucci, J.-L. Bertaux, I. Bertini, S. Boudreault, G. Cremonese, V. Da Deppo, B. Davidsson, S. Debei, M. De Cecco, J. Deller, S. Fornasier, M. Fulle, O. Groussin, P.J. Gutiérrez, C. Güttler, M. Hofmann, S.F. Hviid, L. Jorda, J. Knollenberg, G. Kovacs, J.-R. Kramm, E. Kührt, M. Küppers, L.M. Lara, M. Lazzarin, J.J. Lopez Moreno, F. Marzari, G. Naletto, N. Oklay, X. Shi, C. Tubiana, J.-B. Vincent, Gas outflow and dust transport of comet 67P/Churyumov–Gerasimenko. Mon. Not. R. Astron. Soc. **462**, 533–546 (2016). https://doi.org/10.1093/mnras/stx332

P.L. Lamy, I. Toth, H.A. Weaver, L. Jorda, M. Kaasalainen, P.J. Gutiérrez, Hubble Space Telescope observations of the nucleus and inner coma of comet 67P/Churyumov–Gerasimenko. Astron. Astrophys. **458**, 669–678 (2006). https://doi.org/10.1051/0004-6361:20065253

Y. Langevin, M. Hilchenbach, N. Ligier, S. Merouane, K. Hornung, C. Engrand, R. Schulz, J. Kissel, J. Rynö, P. Eng, Typology of dust particles collected by the COSIMA mass spectrometer in the inner coma of 67P/Churyumov Gerasimenko. Icarus **271**, 76–97 (2016). https://doi.org/10.1016/j.icarus.2016.01.027

L. Le Roy, K. Altwegg, H. Balsiger, J.-J. Berthelier, A. Bieler, C. Briois, U. Calmonte, M.R. Combi, J. De Keyser, F. Dhooghe, B. Fiethe, S.A. Fuselier, S. Gasc, T.I. Gombosi, M. Hässig, A. Jäckel, M. Rubin, C.-Y. Tzou, Inventory of the volatiles on comet 67P/Churyumov–Gerasimenko from Rosetta/ROSINA. Astron. Astrophys. **583**, 1 (2015). https://doi.org/10.1051/0004-6361/201526450

S. Lee, P. von Allmen, M. Allen, G. Beaudin, N. Biver, D. Bockelée-Morvan, M. Choukroun, J. Crovisier, P. Encrenaz, M. Frerking, S. Gulkis, P. Hartogh, M. Hofstadter, W.-H. Ip, M. Janssen, C. Jarchow, S.

Keihm, E. Lellouch, C. Leyrat, L. Rezac, F.P. Schloerb, T. Spilker, B. Gaskell, L. Jorda, H.U. Keller, H. Sierks, Spatial and diurnal variation of water outgassing on comet 67P/Churyumov–Gerasimenko observed from Rosetta/MIRO in August 2014. Astron. Astrophys. **583**, 5 (2015). https://doi.org/10.1051/0004-6361/201526155

H. Leroux, P. Cuvillier, B. Zanda, R.H. Hewins, GEMS-like material in the matrix of the Paris meteorite and the early stages of alteration of CM chondrites. Geochim. Cosmochim. Acta **170**, 247–265 (2015). https://doi.org/10.1016/j.gca.2015.09.019

J.-Y. Li, M.F. A'Hearn, M.J.S. Belton, C.J. Crockett, T.L. Farnham, C.M. Lisse, L.A. McFadden, K.J. Meech, J.M. Sunshine, P.C. Thomas, J. Veverka, Deep Impact photometry of Comet 9P/Tempel 1. Icarus **187**, 41–55 (2007a). https://doi.org/10.1016/j.icarus.2006.09.018

J.-Y. Li, M.F. A'Hearn, L.A. McFadden, M.J.S. Belton, Photometric analysis and disk-resolved thermal modeling of Comet 19P/Borrelly from Deep Space 1 data. Icarus **188**, 195–211 (2007b). https://doi.org/10.1016/j.icarus.2006.11.015

J.-Y. Li, M.F. A'Hearn, T.L. Farnham, L.A. McFadden, Photometric analysis of the nucleus of Comet 81P/Wild 2 from Stardust images. Icarus **204**, 209–226 (2009). https://doi.org/10.1016/j.icarus.2009.06.002

J.-Y. Li, S. Besse, M.F. A'Hearn, M.J.S. Belton, D. Bodewits, T.L. Farnham, K.P. Klaasen, C.M. Lisse, K.J. Meech, J.M. Sunshine, P.C. Thomas, Photometric properties of the nucleus of Comet 103P/Hartley 2. Icarus **222**, 559–570 (2013). https://doi.org/10.1016/j.icarus.2012.11.001

A. Longobardo, E. Palomba, F. Capaccioni, M. Ciarniello, F. Tosi, S. Mottola, L.V. Moroz, G. Filacchione, A. Raponi, E. Quirico, A. Zinzi, M.T. Capria, D. Bockelee-Morvan, S. Erard, C. Leyrat, G. Rinaldi, F. Dirri, Photometric behaviour of 67P/Churyumov–Gerasimenko and analysis of its pre-perihelion diurnal variations. Mon. Not. R. Astron. Soc. **469**, 346–356 (2017). https://doi.org/10.1093/mnras/stx1803

S.C. Lowry, A. Fitzsimmons, S. Collander-Brown, CCD photometry of distant comets. III. Ensemble properties of Jupiter-family comets. Astron. Astrophys. **397**, 329–343 (2003). https://doi.org/10.1051/0004-6361:20021486

A. Lucchetti, G. Cremonese, L. Jorda, F. Poulet, J.-P. Bibring, M. Pajola, F. La Forgia, M. Massironi, M.R. El-Maarry, N. Oklay, H. Sierks, C. Barbieri, P. Lamy, R. Rodrigo, D. Koschny, H. Rickman, H.U. Keller, J. Agarwal, M.F. A'Hearn, M.A. Barucci, J.-L. Bertaux, I. Bertini, V. Da Deppo, B. Davidsson, S. Debei, M. De Cecco, S. Fornasier, M. Fulle, O. Groussin, P.J. Gutierrez, C. Güttler, S.F. Hviid, W.-H. Ip, J. Knollenberg, J.-R. Kramm, E. Kührt, M. Küppers, L.M. Lara, M. Lazzarin, J.J. Lopez Moreno, F. Marzari, S. Mottola, G. Naletto, F. Preusker, F. Scholten, N. Thomas, C. Tubiana, J.-B. Vincent, Characterization of the Abydos region through OSIRIS high-resolution images in support of CIVA measurements. Astron. Astrophys. **585**, 1 (2016). https://doi.org/10.1051/0004-6361/201527330

A. Lucchetti, M. Pajola, S. Fornasier, S. Mottola, L. Penasa, L. Jorda, G. Cremonese, C. Feller, P.H. Hasselmann, M. Massironi, S. Ferrari, G. Naletto, N. Oklay, H. Sierks, C. Barbieri, P.L. Lamy, R. Rodrigo, D. Koschny, H. Rickman, H.U. Keller, J. Agarwal, M.F. A'Hearn, M.A. Barucci, J.L. Bertaux, I. Bertini, S. Boudreault, V. Da Deppo, B. Davidsson, S. Debei, M. De Cecco, J. Deller, M. Fulle, O. Groussin, P.J. Gutierrez, C. Güttler, M. Hoffman, S.F. Hviid, W.H. Ip, J. Knollenberg, J.R. Kramm, E. Kührt, M. Küppers, L.M. Lara, M. Lazzarin, F. La Forgia, L.Z. Lin, J.J. Lopez Moreno, F. Marzari, F. Preusker, F. Scholten, X. Shi, N. Thomas, C. Tubiana, J.B. Vincent, Geomorphological and spectrophotometric analysis of Seth's circular niches on comet 67P/Churyumov–Gerasimenko using OSIRIS images. Mon. Not. R. Astron. Soc. **469**, 238–251 (2017). https://doi.org/10.1093/mnras/stx1590

A. Luspay-Kuti, M. Hässig, S.A. Fuselier, K.E. Mandt, K. Altwegg, H. Balsiger, S. Gasc, A. Jäckel, L. Le Roy, M. Rubin, C.-Y. Tzou, P. Wurz, O. Mousis, F. Dhooghe, J.J. Berthelier, B. Fiethe, T.I. Gombosi, U. Mall, Composition-dependent outgassing of comet 67P/Churyumov–Gerasimenko from ROSINA/DFMS. Implications for nucleus heterogeneity? Astron. Astrophys. **583**, 4 (2015). https://doi.org/10.1051/0004-6361/201526205

A. Luspay-Kuti, O. Mousis, M. Hässig, S.A. Fuselier, J.I. Lunine, B. Marty, K.E. Mandt, P. Wurz, M. Rubin, The presence of clathrates in comet 67P/Churyumov–Gerasimenko. Sci. Adv. **2**, 1501781 (2016). https://doi.org/10.1126/sciadv.1501781

T. Mannel, M.S. Bentley, R. Schmied, H. Jeszenszky, A.C. Levasseur-Regourd, J. Romstedt, K. Torkar, Fractal cometary dust—a window into the early Solar system. Mon. Not. R. Astron. Soc. **462**, 304–311 (2016). https://doi.org/10.1093/mnras/stw2898

U. Marboeuf, B. Schmitt, How to link the relative abundances of gas species in coma of comets to their initial chemical composition? Icarus **242**, 225–248 (2014). https://doi.org/10.1016/j.icarus.2014.07.001

N. Masoumzadeh, N. Oklay, L. Kolokolova, H. Sierks, S. Fornasier, M.A. Barucci, J.-B. Vincent, C. Tubiana, C. Güttler, F. Preusker, F. Scholten, S. Mottola, P.H. Hasselmann, C. Feller, C. Barbieri, P.L. Lamy, R. Rodrigo, D. Koschny, H. Rickman, M.F. A'Hearn, J.-L. Bertaux, I. Bertini, G. Cremonese, V. Da Deppo, B.J.R. Davidsson, S. Debei, M. De Cecco, M. Fulle, A. Gicquel, O. Groussin, P.J. Gutiérrez, I. Hall, M. Hofmann, S.F. Hviid, W.-H. Ip, L. Jorda, H.U. Keller, J. Knollenberg, G. Kovacs, J.-R. Kramm,

E. Kührt, M. Küppers, L.M. Lara, M. Lazzarin, J.J. Lopez Moreno, F. Marzari, G. Naletto, X. Shi, N. Thomas, Opposition effect on comet 67P/Churyumov–Gerasimenko using Rosetta-OSIRIS images. Astron. Astrophys. **599**, 11 (2017). https://doi.org/10.1051/0004-6361/201629734

J. Mayo Greenberg, C.X. Mendoza-Gómez, The seeding of life by comets. Adv. Space Res. **12**, 169–180 (1992). https://doi.org/10.1016/0273-1177(92)90170-3

D.S. McKay, J.L. Carter, W.W. Boles, C.C. Allen, J.H. Allton, JSC-1: a new lunar regolith simulant, in *Lunar and Planetary Science Conference*. Lunar and Planetary Inst. Technical Report, vol. 24 (1993)

G. Moreels, J. Clairemidi, P. Hermine, P. Brechignac, P. Rousselot, Detection of a polycyclic aromatic molecule in comet P/Halley. Astron. Astrophys. **282**, 643–656 (1994)

L.V. Moroz, G. Arnold, A.V. Korochantsev, R. Wäsch, Natural solid bitumens as possible analogs for cometary and asteroid organics: 1. Reflectance spectroscopy of pure bitumens. Icarus **134**, 253–268 (1998). https://doi.org/10.1006/icar.1998.5955

L.V. Moroz, K. Markus, G. Arnold, D. Henkel, D. Kappel, U. Schade, M. Ciarniello, B. Rousseau, E. Quirico, B. Schmitt, F. Capaccioni, D. Bockelee-Morvan, G. Filacchione, S. Erard, C. Leyrat, A. Longobardo, Laboratory spectral reflectance studies aimed at providing clues to composition of refractory phases of comet 67P/CG's nucleus, in *European Planetary Science Congress*, vol. 11 (2017). EPSC2017-266

S. Mottola, S. Lowry, C. Snodgrass, P.L. Lamy, I. Toth, A. Rożek, H. Sierks, M.F. A'Hearn, F. Angrilli, C. Barbieri, M.A. Barucci, J.-L. Bertaux, G. Cremonese, V. Da Deppo, B. Davidsson, M. De Cecco, S. Debei, S. Fornasier, M. Fulle, O. Groussin, P. Gutiérrez, S.F. Hviid, W. Ip, L. Jorda, H.U. Keller, J. Knollenberg, D. Koschny, R. Kramm, E. Kührt, M. Küppers, L. Lara, M. Lazzarin, J.J. Lopez Moreno, F. Marzari, H. Michalik, G. Naletto, H. Rickman, R. Rodrigo, L. Sabau, N. Thomas, K.-P. Wenzel, J. Agarwal, I. Bertini, F. Ferri, C. Güttler, S. Magrin, N. Oklay, C. Tubiana, J.-B. Vincent, The rotation state of 67P/Churyumov–Gerasimenko from approach observations with the OSIRIS cameras on Rosetta. Astron. Astrophys. **569**, 2 (2014). https://doi.org/10.1051/0004-6361/201424590

S. Mottola, G. Arnold, H.-G. Grothues, R. Jaumann, H. Michaelis, G. Neukum, J.-P. Bibring, S.E. Schröder, M. Hamm, K.A. Otto, I. Pelivan, G. Proffe, F. Scholten, D. Tirsch, M. Kreslavsky, E. Remetean, F. Souvannavong, B. Dolives, The structure of the regolith on 67P/Churyumov–Gerasimenko from ROLIS descent imaging. Science **349**(2), (2015). https://doi.org/10.1126/science.aab0232

O. Mousis, A. Guilbert-Lepoutre, B. Brugger, L. Jorda, J.S. Kargel, A. Bouquet, A.-T. Auger, P. Lamy, P. Vernazza, N. Thomas, H. Sierks, Pits formation from volatile outgassing on 67P/Churyumov–Gerasimenko. Astrophys. J. Lett. **814**, 5 (2015). https://doi.org/10.1088/2041-8205/814/1/L5

M.J. Mumma, S.B. Charnley, The chemical composition of comets—emerging taxonomies and natal heritage. Annu. Rev. Astron. Astrophys. **49**, 471–524 (2011). https://doi.org/10.1146/annurev-astro-081309-130811

G.M. Muñoz Caro, W.A. Schutte, UV-photoprocessing of interstellar ice analogs: new infrared spectroscopic results. Astron. Astrophys. **412**, 121–132 (2003). https://doi.org/10.1051/0004-6361:20031408

R.M. Nelson, M.D. Rayman, H.A. Weaver, The Deep Space 1 encounter with Comet 19P/Borrelly. Icarus **167**, 1–3 (2004). https://doi.org/10.1016/j.icarus.2003.10.008

A. Oehler, G. Neukum, Visible and near IR albedo measurements of ice/dust mixtures. Geophys. Res. Lett. **18**, 253–256 (1991). https://doi.org/10.1029/91GL00174

N. Oklay, J.M. Sunshine, M. Pajola, A. Pommerol, J.-B. Vincent, S. Mottola, H. Sierks, S. Fornasier, M.A. Barucci, F. Preusker, F. Scholten, L.M. Lara, C. Barbieri, P.L. Lamy, R. Rodrigo, D. Koschny, H. Rickman, M.F. A'Hearn, J.-L. Bertaux, I. Bertini, D. Bodewits, G. Cremonese, V. Da Deppo, B.J.R. Davidsson, S. Debei, M. De Cecco, J. Deller, M. Fulle, A. Gicquel, O. Groussin, P.J. Gutiérrez, C. Güttler, I. Hall, M. Hofmann, S.F. Hviid, W.-H. Ip, L. Jorda, H.U. Keller, J. Knollenberg, G. Kovacs, J.-R. Kramm, E. Kührt, M. Küppers, M. Lazzarin, Z.-Y. Lin, J.J. Lopez Moreno, F. Marzari, G. Naletto, X. Shi, N. Thomas, C. Tubiana, Comparative study of water ice exposures on cometary nuclei using multispectral imaging data. Mon. Not. R. Astron. Soc. **462**, 394–414 (2016a). https://doi.org/10.1093/mnras/stw2918

N. Oklay, J.-B. Vincent, S. Fornasier, M. Pajola, S. Besse, B.J.R. Davidsson, L.M. Lara, S. Mottola, G. Naletto, H. Sierks, A.M. Barucci, F. Scholten, F. Preusker, A. Pommerol, N. Masoumzadeh, M. Lazzarin, C. Barbieri, P.L. Lamy, R. Rodrigo, D. Koschny, H. Rickman, M.F. A'Hearn, J.-L. Bertaux, I. Bertini, D. Bodewits, G. Cremonese, V. Da Deppo, S. Debei, M. De Cecco, M. Fulle, O. Groussin, P.J. Gutiérrez, C. Güttler, I. Hall, M. Hofmann, S.F. Hviid, W.-H. Ip, L. Jorda, H.U. Keller, J. Knollenberg, G. Kovacs, J.-R. Kramm, E. Kührt, M. Küppers, Z.-Y. Lin, J.J. Lopez Moreno, F. Marzari, F. Moreno, X. Shi, N. Thomas, I. Toth, C. Tubiana, Variegation of comet 67P/Churyumov–Gerasimenko in regions showing activity. Astron. Astrophys. **586**, 80 (2016b). https://doi.org/10.1051/0004-6361/201527369

N. Oklay, S. Mottola, J.-B. Vincent, M. Pajola, S. Fornasier, S.F. Hviid, D. Kappel, E. Kührt, H.U. Keller, M.A. Barucci, C. Feller, F. Preusker, F. Scholten, I. Hall, H. Sierks, C. Barbieri, P.L. Lamy, R. Rodrigo, D. Koschny, H. Rickman, M.F. A'Hearn, J.-L. Bertaux, I. Bertini, D. Bodewits, G. Cremonese, V. Da Deppo, B.J.R. Davidsson, S. Debei, M. De Cecco, J. Deller, J.D.P. Deshapriya, M. Fulle, A. Gicquel, O. Groussin, P.J. Gutiérrez, C. Güttler, P.H. Hasselmann, M. Hofmann, W.-H. Ip, L. Jorda, J. Knollenberg, G. Kovacs, J.-R. Kramm, M. Küppers, L.M. Lara, M. Lazzarin, Z.-Y. Lin, J.J.L. Moreno, A. Lucchetti, F. Marzari, N. Masoumzadeh, G. Naletto, A. Pommerol, X. Shi, N. Thomas, C. Tubiana, Long-term survival of surface water ice on comet 67P. Mon. Not. R. Astron. Soc. **469**, 582–597 (2017). https://doi.org/10.1093/mnras/stx2298

M. Pajola, S. Höfner, J.B. Vincent, N. Oklay, F. Scholten, F. Preusker, S. Mottola, G. Naletto, S. Fornasier, S. Lowry, C. Feller, P.H. Hasselmann, C. Güttler, C. Tubiana, H. Sierks, C. Barbieri, P. Lamy, R. Rodrigo, D. Koschny, H. Rickman, H.U. Keller, J. Agarwal, M.F. A'Hearn, M.A. Barucci, J.-L. Bertaux, I. Bertini, S. Besse, S. Boudreault, G. Cremonese, V. de Deppo, B. Davidsson, S. Debei, M. de Cecco, J. Deller, J.D.P. Deshapriya, M.R. El-Maarry, S. Ferrari, F. Ferri, M. Fulle, O. Groussin, P. Gutierrez, M. Hofmann, S.F. Hviid, W.-H. Ip, L. Jorda, J. Knollenberg, G. Kovacs, J.R. Kramm, E. Kührt, M. Küppers, L.M. Lara, Z.-Y. Lin, M. Lazzarin, A. Lucchetti, J.J. Lopez Moreno, F. Marzari, M. Massironi, H. Michalik, L. Penasa, A. Pommerol, E. Simioni, N. Thomas, I. Toth, E. Baratti, The pristine interior of comet 67P revealed by the combined Aswan outburst and cliff collapse. Nat. Astron. **1**, 0092 (2017). https://doi.org/10.1038/s41550-017-0092

R. Papoular, The use of kerogen data in understanding the properties and evolution of interstellar carbonaceous dust. Astron. Astrophys. **378**, 597–607 (2001). https://doi.org/10.1051/0004-6361:20011224

Y.J. Pendleton, L.J. Allamandola, The organic refractory material in the diffuse interstellar medium: mid-infrared spectroscopic constraints. Astrophys. J. Suppl. Ser. **138**, 75–98 (2002). https://doi.org/10.1086/322999

D. Perna, M. Fulchignoni, M.A. Barucci, S. Fornasier, C. Feller, J.D.P. Deshapriya, P.H. Hasselmann, H. Sierks, C. Barbieri, P.L. Lamy, R. Rodrigo, D. Koschny, H. Rickman, M. A'Hearn, J.-L. Bertaux, I. Bertini, G. Cremonese, V. Da Deppo, B. Davidsson, S. Debei, J. Deller, M. De Cecco, M.R. El-Maarry, M. Fulle, O. Groussin, P.J. Gutierrez, C. Güttler, M. Hofmann, S.F. Hviid, W.-H. Ip, L. Jorda, H.U. Keller, J. Knollenberg, R. Kramm, E. Kührt, M. Küppers, L.M. Lara, M. Lazzarin, J.J. Lopez Moreno, F. Marzari, G. Naletto, N. Oklay, N. Thomas, C. Tubiana, J.-B. Vincent, Multivariate statistical analysis of OSIRIS/Rosetta spectrophotometric data of comet 67P/Churyumov–Gerasimenko. Astron. Astrophys. **600**, 115 (2017). https://doi.org/10.1051/0004-6361/201630015

S. Pizzarello, G.W. Cooper, G.J. Flynn, in *The Nature and Distribution of the Organic Material in Carbonaceous Chondrites and Interplanetary Dust Particles*, ed. by D.S. Lauretta, H.Y. McSween (2006), pp. 625–651

O. Poch, A. Pommerol, B. Jost, N. Carrasco, C. Szopa, N. Thomas, Sublimation of ice-tholins mixtures: a morphological and spectro-photometric study. Icarus **266**, 288–305 (2016a). https://doi.org/10.1016/j.icarus.2015.11.006

O. Poch, A. Pommerol, B. Jost, N. Carrasco, C. Szopa, N. Thomas, Sublimation of water ice mixed with silicates and tholins: evolution of surface texture and reflectance spectra, with implications for comets. Icarus **267**, 154–173 (2016b). https://doi.org/10.1016/j.icarus.2015.12.017

A. Pommerol, N. Thomas, M.R. El-Maarry, M. Pajola, O. Groussin, A.-T. Auger, N. Oklay, S. Fornasier, C. Feller, B. Davidsson, A. Gracia-Berná, B. Jost, R. Marschall, O. Poch, M.A. Barucci, J.-L. Bertaux, F. La Forgia, H.U. Keller, E. Kührt, S.C. Lowry, S. Mottola, G. Naletto, H. Sierks, C. Barbieri, P.L. Lamy, R. Rodrigo, D. Koschny, H. Rickman, J. Agarwal, M.F. A'Hearn, I. Bertini, S. Boudreault, G. Cremonese, V. Da Deppo, M. De Cecco, S. Debei, C. Güttler, M. Fulle, P.J. Gutierrez, S.F. Hviid, W.-H. Ip, L. Jorda, J. Knollenberg, G. Kovacs, J.-R. Kramm, E. Küppers, L. Lara, M. Lazzarin, J.L. Lopez Moreno, F. Marzari, H. Michalik, F. Preusker, F. Scholten, C. Tubiana, J.-B. Vincent, OSIRIS observations of meter-sized exposures of $H_2O$ ice at the surface of 67P/Churyumov–Gerasimenko and interpretation using laboratory experiments. Astron. Astrophys. **583**, 25 (2015). https://doi.org/10.1051/0004-6361/201525977

F. Preusker, F. Scholten, K.-D. Matz, T. Roatsch, K. Willner, S.F. Hviid, J. Knollenberg, L. Jorda, P.J. Gutiérrez, E. Kührt, S. Mottola, M.F. A'Hearn, N. Thomas, H. Sierks, C. Barbieri, P. Lamy, R. Rodrigo, D. Koschny, H. Rickman, H.U. Keller, J. Agarwal, M.A. Barucci, J.-L. Bertaux, I. Bertini, G. Cremonese, V. Da Deppo, B. Davidsson, S. Debei, M. De Cecco, S. Fornasier, M. Fulle, O. Groussin, C. Güttler, W.-H. Ip, J.R. Kramm, M. Küppers, L.M. Lara, M. Lazzarin, J.J. Lopez Moreno, F. Marzari, H. Michalik, G. Naletto, N. Oklay, C. Tubiana, J.-B. Vincent, Shape model, reference system definition, and cartographic mapping standards for comet 67P/Churyumov–Gerasimenko–Stereo-photogrammetric analysis of Rosetta/OSIRIS image data. Astron. Astrophys. **583**, 33 (2015). https://doi.org/10.1051/0004-6361/201526349

D. Prialnik, M.F. A'Hearn, K.J. Meech, A mechanism for short-lived cometary outbursts at sunrise as observed by Deep Impact on 9P/Tempel 1. Mon. Not. R. Astron. Soc. **388**, 20–23 (2008). https://doi.org/10.1111/j.1745-3933.2008.00491.x

E. Quirico, G. Montagnac, V. Lees, P.F. McMillan, C. Szopa, G. Cernogora, J.-N. Rouzaud, P. Simon, J.-M. Bernard, P. Coll, N. Fray, R.D. Minard, F. Raulin, B. Reynard, B. Schmitt, New experimental constraints on the composition and structure of tholins. Icarus **198**, 218–231 (2008). https://doi.org/10.1016/j.icarus.2008.07.012

E. Quirico, L.V. Moroz, B. Schmitt, G. Arnold, M. Faure, P. Beck, L. Bonal, M. Ciarniello, F. Capaccioni, G. Filacchione, S. Erard, C. Leyrat, D. Bockelée-Morvan, A. Zinzi, E. Palomba, P. Drossart, F. Tosi, M.T. Capria, M.C. De Sanctis, A. Raponi, S. Fonti, F. Mancarella, V. Orofino, A. Barucci, M.I. Blecka, R. Carlson, D. Despan, A. Faure, S. Fornasier, M.S. Gudipati, A. Longobardo, K. Markus, V. Mennella, F. Merlin, G. Piccioni, B. Rousseau, F. Taylor, Refractory and semi-volatile organics at the surface of comet 67P/Churyumov–Gerasimenko: Insights from the VIRTIS/Rosetta imaging spectrometer. Icarus **272**, 32–47 (2016). https://doi.org/10.1016/j.icarus.2016.02.028

A. Raponi, M. Ciarniello, F. Capaccioni, G. Filacchione, F. Tosi, M.C. De Sanctis, M.T. Capria, M.A. Barucci, A. Longobardo, E. Palomba, D. Kappel, G. Arnold, S. Mottola, B. Rousseau, E. Quirico, G. Rinaldi, S. Erard, D. Bockelee-Morvan, C. Leyrat, The temporal evolution of exposed water ice-rich areas on the surface of 67P/Churyumov–Gerasimenko: spectral analysis. Mon. Not. R. Astron. Soc. **462**, 476–490 (2016). https://doi.org/10.1093/mnras/stw3281

A. Raponi, M. Ciarniello, G. Filacchione, F. Capaccioni, M.C. De Sanctis, L.V. Moroz, V. Vinogradoff, F. Tosi, G. Arnold, E. Quirico, V. Mennella, P. Beck, C. Pilorget, A. Pommerol, S. Schroder, D. Kappel, I. Istiqomah, B. Rousseau, O. Poch, Organic features on the spectrum of 67P Churyumov–Gerasimenko from the improved calibration of VIRTIS-M-IR, in *European Planetary Science Congress 2018*, 16–21 September 2018, vol. EPSC2018-865 (2018)

R. Reinhard, The Giotto mission to Halley's comet. Adv. Space Res. **2**, 97–107 (1982). https://doi.org/10.1016/0273-1177(82)90293-9

W. Riedler, K. Torkar, H. Jeszenszky, J. Romstedt, H.S.C. Alleyne, H. Arends, W. Barth, J.V.D. Biezen, B. Butler, P. Ehrenfreund, M. Fehringer, G. Fremuth, J. Gavira, O. Havnes, E.K. Jessberger, R. Kassing, W. Klöck, C. Koeberl, A.C. Levasseur-Regourd, M. Maurette, F. Rüdenauer, R. Schmidt, G. Stangl, M. Steller, I. Weber, MIDAS the Micro-Imaging Dust Analysis System for the Rosetta mission. Space Sci. Rev. **128**, 869–904 (2007). https://doi.org/10.1007/s11214-006-9040-y

A. Rotundi, H. Sierks, V. Della Corte, M. Fulle, P.J. Gutierrez, L. Lara, C. Barbieri, P.L. Lamy, R. Rodrigo, D. Koschny, H. Rickman, H.U. Keller, J.J. López-Moreno, M. Accolla, J. Agarwal, M.F. A'Hearn, N. Altobelli, F. Angrilli, M.A. Barucci, J.-L. Bertaux, I. Bertini, D. Bodewits, E. Bussoletti, L. Colangeli, M. Cosi, G. Cremonese, J.-F. Crifo, V. Da Deppo, B. Davidsson, S. Debei, M. De Cecco, F. Esposito, M. Ferrari, S. Fornasier, F. Giovane, B. Gustafson, S.F. Green, O. Groussin, E. Grün, C. Güttler, M.L. Herranz, S.F. Hviid, W. Ip, S. Ivanovski, J.M. Jerónimo, L. Jorda, J. Knollenberg, R. Kramm, E. Kührt, M. Küppers, M. Lazzarin, M.R. Leese, A.C. López-Jiménez, F. Lucarelli, S.C. Lowry, F. Marzari, E.M. Epifani, J.A.M. McDonnell, V. Mennella, H. Michalik, A. Molina, R. Morales, F. Moreno, S. Mottola, G. Naletto, N. Oklay, J.L. Ortiz, E. Palomba, P. Palumbo, J.-M. Perrin, J. Rodríguez, L. Sabau, C. Snodgrass, R. Sordini, N. Thomas, C. Tubiana, J.-B. Vincent, P. Weissman, K.-P. Wenzel, V. Zakharov, J.C. Zarnecki, Dust measurements in the coma of comet 67P/Churyumov–Gerasimenko inbound to the Sun. Science **347**(1), 3905 (2015). https://doi.org/10.1126/science.aaa3905

B. Rousseau, S. Érard, P. Beck, É. Quirico, B. Schmitt, O. Brissaud, G. Montes-Hernandez, F. Capaccioni, G. Filacchione, D. Bockelée-Morvan, C. Leyrat, M. Ciarniello, A. Raponi, D. Kappel, G. Arnold, L.V. Moroz, E. Palomba, F. Tosi, Laboratory simulations of the Vis-NIR spectra of comet 67P using sub-$\mu$m sized cosmochemical analogues. Icarus **306**, 306–318 (2018)

M. Rubin, N. Fougere, K. Altwegg, M.R. Combi, L. Le Roy, V.M. Tenishev, N. Thomas, Mass transport around comets and its impact on the seasonal differences in water production rates. Astrophys. J. **788**, 168 (2014). https://doi.org/10.1088/0004-637X/788/2/168

M. Rubin, K. Altwegg, H. Balsiger, A. Bar-Nun, J.-J. Berthelier, A. Bieler, P. Bochsler, C. Briois, U. Calmonte, M. Combi, J. De Keyser, F. Dhooghe, P. Eberhardt, B. Fiethe, S.A. Fuselier, S. Gasc, T.I. Gombosi, K.C. Hansen, M. Hässig, A. Jäckel, E. Kopp, A. Korth, L. Le Roy, U. Mall, B. Marty, O. Mousis, T. Owen, H. Rème, T. Sémon, C.-Y. Tzou, J.H. Waite, P. Wurz, Molecular nitrogen in comet 67P/Churyumov–Gerasimenko indicates a low formation temperature. Science **348**, 232–235 (2015). https://doi.org/10.1126/science.aaa6100

C. Sagan, B.N. Khare, Tholins—organic chemistry of interstellar grains and gas. Nature **277**, 102–107 (1979). https://doi.org/10.1038/277102a0

R.Z. Sagdeev, G.A. Avanesov, Y.L. Ziman, V.I. Moroz, V.I. Tarnopolsky, B.S. Zhukov, V.A. Shamis, B. Smith, I. Toth, TV experiment of the Vega mission: photometry of the nucleus and the inner coma, in *ESA Proceedings of the 20th ESLAB Symposium on the Exploration of Halley's Comet. Dust and Nucleus*, vol. 2 (1986), pp. 317–326. SEE N87-25908 19-90

S.E. Schröder, Y. Grynko, A. Pommerol, H.U. Keller, N. Thomas, T.L. Roush, Laboratory observations and simulations of phase reddening. Icarus **239**, 201–216 (2014). https://doi.org/10.1016/j.icarus.2014.06.010

M.A. Sephton, I.P. Wright, I. Gilmour, J.W. de Leeuw, M.M. Grady, C.T. Pillinger, High molecular weight organic matter in martian meteorites. Planet. Space Sci. **50**, 711–716 (2002). https://doi.org/10.1016/S0032-0633(02)00053-3

H. Sierks, C. Barbieri, P.L. Lamy, R. Rodrigo, D. Koschny, H. Rickman, H.U. Keller, J. Agarwal, M.F. A'Hearn, F. Angrilli, A.-T. Auger, M.A. Barucci, J.-L. Bertaux, I. Bertini, S. Besse, D. Bodewits, C. Capanna, G. Cremonese, V. Da Deppo, B. Davidsson, S. Debei, M. De Cecco, F. Ferri, S. Fornasier, M. Fulle, R. Gaskell, L. Giacomini, O. Groussin, P. Gutierrez-Marques, P.J. Gutiérrez, C. Güttler, N. Hoekzema, S.F. Hviid, W.-H. Ip, L. Jorda, J. Knollenberg, G. Kovacs, J.R. Kramm, E. Kührt, M. Küppers, F. La Forgia, L.M. Lara, M. Lazzarin, C. Leyrat, J.J. Lopez Moreno, S. Magrin, S. Marchi, F. Marzari, M. Massironi, H. Michalik, R. Moissl, S. Mottola, G. Naletto, N. Oklay, M. Pajola, M. Pertile, F. Preusker, L. Sabau, F. Scholten, C. Snodgrass, N. Thomas, C. Tubiana, J.-B. Vincent, K.-P. Wenzel, M. Zaccariotto, M. Pätzold, On the nucleus structure and activity of comet 67P/Churyumov–Gerasimenko. Science **347**(1), 1044 (2015). https://doi.org/10.1126/science.aaa1044

Y. Skorov, J. Blum, Dust release and tensile strength of the non-volatile layer of cometary nuclei. Icarus **221**, 1–11 (2012). https://doi.org/10.1016/j.icarus.2012.01.012

G. Socrates, *Infrared and Raman Characteristic Group Frequencies—Tables and Charts* (Wiley, New York, 2001)

L.A. Soderblom, T.L. Becker, G. Bennett, D.C. Boice, D.T. Britt, R.H. Brown, B.J. Buratti, C. Isbell, B. Giese, T. Hare, M.D. Hicks, E. Howington-Kraus, R.L. Kirk, M. Lee, R.M. Nelson, J. Oberst, T.C. Owen, M.D. Rayman, B.R. Sandel, S.A. Stern, N. Thomas, R.V. Yelle, Observations of Comet 19P/Borrelly by the miniature integrated camera and spectrometer aboard Deep Space 1. Science **296**, 1087–1091 (2002). https://doi.org/10.1126/science.1069527

S.A. Stern, D.C. Slater, J. Scherrer, J. Stone, M. Versteeg, M.F. A'Hearn, J.L. Bertaux, P.D. Feldman, M.C. Festou, J.W. Parker, O.H.W. Siegmund, Alice: the Rosetta ultraviolet imaging spectrograph. Space Sci. Rev. **128**, 507–527 (2007). https://doi.org/10.1007/s11214-006-9035-8

J.M. Sunshine, M.F. A'Hearn, O. Groussin, J.-Y. Li, M.J.S. Belton, W.A. Delamere, J. Kissel, K.P. Klaasen, L.A. McFadden, K.J. Meech, H.J. Melosh, P.H. Schultz, P.C. Thomas, J. Veverka, D.K. Yeomans, I.C. Busko, M. Desnoyer, T.L. Farnham, L.M. Feaga, D.L. Hampton, D.J. Lindler, C.M. Lisse, D.D. Wellnitz, Exposed water ice deposits on the surface of Comet 9P/Tempel 1. Science **311**, 1453–1455 (2006). https://doi.org/10.1126/science.1123632

M.G.G.T. Taylor, N. Altobelli, B.J. Buratti, M. Choukroun, The Rosetta mission orbiter science overview: the comet phase. Philos. Trans. R. Soc. Lond. Ser. A **375**, 20160262 (2017). https://doi.org/10.1098/rsta.2016.0262

N. Thomas, H.U. Keller, The colour of Comet P/Halley's nucleus and dust. Astron. Astrophys. **213**, 487–494 (1989)

N. Thomas, H. Sierks, C. Barbieri, P.L. Lamy, R. Rodrigo, H. Rickman, D. Koschny, H.U. Keller, J. Agarwal, M.F. A'Hearn, F. Angrilli, A.-T. Auger, M.A. Barucci, J.-L. Bertaux, I. Bertini, S. Besse, D. Bodewits, G. Cremonese, V. Da Deppo, B. Davidsson, M. De Cecco, S. Debei, M.R. El-Maarry, F. Ferri, S. Fornasier, M. Fulle, L. Giacomini, O. Groussin, P.J. Gutierrez, C. Güttler, S.F. Hviid, W.-H. Ip, L. Jorda, J. Knollenberg, J.-R. Kramm, E. Kührt, M. Küppers, F. La Forgia, L.M. Lara, M. Lazzarin, J.J.L. Moreno, S. Magrin, S. Marchi, F. Marzari, M. Massironi, H. Michalik, R. Moissl, S. Mottola, G. Naletto, N. Oklay, M. Pajola, A. Pommerol, F. Preusker, L. Sabau, F. Scholten, C. Snodgrass, C. Tubiana, J.-B. Vincent, K.-P. Wenzel, The morphological diversity of comet 67P/Churyumov–Gerasimenko. Science **347**(1), 0440 (2015). https://doi.org/10.1126/science.aaa0440

C. Tubiana, H. Böhnhardt, J. Agarwal, M. Drahus, L. Barrera, J.L. Ortiz, 67P/Churyumov–Gerasimenko at large heliocentric distance. Astron. Astrophys. **527**, 113 (2011). https://doi.org/10.1051/0004-6361/201016027

J.-B. Vincent, D. Bodewits, S. Besse, H. Sierks, C. Barbieri, P. Lamy, R. Rodrigo, D. Koschny, H. Rickman, H.U. Keller, J. Agarwal, M.F. A'Hearn, A.-T. Auger, M.A. Barucci, J.-L. Bertaux, I. Bertini, C. Capanna, G. Cremonese, V. da Deppo, B. Davidsson, S. Debei, M. de Cecco, M.R. El-Maarry, F. Ferri, S. Fornasier, M. Fulle, R. Gaskell, L. Giacomini, O. Groussin, A. Guilbert-Lepoutre, P. Gutierrez-Marques, P.J. Gutiérrez, C. Güttler, N. Hoekzema, S. Höfner, S.F. Hviid, W.-H. Ip, L. Jorda, J. Knollenberg, G. Kovacs, R. Kramm, E. Kührt, M. Küppers, F. La Forgia, L.M. Lara, M. Lazzarin, V. Lee, C. Leyrat, Z.-Y. Lin, J.J. Lopez Moreno, S. Lowry, S. Magrin, L. Maquet, S. Marchi, F. Marzari, M. Massironi, H. Michalik, R. Moissl, S. Mottola, G. Naletto, N. Oklay, M. Pajola, F. Preusker, F. Scholten, N. Thomas, I. Toth, C. Tubiana, Large heterogeneities in comet 67P as revealed by active pits from sinkhole collapse. Nature **523**, 63–66 (2015). https://doi.org/10.1038/nature14564

J.-B. Vincent, N. Oklay, M. Pajola, S. Höfner, H. Sierks, X. Hu, C. Barbieri, P.L. Lamy, R. Rodrigo, D. Koschny, H. Rickman, H.U. Keller, M.F. A'Hearn, M.A. Barucci, J.-L. Bertaux, I. Bertini, S. Besse, D. Bodewits, G. Cremonese, V. Da Deppo, B. Davidsson, S. Debei, M. De Cecco, M.R. El-Maarry, S. Fornasier, M. Fulle, O. Groussin, P.J. Gutiérrez, P. Gutiérrez-Marquez, C. Güttler, M. Hofmann, S.F. Hviid, W.-H. Ip, L. Jorda, J. Knollenberg, G. Kovacs, J.-R. Kramm, E. Kührt, M. Küppers, L.M. Lara, M. Lazzarin, Z.-Y. Lin, J.J. Lopez Moreno, S. Lowry, F. Marzari, M. Massironi, F. Moreno, S. Mottola, G. Naletto, F. Preusker, F. Scholten, X. Shi, N. Thomas, I. Toth, C. Tubiana, Are fractured cliffs the source of cometary dust jets? Insights from OSIRIS/Rosetta at 67P/Churyumov–Gerasimenko. Astron. Astrophys. **587**, 14 (2016a). https://doi.org/10.1051/0004-6361/201527159

J.-B. Vincent, M.F. A'Hearn, Z.-Y. Lin, M.R. El-Maarry, M. Pajola, H. Sierks, C. Barbieri, P.L. Lamy, R. Rodrigo, D. Koschny, H. Rickman, H.U. Keller, J. Agarwal, M.A. Barucci, J.-L. Bertaux, I. Bertini, S. Besse, D. Bodewits, G. Cremonese, V. Da Deppo, B. Davidsson, S. Debei, M. De Cecco, J. Deller, S. Fornasier, M. Fulle, A. Gicquel, O. Groussin, P.J. Gutiérrez, P. Gutiérrez-Marquez, C. Güttler, S. Höfner, M. Hofmann, S.F. Hviid, W.-H. Ip, L. Jorda, J. Knollenberg, G. Kovacs, J.-R. Kramm, E. Kührt, M. Küppers, L.M. Lara, M. Lazzarin, J.J. Lopez Moreno, F. Marzari, M. Massironi, S. Mottola, G. Naletto, N. Oklay, F. Preusker, F. Scholten, X. Shi, N. Thomas, I. Toth, C. Tubiana, Summer fireworks on comet 67P. Mon. Not. R. Astron. Soc. **462**, 184–194 (2016b). https://doi.org/10.1093/mnras/stw2409

F.L. Whipple, A comet model. I. The acceleration of Comet Encke. Astrophys. J. **111**, 375–394 (1950). https://doi.org/10.1086/145272

D.H. Wooden, S.B. Charnley, P. Ehrenfreund, in *Composition and Evolution of Interstellar Clouds*, ed. by G.W. Kronk (2004), pp. 33–66

I.P. Wright, S.J. Barber, G.H. Morgan, A.D. Morse, S. Sheridan, D.J. Andrews, J. Maynard, D. Yau, S.T. Evans, M.R. Leese, J.C. Zarnecki, B.J. Kent, N.R. Waltham, M.S. Whalley, S. Heys, D.L. Drummond, R.L. Edeson, E.C. Sawyer, R.F. Turner, C.T. Pillinger, Ptolemy an instrument to measure stable isotopic ratios of key volatiles on a cometary nucleus. Space Sci. Rev. **128**, 363–381 (2007). https://doi.org/10.1007/s11214-006-9001-5

I.P. Wright, S. Sheridan, S.J. Barber, G.H. Morgan, D.J. Andrews, A.D. Morse, CHO-bearing organic compounds at the surface of 67P/Churyumov–Gerasimenko revealed by Ptolemy. Science **349**(2) (2015). https://doi.org/10.1126/science.aab0673

I.P. Wright, S. Sheridan, G.H. Morgan, S.J. Barber, A.D. Morse, On the attempts to measure water (and other volatiles) directly at the surface of a comet. Philos. Trans. R. Soc. Lond. Ser. A **375**, 20150385 (2017). https://doi.org/10.1098/rsta.2015.0385

Z. Yoldi, A. Pommerol, B. Jost, O. Poch, J. Gouman, N. Thomas, VIS-NIR reflectance of water ice/regolith analogue mixtures and implications for the detectability of ice mixed within planetary regoliths. Geophys. Res. Lett. **42**, 6205–6212 (2015). https://doi.org/10.1002/2015GL064780

Space Sci Rev (2019) 215:30
https://doi.org/10.1007/s11214-019-0596-8

# Local Manifestations of Cometary Activity

**Jean-Baptiste Vincent[1] · Tony Farnham[2] · Ekkehard Kührt[1] · Yuri Skorov[3] · Raphael Marschall[4] · Nilda Oklay[1] · Mohamed Ramy El-Maarry[5] · Horst Uwe Keller[1,6]**

Received: 19 November 2018 / Accepted: 9 April 2019 / Published online: 2 May 2019


**Abstract** Comets are made of volatile and refractory material and naturally experience various degrees of sublimation as they orbit around the Sun. This gas release, accompanied by dust, represents what is traditionally described as *activity*. Although the basic principles are well established, most details remain elusive, especially regarding the mechanisms by which dust is detached from the surface and subsequently accelerated by the gas flows surrounding the nucleus.

During its 2 years rendez-vous with comet 67P/Churyumov-Gerasimenko, ESA's Rosetta has observed cometary activity with unprecedented details, in both the inbound and outbound legs of the comet's orbit. This trove of data provides a solid ground on which new models of activity can be built. In this chapter, we review how activity manifests at close distance from the surface, establish a nomenclature for the different types of observed features, discuss how activity is at the same time transforming and being shaped by the topography, and finally address several potential mechanisms.

**Keywords** Comets · Activity · Rosetta

Comets: Post 67P/Churyumov-Gerasimenko Perspectives
Edited by Nicolas Thomas, Björn Davidsson, Laurent Jorda, Ekkehard Kührt, Raphael Marschall, Colin Snodgrass and Rafael Rodrigo

✉ J.-B. Vincent
jean-baptiste.vincent@dlr.de

1 DLR Institute of Planetary Research, Rutherfordstrasse, 2, 12489 Berlin, Germany

2 Department of Astronomy, University of Maryland, College Park, MD 20742, USA

3 Institut für Geophysik und extraterrestrische Physik, Technische Universität Braunschweig, Mendelssohnstr. 3, 38106 Braunschweig, Germany

4 Physikalisches Institut, University of Bern, Sidlerstr. 5, 3012 Bern, Switzerland

5 Birkbeck College, University of London, WC1E 7HX, London, UK

6 Institut fur Geophysik und extraterrestrische Physik, TU Braunschweig, 38106 Braunschweig, Germany

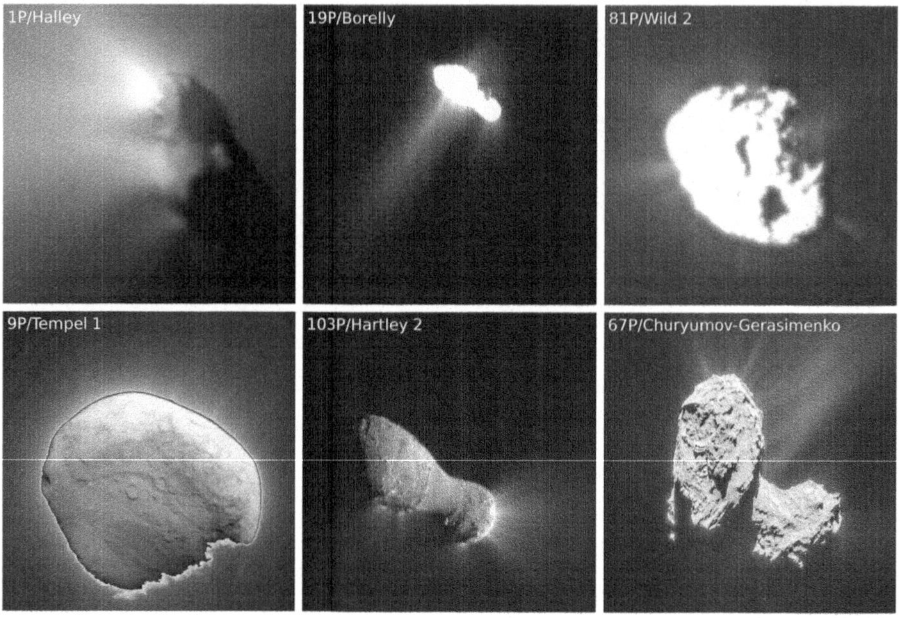


**Fig. 1** Cometary activity observed by space missions, in chronological order from left to right, top to bottom

# 1 Introduction

When the Giotto mission flew by comet 1P/Halley in March 1986, it completely changed our understanding of comets and their activity. Among many discoveries, the mission revealed a dark nucleus with little or no exposed ice and a non-isotropic coma. While most of the surface appeared inactive, a few areas gave rise to strong collimated flows of gas and dust, commonly referred to as "jets".

In the following 30 years, five additional cometary nuclei have been imaged by space probes, confirming the picture revealed by Giotto: cometary nuclei are among the darkest objects in our Solar System, and the distribution of gas and dust density above the surface is highly non-isotropic, with ubiquitous collimated features.

While there is no doubt that coma features are collimated streams of dust and gas (for only the density contrast with the ambient coma makes them detectable in images), the cause of this collimation is challenging to establish, and has led to many publications with conflicting definitions for similar observations. One issue, to be discussed in Sect. 2, is that physical processes that are completely unrelated can give rise to the same type of morphological features. It is for instance well established that topographic variations will collimate gas flows and create regions of higher density where the dust can be more easily accelerated. This phenomenon has been detected on various objects. On the other hand, observations also suggest that some localized areas are more volatile-rich than others, with the ice being sometimes directly exposed on the surface. Consistently, this larger amount of volatile material will sustain activity for a longer time that depleted regions, thus leading to stronger gas and dust flows over those areas, and jet-like features. Figure 1 shows how activity looked like for the 6 cometary nuclei images in situ so far.

Several attempts have been made to classify these features and connect them to processes which can then inform us about the physical and chemical properties of cometary nuclei. A comprehensive review by Belton (2010) proposes the following nomenclature:

- *Type I* describes dust release dominated by the sublimation of $H_2O$ through the porous mantle;
- *Type II* is controlled by the localized and persistent effusion of super-volatiles from the interior;
- *Type III* is characterized by episodic releases of super-volatiles.

In this description, *Type I* jet-like features do not appear to be associated to specific morphology and are generally broader and more diffuse than other dust streams. *Type II* features, also called *filaments* in the literature display a much more collimated structure and have been traced back to specific regions of cometary nuclei. Finally, *Type III* are more sporadic events, probably related to micro-outbursts or other explosive processes.

This description was well suited to describe observations acquired from the first flybys (comets 1P, 9P, 19P, and 81P), but does not apply to all features observed by later missions (comets 103P and 67P). In addition, this former nomenclature imposes an interpretation by associating morphological description of jet-like features with some physical processes which are still a matter of debate.

In this paper, we chose to dissociate morphology and physics: we propose a nomenclature that is purely based on observations and define terms that can remain future proof. We will show how our definitions relate to Belton's. In Sect. 5, we will review the potential mechanisms of cometary activity, and discuss whether some processes relate more closely to one specific manifestation of activity or another.

## 2 Nomenclature of Cometary Activity

A spacecraft orbiting a cometary nucleus will typically observe activity: release of volatile and refractory material, as producing an ambient coma and localized regions of enhanced gas/dust density.

### 2.1 Ambient Coma

The ambient coma is also known as "coma background" is present in all observations and may have to be subtracted from the data in order to detect anisotropies. It is a diffuse background of gas and dust, typically enhanced on the day side of the nucleus where sublimation is stronger.

Different volatile species will display different distributions, depending on their sublimation temperature. For instance $H_2O$ is much more abundant on the day side, while $CO_2$ expands in a more isotropic coma. Of course, this distribution is somewhat affected by local variations of composition on the nucleus' surface, but the diffuse nature of such activity makes it often difficult to associate a specific region with a given gas species abundance. Secondary sources, e.g. ice grains ejected from the nucleus and sublimating at some distance from the surface, may introduce some anisotropy in the distribution. Example of such phenomena can be found in A'Hearn et al. (2011) for comet 103P/Hartley 2 or Bodewits et al. (2016) for comet 67P/Churyumov-Gerasimenko

Likewise, dust also expands in a coma background, accelerated by the gas flows for a few nucleus radii until the gas density becomes too low to provide further acceleration. Beyond that limit, dust motion is mostly controlled by solar gravity and radiation pressure, and the expands into general coma, tail and trail.

A more thorough description of the coma background for dust and gas, as well as the prime mechanisms for expansion, and a summary of numerical simulations efforts on this topic can be found in the "Coma" chapter of this book.

## 2.2 Collimated Streams

Collimated streams have been observed around all cometary nuclei, and are always identified as narrow regions of higher density in gas/dust with respect to the ambient coma. At the distances and resolutions considered in the paper (spacecrafts flying by or orbiting cometary nuclei), these features are typically detected by stretching the contrast of images. This is different than for ground based observation which often require special image filtering in order to reveal coma structures (Samarasinha and Larson 2014). The absence of features in our data means that no variations in coma brightness can be detected against the noise in the background, indicating a very low level of activity, or none at all.

While they are easy to detect, naming them and understanding the physics of their formation has been a challenge ever since the first observations (either ground based or in situ). Many authors have described these collimated streams using different names: e.g. "coma structures", "jets", "jet-like features", "plumes", or "filaments". Often the same words are used to describe collimated streams at very different scales, from spacecraft observations (a few tens of km from the surface) to ground based observations (a few tens of thousands kilometers from the nucleus). In addition, some of these words carry a physical meaning which may not be the correct interpretation of the nature of this phenomenon. A "jet", for instance, implies that the gas/dust stream is being forced out of a small orifice. Although this is a working hypotheses, it has never been observed as such so far. We should also be wary of misinterpretation by people with different knowledge background: "plume" describes accurately a cloud of dust/gas but also carry the general meaning of having different properties (composition, temperature) than its surrounding, and being potentially buoyant. Neither case is necessarily applicable to the case of comets. "Filament" is a better word, and already mentioned in solar physics to describe coronal streamers. It is however seldom used in the cometary literature, where "jet" (although ambiguous) has been the dominating word.

In this paper, we advocate the usage of neutral vocabulary and recommend the usage of "collimated streams" when describing the observations. However, we also recognize that some words are well established in the community and cannot be dismissed totally. We do strongly suggest, though, to restrain from using the single word "jet" and instead prefer the longer version: "jet-like feature". This provides a visual description of the observations that is close enough to what has been already published, while at the same time not claiming definitely that these features are produced by a jetting process.

Note that this definition is based on the morphology of collimated streams themselves and does not assume any relation between a jet-like feature and nucleus surface morphology/composition. This is an important point which will be developed in Sect. 4.

## 2.3 Temporality

All missions prior to Rosetta have been single fly-bys encounters. This means that they provided a snapshot of cometary behaviour at a fixed epoch, with high resolution observations lasting only a couple of hours at best. By orbiting comet 67P for over 2 years, throughout the perihelion passage, Rosetta gave us an unprecedented view into the temporality of activity

**Table 1** Nomenclature of local manifestations of cometary activity

| Description | | Classification |
| --- | --- | --- |
| Long-lasting events, predictable? | ⇒ | Perennial collimated streams (jet-like features) |
| Short-lasting events, sporadic? | ⇒ | Transient plumes |

phenomena. In particular, it was possible to observe some jet-like features repeating from one rotation to the next, seemingly arising from the same area, while others were sporadic events appearing only once.

The temporal factor turns out to be extremely important when trying to infer the mechanisms leading to the release of gas and dust, and to the collimates streams we observe. In fact, a single image can often be misleading. Out of the many jet-like features visible at any time, some may last for many hours—rotating with the nucleus, switching off at night and waking up the next morning—while others may only be detected in that single frame. Yet both types of features may appear morphologicaly identical.

Some attempts have been made to include temporality in the nomenclature, e.g. Belton (2010, 2013), Vincent et al. (2016). Unfortunately, they often mix observation and interpretation. This should be taken with caution as establishing the processes behind cometary outbursts remains a challenge. Yet, a clear distinction can be made between different types of activity based on the duration of each event. Our proposed nomenclature is described hereafter and summarized in Table 1.

Cometary missions have observed two categories of jet-like features. While both share the same morphology and can be traced down to specific areas of the nucleus, their dynamics are quite different.

In the most common case, *collimated streams* appear to last for many hours: they rise as soon as the local temperature enables sublimation, rotate with the nucleus, and wane some time after the local sunset. They will often be reactivated on the next morning, and repeat this cycle for many rotations. For a source more volatile than water, such features has been observed to last for several full rotations, with little or no decrease in intensity during the night (Feaga et al. 2007). We will refer to these features as perennial or long-lasting. Observations suggest that these common streams contain more dust than the surrounding coma (that's why we can detect them) but not necessarily a different speed or composition. A particular case of perennial activity are curtain-shaped streams (i.e. collimated only in one direction) which closely follow the morning terminator and are associated to the sublimation of frost in the early local morning (De Sanctis et al. 2015; Shi et al. 2018). While each source area will only be active for a short time, the active foot print slowly swipes a large fraction of the morning surface and is reactivated every morning, thus we consider them perennial as well.

The second type of jet-like features are transient, or short-lived. They arise suddenly for a few minutes, and may never be detected again from that location. These are traditionally associated to outbursts, i.e. the sudden release of a large quantity of gas and dust, driven by mechanisms that are still debated. While outbursts have usually been interpreted as akin to an explosive decompression, this interpretation is challenged by EPOXI and Rosetta data and the associated models which suggest that the transient release of material can be well explained by avalanches (Steckloff et al. 2016) or topographic collapses (pits, cliffs, Vincent et al. 2016). Spacecraft observations have shown that such events are more likely to occur around perihelion and in the outbound orbit, but not always. They may arise from very different terrains and illumination conditions on the nucleus (Belton 2013; Knollenberg 2015; Vincent et al. 2016; Grün et al. 2016). Some events may also repeat from the same location,

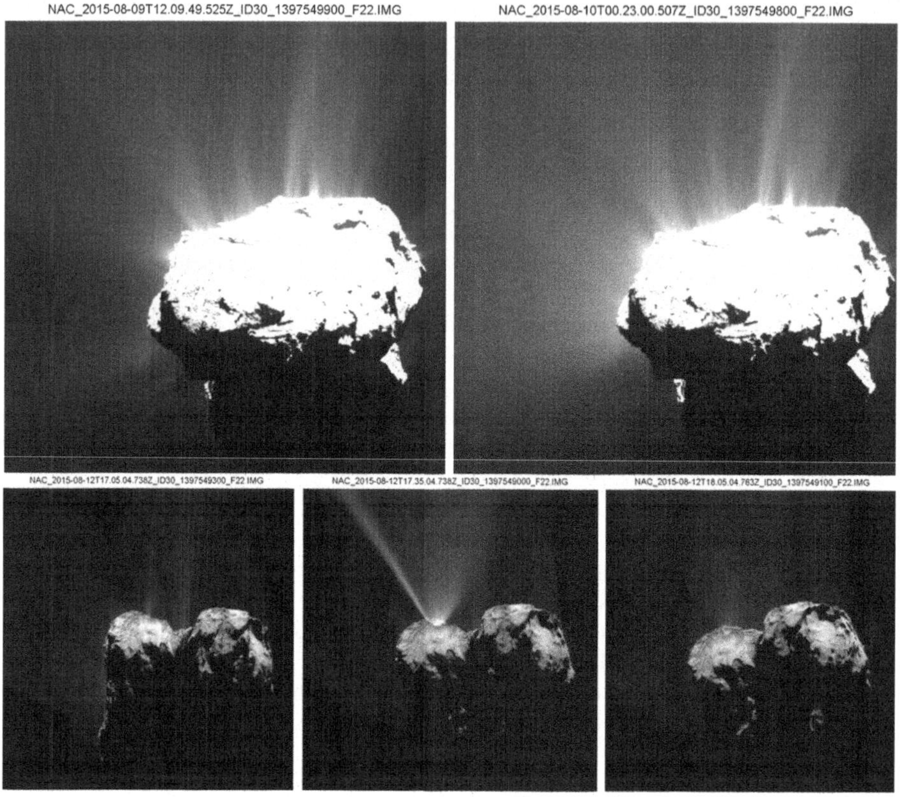


**Fig. 2** The two main local manifestations of cometary activity, as observed by the OSIRIS Narrow Angle Camera on board Rosetta, at comet 67P. The top panels show contrast enhanced views of the comet nucleus acquired one rotation apart. One can see the exact same *perennial collimated streams* consistently repeating from one comet day to the next. The bottom row shows an outburst event, in which a major *transient plume appears* only in the middle frame, and is not observed 1/2 h earlier or later. Adapted from Vincent et al. (2016)

at different frequencies (from days to months, Agarwal et al. 2016; Oklay 2018). Transient events may lead to different morphologies for the associated dust feature. Vincent et al. (2016) distinguish three categories: *Type A* are narrow, strongly collimated streams, *Type B* are broad plumes, and *Type C* are complex morphologies with several streams of types A and B at once. The term *plume* is quite relevant here as the material associated with transient events has often different properties than the ambient coma (an order of magnitude faster, smaller grains, enriched in water ice, see papers cited above).

To summarize, we classify all local manifestations of cometary activity in two major classes: *perennial collimated streams* and *transient plumes*, illustrated in Fig. 2. Sub-classes may be introduced when required by the observations, or when the physical process can be clearly established.

Going back to Belton (2010)'s nomenclature, our *perennial collimated streams* effectively merges the types I & II, as we seen with Rosetta that we often cannot distinguish between jet-like features related to $H_2O$ sublimation alone, or other volatiles. *Transient plumes* are equivalent to Belton's type III, although once again detached from any physical interpretation.

**Table 2** Spacecraft missions to comets

| Mission | Target | Date | Close approach distance |
|---|---|---|---|
| Vega 1 | 1P/Halley | 6 Mar. 1986 | 8889 km |
| Vega 2 | 1P/Halley | 9 Mar. 1986 | 8030 km |
| Giotto | 1P/Halley | 14 Mar. 1986 | 596 km |
| Deep Space 1 | 19P/Borrelly | 22 Sep. 2001 | 2171 km |
| Stardust | 81P/Wild 2 | 2 Jan. 2004 | 237 km |
| Deep Impact | 9P/Tempel 1 | 4 Jul. 2005 | 500 km |
| Deep Impact (EPOXI) | 103P/Hartley 2 | 4 Nov. 2010 | 694 km |
| Stardust NExT | 9P/Tempel 1 | 15 Feb. 2011 | 182 km |
| Rosetta | 67P/C-G | 2014–2016 | Orbital |

# 3 History of Observations

## 3.1 General Introduction

Consideration of studies of cometary activity in a historical context shows a progression in which improvements in technology and data acquisition have built upon previous knowledge to further our understanding of the physical processes at work on cometary nuclei. It has been known from many decades of ground-based observations that active areas on the nucleus can produce features in the coma. Linear filaments, corkscrews, arcs, and spirals are detected in images of both dust and gases, with the different morphologies reflecting the comet's activity, the dynamics of the nucleus and the viewing geometry. These features are often used to infer the nucleus' physical properties and to constrain characteristics of the comet's emission (e.g., features extending thousands of kilometers from the nucleus indicate that the material must be highly collimated). Unfortunately, the large scales of ground-based observations shed little light on the interaction between the nucleus and coma or about the mechanisms producing the observed features. With the advent of spacecraft missions that could obtain images of the resolved nucleus, it became possible to start investigating the nature of the activity itself. Although most missions were simple flybys that captured only "snapshots" of the nucleus' behavior, each successive visit revealed additional details about the physical characteristics of coma structures. From the very first images of 1P/Halley, through the long-term observations of 67P/Churyumov-Gerasimenko, individual comets show different types and scales of activity changing on a variety of timescales, and the aggregate of these measurements has provided the necessary background to better understand the mechanisms that drive cometary activity. We outline here a brief history of the different comet missions, with a summary of some of the highlights of each. A timeline of all missions is given in Table 2.

## 3.2 1P/Halley Flybys (Vega 1 & 2 and Giotto Spacecraft)

1P/Halley, the only cometary mission target to date that is not a Jupiter Family Comet, was the first comet to have its nucleus resolved in spacecraft images (Sagdeev et al. 1987). An important conclusion from these observations was the confirmation that the nucleus did indeed exhibit isolated active areas on its surface that produce features in the coma (Keller et al. 1987; Larson et al. 1987). Unlike objects visited later, however, Halley's inner coma was so dense that it masked the details of the nucleus' surface. This complicated any detailed analyses involving the interaction between the surface and coma and the processes that generated the activity. A number of attempts were made to use the spacecraft images to extract information about the locations and other properties of the active areas on the surface (e.g., Celnik and Schmidt-Kaler 1987), but ambiguities in the spacecraft viewing geometry, nucleus shape and complex rotation state (e.g., Belton et al. 1991; Stooke and Abergel 1991) made it difficult to study the coma-nucleus interactions. There is a growing interest in revisiting the Halley spacecraft observations, using modern image processing and analysis techniques, to maximize the return from these missions and provide more up-to-date comparisons to the more recently studied comets.

### 3.3 19P/Borrelly (DS1)

Because DS1 was a technology demonstration flight redirected to comet Borrelly after the end of its primary mission, it obtained only a few high resolution images near close approach. However, this limited dataset represented the best images of a comet's nucleus available at the time and showing a variety of different coma features, including a highly-collimated linear stream, a broad fan, and a small rapidly changing loop feature (Soderblom et al. 2002). Although the collimated streams exhibited little detail, they provided important insight that foreshadowed what was to come in future missions.

During its approach, DS1 tracked the strong, highly collimated jet-like feature for some time. This stream was unambiguously connected to a long-term coma structure that had been detected in ground-based observations for decades, and whose constant orientation in inertial space indicates that it is aligned with the nucleus' spin axis (Farnham and Cochran 2002). DS1 close approach data showed that this feature arises from a central basin at the waist of the elongated nucleus, but the limited spatial resolution of the comet's surface make it impossible to identify the source with any specific geologic structures on the surface. Even so, this was the first measurement of a common feature that can be traced from its origin on the nucleus to $> 10\,000$ km from the nucleus, and it confirmed that features seen in ground-based observations do indeed reflect the comet's surface activity (though the comet's rotation and observing geometry can obscure the connections).

DS1 also detected the first rapid changes in a comet's innermost coma, recording a "loop" structure at the terminator where dust produced from an unilluminated portion of the surface is emerging into sunlight (Boice et al. 2002). The appearance of this structure changed on timescales of less than a minute, and it was believed that the rapid variability is related to the shutdown in activity after the source had recently entered darkness. Again, the restricted dataset limits any detailed analysis of this phenomenon, but it represents the first occasion of features changing on very short timescales.

### 3.4 81P/Wild 2 (Stardust)

The coma of Comet Wild 2 around perihelion is dominated by a persistent, large-scale sunward fan that is centered on the nucleus' spin axis (Schulz et al. 2003; Farnham and Schleicher 2005). Unfortunately, the Stardust flyby occurred late in the apparition (98 days after perihelion), after this feature had disappeared due to the changing seasonal variations. Thus, not only was the primary fan absent during the encounter, but the surface of the nucleus where it originated was unilluminated and could not be studied. Stardust did record a number of small, isolated jets around close approach, however, and Sekanina et al. (2004)

performed an analysis of these structures. Their results showed the features were widely distributed around the surface, with no obvious correlations in their source regions. It should be noted that this work only projected the jet-like features back to their intersection with the surface of the triaxial ellipsoid that was fit to the nucleus, and a more complete analysis that investigates the source regions with respect to the actual shape model of the nucleus might reveal more meaningful information about specific correlations to the topography at the origins of the activity.

Sekanina et al. (2004) also found that several collimated streams arose from the night side of the nucleus, confirming that activity can persist for some time after the site is no longer illuminated. Given the locations of these features and Wild 2's 13.5-hr rotation period (Farnham et al. 2010) these sources may have remained active for hours after local sunset. Unlike in comet Borrelly, the jet-like features displayed no temporal variations that would suggest the activity might be shutting down.

## 3.5 9P/Tempel 1 (Deep Impact)

Comet 9P/Tempel 1 is unique in two respects: First, it was the target of a large-scale impact experiment designed to investigate the nucleus' sub-surface characteristics. This experiment offered the unprecedented opportunity to explore the conditions that contribute to cometary activity. Second, it was a spacecraft target on two consecutive apparitions, offering the first opportunity to detect changes in the nucleus produced by the comet's long-term activity.

As with comet Wild 2, Tempel 1 is known to exhibit a persistent sunward fan in its coma during the pre-perihelion portion of its orbit. Although continuum images of this feature show no short-term changes (e.g., Lara et al. 2006), CN observations resolved it into a corkscrew that oscillated on timescales consistent with the comet's 41-hr rotation period (Farnham et al. 2005; Schleicher 2006; Boehnhardt et al. 2007). Both observations indicate that the feature was produced by an active area located near the nucleus' south pole. This primary fan had faded in ground-based observations by the time of the Deep Impact (DI) flyby (1 day before perihelion), the spacecraft recorded a diffuse jet-like feature arising from the south pole that likely represents residual activity from the primary fan. This stream is too diffuse to be visible against the illuminated portions of the nucleus, limiting any direct connection to surface features that might indicate its origin. However, projecting the linear structure seen beyond the limb back toward the surface suggests it is associated with the large smooth flow that encompasses the nucleus' south pole (Farnham et al. 2007; Vincent et al. 2010).

The connection between the activity and the smooth terrain was confirmed by other features seen in the DI images. Additional polar activity was observed on the nightside beyond the terminator, where dust is seen in a string of connected sources against the dark surface. Fortuitously, after the impact experiment, this region became illuminated by sunlight reflected off the rising ejecta cloud, and the indirect lighting revealed another, previously undetected, smooth patch on the surface, with the string of dust features aligned along its edge (Farnham et al. 2007; Thomas et al. 2007). This was the first direct observation of cometary activity arising from a scarp or vertical surface, a phenomenon that has been observed on many occasions since. The relationship between active areas and vertical faces provides a natural explanation for the persistence of cometary activity, in that a scarp can erode backward, continuously revealing fresh volatiles, while the vertical surface prevents insulating dust from accumulating and choking off the activity.

The Deep Impact IR Spectrometer obtained the first high quality measurements of volatiles in the coma, hinting at the processes that drive a comet's activity (Feaga et al.

2007). Notably, the southern polar region of Tempel 1 is dominated by $CO_2$ emission, while $H_2O$ is most prominent near the sub-solar point. From these observations, it is clear that the gas production is heterogeneous, with $CO_2$ concentrated in the region where the highest activity levels are seen. This suggests that the distribution of highly volatile material could play as significant a role in the activity as the illumination conditions. Furthermore, the presence of $CO_2$ near the south pole suggests that it is the driver of the activity seen on the nightside of Tempel 1 (and by extension, on Wild 2 and other comets). Because $CO_2$ is more volatile than $H_2O$, even low levels of lingering heat can apparently continue to generate activity long after the region is in darkness.

DI also detected ice on the surface of Tempel 1, in the form of several small patches on the "flat top" that had just rotated into sunlight (Sunshine et al. 2006). These patches were associated with small, narrow streams whose proximity indicates they are related to the ice, but whose morphology suggests that they do not actually arise from it. Instead, the jet-like feature origins were found to coincide with small dark features adjacent to the ice fields, and it was interpreted that those jets were the result of volatile sublimation which dragged sub-surface ice and dust into the coma (Farnham et al. 2007). Some of this material then falls back to the surface in the vicinity of the jets, producing the ice patches that were observed. It is notable that an increased concentration of $CO_2$ was detected in the coma around these jets, possibly permitting nighttime activity that would result in the ice fields (which are likely short-lived) being seen shortly after local sunrise.

Another constraint on the mechanisms that contribute to cometary activity comes from the DI impact experiment. Using IR sequences of the ejecta and modeling of the excavation event, Sunshine et al. (2007) showed that water ice exists within a meter of the surface, even at a site that revealed little evidence for activity in its pre-impact state. This suggests that a thermal wave doesn't need to penetrate very deeply to sublimate volatiles. Furthermore, on vertical surfaces that don't build up an insulating layer, ices are likely to be much closer to the surface, enabling the formation of isolated gas and dust streams.

In 2011, comet Tempel 1 was visited by the Stardust spacecraft on its extended mission (Stardust NExT; SDN), providing the first comparisons of a comet on two subsequent apparitions. Combining the observations of the regions observed by both DI and SDN provides a means of investigating the surface evolution over the course of slightly more than one orbit. Thomas et al. (2013) discuss these changes, the most prominent of which is the recession, by more than 50 m in places, of the leading edge of the south polar smooth flow. The loss of as much as 108 kg of material from this region confirms that the scarp is a significant contributor to the activity arising from the polar region.

As far as activity, the SDN flyby occurred later in the apparition (+33 day), so the nucleus' production rates were much lower than during the DI flyby. Even so, it is notable that no activity was detected from the sources seen in the DI observations, even though the viewing geometry was such that the coma features should have been seen if they were present. The "flat top" region of the surface containing the ice patches was on the limb during the SDN departure, but it was also in darkness, so it is not known whether the small jets were still active at this point in the orbit. The conspicuous differences in this respect sets a tight constraint on how the diurnal and seasonal activity evolves throughout the orbit.

SDN did observe a number of isolated active areas (Farnham et al. 2013), some of which arise from unilluminated regions, so it is not possible to evaluate their origins. The most striking features, however, are a cluster of narrow jets seen on the limb near close approach. These are traceable back to their sources on the surface, and Farnham et al. (2013) showed that they all arise from a terraced structure that forms the boundary between a rough highland area and a smooth lowland region. This further confirms that jet activity is frequently associated with vertical topography.

### 3.6 103P/Hartley 2 (Deep Impact Extended Investigation)

Little was known about comet Hartley 2 before the encounter, other than it is a member of the rare family of hyperactive comets (those that emit more water for their size than would be expected from water production models). As with comet Tempel 1, the IR spectrometer observations show $H_2O$ and $CO_2$ emissions in the inner coma, with links to the continuum features seen in images at visible wavelengths (Protopapa et al. 2014). Unlike Tempel 1, however, the continuum seems to be exclusively associated with the $CO_2$ in the coma, with an almost anti-correlation to the $H_2O$. This strongly indicates that hypervolatiles are driving the jet activity in Hartley 2. There is also a thin layer of water ice seen on the morning terminator, visible in the approach images, that might indicate that water is recondensing or ice grains are falling back to the surface during the night (Sunshine et al. 2011). The rest of the surface appears to be free of ice, suggesting that the recondensed material rapidly sublimes away when in sunlight.

Hartley 2 exhibits numerous isolated active regions, with a strong concentration on the end of the smaller of the two lobes of the nucleus. A second concentration is located along the evening terminator on the side of the larger lobe, and a number of more isolated jet-like features are distributed around the rest of the nucleus, including some that arise from the night side (Farnham et al. 2011). The observed streams range from narrow, highly collimated structures to broader, diffuse fans. Many of the jets can be traced back to their origins at scarps (including the one at the edge of the nucleus' smooth neck region), rimless depressions, and dark features that may be holes or vents (Bruck Syal et al. 2013). Modeling of the activity during approach and departure indicates that the active areas seen at close approach are the main ones on the nucleus. The activity from these sources increases and decreases with rotation/insolation, but doesn't tend to turn off completely. Furthermore, there is little evidence that there are large sources that are inactive at the time of close approach but turn on under different illumination conditions (Farnham 2012).

In addition to the ice seen on the surface, the DI IR spectrometer also identified water ice in the coma of Hartley 2 (Protopapa et al. 2014). Much of this ice is in the form of micron-sized particles located in the jets emanating from the small-lobe. Their correlation with the $CO_2$ emission indicates that the icy grains are being dragged from the nucleus' subsurface by the $CO_2$. The grains sublime in the coma, producing a secondary, extended source of water that explains the comet's perceived hyperactivity.

A second population of grains, never seen in previous comet encounters, was also discovered around the nucleus of Hartley 2. These large, individual entities are 10s of centimeters in size, and likely consist of fluffy aggregates of mixed dust and icy particles (Kelley et al. 2013). Although Hermalyn et al. (2013) mapped the positions and motions of a selection of these grains, their dynamics are not well understood. Some show no motion, and will likely return to the nucleus, others are exceeding the escape velocity but on paths that are not radial, while others are clearly escaping from the nucleus. Because of the oddities of the particles' motions, there is no information about the origin of these grains, or whether the aggregates came directly from the nucleus or formed in the coma.

A final phenomenon of interest observed by the DI IR spectrometer was a water feature arising from the smooth neck region. What makes this feature unusual is that it has no continuum associated with it. The smooth neck has been interpreted as a gravitational low, where dust fallback has filled in the depressions, which may be producing an insulating layer that chokes off emission from the underlying surface. Presumably, part of the fallback includes icy grains and these are likely subliming away from the surface to produce the water feature above the neck. The fact that the escaping water does not entrain any dust grains should provide information about the mechanisms by which dust features are produced (e.g., gas arising directly from the surface may not provide a means of lifting or accelerating the dust grains).

### 3.7 67P/Churyumov-Gerasimenko

Comet 67P/Churyumov-Gerasimenko has been observed by Rosetta from end of March 2014 to end of September 2016, from 4.3 AU inbound to 3.8 AU outbound. In this period, the distance spacecraft-nucleus varied from $5.10^6$ km down to the surface. As this paper deals with local manifestations of activity, we will focus on the epoch August 2014–September 2016, during which the spacecraft orbited mostly between 10 km and 300 km from the nucleus (with the exception of two far away excursions in the sunward and anti-sunward directions).

Activity is a continuous process on comets, and its intensity is strongly driven by the amount of input energy (heliocentric distance). From Rosetta observations, we can roughly define four main epochs showing significant variations in activity pattern: wake-up, equinox and perihelion, post-perihelion.

#### *3.7.1 Wake-Up*

The beginning of the wake-up phase is not known, as the sensitivity of distant observations is limited. OSIRIS/Rosetta observations show that the nucleus was already faintly active in our first observations (4.3 AU), displaying a brightness profile slightly broader than star-like, a signature of the developing coma (Tubiana et al. 2015). We do not know if this comet is active at larger heliocentric distance, but it can be investigated by a sample return mission such as CAESAR, which will approach the comet at its aphelion.

From onset of activity to arrival at the comet in August 2014, Rosetta observed the gradual development of the ambient coma, and started to resolve collimated streams at various scales. Among these features we detected several streams which appear to originate from the bright, ice-rich concavity in between the two lobes of the nucleus. These jet-like features are interpreted as the combination of enhanced volatile content in this specific area and a focusing effect by the topography (Lara et al. 2015; Lin et al. 2015).

It is important to note that the wake-up of activity was not only a slow and quiet process. Tubiana et al. (2015) reported an outburst at the end of April 2014, which increased the brightness by 65%. Unfortunately, the nucleus was unresolved in OSIRIS data at that time, and we do not know how the surface was modified by this event.

After Rosetta arrived at 67P on 06 August 2014, the spacecraft remained about 6 months at distances lesser than 100 km from the nucleus. Thus Rosetta was deeply into the coma, in a prime location to observe the developing activity close to the surface. Vincent et al. (2015a) report on this evolution from arrival (August 2014, 3.6 AU) until equinox (May 2015, 1.6 AU). They observed that the footprints of jet-like features appear to follow the subsolar latitude, and originate from an ever larger area of the nucleus as the heliocentric distance decreases, in agreement with prediction from thermal models such as Keller et al. (2015). They also note that activity seems to switch on and off with the Sun rising and setting over a given region. This suggests that such activity is driven by the diurnal heat wave sublimating water ice, and must originate from the surface or the very shallow subsurface.

Three-dimensional inversion of jet-like features to find their footprints show a correlation with topographically rougher areas, most notably taluses of receding cliffs and large pits (Vincent et al. 2015b,a), as it had been observed previously on other comets (e.g.

9P/Tempel 1, Farnham et al. 2013). One should note, however, that the link between morphology and jet-like features is very complex. As described by Crifo et al. (2005) and many later authors, collimated streams of gas and dust may seem to arise from well defined location which are not intrinsically different from any other place. The collimation is mostly due to interactions of the flow with the topography, which explains why one is more likely to observe jets from cliff and pits rather than smooth, flat terrains. A more recent example including numerical simulations of this effect can be found in Shi et al. (2018).

It is nonetheless important to stress that the footprints of most jet-like features observed on 67P can be related to the remains of collapsed cliffs, where regressive erosion exposes icy layer previously insulated by a layer of dessicated material (Vincent et al. 2015a; Oklay et al. 2016). It seems that both topography and availability of volatile material play an important role in forming the observed features.

### *3.7.2 Equinox and Perihelion*

The equinox crossing in May 2015 triggered a increase in cometary activity which cannot be solely explained by the reduced heliocentric distance. Due to the peculiar shape of the nucleus, and the high obliquity (52°), 67P experiences strong seasonal effects. One important consequence is that the southern hemisphere remains in quasi polar night 80% of the orbit, being illuminated only for about 1 year from one equinox to the next (Keller et al. 2015). This means that a large amount of ice can build up and be preserved on this dark side until the inbound equinox, when it is suddenly illuminated for the first time in several years (Choukroun et al. 2015). This translates into an increase in the production rate (Hansen et al. 2016) and in the number of jet-like features observed from Spring 2015 onward (Lai et al. 2016).

In addition to the increase in number of observed jet-like features, we also note that activity start to persist for some time beyond the evening terminator. Shi et al. (2016) have shown that this time lag can be used to derive a thickness of the dust layer covering the buried volatile material. It typically corresponds to a few mm.

This epoch is also when Rosetta started to observe significant changes on the surface (transient pitted patterns, regressive erosion, boulder movement, fracture extension, see El-Maarry et al. 2017).

This increase of activity can also be traced by the numerous transient event detected after July 2015 onward. Despite a quasi constant monitoring, no outbursts were detected after the distant one caught in April 2014 (see previous subsection). Vincent et al. (2016) show that many major events were observed (roughly one every other comet rotation), each releasing several tens of tons of material in a few minutes. We found that outbursts occur about every 2.4 nucleus rotations. They are comparable to what had been observed on previous comets. Like for perennial collimated streams, transient plume also appear to originate from areas close to the subsolar latitude, although not always. The spatial distribution of their footprints on the nucleus correlates well with morphological region boundaries, especially areas marked by steep scarps or cliffs. In some cases, the plumes could be linked to the sudden collapse of a cliff, but many other events remained to be explained.

### *3.7.3 Post-perihelion*

Further work on outbursts (e.g. Lin et al. 2016; Agarwal et al. 2016; Oklay 2018) show that their frequency decreased after perihelion, but did not stop and transient events could be observed until the end of the mission, more than one year after perihelion. In fact, this type of activity turns out to be extremely asymmetric: Rosetta recorded one single event in the inbound phase down to 1.3 AU, and several hundred from one month before perihelion to the end of the mission (outbound arc, 3.8 AU). This hints at outbursts being driven not by daily processes, but rather by the seasonal heat wave reaching deeper regions of the nucleus, and perhaps more volatile ices.

This asymmetry can also be found in the overall production rate (Hansen et al. 2016) and the number of jets seen by the cameras on board Rosetta (Lai et al. 2016; Kramer et al. 2017). It is compatible with our understanding of the thermal processes involved (Keller et al. 2015).

## 4 Active Sources and Morphological Changes

### 4.1 Active Sources

If the coma present anisotropies, and activity manifests itself as collimated streams of gas and dust, it is logical to assume an anisotropic distribution of sources on the nucleus. That is, areas where conditions are more favourable to the release of material, e.g. a local increase in volatile material, fine dust more easily liftable, or concavities leading to a heat trap. But such active regions are more elusive than it seems, and have been debated ever since the first Giotto images of comet 1P/Halley in 1984. Indeed, this data revealed a relatively homogeneous nucleus surface with almost no ice, although many collimated streams were detected in the coma (Keller et al. 1987). This led the community wonder whether we actually need active and inactive areas to account for such activity (see Crifo and Rodionov 1997 and Crifo et al. 2002 for an exhaustive discussion). The current consensus is that active regions do exist but the topography is a major player in focusing the gas/dust flows.

Analytical examples and numerical simulations show that it is possible to create a jet-like feature from a flat surface if the ice distribution is not homogeneous (Knollenberg et al. 1996), but observations show that this situation is seldom encountered on cometary nuclei. To understand the link between activity and surface morphology, we typically measure the position of the same jet-like features across several images, and use this information to perform a tri-dimensional stereo reconstruction of the stream geometry. Then one can calculate the intersection between the jet and the surface, keeping in mind that jets give only a faint signal and are mostly seen in limb pointing observations which may hide first few meters (or tens of meters) of the stream if the source lies beyond the local horizon. Inversion techniques like these, from spacecraft observations at close distance, are described in details in Farnham et al. (2007, 2013), Vincent et al. (2015a), Shi et al. (2016) and references therein. We find that jets are more typically linked to rough terrains, where the topography present many small cavities that are very efficient at collimated surrounding streams of gas.

On comet 67P, the rendez-vous nature of the Rosetta mission allowed the teams to investigate further the potential source locations inferred from 3d-inversion of jets. We found that many features appear to arise from cliffs, with the topography naturally collimating flows of gas and dust created in the vicinity (Vincent et al. 2015a). It is also important to note that the cliffs almost always display a talus enriched in water ice, thus providing a volatile input, and also indicating a recent collapse—signature of ongoing erosion (Oklay et al. 2016; Fornasier et al. 2017). Transient events associated to such collapse have also been observed on 67P (Vincent et al. 2016; Grün et al. 2016; Pajola et al. 2017). Other comets show similar link between activity and morphology (Farnham et al. 2007, 2013), and also with landslides in areas covered with fine dust like on comet 103P (Steckloff et al. 2016).

We also have evidence for jet-like features arising from smoother terrains Shi et al. (2016, 2018) although modeling shows that one still need some topographic variation to support the collimation (Shi et al. 2018).

Overall, it seems like perennial streams preferentially arise from areas where the topography can collimate the flows of gas and dust, with a preference for places where there is more volatile material available. We note that this does preclude activity to arise elsewhere, only that it is clearly where jet-like features are the most visible.

Forward modeling of cometary activity, and link to topography, and particularly the interaction of dust and gas flows in "jets", have been published by many authors. Prominent results can be found for instance in Crifo et al. (2005), Kramer and Noack (2016), Marschall et al. (2017), Shi et al. (2018). Such simulations provide a strong support for the topography driven model with a very repetitive coma structure with each nucleus rotation, with 90% correlation between simulation and measured intensity around the nucleus (Kramer and Noack 2016).

Still, topography does not explain everything. The transient nature of outbursts suggest that some areas are likely to contain reservoirs of volatile material, released in a sporadic events when the right conditions are met. This phenomenon has been well characterized by the gas related instruments on board Rosetta. For instance, the ROSINA (Rosetta Spectrometer for Ion and Neutral Analysis) analysis has revealed that there is a strong correlation [29 out of 36] between $CO_2$/$H_2O$ ice patches and outburst locations (Kramer et al. 2017): $CO_2$ spots are likely candidates for later outbursts and remain active many months after the first events have been observed, indicating the presence of a reservoir extending into the subsurface. Similarly, Oklay (2018) have reported more than 20 outbursts originating from the same location in the Imhotep region of 67P over a period of 1 year, in the outbound arc of 67P's orbit. This once again suggests that flows from this area relate to a large buried reservoir of ices.

## 4.2 Morphological Changes

If topography affects activity, one may also ask about the reciprocal. As cometary material is ejected away from the nucleus, this must change the surface.

First evidence for topographic changes in relation to activity were reported by Thomas et al. (2013) for comet 9P. After one orbit, they observed the retreat of a scarp with downward erosion of $> 20$ m, and morphological changes of depressions in smooth terrains. Similar large scale changes were observed on comet 67P (Groussin et al. 2015). In addition, Rosetta also discovered many other types of changes, like fault extension, cliff collapse, transport of large boulders, summarized in El-Maarry et al. (2017) and Chapter "Morphology" of this book.

Dust transport from activity (Thomas et al. 2015; Lai et al. 2016) also play a major role in blanketing and unveiling the topography.

In addition to these large scale changes, many small scale ($< 10$ m) surface modifications have been detected and can often be associated to the sudden release of material, for instance forming small pits (Agarwal et al. 2016; Vincent 2018; Oklay 2018). A prominent example from comet 67P is shown in Fig. 3.

Overall, it is clear that activity affects the evolution of the surface. Its long term effects on the global shape of the nucleus starts to be understood (see for instance Vincent et al. 2017) but most of the small scale effects have only been revealed by Rosetta and remain to be explained.

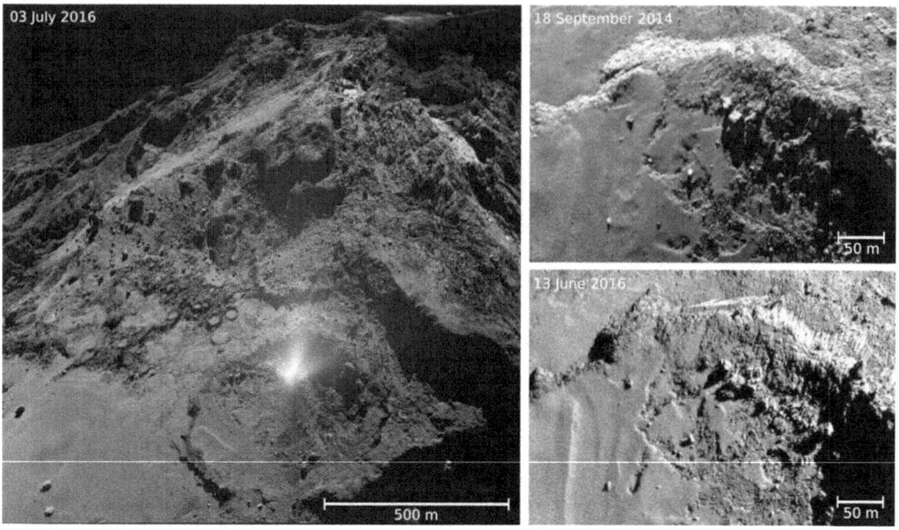


**Fig. 3** Left: Rosetta/OSIRIS Wide Angle Camera image of the outburst plume obtained on 03 July 2016. Right: zoom in the same region before and after perihelion. Although the 03 July 2016 outburst is the only one observed at close distance, Rosetta has detected more than 20 events from this same area during the two years of the mission. The surface has changed significantly and numerous pits have been formed Oklay (2018), Vincent (2018)

## 5 Nature of Activity

A basic goal of the Rosetta mission was to learn more about the nature of cometary activity during the long observation time. Cometary activity is the escape of gas molecules by sublimation of ice and of dust grains from sub- µm to dm-size from the nucleus. It is widely assumed that the dust is accelerated by the gas drag (Finson and Probstein 1968); alternative explanations consider electrostatic forces (Horányi et al. 2015). While the dust in perennial jets seems to depart from the surface with almost zero velocity, material ejected in transient events escapes much faster and the trajectory may be controlled by the initial impulse.

The gas pressure provided by sublimating water molecules at perihelion temperatures is in the order of 1 Pa (Skorov et al. 2017). This is enough to accelerate grains up to a few decimeters in size against the weak gravity but may not be sufficient to overcome the cohesion of cometary matter and actually detach the grains of the surface. This cohesion can be due to van der Waals attraction (Chokshi et al. 1993) and stronger forces like sintering processes (Thomas et al. 1994; Kossacki et al. 2015) or chemical interactions.

According to measurements (Attree et al. 2018a; Thomas et al. 2015; Mannel et al. 2016), the tensile strength of cometary matter reaches several Pa to kPa depending on the scale of observations. In such a scenario of strength higher than the gas pressure one has to explain how dust can be lifted off and how activity is maintained since an isolating dust crust may be formed. As a consequence the comet should become inactive after short time (Kuehrt and Keller 1994). This phenomenon of self-extinction of activity was observed in comet simulation experiments (Grün et al. 1993) but obviously not at comets. This conundrum is sometimes referred to as the activity paradox (Blum et al. 2014).

There are several ways out of this paradox that will be discussed in the following. When the dust to ice volume ratio is $\ll 1$ dust grains do not touch each other and cannot form a cohesive lattice structure. They escape naturally with the subliming gas. However, according to observations (Rotundi et al. 2015) this is an unlikely scenario since dust seems to be the major component in comets. An enhanced gas pressure to crack the cohesive forces would also help. It could be achieved by considering ices with a low sublimation enthalpy. However, such supervolatiles like $CO_2$ and CO do not dominate the volatile composition (Hässig et al. 2015) and are probably present only in deeper layers where the heat flux from Sun is weak. Additional heat sources in the nucleus, e.g. by chemical reactions (Miles 2016) or phase transition from amorphous to crystalline water ice (Marboeuf et al. 2012) can also raise the gas pressure but there are no direct observations that prove the existence of such exothermic processes. A high pressure can also be achieved according to the pressure cooker effect (Kuehrt and Keller 1994). This needs basically closed pores with a good thermal contact to the surface but is difficult to believe in material with 70% porosity (Pätzold et al. 2016). Finally, Vincent et al. (2012, 2015a), Skorov et al. (2017) proposed that fractures, ubiquitous in cometary material, may act as nozzles through which the gas is accelerated. This would increase the gas velocity, thus the drag force applied to the dust.

An other way out of the paradox is a local weakening of the cohesive forces that can be broken by the gas flux. Thermal fatigue is too slow (Delbo et al. 2014; Attree et al. 2018b) to have such an effect but locally weak binding forces can be reached by other effects that were not always considered in models or lab experiments.

When assuming that cohesion is mainly caused by van der Waals forces (Chokshi et al. 1993; Kuehrt and Keller 1994; Güttler et al. 2009; Skorov and Blum 2012) between spherical grains with radii R1 and R2 one can derive the contact force (Chokshi et al. 1993) as

$$F_c = 3\pi R\gamma \tag{1}$$

where $\gamma$ is the surface energy per unit area and $R$ is the reduced radius

$$R = \frac{R_1 R_2}{R_1 + R_2} \tag{2}$$

According to the model of Johnson et al. (1971) the cohesive strength $Y$ of a monodisperse medium consisting of spheres of radius $R$ is given by:

$$Y = \frac{3(1-p)N_c\gamma}{4R} \tag{3}$$

where $p$ is the porosity, $N_c$ is the average number of contact points per particle. Interestingly, this strength does not depend on elastic parameters of the grains.

In the following the importance of several parameters that describe cometary surface layers is discussed to figure out how dust clusters of low of cohesiveness can be formed and, finally, be eroded by the gas.

## 5.1 Material Properties

The stickiness, given by the surface energy per unit area (see table in Chokshi et al. 1993) can vary by orders of magnitude. Water ice is a good glue ($\gamma = 0.37\ \mathrm{J\,m^{-2}}$), quartz is not ($\gamma = 0.025\ \mathrm{J\,m^{-2}}$). Therefore, the composition of the surface layer is of importance. However, at low temperatures, typical for cometary ices, the stickiness of ice and dust are more similar than at higher ones as they occur on Earth (Gundlach et al. 2015).

## 5.2 Dust/Ice Ratio

The dust/ice *mass ratio* $\chi_m$ derived for comet 67P is not completely understood but Rotundi et al. (2015) derived a value in the range 2 to 6. The dust/ice *volume ratio* $\chi_v$, that is more important for the geometric structure (and thus for $N_c$) if we assume that small and large dust grains have the same porosity, results by multiplying $\chi_m$ with the density ratio of ice and dust. It is, therefore, considerably smaller (by about a factor 3) than $\chi_m$. A $\chi_v$ that is not too large (in the order of one or lower) gives a small scale mixture of dust and ice grains where the dust doesn't get the chance to intensely interlink over large distances and to form stable clusters to some depth. Thus the surface can be easily cleared from dust in the sublimation process independently on the cohesion. At the surface this ratio may be enhanced compared to the bulk by ice depletion due to sublimation process.

Deep Impact and Rosetta measurements (Sunshine et al. 2006; Capaccioni et al. 2015) conclude that there are only small amounts of ice on the surface of comets. However, lab results (Yoldi et al. 2015) demonstrate that ice can be masked in IR spectra. Up to now, it is not completely understood to what extend and to what depth the superficial dust/ice ratio is higher than that of the bulk. The problem of determining the ratio of refractory to icy material is complex and often under-constrained. This leads to competing interpretations of the same data set. One can find the most recent discussions on this topic in Levasseur-Regourd et al. (2018), Fulle et al. (2019), and Choukroun et al. (this book).

## 5.3 Porosity

According to Eq. (3), high porosity reduces the strength with $(1 - p)$. Measurements on comet 67P (Pätzold et al. 2016) show a high porosity $p > 70\%$. Thus the strength is reduced to less than 30% of nonporous granular matter.

## 5.4 Grains Sizes

The cohesive strength $Y$ in a layer of equally sized spheres goes down with increasing radius of the grains (Eq. (3)). Thus an arrangements of larger grains can be more effectively cracked by the gas than a structure of small grains. This was already pointed out by Kuehrt and Keller (1994) and discussed later in much more detail by Gundlach et al. (2015). When assuming a surface energy of $0.025\ \mathrm{J\,m^{-2}}$, valid for quartz, a porosity of 75% and four contacts per grain Eq. (3) provides a strength $Y \simeq 0.02/R$ Pa. The cohesion between 1 cm grains is in this model 2 Pa, between dm-sized grains 0.2 Pa, which may be low enough to be torn off by gas pressure. For such large grains—that in terms of cometary structures will present hierarchically built clusters of grains—gravity is the limiting factor. In situ observations (Mottola et al. 2015; Ott et al. 2017) and laboratory measurements (Gundlach et al. 2015) have confirmed that mm to dm grains that can be easily removed against cohesion and gravity and are common on the surface of comet 67P.

## 5.5 Grain Size Distribution

A further way to build clusters of low cohesion is a mixture of grains of different sizes. In such a structure (Fig. 4) with $R_2 \ll R_1$ the cohesive forces are controlled by the small (black) grains but the gas pressure can act on the cross section of the large (blue) spheres with $R_2$. The red dashed line marks a weakly bound cluster and Eqs. (1) and (2) yield:

$$F_c = 3\pi\gamma R_2 \tag{4}$$

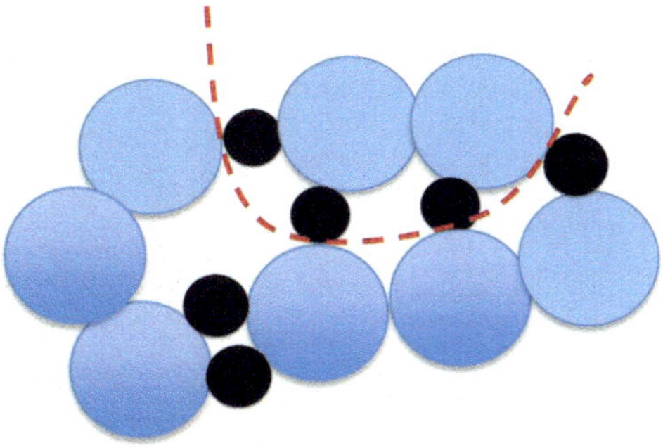

**Fig. 4** Example arrangement of spheres of different sizes in a porous structure. The dashed red line shows a case of a cluster with a weak relative strength

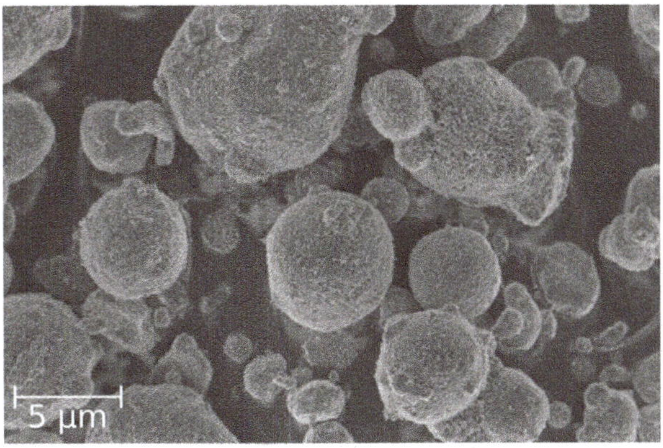


**Fig. 5** SEM images at magnification of 10 000 of a sample with µm-sized Al particles and R972 nanoparticles for reduction of cohesiveness. Adapted from Chen et al. (2010)

The drag force is given by

$$F_d = \frac{1}{2}\rho v^2 C_d \pi R_1^2 \tag{5}$$

where $C_d$ is the drag coefficient, $v$ the gas velocity, $\rho$ the gas density. Consequently, the ratio of drag force and cohesive binding force is $F_d/F_c \propto R_1^2/R_2$. It is enhanced by a factor of 1000 when having a mixture of mm- with µm- or µm- with nm-particles compared to only monodisperse mm- or µm-grains, respectively.

When there is a closed line (or surface in 3D) where the local ratio $F_d/F_c$ is larger than 1 the cluster that is coincidentally formed by this surface can escape by the gas pressure.

An analogous effect can be reached by particles with a rough surface. It should be noted that the described method to reduce the cohesion is known as "dry particle coating" (Fig. 5) in technical applications on Earth (Chen et al. 2010) and is e.g. applied to improve the flow properties of printer toner. More detailed numerical simulations and lab measurements are necessary to quantify this scenario.

For sure the real comet is much more complex in structure than what is shown in Fig. 4 but this is valid approximation. Indeed, atomic force microscope images from the MIDAS instrument on board Rosetta revealed grains with an irregular structure (Fig. 6), which can produce quite weak cohesive bounds, locally, as discussed above.

It should be mentioned that the conditions of such weak structures are mainly given near the surface where most of the ice that clues the dust effectively together by sintering effects has already gone. Therefore, the measured strengths of some Pa to kPa must not be in contradiction to a very low cohesive path ($< 1$ Pa) along a given area in a surface layer.

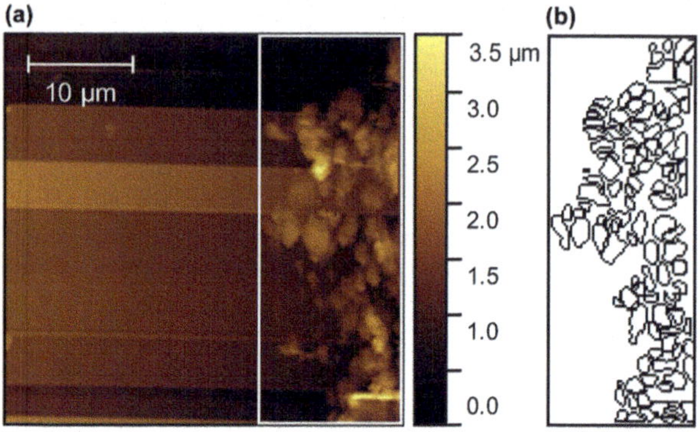


**Fig. 6** (**a**) Crop of a MIDAS post-processed topographic image (**b**) Corresponding area to the white frame in (**a**), showing all identified subunits of particle as 2D projection. Adapted from Mannel et al. (2016)

### 5.6 Escaping the "Activity Paradox"

In summary, based on the concepts described here, we argue that the activity paradox can be addressed in two ways:

1. If the dust to ice volume ratio is < 1 dust grains are isolated in the ice matrix and cannot touch each other to form a cohesive and maybe permanent dust mantle that quenches activity. Independently on the strength of the layer the dust can be removed when the ice is sublimating. However this scenario can only work when the dust/ice mass ratio is small and each dust grain is immediately removed by the gas.
2. Even at a high dust/ice ration physical properties and the microstructure of the cometary surface layer result in local cohesive weaknesses that allow the gas pressure to clear the surface from dust. If irregular clusters at the surface are formed by chance that can be intrinsically stronger than the gas pressure but have weak bonds among themselves they can be lifted off even by low gas pressure. Besides such a zone of low local strength the clusters must be small enough to be removable against gravity. However, this limit is near perihelion in the meter range.

More detailed numerical simulations and lab measurements are necessary to quantify these scenarios. In that context, future sample return missions like CAESAR (NASA New Frontier proposal) will be critical in assessing the real nature of cometary grains and their cohesion.

## 6 Conclusion

This chapter has reviewed what is currently known about local manifestations of cometary activity. We presented a revised nomenclature of active features, recommending the usage of descriptive but neutral terms to describe physical phenomenon we observe but do not yet understand. Throughout the paper, we use *jet-like feature* when referring to perennial collimated streams, and *transient plumes* for sudden events.

We listed historical in-situ observations of cometary activity over a 30 years period, from the Giotto mission in 1984 to Rosetta in 2014–2016.

Finally, we dwelved into the "activity paradox", which is that current models of dust release and subsequent acceleration by gas streams formed by sublimating volatiles are not sufficient to explain our observations. While we do not have an answer to this conundrum, we do propose several ways forward, to be investigated further.

**Acknowledgements** This project has received funding from the European Union's Horizon 2020 research and innovation programme under grant agreement No. 686709 (MiARD). This work was supported by the Swiss State Secretariat for Education, Research and Innovation (SERI) under contract number 16.0008-2. The opinions expressed and arguments employed herein do not necessarily reflect the official view of the Swiss Government. This research has made use of NASA's Astrophysics Data System Bibliographic Services.

**Publisher's Note** Springer Nature remains neutral with regard to jurisdictional claims in published maps and institutional affiliations.

## References

J. Agarwal, M.F. A'Hearn, J.B. Vincent, C. Güttler, S. Höfner, H. Sierks, C. Tubiana, C. Barbieri, P.L. Lamy, R. Rodrigo, D. Koschny, H. Rickman, M.A. Barucci, J.L. Bertaux, I. Bertini, S. Boudreault, G. Cremonese, V. Da Deppo, B. Davidsson, S. Debei, M. De Cecco, J. Deller, S. Fornasier, M. Fulle, A. Gicquel, O. Groussin, P.J. Gutiérrez, M. Hofmann, S.F. Hviid, W.H. Ip, L. Jorda, H.U. Keller, J. Knollenberg, J.R. Kramm, E. Kührt, M. Küppers, L.M. Lara, M. Lazzarin, J.J. Lopez Moreno, F. Marzari, G. Naletto, N. Oklay, X. Shi, N. Thomas, Acceleration of individual, decimetre-sized aggregates in the lower coma of comet 67P/Churyumov-Gerasimenko. Mon. Not. R. Astron. Soc. **462**, S78–S88 (2016). https://doi.org/10.1093/mnras/stw2179. arXiv:1608.07933

M.F. A'Hearn, M.J.S. Belton, W.A. Delamere, L.M. Feaga, D. Hampton, J. Kissel, K.P. Klaasen, L.A. McFadden, K.J. Meech, H.J. Melosh, P.H. Schultz, J.M. Sunshine, P.C. Thomas, J. Veverka, D.D. Wellnitz, D.K. Yeomans, S. Besse, D. Bodewits, T.J. Bowling, B.T. Carcich, S.M. Collins, T.L. Farnham, O. Groussin, B. Hermalyn, M.S. Kelley, M.S. Kelley, J.Y. Li, D.J. Lindler, C.M. Lisse, S.A. McLaughlin, F. Merlin, S. Protopapa, J.E. Richardson, J.L. Williams, EPOXI at comet Hartley 2. Science **332**, 1396 (2011). https://doi.org/10.1126/science.1204054

N. Attree, O. Groussin, L. Jorda, D. Nébouy, N. Thomas, Y. Brouet, E. Kührt, F. Preusker, F. Scholten, J. Knollenberg, P. Hartogh, H. Sierks, C. Barbieri, P. Lamy, R. Rodrigo, D. Koschny, H. Rickman, H.U. Keller, M.F. A'Hearn, A.T. Auger, M.A. Barucci, J.L. Bertaux, I. Bertini, D. Bodewits, S. Boudreault, G. Cremonese, V. Da Deppo, B. Davidsson, S. Debei, M. De Cecco, J. Deller, M.R. El-Maarry, S. Fornasier, M. Fulle, P.J. Gutiérrez, C. Güttler, S. Hviid, W.H. Ip, G. Kovacs, J.R. Kramm, M. Küppers, L.M. Lara, M. Lazzarin, J.J. Lopez Moreno, S. Lowry, S. Marchi, F. Marzari, S. Mottola, G. Naletto, N. Oklay, M. Pajola, I. Toth, C. Tubiana, J.B. Vincent, X. Shi, Tensile strength of 67P/Churyumov-Gerasimenko nucleus material from overhangs. Astron. Astrophys. **611**, A33 (2018a). https://doi.org/10.1051/0004-6361/201732155. arXiv:1712.07508

N. Attree, O. Groussin, L. Jorda, S. Rodionov, A.T. Auger, N. Thomas, Y. Brouet, O. Poch, E. Kührt, M. Knapmeyer, F. Preusker, F. Scholten, J. Knollenberg, S. Hviid, P. Hartogh, Thermal fracturing on comets. Applications to 67P/Churyumov-Gerasimenko. Astron. Astrophys. **610**, A76 (2018b). https://doi.org/10.1051/0004-6361/201731937. arXiv:1711.09809

M.J.S. Belton, Cometary activity, active areas, and a mechanism for collimated outflows on 1P, 9P, 19P, and 81P. Icarus **210**, 881–897 (2010)

M.J.S. Belton, The origin of pits on 9P/Tempel 1 and the geologic signature of outbursts in Stardust-NExT images. Icarus **222**, 477–486 (2013)

M.J.S. Belton, B.E.A. Mueller, W.H. Julian, A.J. Anderson, The spin state and homogeneity of comet Halley's nucleus. Icarus **93**, 183–193 (1991)

J. Blum, B. Gundlach, S. Mühle, J.M. Trigo-Rodriguez, Comets formed in solar-nebula instabilities!—An experimental and modeling attempt to relate the activity of comets to their formation process. Icarus **235**, 156–169 (2014). https://doi.org/10.1016/j.icarus.2014.03.016. arXiv:1403.2610

D. Bodewits, L.M. Lara, M.F. A'Hearn, F. La Forgia, A. Gicquel, G. Kovacs, J. Knollenberg, M. Lazzarin, Z.Y. Lin, X. Shi, C. Snodgrass, C. Tubiana, H. Sierks, C. Barbieri, P.L. Lamy, R. Rodrigo, D. Koschny, H. Rickman, H.U. Keller, M.A. Barucci, J.L. Bertaux, I. Bertini, S. Boudreault, G. Cremonese, V. Da Deppo, B. Davidsson, S. Debei, M. De Cecco, S. Fornasier, M. Fulle, O. Groussin, P.J. Gutiérrez, C. Güttler, S.F. Hviid, W.H. Ip, L. Jorda, J.R. Kramm, E. Kührt, M. Küppers, J.J. López-Moreno, F. Marzari, G. Naletto, N. Oklay, N. Thomas, I. Toth, J.B. Vincent, Changes in the physical environment of the inner coma of 67P/Churyumov-Gerasimenko with decreasing heliocentric distance. Astron. J. **152**, 130 (2016). https://doi.org/10.3847/0004-6256/152/5/130. arXiv:1607.05632

H. Boehnhardt, E. Pompei, G.P. Tozzi, O. Hainaut, N. Ageorges, S. Bagnulo, L. Barrera, T. Bonev, H.U. Käufl, F. Kerber, G. Locurto, O. Marco, E. Pantin, H. Rauer, I. Saviane, F. Selman, C. Sterken, M. Weiler, Broad- and narrowband visible imaging of comet 9P/Tempel 1 at ESO around the time of the Deep Impact event. Astron. Astrophys. **470**, 1175–1183 (2007). https://doi.org/10.1051/0004-6361:20066577

D.C. Boice, L.A. Soderblom, D.T. Britt, R.H. Brown, B.R. Sandel, R.V. Yelle, B.J. Buratti, M.D. Hicks, R.M. Nelson, M.D. Rayman, J. Oberst, N. Thomas, The Deep Space 1 encounter with comet 19p/Borrelly. Earth Moon Planets **89**, 301–324 (2002). https://doi.org/10.1023/A:1021519124588

M. Bruck Syal, P.H. Schultz, J.M. Sunshine, M.F. A'Hearn, T.L. Farnham, D.S.P. Dearborn, Geologic control of jet formation on comet 103P/Hartley 2. Icarus **222**, 610 (2013)

F. Capaccioni, A. Coradini, G. Filacchione, S. Erard, G. Arnold, P. Drossart, M.C. De Sanctis, D. Bockelee-Morvan, M.T. Capria, F. Tosi, C. Leyrat, B. Schmitt, E. Quirico, P. Cerroni, V. Mennella, A. Raponi, M. Ciarniello, T. McCord, L. Moroz, E. Palomba, E. Ammannito, M.A. Barucci, G. Bellucci, J. Benkhoff, J.P. Bibring, A. Blanco, M. Blecka, R. Carlson, U. Carsenty, L. Colangeli, M. Combes, M. Combi, J. Crovisier, T. Encrenaz, C. Federico, U. Fink, S. Fonti, W.H. Ip, P. Irwin, R. Jaumann, E. Kuehrt, Y. Langevin, G. Magni, S. Mottola, V. Orofino, P. Palumbo, G. Piccioni, U. Schade, F. Taylor, D. Tiphene, G.P. Tozzi, P. Beck, N. Biver, L. Bonal, J.P. Combe, D. Despan, E. Flamini, S. Fornasier, A. Frigeri, D. Grassi, M. Gudipati, A. Longobardo, K. Markus, F. Merlin, R. Orosei, G. Rinaldi, K. Stephan, M. Cartacci, A. Cicchetti, S. Giuppi, Y. Hello, F. Henry, S. Jacquinod, R. Noschese, G. Peter, R. Politi, J.M. Reess, A. Semery, The organic-rich surface of comet 67P/Churyumov-Gerasimenko as seen by VIRTIS/Rosetta. Science **347**(1), aaa0628 (2015). https://doi.org/10.1126/science.aaa0628

W.E. Celnik, T. Schmidt-Kaler, Structure and dynamics of plasma-tail condensations of comet P/Halley 1986 and inferences on the structure and activity of the cometary nucleus. Astron. Astrophys. **187**, 233–248 (1987)

Y. Chen, L. Jallo, M.A. Quintanilla, R. Dave, Characterization of particle and bulk level cohesion reduction of surface modified fine aluminum powders. Colloids Surf. A, Physicochem. Eng. Asp. **361**(1), 66–80 (2010). https://doi.org/10.1016/j.colsurfa.2010.03.015. http://www.sciencedirect.com/science/article/pii/S0927775710001895

A. Chokshi, A.G.G.M. Tielens, D. Hollenbach, Dust coagulation. Astrophys. J. **407**, 806–819 (1993). https://doi.org/10.1086/172562

M. Choukroun, S. Keihm, F.P. Schloerb, S. Gulkis, E. Lellouch, Dark side of comet 67P/Churyumov-Gerasimenko in Aug.-Oct. 2014. MIRO/Rosetta continuum observations of polar night in the southern regions. Astron. Astrophys. **583**, A28 (2015). https://doi.org/10.1051/0004-6361/201526181

J.F. Crifo, A.V. Rodionov, The dependence of the circumnuclear coma structure on the properties of the nucleus. Icarus **129**, 72–93 (1997). https://doi.org/10.1006/icar.1997.5714

J.F. Crifo, A.V. Rodionov, K. Szegö, M. Fulle, Challenging a paradigm: do we need active and inactive areas to account for near-nuclear jet activity? Earth Moon Planets **90**, 227–238 (2002)

J.F. Crifo, G.A. Loukianov, A.V. Rodionov, V.V. Zakharov, Direct Monte Carlo and multifluid modeling of the circumnuclear dust coma. Spherical grain dynamics revisited. Icarus **176**, 192–219 (2005)

M.C. De Sanctis, F. Capaccioni, M. Ciarniello, G. Filacchione, M. Formisano, S. Mottola, A. Raponi, F. Tosi, D. Bockelée-Morvan, S. Erard, C. Leyrat, B. Schmitt, E. Ammannito, G. Arnold, M.A. Barucci, M. Combi, M.T. Capria, P. Cerroni, W.H. Ip, E. Kuehrt, T.B. McCord, E. Palomba, P. Beck, E. Quirico (VIRTIS Team), G. Piccioni, G. Bellucci, M. Fulchignoni, R. Jaumann, K. Stephan, A. Longobardo, V. Mennella, A. Migliorini, J. Benkhoff, J.P. Bibring, A. Blanco, M. Blecka, R. Carlson, U. Carsenty, L. Colangeli, M. Combes, J. Crovisier, P. Drossart, T. Encrenaz, C. Federico, U. Fink, S. Fonti, P. Irwin, Y. Langevin, G. Magni, L. Moroz, V. Orofino, U. Schade, F. Taylor, D. Tiphene, G.P. Tozzi, N. Biver, L. Bonal, J.P. Combe, D. Despan, E. Flamini, S. Fornasier, A. Frigeri, D. Grassi, M.S. Gudipati, F. Mancarella, K. Markus, F. Merlin, R. Orosei, G. Rinaldi, M. Cartacci, A. Cicchetti, S. Giuppi, Y. Hello, F. Henry, S. Jacquinod, J.M. Rees, R. Noschese, R. Politi, G. Peter, The diurnal cycle of water ice on comet 67P/Churyumov-Gerasimenko. Nature **525**, 500 (2015)

M. Delbo, G. Libourel, J. Wilkerson, N. Murdoch, P. Michel, K.T. Ramesh, C. Ganino, C. Verati, S. Marchi, Thermal fatigue as the origin of regolith on small asteroids. Nature **508**, 233–236 (2014). https://doi.org/10.1038/nature13153

M.R. El-Maarry, O. Groussin, N. Thomas, M. Pajola, A.T. Auger, B. Davidsson, X. Hu, S.F. Hviid, J. Knollenberg, C. Güttler, C. Tubiana, S. Fornasier, C. Feller, P. Hasselmann, J.B. Vincent, H. Sierks, C. Barbieri, P. Lamy, R. Rodrigo, D. Koschny, H.U. Keller, H. Rickman, M.F. A'Hearn, M.A. Barucci, J.L. Bertaux, I. Bertini, S. Besse, D. Bodewits, G. Cremonese, V. Da Deppo, S. Debei, M. De Cecco, J. Deller, J.D.P. Deshapriya, M. Fulle, P.J. Gutierrez, M. Hofmann, W.H. Ip, L. Jorda, G. Kovacs, J.R. Kramm, E. Kührt, M. Küppers, L.M. Lara, M. Lazzarin, Surface changes on comet 67P/Churyumov-Gerasimenko suggest a more active past. Science **355**, 1392–1395 (2017). https://doi.org/10.1126/science.aak9384

T.L. Farnham, Dust jet activity in comet 103P/Hartley 2, in *Asteroids, Comets, Meteors*, vol. 1667 (2012), p. 6241

T.L. Farnham, A.L. Cochran, A McDonald Observatory study of comet 19P/Borrelly: placing the Deep Space 1 observations into a broader context. Icarus **160**, 398–418 (2002). https://doi.org/10.1006/icar.2002.6969. arXiv:astro-ph/0208445

T.L. Farnham, D.G. Schleicher, Physical and compositional studies of comet 81P/Wild 2 at multiple apparitions. Icarus **173**, 533–558 (2005). https://doi.org/10.1016/j.icarus.2004.08.021

T.L. Farnham, B.E.A. Mueller, M.M. Knight, N.H. Samarasinha, Narrowband observations of comet Tempel 1 in support of the Deep Impact mission, in *AAS/Division for Planetary Sciences Meeting Abstracts*, vol. 37 (2005) p. 43.05

T.L. Farnham, D.D. Wellnitz, D.L. Hampton, J.Y. Li, J.M. Sunshine, O. Groussin, L.A. McFadden, C.J. Crockett, M.F. A'Hearn, M.J.S. Belton, P. Schultz, C.M. Lisse, Dust coma morphology in the Deep Impact images of comet 9P/Tempel 1. Icarus **187**, 26–40 (2007). 10.1016/j.icarus.2006.10.036

T.L. Farnham, N.H. Samarasinha, B.E.A. Mueller, Comet 81P/Wild, in *International Astronomical Union Circular*, vol. 9138 (2010), p. 2

T.L. Farnham, S. Besse, L.M. Feaga, J.M. Sunshine, M.F. A'Hearn, D. Lindler, D. Bodewits, C.M. Lisse, M.J.S. Belton (Dixi Team), Jet activity in comet 103P/Hartley 2 as observed by the Deep Impact Spacecraft, in *Lunar and Planetary Science Conference* (2011), p. 2160

T.L. Farnham, D. Bodewits, J.Y. Li, J. Veverka, P. Thomas, M.J.S. Belton, Connections between the jet activity and surface features on comet 9P/Tempel 1. Icarus **222**, 540–549 (2013). https://doi.org/10.1016/j.icarus.2012.06.019

L.M. Feaga, M.F. A'Hearn, J.M. Sunshine, O. Groussin, T.L. Farnham, Asymmetries in the distribution of $H_2O$ and $CO_2$ in the inner coma of comet 9P/Tempel 1 as observed by Deep Impact. Icarus **190**, 345–356 (2007)

M.L. Finson, R.F. Probstein, A theory of dust comets. I. Model and equations. Astrophys. J. **154**, 353–380 (1968)

S. Fornasier, C. Feller, J.C. Lee, S. Ferrari, M. Massironi, P.H. Hasselmann, J.D.P. Deshapriya, M.A. Barucci, M.R. El-Maarry, L. Giacomini, S. Mottola, H.U. Keller, W.H. Ip, Z.Y. Lin, H. Sierks, C. Barbieri, P.L. Lamy, R. Rodrigo, D. Koschny, H. Rickman, J. Agarwal, M. A'Hearn, J.L. Bertaux, I. Bertini, G. Cremonese, V. Da Deppo, B. Davidsson, S. Debei, M. De Cecco, J. Deller, M. Fulle, O. Groussin, P.J. Gutierrez, C. Güttler, M. Hofmann, S.F. Hviid, L. Jorda, J. Knollenberg, G. Kovacs, R. Kramm, E. Kührt, M. Küppers, M.L. Lara, M. Lazzarin, J.J. Lopez Moreno, F. Marzari, G. Naletto, N. Oklay, M. Pajola, X. Shi, N. Thomas, I. Toth, C. Tubiana, J.B. Vincent, The highly active Anhur-Bes regions in the 67P/Churyumov-Gerasimenko comet: results from OSIRIS/ROSETTA observations. Mon. Not. R. Astron. Soc. **469**, S93–S107 (2017). https://doi.org/10.1093/mnras/stx1275. arXiv:1707.02945

M. Fulle, J. Blum, S.F. Green, B. Gundlach, A. Herique, F. Moreno, S. Mottola, A. Rotundi, C. Snodgrass, The refractory-to-ice mass ratio in comets. Mon. Not. R. Astron. Soc. **482**, 3326–3340 (2019). https://doi.org/10.1093/mnras/sty2926

O. Groussin et al., Gravitational slopes, geomorphology and material strengths of the nucleus of comet 67P/Churyumov-Gerasimenko from OSIRIS observations. Astron. Astrophys. (2015). https://doi.org/10.1051/0004-6361/201526379

E. Grün, J. Gebhard, A. Bar-Nun, J. Benkhoff, H. Düren, G. Eich, R. Hische, W.F. Huebner, H.U. Keller, G. Klees, H. Kochan, G. Kölzer, H. Kroker, E. Kührt, P. Lämmerzahl, E. Lorenz, W.J. Markiewicz, D. Möhlmann, A. Oehler, J. Scholz, K.J. Seidensticker, K. Roessler, G. Schwehm, G. Steiner, K. Thiel, H. Thomas, Development of a dust mantle on the surface of an insolated ice-dust mixture—results from the KOSI-9 experiment. J. Geophys. Res. **98**, 15 (1993). https://doi.org/10.1029/93JE01134

E. Grün, J. Agarwal, N. Altobelli, K. Altwegg, M.S. Bentley, N. Biver, V. Della Corte, N. Edberg, P.D. Feldman, M. Galand, B. Geiger, C. Götz, B. Grieger, C. Güttler, P. Henri, M. Hofstadter, M. Horanyi, E. Jehin, H. Krüger, S. Lee, T. Mannel, E. Morales, O. Mousis, M. Müller, C. Opitom, A. Rotundi, R. Schmied, F. Schmidt, H. Sierks, C. Snodgrass, R.H. Soja, M. Sommer, R. Srama, C.Y. Tzou, J.B. Vincent, P. Yanamandra-Fisher, M.F. A'Hearn, A.I. Erikson, C. Barbieri, M.A. Barucci, J.L. Bertaux, I. Bertini, J. Burch, L. Colangeli, G. Cremonese, V. Da Deppo, B. Davidsson, S. Debei, M. De Cecco, J. Deller, L.M. Feaga, M. Ferrari, S. Fornasier, M. Fulle, A. Gicquel, M. Gillon, S.F. Green, O. Groussin, P.J. Gutiérrez, M. Hofmann, S.F. Hviid, W.H. Ip, S. Ivanovski, L. Jorda, H.U. Keller, M.M. Knight, J. Knollenberg, D. Koschny, J.R. Kramm, E. Kührt, M. Küppers, P.L. Lamy, L.M. Lara, M. Lazzarin, J.J. Lòpez-Moreno, J. Manfroid, E.M. Epifani, F. Marzari, G. Naletto, N. Oklay, P. Palumbo, J.W. Parker, H. Rickman, R. Rodrigo, J. Rodrìguez, E. Schindhelm, X. Shi, R. Sordini, A.J. Steffl, S.A. Stern, N. Thomas, C. Tubiana, H.A. Weaver, P. Weissman, V.V. Zakharov, M.G.G.T. Taylor, The 2016 Feb 19 outburst of comet 67P/CG: an ESA Rosetta multi-instrument study. Mon. Not. R. Astron. Soc. **462**, S220–S234 (2016). https://doi.org/10.1093/mnras/stw2088

B. Gundlach, J. Blum, H.U. Keller, Y.V. Skorov, What drives the dust activity of comet 67P/Churyumov-Gerasimenko? Astron. Astrophys. **583**, A12 (2015). https://doi.org/10.1051/0004-6361/201525828. arXiv:1506.08545

C. Güttler, M. Krause, R.J. Geretshauser, R. Speith, J. Blum, The physics of protoplanetesimal dust agglomerates. IV. Toward a dynamical collision model. Astrophys. J. **701**, 130–141 (2009). https://doi.org/10.1088/0004-637X/701/1/130. arXiv:0906.0088

K.C. Hansen, K. Altwegg, J.J. Berthelier, A. Bieler, N. Biver, D. Bockelée-Morvan, U. Calmonte, F. Capaccioni, M.R. Combi, J. de Keyser, B. Fiethe, N. Fougere, S.A. Fuselier, S. Gasc, T.I. Gombosi, Z. Huang, L. Le Roy, S. Lee, H. Nilsson, M. Rubin, Y. Shou, C. Snodgrass, V. Tenishev, G. Toth, C.Y. Tzou, C.S. Wedlund (Rosina Team), Evolution of water production of 67P/Churyumov-Gerasimenko: an empirical model and a multi-instrument study. Mon. Not. R. Astron. Soc. **462**, S491–S506 (2016). https://doi.org/10.1093/mnras/stw2413

M. Hässig, K. Altwegg, H. Balsiger, A. Bar-Nun, J.J. Berthelier, A. Bieler, P. Bochsler, C. Briois, U. Calmonte, M. Combi, J. De Keyser, P. Eberhardt, B. Fiethe, S.A. Fuselier, M. Galand, S. Gasc, T.I. Gombosi, K.C. Hansen, A. Jäckel, H.U. Keller, E. Kopp, A. Korth, E. Kührt, L. Le Roy, U. Mall, B. Marty, O. Mousis, E. Neefs, T. Owen, H. Rème, M. Rubin, T. Sémon, C. Tornow, C.Y. Tzou, J.H. Waite, P. Wurz, Time variability and heterogeneity in the coma of 67P/Churyumov-Gerasimenko. Science **347**, aaa0276 (2015). https://doi.org/10.1126/science.aaa0276

B. Hermalyn, T.L. Farnham, S.M. Collins, M.S. Kelley, M.F. A'Hearn, D. Bodewits, B. Carcich, D.J. Lindler, C. Lisse, K. Meech, P.H. Schultz, P.C. Thomas, The detection, localization, and dynamics of large icy particles surrounding comet 103P/Hartley 2. Icarus **222**, 625–633 (2013). https://doi.org/10.1016/j.icarus.2012.09.030

M. Horányi, J.R. Szalay, S. Kempf, J. Schmidt, E. Grün, R. Srama, Z. Sternovsky, A permanent, asymmetric dust cloud around the Moon. Nature **522**, 324–326 (2015). https://doi.org/10.1038/nature14479

K.L. Johnson, K. Kendall, A.D. Roberts, Surface energy and the contact of elastic solids. Proc. R. Soc. Lond. Ser. A **324**, 301–313 (1971). https://doi.org/10.1098/rspa.1971.0141

H.U. Keller, W.A. Delamere, H.J. Reitsema, W.F. Huebner, H.U. Schmidt, Comet P/Halley's nucleus and its activity. Astron. Astrophys. **187**, 807–823 (1987)

H.U. Keller, S. Mottola, B. Davidsson, S.E. Schröder, Y. Skorov, E. Kührt, O. Groussin, M. Pajola, S.F. Hviid, F. Preusker, F. Scholten, M.F. A'Hearn, H. Sierks, C. Barbieri, P. Lamy, R. Rodrigo, D. Koschny, H. Rickman, M.A. Barucci, J.L. Bertaux, I. Bertini, G. Cremonese, V. Da Deppo, S. Debei, M. De Cecco, S. Fornasier, M. Fulle, P.J. Gutiérrez, W.H. Ip, L. Jorda, J. Knollenberg, J.R. Kramm, M. Küppers, L.M. Lara, M. Lazzarin, J.J. Lopez Moreno, F. Marzari, H. Michalik, G. Naletto, L. Sabau, N. Thomas, J.B. Vincent, K.P. Wenzel, J. Agarwal, C. Güttler, N. Oklay, C. Tubiana, Insolation, erosion, and morphology of comet 67P/Churyumov-Gerasimenko. Astron. Astrophys. **583**, A34 (2015). https://doi.org/10.1051/0004-6361/201525964

M.S. Kelley, D.J. Lindler, D. Bodewits, M.F. A'Hearn, C.M. Lisse, L. Kolokolova, J. Kissel, B. Hermalyn, A distribution of large particles in the coma of comet 103P/Hartley 2. Icarus **222**, 634–652 (2013). https://doi.org/10.1016/j.icarus.2012.09.037. arXiv:1304.4204

J. Knollenberg, A mini outburst from the nightside of comet P67/Churyumov-Gerasimenko observed by the OSIRIS camera on Rosetta. Astron. Astrophys. (2015)

J. Knollenberg, E. Kührt, H.U. Keller, Interpretation of HMC images by a combined thermal and gasdynamic model. Earth Moon Planets **72**, 103–112 (1996). https://doi.org/10.1007/BF00117509

K.J. Kossacki, T. Spohn, A. Hagermann, E. Kaufmann, E. Kührt, Comet 67P/Churyumov-Gerasimenko: hardening of the sub-surface layer. Icarus **260**, 464–474 (2015). https://doi.org/10.1016/j.icarus.2015.07.024

T. Kramer, M. Noack, On the origin of inner coma structures observed by Rosetta during a diurnal rotation of comet 67P/Churyumov-Gerasimenko. Astrophys. J. Lett. **823**, L11 (2016). https://doi.org/10.3847/2041-8205/823/1/L11. arXiv:1607.03825

T. Kramer, M. Läuter, M. Rubin, K. Altwegg, Seasonal changes of the volatile density in the coma and on the surface of comet 67P/Churyumov-Gerasimenko. Mon. Not. R. Astron. Soc. **469**, S20–S28 (2017). https://doi.org/10.1093/mnras/stx866. arXiv:1704.03874

E. Kuehrt, H.U. Keller, The formation of cometary surface crusts. Icarus **109**, 121–132 (1994). https://doi.org/10.1006/icar.1994.1080

I.L. Lai, W.H. Ip, J.C. Lee, Z.Y. Lin, J.B. Vincent, P. Hartogh, H. Sierks, C. Barbieri, P.L. Lamy, H. Rickman, H.U. Keller, Mapping of the source regions of the dust jets on comet 67P/Churyumov-Gerasimenko, in *AGU Fall Meeting Abstracts P43A-2087* (2016)

L.M. Lara, H. Böhnhardt, R. Gredel, P.J. Gutiérrez, J.L. Ortiz, R. Rodrigo, M.J. Vidal-Nuñez, Pre-impact monitoring of comet 9P/Tempel 1, the Deep Impact target. Astron. Astrophys. **465**, 1151–1157 (2006)

L.M. Lara et al., Large-scale dust jets in the coma of 67P/Churyumov-Gerasimenko as seen by the OSIRIS instrument onboard Rosetta. Astron. Astrophys. (2015). https://doi.org/10.1051/0004-6361/201526103

S. Larson, Z. Sekanina, D. Levy, S. Tapia, M. Senay, Comet p/Halley near nucleus phenomena in 1986. Astron. Astrophys. **187**, 639 (1987)

A.C. Levasseur-Regourd, J. Agarwal, H. Cottin, C. Engrand, G. Flynn, M. Fulle, T. Gombosi, Y. Langevin, J. Lasue, T. Mannel, S. Merouane, O. Poch, N. Thomas, A. Westphal, Cometary dust. Space Sci. Rev. **214**, 64 (2018). https://doi.org/10.1007/s11214-018-0496-3

Z.Y. Lin et al., Morphology and dynamics of jets of comet 67P/Churyumov-Gerasimenko: early phase development. Astron. Astrophys. (2015). https://doi.org/10.1051/0004-6361/201525961

Z.Y. Lin, I.L. Lai, C.C. Su, W.H. Ip, J.C. Lee, J.S. Wu, J.B. Vincent, F. La Forgia, H. Sierks, C. Barbieri, P.L. Lamy, R. Rodrigo, D. Koschny, H. Rickman, H.U. Keller, J. Agarwal, M.F. A'Hearn, M.A. Barucci, J.L. Bertaux, I. Bertini, D. Bodewits, G. Cremonese, V. Da Deppo, B. Davidsson, S. Debei, M. De Cecco, S. Fornasier, M. Fulle, O. Groussin, P.J. Gutiérrez, C. Güttler, S.F. Hviid, L. Jorda, J. Knollenberg, G. Kovacs, J.R. Kramm, E. Kührt, M. Küppers, L.M. Lara, M. Lazzarin, J.J. López-Moreno, S. Lowry, F. Marzari, H. Michalik, S. Mottola, G. Naletto, N. Oklay, M. Pajola, A. Rozek, N. Thomas, C. Tubiana, Observations and analysis of a curved jet in the coma of comet 67P/Churyumov-Gerasimenko. Astron. Astrophys. **588**, L3 (2016). https://doi.org/10.1051/0004-6361/201527784

T. Mannel, M.S. Bentley, R. Schmied, H. Jeszenszky, A.C. Levasseur-Regourd, J. Romstedt, K. Torkar, Fractal cometary dust—a window into the early Solar system. Mon. Not. R. Astron. Soc. **462**, S304–S311 (2016). https://doi.org/10.1093/mnras/stw2898

U. Marboeuf, B. Schmitt, J.M. Petit, O. Mousis, N. Fray, A cometary nucleus model taking into account all phase changes of water ice: amorphous, crystalline, and clathrate. Astron. Astrophys. **542**, A82 (2012). https://doi.org/10.1051/0004-6361/201118176

R. Marschall, S. Mottola, C.C. Su, Y. Liao, M. Rubin, J.S. Wu, N. Thomas, K. Altwegg, H. Sierks, W.H. Ip, H.U. Keller, J. Knollenberg, E. Kührt, I.L. Lai, Y. Skorov, L. Jorda, F. Preusker, F. Scholten, J.B. Vincent (Osiris Team Rosina Team), Cliffs versus plains: can ROSINA/COPS and OSIRIS data of comet 67P/Churyumov-Gerasimenko in autumn 2014 constrain inhomogeneous outgassing? Astron. Astrophys. **605**, A112 (2017). https://doi.org/10.1051/0004-6361/201730849

R. Miles, Heat of solution: a new source of thermal energy in the subsurface of cometary nuclei and the gas-exsolution mechanism driving outbursts of comet 29P/Schwassmann-Wachmann and other comets. Icarus **272**, 356–386 (2016). https://doi.org/10.1016/j.icarus.2015.12.053

S. Mottola, G. Arnold, H.G. Grothues, R. Jaumann, H. Michaelis, G. Neukum, J.P. Bibring, S.E. Schröder, M. Hamm, K.A. Otto, I. Pelivan, G. Proffe, F. Scholten, D. Tirsch, M. Kreslavsky, E. Remetean, F. Souvannavong, B. Dolives, The structure of the regolith on 67P/Churyumov-Gerasimenko from ROLIS descent imaging. Science **349**(2), aab0232 (2015). https://doi.org/10.1126/science.aab0232

N. Oklay, Large sub-surface volatile reservoirs of comet 67P, in *Lunar and Planetary Science Conference*, vol. 49 (2018), p. 1282

N. Oklay, J.B. Vincent, S. Fornasier, M. Pajola, S. Besse, B.J.R. Davidsson, L.M. Lara, S. Mottola, G. Naletto, H. Sierks, A.M. Barucci, F. Scholten, F. Preusker, A. Pommerol, N. Masoumzadeh, M. Lazzarin, C. Barbieri, P.L. Lamy, R. Rodrigo, D. Koschny, H. Rickman, M.F. A'Hearn, J.L. Bertaux, I. Bertini, D. Bodewits, G. Cremonese, V. Da Deppo, S. Debei, M. De Cecco, M. Fulle, O. Groussin, P.J. Gutiérrez, C. Güttler, I. Hall, M. Hofmann, S.F. Hviid, W.H. Ip, L. Jorda, H.U. Keller, J. Knollenberg, G. Kovacs, J.R. Kramm, E. Kührt, M. Küppers, Z.Y. Lin, J.J. Lopez Moreno, F. Marzari, F. Moreno, X. Shi, N. Thomas, I. Toth, C. Tubiana, Variegation of comet 67P/Churyumov-Gerasimenko in regions showing activity. Astron. Astrophys. **586**, A80 (2016). https://doi.org/10.1051/0004-6361/201527369

T. Ott, E. Drolshagen, D. Koschny, C. Güttler, C. Tubiana, E. Frattin, J. Agarwal, H. Sierks, I. Bertini, C. Barbieri, P.I. Lamy, R. Rodrigo, H. Rickman, M.F. A'Hearn, M.A. Barucci, J.L. Bertaux, S. Boudreault, G. Cremonese, V. Da Deppo, B. Davidsson, S. Debei, M. De Cecco, J. Deller, C. Feller, S. Fornasier, M. Fulle, B. Geiger, A. Gicquel, O. Groussin, P.J. Gutiérrez, M. Hofmann, S.F. Hviid, W.H. Ip, L. Jorda, H.U. Keller, J. Knollenberg, G. Kovacs, J.R. Kramm, E. Kührt, M. Küppers, L.M. Lara, M. Lazzarin, Z.Y. Lin, J.J. López-Moreno, F. Marzari, S. Mottola, G. Naletto, N. Oklay, M. Pajola, X. Shi, N. Thomas, J.B. Vincent, B. Poppe, Dust mass distribution around comet 67P/Churyumov-Gerasimenko determined via parallax measurements using Rosetta's OSIRIS cameras. Mon. Not. R. Astron. Soc. **469**, S276–S284 (2017). https://doi.org/10.1093/mnras/stx1419

M. Pajola, S. Hoefner, J.B. Vincent, O. Nea, The pristine interior of comet 67P revealed by the combined Aswan outburst and cliff collapse. Nature **1**, 0092 (2017)

M. Pätzold, T. Andert, M. Hahn, S.W. Asmar, J.P. Barriot, M.K. Bird, B. Häusler, K. Peter, S. Tellmann, E. Grün, P.R. Weissman, H. Sierks, L. Jorda, R. Gaskell, F. Preusker, F. Scholten, A homogeneous nucleus for comet 67P/Churyumov-Gerasimenko from its gravity field. Nature **530**, 63–65 (2016). https://doi.org/10.1038/nature16535

S. Protopapa, J.M. Sunshine, L.M. Feaga, M.S.P. Kelley, M.F. A'Hearn, T.L. Farnham, O. Groussin, S. Besse, F. Merlin, J.Y. Li, Water ice and dust in the innermost coma of comet 103P/Hartley 2. Icarus **238**, 191–204 (2014). https://doi.org/10.1016/j.icarus.2014.04.008. arXiv:1406.3382

A. Rotundi, H. Sierks, V. Della Corte, M. Fulle, P.J. Gutierrez, L. Lara, C. Barbieri, P.L. Lamy, R. Rodrigo, D. Koschny, H. Rickman, H.U. Keller, J.J. López-Moreno, M. Accolla, J. Agarwal, M.F. A'Hearn, N. Altobelli, F. Angrilli, M.A. Barucci, J.L. Bertaux, I. Bertini, D. Bodewits, E. Bussoletti, L. Colangeli, M. Cosi, G. Cremonese, J.F. Crifo, V. Da Deppo, B. Davidsson, S. Debei, M. De Cecco, F. Esposito, M. Ferrari, S. Fornasier, F. Giovane, B. Gustafson, S.F. Green, O. Groussin, E. Grün, C. Güttler, M.L. Herranz, S.F. Hviid, W. Ip, S. Ivanovski, J.M. Jerónimo, L. Jorda, J. Knollenberg, R. Kramm, E. Kührt, M. Küppers, M. Lazzarin, M.R. Leese, A.C. López-Jiménez, F. Lucarelli, S.C. Lowry, F. Marzari, E.M.

Epifani, J.A.M. McDonnell, V. Mennella, H. Michalik, A. Molina, R. Morales, F. Moreno, S. Mottola, G. Naletto, N. Oklay, J.L. Ortiz, E. Palomba, P. Palumbo, J.M. Perrin, J. Rodríguez, L. Sabau, C. Snodgrass, R. Sordini, N. Thomas, C. Tubiana, J.B. Vincent, P. Weissman, K.P. Wenzel, V. Zakharov, J.C. Zarnecki, Dust measurements in the coma of comet 67P/Churyumov-Gerasimenko inbound to the Sun. Science **347**, aaa3905 (2015). https://doi.org/10.1126/science.aaa3905

R.Z. Sagdeev, G.A. Avanesov, B.S. Zhukov, I.V. Zenkin, Y.L. Ziman, V.I. Moroz, V.I. Tarnopol'Skij, Photometric characteristics of the nucleus and inner coma of comet Halley. Kosm. Issled. **25**, 831–839 (1987)

N.H. Samarasinha, S.M. Larson, Image enhancement techniques for quantitative investigations of morphological features in cometary comae: a comparative study. Icarus **239**, 168 (2014)

D.G. Schleicher, Compositional and physical results for Rosetta's new target comet 67P/Churyumov Gerasimenko from narrowband photometry and imaging. Icarus **181**, 442–457 (2006). https://doi.org/10.1016/j.icarus.2005.11.014

R. Schulz, J.A. Stüwe, H. Boehnhardt, W. Gaessler, G.P. Tozzi, Characterization of STARDUST target comet 81P/Wild 2 from 1996 to 1998. Astron. Astrophys. **398**, 345–352 (2003). https://doi.org/10.1051/0004-6361:20021673

Z. Sekanina, D.E. Brownlee, T.E. Economou, A.J. Tuzzolino, S.F. Green, Modeling the nucleus and jets of comet 81P/Wild 2 based on the Stardust encounter data. Science **304**, 1769–1774 (2004)

X. Shi, X. Hu, H. Sierks, C. Güttler, M. A'Hearn, J. Blum, M.R. El-Maarry, E. Kührt, S. Mottola, M. Pajola, N. Oklay, S. Fornasier, C. Tubiana, H.U. Keller, J.B. Vincent, D. Bodewits, S. Höfner, Z.Y. Lin, A. Gicquel, M. Hofmann, C. Barbieri, P.L. Lamy, R. Rodrigo, D. Koschny, H. Rickman, M.A. Barucci, J.L. Bertaux, I. Bertini, G. Cremonese, V. Da Deppo, B. Davidsson, S. Debei, M. De Cecco, M. Fulle, O. Groussin, P.J. Gutiérrez, S.F. Hviid, W.H. Ip, L. Jorda, J. Knollenberg, G. Kovacs, J.R. Kramm, M. Küppers, L.M. Lara, M. Lazzarin, J.J. Lopez-Moreno, F. Marzari, G. Naletto, N. Thomas, Sunset jets observed on comet 67P/Churyumov-Gerasimenko sustained by subsurface thermal lag. Astron. Astrophys. **586**, A7 (2016). https://doi.org/10.1051/0004-6361/201527123

X. Shi, X. Hu, S. Mottola, H. Sierks, H.U. Keller, M. Rose, C. Güttler, M. Fulle, S. Fornasier, J. Agarwal, M. Pajola, C. Tubiana, D. Bodewits, C. Barbieri, P.L. Lamy, R. Rodrigo, D. Koschny, M.A. Barucci, J.L. Bertaux, I. Bertini, S. Boudreault, G. Cremonese, V. Da Deppo, B. Davidsson, S. Debei, M. De Cecco, J. Deller, O. Groussin, P.J. Gutiérrez, S.F. Hviid, W.H. Ip, L. Jorda, J. Knollenberg, G. Kovacs, J.R. Kramm, E. Kührt, M. Küppers, L.M. Lara, M. Lazzarin, J.J. Lopez-Moreno, F. Marzari, G. Naletto, N. Oklay, I. Toth, J.B. Vincent, Coma morphology of comet 67P controlled by insolation over irregular nucleus. Nat. Astron. **2**, 562–567 (2018). https://doi.org/10.1038/s41550-018-0481-5

Y. Skorov, J. Blum, Dust release and tensile strength of the non-volatile layer of cometary nuclei. Icarus **221**, 1 (2012)

Y.V. Skorov, L. Rezac, P. Hartogh, H.U. Keller, Is near-surface ice the driver of dust activity on 67P/Churyumov-Gerasimenko. Astron. Astrophys. **600**, A142 (2017). https://doi.org/10.1051/0004-6361/201630000

L.A. Soderblom, T.L. Becker, G. Bennett, D.C. Boice, D.T. Britt, R.H. Brown, B.J. Buratti, C. Isbell, B. Giese, T. Hare, M.D. Hicks, E. Howington-Kraus, R.L. Kirk, M. Lee, R.M. Nelson, J. Oberst, T.C. Owen, M.D. Rayman, B.R. Sandel, S.A. Stern, N. Thomas, R.V. Yelle, Observations of comet 19P/Borrelly by the miniature integrated camera and spectrometer aboard Deep Space 1. Science **296**, 1087–1091 (2002). https://doi.org/10.1126/science.1069527

J.K. Steckloff, K. Graves, M. Hirabayashi, H.J. Melosh, J.E. Richardson, Rotationally induced surface slope-instabilities and the activation of $CO_2$ activity on comet 103P/Hartley 2. Icarus **272**, 60–69 (2016). https://doi.org/10.1016/j.icarus.2016.02.026. arXiv:1602.08518

P.J. Stooke, A. Abergel, Morphology of the nucleus of comet P/Halley. Astron. Astrophys. **248**, 656–668 (1991)

J.M. Sunshine, M.F. A'Hearn, O. Groussin, J.Y. Li, M.J.S. Belton, W.A. Delamere, J. Kissel, K.P. Klaasen, L.A. McFadden, K.J. Meech, H.J. Melosh, P.H. Schultz, P.C. Thomas, J. Veverka, D.K. Yeomans, I.C. Busko, M. Desnoyer, T.L. Farnham, L.M. Feaga, D.L. Hampton, D.J. Lindler, C.M. Lisse, D.D. Wellnitz, Exposed water ice deposits on the surface of comet 9P/Tempel 1. Science **311**, 1453–1455 (2006). https://doi.org/10.1126/science.1123632

J.M. Sunshine, O. Groussin, P.H. Schultz, M.F. A'Hearn, L.M. Feaga, T.L. Farnham, K.P. Klaasen, The distribution of water ice in the interior of comet Tempel 1. Icarus **190**, 284–294 (2007). https://doi.org/10.1016/j.icarus.2007.04.024

J.M. Sunshine, L.M. Feaga, O. Groussin, S. Protopapa, M.F. A'Hearn, T.L. Farnham, F. Merlin, J.Y. Li, S. Besse, Water ice on comet 103P/Hartley 2, in *EPSC-DPS Joint Meeting 2011*, vol. 2011 (2011), p. 1345

H. Thomas, L. Ratke, H. Kochan, Crushing strength of porous ice-mineral bodies-relevance for comets. Adv. Space Res. **14**, 207–216 (1994). https://doi.org/10.1016/0273-1177(94)90271-2

P.C. Thomas, J. Veverka, M.J.S. Belton, A. Hidy, M.F. A'Hearn, T.L. Farnham, O. Groussin, J.Y. Li, L.A. McFadden, J. Sunshine, D. Wellnitz, C. Lisse, P. Schultz, K.J. Meech, W.A. Delamere, The shape, topography, and geology of Tempel 1 from Deep Impact observations. Icarus **187**, 4–15 (2007)

P. Thomas et al., The nucleus of comet 9P/Tempel 1: shape and geology from two flybys. Icarus **222**, 453–466 (2013)

N. Thomas, B. Davidsson, M.R. El-Maarry, S. Fornasier, L. Giacomini, A.G. Gracia-Berná, S.F. Hviid, W.H. Ip, L. Jorda, H.U. Keller, J. Knollenberg, E. Kührt, F. La Forgia, I.L. Lai, Y. Liao, R. Marschall, M. Massironi, S. Mottola, M. Pajola, O. Poch, A. Pommerol, F. Preusker, F. Scholten, C.C. Su, J.S. Wu, J.B. Vincent, H. Sierks, C. Barbieri, P.L. Lamy, R. Rodrigo, D. Koschny, H. Rickman, M.F. A'Hearn, M.A. Barucci, J.L. Bertaux, I. Bertini, G. Cremonese, V. Da Deppo, S. Debei, M. de Cecco, M. Fulle, O. Groussin, P.J. Gutierrez, J.R. Kramm, M. Küppers, L.M. Lara, M. Lazzarin, J.J. Lopez Moreno, F. Marzari, H. Michalik, G. Naletto, J. Agarwal, C. Güttler, N. Oklay, C. Tubiana, Redistribution of particles across the nucleus of comet 67P/Churyumov-Gerasimenko. Astron. Astrophys. **583**, A17 (2015). https://doi.org/10.1051/0004-6361/201526049

C. Tubiana et al., 67P/Churyumov-Gerasimenko: activity between March and June 2014 as observed from Rosetta/OSIRIS. Astron. Astrophys. **573**, A62 (2015)

J.B. Vincent, Evolution of cometary surfaces, in *Lunar and Planetary Science Conference*, vol. 49 (2018), p. 1280

J.B. Vincent, H. Böhnhardt, I. Bertini, L. Lara, M. Küppers, R. Rodrigo, Coma structures in comet 73P/Schwassmann-Wachmann 3, components B and C, between January and May 2006. Earth Moon Planets **106**, 27 (2010)

J.B. Vincent, H. Sierks, M. Rose, Y. Skorov, 3D modeling of cometary dust jets, from the nucleus surface to infinity, in *EPSC*, vol. 2012 (2012), p. 358

J.B. Vincent, D. Bodewits, S. Besse, H. Sierks, C. Barbieri, P. Lamy, R. Rodrigo, D. Koschny, H. Rickman, H.U. Keller, J. Agarwal, M.F. A'Hearn, A.T. Auger, M.A. Barucci, J.L. Bertaux, I. Bertini, C. Capanna, G. Cremonese, V. da Deppo, B. Davidsson, S. Debei, M. de Cecco, M.R. El-Maarry, F. Ferri, S. Fornasier, M. Fulle, R. Gaskell, L. Giacomini, O. Groussin, A. Guilbert-Lepoutre, P. Gutierrez-Marques, P.J. Gutiérrez, C. Güttler, N. Hoekzema, S. Höfner, S.F. Hviid, W.H. Ip, L. Jorda, J. Knollenberg, G. Kovacs, R. Kramm, E. Kührt, M. Küppers, F. La Forgia, L.M. Lara, M. Lazzarin, V. Lee, C. Leyrat, Z.Y. Lin, J.J. Lopez Moreno, S. Lowry, S. Magrin, L. Maquet, S. Marchi, F. Marzari, M. Massironi, H. Michalik, R. Moissl, S. Mottola, G. Naletto, N. Oklay, M. Pajola, F. Preusker, F. Scholten, N. Thomas, I. Toth, C. Tubiana, Large heterogeneities in comet 67P as revealed by active pits from sinkhole collapse. Nature **523**, 63–66 (2015a). https://doi.org/10.1038/nature14564

J.B. Vincent, N. Oklay, S. Marchi, S. Höfner, H. Sierks, Craters on comets. Planet. Space Sci. **107**, 53–63 (2015b). https://doi.org/10.1016/j.pss.2014.06.008

J.B. Vincent, M.F. A'Hearn, Z.Y. Lin, M.R. El-Maarry, M. Pajola, H. Sierks, C. Barbieri, P.L. Lamy, R. Rodrigo, D. Koschny, H. Rickman, H.U. Keller, J. Agarwal, M.A. Barucci, J.L. Bertaux, I. Bertini, S. Besse, D. Bodewits, G. Cremonese, V. Da Deppo, B. Davidsson, S. Debei, M. De Cecco, J. Deller, S. Fornasier, M. Fulle, A. Gicquel, O. Groussin, P.J. Gutiérrez, P. Gutiérrez-Marquez, C. Güttler, S. Höfner, M. Hofmann, S.F. Hviid, W.H. Ip, L. Jorda, J. Knollenberg, G. Kovacs, J.R. Kramm, E. Kührt, M. Küppers, L.M. Lara, M. Lazzarin, J.J. Lopez Moreno, F. Marzari, M. Massironi, S. Mottola, G. Naletto, N. Oklay, F. Preusker, F. Scholten, X. Shi, N. Thomas, I. Toth, C. Tubiana, Summer fireworks on comet 67P. Mon. Not. R. Astron. Soc. **462**, S184–S194 (2016). https://doi.org/10.1093/mnras/stw2409. arXiv:1609.07743

J.B. Vincent, S.F. Hviid, S. Mottola, E. Kuehrt, F. Preusker, F. Scholten, H.U. Keller, N. Oklay, D. de Niem, B. Davidsson, M. Fulle, M. Pajola, M. Hofmann, X. Hu, H. Rickman, Z.Y. Lin, C. Feller, A. Gicquel, S. Boudreault, H. Sierks, C. Barbieri, P.L. Lamy, R. Rodrigo, D. Koschny, M.F. A'Hearn, M.A. Barucci, J.L. Bertaux, I. Bertini, G. Cremonese, V. Da Deppo, S. Debei, M. De Cecco, J. Deller, S. Fornasier, O. Groussin, P.J. Gutiérrez, P. Gutiérrez-Marquez, C. Güttler, W.H. Ip, L. Jorda, J. Knollenberg, G. Kovacs, J.R. Kramm, M. Küppers, L.M. Lara, M. Lazzarin, J.J. Lopez Moreno, F. Marzari, G. Naletto, L. Penasa, X. Shi, N. Thomas, I. Toth, C. Tubiana, Constraints on cometary surface evolution derived from a statistical analysis of 67P's topography. Mon. Not. R. Astron. Soc. **469**, S329–S338 (2017). https://doi.org/10.1093/mnras/stx1691. arXiv:1707.00734

Z. Yoldi, A. Pommerol, B. Jost, O. Poch, J. Gouman, N. Thomas, VIS-NIR reflectance of water ice/regolith analogue mixtures and implications for the detectability of ice mixed within planetary regoliths. Geophys. Res. Lett. **42**, 6205–6212 (2015). https://doi.org/10.1002/2015GL064780

Space Sci Rev (2019) 215:29
https://doi.org/10.1007/s11214-019-0594-x

# The Thermal, Mechanical, Structural, and Dielectric Properties of Cometary Nuclei After Rosetta

**O. Groussin[1] · N. Attree[1,2] · Y. Brouet[3] · V. Ciarletti[4] · B. Davidsson[5] · G. Filacchione[6] · H.-H. Fischer[7] · B. Gundlach[8] · M. Knapmeyer[9] · J. Knollenberg[9] · R. Kokotanekova[10,11,12] · E. Kührt[9] · C. Leyrat[13] · D. Marshall[10,14] · I. Pelivan[9,15] · Y. Skorov[8,10] · C. Snodgrass[11,16] · T. Spohn[9] · F. Tosi[6]**

Received: 14 August 2018 / Accepted: 1 April 2019 / Published online: 2 May 2019


**Abstract** The physical properties of cometary nuclei observed today relate to their complex history and help to constrain their formation and evolution. In this article, we review some of the main physical properties of cometary nuclei and focus in particular on the thermal, mechanical, structural and dielectric properties, emphasising the progress made during the Rosetta mission. Comets have a low density of $480 \pm 220\ \mathrm{kg\,m^{-3}}$ and a low permittivity of 1.9–2.0, consistent with a high porosity of 70–80%, are weak with a very low global tensile strength $< 100$ Pa, and have a low bulk thermal inertia of 0–60 $\mathrm{J\,K^{-1}\,m^{-2}\,s^{-1/2}}$ that allowed them to preserve highly volatiles species (e.g. $CO$, $CO_2$, $CH_4$, $N_2$) into their interior

Comets: Post 67P / Churyumov-Gerasimenko Perspectives
Edited by Nicolas Thomas, Björn Davidsson, Laurent Jorda, Ekkehard Kührt, Raphael Marschall, Colin Snodgrass and Rafael Rodrigo

✉ O. Groussin
olivier.groussin@lam.fr

1 Aix Marseille Univ, CNRS, CNES, LAM, Laboratoire d'Astrophysique de Marseille, Marseille, France

2 Earth and Planetary Observation Centre, University of Stirling, Stirling, UK

3 Physics Institute, University of Bern, Bern, Switzerland

4 LATMOS/IPSL, UVSQ (Université Paris-Saclay), UPMC (Sorbonne Univ.), CNRS, 78280 Guyancourt, France

5 Jet Propulsion Laboratory, M/S 183-401, 4800 Oak Grove Drive, Pasadena, CA, 91109, USA

6 INAF-IAPS, Istituto di Astrofisica e Planetologia Spaziali, Rome, Italy

7 DLR Microgravity User Support Centre, Linder Höhe, 51147 Cologne, Germany

8 Institut für Geophysik und extraterrestrische Physik, Technische Universität Braunschweig, Braunschweig, Germany

9 DLR Institute of Planetary Research, Rutherfordstr. 2, 12489 Berlin, Germany

10 Max Planck Institute for Solar System Research, Justus-von-Liebig-Weg 3, 37077, Göttingen, Germany

11 Planetary and Space Sciences, School of Physical Sciences, The Open University, Milton Keynes, MK7 6AA, UK

since their formation. As revealed by 67P/Churyumov-Gerasimenko, the above physical properties vary across the nucleus, spatially at its surface but also with depth. The broad picture is that the bulk of the nucleus consists of a weakly bonded, rather homogeneous material that preserved primordial properties under a thin shell of processed material, and possibly covered by a granular material; this cover might in places reach a thickness of several meters. The properties of the top layer (the first meter) are not representative of that of the bulk nucleus. More globally, strong nucleus heterogeneities at a scale of a few meters are ruled out on 67P's small lobe.

**Keywords** Solar system · Comets · Rosetta · Space mission

## 1 Introduction

Comets were formed during the early stages of the solar system, and have been stored in the Kuiper belt and Oort cloud for billions of years, before they got injected into the inner solar system and ultimately became visible. Many processes have altered their nucleus since their formation, including: collisions, radiogenic and solar heating, encounters with giant planets, space weathering, phase transitions and erosion. A key question is to understand how the physical properties of cometary nuclei observed today relate to their complex history and more generally what it tells us about the formation and evolution of our solar system.

The physical properties of cometary nuclei are numerous, from basic attributes such as size, shape, albedo, or rotation period, to more complex properties such as the nature of the cometary material and its strength, porosity, or conductivity. Measuring these properties with a ground- or spaced-based telescope is challenging due to the small size of the nucleus, which is spatially unresolved, and its surrounding coma, which masks it. It has however been possible to derive the basic properties of more than a hundred cometary nuclei, from which the picture emerges of a nucleus with a kilometric size, a non-spherical shape, a low albedo (e.g. Lamy et al. 2004), a low density (e.g. Weissman et al. 2004) and a low conductivity (e.g. Fernández et al. 2013). To derive more accurate parameters, including surface heterogeneities, a dedicated space mission with a flyby or better a rendezvous offers the best solution. However, as shown in Fig. 1, this has only been possible for 6 cometary nuclei so far: 1P/Halley, 19P/Borrelly, 81P/Wild 2, 9P/Tempel 1, 103P/Hartley 2, and 67P/Churyumov-Gerasimenko (not mentioning the "blind" flybys of comets 21P/Giacobini-Zinner in 1985 and 26P/Grigg-Skjellerup in 1992).

In this paper, we review some of the main physical properties of cometary nuclei and focus in particular on the thermal properties (Sect. 2), the mechanical properties (Sect. 3), and the structural and dielectric properties (Sect. 4), emphasising the progress made during the Rosetta mission of the European Space Agency.

[12] European Southern Observatory, Karl-Schwarzschild-Strasse 2, 85748 Garching bei München, Germany

[13] Laboratoire d'Etudes Spatiales et d'Instrumentation en Astrophysique (LESIA), Observatoire de Paris, CNRS, UPMC, Paris Diderot Observatoire de Paris, 5 place Jules Janssen, 92195 Meudon, France

[14] Universität Göttingen, Friedrich-Hund-Platz 1, 37077 Göttingen, Germany

[15] Helmholtz Centre Potsdam, GFZ, German Research Center For Geosciences, 14473 Potsdam, Germany

[16] Institute for Astronomy, Royal Observatory, University of Edinburgh, Edinburgh EH9 3HJ, UK

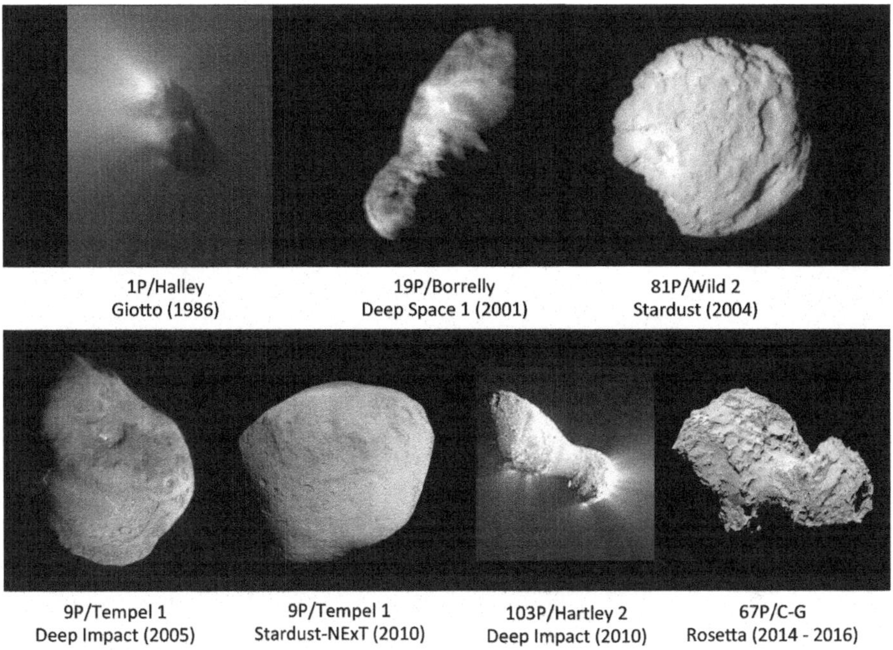


**Fig. 1** Images of the 6 cometary nuclei visited by spacecraft: 1P/Halley (credits ESA-MPAE), 19P/Borrelly (credits NASA/JPL), 81P/ Wild 2 (credits NASA/JPL), 9P/Tempel 1 (credits NASA/JPL/UMD and NASA/JPL/Caltech-Cornell), 103P/Hartley 2 (credits NASA/JPL/UMD), and 67P/Churyumov-Gerasimenko (credits ESA/Rosetta/MPS for OSIRIS Team MPS/UPD/LAM/IAA/SSO/INTA/UPM/DASP/IDA)

## 2 The Nucleus Thermal Properties

The thermal properties drive the ability of the comet nucleus to response to solar illumination and define how heat is transported inside the nucleus material. They define the temperature of the nucleus and are therefore key to properly describe many physical processes occurring on it, including the surface thermal emission, the sublimation of ices and their recondensation, the conduction of heat into the nucleus porous media, phase transitions and the transport of gas. The thermal properties are derived from temperature measurements (Sect. 2.1), and best described by the thermal inertia (Sect. 2.2).

### 2.1 Temperature

#### *2.1.1 State of the Art Before Rosetta*

Before the spacecraft flybys at comet 1P in 1986, the nucleus surface temperature was mainly constrained by modelling, assuming a mixture of ice and dust, possibly covered by a dust mantle of varying thickness (e.g. Whipple 1950; Mendis and Brin 1977; Fanale and Salvail 1984). At 1 au from the Sun, the typical surface temperature at the subsolar point was therefore between ~200 K, the free sublimation temperature of water ice, and ~400 K, the black body temperature, without being able to favour one case to the other.

Following the 1P flyby, the picture emerged of a monolithic and dark nucleus, with active and inactive regions, each of them with a different temperature (Julian et al. 2000). The

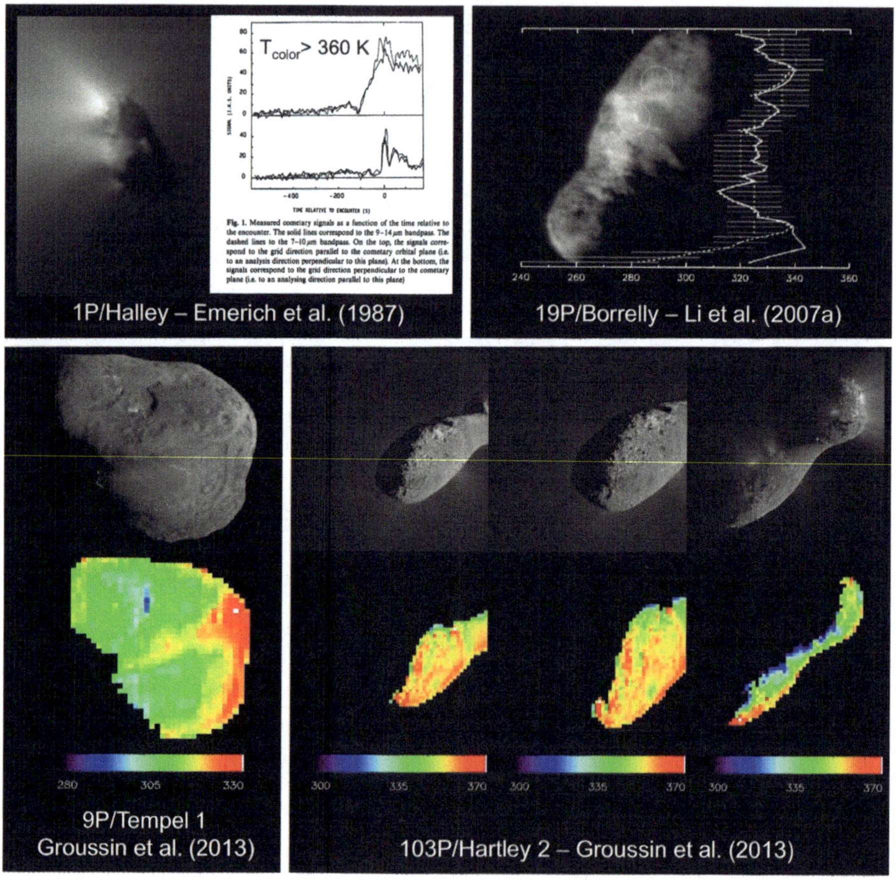


**Fig. 2** Spacecrafts colour temperature measurements of four cometary nuclei, before the Rosetta mission

temperature measured at the surface of 1P by the IKS instrument onboard the VEGA mission was larger than 360 K at 0.8 au from the Sun (Fig. 2), suggesting that the surface is mainly covered by hot inactive materials, depleted in volatiles, and with a low thermal conductivity because of the temperature close to the black body temperature (Emerich et al. 1987).

Progress occurred in the 2000's with the in-situ temperature measurements of four cometary nuclei, moreover spatially resolved (Fig. 2). Results indicate that the nucleus reaches its maximum temperature close to the subsolar point, and that the maximum temperature is close to the black body temperature; these two effects show that the thermal inertia of the nucleus is small enough not to cool the surface temperature significantly or to shift it temporally relatively to the subsolar point (Sect. 2.2). For comets 9P and 103P, the temperature spatial distribution shows no correlation with exposed water ice on the surface, which is therefore not present in sufficient amount to cool it (Sunshine et al. 2006). Finally, the colour temperature decreases by only a few tens of Kelvin from the subsolar point to solar incident angles exceeding 60°, which demonstrates that the surface is rough at the instrument sub-pixel scale ($<10$ m) (Sect. 4.1; Groussin et al. 2013).

Concerning the temperature inside the nucleus, it is only constrained by modelling. Assuming that the comet nucleus is characterised by a low thermal conductivity (Sect. 2.2) and high porosity (Sect. 4.3), temperature decreases rapidly with depth on the day side and only a small fraction of the surface solar energy is transported into the nucleus at depths exceeding 1 m. As a consequence, simple thermodynamic estimates show that the Jupiter Family Comets (JFCs) observed today are out of thermal equilibrium, with a thermal relaxation time ($\sim 10^5$ yr; see Prialnik et al. 2004, with the parameters of 67P) almost one order of magnitude larger than their mean lifetime in the inner solar system ($\sim 1.2 \times 10^4$ yr; Levison and Duncan 1997). On the contrary, their expected lifetime of $\sim 4 \times 10^8$ yr in the primordial disk around 15–30 au (Morbidelli et al. 2012; Marchi et al. 2013) and of $\sim 4.5 \times 10^7$ yr in the Kuiper belt around 40 au (Levison and Duncan 1997), is more than two orders of magnitude larger than the thermal relaxation time, which means that they should be at thermal equilibrium at these distances. It is therefore expected that the core temperature might be low, explaining the presence of volatile species (e.g. $H_2O$, $CO_2$) in comets, but possibly above the sublimation temperature of super volatiles (e.g. CO, $N_2$), which could therefore be trapped in amorphous ice or clathrates (Mousis et al. 2015).

### *2.1.2 Progress During Rosetta*

**VIRTIS Temperature Measurements** Temperature maps of the nucleus of comet 67P have been derived by modelling its thermal emission on spectral radiance data measured by VIRTIS-M (Visible InfraRed and Thermal Imaging Spectrometer, Mapping channel), the VIS-IR hyperspectral imaging channel aboard Rosetta (Coradini et al. 2007). The instrument performs 0.25–5.1 μm imaging spectroscopy on two separate channels, covering the 0.25–1.0 μm and 1.0–5.1 μm ranges respectively, by means of two bidimensional detectors, a CCD and an MCT array, sharing the same telescope equipped with a scan mirror to build hyperspectral images (so-called "cubes"). The average spectral sampling is equal to 1.8 nm and 9.7 nm per band for the two channels. On 3 May 2015, a sudden malfunction of the VIRTIS-M infrared focal plane cryocooler inhibited the use of infrared data acquired by the mapping channel, for which only the visible channel (0.25–1 μm) was still operating. The infrared spectral range 2–5 μm was covered by the high-spectral resolution channel VIRTIS-H, with however no imaging capability, for the rest of the mission. VIRTIS-H is mainly devoted to observe the coma.

VIRTIS could sample temperatures within the uppermost radiatively active surface layer, typically tens of microns due to the spectral range being limited to $\leq 5.1$ μm, the very low surface albedo and the high thermal emissivity. Within this wavelength range, the grey-body radiance is a non-linear function of temperature. Moreover, due to the typical VIRTIS pixel resolution values achieved on the nucleus (2–25 $\mathrm{m\,px^{-1}}$), within a given resolution cell, the measured radiance is preferentially weighted by the warmest sub-pixel areas. This effect is relevant for the case of 67P due to the rough morphology of the surface (Sect. 4.1).

Surface temperatures, as retrieved by VIRTIS-M and covering the early pre-perihelion period from 1 August to 23 September 2014, are widely reported by Tosi et al. (2019). In this period, 67P was rapidly approaching the frost line, with the heliocentric distance decreasing from 3.62 to 3.31 au, and the spacecraft was in the altitude range 61–13 km above the surface, resulting in a spatial resolution from $\sim 15$ to $\sim 3$ $\mathrm{m\,px^{-1}}$ (most data showing a resolution of 13 to 15 $\mathrm{m\,px^{-1}}$). The solar phase angle ranged from 17° to 93°, which also allowed variable illumination and local time conditions to be explored. Due to the large obliquity of the comet (52°, Jorda et al. 2016) that causes pronounced seasonal effects, at the time of these observations Northern latitudes were in the summer season, while Southern latitudes below 45° S were in the winter season and experienced polar night (Tosi et al. 2019).

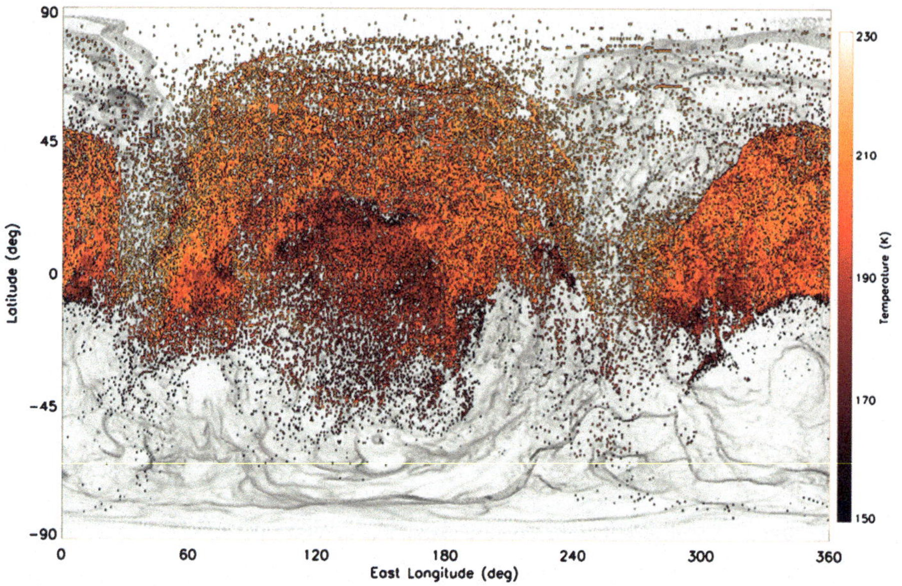


**Fig. 3** Temperature map of the nucleus of comet 67P obtained by VIRTIS between 10 h and 12 h true local solar time from 1 August to 23 September 2014, in cylindrical projection and with a fixed angular binning of 0.5° by 0.5° in both latitude and longitude. Each bin collects the average temperature recorded during a full diurnal rotation period. We display only illuminated points that have at the same time: solar incidence angle $i < 80°$, emission angle $e < 80°$, and accuracy better than 2 K. In this period, circumpolar regions experienced permanent lighting. For comparison, at that heliocentric distance, the black body temperature is around 215 K

The lower limit of temperatures sensed by VIRTIS-M is controlled by the instrumental noise-equivalent spectral radiance (NESR), which may vary over time depending on some instrumental parameters. In the above mentioned time period, this lower bound was 156 K on average, which in fact restricts the VIRTIS-M coverage to the dayside of the nucleus, including areas that recently underwent shadowing or recently exited from shade (Tosi et al. 2019).

When filtering VIRTIS-derived thermal maps into different local time intervals, each covering two hours in true local solar time, the surface of comet 67P does not exhibit outstanding thermal anomalies. These maps show a fairly uniform distribution of surface temperatures, mildly dependent on the latitude. The big lobe and the "neck", i.e. the surface area connecting the two lobes display regional differences despite their mutual proximity, with the neck exhibiting a faster temperature change than the big lobe, both in the morning and in the evening hours. Finally, the regions in the big lobe that experienced grazing sunlight around local noon during this season show the lowest values of maximum daytime temperature. Figure 3 shows one example of such a map obtained for the 10–12 h true local solar time interval (Tosi et al. 2019).

Different morphological regions (Fig. 4; Thomas et al. 2018) show peculiar thermal behaviour: Ash, Babi, Hapi, and Seth regions, located near the neck, and the Ma'at region on the small lobe close to the neck, are the locations where maximum surface temperatures were detected, whereas other regions at comparable latitudes reached lower temperatures. This behaviour is correlated to favourable illumination conditions occurring at the time of VIRTIS observations. Besides increasing diurnal surface temperature, solar illumination drives

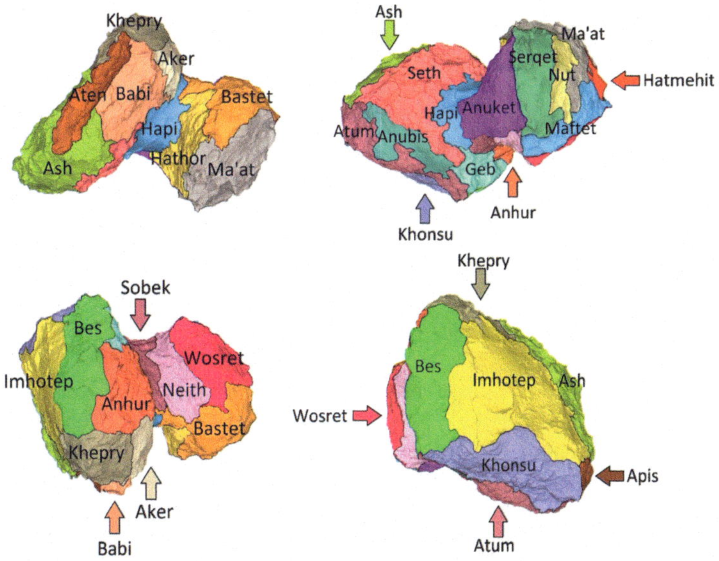


**Fig. 4** Regional units of the nucleus of 67P, from Thomas et al. (2018)

the activity occurring in some of those areas (Keller et al. 2017), and particularly in the Hapi region (De Sanctis et al. 2015).

The temperature recorded on the dayside of the nucleus, for solar incidence angles $<20^\circ$ (i.e. around local noon) and emission angles $<80^\circ$, is $213 \pm 3$ K on average (Tosi et al. 2019). No correlation arises on a global scale between VIRTIS-derived temperature and single scattering albedo, both at visible and near-infrared wavelengths (Ciarniello et al. 2015). Extreme values, reaching up to 230 K, were recorded by VIRTIS at Northern latitudes in the Seth region corresponding to the pit area named "Seth_01" (Vincent et al. 2015), and are most likely due to the peculiar topography of the pit combined with the instantaneous solar illumination shortly after local noon, which enhances the self-heating effect (Tosi et al. 2019). Self-heating results from the irregular shape and topography, which induces mutual heating of regions (or facets on a shape model) seeing each other.

When correlating the measured surface temperature values for all of the morphological regions in the sunlit hemisphere with the solar incidence angle, a strong dispersion is observed. This result indicates that most data points do not follow a black body behaviour. In the case of regions located in the neck area, such deviations can be attributed to macroscopic self-heating effects caused by the presence of large concavities. Such morphologies, along with the low surface albedo, causes local infrared thermal flux enhancement by repeated emissions from mutually facing surface areas. On the other hand, in the equatorial region the presence of several distinct trends in temperature vs. solar incidence angle could be ascribed to the non-uniformity of the morphology within a given region, and thus to a variable role of the mutually illuminating areas. Finally, temperature values higher than the ideal case of a black body, observed in all of the explored morphological regions for large solar incidence angles, are evidence for small-scale roughness (Sect. 4.1) (Tosi et al. 2019).

On 22 August 2014, VIRTIS obtained 7 consecutive snapshots of the nucleus, from an altitude of 60 km above the surface (spatial resolution 15 $\mathrm{m\,px^{-1}}$). This sequence covers large areas of the neck and the two lobes, following them during $\sim$15% of the rotational period, which allowed direct computation of thermal gradients throughout the scene. From these data, it emerges that the dayside of the nucleus shows typical temporal thermal gradients of 0.1 $K\,min^{-1}$, except in the neck area where sudden daytime shadowing takes place. In this latter case, the temporal gradient increases up to 2.0 $K\,min^{-1}$, i.e. twenty times larger than the typical value. However, the real temporal gradient in the neck is likely several times larger than this value, due to the poor ability of VIRTIS in retrieving surface temperatures $<160$ K (Tosi et al. 2019).

**MIRO Temperature Measurements** The Microwave Instrument for the Rosetta Orbiter (MIRO) is a microwave spectrometer with two heterodyne receivers operating at 190 GHz (mm channel) and 560 GHz (sub-mm channel) (Gulkis et al. 2007). Each receiver measures an antenna temperature, which corresponds to the temperature from a sub-surface depth. The sub-mm channel typically probes the first cm, while the mm channel probes the first 4 cm (Schloerb et al. 2015). The measured antenna temperatures are converted to a brightness temperature, which is defined as the required temperature of a black body that fills the observed area to produce the observed power. In this section, whenever we refer to a temperature measured by MIRO, it is therefore a *brightness temperature*.

Measured temperatures from MIRO have been quoted in a variety of sources, although many of them only cover the very beginning of the mission, when 67P was beyond 3 au from the Sun. At the MIRO wavelengths, the measured temperatures originate from a subsurface (a few cm) volume element, not from the surface itself.

The earliest temperature measurements come from Gulkis et al. (2015), who reported results from 2014, showing diurnal and seasonal temperature variations. The data were restricted to a longitude band between 100° and 200° in order to avoid the neck region, where shadowing makes it difficult to interpret the diurnal phase curves. They show the temperature variation in latitude bands at 20°S–30°S and 20°N–30°N. During this early part of the mission, the Northern side was mainly illuminated. In the sub-mm channel, on the Southern side, the sub-surface temperature varied between approximately 60–100 K at midnight but between approximately 50–160 K at midday. In the well illuminated Northern hemisphere, the variation was far less dramatic, varying between 115–125 K at midnight and between 160–190 K at midday. In the mm channel, low sub-surface temperatures of 60 K were recorded at all times in the Southern hemisphere with highs of 110 K at midnight and 155 K at midday. In the Northern hemisphere, the sub-surface temperature varied between 140–155 K at midnight and 160–185 K at midday.

Temperatures from August 2014 are also reported by Lee et al. (2015) and show a similar behaviour: the sub-surface temperature on the well illuminated dayside reaches over 150 K, but the in less illuminated regions, it goes down to about 75 K.

Additionally, Gulkis et al. (2015) suggest that there is a slight temporal lag, with the highest sub-surface temperatures in the Southern hemisphere observed in the early or mid-afternoon, revealing the sensitivity to the diurnal heat cycle. Furthermore, the reduction in temperatures on the Southern side compared to the Northern side shows the seasonal effect on the comet.

The diurnal variation in sub-surface temperature in September 2014 is further investigated by Schloerb et al. (2015). In this month, the comet was at heliocentric distances of 3.45–3.27 au and the sub-solar point in mid-September was at 43°N. The observations are mainly of the Imhotep and Ash regions, and the measurements are binned by latitude and local solar time. Schloerb et al. (2015) find that the sub-surface temperature is a function of latitude, arising from the fact that solar flux is a function of latitude. The diurnal temperature curve in the sub-mm channel has a greater amplitude than the temperature curve of the mm channel. In both channels, the maximum is shifted after midday, with a larger shift for the mm channel compared to the sub-mm channel. The shift is greater for latitudes farther from the sub-solar point. These results indicate that the mm emission arises from greater depths than the sub-mm emission, and Schloerb et al. (2015) find that for the mm channel, the sub-surface temperature is measured over the first 4 cm of the nucleus surface, whereas for the sub-mm channel, the measured sub-surface temperature typically originates from the first centimetre.

A dedicated mapping scan of 67P was performed on 7 September 2014, allowing the temperature of the subsurface to be mapped across the whole nucleus. The scanning procedure took three hours and the results of the sub-mm mapping are shown in Biver et al. (2015). At this point in time, the variation in temperature across the comet is clearly evident, with the illuminated dayside having temperatures of approximately 160 K and as low as 20 K on the night side. In their work, Biver et al. (2015) also calculate the water column densities around the comet, finding the highest densities occur around the regions with the highest temperature, although the neck region is distinctly more productive than the hotter regions surrounding it (Marschall et al. 2019). This suggests that topography plays a key role in outgassing activity, in addition to solar insolation (Vincent et al. 2019).

Choukroun et al. (2015) used MIRO to assess the night-time temperatures on 67P from data taken between August 2014 and October 2014. The night-time sub-surface temperatures varied from 17–40 K in the sub-mm channel and 30–50 K in the mm channel. The mm channel probes deeper than the sub-mm channel, so the measured brightness temperature is a little higher than in the sub-mm channel, due to thermal inversion curve on the night side.

Beyond the early phase of the mission, continuum temperatures from MIRO covering August 2014 to April 2016 are given in Marshall et al. (2017). The observed sub-surface temperatures vary from 70 K to 255 K, with the highest temperatures recorded around perihelion when 67P was at 1.24 au from the Sun. In these measurements, it is unlikely that MIRO observed the hottest regions on the fully illuminated dayside as the spacecraft was in a terminator orbit during perihelion and so the maximum sub-surface temperature of the nucleus is probably higher than these recorded values.

**Philae Temperature Measurements** Philae came to rest at the final landing site Abydos on 12 November 2014, 17:31:17 UTC ($t = 0$ in the following text). Nine months before perihelion, the distance to the Sun was 3 au Several experiments and sensors on Philae provide information about the thermal environment at Abydos. These are the MUPUS (MUlti-PUrpose Sensors for surface and subsurface science) sensors MUPUS-TM (Thermal Mapper) and MUPUS-PEN (PENetrator) (Spohn et al. 2007), the SESAME (Surface Electric Sounding and Acoustic Monitoring Experiment) (Seidensticker et al. 2007) temperature sensors located in the feet of the lander, and Philae's housekeeping sensors and solar array data. Temperature data from the scientific sensors were not continuously acquired but only during certain operational phases. The most extended data set comes from MUPUS-TM that was operational during Block-1 ($t = 0$–11 h), the 4 safe blocks ($t = 13$–21 h), and the final operational Block-6 ($t = 29$–41 h), whereas the PEN was only operated in Block-6 (for a description of the operational blocks on the comet see Ulamec et al. 2016). SESAME temperature data were collected during the four safe blocks and in Block-6.

Figure 5 shows the diurnal cycle of the acquired temperatures after landing in Abydos. Data from different rotations are plotted as a function of local time. Here, we only show data that were not significantly affected by operational issues (i.e. due to self-heating) of the sensors and, therefore, provide real information about the thermal environment including the illumination history. This excludes SESAME data recorded with active accelerometers/transmitters and PEN data acquired before hammering was finished and the deployment

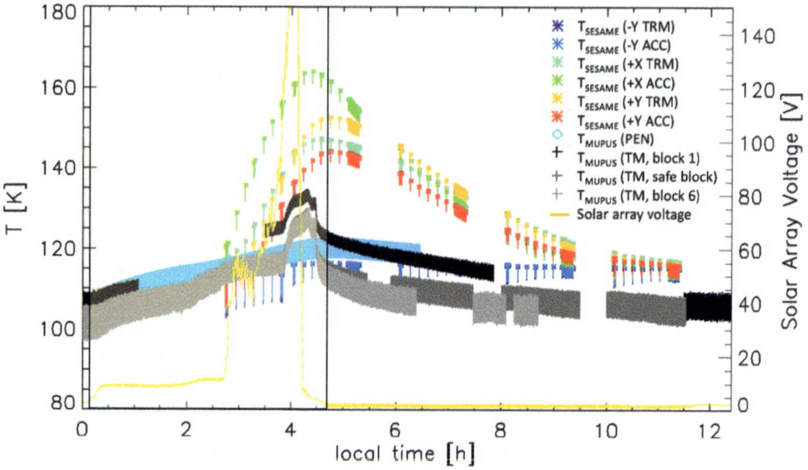


**Fig. 5** Temperatures of MUPUS-TM (brightness temperature of 6–25 μm broadband channel), MUPUS-PEN, and SESAME transmitters (TRM) and accelerometers (ACC) located in the three feet of Philae (designated as +X, −Y and +Y). Additionally, the solar array voltage (yellow line) is plotted as function of local time. The semi-transparent yellow box indicates the length of the local day (∼4.7 hours)

boom retracted as well as all PEN sensors located in the upper 2/3 of the PEN. Because the PEN was not able to penetrate into the surface (Spohn et al. 2015), it was either sticking with its tip in the ground or lying somewhere on the ground close to the big boulder behind the Philae balcony. Therefore, the measured PEN temperatures are mainly reflecting an average of its radiative environment. The SESAME sensors provide good information about the illumination at Abydos but it is difficult to judge how well they measure the temperature of the soil beneath the soles. Knapmeyer et al. (2018) showed that all lander feet were in contact with the ground at Abydos but the reaction to illumination is certainly different between the feet and the cometary ground. Detailed modelling of this problem has not yet been done.

With respect to the MUPUS-TM temperatures it should be noted that the temperatures shown in Fig. 5 are somewhat higher than those given by Spohn et al. (2015). The reason is a mistake in Eq. (2) of the supplementary material of Spohn et al. (2015) that relates the measured thermopile signal to the kinetic temperature. A corrected version of this equation and a discussion of the underlying assumptions can be found in Grott et al. (2017). The consequences of this correction are that no estimate of the comet emissivity is possible from a single TM channel, and that the plotted brightness temperature is a lower bound for the surface temperature. The uncertainty introduced by the unknown emissivity is small for the low target temperatures at Abydos. For the broadband TM channel used here deviations are below 2.2 K if the emissivity is larger than 0.9. This uncertainty is incorporated in the error bars shown in Fig. 5.

The measured temperatures are consistent with surface temperatures between about 100 K and at least 165 K varying over time and with the exact position in Abydos. The temperatures of the feet show that first the +X-foot (towards the observer in the OSIRIS image of the lander at Abydos, Ulamec et al. 2017) of the lander is illuminated, then the +Y foot, and that the −Y foot does not receive any direct illumination at all. The latter sensor gives nearly constant temperature readings of about 115 K, which should reflect the average temperature of the immediate environment of the −Y foot. It is further noteworthy that there is no discontinuity in the rising temperature curve of the SESAME sensors although different parts of the curve were recorded about 2 comet rotations apart, during Block-6 (Nov. 14,

$t_{loc} = 2.7$–$3.7$ h) and the first safe block (Nov. 13, $t_{loc} = 3.7$–$4.7$ h). This is an indication that no significant movement of the lander occurred in between. The temperatures measured by TM and PEN start slightly above 100 K and then rise slowly over 4 h to reach about 120 K. This temperature rise is probably due to diffuse radiation (reflected visible and thermal infrared) reaching the TM target area and the PEN surroundings after sunrise in the Abydos region. This interpretation is supported also by lander solar array data. Here, it should be noted that due to the design of the Philae solar power system the output voltage of the solar cells is more sensitive to weak illumination levels than the generated current. The output voltage summed up over all solar panels rises at $t_{loc} \approx 0$ to approximately 10 V indicating that scattered visible light reaches the Philae hood.

At $t_{loc} \approx 4$ h the TM signal shows a steep increase to a maximum of 130 K whereas the PEN temperature continues to rise slowly. This was interpreted by Spohn et al. (2015) to indicate that a part of the TM FOV receives direct sunlight at this point in time whereas the PEN remains in shadow. At $t_{loc} \approx 4.7$ h (dotted vertical line in Fig. 5) the night in Abydos begins as indicated by the solar array voltage dropping to zero. This is also reflected in the sharp decrease of the TM signal and in the temperature maximum reached at the +Y foot and the subsequent decay of all temperatures over the remaining part of the rotational period.

Independent of the diurnal cycle, there seems to exist a trend to lower temperatures with increasing time. TM temperatures recorded directly after landing (Block-1) are higher than the values measured one (safe block) and three (Block-6) rotational periods later, and also the temperature of the −Y foot recorded late during the first science sequence (in the diurnal curve between $t_{loc} = 2.7$ h and 3.7 h) is lower than during earlier measurements (though this is not fully conclusive due to the large error bar). Such a trend to lower temperatures could be a consequence of additional shadowing of Abydos by the lander itself.

## 2.2 Thermal Inertia and Conductivity

### *2.2.1 State of the Art Before Rosetta*

The thermal inertia drives the ability of a material to adapt its temperature to a change in local insulation, either resulting from day/night variations or shadowing effects. A material with a large thermal inertia takes longer to adapt its temperature compared to a low thermal inertia material. The thermal inertia $I$ (in $\mathrm{J\,K^{-1}\,m^{-2}\,s^{-1/2}}$) is defined as $I = \sqrt{\kappa\rho c}$, where $\kappa$ is the thermal conductivity (in $\mathrm{W\,K^{-1}\,m^{-1}}$), $\rho$ the density (in $\mathrm{kg\,m^{-3}}$), and $c$ the heat capacity (in $\mathrm{J\,K^{-1}}$). Since the thermal conductivity and the heat capacity are temperature dependent, the thermal inertia also depends on temperature; the thermal inertia of a given body changes depending on its heliocentric distance, which is particularly relevant for comets with highly elliptical orbits.

Estimating the thermal inertia has always been challenging, and until the flyby of comet 9P in 2005, mainly relied on in-situ (Vega) or remote infrared (ISO and Spitzer) and millimetre (IRAM) spatially unresolved observations of the nucleus to derive, at minimum its temperature, better the deviation of its spectral energy distribution from that of a black body, and even better a temporal shift in its thermal infrared light curve.

On comet 1P, the hot nucleus (Sect. 2.1.1), with a temperature close to that of a black body, suggested a low thermal inertia with negligible energy penetrating into the nucleus by heat conduction. From ISO observations at four wavelengths in the range 7–15 µm, Jorda et al. (2000) derived a negligible thermal inertia for comet C/1991 O1 Hale-Bopp, as well as Boissier et al. (2011) who derived a thermal inertia of $10\ \mathrm{J\,K^{-1}\,m^{-2}\,s^{-1/2}}$ for comet 8P/Tuttle using Plateau de Bure observations. Fernández et al. (2013) performed a survey of

57 cometary nuclei, using Spitzer two-wavelengths observations, and also concluded that the nucleus thermal inertia is negligible. More precisely, we converted their measured beaming parameter of $1.03 \pm 0.11$ to a thermal inertia of $30 \pm 20\ \mathrm{J\,K^{-1}\,m^{-2}\,s^{-1/2}}$, assuming a reasonable small-scale roughness ($\eta = 0.8$; see Lebofsky et al. 1986 and references therein for a discussion of the $\eta$ parameter). For this purpose, we assumed a spherical nucleus at 4.5 au from the Sun (where most Spitzer data were acquired) with no thermal inertia and $\eta = 1.03$, and computed its flux at 16 µm and 22 µm (Spitzer observational wavelengths); we then looked for the thermal inertia that best reproduces these fluxes, for the same nucleus but with $\eta = 0.8$. Finally, from Spitzer thermal infrared light curves of 67P and 9P, Lamy et al. (2008) and Lisse et al. (2009) respectively, came to the same conclusion, with a thermal inertia $<50\ \mathrm{J\,K^{-1}\,m^{-2}\,s^{-1/2}}$.

A second, less direct, approach has also been used, which is to derive the thermal inertia from the porosity, the latter being derived from density estimates (Sect. 4.2). Since the nucleus density can be inferred from non-gravitational or tidal perturbations, this method is the oldest one, but the least accurate. For example, in the 80's and 90's, the density was estimated to $600\ \mathrm{kg\,m^{-3}}$ for the nucleus of 1P (Sagdeev et al. 1988) and to 300–700 $\mathrm{kg\,m^{-3}}$ for that of D/1993 F2 Shoemaker-Levy 9 (SL9) (Asphaug and Benz 1994). This low density suggested a low conductivity, therefore a low thermal inertia, but the analysis is model dependent, therefore more qualitative than quantitative.

In parallel to the above observations, an effort has been made to constrain the thermal inertia of porous icy materials from laboratory experiments and modelling. From the experimental point of view, we can mention the KOSI experiments (Huebner 1991; Kochan et al. 1998) on porous ice/dust samples with a measured thermal conductivity of 0.3–0.6 $\mathrm{W\,m^{-1}\,K^{-1}}$, ten times lower than that of compact ice (Spohn et al. 1989). More recently, Krause et al. (2011) built an experiment, combined with numerical simulations, to measure the thermal conductivity of highly porous dust aggregates and obtained very low value of 0.002–0.02 $\mathrm{W\,m^{-1}\,K^{-1}}$. Assuming a nucleus with a density of $532\ \mathrm{kg\,m^{-3}}$ (Jorda et al. 2016) and a heat capacity of $770\ \mathrm{J\,kg^{-1}\,K^{-1}}$ (dust at 300 K; Winter and Saari 1969), the thermal conductivity range of 0.002–0.6 $\mathrm{W\,m^{-1}\,K^{-1}}$ translates into a thermal inertia in the range 28–495 $\mathrm{J\,K^{-1}\,m^{-2}\,s^{-1/2}}$. Models also indicate that the thermal conductivity decreases with increasing porosity, with several orders of magnitude difference between low ($<20\%$) and high ($>70\%$) porosity media (Prialnik et al. 2004, and references therein). While the above methods do not provide a direct constraint for the thermal inertia of the nucleus, they demonstrate however that a low thermal inertia is fully compatible with both experimental and theoretical studies on porous aggregates, from which comets are most likely made of.

The most recent estimates are provided by the spatially resolved data of the Deep Impact spacecraft, with the temperature measurements of comets 9P and 103P. From the observed temperature map of Fig. 2, Davidsson et al. (2013) estimated the thermal inertia of 9P from less than $50\ \mathrm{J\,K^{-1}\,m^{-2}\,s^{-1/2}}$ in rough pitted terrains to $200\ \mathrm{J\,K^{-1}\,m^{-2}\,s^{-1/2}}$ in smooth terrains, while Groussin et al. (2013) estimated the thermal inertia of 103P to less than $250\ \mathrm{J\,K^{-1}\,m^{-2}\,s^{-1/2}}$. The lack of temporal coverage (i.e. flyby) is the main limitation for more accurate estimates.

### 2.2.2 *Progress During Rosetta*

**VIRTIS Thermal Inertia Measurements** The huge amount of infrared spectra collected by the −M and −H channels of the VIRTIS instruments over the full Rosetta mission, has been used to investigate the surface thermal inertia. As VIRTIS was sensitive from 0.25 to

5 micron, the thermal radiation measured comes essentially from the first layers of the regolith (in principle, one millimetre maximum). Thus any VIRTIS temperature measurement can be associated to the surface thermal properties themselves.

One big advantage of the VIRTIS Rosetta observations over the previous space missions dedicated to comets studies is the large range of phases angles and local times covered by VIRTIS thanks to the complex orbit of Rosetta. This allows disentangling the surface thermal inertia effects from the surface roughness effects. The thermal emission of a rough surface is not isotropic (not Lambertian) and the same area observed at different phases angles presents different thermal observed fluxes that can be misinterpreted as a variation of the thermal conductivity. At very short wavelengths, typically the sensitivity range of VIRTIS, this effect is critical and affects significantly the Wien's part of the Planck function.

The surface thermal inertia was estimated across the whole surface using VIRTIS data acquired between September 2014 and December 2014, between 3.5 and 3.0 au heliocentric distance, when the activity was still relatively low. At that time, most of the Northern hemisphere was illuminated by the Sun while the Southern part was experiencing winter polar night. The complex geometry of the nucleus required considering topographic effects, which include mutual shadowing, mutual heating, spin rate changes over time. Synthetic thermal spectra were generated assuming heat transfer through the subsurface by conduction, and including some roughness properties both at the nucleus shape model resolution scale (few meters) and at sub-pixel scale (Kuehrt et al. 1992). For about 20 geomorphological regions, 200 spectra were randomly selected and compared to the synthetic ones.

In general, the thermal inertia remains in the range 10–170 $\mathrm{J\,K^{-1}\,m^{-2}\,s^{-1/2}}$ (Leyrat et al. 2015), similar to values found on other comets (see Sect. 2.1.1). The surface temperature increases in average by 0.5 $\mathrm{K\,min^{-1}}$ just after being illuminated by the Sun. Interestingly, two groups of surfaces were identified: the smooth terrains (dusty units like Imhotep, Hatmehit, etc...) present very low thermal inertia ($<30\ \mathrm{J\,K^{-1}\,m^{-2}\,s^{-1/2}}$) while rough consolidated terrains with apparent fractures seem to conduct heat more efficiently, with higher thermal inertia values ($>110\ \mathrm{J\,K^{-1}\,m^{-2}\,s^{-1/2}}$). Since the thermal inertia depends on density and porosity (e.g. for the thermal conductivity), the change in thermal inertia could result mainly from the change in the material properties, from consolidated to unconsolidated, rather than from a change in material composition.

Despite the very smooth texture of the Hapi region, this area presents two interesting "anomalies": first, the Hapi region is about 30 to 40 K warmer than expected, and second, its thermal inertia reaches intermediate values (60 $\mathrm{J\,K^{-1}\,m^{-2}\,s^{-1/2}}$). The high temperatures can be easily explained by the mutual heating due neighbours regions Hathor and Seth that heat Hapi even in the night. A possible explanation for the medium thermal inertia is the role played by the ice trapped inside the regolith and detected by VIRTIS (De Sanctis et al. 2015). Water gas has condensed near the cold surface during night, which increases the effective thermal inertia. Ice favours a better contact between grains at the surface, and thus a better heat conduction.

A clear correlation exists between the distribution of local gravitational slopes (Groussin et al. 2015) and the thermal inertia (Leyrat et al. 2015). Dusty units with very flat surfaces present in general low thermal inertia, while consolidated terrains with very steep slopes are consistent with very high thermal inertia values. This behaviour indicates that low thermal inertia areas consist of fluffy debris accumulated that come from other sources: the Southern hemisphere (Keller et al. 2017) during its very short and strong summer close to perihelion, and the ejecta of neighbouring units created by crack and thermal stress (Groussin et al. 2015). On the highly inclined surfaces, the regolith cannot remain stable and it is moved away, allowing the denser sub-surfaces to be exposed directly to solar light.

More recently, a more detailed analysis was initiated using an accurate shape model of the nucleus and dividing each geomorphological area in sub-regions, in particular to better understand the variations of thermal properties within the large units (Leyrat et al., private communication). Preliminary results confirm the averages values already found for the thermal inertia, with local variations from 5 to 350 $\mathrm{J\,K^{-1}\,m^{-2}\,s^{-1/2}}$.

**MIRO Thermal Inertia Measurements** As described in Sect. 2.1.2, MIRO is capable of measuring two subsurface temperatures from different depths. Schloerb et al. (2015) estimated that the measured sub-mm and mm temperatures arise from depths of 1 cm and 4 cm below the surface, using a model of the nucleus thermal emission that takes into account thermal inertia and absorption properties at the MIRO wavelengths (for lunar soils and cometary analogs). With the MIRO dataset, several attempts have been made to estimate the thermal inertia of the comet subsurface.

The most commonly used method to calculate the thermal inertia from MIRO measurements has two steps. First, the temperature profile in the nucleus subsurface must be found, by solving the 1D heat transfer equation for the propagation of energy into the nucleus. The temperature profiles are then found for a range of thermal inertia values. In the second step, the resulting profiles serve as the input to a radiative transfer model that computes the simulated brightness temperatures. Comparing the simulated brightness temperatures to the measured brightness temperatures from MIRO enables the best estimate for thermal inertia to be found from the range of values that went into solving the heat conduction equation. This method is used by Gulkis et al. (2015), Choukroun et al. (2015), Schloerb et al. (2015) and Marshall et al. (2018).

Using MIRO measurements from August 2014, Gulkis et al. (2015) found the thermal inertia to be low, with values in the range 10–50 $\mathrm{J\,K^{-1}\,m^{-2}\,s^{-1/2}}$. Choukroun et al. (2015) analysed data from the polar night side obtained between August–October 2014 and also found low thermal inertias, 10–40 $\mathrm{J\,K^{-1}\,m^{-2}\,s^{-1/2}}$ in the sub-mm channel and 20–60 $\mathrm{J\,K^{-1}\,m^{-2}\,s^{-1/2}}$ in the mm channel. To explain the difference in thermal inertia between the two channels, the authors suggest that ice may be present near the surface layer, which would extend the electrical penetration depth for each channel. This brings the data and best fitting values into better agreement. Work by Schloerb et al. (2015) with data from September 2014, gives 10–30 $\mathrm{J\,K^{-1}\,m^{-2}\,s^{-1/2}}$ for the thermal inertia. With a value of 22 $\mathrm{J\,K^{-1}\,m^{-2}\,s^{-1/2}}$, they check their modelled value against diurnal phase curves for the temperature and find good agreement between the data and model. Finally, Marshall et al. (2018) also analyse data from September 2014 and calculate bounds for the thermal inertia in five spots on the nucleus surface. Results in the mm channel imply that the thermal inertia is $<80$ $\mathrm{J\,K^{-1}\,m^{-2}\,s^{-1/2}}$. The sub-mm channel gives similar results to the mm channel, but also imply that the thermal inertia could be higher, although this may be an artefact due to the limited available constraints on the electrical skin depth.

**Philae Thermal Inertia Measurements** The first estimation of the local thermal inertia at Philae's final landing site was given by Spohn et al. (2015), based on MUPUS-TM measurements and thermophysical modelling. Without the availability of a local digital terrain model (DTM) at the time of data processing, information about illumination conditions had to be inferred from other sources. While the MUPUS-PEN did not intrude into the cometary material, above-surface measurements could be used for comparison with TM measurements. The fact that the PEN did not capture a peak in temperature, evident in TM data, led to the conclusion that only a fraction of the surface in the field-of-view of the TM was illuminated by direct sunlight for a short duration of about 40 min, whereas indirect illumination by scattered visible light and thermal re-radiation from the environment was assumed to be responsible for the observed slow increase in brightness temperature after sunrise in the Abydos region (see also the discussion in Sect. 2.1.2). This interpretation has later been confirmed by Kömle et al. (2017), upon the availability of a local DTM in combination with the coarser global shape model of the comet.

In their modelling, Spohn et al. (2015) further assumed that the indirect fraction of the total illumination was proportional to the solar illumination of the anticipated landing region, which was modelled by the orientation of a single facet of the global shape model SHAP4 (Jorda et al. 2016) available at that time. Using these assumptions, Spohn et al. (2015) inferred a local thermal inertia at Abydos of $85 \pm 35\ \mathrm{J\,K^{-1}\,m^{-2}\,s^{-1/2}}$. Unfortunately, the derived temperature curve used for this analysis suffers from an error in the conversion between measured TM brightness temperature and kinetic surface temperature as already discussed in Sect. 2.1.2. This led to a distortion of the diurnal temperature curve, more pronounced in the low temperature regime. Therefore, the corrected temperatures shown in Fig. 5 are now about 8–10 K higher for the lowest temperatures encountered, whereas the difference is only about 2 K for the peak temperatures of approximately 130 K. This flattening of the temperature curve results in a higher thermal inertia. By comparison with the temperature curve and the error bars in Fig. 2 of Spohn et al. (2015), one gets a corrected thermal inertia of at least $120\ \mathrm{J\,K^{-1}\,m^{-2}\,s^{-1/2}}$. Assuming that the surface is a dust/ice mixture with a density of $500\ \mathrm{kg\,m^{-3}}$ and a heat capacity of $<540\ \mathrm{J\,kg^{-1}\,K^{-1}}$ (corresponding to a dust/ice mass ratio $>2$ at 120 K, see Winter and Saari 1969; Herman and Weissman 1987), this value for the thermal inertia implies a thermal conductivity $>0.05\ \mathrm{W\,m^{-1}\,K^{-1}}$.

The local value determined at Abydos seems compatible with those deduced from remote sensing observations (Sect. 2.2.2) though it is more on the high side of all derived values. In this respect it could play a role that Abydos lies on the consolidated region of Wosret and is part of a rough talus with numerous boulders (Poulet et al. 2016). The landing site seems to be free of dust (at least down to the 1 mm resolution of the cameras), which could mean that the determined thermal inertia value is representative of that of boulders. Poulet et al. (2016) also found that the texture at Abydos falls into two classes, pebbles in the size range 3–6 mm, and rough material (unresolved, size $<1$ mm) in between. These findings fit well with the updated lower bound for the thermal conductivity derived from the MUPUS-TM data. Applying the theory of Sakatani et al. (2017, 2018) for the thermal conductivity of powdered material, it is found that for the very low temperatures encountered at Abydos, grain radii $<1$ cm, and reasonable comet porosities $>70\%$, the surface energy driven adhesion and radiation are not sufficient to explain the effective thermal conductivity $>0.05\ \mathrm{W\,m^{-1}\,K^{-1}}$. Some sort of cementation or sintering seems to be required to explain the relatively high thermal conductivity derived for Abydos.

It should be noted that the derivation of the Abydos thermal inertia described above is based on rather simplifying assumptions, the strongest one being the assumption that the indirect energy input is directly proportional to the average solar illumination of the environment in Abydos. This may be an oversimplification of the complex local topography, and, furthermore, it requires a low thermal inertia for the surroundings to be realistic (since the method implicitly assumes that not only the visible but also the infrared energy input to the TM measurement spot ceases after sunset). Attempts to apply more realistic and sophisticated models to the problem were performed more recently. Kömle et al. (2017) investigated the illumination and thermal environment at Abydos using a 3D thermal model and geometry based on a combination of (at that time) available global and local digital terrain models. Pelivan (2018) coupled a 1D thermal model with a high-resolution geometrical model and applied it to the Abydos region. These two models can reproduce some of the observed features like the approximate length of illumination in the Abydos region and the duration

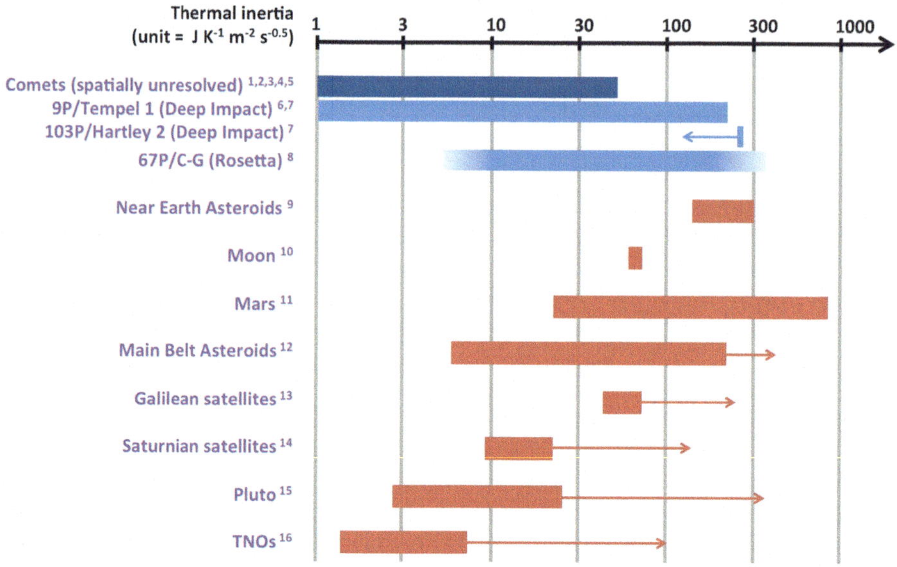


**Fig. 6** Synthesis of the thermal inertia of cometary nuclei compared to other solar system bodies. For 103P, the blue arrow indicates the upper limit ($<250\ \mathrm{J\,K^{-1}\,m^{-2}\,s^{-1/2}}$). For 67P, the blue bar indicates most estimates ($10$–$170\ \mathrm{J\,K^{-1}\,m^{-2}\,s^{-1/2}}$) with extreme values in colour gradient ($<10$ or $>170\ \mathrm{J\,K^{-1}\,m^{-2}\,s^{-1/2}}$). The red arrows indicate the corrected value at 1 au for bodies beyond Mars, using power law dependence with heliocentric distance $r_h^{-0.75}$ (Delbó et al. 2007). References: (1) Jorda et al. (2000), (2) Lamy et al. (2008), (3) Lisse et al. (2009), (4) Boissier et al. (2011), (5) Fernández et al. (2013), (6) Davidsson et al. (2013), (7) Groussin et al. (2013), (8) Marshall et al. (2018), (9) Delbó et al. (2007), (10) Hayne et al. (2017), (11) Mellon et al. (2000), (12) Delbó and Tanga (2009), (13) Morrison and Cruikshank (1973), (14) Howett et al. (2010), (15) Lellouch et al. (2017), (16) Lellouch et al. (2013)

of the direct illumination of the TM spot but so far the measurements could not be reproduced quantitatively. This can mostly be attributed to the imperfections in the detailed local geometry of the complex topography at Abydos that was used in these works.

To conclude, the value of the thermal inertia at Abydos is not yet finally determined, and more modelling work should be done to derive accurate thermal properties at Abydos in the future, in particular if better local DTMs become available. Furthermore, in future models, the constraints from all Philae data should also be taken into account. This comprises the available imaging and DTM information from CIVA and ROLIS cameras, but also the housekeeping data from the lander (Sect. 2.1.2).

## 2.3 Synthesis on Nucleus Thermal Properties

The thermal properties of cometary nuclei are characterised by a thermal inertia in the range 0–350 $\mathrm{J\,K^{-1}\,m^{-2}\,s^{-1/2}}$. When the nucleus is unresolved, the thermal inertia is estimated to less than 50 $\mathrm{J\,K^{-1}\,m^{-2}\,s^{-1/2}}$, which corresponds to a low thermal conductivity of 0–0.006 $\mathrm{W\,m^{-1}\,K^{-1}}$ assuming a nucleus with a density of 532 $\mathrm{kg\,m^{-3}}$ (Jorda et al. 2016) and a heat capacity of 770 $\mathrm{J\,kg^{-1}\,K^{-1}}$ (dust at 300 K; Winter and Saari 1969). Spatially resolved observations show variations across the surface, depending on the nature of the terrain: on 67P, smooth terrains covered by deposits usually have a lower thermal inertia ($<30\ \mathrm{J\,K^{-1}\,m^{-2}\,s^{-1/2}}$) than terrains exposing consolidated materials ($>110\ \mathrm{J\,K^{-1}\,m^{-2}\,s^{-1/2}}$). As illustrated by Fig. 6, comets are among the bodies of the solar system with the lowest thermal inertia, taking into account the fact that the thermal inertia is a function of temperature and therefore decreases with heliocentric distance $r_h$ for a given body (Rozitis et al. 2018) with a power law of $r_h^{-0.75}$ (Delbó et al. 2007).

The values of thermal inertia derived from MIRO data are globally lower than those derived from VIRTIS (Sect. 2.2.2) and MUPUS data (Sect. 2.2.2). This apparent discrepancy could be real, in which case there would be a decrease of the thermal inertia with depth in the first centimetres. The discrepancy may also result from the thermal models used by the different authors, which neglect the sublimation of ices and the cooling by outgassing (i.e., they use "asteroid-like" thermal models) and therefore overestimate the temperature inside the nucleus, resulting in an underestimation of the thermal inertia. This effect is more pronounced for MIRO than for VIRTIS and MUPUS, since MIRO probes greater depths. Finally, it is worth mentioning that thermal inertia is temperature dependant and is therefore expected to vary diurnally and seasonally across the surface for a given area.

Overall, a thermal inertia of 0–60 $\mathrm{J\,K^{-1}\,m^{-2}\,s^{-1/2}}$ is probably the best estimate for the bulk value, which corresponds to a low thermal conductivity of 0–0.009 $\mathrm{W\,m^{-1}\,K^{-1}}$ with our previous assumptions on density and heat capacity. This bulk value is consistent with the range of 0–50 $\mathrm{J\,K^{-1}\,m^{-2}\,s^{-1/2}}$ derived from spatially unresolved observations, which by definition is an average over the nucleus surface, and moreover comes from a large sample of ~60 cometary nuclei (see references in Table 1). This bulk value is also consistent with the range of 10–60 $\mathrm{J\,K^{-1}\,m^{-2}\,s^{-1/2}}$ derived from MIRO observations of the night side of 67P (Choukroun et al. 2015); we consider this estimate as particularly reliable since the nucleus is in a purely cooling conductive regime at that time (no sun light), which simplifies the number of hypothesis in the analysis (e.g. no issues with solar insulation, roughness, surface reflection and multiple scattering). Finally, larger values are found on exposed consolidated terrains, which are usually harder, denser and less porous than the bulk nucleus (Sects. 3 and 4).

Because of the low thermal inertia, the nucleus temperature decreases very rapidly with depth, from hundreds of Kelvin to tens of Kelvin in the first meter. In addition, the erosion rate of the nucleus is, on average, comparable to the propagation velocity of the heat wave, i.e. typically decimetres to meters at each perihelion passage (Prialnik et al. 2004; Gortsas et al. 2011; Keller et al. 2015). The low thermal inertia and rapid erosion provide a good thermal insulation of the nucleus interior, which remains unaffected by solar insultation below the first meter (Capria et al. 2017), and leads to the conclusion that comets could still hold primordial materials, unaffected since their formation billions of years ago. This is mostly relevant for small comets (diameter <10 km) formed by the slow (~3 Myr) hierarchical agglomeration of materials, likely those of Fig. 1, whose internal temperature never exceeded 100 K since their formation, even in the presence of radioactive heating, allowing them to keep their super volatiles species (e.g. CO) trapped in the amorphous water ice (Davidsson et al. 2016).

## 3 The Nucleus Mechanical Properties

The mechanical properties of a comet nucleus drive its ability to respond to an internal or external stress. They are key to properly describing many processes that occurred during the history of a comet, including accretion, encounters with giant planet, or collisions. The mechanical properties can be modified by internal heating and phase transitions, and therefore also provide constraints on these processes. The mechanical properties are described by the tensile, shear and compressive strengths (Sect. 3.1), and by the elastic properties (Sect. 3.2).

By definition, in solid-state physics, the tensile strength ($\sigma_T$ in Pa) is the maximal mechanical tension that can build up inside a material before it breaks, the shear strength ($\sigma_S$ in Pa) is the maximum shear load that can be applied to a material without failure, and the compressive strength ($\sigma_C$ in Pa) is the maximum load that can be applied to a material without changing the size of the sample. Usually, $\sigma_T < \sigma_S < \sigma_C$ for ductile and brittle materials.

## 3.1 Strengths

### *3.1.1 State of the Art Before Rosetta*

Before the Rosetta mission, the tensile, shear and compressive strengths of the cometary material had been determined using observational constraints, laboratory experiments and modelling.

From the observational point of view, it is, a priori, possible to estimate the strength of the nucleus when it experiences a mechanical stress. This stress can be of various origins, including the nucleus rotation (i.e. resistance to centrifugal forces), the encounter with a giant planet or the Sun (i.e. resistance to tidal forces), or an impact (i.e. resistance to external mechanical forces). From the rotational period of several nuclei, Davidsson (2001) and Toth and Lisse (2006) estimated the nucleus tensile strength to be 1–100 Pa. From the close encounter of comet SL9 with Jupiter, Asphaug and Benz (1996) estimated its tensile strength to 5 Pa, while Klinger et al. (1989) estimated that of a 1 km radius sungrazing comet to 100 Pa. The Deep Impact mission, with a controlled impact of a 364 kg impactor at 10.3 $\mathrm{km\,s^{-1}}$ on comet 9P, resulted in a measured shear strength of <65 Pa (A'Hearn et al. 2005). This value is however uncertain since Holsapple and Housen (2007) showed that the Deep impact experiment is compatible with any shear strength in the range 0–12 kPa. From all the above constraints, it however emerges the picture of a very weak nucleus, with a bulk tensile strength most probably lower than 100 Pa. As we will see in the next two paragraphs, this low strength is supported by both experimental and theoretical studies.

From laboratory experiments, it is possible to measure the strengths of cometary material analogs, with the strong limitation that we do not know well the nature of the cometary material, therefore the ground truth of the studied analogs. Various analogs made of water ice grains and/or dust grains have been studied, with an estimated tensile strength in the range 2–4 kPa for pure 200 micron size water ice grains (Bar-Nun et al. 2007) and in the range 1–10 kPa for micrometre size siliceous grains (Blum et al. 2006). The compressive strength has also been estimated to 0.3–1 MPa by Jessberger and Kotthaus (1989) for micron size low density aggregates of water ice and dust grains. Recent experiments are discussed in Sect. 3.1.2 and show that dust aggregates can have a tensile strength of only 1 Pa (e.g. Blum et al. 2014).

Theoretically, it is possible to calculate the strength of an aggregate from the van der Waals forces, depending on the physical properties of the aggregate grains (size, shape, porosity, contact area, ...). Using this method, Greenberg et al. (1995) derived a tensile strength of 270 Pa for a highly porous (80%) aggregate of micron size dust/ice grains, while Kuehrt and Keller (1994) derived a tensile strength of ~10 kPa between two micron size grains. Finally, we can mention the work of Biele et al. (2009) who derived a tensile strength of 5 kPa for a material similar to that of Greenberg et al. (1995). More details on modelling and results from recent works are presented in Sect. 3.1.2.

While the bulk tensile strength of the nucleus is likely low (<100 Pa), there exist variations across the surface and with depth. The unconsolidated materials and fine deposits that cover a large fraction of the surface have, by definition, a lower strength than the consolidated materials from which, for example, boulders and cliffs are made of. Moreover, Jessberger and Kotthaus (1989), Grün et al. (1991), Thomas et al. (1994), and Pommerol et al. (2015) have shown experimentally that a hard (typically 1 MPa) layer of water ice can be produced by sublimation/redeposition cycles and/or sintering of water ice close to the surface, a result consistent with a modelling performed by Kossacki et al. (2015). This processed layer is discussed in more detail below (Sect. 3.1.2), in particular for the interpretation of the Philae measurements.

Finally, it is important to note that the strength is a scale dependent parameter, i.e. the same material has a different strength at the macroscopic and microscopic scale. The general trend is a decreasing of the strength with increasing scale, so that, for example, micron size aggregates have a larger strength than a kilometre body made of such aggregates. The scaling law is material dependent, which makes the comparison between the strengths determined from observations, laboratory experiments and modelling difficult.

### *3.1.2 Progress During Rosetta*

**Strengths from Remote OSIRIS Observations** Images of the nucleus of 67P acquired by the OSIRIS cameras (Keller et al. 2007) show a varied and complex morphology including surface features with sharp topography, such as pits, cliffs, overhangs and fractured consolidated material, suggestive of non-zero material strength.

Groussin et al. (2015) estimated the strength of collapsed surface features, i.e. the tensile strength that was presumably overcome by the weight of overhanging material in order to collapse, and a compressive strength scaled from this. They estimate the consolidated material to have tensile strength 3–15 Pa and compressive strength 30–150 Pa at 5–30 m scales. They also estimate the shear strength or cohesion needed to hold meter sized boulders on slopes as 4–30 Pa, and to resist the lateral pressure at the bottom of the 900 m high Hathor cliffs as >30 Pa.

Basilevsky et al. (2016) performed a similar analysis of overhangs and landslides, resulting in comparable strength estimates. Vincent et al. (2017) also measured the cohesion needed to prevent the collapse of cliffs, finding 1–2 Pa strengths at tens of metre scales globally. They then suggest weakening by sublimation effects dominate over collapse under self-gravity.

Attree et al. (2018a) performed a survey of overhanging cliffs, using a full 67P shape model (Preusker et al. 2017) with calculated gravity vectors to identify facets with >90° local slope. They then examined twenty features in detail, measuring a profile of each and deriving a minimum tensile strength estimate. Strengths ranged from 0.02–1.02 Pa, with no apparent correlation with position on the nucleus head or body. Attree et al. (2018a) argue that the presence of debris at the base of nearly all these overhangs and the observed collapse of two, as well as their depths exceeding likely zones of thermal processing (based on thermal modelling), confirms the comet's low global bulk tensile strength.

Despite these low strengths, the prevalence of fractures on 67P indicates a material with sufficient strength or stiffness to behave in a brittle way. Fractures are present at all scales in OSIRIS images, from hundreds of metres down to tens of cm (Thomas et al. 2015; El-Maarry et al. 2015; Pajola et al. 2015; Matonti et al. 2019); and cm and below scale in images from the Philae lander (Poulet et al. 2016). They may be caused by stresses associated with rotational and shape effects, tidal forces, collisions and thermal cycling. Metre-scale polygonal fracture networks, in particular, resemble similar polygon features on Earth and Mars (Auger et al. 2018) and are probably associated with thermal stresses, as discussed below.

Reconciling low bulk strengths and granular materials with stiff/brittle surface features is a difficult challenge that most likely has its answer in the stratigraphy and hardening and sintering processes described in the next sections. Groussin et al. (2015) also note, finally, that the ratio of the derived material strengths to surface gravity for 67P is similar to that of Earth, which might explain the perceived morphological similarities.

**Strengths from in-Situ Philae Measurements** Despite the unintended excursion of Philae after its touchdown at the Agilkia landing site, the lander could contribute to the investigation of cometary strength in several ways—it was even possible to instrumentalise the bouncing itself in this regard. The associated measurements and observations occurred sequentially, thus we summarise the results in chronological order.

The images of Agilkia taken by the ROLIS camera (Mottola et al. 2007) confirmed the interpretation of earlier OSIRIS images that many subhorizontal surfaces of 67P are covered by a granular material, made of particles bonded so weakly that moat- and wind-tail-like structures can form around boulders under airfall of particles ejected from elsewhere (Mottola et al. 2015). The presence of boulders of several meters on the other hand shows that the granular medium is a mere cover.

All three accelerometers of CASSE were set to triggered mode for the landing and recorded the individual touchdowns of Philae's three feet at Agilkia. Applying Hertzian contact mechanics to the deceleration time series, a compressive strength between 3.5 kPa and 12 kPa could be derived from +Y-foot data, which hit the ground first, likely on one of the boulders at the Agilkia site (Möhlmann et al. 2018).

After touchdown, Philae bounced off and started drifting across the comet. Images taken by OSIRIS before and after the touchdown show fresh excavations up to 20 cm deep in the granular medium, resulting from the bouncing. Biele et al. (2015) find from a parameter study that the uniaxial compressive strength of the material is unlikely to exceed $\sim$10 kPa, since otherwise the depth of the excavations would be reduced to millimetres. They also argue that a layer of much larger strength is located below these surface materials. The depth of the footprints at the same time poses a lower limit for the thickness of the granular layer. Roll et al. (2016) further elaborate this evaluation and find an average compressive strength of 2 kPa, or an increase of 3 $\mathrm{kPa\,m^{-1}}$, both results being valid for the depth the soles penetrated into the ground.

At the final Abydos landing site, a non-granular, fractured surface was found (Bibring et al. 2015). MUPUS attempted to hammer its thermal probe, a 33 cm carbon rod with metal tip, into this material but failed to penetrate. A comparison with laboratory tests shows that an uniaxial compressive strength of the material of at least 2 MPa is required to frustrate the penetration of MUPUS (Spohn et al. 2015).

MUPUS nevertheless hammered for more than three hours, and CASSE recorded 14 of its hammer strokes (Knapmeyer et al. 2018, and references therein). The dispersion of recorded surface waves indicates a decrease in shear wave velocity with depth, suggesting that strength is depth dependent (see also Sect. 3.2).

**Strength from Laboratory Experiments** Due to their formation history, cometary surfaces are believed to consist of granular material possessing very high porosities ($\sim$80%; Sierks et al. 2015) and a composition made of ices, organic materials and minerals (Filacchione et al. 2019). Two different configurations of the cometary material as a result of the formation and evolution of the cometary nucleus are currently under debate (Fig. 7): homogeneous dust layers (see, e.g., Davidsson et al. 2016) versus aggregate layers (see, e.g., Blum et al. 2017). These two different configurations have a strong influence on the strength of

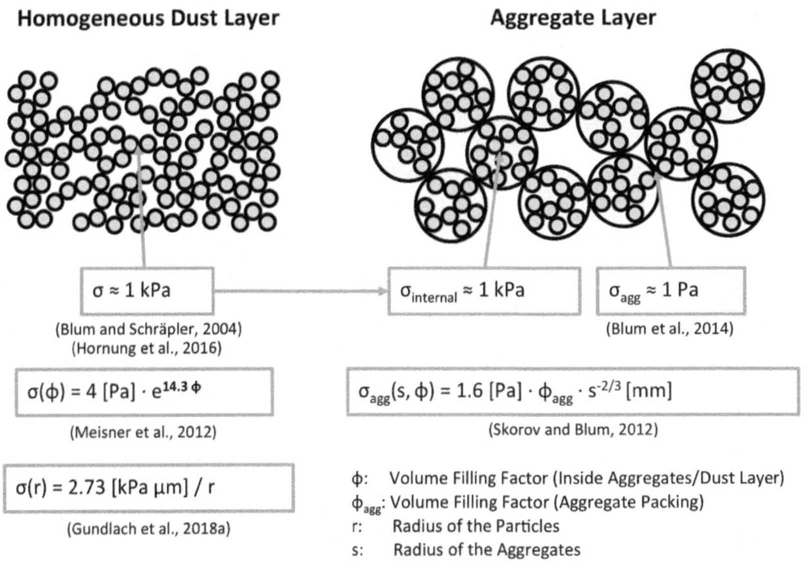


**Fig. 7** Summary of the main laboratory results (including the COSIMA findings). The corresponding references are provided below the equations

the surface material, such as the tensile and the compressive strength. In the past, laboratory experiments were utilised to investigate the strength of granular materials under cometary conditions.

The tensile strength of homogeneous dust layers (composed of silica particles with a diameter of 1.5 μm) was measured by Blum and Schräpler (2004). They found that the dust layers (porosity ~0.8) possess a tensile strength of ~1 kPa. Meisner et al. (2012) utilised the Brazilian Disc Test to measure the tensile strength of centimetre-sized dust samples (composed of irregular silica particles in the 0.1 to 10 μm-size regime) and estimated values in the range from 1 to 5 kPa. Furthermore, a relationship between the tensile strength, $\sigma_T$, and the volume filling factor, $\varphi$, was derived, which reads, $\sigma_T = 4\ \text{Pa} \cdot \exp(14.3\varphi)$. The same approach was used by Gundlach et al. (2018b; their Fig. 4) to study the influence of the grain size on the tensile strength of granular materials. The result was that the tensile strength is decreasing linearly with increasing grain size. Furthermore, the tensile strength of granular water ice was estimated to $0.9 \pm 0.7$ kPa, for samples consisting of spherical poly-disperse particles with a mean radius of $2.38 \pm 1.11$ μm and a porosity of 0.5. This value is ten times lower than expected based on the increased specific surface energy of water ice compared to silica (Gundlach et al. 2011). The only explanation for this behaviour is that water ice possesses a similar specific surface energy as silica at low temperatures ($0.02\ \text{J}\,\text{m}^2$). The results of laboratory experiments investigating the tensile strength of homogeneous granular materials are in good agreement with astrophysical measurements such as the breakup of cometary dust particles observed with the COSIMA experiment onboard the Rosetta spacecraft (~1 kPa; Hornung et al. 2016) and with the derived tensile strength from meteor streams breakup in the Earth atmosphere (0.4–150 kPa by Blum et al. 2014; 40–1000 Pa by Trigo-Rodríguez and Llorca 2006).

Thus, the tensile strength of homogeneous dust layers can be used to describe the internal tensile strength of dust aggregates (sub-decimetre-sized objects composed of micrometre-sized particles). Measurements of the tensile strength of packings of aggregates have, how-

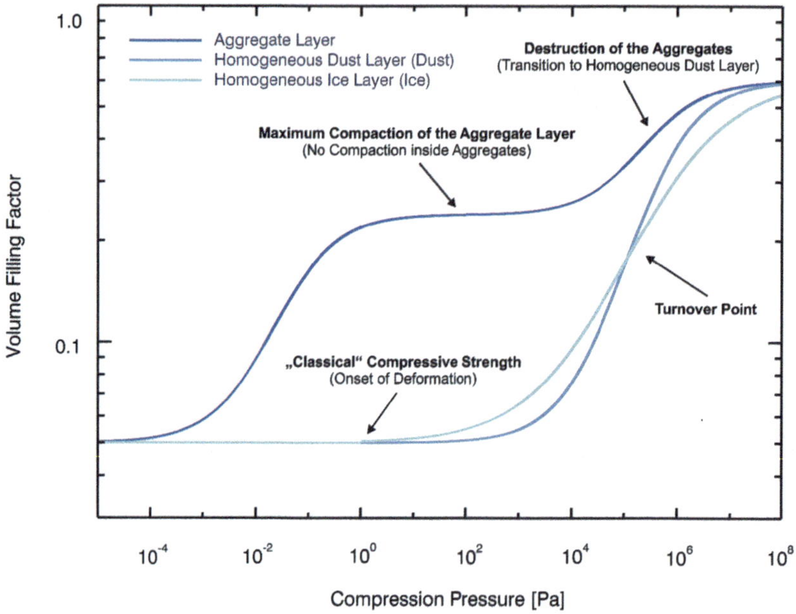


**Fig. 8** Compressive strength of granular materials measured in the laboratory: aggregate layers (Schräpler et al. 2015), homogeneous dust layers (Güttler et al. 2009) and homogeneous water-ice layers (Lorek et al. 2016). Since all the samples start with a different initial volume filling factor, the compression curves were normalised to an initial and a maximum volume filling factor of 0.05 and 0.6, respectively. The first compaction is due to the rearrangement of the structure, aggregates are rolling, etc... Later the aggregates are destroyed as indicated in the figure

ever, revealed much lower values, in the order of only ∼1 Pa (Blum et al. 2014). Furthermore, the measurements have shown that the tensile strength of aggregate packing, $\sigma_{agg}$, decreases with aggregate size, $s$, with the relation, $\sigma_{agg} \sim s^{-2/3}$, as predicted by Skorov and Blum (2012). It is therefore important to note that the tensile strength of a granular material can vary by at least three orders of magnitude just by different arrangements of the material (homogeneous dust layers versus aggregate packings).

In contrast, the compressive behaviour of granular materials cannot be expressed by just a single value. In solid-state physics the compressive strength is a measure for the maximum load that can be applied to a material without changing the size of the sample. However, granular materials have the tendency to react with deformation also if very small loads are applied. Thus, a better description of the behaviour of granular materials under compression is given by "compression curves". While compressed, the size change leads to an increase of the volume filling factor of the material. Figure 8 shows the compression curves of different granular materials measured in the laboratory (Güttler et al. 2009; Schräpler et al. 2015; Lorek et al. 2016). The onset of deformation and the turnover point are characteristics of S-type functions that can be used to derive material properties such as the rolling friction force and, therewith, the specific surface energy, or the particle radius (see Eqs. 5–7 in Schräpler et al. 2015). In order to derive material properties from laboratory, or spacecraft measurements it is mandatory to either derive the entire compression curve, or to estimate the change of the volume filling factor of the material by, e.g., a measurement of the intrusion depth, if only single data points can be acquired. In this context, an application to

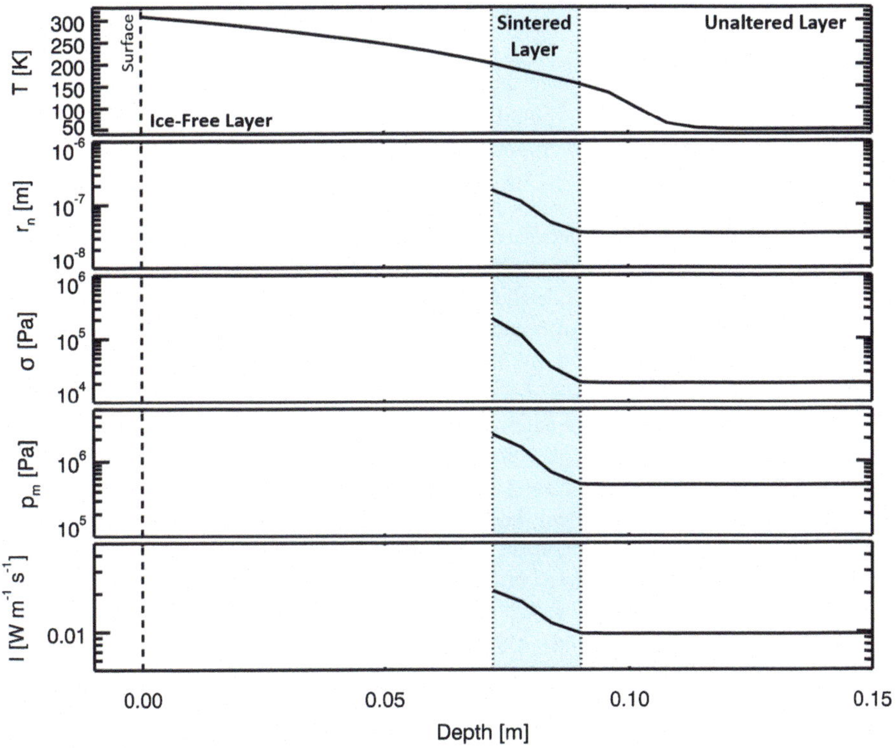


**Fig. 9** Hardening of the cometary surface by sintering of micrometre-sized water-ice particles. The first panel shows a typical temperature distribution (T) inside the surface layers of the cometary nucleus derived from a thermophysical model developed by Blum et al. (2017). In the second panel, the evolved sinter neck ($r_n$) between the ice particles is shown (based on the sinter model described in Gundlach et al. 2018a). The next three panels are visualising the change of the tensile strength ($\sigma$), the change of the turnover point of the compression curve ($p_m$) and the change of the thermal conductivity (l)

the MUPUS findings (>2 MPa; Spohn et al. 2015) to derive cometary material properties is very challenging, not mentioning that the compression curve is also scaled dependent.

Furthermore, hardening of the cometary surface layers can occur by, e.g., sintering of water ice, a process that transports molecules from the particles in contact into their neck region (Kossacki et al. 1994). The growth of the sinter neck can significantly increase the strength of the material as shown by Fig. 9 (derived from the sinter model provided by Gundlach et al. 2018a). However, it is important to note that the sintering is only a short-term effect because sublimation will always lead to evaporation of the contact area. For comets, the sinter timescale is faster than the erosion process, but the sinter neck has a lifetime of $10^{-2}$–$10^3$ s, depending on the ice temperature (220–160 K, respectively; Fig. 12 in Gundlach et al. 2018a). Thus, the sinter neck can form in ice-rich areas at these temperatures, but the increase of the tensile and the compressive strength are only temporary.

Such a hard layer should experience significant thermo-mechanical stresses due to seasonal and diurnal temperature changes, a process which may be an important source of weathering on asteroids, the moon and other airless bodies (Delbó et al. 2014). The large temperature variations experienced by cometary surfaces can induce stresses exceeding the tensile strength of solid water ice (Kuehrt 1984; Tauber and Kuhrt 1987), suggesting they are responsible for at least some of the fractures observed on the surface of 67P (Auger et al. 2018).

Attree et al. (2018b) investigated this with a thermo-viscoelastic model, based on that used for frozen Mars soil. They found the seasonal temperature cycle on 67P to induce large thermal stresses, of up to several tens of MPa, down to tens of centimetres to ~metres depths, proportional to soil thermal inertia and ice content. This assumed a relatively stiff ice-bonded layer (see Sect. 3.2, below, for a discussion of elastic properties) but is entirely consistent with the observed ~metre-scale thermal contraction crack polygons (Auger et al. 2018). These polygons are detected at all latitudes, suggesting that a sufficiently hard layer is globally present. Thermal fracturing should, therefore, be an important erosion mechanism, contributing to the breakdown of boulders, consolidated material and cliffs on cometary surfaces.

**Strength from Modelling** Estimates of the strength of cometary material are inherently related to the evaluation of its cohesion. Since the comet nucleus undoubtedly has a high porosity, plausible estimates of cohesion can be obtained on the basis of the well-known theory (Rumpf 1962, 1970), where the tensile strength is calculated via the cohesion expressed by forces acting at the contact between the neighbouring particles. These forces may form with and without material bridges. In the first case the Van der Waals or electrostatic forces cause the adhesion. In the second case the adhesion arises from the presence of different bridges (e.g. sintering). Speaking of particle-surface contacts and their force-response behaviour one can select the following basic approaches: (1) elastic contact deformation, (2) plastic contact deformation with adhesion, (3) visco-elastic and visco-plastic contact deformations. For cometary conditions, only the first approach seems to be relevant. The theoretical basis of this approach was developed by Hertz (1882) and developed in Derjaguin et al. (1975) and Johnson et al. (1971).

For a homogeneous random porous media composed by spherical grains the tensile strength $\sigma_T$ of a porous bed can be expressed as (Rumpf 1970):

$$\sigma_T = \frac{(1-\varphi)N_c F_c}{4\pi R^2} \tag{1}$$

where $\varphi$, $N_c$, $R$ and $F_c$ are the porosity, the average coordination number, the effective particle radius (for two spheres with radii $R_1$ and $R_2$ $1/R = 1/R_1 + 1/R_2$) and the adhesion force between particles. We note that without a significant loss of accuracy, one can assume that $\varphi N_c \approx \pi$ (Rumpf 1974; Rumpf 1990), so:

$$\sigma_T = \frac{(1-\varphi)/\varphi F_c}{4R^2} \tag{2}$$

The adhesive force between two spheres is directly proportional to the specific surface energy $\gamma$ and $R$. For the elastic spheres two well-known expressions are usually applied:

- the so called JKR model (Johnson et al. 1971):

$$F_c = 3\pi R\gamma \tag{3}$$

- and the DMT model (Derjaguin et al. 1975):

$$F_c = 4\pi R\gamma \tag{4}$$

As it was shown by Tabor (1977), the JKR model is appropriate for relatively large soft particles with high surface energies, whereas the DMT model works well for small hard particles with low surface energies. For our purposes, the difference between these expressions is unimportant. We have note that both JKR and DMT models are limited to spheres with a smooth surface. From the above it is evident that tensile strength is inversely proportional to the effective particle size. Thus, for the porous medium consisting of mono-disperse micron size spheres the tensile strength is about several kPa, however, for the millimetre-sized particles the tensile strength is about few Pascal only. We remind that it only considers Van der Waals forces (no sintering).

The generalisation of this conclusion to a poly-disperse medium is nontrivial. It is well known that a substantial additive of fine particles can increase effective medium cohesion. At the same time, one can imagine a situation (for example, for a bi-disperse mixture) where the number of contacts per square meter is controlled by large particles, while the strength of individual contact between particles is determined by small ones. In this case, small grains perform the same role as the surface roughness of large particles, i.e. they reduce the strength.

An interesting attempt to estimate the effective strength of a hierarchical porous medium was made in Skorov and Blum (2012). Assuming that cometesimals formed by gravitational instability of a cloud of porous dust and ice aggregates in a gentle environment, the authors presented a consistent model of a porous medium having hierarchical structure. Using theoretical estimates and experimental results, they found that the effective tensile strength $\sigma_{eff}$ of the dust layer could be expressed as:

$$\sigma_{eff} = \sigma_a (1 - \varphi) \frac{A}{A_0} \tag{5}$$

with $A$ and $A_0$ being the contact area between two dust aggregates and the cross-sectional area of the aggregates, respectively. The intrinsic tensile strength of the aggregates $\sigma_a$ is estimated by using Eq. (8) of Güttler et al. (2009). Finally, the effective strength is expressed via

$$\sigma_{eff} = \sigma_I (1 - \varphi) \left( \frac{R}{1\ \text{mm}} \right)^{-2/3} \tag{6}$$

with $\sigma_I = 1.6$ Pa a numerical factor obtained for the specific values of the model parameters given in Skorov and Blum (2012). In this model the effective strength is a function of surface energy also (this characteristics is hidden in $\sigma_I$). The predicted theoretical tensile strength of aggregates is in good agreement with the recent experimental results (Blum et al. 2014; Brisset et al. 2016). It was shown that the hierarchical structure of the surface layer formed by porous dust aggregates dramatically reduces the tensile strength of the layer to only a few Pascal.

## 3.2 Elastic Properties

### *3.2.1 State of the Art Before Rosetta*

Tidal breakup of cometary nuclei requires the tidal stress resulting from close encounters with other bodies to exceed the tensile strength of the nucleus (Sect. 3.1.1). This stress depends not only on masses, distances and dimensions of the two bodies, but, as Greenberg et al. (1995) pointed out, also on the Poisson ratio of the cometary material, although the dependency is weak.

Of greater interest is the possibility to explore the subsurface structure and stratification of the comet using the propagation of elastic waves. The CASSE instrument of Philae (Comet Acoustic Surface Sounding Experiment, Seidensticker et al. 2007) was actually designed to this end. It is expected that the elastic moduli of cometary material depend on the degree of sintering and will thus allow investigating the formation of layering predicted by several authors in the past as explained earlier (Sect. 3.1).

Before Rosetta, little was known about the elastic properties of cometary materials, while several models for composition and structure, especially layering were published. Material properties had to be derived from assumed compositions.

Stoffler et al. (1991) describe the composition and preparation of a cometary analog material that was used in the KOSI experiments at DLR Cologne in the early 1990s. Estimations of elastic moduli or elastic wave velocities can be derived from the composition using heuristic approaches based on snow, or using mixture theories. It must however be noted that the elastic moduli of granular material are dominantly defined by grain contacts rather than by grain composition (e.g. Mavko et al. 2009). A common property of granular media is indeed very low seismic velocities, even below the speed of sound in air. Shear wave velocities found in lunar regolith vary over a wider range than those in snow, and actually values as low as 30 m/s are even possible (Knapmeyer et al. 2018; their Fig. 2).

The formation of layers can be considered and is linked directly to the idea that comets are a mixture of volatile and non-volatile constituents. Whipple (1950) not only suggested the dirty snowball concept, he also concluded from this concept that a size-dependent ejection of particles, and the different sublimation properties of $H_2O$ and $CO_2$, may result in the formation of subsurface layering. The observation that most of 1P surface is inactive gave rise to models for the formation of inert crusts (see Kuehrt and Keller 1994 for the post-Halley state of the discussion). Kuehrt and Keller (1994) also discuss the importance of cohesive forces between particles, in addition to sintering and cementing. The time necessary to produce sinter bridges of a significant volume between ice grains is strongly dependent on particle size. Thus, subsurface structure is the result of interplay between accelerated sintering of surface materials, but also abrasion close to perihelion, the inward propagation of sublimation fronts of different volatiles, and airfall of ejected dust. Significant discrepancies between proposed layer thicknesses (e.g. Prialnik and Mekler 1991 versus Kuehrt and Keller 1994) or ablation rates (e.g. Keller et al. 2015 versus Brugger et al. 2016) indicate that ground truth even from a single point would be a valuable constraint to improve the existing models.

### *3.2.2 Progress During Rosetta*

Even Philae's brief contact with the surface of 67P at Agilkia allowed a determination of the material's Young's modulus. The six cup-shaped soles of the three feet of Philae are capable of several modes of free resonances. When in contact with an elastic surface, the coupled oscillator shows modified resonances from which the Young's modulus of the surface can be derived. Using this approach, Möhlmann et al. (2018) found that the material in contact with Philae had a Young's modulus of a few MPa. This is much lower than the 10 GPa found in bubble-free water ice, but compares well to snow with porosity beyond 0.75 (Knapmeyer et al. 2018, and references therein).

The penetration phase of MUPUS at the Abydos site provided an opportunity for further analysis (Knapmeyer et al. 2018). A comparison of recorded CASSE accelerometer signals from the different feet of Philae, as well as comparison with laboratory experiments, indicates that all three feet were in contact with the comet at least during some of MUPUS hammer strokes. Several strokes were recorded on more than one foot, allowing travel times to be compared. Since the topography at Abydos is not known with sufficient accuracy (ROLIS descent images were shot at Agilkia, not Abydos), only an interval of possible propagation velocities (and thus shear modulus) could be derived. In terms of the Young's modulus, 7.2 MPa $< E <$ 980 MPa was found by Knapmeyer et al. (2018) with the lower bound being a firm limit, while the upper bound generously accounts for uncertainties in the path length. Typical correlations between Young's modulus ($E$) and compressive strength ($\sigma_c$) indicate a ratio $E/\sigma_c \sim$ 100–400 for ice or dust/ice mixtures (Jessberger and Kotthaus 1989; Möhlmann et al. 2018 and references therein), which translates to a compressive strength of 18 kPa $< \sigma_c <$ 9800 kPa. Finally, assuming a typical ratio $\sigma_c/\sigma_T \sim 10$ for brittle material, it gives a tensile strength in the range 1.8–980 kPa for the layer that stopped MUPUS.

An independent result is the layering indicated by the frequency dependency of propagation velocity. The presence of dispersion can be established without any assumptions. Surface waves at frequencies below 250 Hz are delayed by about 10 ms with respect to higher frequency waves. Given the short distances between MUPUS and the three feet, this amounts to a velocity reduction of at least 32%. Since the penetration depth of surface waves depends on wavelength, velocity reduction and estimated wavelengths mean that the above Young's modulus applies to a top layer with a thickness of 0.1 m to 0.5 m, while the material below shows reduced Rayleigh wave velocities, probably due to a reduced Young's modulus (density variations might also contribute, but are less plausible, given the available material).

MUPUS non-penetration and thermal inertia measurements (Sect. 2.2.2; Spohn et al. 2015) suggest that MUPUS was stopped on a hard layer, where the Young's modulus is likely increased as well. Given the uncertainty in the CASSE Young's modulus and the MUPUS compressive strength estimates, it cannot be excluded that the fast near surface layer found by CASSE is indeed the sintered layer that stopped MUPUS PEN.

In summary, the combination of MUPUS and CASSE measurements indicates the presence of two or three subsurface layers, with a Young's modulus that decreases with depth.

### 3.3 Synthesis on Mechanical Properties

The mechanical properties of cometary nuclei are characterised by a low bulk tensile strength of typically 100 Pa or less, with spatial variations across the nucleus, and locally the presence of a shallow (10's of cm) sub-surface layer of harder materials with a Young's modulus in the range 7.2–980 MPa, i.e. a tensile strength of typically 1.8–980 kPa. To put these numbers in perspective, the harder material compares well to porous (75%) snow (Knapmeyer et al. 2018) or silica aerogel, while the bulk nucleus material ($\sigma_T$ of 100 Pa) has no analog on Earth among consolidated materials and only exists on comets because of the very low gravity (typically $10^{-4}$ m s$^{-2}$).

Figure 10 shows a synthesis of the strength values for cometary nuclei measured by different techniques. As explained in Sect. 3.1.1, the strength is a scale dependent parameter, which makes the comparison between the different measurements difficult and somewhat speculative. As an indication for Fig. 10, we used the scaling law of water ice, which is a power law $d^{-q}$, where $d$ is the scale and $q \sim 0.6$ for water ice (Petrovic 2003). This scaling law should be taken with caution, since it is material dependent (Bažant 1999). The strength is also temperature dependent, for example water ice is 3 to 4 times stronger at 120 K than at 220 K (Schulson and Duval 2009), but this is of second order effect compared the scale changes of Fig. 10. From Fig. 10, bearing in mind the above limitations on scaling laws, we see that three out of four strengths values are <100 Pa at the meter scale. Only three estimates are incompatible with a tensile strength <100 Pa, including two from Philae

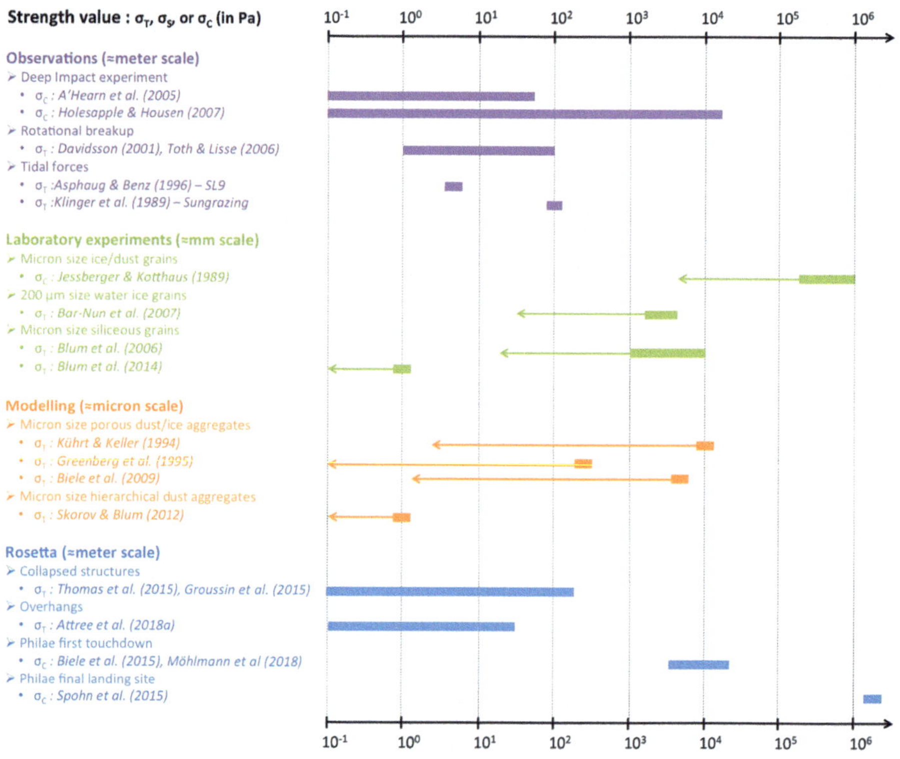


**Fig. 10** Synthesis of strength values for cometary nuclei measured by different techniques and authors, including Rosetta estimates (non exhaustive list). For each strength value, we specify whether it is the tensile ($\sigma_T$), shear ($\sigma_S$) or compressive ($\sigma_C$) strength. For laboratory experiments and modelling, the arrows indicate the strength scaled to the meter scale, using a power law with a power exponent of $-0.6$ (see text for details). This scaling law should be taken with caution, since it is material dependent (Bažant 1999)

surface measurements and resulting from the presence of a processed layer below the surface (Sects. 3.1 and 3.2).

As shown in Fig. 11, a bulk tensile strength of only 15 Pa is sufficient to keep comets stable against rotational splitting, under the assumption that they have densities and shapes similar to 67P. The above value of 100 Pa for the tensile strength of cometary nuclei likely provides sufficient margins against density and shape variations among comets.

Finally, the low bulk tensile strength of cometary nuclei provides an important constraint for their formation and evolution. In a first scenario, one can argue that to keep such a low tensile strength over 4.6 Gy, the formation and evolution must be gentle, favouring a primordial rubble pile origin (Davidsson et al. 2016), with a formation by low velocity ($1\ \mathrm{m\,s^{-1}}$) accretion of pebbles (Blum et al. 2014) for a resulting tensile strength of typically 1 Pa (Skorov and Blum 2012), rather than a collisional scenario between two 50–100 km size bodies at $\sim 0.5\ \mathrm{km\,s^{-1}}$ (Davis and Farinella 1997). In a second scenario, one can argue that the low bulk tensile strength is compatible with comets being collisional fragments of larger bodies (Morbidelli and Rickman 2015), in which case the nucleus could be made of numerous mechanically strong (kPa–MPa) blocks and boulders, assembled together for a resulting low bulk tensile strength (Pa), similar to rubble piles asteroids (Sánchez and Scheeres 2014). Disentangling between these two scenarios certainly requires more constraints than

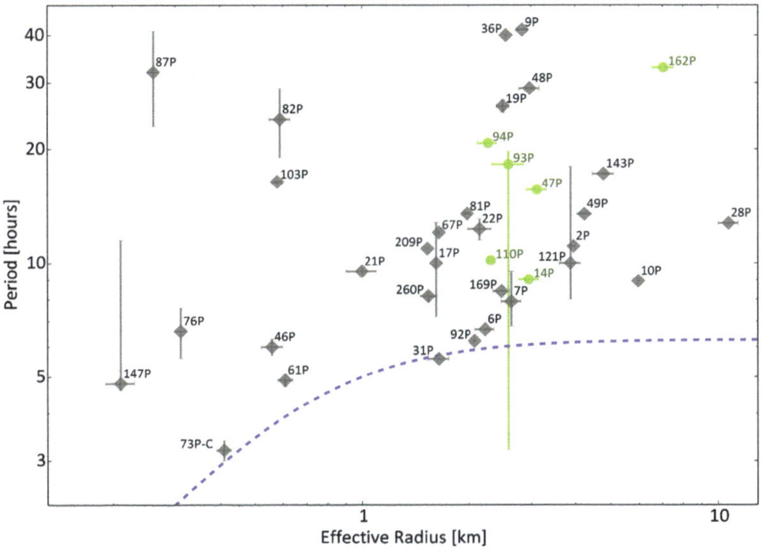


**Fig. 11** Rotation period against effective radius for JFC nuclei (adapted from Kokotanekova et al. 2017). The grey diamonds correspond to comets observed with ground- and space-based telescopes. The green circles denote the comets visited by spacecraft. The dashed curve corresponds to the model for prolate ellipsoids stable against rotational instability by Davidsson (2001), plotted for the parameters derived for 67P, i.e. a density of 532 $\mathrm{kg\,m^{-3}}$, an axis ratio $a/b = 2$ and a bulk tensile strength of 15 Pa

the strength itself, but it emphasises the importance of understanding the scale dependence of the strength for cometary materials, from the microscopic to the macroscopic scale.

## 4 The Nucleus Structural and Dielectric Properties

The structural and dielectric properties of the comet nucleus provide information on the nature of the cometary material and in particular its roughness (Sect. 4.1), density (Sect. 4.2), permittivity and porosity (Sect. 4.3). Excepted for surface roughness, these properties are best estimated by radar and permittivity probes, which operate in the radio wavelength range, therefore providing information on the interior of the nucleus, inaccessible to visible or infrared instruments observing only the surface or sub-surface. Overall, they help to constrain and understand the nucleus vertical stratigraphy from its surface to its centre (Sect. 4.4).

### 4.1 Surface Roughness

Comet nucleus surfaces are rough on different size scales. For practical reasons we here distinguish between two size regimes: (1) *large-scale roughness* ($>10$ m) is resolved by cameras on orbiting or flyby spacecraft, while (2) *small-scale roughness* ($<10$ m) is often only detectable through its shadowing effects on reflected or emitted radiation.

#### *4.1.1 State of the Art Before Rosetta*

Pre-Rosetta spacecraft imaging established the existence of rough terrain containing pits and circular depressions with a cumulative size distribution power law index $p = -1.7$ for comet 81P and $p = -2.1$ for 9P (Thomas et al. 2013a), boulders with $p = -2.7$ on 103P (Pajola et al. 2016), and smooth terrain proposed to be flows on 9P (Thomas et al. 2013a) or fallback on 103P (Thomas et al. 2013b).

Small-scale roughness obtained from optical photometric phase functions and measured in terms of the Hapke (1984) mean slope angle $\theta$ yielded $\theta = 13$–$25°$ for 19P (Buratti et al. 2004), $\theta = 16 \pm 8°$ for 9P (Li et al. 2007), $\theta = 27 \pm 5°$ for 81P (Li et al. 2009), and $\theta = 15 \pm 10°$ for 103P (Li et al. 2013).

Analysis of infrared radiation, that allows for the simultaneous determination of thermal inertia and small-scale roughness yielded $\theta = 45°$ for rough terrain on 9P (Davidsson et al. 2013). Roughness effects are also evident in infrared spectra of 103P (Groussin et al. 2013).

#### 4.1.2 Progress During Rosetta

The unprecedented resolution and complete nucleus coverage provided by the OSIRIS cameras has allowed for the construction of accurate geometric shape models for 67P (Jorda et al. 2016; Preusker et al. 2015, 2017). Combined with methods for accurately calculating the gravity field (Werner and Scheeres 1997), this has allowed for the identification of cliffs, that is one manifestation of large-scale roughness. Vincent et al. (2017) studied the cumulative cliff height distribution and found that regions subjected to a higher level of insolation and sublimation-driven erosion have more negative $p$-values. They therefore suggest that 81P ($p = -1.7$) and 67P ($p = -1.69$ on average) have more pristine surfaces than 9P ($p = -2.1$) and 103P ($p = -2.7$). This exemplifies the importance of measuring comet nucleus surface roughness in order to better understand comet evolution.

Small-scale roughness from optical photometry yielded $\theta = 22 \pm 7°$ (Fornasier et al. 2015) and $\theta = 19\ (+4/-9)°$ (Ciarniello et al. 2015) for 67P. Analysis of roughness derived from infrared emission is on-going (e.g., Leyrat et al. 2014; Marshall et al. 2018). Images with extremely high resolution obtained by CIVA (1 $\mathrm{mm\,px^{-1}}$) of consolidated terrain at Abydos (Bibring et al. 2015), and of smooth terrain by ROLIS (1 $\mathrm{cm\,px^{-1}}$) at Agilkia (Mottola et al. 2015) and by OSIRIS (1.4 $\mathrm{cm\,px^{-1}}$) at Sais (Pajola et al. 2017) have provided rare glimpses of the true nature of small-scale roughness. Smooth terrains consist of facetted and angularly shaped pebbles with sizes in the cm-dm range. Consolidated terrains, on the other hand, consist of a mixture of cemented/sintered agglomerates of mm-cm sized grains, and surfaces of fine-grained material that are smooth on decametric size scales (Bibring et al. 2015). However, these textures sit on structures that are irregular on the meter-scale and upwards, while smooth terrains are essentially flat on such size scales.

The resulting richness in shadows on the cm-dm scale in both types of terrains may explain why $q$-values from optical photometry are consistently high and do not seem to vary much from place to place on one target, or among targets. However, the heat conduction may be sufficiently effective on cm-dm size scales to smooth out the lateral temperature gradients that otherwise would form because of the shadows that the small pebbles cast (e.g., Davidsson and Rickman 2014). If so, self-heating effects (e.g., Hansen 1977) may be rather weak in smooth terrain that lacks larger-scale shadows, while it should be stronger in consolidated terrains where irregularities on the meter-scale and upward are common. Testing this hypothesis through continued analysis of VIRTIS data appears important. If it can be demonstrated that self-heating effects differs strongly between smooth and consolidated terrains, it would offer the possibility of determining the area fractions of both terrain types for a large number of unresolved comet nuclei, e.g., through observations with the James Webb Space Telescope.

**Table 1** Bulk density estimates for cometary nuclei, using different methods (see text for details)

| Comet | Preferred value ($kg\,m^{-3}$) | Range ($kg\,m^{-3}$) | Method | Reference |
|---|---|---|---|---|
| 1P | | 100–200 | NGF | Rickman (1986) |
| | 600 | 200–1500 | NGF | Sagdeev et al. (1988) |
| | | 500–1200 | NGF | Skorov and Rickman (1999) |
| | 1000 | 30–4900 | NGF | Peale (1989) |
| | 500 | 200–800 | NGF | Sosa and Fernández (2009) |
| 2P | 800 | ≤1600 | NGF | Sosa and Fernández (2009) |
| 6P | 150 | 100–200 | NGF | Sosa and Fernández (2009) |
| 8P | | 900–1900 | Radar | Harmon et al. (2010) |
| 9P | 400 | 200–1000 | Deep impact ejecta | Richardson et al. (2007) |
| | 450 | 200–700 | NGF | Davidsson et al. (2007) |
| | 200 | 100–300 | NGF | Sosa and Fernández (2009) |
| | 470 | 240–1250 | Deep impact ejecta | Thomas et al. (2013a) |
| 10P | 700 | 300–1100 | NGF | Sosa and Fernández (2009) |
| 19P | 490 | 290–830 | NGF | Farnham and Cochran (2002) |
| | | 100–300 | NGF | Davidsson and Gutiérrez (2004) |
| 22P | 200 | 100–300 | NGF | Sosa and Fernández (2009) |
| 46P | 400 | 100–700 | NGF | Sosa and Fernández (2009) |
| 67P | ≤500 | 100–370 | NGF | Davidsson and Gutiérrez (2005) |
| | | 600–1000 | Radar | Kamoun et al. (2014) |
| | 535 | 500–570 | In-situ mass and volume | Preusker et al. (2015) |
| | 532 | 525–539 | In-situ mass and volume | Jorda et al. (2016) |
| 81P | | ≤600–800 | NGF | Davidsson and Gutiérrez (2006) |
| | 300 | ≤800 | NGF | Sosa and Fernández (2009) |
| 103P | | 200–400 | Shape and rotation | Thomas et al. (2013b) |
| | 220 | 140–520 | Erosion rate | Richardson and Bowling (2014) |
| D/S-L 9 | 550 | 500–600 | Tidal forces | Solem (1995) |
| | 600 | 500–1000 | Tidal forces | Asphaug and Benz (1996) |

## 4.2 Bulk Density

### *4.2.1 State of the Art Before Rosetta*

Before Rosetta, the density of comet nuclei could only be measured indirectly (Table 1). Solem (1995) and Asphaug and Benz (1996) modelled the tidal disruption of comet SL9 during its close approach to Jupiter, and derived densities of $550 \pm 50\ kg\,m^{-3}$ and $\sim 600\ kg\,m^{-3}$, respectively. The mass of nuclei can also be estimated based on their acceleration due to non-gravitational forces (NGF), caused by the rocket-like effect of outgassing while the comet passes through perihelion, which produce measurable changes in their orbital parameters. The density can then be calculated from this mass and the volume, the latter being estimated from resolved images for spacecraft targets or from photometry for a larger number of comets (Sosa and Fernández 2009). This method based on NGF was first used by Rickman (1986) to derive a surprisingly low density of 100–200 $kg\,m^{-3}$ for comet 1P, and after by several authors (e.g. Davidsson and Gutiérrez 2005, 2006).

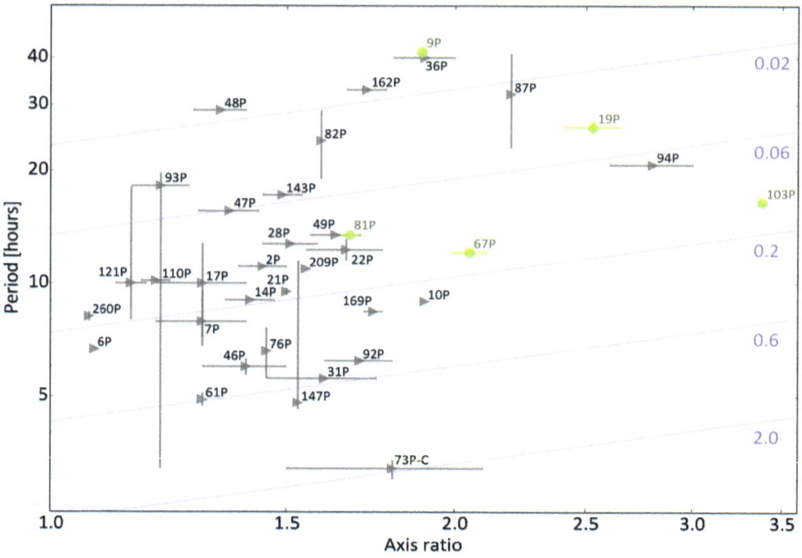


**Fig. 12** Rotation period against axis ratio for JFC nuclei (adapted from Kokotanekova et al. 2017). The triangles correspond to the lower limit of the axis ratio determined from light-curve observations. The circles denote comets visited by spacecraft. The diagonal curves show the minimum density (in $g\,cm^{-3}$) required to keep a strengthless body with the given axis ratio and spin period stable against rotational splitting. Apart from 73P, which has recently undergone a splitting event, no other comets require densities larger than $\sim 0.6\ g\,cm^{-3}$

Flyby missions to comets provided additional constraints on the density. The shape of the ejecta plume resulting from the Deep Impact experiment on comet 9P was used to estimate a bulk density of $470\ kg\,m^{-3}$, with a range of possible values between 240 and $1250\ kg\,m^{-3}$ (Richardson et al. 2007; Thomas et al. 2013a). Using the characteristic shape of the nucleus of comet 103P, and in particular the "waist" connecting the two lobes, A'Hearn et al. (2011) estimated a mean value for the density of $220\ kg\,m^{-3}$ with possible values between 140 and $520\ kg\,m^{-3}$ (Richardson and Bowling 2014).

Radar observations can be used to constrain the density of comets. However, it is important to be aware that the radar measurements are only sensitive to the top layer of the comets (down to the penetration depth of the radar wave; $\sim$10 wavelengths for packed soils, i.e. up to $\sim$10 m). The density of the surface layers can be estimated from the radar albedo if the nucleus surface is covered by a thick homogeneous layer (e.g. Harmon et al. 2004). The radar surface density estimates range between 500 and $1500\ kg\,m^{-3}$ (Harmon et al. 2004) and are generally larger than bulk density estimates derived from NGF, reinforces the idea that harder (denser) materials are present in the top layer (Sect. 3 and Davidsson et al. 2009).

Finally, attempts to constrain the bulk density of JFC nuclei from visible photometry (light-curves) have also been made. If the nuclei are modelled as strengthless prolate ellipsoids that are held together only by gravity, a minimum density is required to balance the centrifugal force for the given rotation rate. Considering all JFCs with available rotation rates and minimum axis ratios, Lowry and Weissman (2003) determined a cut-off in density at $600\ kg\,m^{-3}$, later confirmed by Snodgrass et al. (2006) and Kokotanekova et al. (2017). The lack of objects that require larger minimum densities implies that these objects have been destroyed by fast rotation (Fig. 12). In analogy with the clear spin barrier seen in asteroids (Harris 1996; Pravec et al. 2002), $600\ kg\,m^{-3}$ is therefore considered to correspond to the average bulk density of JFCs. While this density estimate is indirect and relies on a simplified model based on assumptions about the material strength and the shape of JFCs, the derived result is in excellent agreement with the densities estimated from other methods.

### *4.2.2 Progress During Rosetta*

Rosetta provided the first direct and precise measurement of a cometary bulk density. The detailed shape model of the comet nucleus allowed its volume to be calculated with great precision (Preusker et al. 2015; Jorda et al. 2016). The mass of the comet was determined by the Radio Science Investigation (RSI) instrument on board Rosetta (Pätzold et al. 2007), where careful tracking of Rosetta's motion, via the Doppler shift in the radio link with Earth, was used to measure the gravitational influence of the nucleus on the spacecraft. Combining the two parameters allowed Jorda et al. (2016) to determine that the bulk density of the nucleus of 67P is $532 \pm 7\ \mathrm{kg\,m^{-3}}$. This study also determined that the nucleus has high porosity of 70–75%.

This method is precise enough to measure higher-order gravitational terms (i.e. the mass distribution within the comet) and the mass loss during the perihelion passage while Rosetta was in operation. These revealed an offset between the centre of figure and centre of mass, indicating subtly different densities between the "head" and "body" of 67P's nucleus (Pätzold et al. 2016).

It is worth mentioning that the bilobate nucleus of 67P would not be stable without any cohesion, and the cohesive strength has been estimated to 10–200 Pa at its current rotation rate of 12.4 hours by Hirabayashi et al. (2016), compatible with the strength estimates reported in Sect. 3.1.

## 4.3 Porosity from Dielectric Properties

The electrical parameters of a medium describe how the electric field interacts with the charges and dipoles present in the material through a number of physical processes (Kingery et al. 1976) such as electronic, atomic, dipolar and space charge polarisations, as well as the motion of free charges in matter. These processes are not instantaneous and the electrical parameters are consequently frequency dependent. At a given frequency, the electrical properties are described by the effective complex permittivity relative to vacuum, $\varepsilon = \varepsilon' - i\varepsilon''$. $\varepsilon'$ is the real part of the relative permittivity (often called the dielectric constant) and drives the velocity of the electromagnetic wave and the refraction phenomenon (in low loss media where $\varepsilon'' \ll \varepsilon'$). $\varepsilon''$ is the imaginary part of the relative permittivity and is mostly responsible for the attenuation of electromagnetic waves propagating through the medium. The complex permittivity value depends on the physical and chemical properties of the medium, as well as on its temperature.

Instruments like radars and permittivity probes make use of electromagnetic fields and aim to determine in a non-destructive way and at their operating frequency, the complex permittivity value of medium inside the sounded volume. In the case of a non-homogeneous medium, the achieved spatial resolution depends on the instrument's technical characteristics (i.e. frequency bandwidth for a radar and distance between the electrodes for a permittivity probe). Within their spatial resolution, the instruments see a homogeneous medium characterised by its bulk properties. The retrieved bulk permittivity values will then provide constraints on some properties of the material such as composition and porosity.

In the case of a comet nucleus made of ices and refractory materials (hereafter dusts), the mixture permittivity is a function of the known temperature ($T$) and frequency ($\nu$), and of the partially unknown volume fraction of ices ($x_{ices}$), dusts ($x_{dusts}$) and vacuum ($x_{vacuum}$), and their respective permittivity values ($\varepsilon_{ices}, \varepsilon_{dusts}, \varepsilon_{vacuum}$):

$$\varepsilon = f(\varepsilon_{vacuum}, \varepsilon_{dusts}, \varepsilon_{ices}, x_{vacuum}, x_{dusts}, x_{ices}, T, \nu) \quad (7)$$

where $\varepsilon_{vacuum} = 1$ and $x_{vacuum} + x_{dusts} + x_{ices} = 1$.

Different mathematical expressions for the *function* in Eq. (7) are provided by a number of so-called mixing laws, which are based on different assumptions on the density, size, shape and orientation of the inclusions (Sihvola 1999; Hashin and Shtrikman 1962; Looyenga 1965; Choy 1999). This, added to the too many unknown parameters, makes the estimation of the porosity from this single equation quite impossible without additional inputs such as the dust to ice ratio, the bulk density, and the refractory materials to be considered.

An effective alternative approach to avoid the use of mixing laws is based on experimental measurements of the permittivity, performed on mixture samples in known conditions (e.g., Campbell and Ulrichs 1969; Heggy et al. 2001; Brouet et al. 2016b). But again, the same additional inputs are also needed to limit the number of samples to characterise.

In any case, knowledge of the permittivity value does provide additional constraints on the porosity value to the one obtained independently by the bulk density value. In conclusion, using both permittivity and density bulk values, and additional hypotheses about dusts to ices ratio and dusts nature, it is possible to get reliable estimates of the porosity.

### 4.3.1 State of the Art Before Rosetta

Earth-based radars, like the Arecibo Observatory S-band radar ($\lambda = 12.6$ cm) and the Goldstone Solar System Radar S-band ($\lambda = 12.9$ cm) and X-band ($\lambda = 3.54$ cm), are used to observe planets and small bodies, including comets from the ground. For comet nuclei, the measured radar cross-section values, together with the physical dimensions of the nucleus (that can be obtained either from radar or optical measurements), provide an estimate of the nucleus's radar albedo, which sets constraints on the dielectric properties of the sounded volume. As explained earlier, it is then possible to obtain limits on the porosity value of the nucleus subsurface. Note that the penetration depth of the radar waves depends on the radar wavelength and on the sounded matter. A rule of thumb is that, for porous media, bulk values estimated from radar data would be characteristic of the subsurface down to a depth that can reach 20 times the wavelength (i.e. 2.5 m for S-band radars).

Unfortunately, cometary nuclei are not easily detected by Earth-based radars (Ostro 1985). Their kilometric-size compared to their distance to Earth is only part of the explanation. In a number of radar observations, the echo from cm-size grains in the coma prevented the detection of the weak nucleus echo. Nevertheless, in most cases, even the non-detection of the nucleus by radar still allowed limits of the nucleus radar albedo to be set, and to constrain the density or porosity of the nucleus top layer.

The first successful radar detection of a comet nucleus was made in 1980 of comet 2P/Encke with the Arecibo S-band radar (Kamoun et al. 1982), followed by a series of observations leading to the detection of more than 14 nuclei and the non-detection of several of them (e.g., Campbell et al. 1989; Harmon et al. 1997, 2011; Harmon and Nolan 2005).

As explained in Sect. 4.2, most of the density values estimated from nucleus radar echoes lie between 0.3 and 0.7 $\text{g}\,\text{cm}^{-3}$ (Table 1), which is roughly consistent with a high porosity of the subsurface. Comet 8P/Tuttle stands out though, with an estimated density at the subsurface significantly higher at between 0.9 and 1.9 $\text{g}\,\text{cm}^{-3}$ (Harmon et al. 2010).

The case of 67P is of upmost interest because its nucleus has been thoroughly analysed later during the Rosetta mission. The 2.380 GHz Arecibo radar has been operated in 1982 to observe 67P at a geocentric distance of 0.4 a.u, but no detection of the nucleus or the coma was achieved then (Kamoun et al. 1998). The lack of radar detection provided an upper limit of the nucleus radius of around 3 km (Kamoun 1983; Kamoun et al. 1998) and an upper value of the nucleus albedo at 0.05. Kamoun et al. (2014) revisited the same radar data set to provide some constraints on the nucleus properties that would be useful for the Philae landing to come. Considering that, from measurements performed on other cometary nuclei, the lower limit of radar albedo values is likely to be 0.04, this led, for 67P, to an estimated permittivity value between 1.9 and 2.1, converted to a 0.6 to 1 $\mathrm{g\,cm^{-3}}$ density at the surface, compatible with 55–65% porosity.

### *4.3.2 Progress During Rosetta*

Two experiments of the Rosetta mission were able to sound beyond the nucleus surface and to provide estimates for the dielectric constant of the nucleus using electromagnetic waves: the Permittivity Probe of the Surface Electric Sounding and Acoustic Monitoring Experiment (SESAME-PP; Seidensticker et al. 2007) and the Comet Nucleus Sounding Experiment by Radiowave Transmission (CONSERT; Kofman et al. 1998, 2007).

SESAME-PP consisted of five electrodes operating between 10 Hz and 10 KHz, with three transmitting electrodes (located on the +X-foot, MUPUS-PEN and APXS), able to inject a current in the material in contact with the electrodes, and two receiving electrodes (located on the +Y-foot and −Y-foot). However, data processing revealed that only the range of frequency between 409 and 804 Hz was usable to derive a constraint on the dielectric constant of the material, and that only one transmitter could be used (+X-foot). The measurements performed by SESAME-PP were restricted approximately to the first meter of cometary material below the final Abydos landing site of Philae (Lethuillier et al. 2016). Lethuillier et al. (2016) determined a lower limit of $2.45 \pm 0.20$ for the dielectric constant.

CONSERT was a bistatic radar instrument. The principle of measurement was based on the transmission and reception of a radio signal between the orbiting Rosetta spacecraft and the Philae lander. It operated at a centre-band frequency of 90 MHz, so that it propagated long-wavelength electromagnetic waves through part of the cometary nucleus. In a low loss medium, the measurement of the signal travel time allows the determination of the wave velocity and thus to estimate the real part of the permittivity value. CONSERT was able to provide a dielectric constant equal to $1.27 \pm 0.05$ from signals that propagated through the small lobe of the nucleus over distances of hundreds of meters (Kofman et al. 2015).

The very low value of CONSERT contrasts with the lower limit determined by SESAME-PP, although both results indicate a dielectric constant representative of a highly porous material (i.e., a volume fraction of voids larger than 50%; e.g., Heggy et al. 2012; Brouet et al. 2014, 2015, 2016b). The temperature of the near-surface material was low enough during the measurement sequences, i.e. lower than ~180 K, to assume a nondispersive behaviour of the dielectric constant (Lethuillier et al. 2016) in the operating frequency range of SESAME-PP and CONSERT, allowing the discrepancy between the near-surface dielectric constant and the internal dielectric constant to be only caused by a variation of the porosity and/or the composition. Before the final results released by the SESAME-PP science team regarding the measurement of the dielectric constant, Ciarletti et al. (2015) had already pointed out that the dielectric constant may vary as a function of depth within the shallow subsurface of the Abydos landing site, suggesting a decrease of its value with depth, and excluding a significant increase of the bulk density or of the dust-to-ice mass ratio with depth in the first meters.

In order to interpret the dielectric constant value in terms of porosity, two independent methods were used:

(1) A first method is based on the use of dielectric mixing laws previously mentioned. Kofman et al. (2015) used a conservative approach by using the Hashin-Shtrikman bounds that encompass all the solutions suggested by the numerous existing dielectric mixing laws. Kofman et al. (2015) also compiled laboratory dielectric data on relevant refractory materials and ices from the existing literature in order to derive the porosity, and finally, found a porosity of 75%–85% with this method. Lethuillier et al. (2016), following the same method, suggested that a lower limit of $2.45 \pm 0.20$ for the dielectric constant is compatible with a porosity lower than 50% if carbonaceous chondrites are taken into account as the dust component of the ternary mixture (we note that ordinary chondrites have been excluded as potential dust analogues to explain the dielectric constant estimated by CONSERT, see Kofman et al. 2015). With the same method of analysis, Herique et al. (2016) refined the possible range of the zero-porosity dielectric constant of the dusty component of 67P/CG.

(2) A second method is based strictly on laboratory dielectric measurements of cometary analogues performed in a frequency range encompassing the CONSERT operating frequency, i.e. between 50 MHz and 2 GHz (Brouet et al. 2015, 2016b). The studied samples included pure laboratory-grown water ice and JSC-1A Lunar simulants, as well as mixtures of these two components, with controlled porosities in the range of 30% to 91% and dust-to-ice mass ratios in the range of 0 to 1. From these measurements, a semi-empirical dielectric law has been derived and used to estimate a porosity lower than 58%, and between 75% and 86%, from the values of the dielectric constant measured by SESAME-PP and CONSERT, respectively (Brouet et al. 2016a). Thus, the results suggest an increase of the porosity with depth, like the first method, and also indicate that this porosity gradient over the small lobe of the nucleus cannot be excluded even when taking into account the variation of the dust-to-ice mass ratio.

Penetrating radars like CONSERT are able to detect structures of spatial scales much larger than the wavelength (i.e. several tens of meters), which do not seem to be present in the small lobe sounded by the CONSERT radar. At a much smaller scale, commensurate with the wavelength, the analysis of the shape of the received pulses provides information about the potential heterogeneities inside the sounded volume. Since no significant modification of the pulse width has been observed (Ciarletti et al. 2017), it is possible to conclude that no volume scattering took place. Based on extensive simulations of propagation through non-homogeneous media with different spatial characteristic scales and permittivity values, the presence of strong heterogeneities at a scale of a few meters has been ruled out.

### 4.4 Synthesis on Structure and Vertical Stratigraphy

Rosetta investigated the interior of 67P's nucleus on scales ranging from millimetres to kilometres, using electromagnetic waves from the kilohertz via the gigahertz and infrared to the visible light range, its gravity field, and elastic waves. Some instruments provided mappings of large areas, others focused on isolated spots. Putting all parts together into a single stratigraphy requires some simplifications, but nevertheless all of the investigations reported here fit together into a coherent picture (Fig. 13).

The broad picture is that the bulk of the comet consists of a weakly bonded, rather homogeneous material that preserved primordial properties under a thin shell of processed material, and possibly covered by a granular material that is able to form pseudo-aeolian

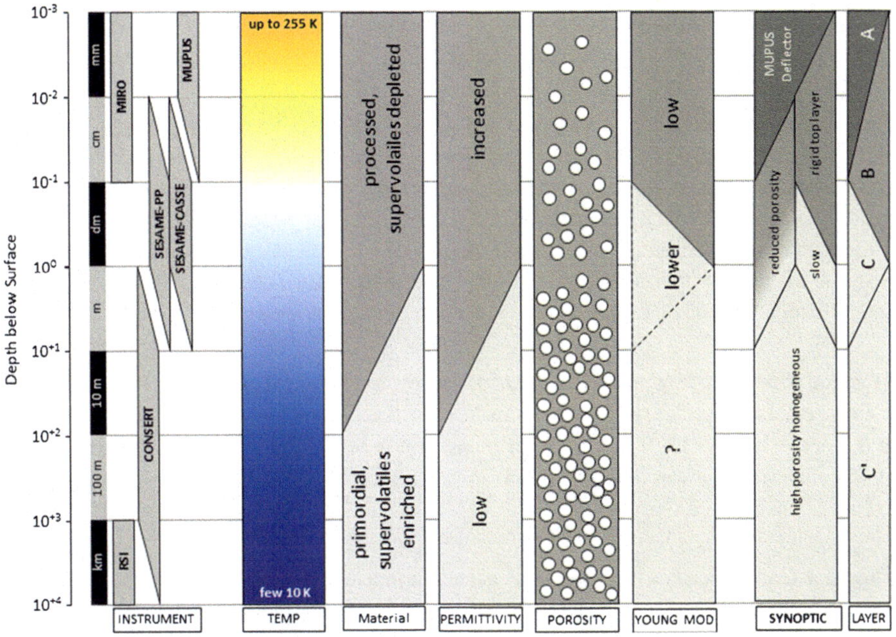


**Fig. 13** Synthesis of the (simplified) stratigraphy of 67P/Churyumov-Gerasimenko on a logarithmic depth scale (granular cover found on subhorizontal surfaces not shown). Columns from left to right: (1) instruments providing observations of the subsurface, (2) temperature range, indicating near-perihelion surface temperatures and estimated temperature of the deep interior, (3) materials, simplified into primordial and not primordial, (4), electrical permittivity resulting from radar and in-situ measurements, (5) porosity, resulting from electrical properties and bulk density, (6) Young's modulus from SESAME/CASSE, (7) and (8) synoptic view—Layer A: MUPUS deflector is a possible thin strong layer that prevented MUPUS from penetrating, but is too thin to be seen with CASSE; Layer B: Rigid top layer; Layer C "slow" (=low shear wave velocity) layer derived from CASSE, porosity gradient in both layers B and C; Layer C' is the deep primordial material, layers C and C' might be identical

morphologies like dunes and moats (El-Maarry et al. 2019). This cover might in places reach thicknesses of many meters.

The orientations of subhorizontal and subvertical surfaces indicates the presence of an onion-like layering in both lobes of 67P, but like layering in snow or sandstone, the associated changes of physical properties are smaller than the resolution of the methods available with Rosetta. There are no indications of strong nucleus heterogeneities at a scale of a few meters on 67P's small lobe.

The processed outer shell is the result of competing effects, resulting from the sublimation of volatiles and super volatiles heated up for the first time since their aggregation, which on the one hand remove and eject material into space, but on the other hand re-condensate between grains and thus cement the shell.

## 5 Conclusions

We have presented a review of the thermal, mechanical, structural, and dielectric properties of cometary nuclei, after Rosetta, including more than 200 references. Our work includes several syntheses on the thermal properties (Sect. 2.3), mechanical properties (Sect. 3.3) and structural and dielectric properties (Sect. 4.4), with key figures (Figs. 6, 10 and 13). Here are the main conclusions for cometary nuclei:

**Temperature** The temperature varies across the surface, following solar insulation to first order. On the dayside, there exist some deviations from the black body temperature due to surface roughness, self-heating, and heat conductivity. On the night side, the temperature is mainly controlled by heat conductivity. At the $\sim$10 $\mathrm{m\,px^{-1}}$ spatial resolution, it is not yet demonstrated that the presence of water ice reduces significantly the surface temperature. Because of the low thermal inertia, the nucleus temperature decreases very rapidly with depth, from hundreds of Kelvin to tens of Kelvin in the first meter.

**Thermal Inertia** (Fig. 6)—The thermal inertia is in the range 0–350 $\mathrm{J\,K^{-1}\,m^{-2}\,s^{-1/2}}$, with a best estimate of 0–60 $\mathrm{J\,K^{-1}\,m^{-2}\,s^{-1/2}}$ for the bulk value. On 67P, most values are in the range 10–170 $\mathrm{J\,K^{-1}\,m^{-2}\,s^{-1/2}}$, with variations across the surface: the thermal inertia of smooth terrains covered by deposits is lower (typically $<$30 $\mathrm{J\,K^{-1}\,m^{-2}\,s^{-1/2}}$) than that of exposed consolidated terrains (typically $>$110 $\mathrm{J\,K^{-1}\,m^{-2}\,s^{-1/2}}$) (Leyrat et al. 2015).

**Thermal Conductivity** We derive a thermal conductivity of 0–0.009 $\mathrm{W\,m^{-1}\,K^{-1}}$, from the above thermal inertia range of 0–60 $\mathrm{J\,K^{-1}\,m^{-2}\,s^{-1/2}}$, assuming a nucleus with a density of 532 $\mathrm{kg\,m^{-3}}$ (Jorda et al. 2016) and a heat capacity of 770 $\mathrm{J\,kg^{-1}\,K^{-1}}$ (dust at 300 K; Winter and Saari 1969).

**Tensile and Compressive Strength** (Fig. 10)—The global tensile strength is $<$100 Pa. This low value is consistent with laboratory experiments and modelling, when scaled to the meter scale. On 67P, the strength varies across the surface, and there exists locally a subsurface layer of "harder" materials with a compressive strength of 3.5–12 kPa at Agilkia (first Philae touchdown) and 18–9800 kPa at Abydos (final Philae landing). At Abydos, this layer is shallow (10's of cm) and compares well with porous (75%) snow or silica aerogel, in terms of mechanical properties.

**Elastic Properties** The Young's modulus is only known for 67P at the Abydos final landing site and amounts to 7.2–980 Mpa (Knapmeyer et al. 2018). This value applies to the top layer, up to a depth of 0.1 m to 0.5 m, below which the Young's modulus is reduced, most likely by the presence of softer, more porous, materials.

**Surface Roughness** At the $>$10 m scale, their exist a large variety of terrains with different degrees of roughness (smooth or irregular), observable visually on spatially resolved images. At the $<$10 m scale, i.e. spatially unresolved excepted for 67P, small-scale roughness yields to Hapke mean slope angles $\theta$ in the range 5–32°, with variations from one comet to the other. On 67P, close-up images reveal that smooth terrains are mostly pebbles (cm-dm scale), while consolidated terrains consist of cemented/sintered agglomerates of mm-cm sized grains and fine-grained material (Bibring et al. 2015).

**Bulk Density** The mean value for the bulk density is $480 \pm 220$ $\mathrm{kg\,m^{-3}}$, based on 20 estimates (Table 1), with a range between 150 $\mathrm{kg\,m^{-3}}$ and 1000 $\mathrm{kg\,m^{-3}}$. The most accurate estimate is for 67P, with $532 \pm 7$ $\mathrm{kg\,m^{-3}}$ (Jorda et al. 2016).

**Dielectric Properties** The permittivity of the nucleus subsurface layer, derived from radar albedo, is most likely in the range 1.9–2.1 (Kamoun et al. 2014). For 67P, at the final landing site, the dielectric constant was estimated to a lower limit of $2.45 \pm 0.20$ by Lethuillier et al. (2016) with SESAME-PP and to $1.27 \pm 0.05$ by Kofman et al. (2015) with CONSERT. This discrepancy suggests a decrease of the dielectric constant with depth, excluding a significant increase of the bulk density or of the dust-to-ice mass ratio with depth in the first meters (Ciarletti et al. 2015).

**Porosity** The low density of cometary nuclei suggests a high porosity (70–80%) for most of them. For 67P, adding the constraint from the dielectric properties, the porosity increases from 58% in the first meter to 75–86% below.

**Vertical Stratigraphy** (Fig. 13)—The broad picture is that the bulk of the comet consists of a weakly bonded, rather homogeneous material that preserved primordial properties under a thin shell of processed material, and possibly covered by a granular material. This cover might in places reach thicknesses of many meters. The processed outer shell is the result of competing effects between (i) the sublimation of volatiles that ejects material into space, and (ii) the recondensation of volatiles between grains that cement the shell. The physical properties of the top layer (the first meter) are not representative of that of the bulk nucleus; the top layer is always thermally processed, and moreover denser, less porous, and harder when made of consolidated materials. Finally, strong nucleus heterogeneities at a scale of a few meters are ruled out in 67P's small lobe.

All these physical properties put constraints on the formation and evolution scenarios of cometary nuclei. Comets are among the least dense bodies of the solar system, highly porous and very weak, which have been able to preserve highly volatiles species (e.g. CO, $CO_2$, $CH_4$, $N_2$) into their interior since their formation, protected from solar heating by a low bulk thermal conductivity. This points toward a formation and evolution scenario that avoids global heating above ~90 K (amorphous/crystalline water ice transition) and global mechanical alteration of the nucleus. As discussed in Weissman et al. (2019), a formation by hierarchical agglomeration (Davidsson et al. 2016) or a formation by pebbles accretion in streaming instabilities (Blum et al. 2014) are both possible. After its formation, 67P either remained intact until today, in which case it would be a primordial object (Davidsson et al. 2016), or it experienced one or several sub-catastrophic collisions (Jutzi and Benz 2017) or catastrophic disruptions (Schwartz et al. 2018) with the re-aggregation of materials as rubble-pile. Alternatively, 67P could also be the fragment of a larger Kuiper belt object (Davis and Farinella 1997; Morbidelli and Rickman 2015), but this hypothesis is likely limited to a parent body not exceeding 10 km in size. Indeed, as explained by Davidsson et al. (2016), only small bodies (diameter <10 km) formed by the slow (~3 Myr) hierarchical agglomeration of materials can preserve a temperature <100 K in their interior, even in the presence of radioactive heating, allowing them to keep their super volatiles, most probably trapped in amorphous ice or clathrates (Mousis et al. 2015). Larger bodies (50–100 km in size) have a lithostatic pressure of several to tens of kPa, much larger than the inferred tensile strength of <100 Pa, which would mean that, in this collisional fragment scenario, 67P would be made of numerous mechanically strong (kPa–MPa) blocks and boulders, assembled together for a resulting low bulk tensile strength (<100 Pa); additionally, preserving the highly volatiles species during the giant catastrophic collision (~0.5 $km\,s^{-1}$) of two large bodies appears challenging. Disentangling between the different above scenarios however goes beyond the scope of this paper and we refer to Weissman et al. (2019) for a more detailed discussion of the pros and cons of each of them.

There is no doubt that space missions and Rosetta in particular have greatly improved our knowledge on comets, thanks to the missions themselves but also to the numerous parallel studies they motivated. For future space missions, the major challenge concerning the nucleus physical properties is probably to interact with the surface in a more controlled way than Philae did, and to accurately probe the deep interior. For ground and space telescopes, especially important due to the scarcity of in-situ space missions, the challenge will be to use the knowledge gained recently to better characterise the thermal, mechanical and structural properties of a large number of cometary nuclei, even possibly resolving spatially the largest and closest ones with forthcoming thirty-meter class telescopes (providing a spatial resolution $<1$ km at 0.3 au from the Earth).

**Acknowledgements** This project has received funding from the European Union's Horizon 2020 research and innovation programme under grant agreement no. 686709. This work was supported by the Swiss State Secretariat for Education, Research and Innovation (SERI) under contract number 16.0008-2. The opinions expressed and arguments employed herein do not necessarily reflect the official view of the Swiss Government. Yuri Skorov thanks the Deutsche Forschungsgemeinschaft (DFG) for support (grant SK 264/2).



## References

M.F. A'Hearn, M.J.S. Belton, W.A. Delamere, J. Kissel, K.P. Klaasen, L.A. McFadden, K.J. Meech, H.J. Melosh, P.H. Schultz, J.M. Sunshine, P.C. Thomas, J. Veverka, D.K. Yeomans, M.W. Baca, I. Busko, C.J. Crockett, S.M. Collins, M. Desnoyer, C.A. Eberhardy, C.M. Ernst, T.L. Farnham, L. Feaga, O. Groussin, D. Hampton, S.I. Ipatov, J.-Y. Li, D. Lindler, C.M. Lisse, N. Mastrodemos, W.M. Owen, J.E. Richardson, D.D. Wellnitz, R.L. White, Deep impact: Excavating comet Tempel 1. Science **310**, 258–264 (2005)

M.F. A'Hearn, M.J.S. Belton, W.A. Delamere, L.M. Feaga, D. Hampton, J. Kissel, K.P. Klaasen, L.A. McFadden, K.J. Meech, H.J. Melosh, P.H. Schultz, J.M. Sunshine, P.C. Thomas, J. Veverka, D.D. Wellnitz, D.K. Yeomans, S. Besse, D. Bodewits, T.J. Bowling, B.T. Carcich, S.M. Collins, T.L. Farnham, O. Groussin, B. Hermalyn, M.S. Kelley, M.S. Kelley, J.-Y. Li, D.J. Lindler, C.M. Lisse, S.A. McLaughlin, F. Merlin, S. Protopapa, J.E. Richardson, J.L. Williams, EPOXI at comet Hartley 2. Science **332**, 1396 (2011)

E. Asphaug, W. Benz, Density of comet Shoemaker-Levy 9 deduced by modelling breakup of the parent 'rubble pile'. Nature **370**, 120–124 (1994)

E. Asphaug, W. Benz, Size, density, and structure of comet Shoemaker-Levy 9 inferred from the physics of tidal breakup. Icarus **121**, 225–248 (1996)

N. Attree, O. Groussin, L. Jorda, D. Nébouy, N. Thomas, Y. Brouet, E. Kührt, F. Preusker, F. Scholten, J. Knollenberg, P. Hartogh, H. Sierks, C. Barbieri, P. Lamy, R. Rodrigo, D. Koschny, H. Rickman, H.U. Keller, M.F. A'Hearn, A.-T. Auger, M.A. Barucci, J.-L. Bertaux, I. Bertini, D. Bodewits, S. Boudreault, G. Cremonese, V. Da Deppo, B. Davidsson, S. Debei, M. De Cecco, J. Deller, M.R. El-Maarry, S. Fornasier, M. Fulle, P.J. Gutiérrez, C. Güttler, S. Hviid, W.-H. Ip, G. Kovacs, J.R. Kramm, M. Küppers, L.M. Lara, M. Lazzarin, J.J. Lopez Moreno, S. Lowry, S. Marchi, F. Marzari, S. Mottola, G. Naletto, N. Oklay, M. Pajola, I. Toth, C. Tubiana, J.-B. Vincent, X. Shi, Tensile strength of 67P/Churyumov-Gerasimenko nucleus material from overhangs. Astron. Astrophys. **611**, A33 (2018a)

N. Attree, O. Groussin, L. Jorda, S. Rodionov, A.-T. Auger, N. Thomas, Y. Brouet, O. Poch, E. Kührt, M. Knapmeyer, F. Preusker, F. Scholten, J. Knollenberg, S. Hviid, P. Hartogh, Thermal fracturing on comets. Applications to 67P/Churyumov-Gerasimenko. Astron. Astrophys. **610**, A76 (2018b)

A.-T. Auger, O. Groussin, L. Jorda, M.R. El-Maarry, S. Bouley, A. Séjourné, R. Gaskell, C. Capanna, B. Davidsson, S. Marchi, S. Höfner, P.L. Lamy, H. Sierks, C. Barbieri, R. Rodrigo, D. Koschny, H. Rickman, H.U. Keller, J. Agarwal, M.F. A'Hearn, M.A. Barucci, J.-L. Bertaux, I. Bertini, G. Cremonese, V. Da Deppo, S. Debei, M. De Cecco, S. Fornasier, M. Fulle, P.J. Gutiérrez, C. Güttler, S. Hviid, W.-H. Ip, J. Knollenberg, J.-R. Kramm, E. Kührt, M. Küppers, L.M. Lara, M. Lazzarin, J.J. Lopez Moreno, F. Marzari, M. Massironi, H. Michalik, G. Naletto, N. Oklay, A. Pommerol, L. Sabau, N. Thomas, C. Tubiana, J.-B. Vincent, K.-P. Wenzel, Meter-scale thermal contraction crack polygons on the nucleus of comet 67P/Churyumov-Gerasimenko. Icarus **301**, 173–188 (2018)

A. Bar-Nun, I. Pat-El, D. Laufer, Comparison between the findings of deep impact and our experimental results on large samples of gas-laden amorphous ice. Icarus **187**, 321–325 (2007)

A.T. Basilevsky, S.S. Krasil'nikov, A.A. Shiryaev, U. Mall, H.U. Keller, Y.V. Skorov, S. Mottola, S.F. Hviid, Estimating the strength of the nucleus material of comet 67P Churyumov-Gerasimenko. Sol. Syst. Res. **50**, 225–234 (2016)

Z.P. Bažant, Size effect on structural strength: A review. Arch. Appl. Mech. **69**, 702–725 (1999)

J.-P. Bibring, Y. Langevin, J. Carter, P. Eng, B. Gondet, L. Jorda, S. Le Mouélic, S. Mottola, C. Pilorget, F. Poulet, M. Vincendon, 67P/Churyumov-Gerasimenko surface properties as derived from CIVA panoramic images. Science **349** (2015)

J. Biele, S. Ulamec, L. Richter, J. Knollenberg, E. Kührt, D. Möhlmann, The putative mechanical strength of comet surface material applied to landing on a comet. Acta Astronaut. **65**, 1168–1178 (2009)

J. Biele, S. Ulamec, M. Maibaum, R. Roll, L. Witte, E. Jurado, P. Muñoz, W. Arnold, H.-U. Auster, C. Casas, C. Faber, C. Fantinati, F. Finke, H.-H. Fischer, K. Geurts, C. Güttler, P. Heinisch, A. Herique, S. Hviid, G. Kargl, M. Knapmeyer, J. Knollenberg, W. Kofman, N. Kömle, E. Kührt, V. Lommatsch, S. Mottola, R. Pardo de Santayana, E. Remetean, F. Scholten, K.J. Seidensticker, H. Sierks, T. Spohn, The landing(s) of Philae and inferences about comet surface mechanical properties. Science **349** (2015)

N. Biver, M. Hofstadter, S. Gulkis, D. Bockelée-Morvan, M. Choukroun, E. Lellouch, F.P. Schloerb, L. Rezac, W.H. Ip, C. Jarchow, P. Hartogh, S. Lee, P. von Allmen, J. Crovisier, C. Leyrat, P. Encrenaz, Distribution of water around the nucleus of comet 67P/Churyumov-Gerasimenko at 3.4 au from the Sun as seen by the MIRO instrument on Rosetta. Astron. Astrophys. **583**, A3 (2015)

J. Blum, R. Schräpler, Structure and mechanical properties of high-porosity macroscopic agglomerates formed by random ballistic deposition. Phys. Rev. Lett. **93**, 115503 (2004)

J. Blum, R. Schräpler, B.J.R. Davidsson, J.M. Trigo-Rodríguez, The physics of protoplanetesimal dust agglomerates. I. Mechanical properties and relations to primitive bodies in the solar system. Astrophys. J. **652**, 1768–1781 (2006)

J. Blum, B. Gundlach, S. Mühle, J.M. Trigo-Rodriguez, Comets formed in solar-nebula instabilities!—An experimental and modeling attempt to relate the activity of comets to their formation process. Icarus **235**, 156–169 (2014)

J. Blum, B. Gundlach, M. Krause, M. Fulle, A. Johansen, J. Agarwal, I. von Borstel, X. Shi, X. Hu, M.S. Bentley, F. Capaccioni, L. Colangeli, V. Della Corte, N. Fougere, S.F. Green, S. Ivanovski, T. Mannel, S. Merouane, A. Migliorini, A. Rotundi, R. Schmied, C. Snodgrass, Evidence for the formation of comet 67P/Churyumov-Gerasimenko through gravitational collapse of a bound clump of pebbles. Mon. Not. R. Astron. Soc. **469**, S755–S773 (2017)

J. Boissier, O. Groussin, L. Jorda, P. Lamy, D. Bockelée-Morvan, J. Crovisier, N. Biver, P. Colom, E. Lellouch, R. Moreno, Earth-based detection of the millimetric thermal emission from the nucleus of comet 8P/Tuttle. Astron. Astrophys. **528**, A54 (2011)

J. Brisset, D. Heißelmann, S. Kothe, R. Weidling, J. Blum, Submillimetre-sized dust aggregate collision and growth properties. Experimental study of a multi-particle system on a suborbital rocket. Astron. Astrophys. **593**, A3 (2016)

Y. Brouet, A.C. Levasseur-Regourd, P. Encrenaz, S. Gulkis, Permittivity of porous granular matter, in relation with Rosetta cometary mission. Planet. Space Sci. **103**, 143–152 (2014)

Y. Brouet, A.C. Levasseur-Regourd, P. Sabouroux, P. Encrenaz, N. Thomas, E. Heggy, W. Kofman, Permittivity measurements of porous matter in support of investigations of the surface and interior of 67P/Churyumov-Gerasimenko. Astron. Astrophys. **583**, A39 (2015)

Y. Brouet, A.C. Levasseur-Regourd, P. Sabouroux, L. Neves, P. Encrenaz, O. Poch, A. Pommerol, N. Thomas, W. Kofman, A. Le Gall, V. Ciarletti, A. Hérique, A. Lethuillier, N. Carrasco, C. Szopa, A porosity gradient in 67P/C-G nucleus suggested from CONSERT and SESAME-PP results: An interpretation based on new laboratory permittivity measurements of porous icy analogues. Mon. Not. R. Astron. Soc. **462**, S89–S98 (2016a)

Y. Brouet, L. Neves, P. Sabouroux, A.C. Levasseur-Regourd, O. Poch, P. Encrenaz, A. Pommerol, N. Thomas, W. Kofman, Characterization of the permittivity of controlled porous water ice-dust mixtures to support the radar exploration of icy bodies. J. Geophys. Res., Planets **121**, 2426–2443 (2016b)

B. Brugger, O. Mousis, A. Morse, U. Marboeuf, L. Jorda, A. Guilbert-Lepoutre, D. Andrews, S. Barber, P. Lamy, A. Luspay-Kuti, K. Mandt, G. Morgan, S. Sheridan, P. Vernazza, I.P. Wright, Subsurface characterization of 67P/Churyumov-Gerasimenko's Abydos site. Astrophys. J. **822**, 98 (2016)

B.J. Buratti, M.D. Hicks, L.A. Soderblom, D. Britt, J. Oberst, J.K. Hillier, Deep space 1 photometry of the nucleus of comet 19P/Borrelly. Icarus **167**, 16–29 (2004)

M.J. Campbell, J. Ulrichs, Electrical properties of rocks and their significance for lunar radar observations. J. Geophys. Res. **74**, 5867–5881 (1969)

D.B. Campbell, J.K. Harmon, I.I. Shapiro, Radar observations of comet Halley. Astrophys. J. **338**, 1094–1105 (1989)

M.T. Capria, F. Capaccioni, G. Filacchione, F. Tosi, M.C. De Sanctis, S. Mottola, M. Ciarniello, M. Formisano, A. Longobardo, A. Migliorini, E. Palomba, A. Raponi, E. Kührt, D. Bockelée-Morvan, S. Erard, C. Leyrat, A. Zinzi, How pristine is the interior of the comet 67P/Churyumov-Gerasimenko? Mon. Not. R. Astron. Soc. **469**, S685–S694 (2017)

M. Choukroun, S. Keihm, F.P. Schloerb, S. Gulkis, E. Lellouch, C. Leyrat, P. von Allmen, N. Biver, D. Bockelée-Morvan, J. Crovisier, P. Encrenaz, P. Hartogh, M. Hofstadter, W.-H. Ip, C. Jarchow, M. Janssen, S. Lee, L. Rezac, G. Beaudin, B. Gaskell, L. Jorda, H.U. Keller, H. Sierks, Dark side of comet 67P/Churyumov-Gerasimenko in Aug.-Oct. 2014. MIRO/Rosetta continuum observations of polar night in the Southern regions. Astron. Astrophys. **583**, A28 (2015)

T.C. Choy, *Effective Medium Theory Principles and Applications* (Oxford University Press, London, 1999)

V. Ciarletti, A.C. Levasseur-Regourd, J. Lasue, C. Statz, D. Plettemeier, A. Hérique, Y. Rogez, W. Kofman, CONSERT suggests a change in local properties of 67P/Churyumov-Gerasimenko's nucleus at depth. Astron. Astrophys. **583**, A40 (2015)

V. Ciarletti, A. Herique, J. Lasue, A.-C. Levasseur-Regourd, D. Plettemeier, F. Lemmonier, C. Guiffaut, P. Pasquero, W. Kofman, CONSERT constrains the internal structure of 67P at a few metres size scale. Mon. Not. R. Astron. Soc. **469**, S805–S817 (2017)

M. Ciarniello, F. Capaccioni, G. Filacchione, A. Raponi, F. Tosi, M.C. De Sanctis, M.T. Capria, S. Erard, D. Bockelee-Morvan, C. Leyrat, G. Arnold, A. Barucci, P. Beck, G. Bellucci, S. Fornasier, A. Longobardo, S. Mottola, E. Palomba, E. Quirico, B. Schmitt, Photometric properties of comet 67P/Churyumov-Gerasimenko from VIRTIS-M onboard Rosetta. Astron. Astrophys. **583**, A31 (2015)

A. Coradini, F. Capaccioni, P. Drossart, G. Arnold, E. Ammannito, F. Angrilli, A. Barucci, G. Bellucci, J. Benkhoff, G. Bianchini, J.P. Bibring, M. Blecka, D. Bockelee-Morvan, M.T. Capria, R. Carlson, U. Carsenty, P. Cerroni, L. Colangeli, M. Combes, M. Combi, J. Crovisier, M.C. De Sanctis, E.T. Encrenaz, S. Erard, C. Federico, G. Filacchione, U. Fink, S. Fonti, V. Formisano, W.H. Ip, R. Jaumann, E. Kuehrt, Y. Langevin, G. Magni, T. McCord, V. Mennella, S. Mottola, G. Neukum, P. Palumbo, G. Piccioni, H. Rauer, B. Saggin, B. Schmitt, D. Tiphene, G. Tozzi, Virtis: An imaging spectrometer for the Rosetta mission. Space Sci. Rev. **128**, 529–559 (2007)

B.J.R. Davidsson, Tidal splitting and rotational breakup of solid biaxial ellipsoids. Icarus **149**, 375–383 (2001)

B.J.R. Davidsson, P.J. Gutiérrez, Estimating the nucleus density of comet 19P/Borrelly. Icarus **168**, 392–408 (2004)

B.J.R. Davidsson, P.J. Gutiérrez, Nucleus properties of comet 67P/Churyumov Gerasimenko estimated from non-gravitational force modeling. Icarus **176**, 453–477 (2005)

B.J.R. Davidsson, P.J. Gutiérrez, Non-gravitational force modeling of comet 81P/Wild 2. I. A nucleus bulk density estimate. Icarus **180**, 224–242 (2006)

B.J.R. Davidsson, H. Rickman, Surface roughness and three-dimensional heat conduction in thermophysical models. Icarus **243**, 58–77 (2014)

B.J.R. Davidsson, P.J. Gutiérrez, H. Rickman, Nucleus properties of comet 9P/Tempel 1 estimated from non-gravitational force modeling. Icarus **187**, 306–320 (2007)

B.J.R. Davidsson, P.J. Gutiérrez, H. Rickman, Physical properties of morphological units on comet 9P/Tempel 1 derived from near-IR deep impact spectra. Icarus **201**, 335–357 (2009)

B.J.R. Davidsson, P.J. Gutiérrez, O. Groussin, M.F. A'Hearn, T. Farnham, L.M. Feaga, M.S. Kelley, K.P. Klaasen, F. Merlin, S. Protopapa, H. Rickman, J.M. Sunshine, P.C. Thomas, Thermal inertia and surface roughness of comet 9P/Tempel 1. Icarus **224**, 154–171 (2013)

B.J.R. Davidsson, H. Sierks, C. Güttler, F. Marzari, M. Pajola, H. Rickman, M.F. A'Hearn, A.-T. Auger, M.R. El-Maarry, S. Fornasier, P.J. Gutiérrez, H.U. Keller, M. Massironi, C. Snodgrass, J.-B. Vincent, C. Barbieri, P.L. Lamy, R. Rodrigo, D. Koschny, M.A. Barucci, J.-L. Bertaux, I. Bertini, G. Cremonese, V. Da Deppo, S. Debei, M. De Cecco, C. Feller, M. Fulle, O. Groussin, S.F. Hviid, S. Höfner, W.-H. Ip, L. Jorda, J. Knollenberg, G. Kovacs, J.-R. Kramm, E. Kührt, M. Küppers, F. La Forgia, L.M. Lara, M. Lazzarin, J.J. Lopez Moreno, R. Moissl-Fraund, S. Mottola, G. Naletto, N. Oklay, N. Thomas, C. Tubiana, The primordial nucleus of comet 67P/Churyumov-Gerasimenko. Astron. Astrophys. **592**, A63 (2016)

D.R. Davis, P. Farinella, Collisional evolution of Edgeworth-Kuiper belt objects. Icarus **125**, 50–60 (1997)

M.C. De Sanctis, F. Capaccioni, M. Ciarniello, G. Filacchione, M. Formisano, S. Mottola, A. Raponi, F. Tosi, D. Bockelée-Morvan, S. Erard, C. Leyrat, B. Schmitt, E. Ammannito, G. Arnold, M.A. Barucci, M. Combi, M.T. Capria, P. Cerroni, W.-H. Ip, E. Kuehrt, T.B. McCord, E. Palomba, P. Beck, E. Quirico, G. Piccioni, G. Bellucci, M. Fulchignoni, R. Jaumann, K. Stephan, A. Longobardo, V. Mennella, A. Migliorini, J. Benkhoff, J.P. Bibring, A. Blanco, M. Blecka, R. Carlson, U. Carsenty, L. Colangeli, M. Combes, J. Crovisier, P. Drossart, T. Encrenaz, C. Federico, U. Fink, S. Fonti, P. Irwin, Y. Langevin, G. Magni, L. Moroz, V. Orofino, U. Schade, F. Taylor, D. Tiphene, G.P. Tozzi, N. Biver, L. Bonal, J.-P. Combe, D. Despan, E. Flamini, S. Fornasier, A. Frigeri, D. Grassi, M.S. Gudipati, F. Mancarella, K. Markus, F. Merlin, R. Orosei, G. Rinaldi, M. Cartacci, A. Cicchetti, S. Giuppi, Y. Hello, F. Henry, S. Jacquinod, J.M. Rees, R. Noschese, R. Politi, G. Peter (VIRTIS Team), The diurnal cycle of water ice on comet 67P/Churyumov-Gerasimenko. Nature **525**, 500–503 (2015)

M. Delbó, P. Tanga, Thermal inertia of main belt asteroids smaller than 100 km from IRAS data. Planet. Space Sci. **57**, 259–265 (2009)

M. Delbó, A. dell'Oro, A.W. Harris, S. Mottola, M. Mueller, Thermal inertia of near-Earth asteroids and implications for the magnitude of the Yarkovsky effect. Icarus **190**, 236–249 (2007)

M. Delbó, G. Libourel, J. Wilkerson, N. Murdoch, P. Michel, K.T. Ramesh, C. Ganino, C. Verati, S. Marchi, Thermal fatigue as the origin of regolith on small asteroids. Nature **508**, 233–236 (2014)

B.V. Derjaguin, V.M. Muller, U.P. Toporov, Effect of contact deformations on the adhesion of particles. J. Colloid Interface Sci. **53**, 314–326 (1975)

M.R. El-Maarry, N. Thomas, A. Gracia-Berná, R. Marschall, A.-T. Auger, O. Groussin, S. Mottola, M. Pajola, M. Massironi, S. Marchi, S. Höfner, F. Preusker, F. Scholten, L. Jorda, E. Kührt, H.U. Keller, H. Sierks, M.F. A'Hearn, C. Barbieri, M.A. Barucci, J.-L. Bertaux, I. Bertini, G. Cremonese, V. Da Deppo, B. Davidsson, S. Debei, M. De Cecco, J. Deller, C. Güttler, S. Fornasier, M. Fulle, P.J. Gutierrez, M. Hofmann, S.F. Hviid, W.-H. Ip, J. Knollenberg, D. Koschny, G. Kovacs, J.-R. Kramm, M. Küppers, P.L. Lamy, L.M. Lara, M. Lazzarin, J.J. Lopez Moreno, F. Marzari, H. Michalik, G. Naletto, N. Oklay, A. Pommerol, H. Rickman, R. Rodrigo, C. Tubiana, J.-B. Vincent, Fractures on comet 67P/Churyumov-Gerasimenko observed by Rosetta/OSIRIS. Geophys. Res. Lett. **42**, 5170–5178 (2015)

M.R. El-Maarry, O. Groussin, H.U. Keller, N. Thomas, J-B. Vincent, S. Mottola, M. Pajola, K. Otto, C. Herny, S. Krasilnikov, Surface morphology of comets and associated evolutionary processes: a review of Rosetta's observations of 67P/Churyumov-Gerasimenko. Space Sci. Rev. (2019), this issue

C. Emerich, J.M. Lamarre, V.I. Moroz, M. Combes, N.F. Sanko, Y.V. Nikolsky, F. Rocard, R. Gispert, N. Coron, J.P. Bibring, T. Encrenaz, J. Crovisier, Temperature and size of the nucleus of comet p/Halley deduced from IKS infrared VEGA-1 measurements. Astron. Astrophys. **187**, 839 (1987)

F.P. Fanale, J.R. Salvail, An idealized short-period comet model—Surface insolation, $H_2O$ flux, dust flux, and mantle evolution. Icarus **60**, 476–511 (1984)

T.L. Farnham, A.L. Cochran, A McDonald observatory study of comet 19P/Borrelly: Placing the deep space 1 observations into a broader context. Icarus **160**, 398–418 (2002)

Y.R. Fernández, M.S. Kelley, P.L. Lamy, I. Toth, O. Groussin, C.M. Lisse, M.F. A'Hearn, J.M. Bauer, H. Campins, A. Fitzsimmons, J. Licandro, S.C. Lowry, K.J. Meech, J. Pittichová, W.T. Reach, C. Snodgrass, H.A. Weaver, Thermal properties, sizes, and size distribution of Jupiter-family cometary nuclei. Icarus **226**, 1138–1170 (2013)

G. Filacchione, O. Groussin, C. Herny, D. Kappel, S. Mottola, N. Oklay, A. Pommerol, I. Wright, Z. Yoldi, M. Ciarniello, L. Moroz, A. Raponi, Comet 67P/CG nucleus composition and comparison to other comets. Space Sci. Rev. **215**, 19 (2019). https://doi.org/10.1007/s11214-019-0580-3

S. Fornasier, P.H. Hasselmann, M.A. Barucci, C. Feller, S. Besse, C. Leyrat, L. Lara, P.J. Gutierrez, N. Oklay, C. Tubiana, F. Scholten, H. Sierks, C. Barbieri, P.L. Lamy, R. Rodrigo, D. Koschny, H. Rickman, H.U. Keller, J. Agarwal, M.F. A'Hearn, J.-L. Bertaux, I. Bertini, G. Cremonese, V. Da Deppo, B. Davidsson, S. Debei, M. De Cecco, M. Fulle, O. Groussin, C. Güttler, S.F. Hviid, W. Ip, L. Jorda, J. Knollenberg, G. Kovacs, R. Kramm, E. Kührt, M. Küppers, F. La Forgia, M. Lazzarin, J.J. Lopez Moreno, F. Marzari, K.-D. Matz, H. Michalik, F. Moreno, S. Mottola, G. Naletto, M. Pajola, A. Pommerol, F. Preusker, X. Shi, C. Snodgrass, N. Thomas, J.-B. Vincent, Spectrophotometric properties of the nucleus of comet 67P/Churyumov-Gerasimenko from the OSIRIS instrument onboard the ROSETTA spacecraft. Astron. Astrophys. **583**, A30 (2015)

N. Gortsas, E. Kührt, U. Motschmann, H.U. Keller, Thermal model of water and CO activity of comet C/1995 O1 (Hale-Bopp). Icarus **212**, 858–866 (2011)

J.M. Greenberg, H. Mizutani, T. Yamamoto, A new derivation of the tensile strength of cometary nuclei: Application to comet Shoemaker-Levy 9. Astron. Astrophys. **295**, L35–L38 (1995)

M. Grott, J. Knollenberg, B. Borgs, F. Hänschke, E. Kessler, J. Helbert, A. Maturilli, N. Müller, The MASCOT radiometer MARA for the Hayabusa 2 mission. Space Sci. Rev. **208**, 413–431 (2017)

O. Groussin, J.M. Sunshine, L.M. Feaga, L. Jorda, P.C. Thomas, J.-Y. Li, M.F. A'Hearn, M.J.S. Belton, S. Besse, B. Carcich, T.L. Farnham, D. Hampton, K. Klaasen, C. Lisse, F. Merlin, S. Protopapa, The temperature, thermal inertia, roughness and color of the nuclei of comets 103P/Hartley 2 and 9P/Tempel 1. Icarus **222**, 580–594 (2013)

O. Groussin, L. Jorda, A.-T. Auger, E. Kührt, R. Gaskell, C. Capanna, F. Scholten, F. Preusker, P. Lamy, S. Hviid, J. Knollenberg, U. Keller, C. Huettig, H. Sierks, C. Barbieri, R. Rodrigo, D. Koschny, H. Rickman, M.F. A'Hearn, J. Agarwal, M.A. Barucci, J.-L. Bertaux, I. Bertini, S. Boudreault, G. Cremonese, V. Da Deppo, B. Davidsson, S. Debei, M. De Cecco, M.R. El-Maarry, S. Fornasier, M. Fulle, P.J. Gutiérrez, C. Güttler, W.-H. Ip, J.-R. Kramm, M. Küppers, M. Lazzarin, L.M. Lara, J.J. Lopez Moreno, S. Marchi, F. Marzari, M. Massironi, H. Michalik, G. Naletto, N. Oklay, A. Pommerol, M. Pajola, N. Thomas, I. Toth, C. Tubiana, J.-B. Vincent, Gravitational slopes, geomorphology, and material strengths of the nucleus of comet 67P/Churyumov-Gerasimenko from OSIRIS observations. Astron. Astrophys. **583**, A32 (2015)

E. Grün, J. Benkhoff, R. Heidrich, P. Hesselbarth, H. Kohl, E. Kührt, Energy balance of the KOSI 4 experiment. Geophys. Res. Lett. **18**, 257–260 (1991)

S. Gulkis, M. Frerking, J. Crovisier, G. Beaudin, P. Hartogh, P. Encrenaz, T. Koch, C. Kahn, Y. Salinas, R. Nowicki, R. Irigoyen, M. Janssen, P. Stek, M. Hofstadter, M. Allen, C. Backus, L. Kamp, C. Jarchow, E. Steinmetz, A. Deschamps, J. Krieg, M. Gheudin, D. Bockelée-Morvan, N. Biver, T. Encrenaz, D. Despois, W. Ip, E. Lellouch, I. Mann, D. Muhleman, H. Rauer, P. Schloerb, T. Spilker, MIRO: Microwave instrument for Rosetta orbiter. Space Sci. Rev. **128**, 561–597 (2007)

S. Gulkis, M. Allen, P. von Allmen, G. Beaudin, N. Biver, D. Bockelée-Morvan, M. Choukroun, J. Crovisier, B.J.R. Davidsson, P. Encrenaz, T. Encrenaz, M. Frerking, P. Hartogh, M. Hofstadter, W.-H. Ip, M. Janssen, C. Jarchow, S. Keihm, S. Lee, E. Lellouch, C. Leyrat, L. Rezac, F.P. Schloerb, T. Spilker, Subsurface properties and early activity of comet 67P/Churyumov-Gerasimenko. Science **347**, aaa0709 (2015)

B. Gundlach, S. Kilias, E. Beitz, J. Blum, Micrometer-sized ice particles for planetary-science experiments—I. Preparation, critical rolling friction force, and specific surface energy. Icarus **214**, 717–723 (2011)

B. Gundlach, J. Ratte, J. Blum, J. Oesert, S.N. Gorb, Sintering and sublimation of micrometre-sized water-ice particles: The formation of surface crusts on icy Solar System bodies. Mon. Not. R. Astron. Soc. **479**, 5272–5287 (2018a)

B. Gundlach, K.P. Schmidt, C. Kreuzig, D. Bischoff, F. Rezaei, S. Kothe, J. Blum, B. Grzesik, E. Stoll, The tensile strength of ice and dust aggregates and its dependence on particle properties. Mon. Not. R. Astron. Soc. **479**, 1273–1277 (2018b)

C. Güttler, M. Krause, R.J. Geretshauser, R. Speith, J. Blum, The physics of protoplanetesimal dust agglomerates. IV. Toward a dynamical collision model. Astrophys. J. **701**, 130–141 (2009)

O.L. Hansen, An explication of the radiometric method for size and albedo determination. Icarus **31**, 456–482 (1977)

B. Hapke, Bidirectional reflectance spectroscopy. III—Correction for macroscopic roughness. Icarus **59**, 41–59 (1984)

J.K. Harmon, M.C. Nolan, Radar observations of comet 2P/Encke during the 2003 apparition. Icarus **176**, 175–183 (2005)

J.K. Harmon, S.J. Ostro, L.A.M. Benner, K.D. Rosema, R.F. Jurgens, R. Winkler, D.K. Yeomans, D. Choate, R. Cormier, J.D. Giorgini, D.L. Mitchell, P.W. Chodas, R. Rose, D. Kelley, M.A. Slade, M.L. Thomas, Radar detection of the nucleus and coma of comet Hyakutake (C/1996 B2). Science **278**, 1921 (1997)

J.K. Harmon, M.C. Nolan, S.J. Ostro, D.B. Campbell, Radar studies of comet nuclei and grain comae, in *Comets II*, ed. by M.C. Festou, H.U. Keller, H.A. Weaver (University of Arizona Press, Tucson, 2004), pp. 265–279

J.K. Harmon, M.C. Nolan, J.D. Giorgini, E.S. Howell, Radar observations of 8P/Tuttle: A contact-binary comet. Icarus **207**, 499–502 (2010)

J.K. Harmon, M.C. Nolan, E.S. Howell, J.D. Giorgini, P.A. Taylor, Radar observations of comet 103P/Hartley 2. Astrophys. J. **734**, L2 (2011)

A.W. Harris, The rotation rates of very small asteroids: Evidence for 'Rubble Pile' structure, in *Lunar and Planetary Science Conference*, vol. 27 (1996)

Z. Hashin, S.A. Shtrikman, Variational approach to the theory of the effective magnetic permeability of multiphase materials. J. Appl. Phys. **33**, 3125–3131 (1962)

P.O. Hayne, J.L. Bandfield, M.A. Siegler, A.R. Vasavada, R.R. Ghent, J.-P. Williams, B.T. Greenhagen, O. Aharonson, C.M. Elder, P.G. Lucey, D.A. Paige, Global regolith thermophysical properties of the Moon from the diviner lunar radiometer experiment. J. Geophys. Res., Planets **122**, 2371–2400 (2017)

E. Heggy, P. Paillou, G. Ruffie, J.M. Malezieux, F. Costard, G. Grandjean, On water detection in the Martian subsurface using sounding radar. Icarus **154**, 244–257 (2001)

E. Heggy, E.M. Palmer, W. Kofman, S.M. Clifford, K. Righter, A. Hérique, Radar properties of comets: Parametric dielectric modeling of comet 67P/Churyumov-Gerasimenko. Icarus **221**, 925–939 (2012)

A. Herique, W. Kofman, P. Beck, L. Bonal, I. Buttarazzi, E. Heggy, J. Lasue, A.C. Levasseur-Regourd, E. Quirico, S. Zine, Cosmochemical implications of CONSERT permittivity characterization of 67P/CG. Mon. Not. R. Astron. Soc. **462**, S516–S532 (2016)

G. Herman, P.R. Weissman, Numerical simulation of cometary nuclei. III—Internal temperatures of cometary nuclei. Icarus **69**, 314–328 (1987)

H. Hertz, Über die Berührung fester elastischer Körper. J. Reine Angew. Math. **92**, 156–171 (1882)

M. Hirabayashi, D.J. Scheeres, S.R. Chesley, S. Marchi, J.W. McMahon, J. Steckloff, S. Mottola, S.P. Naidu, T. Bowling, Fission and reconfiguration of bilobate comets as revealed by 67P/Churyumov-Gerasimenko. Nature **534**, 352–355 (2016)

K.A. Holsapple, K.R. Housen, A crater and its ejecta: An interpretation of deep impact. Icarus **187**, 345–356 (2007)

K. Hornung, S. Merouane, M. Hilchenbach, Y. Langevin, E.M. Mellado, V. Della Corte, J. Kissel, C. Engrand, R. Schulz, J. Ryno, J. Silen (COSIMA Team), A first assessment of the strength of cometary particles collected in-situ by the COSIMA instrument onboard ROSETTA. Planet. Space Sci. **133**, 63–75 (2016)

C.J.A. Howett, J.R. Spencer, J. Pearl, M. Segura, Thermal inertia and bolometric Bond albedo values for Mimas, Enceladus, Tethys, Dione, Rhea and Iapetus as derived from Cassini/CIRS measurements. Icarus **206**, 573–593 (2010)

W.F. Huebner, The KOSI experiments. Geophys. Res. Lett. **18**, 243 (1991)

H.L. Jessberger, M. Kotthaus, Compressive strength of synthetic comet nucleus samples. Phys. Mech. Cometary Mater. **302** (1989)

K.L. Johnson, K. Kendall, A.D. Roberts, Surface energy and the contact of elastic solids. Proc. R. Soc. Lond. A **324**, 301 (1971)

L. Jorda, P. Lamy, O. Groussin, I. Toth, M.F. A'Hearn, S. Peschke, ISOCAM observations of cometary nuclei, in *ISO Beyond Point Sources: Studies of Extended Infrared Emission*, vol. 455 (2000), p. 61

L. Jorda, R. Gaskell, C. Capanna, S. Hviid, P. Lamy, J. Ďurech, G. Faury, O. Groussin, P. Gutiérrez, C. Jackman, S.J. Keihm, H.U. Keller, J. Knollenberg, E. Kührt, S. Marchi, S. Mottola, E. Palmer, F.P. Schloerb, H. Sierks, J.-B. Vincent, M.F. A'Hearn, C. Barbieri, R. Rodrigo, D. Koschny, H. Rickman, M.A. Barucci, J.L. Bertaux, I. Bertini, G. Cremonese, V. Da Deppo, B. Davidsson, S. Debei, M. De Cecco, S. Fornasier, M. Fulle, C. Güttler, W.-H. Ip, J.R. Kramm, M. Küppers, L.M. Lara, M. Lazzarin, J.J. Lopez Moreno, F. Marzari, G. Naletto, N. Oklay, N. Thomas, C. Tubiana, K.-P. Wenzel, The global shape, density and rotation of comet 67P/Churyumov-Gerasimenko from preperihelion Rosetta/OSIRIS observations. Icarus **277**, 257–278 (2016)

W.H. Julian, N.H. Samarasinha, M.J.S. Belton, Thermal structure of cometary active regions: Comet 1P/Halley. Icarus **144**, 160–171 (2000)

M. Jutzi, W. Benz, Formation of bi-lobed shapes by sub-catastrophic collisions. A late origin of comet 67P's structure. Astron. Astrophys. **597**, A62 (2017)

P.G.D. Kamoun, Radar observations of cometary nuclei. Ph.D. thesis (1983)

P.G. Kamoun, D.B. Campbell, S.J. Ostro, G.H. Pettengill, I.I. Shapiro, Comet Encke—Radar detection of nucleus. Science **216**, 293–295 (1982)

P. Kamoun, D. Campbell, G. Pettengill, I. Shapiro, Radar observations of three comets and detection of echoes from one: P/Grigg-Skjellerup. Planet. Space Sci. **47**, 23–28 (1998)

P. Kamoun, P.L. Lamy, I. Toth, A. Herique, Constraints on the subsurface structure and density of the nucleus of comet 67P/Churyumov-Gerasimenko from Arecibo radar observations. Astron. Astrophys. **568**, A21 (2014)

H.U. Keller, C. Barbieri, P. Lamy, H. Rickman, R. Rodrigo, K.-P. Wenzel, H. Sierks, M.F. A'Hearn, F. Angrilli, M. Angulo, M.E. Bailey, P. Barthol, M.A. Barucci, J.-L. Bertaux, G. Bianchini, J.-L. Boit, V. Brown, J.A. Burns, I. Büttner, J.M. Castro, G. Cremonese, W. Curdt, V. da Deppo, S. Debei, M. de Cecco, K. Dohlen, S. Fornasier, M. Fulle, D. Germerott, F. Gliem, G.P. Guizzo, S.F. Hviid, W.-H. Ip, L. Jorda, D. Koschny, J.R. Kramm, E. Kührt, M. Küppers, L.M. Lara, A. Llebaria, A. López, A. López-Jimenez, J. López-Moreno, R. Meller, H. Michalik, M.D. Michelena, R. Müller, G. Naletto, A. Origné, G. Parzianello, M. Pertile, C. Quintana, R. Ragazzoni, P. Ramous, K.-U. Reiche, M. Reina, J. Rodríguez, G. Rousset, L. Sabau, A. Sanz, J.-P. Sivan, K. Stöckner, J. Tabero, U. Telljohann, N. Thomas, V. Timon, G. Tomasch, T. Wittrock, M. Zaccariotto, OSIRIS the scientific camera system onboard Rosetta. Space Sci. Rev. **128**, 433–506 (2007)

H.U. Keller, S. Mottola, B. Davidsson, S.E. Schröder, Y. Skorov, E. Kührt, O. Groussin, M. Pajola, S.F. Hviid, F. Preusker, F. Scholten, M.F. A'Hearn, H. Sierks, C. Barbieri, P. Lamy, R. Rodrigo, D. Koschny, H. Rickman, M.A. Barucci, J.-L. Bertaux, I. Bertini, G. Cremonese, V. Da Deppo, S. Debei, M. De Cecco, S. Fornasier, M. Fulle, P.J. Gutiérrez, W.-H. Ip, L. Jorda, J. Knollenberg, J.R. Kramm, M. Küppers, L.M. Lara, M. Lazzarin, J.J. Lopez Moreno, F. Marzari, H. Michalik, G. Naletto, L. Sabau, N. Thomas,

J.-B. Vincent, K.-P. Wenzel, J. Agarwal, C. Güttler, N. Oklay, C. Tubiana, Insolation, erosion, and morphology of comet 67P/Churyumov-Gerasimenko. Astron. Astrophys. **583**, A34 (2015)

H.U. Keller, S. Mottola, S.F. Hviid, J. Agarwal, E. Kührt, Y. Skorov, K. Otto, J.-B. Vincent, N. Oklay, S.E. Schröder, B. Davidsson, M. Pajola, X. Shi, D. Bodewits, I. Toth, F. Preusker, F. Scholten, H. Sierks, C. Barbieri, P. Lamy, R. Rodrigo, D. Koschny, H. Rickman, M.F. A'Hearn, M.A. Barucci, J.-L. Bertaux, I. Bertini, G. Cremonese, V. Da Deppo, S. Debei, M. De Cecco, J. Deller, S. Fornasier, M. Fulle, O. Groussin, P.J. Gutiérrez, C. Güttler, M. Hofmann, W.-H. Ip, L. Jorda, J. Knollenberg, J.R. Kramm, M. Küppers, L.-M. Lara, M. Lazzarin, J.J. Lopez-Moreno, F. Marzari, G. Naletto, C. Tubiana, N. Thomas, Seasonal mass transfer on the nucleus of comet 67P/Chuyumov-Gerasimenko. Mon. Not. R. Astron. Soc. **469**, S357–S371 (2017)

W.D. Kingery, H.K. Bowen, D.R. Uhlmann, *Introduction to Ceramics* (Wiley, New York, 1976)

J. Klinger, S. Espinasse, B. Schmidt, Some considerations on cohesive forces in sun-grazing comets. Phys. Mech. Cometary Mater. **302** (1989)

M. Knapmeyer, H.-H. Fischer, J. Knollenberg, K.J. Seidensticker, K. Thiel, W. Arnold, C. Faber, D. Möhlmann, Structure and elastic parameters of the near surface of Abydos site on comet 67P/Churyumov-Gerasimenko, as obtained by SESAME/CASSE listening to the MUPUS insertion phase. Icarus **310**, 165–193 (2018)

H.W. Kochan, W.F. Huebner, D.W.G. Sears, Simulation experiments with cometary analogous material. Earth Moon Planets **80**, 369–411 (1998)

W. Kofman, Y. Barbin, J. Klinger, A.-C. Levasseur-Regourd, J.-P. Barriot, A. Hérique, T. Hagfors, E. Nielsen, E. Grün, P. Edenhofer, H. Kochan, G. Picardi, R. Seu, J. van Zyl, C. Elachi, J. Melosh, J. Veverka, P. Weissman, L.H. Svedhem, S.E. Hamran, I.P. Williams, Comet nucleus sounding experiment by radiowave transmission. Adv. Space Res. **21**, 1589–1598 (1998)

W. Kofman, A. Herique, J.-P. Goutail, T. Hagfors, I.P. Williams, E. Nielsen, J.-P. Barriot, Y. Barbin, C. Elachi, P. Edenhofer, A.-C. Levasseur-Regourd, D. Plettemeier, G. Picardi, R. Seu, V. Svedhem, The Comet Nucleus Sounding Experiment by Radiowave Transmission (CONSERT): A short description of the instrument and of the commissioning stages. Space Sci. Rev. **128**, 413–432 (2007)

W. Kofman, A. Herique, Y. Barbin, J.-P. Barriot, V. Ciarletti, S. Clifford, P. Edenhofer, C. Elachi, C. Eyraud, J.-P. Goutail, E. Heggy, L. Jorda, J. Lasue, A.-C. Levasseur-Regourd, E. Nielsen, P. Pasquero, F. Preusker, P. Puget, D. Plettemeier, Y. Rogez, H. Sierks, C. Statz, H. Svedhem, I. Williams, S. Zine, J. Van Zyl, Properties of the 67P/Churyumov-Gerasimenko interior revealed by CONSERT radar. Science **349** (2015)

R. Kokotanekova, C. Snodgrass, P. Lacerda, S.F. Green, S.C. Lowry, Y.R. Fernández, C. Tubiana, A. Fitzsimmons, H.H. Hsieh, Rotation of cometary nuclei: New light curves and an update of the ensemble properties of Jupiter-family comets. Mon. Not. R. Astron. Soc. **471**, 2974–3007 (2017)

N.I. Kömle, W. Macher, P. Tiefenbacher, G. Kargl, I. Pelivan, J. Knollenberg, T. Spohn, L. Jorda, C. Capanna, V. Lommatsch, B. Cozzoni, F. Finke, Three-dimensional illumination and thermal model of the Abydos region on comet 67P/Churyumov-Gerasimenko. Mon. Not. R. Astron. Soc. **469**, S2–S19 (2017)

K.J. Kossacki, N.I. Koemle, G. Kargl, G. Steiner, The influence of grain sintering on the thermoconductivity of porous ice. Planet. Space Sci. **42**, 383–389 (1994)

K.J. Kossacki, T. Spohn, A. Hagermann, E. Kaufmann, E. Kührt, Comet 67P/Churyumov-Gerasimenko: Hardening of the sub-surface layer. Icarus **260**, 464–474 (2015)

M. Krause, J. Blum, Y.V. Skorov, M. Trieloff, Thermal conductivity measurements of porous dust aggregates: I. Technique, model and first results. Icarus **214**, 286–296 (2011)

E. Kuehrt, Temperature profiles and thermal stresses in cometary nuclei. Icarus **60**, 512–521 (1984)

E. Kuehrt, H.U. Keller, The formation of cometary surface crusts. Icarus **109**, 121–132 (1994)

E. Kuehrt, B. Giese, H.U. Keller, L.V. Ksanfomality, Interpretation of the KRFM-infrared measurements of PHOBOS. Icarus **96**, 213–218 (1992)

P.L. Lamy, I. Toth, Y.R. Fernandez, H.A. Weaver, The sizes, shapes, albedos, and colors of cometary nuclei, in *Comets II* (2004), pp. 223–264

P.L. Lamy, I. Toth, O. Groussin, L. Jorda, M.S. Kelley, J.A. Stansberry, Spitzer Space Telescope observations of the nucleus of comet 67P/Churyumov-Gerasimenko. Astron. Astrophys. **489**, 777–785 (2008)

L.A. Lebofsky, M.V. Sykes, E.F. Tedesco, G.J. Veeder, D.L. Matson, R.H. Brown, J.C. Gradie, M.A. Feierberg, R.J. Rudy, A refined 'standard' thermal model for asteroids based on observations of 1 Ceres and 2 Pallas. Icarus **68**, 239–251 (1986)

S. Lee, P. von Allmen, M. Allen, G. Beaudin, N. Biver, D. Bockelée-Morvan, M. Choukroun, J. Crovisier, P. Encrenaz, M. Frerking, S. Gulkis, P. Hartogh, M. Hofstadter, W.-H. Ip, M. Janssen, C. Jarchow, S. Keihm, E. Lellouch, C. Leyrat, L. Rezac, F.P. Schloerb, T. Spilker, B. Gaskell, L. Jorda, H.U. Keller, H. Sierks, Spatial and diurnal variation of water outgassing on comet 67P/Churyumov-Gerasimenko observed from Rosetta/MIRO in August 2014. Astron. Astrophys. **583**, A5 (2015)

E. Lellouch, P. Santos-Sanz, P. Lacerda, M. Mommert, R. Duffard, J.L. Ortiz, T.G. Müller, S. Fornasier, J. Stansberry, C. Kiss, E. Vilenius, M. Mueller, N. Peixinho, R. Moreno, O. Groussin, A. Delsanti, A.W. Harris, "TNOs are Cool": A survey of the trans-Neptunian region. IX. Thermal properties of Kuiper belt objects and Centaurs from combined Herschel and Spitzer observations. Astron. Astrophys. **557**, A60 (2013)

E. Lellouch, R. Moreno, T. Müller, S. Fornasier, P. Santos-Sanz, A. Moullet, M. Gurwell, J. Stansberry, R. Leiva, B. Sicardy, B. Butler, J. Boissier, The thermal emission of Centaurs and trans-Neptunian objects at millimeter wavelengths from ALMA observations. Astron. Astrophys. **608**, A45 (2017)

A. Lethuillier, A. Le Gall, M. Hamelin, W. Schmidt, K.J. Seidensticker, R. Grard, V. Ciarletti, S. Caujolle-Bert, H.-H. Fischer, R. Trautner, Electrical properties and porosity of the first meter of the nucleus of 67P/Churyumov-Gerasimenko. As constrained by the permittivity probe SESAME-PP/Philae/Rosetta. Astron. Astrophys. **591**, A32 (2016)

H.F. Levison, M.J. Duncan, From the Kuiper belt to Jupiter-family comets: The spatial distribution of ecliptic comets. Icarus **127**, 13–32 (1997)

C. Leyrat, S. Erard, M. Capria, F. Capaccioni, Effects of surface roughness on the VIRTIS/Rosetta thermal measurements, in *Asteroids, Comets, Meteors*, 2014 (2014)

C. Leyrat, D. Blain, E. Lellouch, P. von Allmen, S. Keihm, M. Choukroun, P. Schloerb, N. Biver, S. Gulkis, M. Hofstadter, Search for regional variations of thermal and electrical properties of comet 67P/CG probed by MIRO/Rosetta, in *AAS/Division for Planetary Sciences Meeting Abstracts #47 47*, 413.05 (2015)

J.-Y. Li, M.F. A'Hearn, L.A. McFadden, M.J.S. Belton, Photometric analysis and disk-resolved thermal modeling of comet 19P/Borrelly from deep space 1 data. Icarus **188**, 195–211 (2007)

J.-Y. Li, M.F. A'Hearn, T.L. Farnham, L.A. McFadden, Photometric analysis of the nucleus of comet 81P/Wild 2 from Stardust images. Icarus **204**, 209–226 (2009)

J.-Y. Li, S. Besse, M.F. A'Hearn, M.J.S. Belton, D. Bodewits, T.L. Farnham, K.P. Klaasen, C.M. Lisse, K.J. Meech, J.M. Sunshine, P.C. Thomas, Photometric properties of the nucleus of comet 103P/Hartley 2. Icarus **222**, 559–570 (2013)

C.M. Lisse, Y.R. Fernandez, W.T. Reach, J.M. Bauer, M.F. A'Hearn, T.L. Farnham, O. Groussin, M.J. Belton, K.J. Meech, C.D. Snodgrass, Spitzer space telescope observations of the nucleus of comet 103P/Hartley 2. Publ. Astron. Soc. Pac. **121**, 968 (2009)

H. Looyenga, Dielectric constants of heterogeneous mixtures. Physica **31**(3), 401–406 (1965)

S. Lorek, B. Gundlach, P. Lacerda, J. Blum, Comet formation in collapsing pebble clouds. What cometary bulk density implies for the cloud mass and dust-to-ice ratio. Astron. Astrophys. **587**, A128 (2016)

S.C. Lowry, P.R. Weissman, CCD observations of distant comets from Palomar and Steward Observatories. Icarus **164**, 492–503 (2003)

S. Marchi, W.F. Bottke, B.A. Cohen, K. Wünnemann, D.A. Kring, H.Y. McSween, M.C. de Sanctis, D.P. O'Brien, P. Schenk, C.A. Raymond, C.T. Russell, High-velocity collisions from the lunar cataclysm recorded in asteroidal meteorites. Nat. Geosci. **6**, 303–307 (2013)

D. Marshall, P. Hartogh, L. Rezac, P. von Allmen, N. Biver, D. Bockelée-Morvan, J. Crovisier, P. Encrenaz, S. Gulkis, M. Hofstadter, W.-H. Ip, C. Jarchow, S. Lee, E. Lellouch, Spatially resolved evolution of the local $H_2O$ production rates of comet 67P/Churyumov-Gerasimenko from the MIRO instrument on Rosetta. Astron. Astrophys. **603**, A87 (2017)

D. Marshall, O. Groussin, J.-B. Vincent, Y. Brouet, D. Kappel, G. Arnold, M.T. Capria, G. Filacchione, P. Hartogh, M. Hofstadter, W.-H. Ip, L. Jorda, E. Kuehrt, E. Lellouch, S. Mottola, L. Rezac, R. Rodrigo, S. Rodionov, P. Schloerb, N. Thomas, Thermal inertia and roughness of the nucleus of comet 67P/Churyumov-Gerasimenko from MIRO and VIRTIS observations. Astron. Astrophys. **616**, A122 (2018)

R. Marschall et al., Coma surface relationships. Space Sci. Rev. (2019), this issue

C. Matonti, N. Attree, O. Groussin, L. Jorda, S. Viseur, et al., Bilobate comet morphology and internal structure controlled by shear deformation. Nat. Geosci. **12**, 157–162 (2019)

G. Mavko, T. Mukerji, J. Dvorkin, *The Rock Physics Handbook* (Cambridge University Press, Cambridge, 2009)

T. Meisner, G. Wurm, J. Teiser, Experiments on centimeter-sized dust aggregates and their implications for planetesimal formation. Astron. Astrophys. **544**, A138 (2012)

M.T. Mellon, B.M. Jakosky, H.H. Kieffer, P.R. Christensen, High-resolution thermal inertia mapping from the Mars global surveyor thermal emission spectrometer. Icarus **148**, 437–455 (2000)

D.A. Mendis, G.D. Brin, Monochromatic brightness variations of comets. II—Core-mantle model. Moon **17**, 359–372 (1977)

D. Möhlmann, K.J. Seidensticker, H.-H. Fischer, C. Faber, A. Flandes, M. Knapmeyer, H. Krüger, R. Roll, F. Scholten, K. Thiel, W. Arnold, Compressive strength and elastic modulus at Agilkia on comet 67P/Churyumov-Gerasimenko derived from the SESAME/CASSE touchdown signals. Icarus **303**, 251–264 (2018)

A. Morbidelli, H. Rickman, Comets as collisional fragments of a primordial planetesimal disk. Astron. Astrophys. **583**, A43 (2015)

A. Morbidelli, S. Marchi, W.F. Bottke, D.A. Kring, A sawtooth-like timeline for the first billion years of lunar bombardment. Earth Planet. Sci. Lett. **355**, 144–151 (2012)

D. Morrison, D.P. Cruikshank, Thermal properties of the Galilean satellites. Icarus **18**, 224–236 (1973)

S. Mottola, G. Arnold, H.-G. Grothues, R. Jaumann, H. Michaelis, G. Neukum, J.-P. Bibring, The Rolis experiment on the Rosetta lander. Space Sci. Rev. **128**, 241–255 (2007)

S. Mottola, G. Arnold, H.-G. Grothues, R. Jaumann, H. Michaelis, G. Neukum, J.-P. Bibring, S.E. Schröder, M. Hamm, K.A. Otto, I. Pelivan, G. Proffe, F. Scholten, D. Tirsch, M. Kreslavsky, E. Remetean, F. Souvannavong, B. Dolives, The structure of the regolith on 67P/Churyumov-Gerasimenko from ROLIS descent imaging. Science **349** (2015)

O. Mousis, A. Guilbert-Lepoutre, B. Brugger, L. Jorda, J.S. Kargel, A. Bouquet, A.-T. Auger, P. Lamy, P. Vernazza, N. Thomas, H. Sierks, Pits formation from volatile outgassing on 67P/Churyumov-Gerasimenko. Astrophys. J. **814**, L5 (2015)

S.J. Ostro, Radar observations of asteroids and comets. Publ. Astron. Soc. Pac. **97**, 877–884 (1985)

M. Pajola, J.-B. Vincent, C. Güttler, J.-C. Lee, I. Bertini, M. Massironi, E. Simioni, F. Marzari, L. Giacomini, A. Lucchetti, C. Barbieri, G. Cremonese, G. Naletto, A. Pommerol, M.R. El-Maarry, S. Besse, M. Küppers, F. La Forgia, M. Lazzarin, N. Thomas, A.-T. Auger, H. Sierks, P. Lamy, R. Rodrigo, D. Koschny, H. Rickman, H.U. Keller, J. Agarwal, M.F. A'Hearn, M.A. Barucci, J.-L. Bertaux, V. Da Deppo, B. Davidsson, M. De Cecco, S. Debei, F. Ferri, S. Fornasier, M. Fulle, O. Groussin, P.J. Gutierrez, S.F. Hviid, W.-H. Ip, L. Jorda, J. Knollenberg, J.-R. Kramm, E. Kürt, L.M. Lara, Z.-Y. Lin, J.J. Lopez Moreno, S. Magrin, S. Marchi, H. Michalik, R. Moissl, S. Mottola, N. Oklay, F. Preusker, F. Scholten, C. Tubiana, Size-frequency distribution of boulders $\geq$7 m on comet 67P/Churyumov-Gerasimenko. Astron. Astrophys. **583**, A37 (2015)

M. Pajola, A. Lucchetti, I. Bertini, F. Marzari, M.F. A'Hearn, F. La Forgia, M. Lazzarin, G. Naletto, C. Barbieri, Size-frequency distribution of boulders $\geq$10 m on comet 103P/Hartley 2. Astron. Astrophys. **585**, A85 (2016)

M. Pajola, A. Lucchetti, M. Fulle, S. Mottola, M. Hamm, V. Da Deppo, L. Penasa, G. Kovacs, M. Massironi, X. Shi, C. Tubiana, C. Güttler, N. Oklay, J.B. Vincent, I. Toth, B. Davidsson, G. Naletto, H. Sierks, C. Barbieri, P.L. Lamy, R. Rodrigo, D. Koschny, H. Rickman, H.U. Keller, J. Agarwal, M.F. A'Hearn, M.A. Barucci, J.L. Bertaux, I. Bertini, G. Cremonese, S. Debei, M. De Cecco, J. Deller, M.R. El Maarry, S. Fornasier, E. Frattin, A. Gicquel, O. Groussin, P.J. Gutierrez, S. Höfner, M. Hofmann, S.F. Hviid, W.H. Ip, L. Jorda, J. Knollenberg, J.R. Kramm, E. Kührt, M. Küppers, L.M. Lara, M. Lazzarin, J.J.L. Moreno, F. Marzari, H. Michalik, F. Preusker, F. Scholten, N. Thomas, The pebbles/boulders size distributions on Sais: Rosetta's final landing site on comet 67P/Churyumov-Gerasimenko. Mon. Not. R. Astron. Soc. **469**, S636–S645 (2017)

M. Pätzold, B. Häusler, K. Aksnes, J.D. Anderson, S.W. Asmar, J.-P. Barriot, M.K. Bird, H. Boehnhardt, W. Eidel, E. Grün, W.H. Ip, E. Marouf, T. Morley, F.M. Neubauer, H. Rickman, N. Thomas, B.T. Tsurutani, M.K. Wallis, N.C. Wickramasinghe, E. Mysen, O. Olson, S. Remus, S. Tellmann, T. Andert, L. Carone, M. Fels, C. Stanzel, I. Audenrieth-Kersten, A. Gahr, A.-L. Müller, D. Stupar, C. Walter, Rosetta Radio Science Investigations (RSI). Space Sci. Rev. **128**, 599–627 (2007)

M. Pätzold, T. Andert, M. Hahn, S.W. Asmar, J.-P. Barriot, M.K. Bird, B. Häusler, K. Peter, S. Tellmann, E. Grün, P.R. Weissman, H. Sierks, L. Jorda, R. Gaskell, F. Preusker, F. Scholten, A homogeneous nucleus for comet 67P/Churyumov-Gerasimenko from its gravity field. Nature **530**, 63–65 (2016)

S.J. Peale, On the density of Halley's comet. Icarus **82**, 36–49 (1989)

I. Pelivan, Thermophysical modelling for high-resolution digital terrain models. Mon. Not. R. Astron. Soc. **478**, 386–398 (2018)

J.J. Petrovic, Mechanical properties of ice and snow. J. Mater. Sci. **38**, 1–6 (2003)

A. Pommerol, N. Thomas, M.R. El-Maarry, M. Pajola, O. Groussin, A.-T. Auger, N. Oklay, S. Fornasier, C. Feller, B. Davidsson, A. Gracia-Berná, B. Jost, R. Marschall, O. Poch, M.A. Barucci, J.-L. Bertaux, F. La Forgia, H.U. Keller, E. Kührt, S.C. Lowry, S. Mottola, G. Naletto, H. Sierks, C. Barbieri, P.L. Lamy, R. Rodrigo, D. Koschny, H. Rickman, J. Agarwal, M.F. A'Hearn, I. Bertini, S. Boudreault, G. Cremonese, V. Da Deppo, M. De Cecco, S. Debei, C. Güttler, M. Fulle, P.J. Gutierrez, S.F. Hviid, W.-H. Ip, L. Jorda, J. Knollenberg, G. Kovacs, J.-R. Kramm, E. Küppers, L. Lara, M. Lazzarin, J.L. Lopez Moreno, F. Marzari, H. Michalik, F. Preusker, F. Scholten, C. Tubiana, J.-B. Vincent, OSIRIS observations of meter-sized exposures of $H_2O$ ice at the surface of 67P/Churyumov-Gerasimenko and interpretation using laboratory experiments. Astron. Astrophys. **583**, A25 (2015)

F. Poulet, A. Lucchetti, J.-P. Bibring, J. Carter, B. Gondet, L. Jorda, Y. Langevin, C. Pilorget, C. Capanna, G. Cremonese, Origin of the local structures at the Philae landing site and possible implications on the formation and evolution of 67P/Churyumov-Gerasimenko. Mon. Not. R. Astron. Soc. **462**, S23–S32 (2016)

P. Pravec, A.W. Harris, T. Michalowski, Asteroid rotations, in *Asteroids III*, ed. by W.F. Bottke Jr., A. Cellino, P. Paolicchi, R.P. Binzel (University of Arizona Press, Tucson, 2002), pp. 113–122
F. Preusker, F. Scholten, K.-D. Matz, T. Roatsch, K. Willner, S.F. Hviid, J. Knollenberg, L. Jorda, P.J. Gutiérrez, E. Kührt, S. Mottola, M.F. A'Hearn, N. Thomas, H. Sierks, C. Barbieri, P. Lamy, R. Rodrigo, D. Koschny, H. Rickman, H.U. Keller, J. Agarwal, M.A. Barucci, J.-L. Bertaux, I. Bertini, G. Cremonese, V. Da Deppo, B. Davidsson, S. Debei, M. De Cecco, S. Fornasier, M. Fulle, O. Groussin, C. Güttler, W.-H. Ip, J.R. Kramm, M. Küppers, L.M. Lara, M. Lazzarin, J.J. Lopez Moreno, F. Marzari, H. Michalik, G. Naletto, N. Oklay, C. Tubiana, J.-B. Vincent, Shape model, reference system definition, and cartographic mapping standards for comet 67P/Churyumov-Gerasimenko—Stereo-photogrammetric analysis of Rosetta/OSIRIS image data. Astron. Astrophys. **583**, A33 (2015)
F. Preusker, F. Scholten, K.-D. Matz, T. Roatsch, S.F. Hviid, S. Mottola, J. Knollenberg, E. Kührt, M. Pajola, N. Oklay, J.-B. Vincent, B. Davidsson, M.F. A'Hearn, J. Agarwal, C. Barbieri, M.A. Barucci, J.-L. Bertaux, I. Bertini, G. Cremonese, V. Da Deppo, S. Debei, M. De Cecco, S. Fornasier, M. Fulle, O. Groussin, P.J. Gutiérrez, C. Güttler, W.-H. Ip, L. Jorda, H.U. Keller, D. Koschny, J.R. Kramm, M. Küppers, P. Lamy, L.M. Lara, M. Lazzarin, J.J. Lopez Moreno, F. Marzari, M. Massironi, G. Naletto, H. Rickman, R. Rodrigo, H. Sierks, N. Thomas, C. Tubiana, The global meter-level shape model of comet 67P/Churyumov-Gerasimenko. Astron. Astrophys. **607**, L1 (2017)
D. Prialnik, Y. Mekler, The formation of an ice crust below the dust mantle of a cometary nucleus. Astrophys. J. **366**, 318–323 (1991)
D. Prialnik, J. Benkhoff, M. Podolak, Modeling the structure and activity of comet nuclei, in *Comets II* (2004), pp. 359–387
J.E. Richardson, T.J. Bowling, Investigating the combined effects of shape, density, and rotation on small body surface slopes and erosion rates. Icarus **234**, 53–65 (2014)
J.E. Richardson, H.J. Melosh, C.M. Lisse, B. Carcich, A ballistics analysis of the deep impact ejecta plume: Determining comet Tempel 1's gravity, mass, and density. Icarus **190**, 357–390 (2007)
H. Rickman, Masses and densities of comets Halley and Kopf, in *ESA Conference Proceedings*, vol. 249 (1986), pp. 195–205
R. Roll, L. Witte, W. Arnold, ROSETTA lander Philae—Soil strength analysis. Icarus **280**, 359–365 (2016)
B. Rozitis, S.F. Green, E. MacLennan, J.P. Emery, Observing the variation of asteroid thermal inertia with heliocentric distance. Mon. Not. R. Astron. Soc. **477**, 1782–1802 (2018)
H.C. Rumpf, The strength of granules and agglomerates, in *Agglomeration* (Interscience, New York, 1962), pp. 379–413
H.C. Rumpf, Zur Theorie der Zugfestigkeit von Agglomeraten bei Kraftübertragung an Kontaktpunkten. Chem. Ing. Tech. **42**, 538–540 (1970)
H.C. Rumpf, Die Wissenschaft des Agglomerierens. Chem. Ing. Tech. **46**, 1–11 (1974)
H.C. Rumpf, *Particle Technology* (Chapman and Hall, London, 1990)
R.Z. Sagdeev, P.E. Elyasberg, V.I. Moroz, Is the nucleus of comet Halley a low density body? Nature **331**, 240–242 (1988)
N. Sakatani, K. Ogawa, Y. Iijima, M. Arakawa, R. Honda, S. Tanaka, Thermal conductivity model for powdered materials under vacuum based on experimental studies. AIP Adv. **7**, 015310 (2017)
N. Sakatani, K. Ogawa, M. Arakawa, S. Tanaka, Thermal conductivity of lunar regolith simulant JSC-1A under vacuum. Icarus **309**, 13–24 (2018)
P. Sánchez, D.J. Scheeres, The strength of regolith and rubble pile asteroids. Meteorit. Planet. Sci. **49**, 788–811 (2014)
F.P. Schloerb, S. Keihm, P. von Allmen, M. Choukroun, E. Lellouch, C. Leyrat, G. Beaudin, N. Biver, D. Bockelée-Morvan, J. Crovisier, P. Encrenaz, R. Gaskell, S. Gulkis, P. Hartogh, M. Hofstadter, W.-H. Ip, M. Janssen, C. Jarchow, L. Jorda, H.U. Keller, S. Lee, L. Rezac, H. Sierks, MIRO observations of subsurface temperatures of the nucleus of 67P/Churyumov-Gerasimenko. Astron. Astrophys. **583**, A29 (2015)
R. Schräpler, J. Blum, I. von Borstel, C. Güttler, The stratification of regolith on celestial objects. Icarus **257**, 33–46 (2015)
E.M. Schulson, P. Duval, *Creep and Fracture of Ice* (Cambridge University Press, Cambridge, 2009)
S.R. Schwartz, P. Michel, M. Jutzi, S. Marchi, Y. Zhang, D.C. Richardson, Catastrophic disruptions as the origin of bilobate comets. Nat. Astron. **2**, 379–382 (2018)
K.J. Seidensticker, D. Möhlmann, I. Apathy, W. Schmidt, K. Thiel, W. Arnold, H.-H. Fischer, M. Kretschmer, D. Madlener, A. Péter, R. Trautner, S. Schieke, Sesame—An experiment of the Rosetta lander Philae: Objectives and general design. Space Sci. Rev. **128**, 301–337 (2007)
H. Sierks, C. Barbieri, P.L. Lamy, R. Rodrigo, D. Koschny, H. Rickman, H.U. Keller, J. Agarwal, M.F. A'Hearn, F. Angrilli, A.-T. Auger, M.A. Barucci, J.-L. Bertaux, I. Bertini, S. Besse, D. Bodewits, C. Capanna, G. Cremonese, V. Da Deppo, B. Davidsson, S. Debei, M. De Cecco, F. Ferri, S. Fornasier, M. Fulle, R. Gaskell, L. Giacomini, O. Groussin, P. Gutierrez-Marques, P.J. Gutiérrez, C. Güttler, N.

Hoekzema, S.F. Hviid, W.-H. Ip, L. Jorda, J. Knollenberg, G. Kovacs, J.R. Kramm, E. Kührt, M. Küppers, F. La Forgia, L.M. Lara, M. Lazzarin, C. Leyrat, J.J. Lopez Moreno, S. Magrin, S. Marchi, F. Marzari, M. Massironi, H. Michalik, R. Moissl, S. Mottola, G. Naletto, N. Oklay, M. Pajola, M. Pertile, F. Preusker, L. Sabau, F. Scholten, C. Snodgrass, N. Thomas, C. Tubiana, J.-B. Vincent, K.-P. Wenzel, M. Zaccariotto, M. Pätzold, On the nucleus structure and activity of comet 67P/Churyumov-Gerasimenko. Science **347**, aaa1044 (2015)

A.H. Sihvola, *Electromagnetic Mixing Formulas and Applications*. Electromagnetic Waves Serie, vol. 47 (IEEE, London, 1999)

Y. Skorov, J. Blum, Dust release and tensile strength of the non-volatile layer of cometary nuclei. Icarus **221**, 1–11 (2012)

Y.V. Skorov, H. Rickman, Gas flow and dust acceleration in a cometary Knudsen layer. Planet. Space Sci. **47**, 935–949 (1999)

C. Snodgrass, S.C. Lowry, A. Fitzsimmons, Photometry of cometary nuclei: Rotation rates, colours and a comparison with Kuiper belt objects. Mon. Not. R. Astron. Soc. **373**, 1590–1602 (2006)

J.C. Solem, Cometary breakup calculations based on a gravitationally-bound agglomeration model: The density and size of Shoemaker-Levy 9. Astron. Astrophys. **302**, 596 (1995)

A. Sosa, J.A. Fernández, Cometary masses derived from non-gravitational forces. Mon. Not. R. Astron. Soc. **393**, 192–214 (2009)

T. Spohn, J. Benkhoff, J. Klinger, E. Grun, H. Kochan, Thermal modelling of two KOSI comet nucleus simulation experiments. Adv. Space Res. **9**, 127–131 (1989)

T. Spohn, K. Seiferlin, A. Hagermann, J. Knollenberg, A.J. Ball, M. Banaszkiewicz, J. Benkhoff, S. Gadomski, W. Gregorczyk, J. Grygorczuk, M. Hlond, G. Kargl, E. Kührt, N. Kömle, J. Krasowski, W. Marczewski, J.C. Zarnecki, Mupus a thermal and mechanical properties probe for the Rosetta lander Philae. Space Sci. Rev. **128**, 339–362 (2007)

T. Spohn, J. Knollenberg, A.J. Ball, M. Banaszkiewicz, J. Benkhoff, M. Grott, J. Grygorczuk, C. Hüttig, A. Hagermann, G. Kargl, E. Kaufmann, N. Kömle, E. Kührt, K.J. Kossacki, W. Marczewski, I. Pelivan, R. Schrödter, K. Seiferlin, Thermal and mechanical properties of the near-surface layers of comet 67P/Churyumov-Gerasimenko. Science **349** (2015)

D. Stoffler, H. Duren, J. Knolker, R. Hische, A. Bischoff, Cometary analogue material—Preparation, composition, and thin section petrography. Geophys. Res. Lett. **18**, 285–288 (1991)

J.M. Sunshine, M.F. A'Hearn, O. Groussin, J.-Y. Li, M.J.S. Belton, W.A. Delamere, J. Kissel, K.P. Klaasen, L.A. McFadden, K.J. Meech, H.J. Melosh, P.H. Schultz, P.C. Thomas, J. Veverka, D.K. Yeomans, I.C. Busko, M. Desnoyer, T.L. Farnham, L.M. Feaga, D.L. Hampton, D.J. Lindler, C.M. Lisse, D.D. Wellnitz, Exposed water ice deposits on the surface of comet 9P/Tempel 1. Science **311**, 1453–1455 (2006)

D. Tabor, Surface forces and surface interactions. J. Colloid Interface Sci. **58**, 2–13 (1977)

F. Tauber, E. Kuhrt, Thermal stresses in cometary nuclei. Icarus **69**, 83–90 (1987)

H. Thomas, L. Ratke, H. Kochan, Crushing strength of porous ice-mineral bodies-relevance for comets. Adv. Space Res. **14**, 207–216 (1994)

P. Thomas, M. A'Hearn, M.J.S. Belton, D. Brownlee, B. Carcich, B. Hermalyn, K. Klaasen, S. Sackett, P.H. Schultz, J. Veverka, S. Bhaskaran, D. Bodewits, S. Chesley, B. Clark, T. Farnham, O. Groussin, A. Harris, J. Kissel, J.-Y. Li, K. Meech, J. Melosh, A. Quick, J. Richardson, J. Sunshine, D. Wellnitz, The nucleus of comet 9P/Tempel 1: Shape and geology from two flybys. Icarus **222**, 453–466 (2013a)

P.C. Thomas, M.F. A'Hearn, J. Veverka, M.J.S. Belton, J. Kissel, K.P. Klaasen, L.A. McFadden, H.J. Melosh, P.H. Schultz, S. Besse, B.T. Carcich, T.L. Farnham, O. Groussin, B. Hermalyn, J.-Y. Li, D.J. Lindler, C.M. Lisse, K. Meech, J.E. Richardson, Shape, density, and geology of the nucleus of comet 103P/Hartley 2. Icarus **222**, 550–558 (2013b)

N. Thomas, H. Sierks, C. Barbieri, P.L. Lamy, R. Rodrigo, H. Rickman, D. Koschny, H.U. Keller, J. Agarwal, M.F. A'Hearn, F. Angrilli, A.-T. Auger, M.A. Barucci, J.-L. Bertaux, I. Bertini, S. Besse, D. Bodewits, G. Cremonese, V. Da Deppo, B. Davidsson, M. De Cecco, S. Debei, M.R. El-Maarry, F. Ferri, S. Fornasier, M. Fulle, L. Giacomini, O. Groussin, P.J. Gutierrez, C. Güttler, S.F. Hviid, W.-H. Ip, L. Jorda, J. Knollenberg, J.-R. Kramm, E. Kührt, M. Küppers, F. La Forgia, L.M. Lara, M. Lazzarin, J.J.L. Moreno, S. Magrin, S. Marchi, F. Marzari, M. Massironi, H. Michalik, R. Moissl, S. Mottola, G. Naletto, N. Oklay, M. Pajola, A. Pommerol, F. Preusker, L. Sabau, F. Scholten, C. Snodgrass, C. Tubiana, J.-B. Vincent, K.-P. Wenzel, The morphological diversity of comet 67P/Churyumov-Gerasimenko. Science **347**, aaa0440 (2015)

N. Thomas, M.R. El Maarry, P. Theologou, F. Preusker, F. Scholten, L. Jorda, S.F. Hviid, R. Marschall, E. Kührt, G. Naletto, H. Sierks, P.L. Lamy, R. Rodrigo, D. Koschny, B. Davidsson, M.A. Barucci, J.L. Bertaux, I. Bertini, D. Bodewits, G. Cremonese, V. Da Deppo, S. Debei, M. De Cecco, S. Fornasier, M. Fulle, O. Groussin, P.J. Gutièrrez, C. Güttler, W.H. Ip, H.U. Keller, J. Knollenberg, L.M. Lara, M. Lazzarin, J.J. Lòpez-Moreno, F. Marzari, C. Tubiana, J.B. Vincent, Regional unit definition for the nucleus of comet 67P/Churyumov-Gerasimenko on the SHAP7 model. Planet. Space Sci. **164**, 19–36 (2018)

F. Tosi, F. Capaccioni, M.T. Capria, S. Mottola, A. Zinzi, M. Ciarniello, G. Filacchione, M. Hofstadter, S. Fonti, M. Formisano, D. Kappel, E. Kührt, C. Leyrat, J.-B. Vincent, G. Arnold, M.C. De Sanctis, A. Longobardo, E. Palomba, A. Raponi, B. Rousseau, B. Schmitt, M.A. Barucci, G. Bellucci, J. Benkhoff, D. Bockelée-Morvan, P. Cerroni, J.-Ph. Combe, D. Despan, S. Erard, F. Mancarella, T.B. McCord, A. Migliorini, V. Orofino, G. Piccioni (the VIRTIS Team), The changing temperature of the nucleus of comet 67P induced by morphological and seasonal effects. Nat. Astron. (2019). https://doi.org/10.1038/s41550-019-0740-0

I. Toth, C.M. Lisse, On the rotational breakup of cometary nuclei and centaurs. Icarus **181**, 162–177 (2006)

J.M. Trigo-Rodríguez, J. Llorca, The strength of cometary meteoroids: Clues to the structure and evolution of comets. Mon. Not. R. Astron. Soc. **372**, 655–660 (2006)

S. Ulamec, C. Fantinati, M. Maibaum, K. Geurts, J. Biele, S. Jansen, O. Küchemann, B. Cozzoni, F. Finke, V. Lommatsch, A. Moussi-Soffys, C. Delmas, L. O'Rourke, Rosetta lander—Landing and operations on comet 67P/Churyumov-Gerasimenko. Acta Astronaut. **125**, 80–91 (2016)

S. Ulamec, L. O'Rourke, J. Biele, B. Grieger, R. Andres, S. Lodiot, P. Munoz, A. Charpentier, S. Mottola, J. Knollenberg, M. Knapmeyer, E. Kührt, F. Scholten, K. Geurts, M. Maibaum, C. Fantinati, O. Küchemann, V. Lommatsch, C. Delmas, E. Jurado, R. Garmier, T. Martin, Rosetta lander—Philae: Operations on comet 67P/Churyumov-Gerasimenko, analysis of wake-up activities and final state. Acta Astronaut. **137**, 38–43 (2017)

J.-B. Vincent, D. Bodewits, S. Besse, H. Sierks, C. Barbieri, P. Lamy, R. Rodrigo, D. Koschny, H. Rickman, H.U. Keller, J. Agarwal, M.F. A'Hearn, A.-T. Auger, M.A. Barucci, J.-L. Bertaux, I. Bertini, C. Capanna, G. Cremonese, V. da Deppo, B. Davidsson, S. Debei, M. de Cecco, M.R. El-Maarry, F. Ferri, S. Fornasier, M. Fulle, R. Gaskell, L. Giacomini, O. Groussin, A. Guilbert-Lepoutre, P. Gutierrez-Marques, P.J. Gutiérrez, C. Güttler, N. Hoekzema, S. Höfner, S.F. Hviid, W.-H. Ip, L. Jorda, J. Knollenberg, G. Kovacs, R. Kramm, E. Kührt, M. Küppers, F. La Forgia, L.M. Lara, M. Lazzarin, V. Lee, C. Leyrat, Z.-Y. Lin, J.J. Lopez Moreno, S. Lowry, S. Magrin, L. Maquet, S. Marchi, F. Marzari, M. Massironi, H. Michalik, R. Moissl, S. Mottola, G. Naletto, N. Oklay, M. Pajola, F. Preusker, F. Scholten, N. Thomas, I. Toth, C. Tubiana, Large heterogeneities in comet 67P as revealed by active pits from sinkhole collapse. Nature **523**, 63–66 (2015)

J.-B. Vincent, S.F. Hviid, S. Mottola, E. Kuehrt, F. Preusker, F. Scholten, H.U. Keller, N. Oklay, D. de Niem, B. Davidsson, M. Fulle, M. Pajola, M. Hofmann, X. Hu, H. Rickman, Z.-Y. Lin, C. Feller, A. Gicquel, S. Boudreault, H. Sierks, C. Barbieri, P.L. Lamy, R. Rodrigo, D. Koschny, M.F. A'Hearn, M.A. Barucci, J.-L. Bertaux, I. Bertini, G. Cremonese, V. Da Deppo, S. Debei, M. De Cecco, J. Deller, S. Fornasier, O. Groussin, P.J. Gutiérrez, P. Gutiérrez-Marquez, C. Güttler, W.-H. Ip, L. Jorda, J. Knollenberg, G. Kovacs, J.-R. Kramm, M. Küppers, L.M. Lara, M. Lazzarin, J.J. Lopez Moreno, F. Marzari, G. Naletto, L. Penasa, X. Shi, N. Thomas, I. Toth, C. Tubiana, Constraints on cometary surface evolution derived from a statistical analysis of 67P's topography. Mon. Not. R. Astron. Soc. **469**, S329–S338 (2017)

J.-B. Vincent et al., Local manifestations of activity. Space Sci. Rev. (2019), this issue

P.R. Weissman, E. Asphaug, S.C. Lowry, Structure and density of cometary nuclei, in *Comets II* (2004), pp. 337–357

P.R. Weissman et al., The origin and evolution of cometary nuclei. Space Sci. Rev. (2019), this issue

R.A. Werner, D.J. Scheeres, Exterior gravitation of a polyhedron derived and compared with harmonic and mascon gravitation representations of asteroid 4769 Castalia. Celest. Mech. Dyn. Astron. **65**, 313–344 (1997)

F.L. Whipple, A comet model. I. The acceleration of comet Encke. Astrophys. J. **111**, 375–394 (1950)

D.F. Winter, J.M. Saari, A particulate thermophysical model of the lunar soil. Astrophys. J. **156**, 1135 (1969)

Space Sci Rev (2020) 216:130
https://doi.org/10.1007/s11214-020-00744-0

# Cometary Comae-Surface Links

## The Physics of Gas and Dust from the Surface to a Spacecraft

**Raphael Marschall[1,2] · Yuri Skorov[3,4] · Vladimir Zakharov[5] · Ladislav Rezac[4] · Selina-Barbara Gerig[6,7] · Chariton Christou[8] · S. Kokou Dadzie[8] · Alessandra Migliorini[5] · Giovanna Rinaldi[5] · Jessica Agarwal[4] · Jean-Baptiste Vincent[9] · David Kappel[9,10]**

Received: 28 August 2019 / Accepted: 28 September 2020 / Published online: 6 November 2020


**Abstract** A comet is a highly dynamic object, undergoing a permanent state of change. These changes have to be carefully classified and considered according to their intrinsic temporal and spatial scales. The Rosetta mission has, through its contiguous in-situ and remote sensing coverage of comet 67P/Churyumov-Gerasimenko (hereafter 67P) over the time span of August 2014 to September 2016, monitored the emergence, culmination, and winding down of the gas and dust comae. This provided an unprecedented data set and has spurred a large effort to connect in-situ and remote sensing measurements to the surface. In this review, we address our current understanding of cometary activity and the challenges involved when linking comae data to the surface. We give the current state of research by describing what we know about the physical processes involved from the surface to a few

Comets: Post 67P / Churyumov-Gerasimenko Perspectives
Edited by Nicolas Thomas, Björn Davidsson, Laurent Jorda, Ekkehard Kührt, Raphael Marschall, Colin Snodgrass and Rafael Rodrigo

✉ R. Marschall
marschall@boulder.swri.edu

1 Southwest Research Institute, 1050 Walnut St, Suite 300, Boulder, CO 80302, USA

2 International Space Science Institute (ISSI), Hallerstrasse 6, 3012 Bern, Switzerland

3 Institut für Geophysik und extraterrestrische Physik, Technische Universität Braunschweig, Mendelssohnstr. 3, 38106 Braunschweig, Germany

4 Max-Planck-Institut für Sonnensystemforschung, Justus-von-Liebig-Weg 3, 37077 Göttingen, Germany

5 IAPS-INAF, via Fosso del Cavaliere, 100, 00133, Rome, Italy

6 Physikalisches Institut, University of Bern, Sidlerstr. 5, 3012 Bern, Switzerland

7 NCCR PlanetS, Sidlerstrasse 5, 3012 Bern, Switzerland

8 School of Engineering and Physical Sciences, Heriot-Watt University, Edinburgh, EH14 4AS, Scotland, UK

9 Deutsches Zentrum für Luft- und Raumfahrt (DLR), Institut für Planetenforschung, Rutherfordstrasse 2, 12489 Berlin, Germany

10 Institute of Physics and Astronomy, University of Potsdam, Potsdam-Golm, Germany

tens of kilometres above it with respect to the gas and dust emission from cometary nuclei. Further, we describe how complex multidimensional cometary gas and dust models have developed from the Halley encounter of 1986 to today. This includes the study of inhomogeneous outgassing and determination of the gas and dust production rates. Additionally, the different approaches used and results obtained to link coma data to the surface will be discussed. We discuss forward and inversion models and we describe the limitations of the respective approaches. The current literature suggests that there does not seem to be a single uniform process behind cometary activity. Rather, activity seems to be the consequence of a variety of erosion processes, including the sublimation of both water ice and more volatile material, but possibly also more exotic processes such as fracture and cliff erosion under thermal and mechanical stress, sub-surface heat storage, and a complex interplay of these processes. Seasons and the nucleus shape are key factors for the distribution and temporal evolution of activity and imply that the heliocentric evolution of activity can be highly individual for every comet, and generalisations can be misleading.

**Keywords** Comets · Coma · Gas · Dust · Dynamics · Modelling · Inversion

## 1 Introduction and Historical Context

The Rosetta mission has, through its contiguous in-situ and remote sensing coverage of comet 67P/Churyumov-Gerasimenko (hereafter 67P) over the time span of August 2014 to September 2016, monitored the emergence, culmination, and winding down of the gas and dust comae. This provided an unprecedented data set. One of five prime goals the mission addressed from the original Rosetta announcement of opportunity (RO-EST-AO-0001, 1st March 1995; see also Keller and Kührt (2020)) was to study "The development of cometary activity and the processes in the surface layer of the nucleus and in the inner coma". The processes involved in the gas and dust release and subsequent expansion have thus been intensely studied. Linking the observations of the comae to the diverse morphology of the nucleus as described in this book in El-Maarry et al. (2019) and the local manifestations of activity as described in this book in Vincent et al. (2019) is therefore of particular interest. Even though understanding this link of the coma to the surface is of vital importance to understanding cometary activity on a more basic level this has turned out to be immensely challenging, even with a large data set as provided by the in-situ and remote sensing instruments. There are fundamental reasons for this. First, remote sensing instruments (e.g. visible to infrared imaging systems and spectrometers) all provide line-of-sight (LOS) measurements of the comae which are thus naturally a convolution of the gas or dust distribution around the nucleus and the viewing geometry. In addition, long integration times are often necessary for a sufficiently high signal-to-noise ratio (S/N). This complicates the interpretation if the integration time exceeds a significant fraction of e.g. the rotation period of the nucleus, therefore additionally convolving with the temporal evolution of the comae. High optical depths along the LOS can be a further challenge. Deconvolving all of these effects is therefore clearly non-trivial. Second, in-situ instruments (e.g. particle detectors, collection plates) provide local measurements – often tens or hundreds of kilometres from the surface. These measurements are not only a product of the surface emission but can be heavily influenced by the dynamics of the gas molecules or dust particles between the surface and the spacecraft. In addition, the location where the measurements were made needs to be taken into consideration. This is due to the fact that the emission and subsequent dynamical effects in the sub-solar direction compared to the one at the terminator or on the anti-solar direction (night side) can be very different. Measurements taken e.g. in a terminator orbit (very common for Rosetta) are likely heavily biased due to the orbit as they are sampling a very specific part of the comae. Both of these fundamental reasons need to be appreciated and dealt with when analysing the data trying to link the comae to the surface, and we will discuss the ways these problems have been tackled.

Before going into more detail we would like to point out a common shortcut that can be taken when trying to link the comae to the surface. In the pursuit to detect surface sources of gas we are faced with the problem that remote sensing instruments that are capable of detecting gas, in particular infrared imagers such as VIRTIS on Rosetta, lack the spatial resolution compared to simultaneous visible imaging to resolve small scale gas coma structures that could be traced to the surface. Scanning instruments such as VIRTIS-H and MIRO can also map the inner gas coma. For such maps, the disentanglement between the spatial and temporal variations can become challenging because it may take hours (a significant fraction of the nucleus rotation of 12.4 h in case of comet 67P (Mottola et al. 2014)) to build up a map. To resolve some of the fine structures of the gas coma a spatial resolution of the order of tens of meters would be desirable. We therefore often use dust as a proxy for the gas because it is easily detected with high resolution visible images such as the ones acquired by OSIRIS on Rosetta. This approach is indeed valid as long as the dust is coupled to the gas which can be safely assumed close to the surface. This assumption does break down for e.g. very dusty outbursts. We will discuss further limitations of this assumption later in the paper. Unfortunately, in the case of 67P the brightness of the dust coma is mostly orders of magnitudes lower than that of the surface and can hence not be seen against the bright backdrop of the illuminated nucleus. There are some cases where the dust brightness is high enough to be detected against the nucleus background, and in these cases the source can be accurately detected. This regrettably is seldom the case. Thus even for this case of linking dust features to the surface a traceback method needs to be applied for most coma structures.

To address the challenges posed above and given the current state of research we will begin in Sect. 2 by describing what we know about the physical processes involved from the surface to a few tens of kilometres above it with respect to the gas and dust emission from cometary nuclei. In Sect. 3 we describe how multidimensional cometary gas and dust models have developed since the Halley encounter of 1986 to today. In Sect. 4 we present the different approaches used and results obtained to link coma data to the surface. We will also discuss the limitations of these approaches.

## 2 The Physics of Developing Cometary Comae

In this section we want to explore the current knowledge of the physical processes that transpire from roughly one metre below the surface of the nucleus to a few tens of kilometres above it. Historically there has been a lot of theoretical work done, and, as computers have emerged, numerical approaches have been developed and become more sophisticated as the computing power has increased over the course of the past decades. In the following we will go through the different stages in natural sequence, from a refractory-ice mixture on the surface that is heated and causes the ice to sublimate, up to the point where the gas molecules and dust particles reach a spacecraft detector tens of kilometres from the surface.

### 2.1 Illumination, Heating, and Sublimation

We start at the origin of the activity which is the close surface layer composed of a mixture of refractory material and ices. For a detailed description of what we have learned about this layer we refer to Groussin et al. (2019) within this book. For our purposes we are interested how the surface responds to heat input from solar illumination, and it is sufficient that we have learned that cometary nuclei have very low thermal inertia, $\Gamma$ (e.g. Gulkis et al. 2015, Schloerb et al. 2015, Choukroun et al. 2015, Marshall et al. 2018 for 67P, and Groussin et al. 2013 for 9P/Tempel 1). This implies that the diurnal heat wave does not penetrate deep into the interior and thus that sublimation takes place relatively close to the surface if we see the sublimation rate responding fast to illumination changes as shown e.g. by Zakharov et al. (2018a). Furthermore, we assume that the considered surface element is flat i.e. must be understood as an effective surface rather than a real rough surface. For the case of a homogeneous opaque continuous medium the heat balance to be solved for each surface element is thus

$$\frac{S(1-A_H)}{r_h^2}\cos(i)\cdot\delta+\Phi_r=\epsilon\sigma T^4+Z(T)\mathcal{L}+\kappa\frac{dT}{dz}, \tag{1}$$

where on the left side we have the solar energy input which is proportional to the solar constant, $S$, at 1 AU, and the cosine of the solar incidence angle with respect to the surface normal, $i$, and inversely proportional to the square of the heliocentric distance, $r_h$, and $A_H$ is the bolometric Bond albedo. $\delta$ is a step function which takes the value 1 if a surface element is sunlit and 0 if it is in shadow or on the night side. A further contribution to the energy input comes from the re-radiation flux, $\Phi_r$, from other surfaces within the respective field of view of the considered facet. On the right side we have first the thermal re-emission term with the temperature, $T$, to the fourth power, the Stefan-Boltzmann constant, $\sigma$, and emissivity, $\epsilon$, then second the sublimation term with the latent heat, $\mathcal{L}$, and net sublimation flux, $Z(T)$. Knowing the surface temperature, it is possible to calculate the gas production using the classical Hertz-Knudsen formula and the velocity distribution function (VDF; in this case semi-Maxwell distribution) for outgoing molecules. The sublimation rate without back flux is $Z(T)=P(T)/(0.5\pi\, v_{th})$, where $P(T)$ is the water saturation vapor pressure and the average thermal speed $v_{th}(T)=\sqrt{8Tk_b/(\pi m_g)}$, where $k_b$ is the Boltzmann constant and $m_g$ is the mass of the molecule. Finally, there is the conduction term with $\frac{dT}{dz}$ being the temperature gradient with depth, $z$, and with the thermal conductivity, $\kappa$, which is related to the thermal inertia, $\Gamma$, defined as

$$\Gamma=\sqrt{\kappa\rho c}, \tag{2}$$

with $\rho$ the effective density and $c$ the specific heat capacity. $\Gamma$ represents the ability of the surface to respond to the temperature forcing provided by the solar illumination, effectively yielding a time lag between the maximum energy input and maximum surface temperature. The case when thermal conductivity is neglected will result in the maximum possible gas production rate (Keller et al. 2015).

Equation (1) thus defines the sublimation temperature and sublimation rate as a functions of the energy input from the Sun and the topographical context of the considered surface area. The complex shape of 67P's nucleus has shown the need to incorporate shadow casting of parts of the nucleus on large areas. Thus, the direct solar input radiation can be zero, even if $\cos(i)>0$ (when $\delta=0$ in Eq. (1)). This is especially pronounced in concave regions, such as the Hapi region on comet 67P (Thomas et al. 2015c) which can experience two phases of complete darkness per comet rotation during parts of the cometary year.

## 2.2 The Dust and Ice Boundary Layer

In the previous section we have discussed the process that leads from the illumination of the surface and the associated heat input to the sublimation of ices. Determining the flux of sublimating gas from a bare ice surface is in general not sufficiently accurate. This is due to the fact that very little to no ice has been found directly at the surface (Pommerol et al. 2015; Capaccioni et al. 2015). There are three possibilities to explain this. First, most of the surface is indeed completely devoid of ice. Second, the ice is masked within a mixture of refractories (Yoldi et al. 2015). Third, the ice-containing material is covered by a thin layer of pure refractory material. The first of these three options seems unlikely because the comet appears to be active everywhere, i.e. no completely inactive areas have been discovered. If the second option is valid sublimation takes place at the surface itself. The implications of the third option need to be discussed, considering that a large fraction of the surface of 67P is covered by dust, resulting in smooth looking surfaces, that has likely been redeposited by airfall (Thomas et al. 2015a). In this case the gas from the sublimated ice will flow through a dry dust layer before being released into the coma. During day time, this dry dust layer is potentially much hotter than the sublimating layer because it has no ice cooling it and has thus the potential to alter the gas properties in at least three ways. First, a porous dust layer weakens the gas production rate, and in the most extreme case quenches it completely. Second, the gas can be heated if the molecules interact sufficiently with the dust layer. Third, the VDF can be altered for a flow with or without a dust layer. In general we assume the angular distribution of the emitted gas obeys a cosine law ($\sim \cos^n i$), with power, $n$, and angle to the surface normal, $i$. For sublimation from a plane $n = 1$ at 1 AU. For more complex cases, $n > 1$ can be used as a good approximation. A dust layer through which the gas needs to flow before escaping can thus alter the amount of gas escaping, the temperature of the gas and thus the energy available for acceleration of the gas due to molecular collisions but also the directionality of the molecules upon emission. We will break down the different possible structures of the surface boundary layer in the following sections going from the simplest to the more complex ones.

In the simplest case, it is assumed that solid ice is located on the surface of the cometary nucleus and sublimates directly into vacuum. Then, the ice temperature is determined from a simple algebraic equation expressing the energy balance on the surface (model A in Keller et al. 2015) and Eq. (1) without the last term. In a more accurate approximation, to calculate the surface temperature, it is necessary to solve the heat transfer equation for the surface layer. The estimates made in Keller et al. (2015) show that the difference in gas production estimates between these two approaches to determining the ice surface temperature is insignificant ($\sim 10\%$), if the sublimation rate is high enough (that is, at heliocentric distances $< 3$ AU).

The first modification of this simplest model is related to the fact that cometary ice (even if it exists on the surface) has, plausibly, considerable porosity. The porosity of the material in the general case leads to the fact that both the process of absorption and release of energy and the process of sublimation acquire a volumetric character, that is, they take place in a certain near-surface layer. Note that, if we neglect the temperature gradient in the layer in question, then porosity does not lead to any new results. Both the rate of sublimation and the VDF of the molecules remain the same as in the model where a solid (i.e. non-porous) slab of ice is considered. It is clear that a layer can be considered as quasi-isothermal only if its thermal conductivity is large. This assumption contradicts the results obtained during the mission (Schloerb et al. 2015; Choukroun et al. 2015). The near-surface layer is obviously non-isothermal, which makes the problem even more difficult. The most complete discussion of a suitable model and results is presented in Davidsson and Skorov (2002) and Davidsson and Skorov (2004). It was shown that in the case of dirty porous ice volumetric effects may lead to a noticeable change in the rate of sublimation. The nature of the VDF for the case of a non-isothermal porous medium has not been studied, but we can assume that the deviation from the semi-Maxwell distribution should be less than for the case when the nucleus surface is covered with a non-volatile porous crust. We now turn to the consideration of the last case.

The results obtained by instruments on board Rosetta (Capaccioni et al. 2015) have shown that the surface of 67P is covered with a non-volatile crust. Ice on the surface appears to be rare. At the same time, there are currently no publications showing the presence of areas on 67P where this crust effectively turns off gas production, i.e., gas activity is observed ubiquitously. This in itself does not yet prove that there are no inert areas but rather that they are difficult or maybe with the current data even impossible to detect because of the rapid lateral expansion of the gas that can hide inert surfaces (see e.g. Marschall et al. 2020a). This could suggests that the dust layer has a high porosity, possibly similar to the macroscopic porosity of the nucleus (Preusker et al. 2017; Kofman et al. 2015; Brouet et al. 2016). The low thermal inertia of the surface layer is also consistent with this conclusion. If gas molecules must pass through a non-isothermal porous crust, then many of the characteristics that are important to us, namely the rate of sublimation (or the density of the gas on the surface), and the VDF of the molecules can change significantly during diffusion.

Considering changes in gas parameters when passing through a porous medium, two cases can be distinguished: (a) the porous medium is described as a system of microchannels (or orifices), that is, the gas flow is regarded as internal with respect to the scattering medium, (b) the porous medium is described as a system of particle-scatters, that is, the gas flow is considered as external to the scattering medium. In both cases, due to collisions and scattering of gas molecules with a non-volatile fraction, the effective sublimation rate is weakened. In addition, temperature and VDF also change.

The analysis of free molecular gas flow through channels has more than a century of history and goes back to the work of Maxwell and Knudsen. Considering the diffuse reflection of molecules from the walls of a cylindrical channel, the latter author obtained a simple relationship in which effective gas production is directly proportional to the radius, $r$, and inversely proportional to the channel length (Knudsen 1909). This formula agrees well with the experimental results only for long tubes (channel length $\gg$ channel radius). Later, Clausing (1932) proposed a more complex formula that is valid for straight channels of arbitrary length. In order to take into account the fact that the shape of the natural channels differs from a straight cylinder, the so-called phenomenological coefficient of tortuosity $\tau$ ($> 1$) is introduced. There are many theoretical and empirical expressions evaluating this parameter (see for example Skorov et al. 2011). However, its specific value cannot be determined without a detailed description of the structure of the porous medium. The Knudsen and Clausing formulas are widely used in theoretical cometary models (capillary models) in order to evaluate the weakening by a porous dust layer.

For the capillary model the angular distribution of the emitted molecules was investigated, for example, by Nanbu (1985) and Skorov and Rickman (1995), using the test-particle Monte Carlo method. They have shown that even for a small layer-thickness to radius ratio ($\sim 1$) the distribution shows a marked deviation from the ideal cosine law. For a larger thickness it is collimated near the axis of the orifice. The difference is about 7–9 degrees. For our purposes, this means that the macroscopic velocity of the gas (and therefore, for example, the reactive force causing non-gravitational perturbations in the motion of the comet) exceeds the velocity calculated for sublimation from the plane.

Experiments (Krause et al. 2011) and theoretical models (Kaviany 1995) show that the thermal conductivity of a porous medium can be two to three orders of magnitude lower than the thermal conductivity of a non-porous solid. In the porous case, the temperature difference at the boundary of the dust layer (having a thickness of only a few millimetres) can be tens of degrees (Skorov et al. 2017). When molecules collide with a scattering medium, thermal accommodation occurs, that is, both the velocity direction and its magnitude are modified. A change in the effective gas temperature during passage through a non-isothermal channel was considered in Skorov and Rickman (1995). It was shown that the temperature of the emitted molecules differs markedly from the temperature of the subliming ice already for a channel length of several channel radii.

The Monte Carlo test-particle method can be effectively used also to analyse a free molecular flow through a porous medium, described as a system of particle scatters. This approach allows us to get all the necessary information about the outgoing molecules (e.g. the resulting gas production and VDF). For example, a microphysical computational model for molecular flow in a random porous medium formed by packed spheres was presented and the weakening effect was investigated in detail in Skorov et al. (2011). The main transport characteristics such as the mean free path distribution and the layer permeability were calculated for a wide range of model parameters varying in the range of values expected for the near-surface regions of a cometary nucleus. The results were compared with those obtained by idealized capillary models, and a practical way was suggested to adjust the algebraic Clausing formula taking into consideration the nonlinear dependence of permeability on layer porosity. Although the gas diffusion through a non-isothermal layer has not been considered in this paper (Skorov et al. 2011), the obtained distribution of the mean free path of molecules as a function of porosity makes it possible to estimate the change in gas temperature. Even at very high porosity ($\sim 85\%$), the mean free path is shorter than approximately two particle sizes. This means that, like for the capillary model, a substantial part of the molecules changes speed and temperature during transport through the hot non-volatile porous dust layer.

An approach where a random porous medium consisting of spheres is generated via either random ballistic deposition or random sequential packing is rather general. While this approach can provide us with any targeted porosity it does not fully represent real porous medium (rock) samples. Feldkamp et al. (1989) first reported micro-CT X-Ray technology in the 1980's, and it was used to study bones. Micro-CT technology offers a three-dimensional representation of a rock with high resolution on the scale of μm. It also offers a highly accurate method to measure material porosity. Micro-CT technology attracted interest in the last decade from the petroleum industry for representation of natural gas reservoir porous media (Okabe and Blunt 2004; Dong and Blunt 2009). It is now the state-of-the-art method to represent porous medium structure with a resolution of a few microns.

Christou et al. (2018) and Christou et al. (2019) have directly applied the Direct Simulation Monte Carlo (DSMC) method to the computational mesh obtained from micro-CT images of highly porous rock samples. The flow of sublimating water ice from beneath the sample through the porous layer was studied. At higher porosities, the outgassing temperature remains nearly constant over a longer distance above the surface than at low porosities, $\Phi < 45\%$. Lower porosity media are associated with higher surface gas temperature as the gas interacts more strongly with the warmer surface layer. This may be explained by the fact that, as low porosity implies a low level of void, gas passing through a medium with low porosity is heated more effectively by the higher temperature and expands quickly as it exits. We note that porosity in the range of 88–90% yields a surface gas temperature below 200 K for $T_{\mathrm{rock}} = 300$ K and just above 200 K in the case of $T_{\mathrm{rock}} = 330$ K. Gas passing through high-porosity media receives less heating and experiences an expansion and therefore cooling before exiting the medium. Similar results were reported by Skorov et al. (2011). This is illustrated in Fig. 1. As the porosity increases the gas temperature decreases. Furthermore, Fig. 2 shows the surface pressure as function of porosity. Higher porosity leads to higher

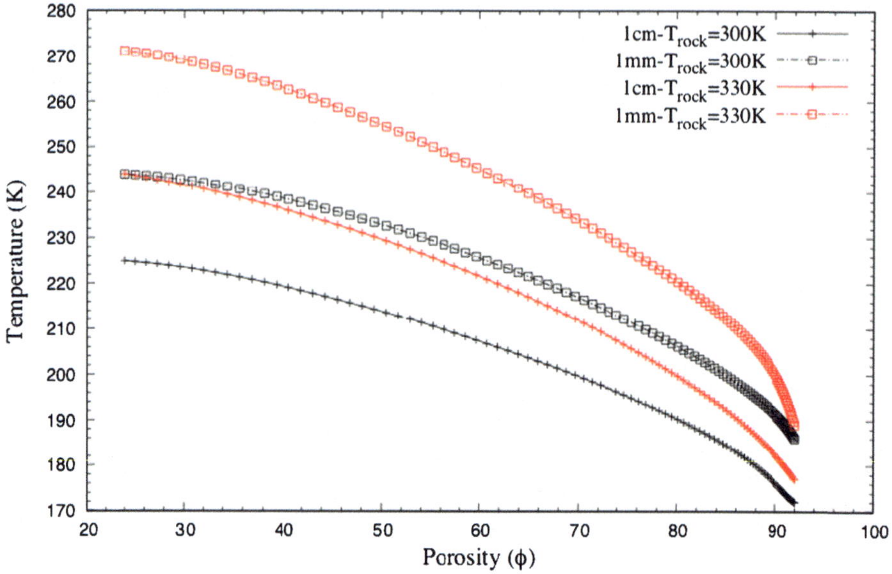


**Fig. 1** Gas temperature as function of porosity, $\Phi$, at 1 mm and 1 cm above the surface for two rock temperatures, $T_{\text{rock}} = 300$ K, 330 K. The graphs were obtained by a polynomial interpolation from six different rock samples. Figure adapted from Christou et al. (2018)

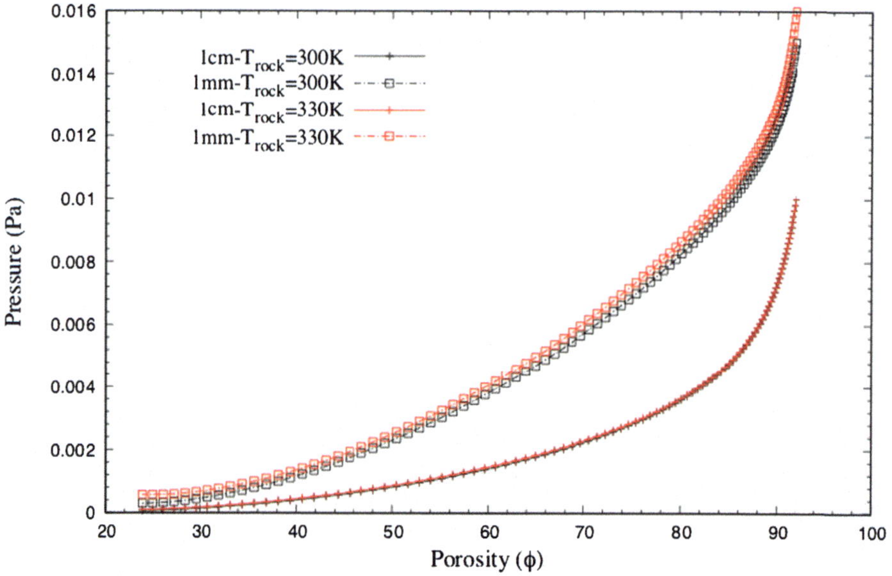


**Fig. 2** Gas pressure as function of porosity at 1 mm and 1 cm above the surface for two rock temperatures, $T_{\text{rock}} = 300$ K, 330 K. Figure adapted from Christou et al. (2018)

pressure. Porosity can thus play an important role in altering gas properties before emission at the surface. Even quenching of the activity can be achieved with a dust layer of very low porosity. Christou et al. (2019) have shown that due to the large variations of observed gas pressures as a function of porosity, lateral gradients in porosity along the surface can produce strong lateral gas flows. Coma inhomogenities can be attributed to non-uniformly distributed sources of $H_2O$ and $CO_2$ over the surface of the nucleus (Fink et al. 2016) or variations in illumination in conjunction with the local topography (Marschall et al. 2016). These lateral flows caused by lateral porosity gradients may be capable of generating dust grain transports. Lara et al. (2015) reported that dust particles on the scale of mm could be ejected. More recently, Lai et al. (2016) calculated for 67P that during peak production, near perihelion passage, grains as large as 10 mm can be ejected. Furthermore, dust particles up to the size of decimetres have been observed flying around 67P (Agarwal et al. 2016). Based on those observations and the lateral velocities predicted by Christou et al. (2019) we can conclude that porosity gradients can generate the necessary lateral flow to transport dust grains after they are ejected from the surface.

These considerations above have covered what may be referred to as the contiguous activity. In addition, short-term activity ("outbursts") has been observed on various time and spatial scales throughout the Rosetta mission (Tubiana et al. 2015; Vincent et al. 2016b; Lin et al. 2017; Bockelée-Morvan et al. 2017; Rinaldi et al. 2018). Outbursts occur not only in case of 67P but are typical for comets. While some of these events can be associated with the erosion of cliffs (Vincent et al. 2016c; Pajola et al. 2017), others were likely driven by either the sublimation of freshly uncovered supervolatiles (Skorov et al. 2016b) or by the release of energy stored in the cometary subsurface (Knollenberg et al. 2016; Agarwal et al. 2017).

## 2.3 Expansion of the Gas

The gas flow in the inner coma of an active cometary nucleus offers severe challenges for numerical simulations because of its complex spatial structure. A macro-scale simulation involves consideration of the complex nucleus shape and topography while a micro-scale simulation will involve a detailed local near surface outgassing modelling. In the most general case, the flow in the coma is a juxtaposition of regions with widely differing conditions – from fully collision-less to fluid. The degree of rarefaction is characterised by the Knudsen number $Kn$:

$$Kn = \lambda / L, \tag{3}$$

where $\lambda$ is the mean free path of the molecules (MFP) and $L$ is the characteristic dimension. To describe the complete flow by a single overall Knudsen number a characteristic scale of the flow (traditionally the equivalent radius of the nucleus) can be used as $L$. To characterise the rarefaction of the flow locally, the scale length of the macroscopic gradient can be used as $L$ (e.g. using the gas density, $n_g$, such that $L = n_g / |\nabla n_g|$). Depending on the local $Kn$ three flow regimes can be roughly distinguished (illustrated in Fig. 3):

1. continuum/fluid, $Kn < 0.01$;
2. transitional, $0.01 \leq Kn \leq 100$;
3. free molecular, $Kn > 100$.

As was shown in the previous sections, the gas coma is formed by the ice sublimation from the nucleus in different ways. The VDF of the sublimating molecules is non-Maxwellian (it is commonly assumed to be half-Maxwellian corresponding to the surface temperature). Therefore, a non-equilibrium layer (Knudsen layer) always occurs adjacent to the surface. Due to inter-molecular collisions the initial VDF evolves (relaxes) and may become at some distance (if and when there are a sufficient number of collisions) completely

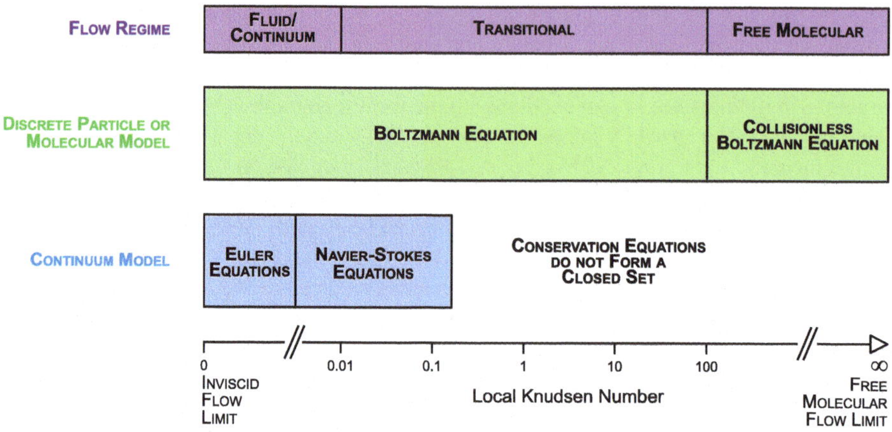


**Fig. 3** Different flow regimes as a function of the Knudsen number as well as the intervals for which different models (continuum and discrete) are valid. This figure has been adapted from its original in Bird (1994)

(or quasi) Maxwellian. The expansion of the flow leads to a decrease of collisions with radial distance and therefore, even if an equilibrium flow was established at the top of the Knudsen layer, the flow becomes non-equilibrium again at larger distances due to an insufficient collision rate to maintain the equilibrium distribution.

The structure of the flow depends considerably on the rarefaction of the flow. For high initial densities of the gas (i.e. high production rates of the surface) three characteristic regions can be distinguished in the flow:

1. subsonic near-surface layer;
2. supersonic continuum flow;
3. supersonic non-continuum, non-equilibrium flow.

The boundary between regions (1) and (2) is the surface where the gas velocity is equal to the local sound velocity. The boundary between regions (2) and (3) is the conditional boundary of the region of the continuum flow (its exact position depends on the formal definition of the continuum breakdown). The part of the subsonic region near the surface is also in non-equilibrium (owing to the Knudsen layer). The thickness of the Knudsen layer depends on the specific conditions, but in general, it is about 10–100 mean free paths (Ytrehus 1975; Davidsson 2008). When $Kn$ decreases, it leads to a decrease of the non-equilibrium subsonic layer and an increase of the continuum supersonic flow (the boundary of the continuous flow is shifted downstream). When $Kn$ increases, the subsonic and the Knudsen layers become thick, and at some rarefaction the Knudsen layer becomes thicker than the subsonic layer. The increase of $Kn$ leads to a decrease of the supersonic continuum flow region because the outer boundary of the continuum flow approaches the surface. At some rarefaction the near-surface non-equilibrium layer (Knudsen layer) and the outer region of non-equilibrium flow merge together, and the Knudsen layer as a relaxation zone of the initial non-equilibrium cannot be distinguished from the outer non-equilibrium region.

The interactions of supersonic fluxes in the coma lead to the appearance of shock structures (with localised enhancements of density, temperature and decrease of velocity). The rarefaction of the flow with such structures is characterised using the scale length of the macroscopic gradients. Therefore, even if the overall flow is dense, locally the flow may be rarefied and non-equilibrium.

In general, the transfer of thermal energy into kinetic energy in a rarefied flow is less efficient than in a fluid flow, therefore the rarefied flow accelerates slower, and the flow structures are more diffused.

The gas dynamics approach uses for its description of the gas only the first moments of the VDF – the number density, bulk velocity, and temperature (Euler equations, see Fig. 3), heat conduction, and viscous dissipation (Navier-Stokes equations, see Fig. 3). This approach is appropriate for those regions where the VDF is either a strict Maxwellian (i.e. the fluid flow) or a moderately distorted Maxwellian. Since this approach can't be used for the simulation in the Knudsen layer, it is necessary to introduce an additional model which links parameters on the surface and on the top of the Knudsen layer – the initial boundary for the gas dynamics simulations. It was shown that the gas dynamics approach with appropriate boundary conditions can provide a physically adequate description of the gas flow even at a moderate degree of rarefaction, but in this case, it is necessary to verify the adequacy of solutions in each particular case (Crifo et al. 2002a, 2003; Lukyanov et al. 2005; Zakharov et al. 2008). It is important to note the difference between "an adequate description" and "a precise description". For example, for the expansion into a vacuum at large distances from the source the Euler solution predicts a continuous acceleration of the gas but the Navier-Stokes solution predicts deceleration due to the excessive viscous dissipation. In this case, the solution of the Euler equations is physically adequate (the gas does accelerate in reality) but it is not precise (e.g. it exaggerates the transfer of thermal energy into kinetic energy). The solution of the Navier-Stokes equations, on the other hand, is physically inadequate. In many cases, the fluid approach keeps physical adequacy of the rarefied flow description – though with different degrees of precision. It is thus necessary to verify the precision for every case.

Gas kinetics methods – based on the collisional Boltzmann equation (see Fig. 3) or the direct simulation Monte Carlo (DSMC) method proposed by Bird (Bird 1994) – are appropriate at any level of distortion of the VDF. The main drawback of these methods is their high computational demands. With the emergence of high-performance computational facilities, the DSMC method became the main method of simulations for rarefied non-equilibrium flows. For example, DSMC was used for simulation of the coma in e.g. Combi (1996), Skorov and Rickman (1999), Crifo et al. (2002a, 2003), Tenishev et al. (2008), Davidsson (2008), Zakharov et al. (2009), Fougere et al. (2013), Finklenburg (2014), Marschall et al. (2016).

As the rate of collisions (if there were any initially) between molecules decreases with distance the momentum and energy exchange decreases as well. This results in the conservation of the flow velocity and internal energy with distance. Also, the kinetic temperature (as a measure of the velocity dispersion with respect to the mean velocity at some point) freezes for the parallel component and slowly decreases for the perpendicular component. Therefore, starting from some distance the gas flow approaches the flow with practically constant temperature and velocity (directed radially from the centre of the nucleus), and the variation of the density, $n_g$, as a function of distance to the comet, $r$, changes according to the "inverse $r^2$" law:

$$n_g \propto \frac{1}{r^2}. \tag{4}$$

The "Haser model" (Haser 1957) uses this property of the flow and includes additionally photo-chemical destruction of molecules. This model is still the most commonly used approximation owing to its apparent simplicity. But it should be noted that it is physically adequate only at a large distance from the nucleus.

Depending on the size of the simulation region the relative importance of physical processes varies. Usually, in the near-nucleus region the collisional processes dominate over photo-dissociation and ionisation, but for large distances and thus long travel times of the molecules the time scales of photo-dissociation and ionisation may become significant (Crovisier 1989; Xie and Mumma 1996; Combi et al. 2004). For the typical conditions, the velocity of gas molecules is of the order of hundreds of meters per second and the typical time scale of processes such as photo-dissociation and ionisation is about $10^5$–$10^7$ seconds. For regions greater than $10^4$ km it is, therefore, necessary to take into account the processes of photo-dissociation and ionisation. The numerical simulations in a large region can be complicated additionally if the settling time of the flow (which is proportional to the size of the region) is longer than the characteristic variation time of boundary conditions (e.g. due to the rotation of the nucleus). In this case, the flow might never reach a steady state.

## 2.4 The Dust Activity Paradox

Up to this point we have mainly focused on the icy constituents and not on the refractory ones. When dust is separated from the surface it can be accelerated in the gas flow. We will discuss this in more detail in the next section. Very little is known about the refractory and ice mixture at the surface (see also Choukroun et al. 2020, in this book), and thus the nature of how dust is released out of this matrix remains an open problem often referred to as the "activity paradox" (Kührt and Keller 1994). The reason for this paradox is the fact that we see the strong cometary dust and gas activity but do not have a working model of how dust is released. The simplest mechanism where the gas pressure of the sublimating ice overcomes the cohesive forces of the material does not work (Skorov et al. 2017). The material strength inferred for cometary material (e.g. Attree et al. 2018) is much larger than the corresponding gas pressures available. For more details and ways out of this paradox as well as our state of knowledge about the physical properties of the surface we refer to Vincent et al. (2019) as well as Groussin et al. (2019) in this book.

## 2.5 The Dust Dynamics in the Coma

Neglecting for the time being the problems associated with ejecting dust from the surface as described in the previous section we can nevertheless continue and discuss the dynamics of the dust particles once they are no longer attached to the surface. At that point dust particles are subject to several forces driving their motion. Figure 4 shows a sketch of the different dynamical regions of the cometary dust environment which we can divide into three regions (in the order from closest to furthest from the surface):

1. **The coupled coma region:** In this part of the coma the dynamics of dust particles is dominated by local forces that are linked to the nucleus. These are the gas drag force and the gravitational force of the nucleus. Because of the non-sphericity of the nucleus the gravitational force from the nucleus is in general non-radial. Likewise because of the complex nucleus shape, local topography, and emission inhomogeneities the gas flow is in general strongly non-radial (beyond the non-radiality that can occur even for a spherical nucleus). Both forces converge to radial forces as the distance to the surface increases.

   The size of this first region extends to from the surface to roughly 10 nucleus radii, $R_N$, and the time particles spend in this region is of the order of seconds to hours depending on the dust size.
2. **The transitional coma region:** In this part of the coma the governing forces dominating the dust dynamics transition from local forces linked to the nucleus to solar forces such as solar gravity and solar radiation pressure. In this region, the gas and dust have completely

| Region | Coupled Coma | Transitional Coma | Tail/Trail |
|---|---|---|---|
| Spatial Scale: | 0 - $10^4$ m | $10^4$ - $10^6$ m | $10^6$ m - ... |
| | 0 - 10 $R_N$ | 10 $R_N$ - 1 $R_{Hill}$ | 1 $R_{Hill}$ - ... |
| Temporal Scale: | sec. - hours | hours - weeks | weeks - years - ... |

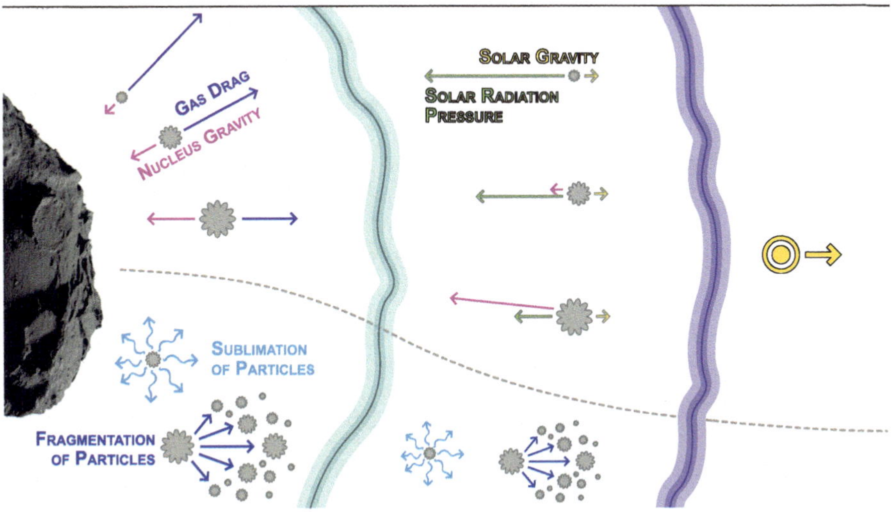


**Fig. 4** Sketch of the different dynamical regions of the cometary dust environment which we can divide into three regions: (1) The coupled coma region where the dust dynamics is dominated by local forces connected to the nucleus (gas drag and nucleus gravity): (2) the transitional coma region where the dust has decoupled from the gas and small particles transition to being dominated by solar forces (gravity and radiation pressure) and large particles are bound in the gravitational field of the nucleus; (3) the dust tail and trail within which the escaping dust particles are purely governed by solar forces. The boundaries between these regions are complex 3D surfaces and not spherical shells and the spatial and temporal scales are rough estimates for a 67P-like comet at 1 AU). $R_N$ is the nucleus radius, and $R_{Hill}$ the Hill radius. See text in this section and Sect. 2.6, 2.8, and 2.7 for more details

decoupled. This is also the region within which the dynamics of very large particles is dominated by the nucleus gravity. Therefore, this region also encompasses the region within which a separation of the coma occurs: small particles escape the gravitational field of the comet and are dominated by solar forces, and large particles remain bound to the comet and eventually re-deposit on the surface.

The size of this second region extends roughly to the comet's Hill radii, $R_{Hill}$, and the time particles spend in this region is of the order of hours to weeks depending on the dust size.

3. **The dust tail and trail:** In this final region of a comet's dust environment the dust particles have left all influence of the nucleus and their dynamics is solely governed by solar forces (gravity and radiation pressure).

Additional processes to the ones named above can alter the dynamics of the dust particles in all three regions of the cometary dust environment. These processes are related to the properties of the dust particles themselves such as sublimation of icy components of the dust particles or fragmentation. See Sect. 2.8 for more details on these processes. We should stress that the picture described above is overly simplified and that each of these regions has a complex 3D structure. Further, the interfaces between these regions are not sharp spherical shells but rather complex 3D surfaces that vary depending on the dust sizes, the heliocentric distance of the comet, overall activity and its distribution on the surface, and more.

Here we focus primarily on the coupled coma region but will comment on some of the above mentioned additional forces and processes in Sects. 2.6, 2.7, and 2.8.

The accelerating force for a spherical dust particle away from the surface is the gas drag (a non-spherical particle will have additional aerodynamic forces). In the immediate vicinity the nucleus gravity is the main opposing force. For longer time scales solar radiation pressure needs to be included as well but can be neglected in the innermost coma (Tenishev et al. 2011; Skorov et al. 2016a, 2018). The dust can be treated either in a test particle approach within the gas flow or as a separate fluid within a gas-dust multi-fluid approach (e.g. Shou et al. 2017). We will discuss the former here in more detail as it is the more common approach used. In this approach it is assumed that dust does not affect the gas flow. In addition, since in most cases the mean free path of the molecules is much larger than the dust particles size, the free molecular aerodynamics is applied. For this approach (Finson and Probstein 1968; Gombosi et al. 1985, 1986; Gombosi 1987; Sengers et al. 2014; Marschall et al. 2016) the equation of motion in the inertial frame for each dust particle is

$$\begin{aligned} m_d \frac{d^2\mathbf{x}}{dt^2} &= \mathbf{F}_G + \mathbf{F}_D \\ &= \mathbf{F}_G + \frac{1}{2} C_D m_g n_g \sigma_d \left|\mathbf{v}_g - \mathbf{v}_d\right| \left(\mathbf{v}_g - \mathbf{v}_d\right), \end{aligned} \tag{5}$$

where $m_d$ is the dust particle mass, and $\sigma_d$ the geometric cross-section. The equation evaluates the dynamic properties of the dust at position $\mathbf{x}$ and with velocity $\mathbf{v}_d = \frac{d\mathbf{x}}{dt}$, $\mathbf{F}_G$ is the gravitational force between the dust particle and the nucleus, $m_g$ the mass of the gas molecule considered (in our case molecular water), and $n_g$ and $\mathbf{v}_g$ are the number density and macroscopic velocity of the gas, respectively. If we assume an equilibrium gas flow and that the mean free path of the molecules is much larger than the dust size, the drag coefficient $C_D$ (Bird 1994) is defined as

$$C_D = \frac{2\zeta^2 + 1}{\sqrt{\pi}\zeta^3} e^{-\zeta^2} + \frac{4\zeta^4 + 4\zeta^2 - 1}{2\zeta^4} \operatorname{erf}(\zeta) + \frac{2(1-\varepsilon)\sqrt{\pi}}{3\zeta} \sqrt{\frac{T_d}{T_g}}, \tag{6}$$

with the gas temperature $T_g$, the dust particle temperature $T_d$, and $\varepsilon$ is the fraction of specular reflection, and

$$\zeta = \frac{\left|\mathbf{v}_g - \mathbf{v}_d\right|}{\sqrt{\frac{2k_b T_g}{m_g}}}. \tag{7}$$

Although we often approximate the dust grains as test particles and consider them to be spherical, especially the larger dust grains are porous and fluffy aggregates (Kolokolova and Kimura 2010; Schulz et al. 2015; Rotundi et al. 2015; Langevin et al. 2016; Bentley et al. 2016; Mannel et al. 2016; Levasseur-Regourd et al. 2018). This affects the dynamics of the particles as shown by Skorov et al. (2016a, 2018) who have found that porous aggregates are accelerated to significantly higher speeds than their compact counterparts. Additionally, rotating particles (Fulle et al. 2015) that have oblate or prolate shapes will also affect the dynamics as studied by Ivanovski et al. (2017a,b). These latter authors have shown that not only may non-spherical particles begin to rotate in the gas flow but they may also accelerate to higher speeds than spherical ones.

## 2.6 Further Forces Acting on Dust

We have already mentioned the two other forces that influence the dynamics of the dust particles namely solar radiation pressure and gravity. The effect of solar radiation pressure

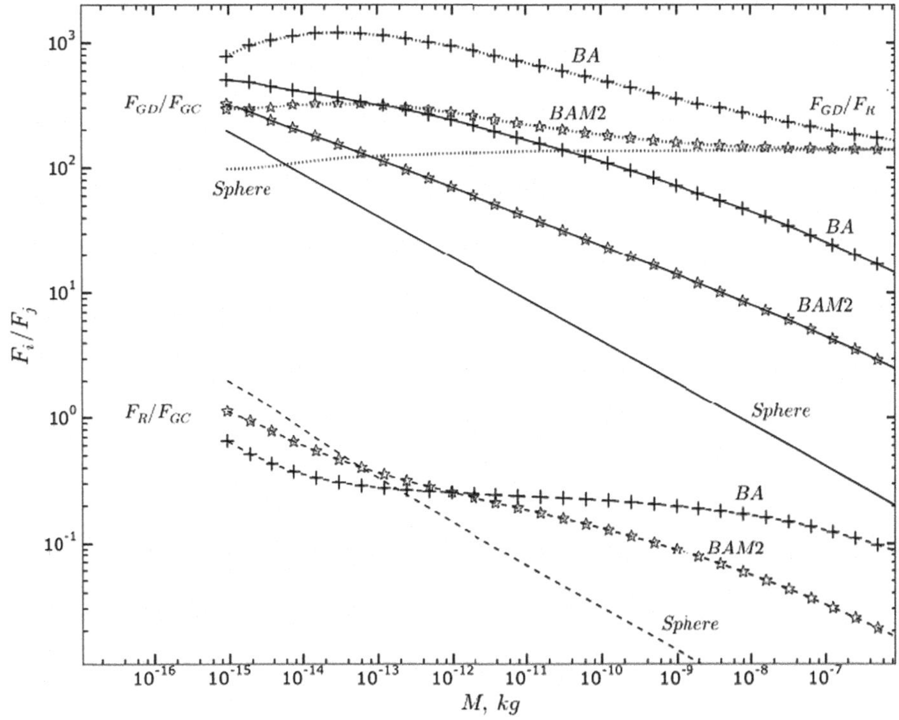


**Fig. 5** Ratios of drag force, $F_{GD}$, gravity, $F_{GC}$, and radiation pressure, $F_R$, on dust particles of different masses: $F_{GD}/F_{GC}$ (solid curves), $F_{GD}/F_R$ (dotted curves), and $F_R/F_{GC}$ (dashed curves) as functions of aggregate mass at 2 km from the comet centre. Results for ballistic aggregates (BA, crosses) and ballistic aggregation with monomer migration (BAM, stars) and non-porous spheres are shown. Euler flow was assumed for the gas flow with a production rate of 2 $\mathrm{kg\,s^{-1}}$ from a spherical nucleus of 2 km radius. Figure adapted from Skorov et al. (2018)

has long been known. Even though dust particles are ejected mainly in the sunward direction cometary tails point in the anti-sunward direction. This is caused by solar radiation pressure decelerating the dust particles as they move towards the Sun until they come to rest and are then pushed in the anti-sunward direction. This back flux towards the comet has, in case of comet 67P, even been found close to the nucleus by GIADA (Colangeli et al. 2007) on board Rosetta.

The effect of gravity is twofold. First, it is an additional opposing force when particles are lifted and thus even in absence of any cohesive forces provide an upper size limit that can be ejected from the surface. Second, large particles which are not accelerated by the gas beyond the escape speed will go into orbit or ballistic trajectories. In the case of 67P its complex nucleus shape leads to a complex gravity field and thus to a non-trivial gravitational influence. Dust deposits from gravitational back-fall on the surface alter the surface properties which in turn influence the nature of activity. In the case of 67P this contributes to the observed seasonality of the activity.

Figure 5 illustrates the ratios of the three major forces acting on a dust particle (Skorov et al. 2018). Different types of particles are shown, from the common assumption of solid spheres to ballistic aggregates (hit and stick of monomers; BA) and ballistic aggregates with monomer migration (hit, stick, and roll; BAM). BA have the highest porosity and the solid spheres per definition the lowest. The figure shows that close to the surface the drag force is the dominant force (by orders of magnitude) for the entire mass range of dust particles. The higher the porosity of the particles the higher also the ratio of the drag to the gravitational force. This is mainly driven by the higher cross-section to mass ratio of the porous particles. Radiation pressure is between two and three orders of magnitude weaker than the gas drag interaction for the given dust mass range and does not depend as strongly on mass as the drag-to-gravity force ratio. For spherical expansion of the gas the drag force remains dominant over gravity even at 20 km from the nucleus centre (Skorov et al. 2018), but radiation pressure becomes more important, and the respective ratio to the drag force can be close to unity. The authors conclude further for the described case that the drag force is indeed dominant for the near-nucleus environment ($< 20$ km) but that beyond that radiation pressure can be a significant factor in the dust dynamics at distances as close as 100 km from the nucleus centre. Finally, the shown results also imply that the assumption of solid spheres leads to the lowest acceleration and thus lowest terminal dust speeds. The higher the porosity the higher the acceleration and thus the higher the terminal dust speed. The solid sphere approximation thus provides for a strict lower boundary for the dust speed.

### 2.7 Redeposition of Dust on the Surface, and Re-Condensation of Gas

So far we have primarily described the mechanisms of mass loss of the comet. But Rosetta has shown that there are two important processes of redeposition on the nucleus which were not well appreciated pre-Rosetta. The first one has been hinted at in the previous section. Due to gravity dust particles – or rather large chunks – might not reach escape speed and are therefore redeposited. This is mainly evident in the morphology of the nucleus surface (Thomas et al. 2015c; El-Maarry et al. 2015, 2016, 2017) showing widespread dusty deposits in particular in the northern hemisphere. This has been described in more detail in Thomas et al. (2015b) who have also shown how particles are gravitationally bound and redeposited. The deposition of particles, especially in gravitational lows, has become quite clear since then. This has been further studied by Lai et al. (2016) who have found that gravitational lows are indeed primary deposition areas. The idea of a seasonal mass transport from the southern to the northern hemisphere has also been pointed out by Keller et al. (2017). This seasonal dust transport can lead to significant altering of the surface properties which in turn define the respective activity. Activity is therefore not simply a function of the local surface properties but can be influenced by non-local long term alterations due to mass transport.

Metre-sized chunks of debris falling back towards the surface have been directly observed in at least one image sequence obtained at 2.6 AU, when comet 67P was outbound from perihelion (Agarwal et al. 2016). Due to the nucleus rotation and the varying trajectories of individual chunks, the material will get distributed widely across the comet (Keller et al. 2017). Generally, the locations of sources and sinks of fall-back debris are determined by the local gas pressure, which in turn is likely mainly determined by solar irradiation and changes with season (Lai et al. 2016; Pajola et al. 2017).

The other possible deposition is that of re-condensation of water vapour. As found by De Sanctis et al. (2015), Filacchione et al. (2016a) a diurnal frost layer has been found on the surface of comet 67P. This surface ice can either be due to re-condensation from below in the near-surface layer due to a temperature inversion caused by thermal lag, or due to re-condensation of gas in the coma from above the surface. Liao et al. (2018) have shown that this second process is also feasible.

Different authors (Fornasier et al. 2016; Filacchione et al. 2016b; Ciarniello et al. 2016; Fornasier et al. 2017) found an increased surface ice content in the southern hemisphere near perihelion, indicating temporary removal or thinning of an otherwise present dust mantle, and the exposure of relatively pristine icy material. These effects were accompanied by significant surface changes such as boulder displacements and vanishing of structures (Fornasier et al. 2019).

## 2.8 Additional Processes Affecting Dust and Gas Flows

Now that we have described the dynamics that govern dust particles, we will also highlight the regime within which this is applicable. There are two main restrictions to the approach described in the previous sections. First, if the flow has a high dust-to-gas mass ratio (mass loading) the two flows cannot be treated independently. Because the dust is accelerated by the gas it draws energy out of it. When there is enough dust in the flow that this ratio becomes significant then the presence of the dust will influence the gas flow. On the one hand the gas will be heated by the usually much hotter dust particles, and on the other hand the dust will decrease the energy in the gas flow and thus tends to slow it down. Additionally, if there is a lot of dust – which is heated very efficiently – it will thermally re-radiate energy that can reach the surface and be an additional source of energy into the surface. The approach above also excludes any icy component in the dust that can sublimate. If ice-bearing dust particles or debris are able to sustain a temperature gradient, e.g. between their illuminated and shadowed sides, the recoil force from asymmetric outgassing can affect their dynamics (Kelley et al. 2013; Agarwal et al. 2016; Güttler et al. 2017). This would also be an additional gas source term – a so called extended source – above the surface and thus alter the gas flow. For comet 67P there has not been any direct evidence for such an extended source but it has been observed for other comets. This process requires coupling of the gas and dust flows and thus a simultaneous numerical treatment of the gas and dust comae. Apart from sublimation, dust particles can also fragment e.g. due to collisions, thermal stresses, or sublimation of icy components. This will alter the size distribution of the dust flow. Further, particles can be charged by the solar wind and their dynamics strongly influenced by electromagnetic forces. This has been first studied by Horanyi and Mendis (1985) and Wallis and Hassan (1983). The latter have found that wavy dust features far down the tail could be explained by charged dust. The charging of dust mainly alters the dynamics of sub-micron to micron particles and acts on long time scales.

## 2.9 Observations of the Dynamics of Dust from the Surface into the Coma

We expect the density of a flow that is radial and collisionless to drop with the inverse square of the distance to the nucleus centre. Similarly to the gas density (Eq. (4)), assuming such a force-free radial outflow, the dust density $n_d$, for a specific dust size, as a function of distance $r$ can be calculated as

$$n_d = \frac{Q_d}{4\pi r^2 \, m_d \, v_d}, \tag{8}$$

with $Q_d$ being the total dust mass production rate, $m_d$ the dust particle mass, and $v_d$ the constant outflow velocity of the dust particles. However, imaging instruments like OSIRIS on Rosetta do not see the density but rather the dust brightness which is the result of the column-integrated density (column density) along the LOS convolved with the scattering properties of the dust particles. In an optically thin medium the observed brightness is proportional to the column density. Due to the integration along the LOS Eq. (8) implies that for a free radial flow the column density should drop by the inverse of the distance. Because the brightness of the dust is proportional to the column density, the brightness multiplied by the distance should thus a conserved quantity as a function of distance. Thomas and Keller

(1990) introduced this quantity as the so-called "azimuthal average" in a study about the dynamics in the inner coma of comet 1P/Halley. They used aperture photometry to study the change in brightness as a function of distance to the comet centre of mass. Deviations in the azimuthal average from this conservation law contain information about additional physical processes involved in the inner coma dynamics compared to free radial outflow. Different processes such as acceleration or fragmentation of the dust particles have different impacts on the azimuthal average. Several of the processes discussed in the literature and their effects on the azimuthal average are revisited below. For one, acceleration of the particles due to gas drag will cause the azimuthal average to drop as a function of distance. This is because the scattering area and thus the brightness of the observed dust disperses into space more quickly than is expected for a free radial outflow due to the increasing velocity (Eq. (8)). When considering fragmentation of dust particles, two different outcomes are possible. On the one hand, for dust particles fragmenting into optically more active daughter particles and thus increasing the scattering area, the azimuthal average will increase with distance from the nucleus (Konno et al. 1993). On the other hand, when a majority of the available dust particles is fragmenting into optically small particles (daughter particles invisible for scattering), the azimuthal average will decrease with distance from the nucleus surface. Sublimation of icy grains in the coma has much the same effect on the azimuthal average like this latter case because here scattering area is lost in the sublimation process. Conversely, optical depth effects can increase the azimuthal average as a function of distance as a larger and larger fraction of the scattering cross section comes into view. The effects of outgassing from localised sources on an extended surface area as modelled by Huebner et al. (1988) and Reitsema et al. (1989) can lead to an increase in azimuthal average with distance from the nucleus. Furthermore, gravitationally bound particles in Keplerian orbits out to the Hill radius can lead to an increase in azimuthal average with distance from the nucleus surface, whereas gravitationally bound particles on ballistic trajectories have the opposite effect as demonstrated by Gerig et al. (2018). In the case of 1P/Halley an increasing azimuthal average curve was found close to the nucleus ($<$ 100 km or 10 nucleus radii) (Thomas and Keller 1990). In several studies a combination of localised sources over an extended active area and particle fragmentation were identified as the dominating processes in the innermost dust coma of 1P/Halley (Huebner et al. 1988; Szego et al. 1989; Thomas et al. 1988; Reitsema et al. 1989; Thomas and Keller 1990; Konno et al. 1993). At larger distances (100–600 km), the profile was flat as expected for a free-radial flow. In contrast, at 67P a decline in the azimuthal average profile as a function of distance was found, indicating that we are most likely seeing the acceleration of the particles. Using models described in Sect. 4.2, Gerig et al. (2018) demonstrated that this trend can be reproduced in a model only taking into account acceleration of the dust particles entrained in the gas outflow and gravity from the nucleus. They also concluded that other processes in the inner coma such as particle sublimation or contributions from gravitationally bound particles might be necessary to fully explain the data. In a study of the radial profile of a jet-like feature above the Imhotep region around spring equinox 2015, Gicquel et al. (2016) concluded that sublimation of small icy grains is needed to explain the observed steep profile decline in the first 10 km above the nucleus surface. Like for comet Halley, the azimuthal average curve for 67P also flattens out at larger distances from the nucleus ($>$ 12 km or 6 nucleus radii; Gerig et al. 2018), and the flow can be treated as a force-free radial outflow beyond that distance. Figure 6 shows a typical example of the drop and then flattening of the curve (in black) with distance. Generalised theoretical considerations as shown by Zakharov et al. (2018b) predict the dust to reach 90% of its terminal velocity at a distance of 6 nucleus radii. This is in remarkable agreement with the results for 67P, and also agrees well with the results found

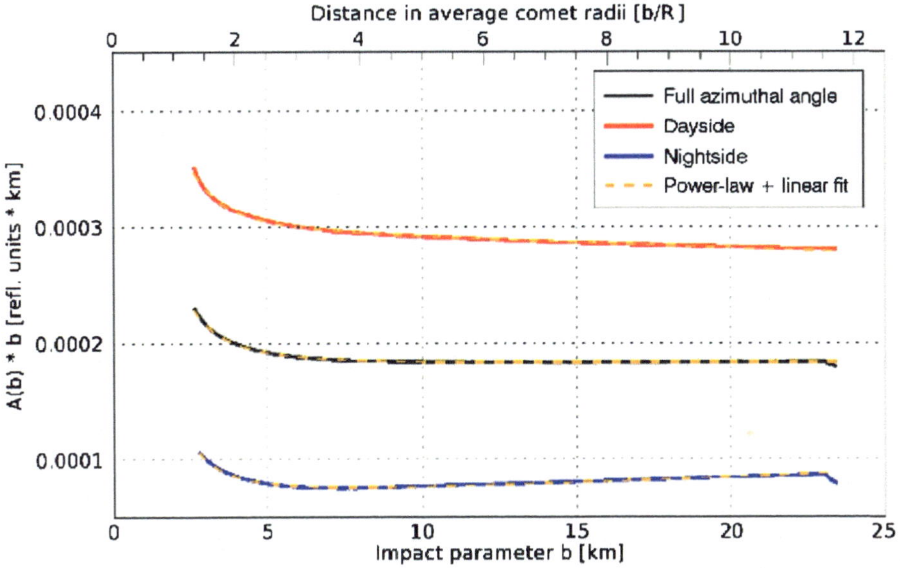


**Fig. 6** Brightness, *A*, for the case of 67P multiplied by the distance, b, as a function of the distance (taken from Gerig et al. 2018). The black curve shows the full result and the red/blue curves are restricted to the day/night side of the comet

for 1P/Halley. During the time of peak activity, about one month after perihelion, Boehnhardt et al. (2016) in ground-based images of 67P found radial brightness profiles flatter than expected for a steady-state coma, which they interpreted as a possible sign of dust fragmentation at nucleus distances > 5000 km from the nucleus. Also ground based polarimetric measurements from several months after perihelion hint at the presence of fragmenting dust in the outer coma (Rosenbush et al. 2017). This illustrates how the combination of theory, complex models, and data analysis have become very successful in the understanding of the dynamics within a cometary coma.

## 3 The Development of Multidimensional Cometary Gas and Dust Models

Since the observations of comet 1P/Halley from a close distance in 1986, it has become evident that the near-nucleus coma has a complex spatial structure (previous observations from large distances just suspected that). Therefore, the simulation of such flows requires multidimensional models of the dusty gas flows. In this section, we will briefly describe the development of such multidimensional models which, triggered by the lead up to the encounter of comet 1P/Halley in 1986, have progressed to this day and applied to Rosetta's data of comet 67P. Though these models build on earlier descriptions in one dimension we will not go into the details of those models.

The first multidimensional models (axially-symmetric and 3D) of a dusty gas coma with the physically consistent description of a flow as a fluid were presented in Kitamura (1986, 1987, 1990), and Korosmezey and Gombosi (1990). These models were based on the numerical solution of the coupled hydrodynamic equations (representing the mass, momentum, and energy conservation). The gas (pure $H_2O$) was treated as viscous in Kitamura (1986, 1987) or inviscid in Kitamura (1990), and Korosmezey and Gombosi (1990). The dust was treated as one of the components of the fluid consisting of single-sized spherical grains (< 1 μm). The studies were performed for a spherical nucleus with (1) an active spot on an inactive surface (i.e. an isolated jet), (2) a spot of enhanced activity on an active surface (i.e. a jet expanding in a co-current flow), and (3) several active spots. The Knudsen number in the jet near the surface was $\sim 10^{-3}$ (estimated by the radius of the nucleus).

These studies showed the formation of the shock structures due to interactions of several jets and that even a single pure gas jet expanding in a co-current flow produces a shock structure. It was shown that the dust structures are more pronounced and persist even at large radial distances.

In all these works the boundary conditions were postulated in a simplified manner (i.e. without a thermo-physical model of the surface) – a given variation of surface flux or stagnation density and constant surface temperature.

These same problems were studied by Itkin et al. (1995), Crifo et al. (1995), Rodionov (1996) with a more efficient computational scheme. These studies revealed that the "single fluid" approach (when all the dust ejected from the nucleus is treated as one single fluid) leads to an unphysical result (artificially strong enhancement of the density) in the region where dust fluxes penetrate each other. To resolve this problem a "multi-fluid" approach, where the dust flow is represented by several fluids originating from different parts of the surface, was applied. Though this model is physically more accurate it is also computationally more expensive.

The studies by Crifo and Rodionov (1997a), Fulle et al. (1999), and Crifo and Rodionov (1999) included a non-spherical shape of the nucleus. A homogeneous "bean" shaped nucleus was used as a model of comet 1P/Halley in Crifo and Rodionov (1997a) and of comet 46P/Wirtanen in Fulle et al. (1999), and Crifo and Rodionov (1999). Also, a realistic model of surface ice sublimation which takes into account the influence of the ambient flow (see Crifo and Rodionov 1997b) was used to get the gas production of the surface. This model was based on the analytical expression for the Knudsen layer obtained by Cercignani (1981). These studies showed for the first time that even in the case of a homogeneous nucleus, the fluxes emanating from different parts of the nucleus can interact and form shock structures i.e. even the absence of any surface inhomogeneity can result in the appearance of inhomogeneity in the cometary coma.

The effect of a realistic shape for comet 1P/Halley – described by spherical harmonics as derived from observational data of Giotto, Vega-1 and Vega-2 – was studied in Crifo et al. (2002c). This study was performed for cases of homogeneous and inhomogeneous (set of spots) surface properties. This study showed that a more detailed description of the shape leads to a considerably more complicated flow structure near the surface. Most notably, geometrical effects of the surface can be stronger than the effect of a surface inhomogeneity.

A model with two gas components – water and CO – was studied in Crifo et al. (2004). It was assumed that the water sublimation at the surface and the CO diffusion from the interior occur simultaneously. This model was used before Rosetta had arrived to study different plausible conditions in the coma of comet 67P at 3 AU. An analytical "starfish" shape with topological similarity to the shape derived from Hubble Space Telescope observations in 2002 was used for the nucleus. This study showed once again that the coma around a non-spherical nucleus is strongly structured by the topography.

Ground-based observations have a large field of view, and therefore to simulate their observed coma a large computational domain is required. In contrast to the simulation of the near-nucleus coma, two additional factors need to be considered when modelling structures which are tens of thousands of kilometres from the surface: (1) processes of photodissociation and (2) the unsteady character of the flow. For example, in Rodionov and Crifo (2006) an attempt was made to explain the nature of the weak spiral structures seen in ground-based observations and, more specifically, in observations of the comet C/1995 O1 Hale-Bopp. It was assumed that water is produced by surface ice sublimation and CO coming from diffusion from the interior of the nucleus. Two processes of photodissociation were taken into account: $H_2O \rightarrow OH + H$ and $OH \rightarrow O + H$. These two processes have characteristic rates of $\sim (1-3) \cdot 10^{-5}/r_h^2$ seconds. At a heliocentric distance of 1 AU their importance thus becomes apparent at a distance of $10^4$ km from the nucleus. The rotational period of Hale-Bopp is 11.4 hours. The gas expanding from the surface travels about $2 \cdot 10^4$ km during one rotation. Therefore the flow in the region of $10^5$ km radius is in an unsteady state. The model demonstrated the capability to generate large-scale gas spirals of the kind observed in comet C/1995 O1 Hale-Bopp, thus demonstrating that the existence of such spirals does not require surface heterogeneity. On the other hand, this result does not in itself exclude heterogeneity either.

Including in the model the photodissociation process may be important not only for a large region of simulation. For example, in observations of comet C/1996 B2 Hyakutake, a bow structure was noticed in spectra showing the minor species (like OH) at the distance $\sim 10^3$ km. Even though at this distance the abundance of OH is much lower than that of $H_2O$, the process of photodissociation is of crucial importance for the spatial distribution of OH.

In the case when activity is strong (usually at small heliocentric distances) and the flow is in the continuum regime in the near-nucleus region, the approach which describes the dusty gas flow as a fluid is justified. But when a comet is weakly active or has less active regions the flow in the coma can be rarefied ($Kn > 0.01$). In the rarefied flow, the collision time between molecules is long compared with transport times. The mean free path between collisions is comparable to or larger than the flow gradient. In these cases the validity of the hydrodynamic approach is questionable (see e.g. Finklenburg et al. 2011). The Direct Simulation Monte Carlo (DSMC) method (Bird 1976, 1994) is based on the kinetic description of the flow and can be applied at any degree of rarefaction.

The first kinetic model of the coma based on the DSMC method was proposed in Combi (1996). This model takes into consideration gravity, intermolecular collisions, photochemistry, and radiative heating/cooling. It was applied for the simulation of an axisymmetric pure water expansion from a non-spherical nucleus. The boundary conditions were postulated in a simplistic manner – it was assumed that the nucleus has a uniform surface temperature, the gas is at rest on the surface with the density varying as the square of the cosine of the sub-solar angle. Gas emission on the night side was set to be zero. The study was carried out for the case $Kn \sim 10^{-4}$ (based on the radius of the nucleus and the mean free path in the sub-solar point). It was shown that kinetic effects become important in the periphery of the jet and on the night side in the immediate vicinity of the nucleus. At the same time, comparisons with hydrodynamic simulations of the same problem show excellent agreement in the dense regions of the flow.

A series of papers (Crifo et al. 2002b; Zakharov et al. 2003; Crifo et al. 2003; Lukyanov et al. 2005; Zakharov et al. 2008; Crifo et al. 2004) was devoted to the comparison of the fluid (the Euler and the Navier-Stokes equations) and the kinetic (DSMC) description of the coma. Several nuclei with some typical geometrical or compositional features were studied: spherically homogeneous and inhomogeneous, aspherical with a cavity fully sunlit or partially shaded. The studies were carried out in the range of $10^{-7} < Kn < 10^{-3}$ (computed for the gas parameters at the sub-solar point and the radius of the nucleus). The results showed that: (1) on the dayside both approaches agree quite well; (2) on the night side, where the local Knudsen number $Kn = \frac{\lambda|\nabla n_g|}{n_g} < 0.1$, the agreement is satisfactory as well; (3) if the night side is active then the agreement is satisfactory even for the local Knudsen number $> 0.1$.

A kinetic model (based on the DSMC method) of the axisymmetric multispecies ($H_2O$, OH, O, $H_2$, H, and CO) coma with a realistic thermophysical model of outgassing from the surface of the nucleus was presented in Tenishev et al. (2008). The surface was assumed as a porous layer of solid grains and ice particles which sublimate under solar illumination. The thermophysical model (Davidsson and Gutiérrez 2004, 2005, 2006) takes into account the processes of the gradual absorption of solar energy in the surface layer due to a finite optical opacity of the ice/dust mixture, the thermal re-radiation, the re-condensation of the gaseous constituents of the coma on the surface of the nucleus, the solid-state heat conduction, the subsurface ice sublimation and re-condensation, and the subsurface transport of mass and energy due to gas diffusion.

This gas coma model (Tenishev et al. 2008) was further completed by the dusty coma model (Tenishev et al. 2011). The numerical model of the dusty coma was constructed in the spirit of DSMC for the gas (i.e. the dust phase in the coma is represented by a large but finite number of model particles that represent real dust grains). It was assumed that the dust motion is governed by the gas drag and nucleus gravity force. The distinctive feature of this model is the self-consistent kinetic treatment of both the gas and the dust phases of the coma. Though, the numerical simulation performed for the comet 67P at heliocentric distances from 1.29 to 3.25 AU and the dust-to-gas mass production rate $\sim 1$ have shown that the back coupling of momentum exchange in the coma and the heating of the gas by collisions with the dust is negligible.

In Fougere et al. (2013) the model of Tenishev et al. (2008) and Tenishev et al. (2011) was applied for modelling the heterogeneous ice and gas coma of comet 103P/Hartley 2. Axisymmetric simulations were performed for a gas coma ($H_2O$ and $CO_2$) using a realistic bi-lobed shaped nucleus with dimensions corresponding to the ones observed by the EPOXI mission (A'Hearn et al. 2011). The model included the possibility that icy grains can sublimate while being carried away from the nucleus and constitute an additional source of water in the coma. A model of pure icy grains ($1\ \mu m \le \cdots \le 8$ cm) was used with equilibrium temperature and sublimation rates which were computed using the energy balance between sublimation of volatiles, and absorption/emission of radiation (see Fougere et al. 2012). The adjustments of the model were made based on comparisons of (1) computed column densities of $H_2O$, $CO_2$ and observed brightness of water, and $CO_2$; (2) computed and measured IR spectra of $H_2O$ (2.7 μm) and $CO_2$ (4.3 μm). The results have shown that most of the water released by Comet Hartley 2 comes from the sublimation from icy grains as they are carried away from the nucleus by the gas drag. A strong correlation was found between the icy grain and the $CO_2$ densities, which raised the hypothesis that the icy grains were lifted by the hyper-volatiles in collimated jets mostly present at the sub-solar lobe of the nucleus.

Up to the rendezvous of Rosetta with comet 67P in 2014, the coma modelling was focused mostly on possible scenarios of cometary activity and trying to study potential influences of different factors.

With Rosetta's data obtained from close distance to the comet by different instruments during a long period of time a new stage of coma modelling has begun. The modelling of 67P can be based on much more precise data and the models are more tightly constrained. New models build on and combine previous studies of particular effects. Now the main problem is not the general simulation of the dusty gas flows but rather the derivation of the boundary conditions via fitting of simulations to the available observational data (e.g. Bieler et al. 2015a; Fougere et al. 2013, 2016b; Marschall et al. 2016, 2017; Zakharov et al. 2018a; Marschall et al. 2019; Combi et al. 2020; Marschall et al. 2020b). We discuss this link of the nucleus surface (i.e. the boundary conditions) to measurements in the next section.

# 4 Linking Coma Data with the Surface

One of the most important goals of Rosetta was to deliver observations which can be used to piece together a consistent picture on the spatial distribution and time evolution of surface sources for gas and dust. It had been envisioned that inverting the remote sensing measurements was going to yield the much needed constraints on the "intrinsic" scales of gas/dust sources which can then be linked with specific surface features identified from the visible wavelength images. We will discuss this approach and its drawbacks in Sect. 4.1. In addition, the computational and numerical modelling has advanced to such a degree by the time Rosetta arrived at the comet that rarefied gas flow calculations in three-dimensional domains have become feasible, simultaneously coupling the modelling of measurements of different instruments. This approach we will label "forward" modelling, and by means of varying the input parameters we can fit the observed data, hence assess the physics of the comae. We will explore this approach and its drawbacks in Sect. 4.2 as well as the nature of how dust filaments (jet-like structures) can arise (Sect. 4.3). In Sect. 4.4 we will discuss the distribution of sources from the different models currently used, how they differ, and how conflicts between them can be resolved. Finally, in Sect. 4.5 we give an overview of how VIRTIS data has informed our inner comae knowledge.

## 4.1 Inversion Techniques for the Inner Gas and Dust Comae

### *4.1.1 Inverting In-Situ Gas Measurements*

On August 6, 2014, the Rosetta spacecraft had for the first time reached a cometocentric distance of about 100 km, which marked the beginning of near-nucleus operation at comet 67P. At this time the ROSINA instrument suite began to record signals consistently above (spacecraft) background (Altwegg et al. 2015). The direct in-situ ROSINA data provided estimates on the local number density of many molecular species. Each measurement was registered on the digital terrain model of 67P at the sub-spacecraft position. The surface outgassing rate may then be obtained, in the case of the simplest inversion approach, from the "Haser model" (Haser 1957) where the density measured at the spacecraft is simply nadir-projected onto the nucleus shape being scaled using the "$r^2$-law" of Eq. (4). It was rather surprising to find this approach to give reasonable results regarding the diurnal evolution of outgassing (Bieler et al. 2015b). Given the large field of view (FOV) of ROSINA and the complex nucleus shape and detailed pattern of illumination it was expected that this kind of nadir projection would not be able to yield a reasonable model of gas coma evolution. In addition the model assumes a collisionless flow, an assumption that can be safely considered only for outgassing conditions at large heliocentric distances $> 3.0$ AU (in the context of 67P). Applying this model to measurements in a given time period it has been also used to estimate the global gas production. The validity of this approach is limited by its assumptions, and accurate global production rates were also not expected from this method. In addition, locations of outgassing on the surface cannot be obtained from this simple inversion approach. This has been shown in Marschall et al. (2017) where emission from cliffs and the Hapi region resulted in equally good fits of ROSINA data as with inhomogeneous outgassing involving the entire surface. In short, the results from a simple inversion must be considered with some care.

A more sophisticated inversion approach was applied to constrain the gas sources at the surface from the ROSINA data in Kramer et al. (2017) and Läuter et al. (2019). In their work they assume gas sources on a shape model with roughly 4′000 facets that emit gas with a certain opening angle with respect to the local surface normal. The gas flow in this model is still assumed to remain collisionless in the entire computation domain, hence, each measurement of ROSINA is simply the (weighted) linear combination of the contribution of all these facets. A global minimisation method of reducing the difference between thousands of measured and simulated ROSINA densities yields the surface facet weights. The facets with the largest weights are the "local" surface sources of gas. The advantage of this approach is its speed, hence it can be applied to data for the entire orbit. The downside of the method is that, even if it yields a good estimate of the global production rate, the resulting source distribution should be treated with care because it is not the outcome of a model that takes into account the necessary physics. For one, the basic assumption that each surface facet contributes linearly is not justified because the gas molecules originating from different areas do interact with each other thus altering the flow. Other challenges of inverse models will be presented in Sect. 4.4.

### *4.1.2 Inverting Remote Sensing Measurements*

At the onset of near-nucleus observations (Aug. 2014) the microwave instrument for the Rosetta orbiter, MIRO, was already monitoring the water production rate for more than two months (Biver et al. 2015), triggered by the first detection of water activity on June 7, 2014 (heliocentric distance of 3.925 AU) of merely 300 g/s (Gulkis et al. 2015). In addition of retrieving molecular densities along the line-of-sight (LOS), the line shapes in the MIRO measurements carry further information on gas velocity (projected on the LOS) as well as kinetic temperature of the gas. The measured photons in the sub-millimetre wavelength range originate from the rotational transitions of molecules which are spaced closely in energy and thus still efficiently thermalised in the inner coma due to inter-molecular collisions. Nevertheless, in the context of cometary comae the population of rotational levels does deviate from local thermodynamical equilibrium (LTE) at some distance from the nucleus (Zakharov et al. 2007; Davidsson et al. 2010), and forward models of MIRO spectra have to generally take this into account (Lee et al. 2011). The velocity profile along the LOS can be estimated from the Doppler shift of the frequency of the measured line shape of a particular rotational transition, as will be discussed later. In this respect, the MIRO observations are truly exceptional in providing far more information constraining the coma's dynamical state than, for instance, the point measurements of density by the ROSINA instrument suite. However, the process of extracting this entangled information from the spectra requires more sophisticated analysis and coma models; this is still a topic of active research.

The first parametric inversion estimating simultaneously the water production rate and terminal velocity along the LOS was carried out by Lee et al. (2015) on a subset of MIRO data taken in nadir geometry in August, 2014. During this period the comet was only weakly active (Q[$H_2O$] usually below $10^{26}$ $s^{-1}$). These conditions in fact provided two advantages for the interpretation of data and played a role in the selection of a particular parametric coma model. First, the main isotopic line $H_2^{16}O$ at 557 GHz was still not so optically thick to lead to saturated line cores, such that the line area ($K\,km\,s^{-1}$) remains quasi-linearly dependent on the water column density along the LOS. Second, due to the relatively low optical thickness the determination of expansion velocity is somewhat easier as the line cores are not saturated. The coma model in the cited work was compressed to five parameters (Q, a, b, $v_0$, c), where Q is the water production rate, (a, b) are related to the power law approximation of the temperature profile required to simulate the MIRO spectra, and ($v_0$, c) are used in a tanh function used to model the velocity along the LOS. The solution of the inverse problem is performed within the framework of optimal estimation (Rodgers 2000)

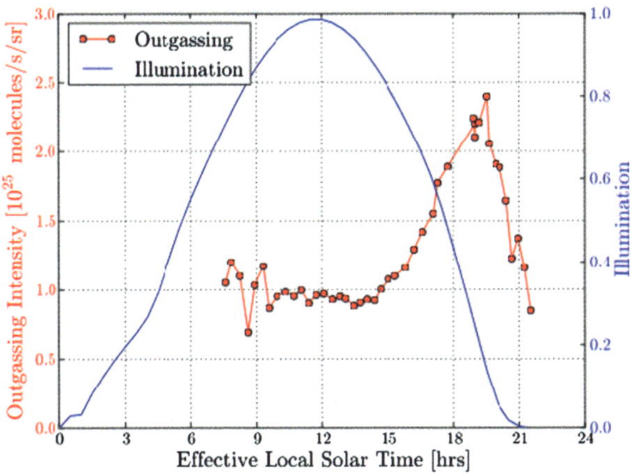


**Fig. 7** Local water production rate versus effective local solar time. The effective local solar time is determined by calculating the angle between the direction of the local normal direction at the location of the MIRO centre beam and the Sun direction. Adapted from Fig. 13 (Lee et al. 2015)

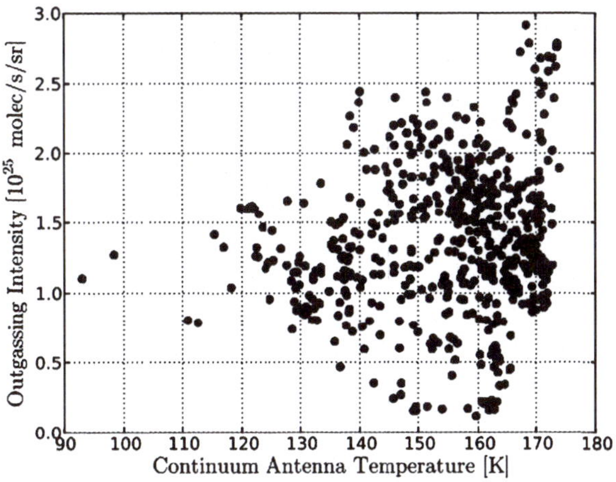


**Fig. 8** Adopted from Fig. 11 in Lee et al. (2015) which shows the distribution of the water outgassing intensity as a function of MIRO sub-millimetre antenna temperature registered at the MIRO beam location on the nucleus. No obvious correlations can be observed

which provides a statistically sound regularisation and error characterisation of the results given that the required assumption can be fulfilled. As acknowledged in the work of Lee et al. (2015) an important but often neglected consideration for non-linear inverse problem solution concerns the degree of uniqueness. Generally, it is not a simple matter to always demonstrate that a global optimum has been found for each processed MIRO measurement. The work is usually tedious and typically relies on a Monte Carlo type sensitivity analysis.

These first MIRO inversions were in an apparent conflict. It was recognised since the earliest observations that the water outgassing originates from the illuminated hemisphere, with the largest water density found when pointing toward the "neck" region of 67P. This was an unambiguous conclusion even with 1D Haser models used in the MIRO measurement reduction. However, when individual MIRO observations were projected onto the surface (at the beam-nucleus intersection location) and correlated with local time (Fig. 7) the expected relationship for column density correlation with mean illumination (or local time) was not found (Lee et al. 2015). Similarly, no obvious correlation was discovered between water flux and the surface continuum temperatures (see Fig. 8). It was speculated that this is a result of heterogeneous subsurface ice distribution, such that the largest contribution of molecules into the MIRO beam comes from other regions than the location of the FOV.

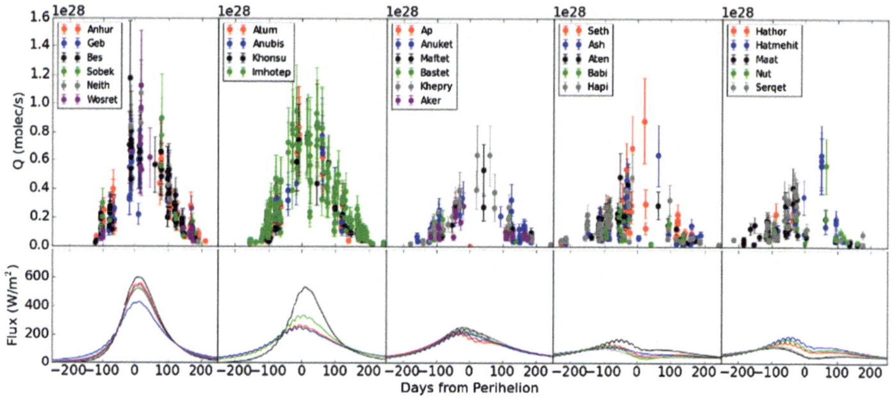


**Fig. 9** (Top panels) Water production rate 250 days before and after perihelion sorted by regions; adopted from Fig. 6 in Marshall et al. (2017). In particular we highlight the northern region results in the fourth panel of the figure where five northern regions are plotted: Seth, Ash, Aten, Babi, and Hapi. These regions receive relatively low mean daily solar flux (bottom panels), however, Seth and Ash are still quite active as determined by MIRO

A different approach to the inversion was taken in the work of Marshall et al. (2017), relying on constructed look-up tables for the line areas of the MIRO main spectroscopic transitions $H_2^{16}O$ and $H_2^{18}O$. In addition, due to the simplicity of modelling the nadir geometry, the limb measurements were not considered in this work. In order to be able to deal with the different optical thicknesses in these transitions a line area ratio was formed and used to invert individual MIRO measurements for water density. As in previous works, all observations were also registered to the surface at the location of the MIRO beam. The look-up table method's main advantage is computational speed, and nearly the entire MIRO data set has been processed such that water production rate can be statistically examined as function of time and space. It is also worth pointing out that, just as in Lee et al. (2015), the water production rate was calculated assuming a 2 km sphere with Haser-like water outgassing, that is, each observation was treated as if a spherical nucleus was observed. It was referred to as "local effective Haser" production rate by Marshall et al. (2017). With the processing of the entire data set available at the time several interesting findings were made. First, the plot of production rates as functions of heliocentric distance showed a delayed peak (20–40 days) with respect to perihelion. This feature was also observed with ROSINA data (Hansen et al. 2016; Läuter et al. 2019; Biver et al. 2019), but it is not yet clearly explained by theoretical thermal models. Second, the calculated slope of the heliocentric water production rate averaged as a function of latitude can be very well described by a purely illumination driven model assuming surface ice activity from each facet of the digital shape model. Third, a relatively high production rate (around perihelion) is implied by the MIRO data in the northern regions of the comet (especially the Seth region), which according to the shape model experienced a rather low illumination (mean daily flux of 50 $Wm^{-2}$), see panel 4 in Fig. 9. Incidentally, the Seth region is adjacent to the Hapi region, both located in the northern hemisphere neck. Despite the successful application of this approach a question remains: Why does the "local effective Haser" coma model work so reasonably well explaining the measurements, although we see such a morphological heterogeneity on the nucleus? And, how accurately can we recover the 3D spatial pattern of activity from 1D models of column density? These issues will be discussed later.

A yet different approach to inverting the nadir MIRO measurements taken around the equinox conditions in May 2015 is described in the multi-instrument study of coma data in Marschall et al. (2019). In the present section we treat only the noteworthy details of the inverse problem related to the MIRO spectra, while the general 3D DSMC modelling and comparisons to multi-instrument data are further discussed in Sect. 4.2 in this chapter. In the cited paper, the simultaneous inversion of the two water transitions ($H_2^{16}O$ and $H_2^{18}O$) was performed in a framework very similar to the one of optimal estimation formalism as used in Lee et al. (2015). However, the vertical profiles of density, temperature, and velocity along the LOS were discretised along the radial direction (along the LOS). The physical and stabilising factor of this non-linear inversion was to start the retrieval from self-consistent a-priori profiles and co-variance matrices constructed from the output of a 3D DSMC model. The retrieval iteration was allowed to rigorously fit the measured signal, which revealed several never before reported features that remain a challenge to explain with the current understanding of gas flow physics near the nucleus. An alternative explanation is that discrepancies in the results are due to some missing physical processes driving the rotational populations of water molecules not taken into account in the current radiative transfer models. The first such surprise was an ubiquitous presence of a "red" emission wing in the nadir absorption spectra at the 557 GHz transition (see e.g. Fig. 10). The MIRO observations record enhanced radiance above the continuum temperatures for red-shifted frequencies of the line, implying a stagnant (not expanding) or tangential flow to the MIRO's LOS of the near-nucleus gas field which is also significantly warmer (by 10–20 K or more) than the surface (the near sub-surface in the case of sub-mm wavelengths). The second unexpected finding was a temperature enhancement for several studied cases in the altitude region of 30–80 km. There were several speculations offered how to explain this feature, but rigorous physical modelling is yet to be constructed. Currently there is no indication that this feature is an artefact of the data calibration. In addition, the retrieved velocity profiles show evidence just above the 2-$\sigma$ uncertainty levels that acceleration continues up to an altitude of 20–40 km, and the radial profiles where the transition into free-molecular flow is expected are not necessarily smooth. Finally, the retrieved water column densities are found at least a factor 2 smaller than the modelled ones, which also seems to be a robust result. Figure 10 illustrates these features.

All nadir remote sensing observations have a limit on the vertical resolution they can achieve. That is dictated by the physics of the line formation and its dependence on the parameters we want to retrieve. In addition, the vertical smoothing also varies for the different observing geometries and the instrument's FOV. The task of inverting measurements to constrain local sources of water activity on the comet surface turns out to be a more complex problem than initially thought as recently reported in Rezac et al. (2019). That study relied on a simplified 3D approach to evaluate the relative water density contribution from facets inside the MIRO beam versus all other visible (and active) facets to a given grid point along the LOS of MIRO. Under the assumption of homogeneous, illumination driven outgassing, they found that less than 1% of the water column density originates from facets directly in the MIRO FOV. This implies that there is little hope that individual inversion can discriminate active from non-active regions at the scale of several FOVs (see Fig. 11). Additionally, it also confirms, or explains, why the beam-averaged illumination, local time or continuum temperatures may not correlate with the observed water density as reported in Lee et al. (2015). However, this result does not support the original speculations that this weak correlation necessarily implies a heterogeneous surface activity. To confirm this a more sophisticated 3D spatial inversion is recommended. And finally, it also provides a qualitative argument why the 1D Haser coma model when applied to MIRO, but also to other instruments, achieves reasonable success in explaining the measurements to first order. It appears

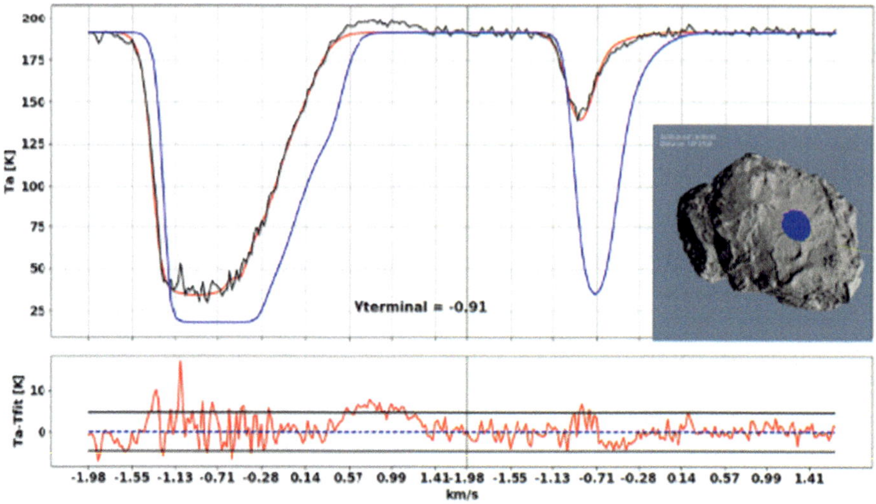


(a)

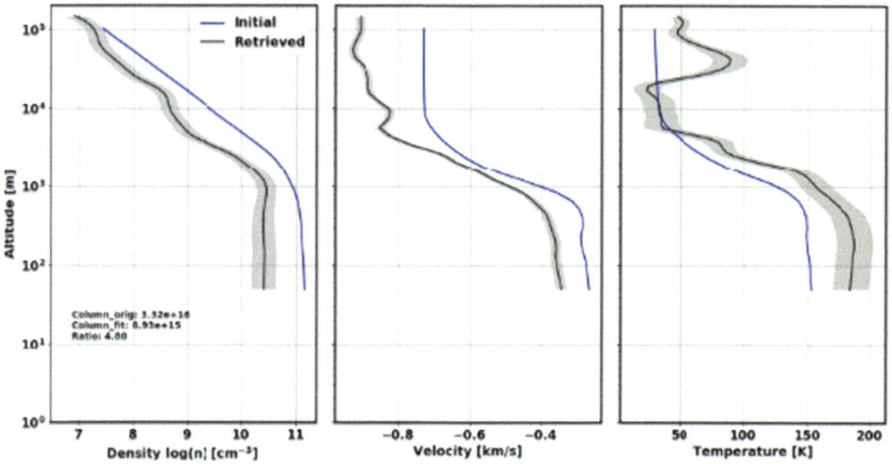


(b)

**Fig. 10** The top panel shows three spectra for the 557 GHz transition of $H_2^{16}O$ (stronger absorption) and 548 GHz $H_2^{18}O$ (weaker absorption) for the case of a nadir viewing geometry (footprint illustrated in inset image). The black line shows a MIRO measurement, red line represents simulated spectra after final retrieval iterations and the blue curve is a synthetic spectra for the a-priori profile supplied by the 3D model. The middle panel depicts the residual between measurement and the best fit (red line). The bottom three-panel plot shows vertical profiles of (left) number density, (middle) expansion velocity, and (right) kinetic temperature for initial conditions as blue lines, and final retrieved profiles as black lines. The shaded region represents a $2\sigma$ component of uncertainty due to measurement random error propagation. Figure adopted from Fig. 12 in Marschall et al. (2019)

that the same physical limitations as to why observations cannot provide a unique solution for the local sources of activity are also the reason why the Haser density distribution is a reasonable approximation for a first rough approach. A similar conclusion on the degener-

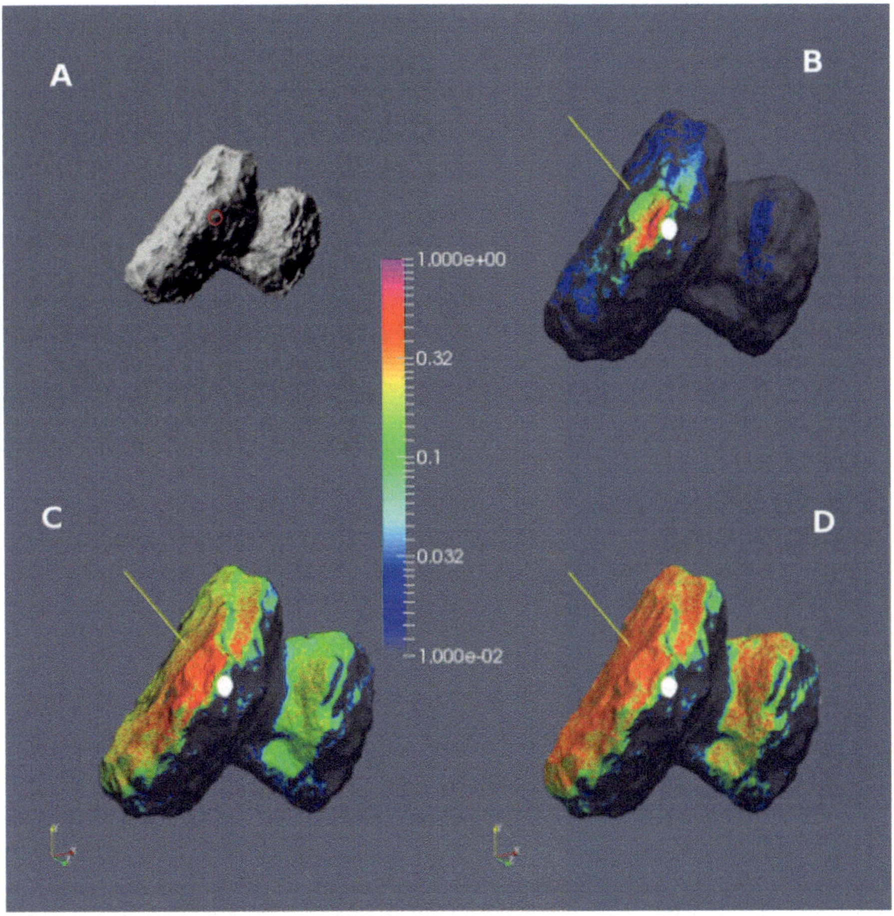


**Fig. 11** Panel (A) shows illumination conditions for a given case of MIRO nadir viewing geometry, with beam footprint located (red circle) at the terminator region on the larger lobe of 67P. Panels B, C, D show horizontal distributions of $H_2O$ density at all the different facets visible from a grid point at 2.5 km (panel B), 5 km (panel C), and 20 km (Panel D). The grid points are located inside the MIRO FOV, and the camera position in this image coincides with the S/C centred on the MIRO beam centre. The colour bar is in relative units (with respect to the facet of maximum contribution to a given grid point). This figure illustrates that even at heights of 5 km and especially 20 km the sample point on the LOS may receive contribution from a large area on the surface. This figure is adopted from Fig. 6 in Rezac et al. (2019)

acy of localising surface sources of gas activity was already drawn in Marschall et al. (2017) in application to ROSINA/COPS data.

### *4.1.3 Inverting Dust Measurements*

The processes of interaction between coma and nucleus surface have been debated since ground based observations revealed anisotropies in cometary comae. It is well known that active comets typically display regions of higher gas and dust density in their coma, often named "coma structures" by astronomers, or sometimes "jet-like" features to highlight their apparent collimated nature. We focus here on the large-scale phenomena which span the full coma and refer the reader to Vincent et al. (2019) in this book for a review of resolved jet-like features and their relation to the surface topography.

While there is no question about the presence of coma structures, extending for tens of thousands of kilometres away from the nucleus, their nature has proven more elusive. Two prominent questions must be answered: Is the collimation real? If yes, are coma structures a signature of compositional heterogeneities on the nucleus surface, or simply due to the topography?

In order to relate nucleus and coma structures, one must first determine whether the "jet-like" appearance actually implies a narrow collimation or if is just an illusion due to the geometry of our observations. It was recognised early by Sekanina (1981, 1987) that a dust/gas source at the surface of a rotating nucleus will automatically lead to a fan-shaped structure in remote observations. Indeed, as the nucleus spins the flow structure emitted from an active source will sweep the surface of a cone, centred on the apparent source and aligned with the spin axis of the nucleus. Because of the scale and resolution of ground based data, this cone cannot be observed fully but only its edges. This is because we are looking at the two dimensional projection of a tri-dimensional structure. As such, the density of gas/dust per pixel appears much higher on surfaces parallel to our line of sight, than on surfaces parallel to the image plane. Hence, the collimation and fan-like appearance of coma structures is more a result of the nucleus rotation than of an inherent mechanism leading to the release of cometary material in a spiral pattern.

This hypothesis has been put to test by the long-term monitoring of many comets. Over the course of months they develop various coma structures which can be used to constrain the nucleus spin axis. This technique has proven to be quite reliable; several authors used coma structures to determine the rotational state of comets 9P (Vincent et al. 2010), 103P (Lin et al. 2012), or 67P (Vincent et al. 2013) with excellent agreement between predictions from ground-based data and in-situ observations by space missions.

One can even go one step further to constrain the activity. Provided monitoring over several comet rotations, it is possible to determine whether the source from where the fan appears to originate is constantly active or only during the local day. This puts constraints on the type of sublimation driving the activity. Such considerations were used for instance to derive the location of a $CO_2$ source on comet 9P (Vincent et al. 2010) and seasonal variations of activity on comet 67P (Vincent et al. 2013).

Additional considerations such as the curvature of coma structures from radiation pressure are also used to retrieve information on the dust present in the coma, for instance grain size and velocity.

These observations tell us that activity appears to be more prominent in certain areas of the nucleus and that the distribution of active regions varies with seasons, typically following the subsolar latitude. Yet, they do not address the nature of those sources. In fact, the inversion of ground based data shows that flows of gas and dust appear to originate from discrete areas but this can be equally explained by real "active regions" (intrinsically different from surrounding areas) or by topographic collimation (large areas are active, but local height variations confine the flows, giving the illusion of a discrete source).

Spacecraft visiting comets have observed jet-like features at high spatial resolution and for many nucleus rotations. They revealed that both explanations are valid and play a role on comets. Regardless of the distribution of ices, local topography will always lead to heterogeneities in the gas flow (e.g Crifo et al. 2002c; Shi et al. 2018). At the same time, it is possible to find areas of the nucleus that are actually releasing more material, without a focusing effect induced by topography. Long-term monitoring at close distances also confirms that the distribution of discrete sources follows the solar insolation, with maximum density of local sources close to the subsolar latitude (Lin et al. 2015; Vincent et al. 2016c; Schmitt et al. 2017; Lai et al. 2016). In the case of 67P these inversions relied on the OSIRIS camera on-board Rosetta that detected the light reflected from the dust coma and jet-like structures within it. If the same dust filament is observed in at least two images a geometric inversion of the jet-like feature can be performed. Each image defines a plane within which the jet-like feature lies, and the intersection of the two planes of the two images gives a single line (except if the planes are parallel), the intersection of which with the surface can be determined, and thus the footprint of the feature on the surface can be found. This has been done by Vincent et al. (2016c) and Lai et al. (2017). They have found that the footprints correlate heavily with cliffs on the comets. Increased activity near freshly eroded cliffs and scarps was also found by Fornasier et al. (2017). The only caveat we would like to point out here is that the method assumes that the dust jets are not curved and are traced straight down to the surface. The actual sources of the jet-like features are mostly not actually seen because, as we mentioned in the introduction, the brightness of the dust is much lower than that of the illuminated surface at least for 67P. So, any curvature of these features close to the surface adds to the uncertainty on the origin of the dust feature. Furthermore the trace-back of these dust features does not yet imply that there is a distinct source at the surface. We will discuss this in more detail in Sect. 4.3.

Finally, the GIADA instrument on Rosetta has measured the momentum and size of dust particles for comet 67P, and thus it can be attempted to trace-back these particles to the surface. Because their speeds are rather low ($\sim$ m/s) the trace-back problem gets challenging, and therefore the uncertainties on the origin of the dust particles on the surface is very large. This has been shown in detail by Ivanovski et al. (2018). Thus, GIADA does not yield the locations of the dust particles' origins to a high resolution, but the particles can rather only be traced to large regions of the comet (Longobardo et al. 2019).

## 4.2 Forward Modelling for Constraining Emission Region

A complimentary approach to the formulation of an explicit inversion is to utilise a self-consistent 3D physics model (or 4D including the time dimension) of the coma coupled with synthetic measurement algorithms for various instruments. Under the assumption that at each narrow time interval we have a unique snapshot of the cometary state such that we can vary the input parameters in order to fit self-consistently all the measurements at the same time accounting properly for gas and dust density, temperature and velocity. We refer to this approach as "forward modelling" (or "forward fitting"). From a strictly mathematical point of view, this approach can be regarded as an incomplete optimisation (inversion) problem, where only a single (or few) iteration(s) are performed, but a convergence is not reached.

The advantage is to engage complimentary information from different instruments on the state of the coma (gas/dust density, temperature, velocity) to guide the full 3D comae physics model. For completeness, and perhaps for stimulation of future work, we note that there is an entire research field devoted to such problems in application to weather forecasting. There, vast real-time data sets of measurements are assimilated into global weather models through so-called 4D-variational assimilation procedures based on the Ensemble Kalman Filtering method (Rabier 1997; Rabier et al. 2000). In this section we will describe efforts applied only to the Rosetta data in terms of forward fitting using 3D DSMC models.

This, what can be considered a common approach, has been applied by several research groups and the results published in many papers (Bieler et al. 2015b,a; Marschall et al. 2016; Fougere et al. 2016b,c; Marschall et al. 2017; Zakharov et al. 2018a). The basic simulation sequence – which is quite similar for all groups – is illustrated in Fig. 12. The different

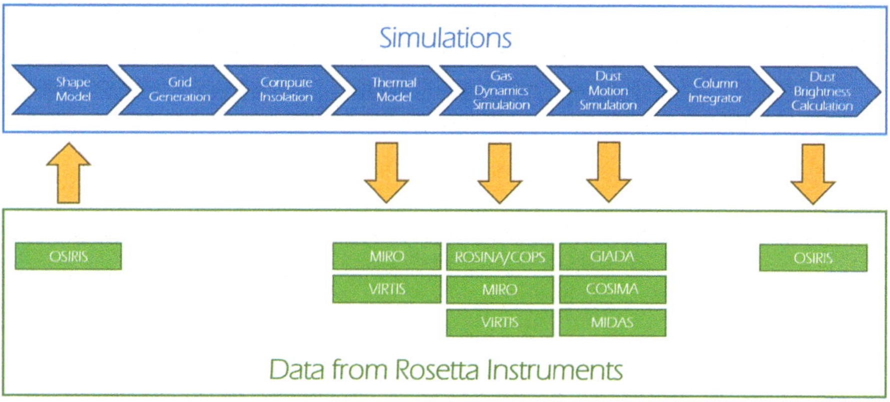


**Fig. 12** Sketch of the simulation sequence representative for the forward models and how the different pieces can be used to compare with the data (adapted from Marschall et al. 2019)

groups use very similar approaches to fit mostly ROSINA data of the local gas density either using a fluid or DSMC approach for modelling the gas dynamics. All groups start with a nucleus shape model upon which the real illumination conditions are calculated taking into account shadow casting. This is used as an input to a thermal model that determines the gas production rates and temperatures at the surface which are the inputs for the gas dynamics programs, the outputs of which can be compared to measurements of the gas instruments ROSINA, VIRTIS, and MIRO. The gas flow field can then be used to simulate the dust flow which can either be compared to results from the in-situ dust experiments GIADA, COSIMA, and MIDAS, or, by means of column integration and scattering theory, to OSIRIS dust coma data. The strong advantage of this approach is that it provides a physical connection between the emission of gas at the surface and measurements in the coma. These models should be favoured over others which run fast but don't include the necessary physical processes. The major drawback of this approach is that it is very computationally expensive and thus very time consuming. This in turn makes a systematic examination of the large parameter space challenging.

Although the basic approach of all groups using these forward models is basically the same they have different optimisation schemes. The group from Michigan, US, (e.g. Fougere et al. 2016b) have optimised the sources of gas at the surface using spherical harmonics. The group in Bern, Switzerland, (e.g. Marschall et al. 2016) assume that the emission strength correlates with morphological regions and optimise the fit according to those regions. Finally, the group in Paris, France, (Zakharov et al. 2018a) run an iterative model where an initial DSMC simulation is run and then the measurements are compared to the model. The deviations of the model to the data adjusts the initial conditions of the source distributions at the surface, by tracing the measurements to the surface within the flow field and changing the respective weights. A new DSMC simulation is run and the iteration continues. We will discuss the similarities and differences of the results of the different groups in Sect. 4.4.

## 4.3 Mechanisms that Result in Filament Structures

In the previous section we have discussed the forward modelling approach for the gas coma. Here we want to summarise some results from forward modelling of the dust coma. The simplest case to produce a dust jet-like structure is having a surface source, i.e. a strong enhancement of the emission in a confined area in contrast to its surrounding. Such a source can occur due to compositional heterogeneities of the nucleus itself or evolutionary differences in surface properties (e.g. thickness of isolating dust layer). But this is not the only way to produce features in the coma, and the interpretation of these features should thus occur with great caution. Crifo et al. (2002a) have pointed out that dust structures in the coma don't require sources on the surface. Rather a non-spherical shape is sufficient to dynamically produce such features by dynamical means. The surface topography focuses gas and thus also dust flows resulting in higher density regions within the coma. A similar case has been pointed out by Shi et al. (2018) where the sublimation along the morning terminator line can produce pronounced features because the terminator is not a straight line. More complex examples of how these dust structures form are illustrated in e.g. Marschall et al. (2017). In addition the viewing geometry plays an important role. Because imaging instruments observe column-integrated quantities, something that might appear to be a fan like structure in one orientation will look like a very fine line in another. Furthermore, variations of the surface illumination due to topography (e.g. due to shadow casting) can produce heterogeneous emission patterns leading to jet-like features that are not linked to any compositional heterogeneity of the nucleus. Jet-like structures in the dust coma can thus arise due to a wealth of reasons. These structures can occur due to genuine surface sources, dynamical focusing caused by topography, emission variations caused by topography-induced illumination conditions, and observational effects caused by the viewing geometry.

## 4.4 Similarities and Differences of Models of 67P

Having discussed the different approaches we can now examine the results of the different models. We will discuss only the maps of sources resulting from the physical models discussed in Sect. 4.2. One important conclusion all research groups arrive at is that the gas – and for that matter also the dust – activity cannot simply be explained by the variations in illumination as the comet rotates and the associated changes in energy input to the surface (Fougere et al. 2016b; Marschall et al. 2016; Zakharov et al. 2018a). The nucleus can therefore not be treated as a homogeneous piece of material. Given the diverse morphology of the surface this might not be too surprising but it is an important result. This does not yet imply though that the bulk nucleus is heterogeneous. Given the strong indications of mass transport and dust deposition, variations in the activity potential of the surface are likely evolutionary rather than compositionally. The strong seasonal differences are therefore mainly the result of the mass transport and not the inherent composition of the nucleus. Similar conclusions can be drawn for the mass transport observed at Hartley 2. The water source distribution derived by the different research groups is shown in Fig. 13. Two things need to be noted. The models don't agree on the inhomogeneity. The models from Michigan and Bern agree qualitatively in that the main source of water emission comes from the north polar region (north pole in Fougere et al. 2016c and Hapi region in Marschall et al. 2016). In contrast to that the Paris group sees enhancements of the emission towards the northern parts of the head and body lobes rather than the neck region (Zakharov et al. 2018a). At this point all groups claim good agreement with the data. This seems like a contradiction but in fact it is not. On the contrary, it shows that the gas flow smears out any inhomogeneities from the surface rather efficiently. This degrades the resolution and thus the possibility of an instrument like ROSINA, measuring the gas density tens of kilometres from the surface, to accurately determine the sources. This also highlights the issues of inverse models as discussed in Sect. 4.1. The physics involved in the expansion of gas allows for multiple surface distributions to result in the same measurements. The solutions are thus naturally degenerate. But where does this leave us in our pursuit of determining the source distribution at

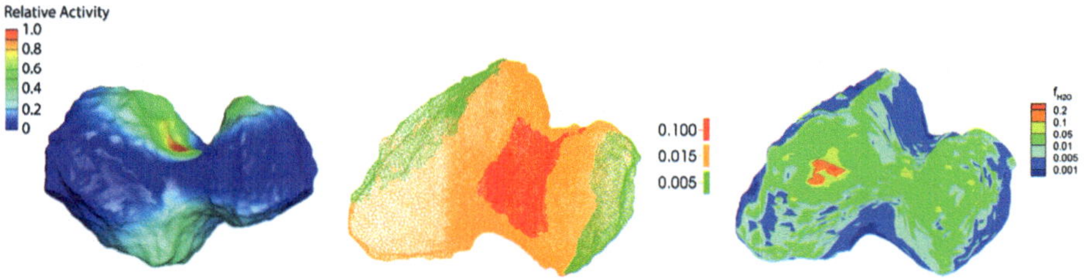


**Fig. 13** These three views of the cometary surface show the relative potential for outgassing all for northern summer, i.e. $\sim 3$ AU according to three research groups (from left to right): Fougere et al. (2016c), Marschall et al. (2016), Zakharov et al. (2018a). While the two right maps show the nucleus in the same orientation the first is shown from the side

the surface? There is a way out because the above argument only holds if we include just ROSINA data. This implies that additional data from other instruments such as VIRTIS and MIRO must be included to break this degeneracy (Marschall et al. 2019). That the solutions of the gas emission distribution are fundamentally ill-constrained below a spatial scale of a few hundred metres was shown in Marschall et al. (2020a).

The Michigan and Paris groups have also provided results for $CO_2$ measurements. Both saw an anti-correlation between the distributions of water and carbon dioxide (Fougere et al. 2016c; Zakharov et al. 2018a; Combi et al. 2020). While water is concentrated to the northern hemisphere carbon dioxide is emitted predominantly from the equatorial region and southern hemisphere. The fact that there seems to be a strong seasonal effect in the emission of water vs. carbon dioxide has already been pointed out by Hässig et al. (2015) who saw strong variations in the species' ratio. In particular they have found that the southern Imhotep region has a high carbon dioxide over water ratio during northern summer.

Finally, linking gas and dust models as done e.g. by Marschall et al. (2020b) for the entire Rosetta mission fitting ROSINA and OSIRIS data allows for a comprehensive look at the dust and gas mass loss. They found that a power-law of $q = 3.7^{+0.57}_{-0.078}$ for the dust size distribution is consistent with the data and in line with independent measurements previously reported by COSIMA (Merouane et al. 2017) and GIADA (Fulle et al. 2016). This resulted in a total (integrated over one orbital period) $5.1^{+6.0}_{-4.9} \cdot 10^9$ kg of dust being ejected from the nucleus surface, of which $4.4^{+4.9}_{-4.2} \cdot 10^9$ kg escaped to space and $6.8^{+11}_{-6.8} \cdot 10^8$ kg (or an equivalent of $14^{+22}_{-14}$ cm over the smooth regions) were re-deposited on the surface. In turn, this led to a dust-to-gas ratio of $0.73^{+1.3}_{-0.70}$ for the escaping material and $0.84^{+1.6}_{-0.81}$ for the ejected material. This is in line with other in-situ and remote sensing estimates (Choukroun et al. 2020) but lower than e.g. Fulle et al. (2019). For a detailed discussion of the dust-to-gas ratio we refer to Choukroun et al. (2020), a dedicated review of the topic.

### 4.5 Rosetta-VIRTIS Perspective on the Gas and Dust Comae

Thanks to the Visible and InfraRed Thermal Imaging Spectrometer (VIRTIS) (Coradini et al. 2007) onboard Rosetta it was possible to identify several molecule species outgassing from comet 67P starting with the first operational mission phase, when the spacecraft approached the comet at about 3 AU (Bockelée-Morvan et al. 2015). The first observed molecules forming its coma were $H_2O$ and $CO_2$ (Bockelée-Morvan et al. 2015; Migliorini et al. 2016; Fink et al. 2016). Fluorescence due the $\nu_3$ band of $H_2O$ and $CO_2$ centred at 2.67 μm and 4.27 μm, respectively, was identified. $H_2O$ emissions were found to be more localised in the comet's neck, where strong dust activity was also reported as well as an enrichment of the surface ice as testified by spectrophotometric properties (De Sanctis et al. 2015; Fornasier et al. 2015; Filacchione et al. 2016b). $CO_2$ emissions were more intense from the southern latitudes (Bockelée-Morvan et al. 2016; Migliorini et al. 2016). During pre-perihelion, $CO_2/H_2O$ ratios of 1–4% were determined above the illuminated regions (Aten, Babi, Seth and Hapi) in the Northern hemisphere (Bockelée-Morvan et al. 2015; Migliorini et al. 2016), with column densities in the order of $(2\text{–}5) \cdot 10^{20}$ and $1 \cdot 10^{19}$ molecules/m$^2$ for $H_2O$ and $CO_2$, respectively. In August 2015, few days after perihelion, an increase of $CO_2/H_2O$ was reported to as high as 30–40% (Bockelée-Morvan et al. 2016). This different distribution is a consequence of the volatile characteristics: water is less volatile than $CO_2$ and hence can be activated by the direct solar illumination, whereas $CO_2$ can sublimate also from regions with a lower insolation. $CO_2$ emission was observed especially from the southern regions, e.g. Imhotep, that were experiencing a lower solar illumination at that observing time with Rosetta (April 2015) due to the season. On the other hand, a $CO_2$ ice rich region 80 m by 60 m wide has been identified on the Southern hemisphere (especially in the Anhur region) at the end of March 2015. The same spot, observed one month later, did not show any $CO_2$ ice on the surface (Filacchione et al. 2016b). Other gaseous species, including $^{13}CO_2$, $CH_4$ and OCS, were also identified with VIRTIS (Bockelée-Morvan et al. 2016). Their abundances close to perihelion showed that the southern latitudes were more productive in $CO_2$ and OCS relative to water while a lower enhancement of $CH_4$ with respect to $H_2O$ was observed, due to the different illumination conditions of the two hemispheres during the different seasons (Bockelée-Morvan et al. 2016). However, the first close approach to the comet showed a $CO_2/H_2O$ ratio in the northern latitudes which is believed not to be representative of the primordial values, but rather a witness of the devolatilisation process occurring in the uppermost surface layers during the several approaches of the comet towards the inner solar system. The $CO/H_2O$ production rate ratio shows an increase during few months close to perihelion and at perihelion as seen with Rosetta/MIRO (Biver et al. 2019), in agreement also with previous measurements (i.e. Bockelée-Morvan et al. 2016; Läuter et al. 2019). Variability with respect to sub-spacecraft latitude is also reported (Biver et al. 2019). Similarly, the $CO/H_2O$ ratio obtained for the southern hemisphere determined close to perihelion is compatible with that of the Jupiter Family Comets (Bockelée-Morvan et al. 2016).

The imaging spectrometer VIRTIS-M, that is, the mapping channel of the VIRTIS instrument aboard Rosetta, allowed observing the full visible and infrared spectrum, from 0.22 to 5.1 μm, acquired almost simultaneously for each pixel in an image. This technique has proven to be very powerful to identify visually on the images and also spectroscopically the emissions due to $H_2O$ and $CO_2$ (Migliorini et al. 2016; Fink et al. 2016; Rinaldi et al. 2016). Spectral continuum radiance, primarily attributed to scattering from dust, can be modelled with a fourth-degree polynomial in the 2.5–3.0 μm range, around the $\nu_3$ fundamental $H_2O$ band, and with a second-degree polynomial for the $CO_2$ band (4.0–4.4 μm range), and can be removed to calculate the gas column densities. Although it is not possible to directly infer the exact spot on the surface where the observed gas emissions originated from, due to difficulties in identifying at least two VIRTIS observations targeting the same region close in time with different geometries to be used for triangulation, some efforts have been spent to address this issue, based on DSMC modelling (Fougere et al. 2016b,a). The model was applied to images acquired by VIRTIS-M when the comet was at a heliocentric distance smaller than 1.9 AU, in order to reproduce the observed emissions due to the two species, and their distributions around the nucleus. The 3D model allowed the conversion of the column density, derived from VIRTIS measurements, into production rates of $CO_2$ ($Q(CO_2) \sim 2 \cdot 10^{25}$ s$^{-1}$) and $H_2O$ ($Q(H_2O) \sim 4 \cdot 10^{26}$ s$^{-1}$) and inferring the more productive areas on the surface from which the emissions originated. As an example, the DSMC result for the most active regions on the 67P surface derived for $H_2O$ is shown in Fig. 13 (left panel), based on VIRTIS-M observations acquired in April 2015. The model confirmed that the $CO_2$ content is higher in the southern hemisphere of comet 67P, with possible sources located in the Imhotep and Khonsu regions. In addition, the evolution of the outgassing as the comet approached the Sun, indicates that water outgassing increased, while a relatively low variation in $CO_2$ outgassing was observed, as indicated by the $CO_2$ to $H_2O$ ratio, which decreases moving closer towards the Sun (Fougere et al. 2016b). Close to perihelion (July 8th, 2015 to September 27th, 2015), the emission location of the major species, $H_2O$, $CO_2$, CO and $O_2$ were more correlated with respect to pre-perihelion observations, as derived from DSMC modelling (Fougere et al. 2016a) applied to observations with the high spectral resolution channel of VIRTIS without imaging capabilities, VIRTIS-H.

Simultaneous observations of the fundamental $\nu_3$ band of $H_2O$ with VIRTIS and the H-Lyman $\beta$ band with Alice (Stern et al. 2007) on board Rosetta showed a good correlation of Lyman $\beta$ brightness and water column densities (Chaufray et al. 2017) and were also used to determine an unexpectedly high intensity of atomic line emissions compared to model predictions based on water column density scattering dissociation. This extra-signal was explained with electron dissociative excitation of water molecules located within few km from the nucleus that contributes to the final atomic emissions (Feldman et al. 2015).

The surface temperatures, derived by inverting the long-wavelength part of VIRTIS-M spectra, presented a latitudinal dependence, being higher than expected in the regions in and close to the neck, like Ash, Babi, Hapi, Seth and Ma'at (Tosi et al. 2019). In contrast, at equatorial and southern latitudes (Aker, Anubis, Atum, Imhotep and Kephry), the retrieved surface temperatures were consistent with a black body behaviour. High temperatures might be explained with small-scale roughness of the surface and indirect radiation contributions in topographically concave regions. By combining VIRTIS and MIRO observations, it clearly appears that the active physical processes occur in the uppermost layers of the nucleus. These include also erosion of the first few cm of the surface layer due to the removal of volatiles and dust caused by solar heating. Further work is required to investigate thermal seasonal effects on regions of the 67P surface and link the retrieved surface temperatures to the dust and gas properties in the coma.

During the inbound phase of 67P the spatial correlation between dust and gas shows that the dust emission in the coma was strongly correlated with the presence of $H_2O$ emissions and was quite different from the distribution of $CO_2$ emissions. This indicates that water is the main driver of dust activity in this time period (Rinaldi et al. 2016). This is also generally consistent with observations from the ground, showing that long-term variations in total water production rate correlate with the total dust brightness (Hansen et al. 2016). Despite water is the main driver of dust activity at this period, no strong temporal correlation between total dust brightness and water production rates is found. As discussed in (Tubiana et al. 2019), when 67P is approaching perihelion, the dust activity cannot be understood based on water-driven activity alone. This is likely to be attributed to regional changes of surface properties, volatile content or access to this. This is in agreement with other modelling results on the seasonal evolution of the near-nucleus coma, which show that the correlation observed earlier in the mission, between the observed dust coma and a modelled water coma from a homogeneously sublimating nucleus (e.g. Shi et al. 2018), is significantly degraded.

During the period of the inbound equinox the inner coma radiance was dominated by particles larger than 10 μm with a power law index larger than −3.1. This is consistent with the result found by GIADA (Fulle et al. 2016). Furthermore, the dust did not change its spectral properties and size distribution from March to April 2015 (Rinaldi et al. 2016,

2017). The colour of the dust in the coma has a fairly steep red slope in the visible of about 9 to $12 \pm 1\%/100$ nm from 0.45 to 0.75 μm both on the sunlit side and on the dark side. The reflectance then transitions into a much shallower slope in the infrared from 1 to 2.5 μm with values of $1.7 \pm 0.2\%/100$ nm on the sunlit side and $3 \pm 1\%100$ nm on the dark side. A possible explanation for these values is the presence of dark particles in the coma. No spectral features attributable to the dust could be detected. The colours of the dust do not show any spatial variation, nor any variation with distance from the nucleus. This indicates that the dust ejected from the nucleus into the coma is quite uniform, although it could also be that any changes in its composition or particle size distribution do not affect its colour.

During perihelion 67P's activity led to rejuvenation of its surface, exposing underlying and mostly primordial material. A comprehensive study of the diffuse coma and outbursts observed with VIRTIS was presented by Bockelée-Morvan et al. (2017, 2019) and Rinaldi et al. (2018). The main results are the particle size distribution in the quiescent coma and the detection of high-temperature grains during outburst. Bockelée-Morvan et al. (2017) modelled the infrared dust emission to explain the quiescent dust coma characterised by a bolometric albedo of 0.13, a colour of 2.5%/100 nm in the IR, and a colour temperature which is 20% in excess with respect to the equilibrium temperature. The best fit for the quiescent coma was achieved with a differential size distribution with a size index of $-2.5$ to $-3$, the same values as obtained during the pre-perihelion period (Rinaldi et al. 2017). The observed colour temperature can be attributed either to the presence of sub-micrometre particles made of absorbing material or, alternatively, to fractal agglomerates with sub-micrometre units. The scattering and thermal properties of 67P's diffuse coma are the same as found for other moderately active comets.

Many outbursts have been observed (e.g. Vincent et al. 2016a; Rinaldi et al. 2018), characterised by a sudden increase of the dust emission, peaking a few minutes later, followed by a smooth decrease of the emission until it returns to the pre-outburst value. The dust material detected during the increase phase of the dust emission is associated with a large increase of the colour temperature (from 300 to 630 K) and a change of the dust colour in the visible and infrared spectra from red to blue, revealing the presence of very small grains ($\leq 100$ nm) with a velocity of 20–40 m/s. In addition, the measured large bolometric albedos ($\sim$0.7) indicate bright grains in the ejecta, which could either be silicatic grains, implying thermal degradation of the carbonaceous material, or icy grains (Bockelée-Morvan et al. 2017; Rinaldi et al. 2018). The far-ultraviolet spectrograph, Alice, onboard Rosetta observed the same outburst in the range between 70 and 205 nm. The spectra show dust in the plume that is brighter than the sunlit nucleus, and which displays a strong absorption feature around 170 nm. This feature is characteristic of water ice (Steffl et al. 2016), suggesting that the observed material is likely to consist primarily of icy grains. The absence of a corresponding feature at 3 μm would be consistent with the presence of very small icy dust particles with radii less than 100 nm (Bockelée-Morvan et al. 2017).

## 5 Summary and Outlook

A comet is a highly dynamic object, undergoing a permanent state of change. These changes have to be carefully classified and considered according to their intrinsic temporal and spatial scales. For instance, one can consider the evolution of micro-physical properties on long time scales (years), or large morphological changes associated with outbursts or similar transient events on the scales of minutes. However, the challenge is still to fully understand the processes we usually associate with "typical" or "normal cometary" activity. On the surface and near sub-surface the changing conditions are clearly demonstrated in optical measurements of light scattered by dust lifted from the surface. The gas, considered to be the driving force liberating dust particles, can be observed remotely in a number of wavelengths. For this purpose the latest comet mission Rosetta had "eyes" in the UV (Alice, VIRTIS-M), VIS (OSIRIS, VIRTIS), IR (VIRTIS, OSIRIS), sub-mm (MIRO), and also featured in-situ instruments for the gas composition (ROSINA) and dust properties (GIADA, COSIMA, MIDAS). This allowed the long-term investigation of the inner coma as never observed before from ground or with other space missions. This enabled the identification of different regimes occurring in the inner coma, compared to the outer. One key type of measurement that could not be performed by Rosetta was the determination of the surface temperature with a high-resolution infrared mapping spectrometer sensitive in the wavelength range of the peak of the thermal emission in the 1–3 AU heliocentric range (10–30 micron).

One of the key questions is the characterisation and understanding on the gas and dust sources on the nucleus. This includes their spatial distribution, whether they posses an intrinsic "size", and whether there is a time scale over which they are activated and/or become dormant. Finally, we want to know the conditions and mechanism of how the dust is actually released from the surface. The final part of a life cycle of a comet, other than e.g. the dynamical ejection from the solar system, is the integral effect of the evolution of such sources.

We have outlined the processes of dust and gas release from the surface and the dynamical aspects of the expansion of the subsequent flows. Driven by the recent Rosetta mission the development of ever more complex and physically precise dynamical models has greatly enriched our understanding of cometary gas and dust dynamics in the innermost coma. The interplay of topography/roughness and global nucleus shape effects, surface compositional inhomogeneities as well as ever-changing illumination and shadowing conditions can no longer be denied. Yet challenges remain. How can we describe the cometary surface such that it can be translated into an adequate 'effective boundary' used in dynamical dusty-gas simulations? Or how do we describe physical properties of the dust considering the diversity of particles observed and the fact that each particle is unique in shape, composition etc.?

Individual remote sensing observations are inherently convolved in space and time. Therefore, solving the inverse problem for the characterisation of local sources is highly challenging. In-situ measurements are inherently local, and inverting for local surface sources is a rather ill-posed problem as shown in several publications. The identification of further active areas on the surface is thus of importance. A non-identification might be a consequence of the limited instrument resolution and natural resolution limits within models. Furthermore, the determination of inactive areas or regions has yet to be successful. Are there indeed no inactive areas on a comet? On what scale does activity occur and on what spatial scales are refractories and ice mixed? It is likely that only a new mission can answer these questions.

There does not seem to be a single uniform process behind cometary activity. Rather, activity seems to be the consequence of a variety of erosion processes, including the sublimation of both water ice and more volatile material, but possibly also more exotic processes such as fracture and cliff erosion under thermal and mechanical stress, sub-surface heat storage, and a complex interplay of these processes.

While Rosetta has provided a a wealth of data for volatile species, observations at larger heliocentric distances would help to investigate some of the super-volatile species (like $CO_2$ and CO) and their interactions with dust. This would also contribute to our understanding of more distant active objects in our solar system such as some of the Centaurs. For this objective again, a new mission would be required.

The future outlook on how to proceed with the existing Rosetta dataset can be simply put as "synergy". What is needed is to combine forward models of the different measurements. This includes explicitly a retrospective assimilation methodology borrowed e.g. from Earth meteorology models. Of course, a step in between might be still trying to combine them implicitly, where measurements of other instruments (or results of inversions thereof) can be put as constraints for one another. The 3D and/or 4D modelling of such constrained modelling is still ahead of us, in part due to its high computational demands.

Another key result of Rosetta has been the importance of seasonal effects. The interplay of redeposition and mass transport at the surface, and cometary activity remains elusive. Mass transport from one region to another provides a feedback effect on the activity observed. Icy spots were identified on the surface and investigated for their seasonal variability. They are in agreement with identified active areas on the surface that are supposed to be the origin of the observed emissions in the coma. However, a direct link is still missing. Furthermore, the degree to which observed inhomogeneities of the surface are the result of evolutionary processes versus reflective of the nucleus composition remains to be explored. Thus a detailed monitoring of some regions to investigate their activity and seasonal variability would hold important information. With Rosetta global coverage was privileged instead of a survey of selected interesting regions. Hence, this task is only partly achievable with the current data sets, and a new mission would be required.

The importance of seasons and the nucleus shape for the distribution and temporal evolution of activity implies that the heliocentric evolution of activity can be highly individual for every comet, and generalisations can be misleading. Finally, the large amount of backfall material suggests that there is no simple relationship between a global gas production rate measured from ground and the "largest liftable grain size" which typically determines the total mass of the ejected dust. This leaves open a big question of how ground based observations can inform our understanding of cometary nuclei.

**Acknowledgements** The project has received funding from the European Union's Horizon 2020 research and innovation programme under grant agreement No. 686709. This work was supported by the Swiss State Secretariat for Education, Research and Innovation (SERI) under contract number 16.0008-2. The opinions expressed and arguments employed herein do not necessarily reflect the official views of the Swiss Government.

Raphael Marschall acknowledges the support of the Swiss National Science Foundation (SNSF) through the grant P2BEP2_184482.

David Kappel acknowledges DFG-grant KA 3757/2-1.



## References

J. Agarwal, M.F. A'Hearn, J.B. Vincent, C. Güttler, S. Höfner, H. Sierks, C. Tubiana, C. Barbieri, P.L. Lamy, R. Rodrigo, D. Koschny, H. Rickman, M.A. Barucci, J.L. Bertaux, I. Bertini, S. Boudreault, G. Cremonese, V. Da Deppo, B. Davidsson, S. Debei, M. De Cecco, J. Deller, S. Fornasier, M. Fulle, A.

Gicquel, O. Groussin, P.J. Gutiérrez, M. Hofmann, S.F. Hviid, W.H. Ip, L. Jorda, H.U. Keller, J. Knollenberg, J.R. Kramm, E. Kührt, M. Küppers, L.M. Lara, M. Lazzarin, J.J. Lopez Moreno, F. Marzari, G. Naletto, N. Oklay, X. Shi, N. Thomas, Acceleration of individual, decimetre-sized aggregates in the lower coma of comet 67P/Churyumov-Gerasimenko. Mon. Not. R. Astron. Soc. **462**, S78–S88 (2016). https://doi.org/10.1093/mnras/stw2179. 1608.07933

J. Agarwal, V. Della Corte, P.D. Feldman, B. Geiger, S. Merouane, I. Bertini, D. Bodewits, S. Fornasier, E. Grün, P. Hasselmann, M. Hilchenbach, S. Höfner, S. Ivanovski, L. Kolokolova, M. Pajola, A. Rotundi, H. Sierks, A.J. Steffl, N. Thomas, M.F. A'Hearn, C. Barbieri, M.A. Barucci, J.L. Bertaux, S. Boudreault, G. Cremonese, V. Da Deppo, B. Davidsson, S. Debei, M. De Cecco, J.F. Deller, L.M. Feaga, H. Fischer, M. Fulle, A. Gicquel, O. Groussin, C. Güttler, P.J. Gutiérrez, M. Hofmann, K. Hornung, S.F. Hviid, W.H. Ip, L. Jorda, H.U. Keller, J. Kissel, J. Knollenberg, A. Koch, D. Koschny, J.R. Kramm, E. Kührt, M. Küppers, P.L. Lamy, Y. Langevin, L.M. Lara, M. Lazzarin, Z.Y. Lin, J.J. Lopez Moreno, S.C. Lowry, F. Marzari, S. Mottola, G. Naletto, N. Oklay, J.W. Parker, R. Rodrigo, J. Rynö, X. Shi, O. Stenzel, C. Tubiana, J.B. Vincent, H.A. Weaver, B. Zaprudin, Evidence of sub-surface energy storage in comet 67P from the outburst of 2016 July 03. Mon. Not. R. Astron. Soc. **469**, s606–s625 (2017). https://doi.org/10.1093/mnras/stx2386. 1710.10235

M.F. A'Hearn, M.J.S. Belton, W.A. Delamere, L.M. Feaga, D. Hampton, J. Kissel, K.P. Klaasen, L.A. McFadden, K.J. Meech, H.J. Melosh, P.H. Schultz, J.M. Sunshine, P.C. Thomas, J. Veverka, D.D. Wellnitz, D.K. Yeomans, S. Besse, D. Bodewits, T.J. Bowling, B.T. Carcich, S.M. Collins, T.L. Farnham, O. Groussin, B. Hermalyn, M.S. Kelley, M.S. Kelley, J.Y. Li, D.J. Lindler, C.M. Lisse, S.A. McLaughlin, F. Merlin, S. Protopapa, J.E. Richardson, J.L. Williams, EPOXI at comet Hartley 2. Science **332**(6036), 1396 (2011). https://doi.org/10.1126/science.1204054

K. Altwegg, H. Balsiger, A. Bar-Nun, J.J. Berthelier, A. Bieler, P. Bochsler, C. Briois, U. Calmonte, M. Combi, J. De Keyser, P. Eberhardt, B. Fiethe, S. Fuselier, S. Gasc, T.I. Gombosi, K.C. Hansen, M. Hässig, A. Jäckel, E. Kopp, A. Korth, L. LeRoy, U. Mall, B. Marty, O. Mousis, E. Neefs, T. Owen, H. Rème, M. Rubin, T. Sémon, C.Y. Tzou, H. Waite, P. Wurz, 67P/Churyumov-Gerasimenko, a Jupiter family comet with a high D/H ratio. Science **347**(6220), 1261,952–1261,952 (2015). https://doi.org/10.1126/science.1261952

N. Attree, O. Groussin, L. Jorda, D. Nébouy, N. Thomas, Y. Brouet, E. Kührt, F. Preusker, F. Scholten, J. Knollenberg, P. Hartogh, H. Sierks, C. Barbieri, P. Lamy, R. Rodrigo, D. Koschny, H. Rickman, H.U. Keller, M.F. A'Hearn, A.T. Auger, M.A. Barucci, J.L. Bertaux, I. Bertini, D. Bodewits, S. Boudreault, G. Cremonese, V. Da Deppo, B. Davidsson, S. Debei, M. De Cecco, J. Deller, M.R. El-Maarry, S. Fornasier, M. Fulle, P.J. Gutiérrez, C. Güttler, S. Hviid, W.H. Ip, G. Kovacs, J.R. Kramm, M. Küppers, L.M. Lara, M. Lazzarin, J.J. Lopez Moreno, S. Lowry, S. Marchi, F. Marzari, S. Mottola, G. Naletto, N. Oklay, M. Pajola, I. Toth, C. Tubiana, J.B. Vincent, X. Shi, Tensile strength of 67P/Churyumov-Gerasimenko nucleus material from overhangs. Astron. Astrophys. **611**, A33 (2018). https://doi.org/10.1051/0004-6361/201732155

M.S. Bentley, R. Schmied, T. Mannel, K. Torkar, H. Jeszenszky, J. Romstedt, A.C. Levasseur-Regourd, I. Weber, E.K. Jessberger, P. Ehrenfreund, C. Koeberl, O. Havnes, Aggregate dust particles at comet 67P/Churyumov-Gerasimenko. Nature **537**, 73–75 (2016). https://doi.org/10.1038/nature19091. 1704.00526

A. Bieler, K. Altwegg, H. Balsiger, J.J. Berthelier, U. Calmonte, M. Combi, J. De Keyser, B. Fiethe, N. Fougere, S. Fuselier, S. Gasc, T. Gombosi, K. Hansen, M. Hässig, Z. Huang, A. Jäckel, X. Jia, L. Le Roy, U.A. Mall, H. Rème, M. Rubin, V. Tenishev, G. Tóth, C.Y. Tzou, P. Wurz, Comparison of 3D kinetic and hydrodynamic models to ROSINA-COPS measurements of the neutral coma of 67P/Churyumov-Gerasimenko. Astron. Astrophys. **583**, A7 (2015a). https://doi.org/10.1051/0004-6361/201526178

A. Bieler, K. Altwegg, H. Balsiger, J.J. Berthelier, J. De Keyser, S. Fuselier, T. Gombosi, H. Reme, M. Rubin, The role of numerical models in data analysis for the Rosetta mission, in *European Planetary Science Congress*, vol. 10 (2015b). EPSC2015

G.A. Bird, Molecular gas dynamics. NASA STI/Recon Technical Report A **76**, 40,225 (1976)

G.A. Bird, *Molecular Gas Dynamics and the Direct Simulation of Gas Flows* (1994)

N. Biver, M. Hofstadter, S. Gulkis, D. Bockelée-Morvan, M. Choukroun, E. Lellouch, F.P. Schloerb, L. Rezac, W.H. Ip, C. Jarchow, P. Hartogh, S. Lee, P. von Allmen, J. Crovisier, C. Leyrat, P. Encrenaz, Distribution of water around the nucleus of comet 67P/Churyumov-Gerasimenko at 3.4 AU from the Sun as seen by the MIRO instrument on Rosetta. Astron. Astrophys. **583**, A3 (2015). https://doi.org/10.1051/0004-6361/201526094

N. Biver, D. Bockelée-Morvan, M. Hofstadter, E. Lellouch, M. Choukroun, S. Gulkis, J. Crovisier, F.P. Schloerb, L. Rezac, P. von Allmen, S. Lee, C. Leyrat, W.H. Ip, P. Hartogh, P. Encrenaz, G. Beaudin (Miro Team), Long-term monitoring of the outgassing and composition of comet 67P/Churyumov-Gerasimenko with the Rosetta/MIRO instrument. Astron. Astrophys. **630**, A19 (2019). https://doi.org/10.1051/0004-6361/201834960

D. Bockelée-Morvan, V. Debout, S. Erard, C. Leyrat, F. Capaccioni, G. Filacchione, N. Fougere, P. Drossart, G. Arnold, M. Combi, First observations of $H_2O$ and $CO_2$ vapor in comet 67P/Churyumov-Gerasimenko made by VIRTIS onboard Rosetta. Astron. Astrophys. **583**, A6 (2015). https://doi.org/10.1051/0004-6361/201526303

D. Bockelée-Morvan, J. Crovisier, S. Erard, F. Capaccioni, C. Leyrat, G. Filacchione, P. Drossart, T. Encrenaz, N. Biver, de M.C. Sanctis, Evolution of $CO_2$, $CH_4$, and OCS abundances relative to $H_2O$ in the coma of comet 67P around perihelion from Rosetta/VIRTIS-H observations. Mon. Not. R. Astron. Soc. **462**, S170–S183 (2016). https://doi.org/10.1093/mnras/stw2428. 1609.07252

D. Bockelée-Morvan, G. Rinaldi, S. Erard, C. Leyrat, F. Capaccioni, P. Drossart, G. Filacchione, A. Migliorini, E. Quirico, S. Mottola, Comet 67P outbursts and quiescent coma at 1.3 au from the Sun: dust properties from Rosetta/VIRTIS-H observations. Mon. Not. R. Astron. Soc. **469**, S443–S458 (2017). https://doi.org/10.1093/mnras/stx1950. 1708.00210

D. Bockelée-Morvan, C. Leyrat, S. Erard, F. Andrieu, F. Capaccioni, G. Filacchione, P.H. Hasselmann, J. Crovisier, P. Drossart, G. Arnold, M. Ciarniello, D. Kappel, A. Longobardo, M.T. Capria, M.C. De Sanctis, G. Rinaldi, F. Taylor, VIRTIS-H observations of the dust coma of comet 67P/Churyumov-Gerasimenko: spectral properties and color temperature variability with phase and elevation. Astron. Astrophys. **630**, A22 (2019). https://doi.org/10.1051/0004-6361/201834764. 1901.03074

H. Boehnhardt, A. Riffeser, M. Kluge, C. Ries, M. Schmidt, U. Hopp, Mt. Wendelstein imaging of the post-perihelion dust coma of 67P/Churyumov-Gerasimenko in 2015/2016. Mon. Not. R. Astron. Soc. **462**, S376–S393 (2016). https://doi.org/10.1093/mnras/stw2859. 1611.03085

Y. Brouet, A.C. Levasseur-Regourd, P. Sabouroux, L. Neves, P. Encrenaz, O. Poch, A. Pommerol, N. Thomas, W. Kofman, A. Le Gall, A porosity gradient in 67P/C-G nucleus suggested from CONSERT and SESAME-PP results: an interpretation based on new laboratory permittivity measurements of porous icy analogues. Mon. Not. R. Astron. Soc. **462**, S89–S98 (2016). https://doi.org/10.1093/mnras/stw2151

F. Capaccioni, A. Coradini, G. Filacchione, S. Erard, G. Arnold, P. Drossart, M.C. De Sanctis, D. Bockelee-Morvan, M.T. Capria, F. Tosi, C. Leyrat, B. Schmitt, E. Quirico, P. Cerroni, V. Mennella, A. Raponi, M. Ciarniello, T. McCord, L. Moroz, E. Palomba, E. Ammannito, M.A. Barucci, G. Bellucci, J. Benkhoff, J.P. Bibring, A. Blanco, M. Blecka, R. Carlson, U. Carsenty, L. Colangeli, M. Combes, M. Combi, J. Crovisier, T. Encrenaz, C. Federico, U. Fink, S. Fonti, W.H. Ip, P. Irwin, R. Jaumann, E. Kuehrt, Y. Langevin, G. Magni, S. Mottola, V. Orofino, P. Palumbo, G. Piccioni, U. Schade, F. Taylor, D. Tiphene, G.P. Tozzi, P. Beck, N. Biver, L. Bonal, J.P. Combe, D. Despan, E. Flamini, S. Fornasier, A. Frigeri, D. Grassi, M. Gudipati, A. Longobardo, K. Markus, F. Merlin, R. Orosei, G. Rinaldi, K. Stephan, M. Cartacci, A. Cicchetti, S. Giuppi, Y. Hello, F. Henry, S. Jacquinod, R. Noschese, G. Peter, R. Politi, J.M. Reess, A. Semery, The organic-rich surface of comet 67P/Churyumov-Gerasimenko as seen by VIRTIS/Rosetta. Science **347**, aaa0628 (2015). https://doi.org/10.1126/science.aaa0628

C. Cercignani, Strong evaporation of a polyatomic gas. Prog. Astronaut. Aeronaut. **74**, 305–320 (1981)

J.Y. Chaufray, D. Bockelée-Morvan, J.L. Bertaux, S. Erard, P.D. Feldman, F. Capaccioni, E. Schindhelm, C. Leyrat, J. Parker, G. Filacchione, Rosetta Alice/VIRTIS observations of the water vapour UV electroglow emissions around comet 67P/Churyumov-Gerasimenko. Mon. Not. R. Astron. Soc. **469**, S416–S426 (2017). https://doi.org/10.1093/mnras/stx1895

M. Choukroun, S. Keihm, F.P. Schloerb, S. Gulkis, E. Lellouch, Dark side of comet 67P/Churyumov-Gerasimenko in Aug.-Oct. 2014. MIRO/Rosetta continuum observations of polar night in the southern regions. Astron. Astrophys. **583**, A28 (2015). https://doi.org/10.1051/0004-6361/201526181

M. Choukroun, K. Altwegg, E. Kührt, N. Biver, D. Bockelée-Morvan, J. Drążkowska, A. Hérique, M. Hilchenbach, R. Marschall, M. Pätzold, M.G.G.T. Taylor, N. Thomas, Dust-to-gas and refractory-to-ice mass ratios of comet 67P/Churyumov-Gerasimenko from Rosetta observations. Space Sci. Rev. **216**(3), 44 (2020). https://doi.org/10.1007/s11214-020-00662-1

C. Christou, S.K. Dadzie, N. Thomas, R. Marschall, P. Hartogh, L. Jorda, E. Kührt, I. Wright, R. Rodrigo, Gas flow in near surface comet like porous structures: application to 67P/Churyumov-Gerasimenko. Planet. Space Sci. **161**, 57–67 (2018). https://doi.org/10.1016/j.pss.2018.06.009

C. Christou, S.K. Dadzie, R. Marschall, N. Thomas, Porosity gradient driving flows at comet near surfaces. Planet. Space Sci. (2019)

M. Ciarniello, A. Raponi, F. Capaccioni, G. Filacchione, F. Tosi, M.C. De Sanctis, D. Kappel, B. Rousseau, G. Arnold, M.T. Capria, M.A. Barucci, E. Quirico, A. Longobardo, E. Kuehrt, S. Mottola, S. Erard, D. Bockelée-Morvan, C. Leyrat, A. Migliorini, A. Zinzi, E. Palomba, B. Schmitt, G. Piccioni, P. Cerroni, W.H. Ip, G. Rinaldi, M. Salatti, The global surface composition of 67P/Churyumov-Gerasimenko nucleus by Rosetta/VIRTIS. II Diurnal and seasonal variability. Mon. Not. R. Astron. Soc. **462**, S443–S458 (2016). https://doi.org/10.1093/mnras/stw3177

P. Clausing, Über die Strömung sehr verdünnter Gase durch Röhren von beliebiger Länge. Ann. Phys. **404**(8), 961–989 (1932). https://doi.org/10.1002/andp.19324040804

L. Colangeli, J.J. Lopez Moreno, P. Palumbo, J. Rodriguez, E. Bussoletti, V. Della Corte, F. Esposito, M. Herranz, J.M. Jerónimo, A. Lopez-Jimenez, E.M. Epifani, R. Morales, E. Palomba, A. Rotundi, S. Vergara (International GIADA Team), GIADA: the grain impact analyser and dust accumulator for the Rosetta space mission. Adv. Space Res. **39**, 446–450 (2007). https://doi.org/10.1016/j.asr.2006.12.048

M.R. Combi, Time-dependent gas kinetics in tenuous planetary atmospheres: the cometary coma. Icarus **123**, 207–226 (1996). https://doi.org/10.1006/icar.1996.0150

M.R. Combi, W.M. Harris, W.H. Smyth, Gas dynamics and kinetics in the cometary coma: theory and observations (2004) pp. 523–552

M. Combi, Y. Shou, N. Fougere, V. Tenishev, K. Altwegg, M. Rubin, D. Bockelée-Morvan, F. Capaccioni, Y.C. Cheng, U. Fink, T. Gombosi, K.C. Hansen, Z. Huang, D. Marshall, G. Toth, The surface distributions of the production of the major volatile species, $H_2O$, $CO_2$, CO and $O_2$, from the nucleus of comet 67P/Churyumov-Gerasimenko throughout the Rosetta mission as measured by the ROSINA double focusing mass spectrometer. Icarus **335**, 113421 (2020). https://doi.org/10.1016/j.icarus.2019.113421. 1909.02082

A. Coradini, F. Capaccioni, P. Drossart, G. Arnold, E. Ammannito, F. Angrilli, A. Barucci, G. Bellucci, J. Benkhoff, G. Bianchini, J.P. Bibring, M. Blecka, D. Bockelee-Morvan, M.T. Capria, R. Carlson, U. Carsenty, P. Cerroni, L. Colangeli, M. Combes, M. Combi, J. Crovisier, M.C. De Sanctis, E.T. Encrenaz, S. Erard, C. Federico, G. Filacchione, U. Fink, S. Fonti, V. Formisano, W.H. Ip, R. Jaumann, E. Kuehrt, Y. Langevin, G. Magni, T. McCord, V. Mennella, S. Mottola, G. Neukum, P. Palumbo, G. Piccioni, H. Rauer, B. Saggin, B. Schmitt, D. Tiphene, G. Tozzi, Virtis: an imaging spectrometer for the Rosetta mission. Space Sci. Rev. **128**, 529–559 (2007). https://doi.org/10.1007/s11214-006-9127-5

J.F. Crifo, A.V. Rodionov, The dependence of the circumnuclear coma structure on the properties of the nucleus. Icarus **129**, 72–93 (1997a). https://doi.org/10.1006/icar.1997.5714

J.F. Crifo, A.V. Rodionov, The dependence of the circumnuclear coma structure on the properties of the nucleus. Icarus **127**, 319–353 (1997b). https://doi.org/10.1006/icar.1997.5690

J.F. Crifo, A.V. Rodionov, Modelling the circumnuclear coma of comets: objectives, methods and recent results. Planet. Space Sci. **47**(6–7), 797–826 (1999). https://doi.org/10.1016/S0032-0633(98)00091-9

J.F. Crifo, A.L. Itkin, A.V. Rodionov, The near-nucleus coma formed by interacting dusty gas jets effusing from a cometary nucleus: I. Icarus **116**(1), 77–112 (1995). https://doi.org/10.1006/icar.1995.1114

J.F. Crifo, G.A. Lukianov, A.V. Rodionov, G.O. Khanlarov, V.V. Zakharov, Comparison between Navier-Stokes and direct Monte-Carlo simulations of the circumnuclear coma. I. Homogeneous, spherical source. Icarus **156**, 249–268 (2002a). https://doi.org/10.1006/icar.2001.6769

J.F. Crifo, G.A. Lukianov, A.V. Rodionov, G.O. Khanlarov, V.V. Zakharov, Comparison between Navier-Stokes and direct Monte-Carlo simulations of the circumnuclear coma. I. Homogeneous, spherical source. Icarus **156**(1), 249–268 (2002b). https://doi.org/10.1006/icar.2001.6769

J.F. Crifo, A.V. Rodionov, K. Szegö, M. Fulle, Challenging a paradigm: do we need active and inactive areas to account for near-nuclear jet activity? Earth Moon Planets **90**, 227–238 (2002c)

J.F. Crifo, G.A. Loukianov, A.V. Rodionov, V.V. Zakharov, Navier-Stokes and direct Monte Carlo simulations of the circumnuclear coma II. Homogeneous, aspherical sources. Icarus **163**, 479–503 (2003). https://doi.org/10.1016/S0019-1035(03)00041-1

J.F. Crifo, G.A. Lukyanov, V.V. Zakharov, A.V. Rodionov, Physical model of the coma of comet 67P/Churyumov-Gerasimenko, the new Rosetta target, in *Observations, Simulations and Instrument Performances*. Astrophysics and Space Science Library, vol. 311 (Kluwer Academic, Dordrecht, 2004)

J. Crovisier, The photodissociation of water in cometary atmospheres. Astron. Astrophys. **213**, 459–464 (1989)

B.J.R. Davidsson, Comet Knudsen layers. Space Sci. Rev. **138**, 207–223 (2008). https://doi.org/10.1007/s11214-008-9305-8

B.J.R. Davidsson, P.J. Gutiérrez, Estimating the nucleus density of Comet 19P/Borrelly. Icarus **168**(2), 392–408 (2004). https://doi.org/10.1016/j.icarus.2003.11.009

B.J.R. Davidsson, P.J. Gutiérrez, Nucleus properties of Comet 67P/Churyumov Gerasimenko estimated from non-gravitational force modeling. Icarus **176**(2), 453–477 (2005). https://doi.org/10.1016/j.icarus.2005.02.006

B.J.R. Davidsson, P.J. Gutiérrez, Non-gravitational force modeling of Comet 81P/Wild 2. I. A nucleus bulk density estimate. Icarus **180**(1), 224–242 (2006). https://doi.org/10.1016/j.icarus.2005.07.023

B.J.R. Davidsson, Y.V. Skorov, On the light-absorbing surface layer of cometary nuclei. II. Thermal modeling. Icarus **159**(1), 239–258 (2002). https://doi.org/10.1006/icar.2002.6912

B.J.R. Davidsson, Y.V. Skorov, A practical tool for simulating the presence of gas comae in thermophysical modeling of cometary nuclei. Icarus **168**(1), 163–185 (2004). https://doi.org/10.1016/j.icarus.2003.11.002

B.J. Davidsson, S. Gulkis, C. von Alexander, P. Allmen, L. Kamp, S. Lee, J. Warell, Gas kinetics and dust dynamics in low-density comet comae. Icarus **210**(1), 455–471 (2010). https://doi.org/10.1016/j.icarus.2010.06.022

M.C. De Sanctis, F. Capaccioni, M. Ciarniello, G. Filacchione, M. Formisano, S. Mottola, A. Raponi, F. Tosi, D. Bockelée-Morvan, S. Erard, The diurnal cycle of water ice on comet 67P/Churyumov-Gerasimenko. Nature **525**(7570), 500–503 (2015). https://doi.org/10.1038/nature14869

H. Dong, M.J. Blunt, Pore-network extraction from micro-computerized-tomography images. Phys. Rev. E **80**, 036,307 (2009). https://doi.org/10.1103/PhysRevE.80.036307

M.R. El-Maarry, N. Thomas, L. Giacomini, M. Massironi, M. Pajola, R. Marschall, A. Gracia-Berná, H. Sierks, C. Barbieri, P.L. Lamy, R. Rodrigo, H. Rickman, D. Koschny, H.U. Keller, J. Agarwal, M.F. A'Hearn, A.-T. Auger, M.A. Barucci, J.-L. Bertaux, I. Bertini, S. Besse, D. Bodewits, G. Cremonese, V. Da Deppo, B. Davidsson, M. De Cecco, S. Debei, C. Güttler, S. Fornasier, M. Fulle, O. Groussin, P.J. Gutierrez, S.F. Hviid, W.-H. Ip, L. Jorda, J. Knollenberg, G. Kovacs, J.-R. Kramm, E. Kührt, M. Küppers, F. La Forgia, L.M. Lara, M. Lazzarin, J.J. Lopez Moreno, S. Marchi, F. Marzari, H. Michalik, G. Naletto, N. Oklay, A. Pommerol, F. Preusker, F. Scholten, C. Tubiana, J.-B. Vincent, Regional surface morphology of comet 67p/Churyumov-Gerasimenko from Rosetta/OSIRIS images. Astron. Astrophys. **583**, A26 (2015). https://doi.org/10.1051/0004-6361/201525723

M.R. El-Maarry, N. Thomas, A. Gracia-Berná, M. Pajola, J.C. Lee, M. Massironi, B. Davidsson, S. Marchi, H.U. Keller, S.F. Hviid, S. Besse, H. Sierks, C. Barbieri, P.L. Lamy, D. Koschny, H. Rickman, R. Rodrigo, M.F. A'Hearn, A.T. Auger, M.A. Barucci, J.L. Bertaux, I. Bertini, D. Bodewits, G. Cremonese, V. Da Deppo, M. De Cecco, S. Debei, C. Güttler, S. Fornasier, M. Fulle, L. Giacomini, O. Groussin, P.J. Gutierrez, W.H. Ip, L. Jorda, J. Knollenberg, G. Kovacs, J.R. Kramm, E. Kührt, M. Küppers, L.M. Lara, M. Lazzarin, J.J. Lopez Moreno, R. Marschall, F. Marzari, G. Naletto, N. Oklay, A. Pommerol, F. Preusker, F. Scholten, C. Tubiana, J.B. Vincent, Regional surface morphology of comet 67P/Churyumov-Gerasimenko from Rosetta/OSIRIS images: the southern hemisphere. Astron. Astrophys. **593**, A110 (2016). https://doi.org/10.1051/0004-6361/201628634

M.R. El-Maarry, N. Thomas, A. Gracia-Berná, M. Pajola, J.C. Lee, M. Massironi, B. Davidsson, S. Marchi, H.U. Keller, S.F. Hviid, S. Besse, H. Sierks, C. Barbieri, P.L. Lamy, D. Koschny, H. Rickman, R. Rodrigo, M.F. A'Hearn, A.T. Auger, M.A. Barucci, J.L. Bertaux, I. Bertini, D. Bodewits, G. Cremonese, V. Da Deppo, M. De Cecco, S. Debei, C. Güttler, S. Fornasier, M. Fulle, L. Giacomini, O. Groussin, P.J. Gutierrez, W.H. Ip, L. Jorda, J. Knollenberg, G. Kovacs, J.R. Kramm, E. Kührt, M. Küppers, L.M. Lara, M. Lazzarin, J.J. Lopez Moreno, R. Marschall, F. Marzari, G. Naletto, N. Oklay, A. Pommerol, F. Preusker, F. Scholten, C. Tubiana, J.B. Vincent, Regional surface morphology of comet 67P/Churyumov-Gerasimenko from Rosetta/OSIRIS images: the southern hemisphere (corrigendum). Astron. Astrophys. **598**, C2 (2017). https://doi.org/10.1051/0004-6361/201628634e

M.R. El-Maarry, O. Groussin, H.U. Keller, N. Thomas, J.B. Vincent, S. Mottola, M. Pajola, K. Otto, C. Herny, S. Krasilnikov, Surface morphology of comets and associated evolutionary processes: a review of Rosetta's observations of 67p/Churyumov–Gerasimenko. Space Sci. Rev. **215**(4), 36 (2019). https://doi.org/10.1007/s11214-019-0602-1

L.A. Feldkamp, S.A. Goldstein, M.A. Parfitt, G. Jesion, M. Kleerekoper, The direct examination of three-dimensional bone architecture in vitro by computed tomography. J. Bone Miner. Res. **4**(1), 3–11 (1989). https://doi.org/10.1002/jbmr.5650040103

P.D. Feldman, M.F. A'Hearn, J.L. Bertaux, L.M. Feaga, J.W. Parker, E. Schindhelm, A.J. Steffl, S.A. Stern, H.A. Weaver, H. Sierks, Measurements of the near-nucleus coma of comet 67P/Churyumov-Gerasimenko with the Alice far-ultraviolet spectrograph on Rosetta. Astron. Astrophys. **583**, A8 (2015). https://doi.org/10.1051/0004-6361/201525925. 1506.01203

G. Filacchione, M.C. de Sanctis, F. Capaccioni, A. Raponi, F. Tosi, M. Ciarniello, P. Cerroni, G. Piccioni, M.T. Capria, E. Palomba, G. Bellucci, S. Erard, D. Bockelee-Morvan, C. Leyrat, G. Arnold, M.A. Barucci, M. Fulchignoni, B. Schmitt, E. Quirico, R. Jaumann, K. Stephan, A. Longobardo, V. Mennella, A. Migliorini, E. Ammannito, J. Benkhoff, J.P. Bibring, A. Blanco, M.I. Blecka, R. Carlson, U. Carsenty, L. Colangeli, M. Combes, M. Combi, J. Crovisier, P. Drossart, T. Encrenaz, C. Federico, U. Fink, S. Fonti, W.H. Ip, P. Irwin, E. Kuehrt, Y. Langevin, G. Magni, T. McCord, L. Moroz, S. Mottola, V. Orofino, U. Schade, F. Taylor, D. Tiphene, G.P. Tozzi, P. Beck, N. Biver, L. Bonal, J.P. Combe, D. Despan, E. Flamini, M. Formisano, S. Fornasier, A. Frigeri, D. Grassi, M.S. Gudipati, D. Kappel, F. Mancarella, K. Markus, F. Merlin, R. Orosei, G. Rinaldi, M. Cartacci, A. Cicchetti, S. Giuppi, Y. Hello, F. Henry, S. Jacquinod, J.M. Reess, R. Noschese, R. Politi, G. Peter, Exposed water ice on the nucleus of comet 67P/Churyumov-Gerasimenko. Nature **529**(7586), 368–372 (2016a). https://doi.org/10.1038/nature16190

G. Filacchione, A. Raponi, F. Capaccioni, M. Ciarniello, F. Tosi, M.T. Capria, M.C. De Sanctis, A. Migliorini, G. Piccioni, P. Cerroni, M.A. Barucci, S. Fornasier, B. Schmitt, E. Quirico, S. Erard, D. Bockelee-Morvan, C. Leyrat, G. Arnold, V. Mennella, E. Ammannito, G. Bellucci, J. Benkhoff, J.P. Bibring,

A. Blanco, M.I. Blecka, R. Carlson, U. Carsenty, L. Colangeli, M. Combes, M. Combi, J. Crovisier, P. Drossart, T. Encrenaz, C. Federico, U. Fink, S. Fonti, M. Fulchignoni, W.H. Ip, P. Irwin, R. Jaumann, E. Kuehrt, Y. Langevin, G. Magni, T. McCord, L. Moroz, S. Mottola, E. Palomba, U. Schade, K. Stephan, F. Taylor, D. Tiphene, G.P. Tozzi, P. Beck, N. Biver, L. Bonal, J.P. Combe, D. Despan, E. Flamini, M. Formisano, A. Frigeri, D. Grassi, M.S. Gudipati, D. Kappel, A. Longobardo, F. Mancarella, K. Markus, F. Merlin, R. Orosei, G. Rinaldi, M. Cartacci, A. Cicchetti, Y. Hello, F. Henry, S. Jacquinod, J.M. Reess, R. Noschese, R. Politi, G. Peter, Seasonal exposure of carbon dioxide ice on the nucleus of comet 67P/Churyumov-Gerasimenko. Science **354**(6319), 1563–1566 (2016b). https://doi.org/10.1126/science.aag3161

U. Fink, L. Doose, G. Rinaldi, A. Bieler, F. Capaccioni, D. Bockelée-Morvan, G. Filacchione, S. Erard, C. Leyrat, M. Blecka, Investigation into the disparate origin of $CO_2$ and $H_2O$ outgassing for Comet 67/P. Icarus **277**, 78–97 (2016). https://doi.org/10.1016/j.icarus.2016.04.040

S. Finklenburg, Investigations of the near nucleus gas and dust coma of comets. PhD thesis, Universität Bern, Switzerland (2014)

S. Finklenburg, N. Thomas, J. Knollenberg, E. Kührt, Comparison of DSMC and Euler equations solutions for inhomogeneous sources on comets, in *American Institute of Physics Conference Series*. American Institute of Physics Conference Series, vol. 1333 (2011), pp. 1151–1156. https://doi.org/10.1063/1.3562799

M.J. Finson, R.F. Probstein, A theory of dust comets. I. Model and equations. Astrophys. J. **154**, 327–352 (1968). https://doi.org/10.1086/149761

S. Fornasier, P.H. Hasselmann, M.A. Barucci, C. Feller, S. Besse, C. Leyrat, L. Lara, P.J. Gutierrez, N. Oklay, C. Tubiana, F. Scholten, H. Sierks, C. Barbieri, P.L. Lamy, R. Rodrigo, D. Koschny, H. Rickman, H.U. Keller, J. Agarwal, M.F. A'Hearn, J.L. Bertaux, I. Bertini, G. Cremonese, V. Da Deppo, B. Davidsson, S. Debei, M. De Cecco, M. Fulle, O. Groussin, C. Güttler, S.F. Hviid, W. Ip, L. Jorda, J. Knollenberg, G. Kovacs, R. Kramm, E. Kührt, M. Küppers, F. La Forgia, M. Lazzarin, J.J. Lopez Moreno, F. Marzari, K.D. Matz, H. Michalik, F. Moreno, S. Mottola, G. Naletto, M. Pajola, A. Pommerol, F. Preusker, X. Shi, C. Snodgrass, N. Thomas, J.B. Vincent, Spectrophotometric properties of the nucleus of comet 67P/Churyumov-Gerasimenko from the OSIRIS instrument onboard the ROSETTA spacecraft. Astron. Astrophys. **583**, A30 (2015). https://doi.org/10.1051/0004-6361/201525901. 1505.06888

S. Fornasier, S. Mottola, H.U. Keller, M.A. Barucci, B. Davidsson, C. Feller, J.D.P. Deshapriya, H. Sierks, C. Barbieri, P.L. Lamy, R. Rodrigo, D. Koschny, H. Rickman, M. A'Hearn, J. Agarwal, J.L. Bertaux, I. Bertini, S. Besse, G. Cremonese, V. Da Deppo, S. Debei, M. De Cecco, J. Deller, M.R. El-Maarry, M. Fulle, O. Groussin, P.J. Gutierrez, C. Güttler, M. Hofmann, S.F. Hviid, W.H. Ip, L. Jorda, J. Knollenberg, G. Kovacs, R. Kramm, E. Kührt, M. Küppers, M.L. Lara, M. Lazzarin, J.J.L. Moreno, F. Marzari, M. Massironi, G. Naletto, N. Oklay, M. Pajola, A. Pommerol, F. Preusker, F. Scholten, X. Shi, N. Thomas, I. Toth, C. Tubiana, J.B. Vincent, Rosetta's comet 67P/Churyumov-Gerasimenko sheds its dusty mantle to reveal its icy nature. Science **354**, 1566–1570 (2016). https://doi.org/10.1126/science.aag2671

S. Fornasier, C. Feller, J.C. Lee, S. Ferrari, M. Massironi, P.H. Hasselmann, J.D.P. Deshapriya, M.A. Barucci, M.R. El-Maarry, L. Giacomini, S. Mottola, H.U. Keller, W.H. Ip, Z.Y. Lin, H. Sierks, C. Barbieri, P.L. Lamy, R. Rodrigo, D. Koschny, H. Rickman, J. Agarwal, M. A'Hearn, J.L. Bertaux, I. Bertini, G. Cremonese, V. Da Deppo, B. Davidsson, S. Debei, M. De Cecco, J. Deller, M. Fulle, O. Groussin, P.J. Gutierrez, C. Güttler, M. Hofmann, S.F. Hviid, L. Jorda, J. Knollenberg, G. Kovacs, R. Kramm, E. Kührt, M. Küppers, M.L. Lara, M. Lazzarin, J.J. Lopez Moreno, F. Marzari, G. Naletto, N. Oklay, M. Pajola, X. Shi, N. Thomas, I. Toth, C. Tubiana, J.B. Vincent, The highly active Anhur-Bes regions in the 67P/Churyumov-Gerasimenko comet: results from OSIRIS/ROSETTA observations. Mon. Not. R. Astron. Soc. **469**, S93–S107 (2017). https://doi.org/10.1093/mnras/stx1275. 1707.02945

S. Fornasier, C. Feller, P.H. Hasselmann, M.A. Barucci, J. Sunshine, J.B. Vincent, X. Shi, H. Sierks, G. Naletto, P.L. Lamy, R. Rodrigo, D. Koschny, B. Davidsson, J.L. Bertaux, I. Bertini, D. Bodewits, G. Cremonese, V. Da Deppo, S. Debei, M. De Cecco, J. Deller, S. Ferrari, M. Fulle, P.J. Gutierrez, C. Güttler, W.H. Ip, L. Jorda, H.U. Keller, M.L. Lara, M. Lazzarin, J.J. Lopez Moreno, A. Lucchetti, F. Marzari, S. Mottola, M. Pajola, I. Toth, C. Tubiana, Surface evolution of the Anhur region on comet 67P/Churyumov-Gerasimenko from high-resolution OSIRIS images. Astron. Astrophys. **630**, A13 (2019). https://doi.org/10.1051/0004-6361/201834824. 1903.09017

N. Fougere, M.R. Combi, V. Tenishev, M. Rubin, B.P. Bonev, M.J. Mumma, Understanding measured water rotational temperatures and column densities in the very innermost coma of Comet 73P/Schwassmann-Wachmann 3 B. Icarus **221**(1), 174–185 (2012). https://doi.org/10.1016/j.icarus.2012.07.019

N. Fougere, M.R. Combi, M. Rubin, V. Tenishev, Modeling the heterogeneous ice and gas coma of Comet 103P/Hartley 2. Icarus **225**, 688–702 (2013). https://doi.org/10.1016/j.icarus.2013.04.031

N. Fougere, K. Altwegg, J.J. Berthelier, A. Bieler, D. Bockelée-Morvan, U. Calmonte, F. Capaccioni, M.R. Combi, J. De Keyser, V. Debout, Direct simulation Monte Carlo modelling of the major species in the coma of comet 67P/Churyumov-Gerasimenko. Mon. Not. R. Astron. Soc. **462**, S156–S169 (2016a). https://doi.org/10.1093/mnras/stw2388

N. Fougere, K. Altwegg, J.J. Berthelier, A. Bieler, D. Bockelée-Morvan, U. Calmonte, F. Capaccioni, M.R. Combi, J. De Keyser, V. Debout, S. Erard, B. Fiethe, G. Filacchione, U. Fink, S.A. Fuselier, T.I. Gombosi, K.C. Hansen, M. Hässig, Z. Huang, L. Le Roy, C. Leyrat, A. Migliorini, G. Piccioni, G. Rinaldi, M. Rubin, Y. Shou, V. Tenishev, G. Toth, C.Y. Tzou, Three-dimensional direct simulation Monte-Carlo modeling of the coma of comet 67P/Churyumov-Gerasimenko observed by the VIRTIS and ROSINA instruments on board Rosetta. Astron. Astrophys. **588**, A134 (2016b). https://doi.org/10.1051/0004-6361/201527889

N. Fougere, K. Altwegg, J.J. Berthelier, A. Bieler, D. Bockelée-Morvan, U. Calmonte, F. Capaccioni, M.R. Combi, J. De Keyser, V. Debout, S. Erard, B. Fiethe, G. Filacchione, U. Fink, S.A. Fuselier, T.I. Gombosi, K.C. Hansen, M. Hässig, Z. Huang, L. Le Roy, C. Leyrat, A. Migliorini, G. Piccioni, G. Rinaldi, M. Rubin, Y. Shou, V. Tenishev, G. Toth, C.Y. Tzou, Three-dimensional direct simulation Monte-Carlo modeling of the coma of comet 67P/Churyumov-Gerasimenko observed by the VIRTIS and ROSINA instruments on board Rosetta. Astron. Astrophys. **588**, A134 (2016c). https://doi.org/10.1051/0004-6361/201527889

M. Fulle, J.F. Crifo, A.V. Rodionov, Numerical simulation of the dust flux on a spacecraft in orbit around an aspherical cometary nucleus – I. Astron. Astrophys. **347**, 1009–1028 (1999)

M. Fulle, V. Della Corte, A. Rotundi, P. Weissman, A. Juhasz, K. Szego, R. Sordini, M. Ferrari, S. Ivanovski, F. Lucarelli, M. Accolla, S. Merouane, V. Zakharov, E. Mazzotta Epifani, J.J. López-Moreno, J. Rodríguez, L. Colangeli, P. Palumbo, E. Grün, M. Hilchenbach, E. Bussoletti, F. Esposito, S.F. Green, P.L. Lamy, J.A.M. McDonnell, V. Mennella, A. Molina, R. Morales, F. Moreno, J.L. Ortiz, E. Palomba, R. Rodrigo, J.C. Zarnecki, M. Cosi, F. Giovane, B. Gustafson, M.L. Herranz, J.M. Jerónimo, M.R. Leese, A.C. López-Jiménez, N. Altobelli, Density and charge of pristine fluffy particles from Comet 67P/Churyumov-Gerasimenko. Astrophys. J. Lett. **802**, L12 (2015). https://doi.org/10.1088/2041-8205/802/1/L12

M. Fulle, F. Marzari, V. Della Corte, S. Fornasier, H. Sierks, A. Rotundi, C. Barbieri, P.L. Lamy, R. Rodrigo, D. Koschny, H. Rickman, H.U. Keller, J.J. López-Moreno, M. Accolla, J. Agarwal, M.F. A'Hearn, N. Altobelli, M.A. Barucci, J.L. Bertaux, I. Bertini, D. Bodewits, E. Bussoletti, L. Colangeli, M. Cosi, G. Cremonese, J.F. Crifo, V. Da Deppo, B. Davidsson, S. Debei, M. De Cecco, F. Esposito, M. Ferrari, F. Giovane, B. Gustafson, S.F. Green, O. Groussin, E. Grün, P. Gutierrez, C. Güttler, M.L. Herranz, S.F. Hviid, W. Ip, S.L. Ivanovski, J.M. Jerónimo, L. Jorda, J. Knollenberg, R. Kramm, E. Kührt, M. Küppers, L. Lara, M. Lazzarin, M.R. Leese, A.C. López-Jiménez, F. Lucarelli, E. Mazzotta Epifani, J.A.M. McDonnell, V. Mennella, A. Molina, R. Morales, F. Moreno, S. Mottola, G. Naletto, N. Oklay, J.L. Ortiz, E. Palomba, P. Palumbo, J.M. Perrin, F.J.M. Rietmeijer, J. Rodríguez, R. Sordini, N. Thomas, C. Tubiana, J.B. Vincent, P. Weissman, K.P. Wenzel, V. Zakharov, J.C. Zarnecki, Evolution of the dust size distribution of comet 67P/Churyumov-Gerasimenko from 2.2 au to perihelion. Astrophys. J. **821**, 19 (2016). https://doi.org/10.3847/0004-637X/821/1/19

M. Fulle, J. Blum, S.F. Green, B. Gundlach, A. Herique, F. Moreno, S. Mottola, A. Rotundi, C. Snodgrass, The refractory-to-ice mass ratio in comets. Mon. Not. R. Astron. Soc. **482**(3), 3326–3340 (2019). https://doi.org/10.1093/mnras/sty2926

S.B. Gerig, R. Marschall, N. Thomas, I. Bertini, D. Bodewits, B. Davidsson, M. Fulle, W.H. Ip, H.U. Keller, M. Küppers, F. Preusker, F. Scholten, C.C. Su, I. Toth, C. Tubiana, J.S. Wu, H. Sierks, C. Barbieri, P.L. Lamy, R. Rodrigo, D. Koschny, H. Rickman, J. Agarwal, M.A. Barucci, J.L. Bertaux, G. Cremonese, V. Da Deppo, S. Debei, M. De Cecco, J. Deller, S. Fornasier, O. Groussin, P.J. Gutierrez, C. Güttler, S.F. Hviid, L. Jorda, J. Knollenberg, J.R. Kramm, E. Kührt, L.M. Lara, M. Lazzarin, J.J. Lopez Moreno, F. Marzari, S. Mottola, G. Naletto, N. Oklay, J.B. Vincent, On deviations from free-radial outflow in the inner coma of comet 67P /Churyumov-Gerasimenko. Icarus **311**, 1–22 (2018). https://doi.org/10.1016/j.icarus.2018.03.010

A. Gicquel, J.B. Vincent, J. Agarwal, M.F. A'Hearn, I. Bertini, D. Bodewits, H. Sierks, Z.Y. Lin, C. Barbieri, P.L. Lamy, Sublimation of icy aggregates in the coma of comet 67P/Churyumov-Gerasimenko detected with the OSIRIS cameras on board Rosetta. Mon. Not. R. Astron. Soc. **462**, S57–S66 (2016). https://doi.org/10.1093/mnras/stw2117. 1608.08774

T.I. Gombosi, Dusty cometary atmospheres. Adv. Space Res. **7**, 137–145 (1987). https://doi.org/10.1016/0273-1177(87)90211-0

T.I. Gombosi, T.E. Cravens, A.F. Nagy, Time-dependent dusty gasdynamical flow near cometary nuclei. Astrophys. J. **293**, 328–341 (1985). https://doi.org/10.1086/163240

T.I. Gombosi, A.F. Nagy, T.E. Cravens, Dust and neutral gas modeling of the inner atmospheres of comets. Rev. Geophys. **24**(3), 667–700 (1986). https://doi.org/10.1029/RG024i003p00667

O. Groussin, J.M. Sunshine, L.M. Feaga, L. Jorda, P.C. Thomas, J.Y. Li, M.F. A'Hearn, M.J.S. Belton, S. Besse, B. Carcich, T.L. Farnham, D. Hampton, K. Klaasen, C. Lisse, F. Merlin, S. Protopapa, The temperature, thermal inertia, roughness and color of the nuclei of Comets 103P/Hartley 2 and 9P/Tempel 1. Icarus **222**, 580–594 (2013). https://doi.org/10.1016/j.icarus.2012.10.003

O. Groussin, N. Attree, Y. Brouet, V. Ciarletti, B. Davidsson, G. Filacchione, H.H. Fischer, B. Gundlach, M. Knapmeyer, J. Knollenberg, R. Kokotanekova, E. Kührt, C. Leyrat, D. Marshall, I. Pelivan, Y. Skorov, C. Snodgrass, T. Spohn, F. Tosi, The thermal, mechanical, structural, and dielectric properties of cometary nuclei after rosetta. Space Sci. Rev. **215**(4), 29 (2019). https://doi.org/10.1007/s11214-019-0594-x

S. Gulkis, M. Allen, P. von Allmen, G. Beaudin, N. Biver, D. Bockelée-Morvan, M. Choukroun, J. Crovisier, B.J.R. Davidsson, P. Encrenaz, T. Encrenaz, M. Frerking, P. Hartogh, M. Hofstadter, W.H. Ip, M. Janssen, C. Jarchow, S. Keihm, S. Lee, E. Lellouch, C. Leyrat, L. Rezac, F.P. Schloerb, T. Spilker, Subsurface properties and early activity of comet 67P/Churyumov-Gerasimenko. Science **347**, aaa0709 (2015). https://doi.org/10.1126/science.aaa0709

C. Güttler, P.H. Hasselmann, Y. Li, M. Fulle, C. Tubiana, G. Kovacs, J. Agarwal, H. Sierks, S. Fornasier, M. Hofmann, P. Gutiérrez Marqués, T. Ott, E. Drolshagen, I. Bertini, C. Barbieri, P.L. Lamy, R. Rodrigo, D. Koschny, H. Rickman, M.F. A'Hearn, M.A. Barucci, D. Bodewits, J.L. Bertaux, S. Boudreault, G. Cremonese, V. Da Deppo, B. Davidsson, S. Debei, M. De Cecco, J. Deller, B. Geiger, O. Groussin, P.J. Gutiérrez, S.F. Hviid, W.H. Ip, L. Jorda, H.U. Keller, J. Knollenberg, J.R. Kramm, E. Kührt, M. Küppers, L.M. Lara, M. Lazzarin, J.J. López-Moreno, F. Marzari, S. Mottola, G. Naletto, N. Oklay, M. Pajola, X. Shi, N. Thomas, J.B. Vincent, Characterization of dust aggregates in the vicinity of the Rosetta spacecraft. Mon. Not. R. Astron. Soc. **469**, S312–S320 (2017). https://doi.org/10.1093/mnras/stx1692

K.C. Hansen, K. Altwegg, J.J. Berthelier, A. Bieler, N. Biver, D. Bockelée-Morvan, U. Calmonte, F. Capaccioni, M.R. Combi, J. de Keyser, B. Fiethe, N. Fougere, S.A. Fuselier, S. Gasc, T.I. Gombosi, Z. Huang, L. Le Roy, S. Lee, H. Nilsson, M. Rubin, Y. Shou, C. Snodgrass, V. Tenishev, G. Toth, C.Y. Tzou, C.S. Wedlund (Rosina Team), Evolution of water production of 67P/Churyumov-Gerasimenko: an empirical model and a multi-instrument study. Mon. Not. R. Astron. Soc. **462**, S491–S506 (2016). https://doi.org/10.1093/mnras/stw2413

L. Haser, Distribution d'intensité dans la tête d'une comète. Bull. Soc. R. Sci. Liège **43**, 740–750 (1957)

M. Hässig, K. Altwegg, H. Balsiger, A. Bar-Nun, J.J. Berthelier, A. Bieler, P. Bochsler, C. Briois, U. Calmonte, M. Combi, J. De Keyser, P. Eberhardt, B. Fiethe, S.A. Fuselier, M. Galand, S. Gasc, T.I. Gombosi, K.C. Hansen, A. Jäckel, H.U. Keller, E. Kopp, A. Korth, E. Kührt, L. Le Roy, U. Mall, B. Marty, O. Mousis, E. Neefs, T. Owen, H. Rème, M. Rubin, T. Sémon, C. Tornow, C.Y. Tzou, J.H. Waite, P. Wurz, Time variability and heterogeneity in the coma of 67P/Churyumov-Gerasimenko. Science **347**, aaa0276 (2015). https://doi.org/10.1126/science.aaa0276

M. Horanyi, D.A. Mendis, Trajectories of charged dust grains in the cometary environment. Astrophys. J. **294**, 357–368 (1985). https://doi.org/10.1086/163303

W.F. Huebner, D.C. Boice, H.J. Reitsema, W.A. Delamere, F.L. Whipple, A model for intensity profiles of dust jets near the nucleus of comet Halley. Icarus **76**(1), 78–88 (1988). https://doi.org/10.1016/0019-1035(88)90140-6

A. Itkin, J. Crifo, A. Rodionov, Study of a structure of the multiphase jets outflowing from the surface of a cometary nucleus. Space Res. **33**, 3 (1995)

S.L. Ivanovski, V. Della Corte, A. Rotundi, M. Fulle, N. Fougere, A. Bieler, M. Rubin, S. Ivanovska, V. Liuzzi, Dynamics of non-spherical dust in the coma of 67P/Churyumov-Gerasimenko constrained by GIADA and ROSINA data. Mon. Not. R. Astron. Soc. **469**, S774–S786 (2017a). https://doi.org/10.1093/mnras/stx3008

S.L. Ivanovski, V.V. Zakharov, V. Della Corte, J.F. Crifo, A. Rotundi, M. Fulle, Dynamics of aspherical dust grains in a cometary atmosphere: I. Axially symmetric grains in a spherically symmetric atmosphere. Icarus **282**, 333–350 (2017b). https://doi.org/10.1016/j.icarus.2016.09.024

S. Ivanovski, V. Zakharov, V. Della Corte, A. Rotundi, M. Fulle, Dust dynamical traceback problem for derivation the surface properties of 67P/Churyumov-Gerasimenko based on the GIADA measurements, in *European Planetary Science Congress, EPSC2018* (2018), p. 1052

M. Kaviany, *Principles of Heat Transfer in Porous Media* (Springer, Berlin, 1995)

H.U. Keller, E. Kührt, Cometary nuclei – from Giotto to Rosetta. Space Sci. Rev. **216**(1), 14 (2020). https://doi.org/10.1007/s11214-020-0634-6

H.U. Keller, S. Mottola, B. Davidsson, S.E. Schröder, Y. Skorov, E. Kührt, O. Groussin, M. Pajola, S.F. Hviid, F. Preusker, F. Scholten, M.F. A'Hearn, H. Sierks, C. Barbieri, P. Lamy, R. Rodrigo, D. Koschny, H. Rickman, M.A. Barucci, J.L. Bertaux, I. Bertini, G. Cremonese, V. Da Deppo, S. Debei, M. De Cecco, S. Fornasier, M. Fulle, P.J. Gutiérrez, W.H. Ip, L. Jorda, J. Knollenberg, J.R. Kramm, M. Küppers, L.M. Lara, M. Lazzarin, J.J. Lopez Moreno, F. Marzari, H. Michalik, G. Naletto, L. Sabau, N. Thomas, J.B. Vincent, K.P. Wenzel, J. Agarwal, C. Güttler, N. Oklay, C. Tubiana, Insolation, erosion, and morphology of comet 67P/Churyumov-Gerasimenko. Astron. Astrophys. **583**, A34 (2015). https://doi.org/10.1051/0004-6361/201525964

H.U. Keller, S. Mottola, S.F. Hviid, J. Agarwal, E. Kührt, Y. Skorov, K. Otto, J.B. Vincent, N. Oklay, S.E. Schröder, B. Davidsson, M. Pajola, X. Shi, D. Bodewits, I. Toth, F. Preusker, F. Scholten, H. Sierks, C. Barbieri, P. Lamy, R. Rodrigo, D. Koschny, H. Rickman, M.F. A'Hearn, M.A. Barucci, J.L. Bertaux,

I. Bertini, G. Cremonese, V. Da Deppo, S. Debei, M. De Cecco, J. Deller, S. Fornasier, M. Fulle, O. Groussin, P.J. Gutiérrez, C. Güttler, M. Hofmann, W.H. Ip, L. Jorda, J. Knollenberg, J.R. Kramm, M. Küppers, L.M. Lara, M. Lazzarin, J.J. Lopez-Moreno, F. Marzari, G. Naletto, C. Tubiana, N. Thomas, Seasonal mass transfer on the nucleus of comet 67P/Chuyumov-Gerasimenko. Mon. Not. R. Astron. Soc. **469**, S357–S371 (2017). https://doi.org/10.1093/mnras/stx1726

M.S. Kelley, D.J. Lindler, D. Bodewits, M.F. A'Hearn, C.M. Lisse, L. Kolokolova, J. Kissel, B. Hermalyn, A distribution of large particles in the coma of Comet 103P/Hartley 2. Icarus **222**, 634–652 (2013). https://doi.org/10.1016/j.icarus.2012.09.037. 1304.4204

Y. Kitamura, Axisymmetric dusty gas jet in the inner coma of a comet. Icarus **66**, 241–257 (1986). https://doi.org/10.1016/0019-1035(86)90155-7

Y. Kitamura, Axisymmetric dusty gas jet in the inner coma of a comet. II – the case of isolated jets. Icarus **72**, 555–567 (1987). https://doi.org/10.1016/0019-1035(87)90052-2

Y. Kitamura, A numerical study of the interaction between two cometary jets: a possibility of shock formation in cometary atmospheres. Icarus **86**(2), 455–475 (1990). https://doi.org/10.1016/0019-1035(90)90229-3

J. Knollenberg, Z.Y. Lin, S.F. Hviid, N. Oklay, J.B. Vincent, D. Bodewits, S. Mottola, M. Pajola, H. Sierks, C. Barbieri, P. Lamy, R. Rodrigo, D. Koschny, H. Rickman, M.F. A'Hearn, M.A. Barucci, J.L. Bertaux, I. Bertini, G. Cremonese, B. Davidsson, V. Da Deppo, S. Debei, M. De Cecco, S. Fornasier, M. Fulle, O. Groussin, P.J. Gutiérrez, W.H. Ip, L. Jorda, H.U. Keller, E. Kührt, J.R. Kramm, M. Küppers, L.M. Lara, M. Lazzarin, J.J. Lopez Moreno, F. Marzari, G. Naletto, N. Thomas, C. Güttler, F. Preusker, F. Scholten, C. Tubiana, A mini outburst from the nightside of comet 67P/Churyumov-Gerasimenko observed by the OSIRIS camera on Rosetta. Astron. Astrophys. **596**, A89 (2016). https://doi.org/10.1051/0004-6361/201527744

M. Knudsen, Die Gesetze der Molekularströmung und der inneren Reibungsströmung der Gase durch Röhren. Ann. Phys. **333**(1), 75–130 (1909). https://doi.org/10.1002/andp.19093330106

W. Kofman, A. Herique, Y. Barbin, J.P. Barriot, V. Ciarletti, S. Clifford, P. Edenhofer, C. Elachi, C. Eyraud, J.P. Goutail, E. Heggy, L. Jorda, J. Lasue, A.C. Levasseur-Regourd, E. Nielsen, P. Pasquero, F. Preusker, P. Puget, D. Plettemeier, Y. Rogez, H. Sierks, C. Statz, H. Svedhem, I. Williams, S. Zine, J. Van Zyl, Properties of the 67P/Churyumov-Gerasimenko interior revealed by CONSERT radar. Science **349**(6247), 2.639 (2015). https://doi.org/10.1126/science.aab0639

L. Kolokolova, H. Kimura, Comet dust as a mixture of aggregates and solid particles: model consistent with ground-based and space-mission results. Earth Planets Space **62**, 17–21 (2010). https://doi.org/10.5047/eps.2008.12.001. 0910.4755

I. Konno, W.F. Huebner, D.C. Boice, A model of dust fragmentation in near-nucleus jet-like features in comet P/Halley. Icarus **101**(1), 84–94 (1993). https://doi.org/10.1006/icar.1993.1008

A. Korosmezey, T.I. Gombosi, A time-dependent dusty gas dynamic model of axisymmetric cometary jets. Icarus **84**(1), 118–153 (1990). https://doi.org/10.1016/0019-1035(90)90162-3

T. Kramer, M. Läuter, M. Rubin, K. Altwegg, Seasonal changes of the volatile density in the coma and on the surface of comet 67P/Churyumov-Gerasimenko. Mon. Not. R. Astron. Soc. **469**, S20–S28 (2017). https://doi.org/10.1093/mnras/stx866

M. Krause, J. Blum, Y.V. Skorov, M. Trieloff, Thermal conductivity measurements of porous dust aggregates: I. Technique, model and first results. Icarus **214**(1), 286–296 (2011). https://doi.org/10.1016/j.icarus.2011.04.024. 1101.2586

E. Kührt, H.U. Keller, The formation of cometary surface crusts. Icarus **109**, 121–132 (1994). https://doi.org/10.1006/icar.1994.1080

I.L. Lai, W.H. Ip, J.C. Lee, Z.Y. Lin, J.B. Vincent, P. Hartogh, H. Sierks, C. Barbieri, P.L. Lamy, H. Rickman, H.U. Keller, Mapping of the source regions of the dust jets on comet 67P/Churyumov-Gerasimenko, in *AGU Fall Meeting Abstracts P43A-2087* (2016)

I.L. Lai, W.H. Ip, J.C. Lee, Z.Y. Lin, J.B. Vincent, N. Oklay, Seasonal variations of the source regions of the dust jets of comet 67P/Churyumov-Gerasimenko, in *European Planetary Science Congress, EPSC2017* (2017), p. 784

Y. Langevin, M. Hilchenbach, N. Ligier, S. Merouane, K. Hornung, C. Engrand, R. Schulz, J. Kissel, J. Rynö, P. Eng, Typology of dust particles collected by the COSIMA mass spectrometer in the inner coma of 67P/Churyumov Gerasimenko. Icarus **271**, 76–97 (2016). https://doi.org/10.1016/j.icarus.2016.01.027

L.M. Lara, S. Lowry, J.B. Vincent, P.J. Gutiérrez, A. Rożek, F. La Forgia, N. Oklay, H. Sierks, C. Barbieri, P.L. Lamy, R. Rodrigo, D. Koschny, H. Rickman, H.U. Keller, J. Agarwal, A.T. Auger, M.F. A'Hearn, M.A. Barucci, J.L. Bertaux, I. Bertini, S. Besse, D. Bodewits, G. Cremonese, B. Davidsson, V. Da Deppo, S. Debei, M. De Cecco, M.R. El-Maarry, F. Ferri, S. Fornasier, M. Fulle, O. Groussin, P. Gutiérrez-Marques, C. Güttler, S.F. Hviid, W.H. Ip, L. Jorda, J. Knollenberg, G. Kovacs, J.R. Kramm, E. Kührt, M. Küppers, M. Lazzarin, Z.Y. Lin, J.J. López-Moreno, S. Magrin, F. Marzari, H. Michalik, R. Moissl-Fraund, F. Moreno, S. Mottola, G. Naletto, M. Pajola, A. Pommerol, N. Thomas, M.D. Sabau, C. Tubiana, Large-scale dust jets in the coma of 67P/Churyumov-Gerasimenko as seen by the OSIRIS instrument onboard Rosetta. Astron. Astrophys. **583**, A9 (2015). https://doi.org/10.1051/0004-6361/201526103

M. Läuter, T. Kramer, M. Rubin, K. Altwegg, Surface localization of gas sources on comet 67P/Churyumov-Gerasimenko based on DFMS/COPS data. Mon. Not. R. Astron. Soc. **483**(1), 852–861 (2019). https://doi.org/10.1093/mnras/sty3103. 1804.06696

S. Lee, P. von Allmen, L. Kamp, S. Gulkis, B. Davidsson, Non-LTE radiative transfer for sub-millimeter water lines in Comet 67P/Churyumov-Gerasimenko. Icarus **215**, 721–731 (2011)

S. Lee, P. von Allmen, M. Allen, G. Beaudin, N. Biver, D. Bockelée-Morvan, M. Choukroun, J. Crovisier, P. Encrenaz, M. Frerking, S. Gulkis, P. Hartogh, M. Hofstadter, W.H. Ip, M. Janssen, C. Jarchow, S. Keihm, E. Lellouch, C. Leyrat, L. Rezac, F.P. Schloerb, T. Spilker, B. Gaskell, L. Jorda, H.U. Keller, H. Sierks, Spatial and diurnal variation of water outgassing on comet 67P/Churyumov-Gerasimenko observed from Rosetta/MIRO in August 2014. Astron. Astrophys. **583**, A5 (2015). https://doi.org/10.1051/0004-6361/201526155

A.C. Levasseur-Regourd, J. Agarwal, H. Cottin, C. Engrand, G. Flynn, M. Fulle, T. Gombosi, Y. Langevin, J. Lasue, T. Mannel, S. Merouane, O. Poch, N. Thomas, A. Westphal, Cometary dust. Space Sci. Rev. **214**, 64 (2018). https://doi.org/10.1007/s11214-018-0496-3

Y. Liao, R. Marschall, C.C. Su, J.S. Wu, I.L. Lai, O. Pinzon, N. Thomas, Water vapor deposition from the inner gas coma onto the nucleus of comet 67P/Churyumov-Gerasimenko. Planet. Space Sci. **157**, 1–9 (2018). https://doi.org/10.1016/j.pss.2018.04.014

Z.Y. Lin, L.M. Lara, J.B. Vincent, W.H. Ip, Physical studies of 81P/Wild 2 from the last two apparitions. Astron. Astrophys. **537** (2012)

Z.-Y. Lin et al., Morphology and dynamics of jets of comet 67P/Churyumov-Gerasimenko: early phase development. Astron. Astrophys. (2015). https://doi.org/10.1051/0004-6361/201525961

Z.Y. Lin, J. Knollenberg, J.B. Vincent, M.F. A'Hearn, W.H. Ip, H. Sierks, C. Barbieri, P.L. Lamy, R. Rodrigo, D. Koschny, H. Rickman, H.U. Keller, S. Mottola, D. Bodewits, J. Agarwal, M.A. Barucci, J.L. Bertaux, I. Bertini, G. Cremonese, J. Deller, V. Da Deppo, B. Davidsson, S. Debei, M. De Cecco, S. Fornasier, M. Fulle, O. Groussin, P.J. Gutiérrez, C. Güttler, M. Hofmann, S.F. Hviid, L. Jorda, G. Kovacs, J.R. Kramm, E. Kührt, M. Küppers, I.L. Lai, L.M. Lara, M. Lazzarin, J.C. Lee, J.J. López-Moreno, F. Marzari, G. Naletto, N. Oklay, T. Ott, E. Drolshagen, X. Shi, N. Thomas, C. Tubiana, Investigating the physical properties of outbursts on comet 67P/Churyumov-Gerasimenko. Mon. Not. R. Astron. Soc. **469**, S731–S740 (2017). https://doi.org/10.1093/mnras/stx2768

A. Longobardo, V. Della Corte, S. Ivanovski, G. Rinaldi, V. Zakharov, A. Rotundi, F. Capaccioni, M. Fulle, G. Filacchione, E. Palomba, 67P/Churyumov-Gerasimenko active areas before perihelion identified by GIADA and VIRTIS data fusion. Mon. Not. R. Astron. Soc. **483**(2), 2165–2176 (2019). https://doi.org/10.1093/mnras/sty3244

G.A. Lukyanov, V.V. Zakharov, A.V. Rodionov, J.F. Crifo, Comparison between Navier-Stokes and DSMC simulations of the rarefied gas flow from model cometary nuclei, in *Rarefied Gas Dynamics: 24th International Symposium on Rarefied Gas Dynamics*, ed. by M. Capitelli. American Institute of Physics Conference Series, vol. 762 (2005), pp. 331–336. https://doi.org/10.1063/1.1941558

T. Mannel, M.S. Bentley, R. Schmied, H. Jeszenszky, A.C. Levasseur-Regourd, J. Romstedt, K. Torkar, Fractal cometary dust – a window into the early Solar system. Mon. Not. R. Astron. Soc. **462**, S304–S311 (2016). https://doi.org/10.1093/mnras/stw2898

R. Marschall, C.C. Su, Y. Liao, N. Thomas, K. Altwegg, H. Sierks, W.H. Ip, H.U. Keller, J. Knollenberg, E. Kührt, I.L. Lai, M. Rubin, Y. Skorov, J.S. Wu, L. Jorda, F. Preusker, F. Scholten, A. Gracia-Berná, A. Gicquel, G. Naletto, X. Shi, J.B. Vincent, Modelling observations of the inner gas and dust coma of comet 67P/Churyumov-Gerasimenko using ROSINA/COPS and OSIRIS data: first results. Astron. Astrophys. **589**, A90 (2016). https://doi.org/10.1051/0004-6361/201628085

R. Marschall, S. Mottola, C.C. Su, Y. Liao, M. Rubin, J.S. Wu, N. Thomas, K. Altwegg, H. Sierks, W.H. Ip, H.U. Keller, J. Knollenberg, E. Kührt, I.L. Lai, Y. Skorov, L. Jorda, F. Preusker, F. Scholten, J.B. Vincent (Osiris Team Rosina Team), Cliffs versus plains: can ROSINA/COPS and OSIRIS data of comet 67P/Churyumov-Gerasimenko in autumn 2014 constrain inhomogeneous outgassing? Astron. Astrophys. **605**, A112 (2017). https://doi.org/10.1051/0004-6361/201730849

R. Marschall, L. Rezac, D. Kappel, C.C. Su, S.B. Gerig, M. Rubin, O. Pinzón-Rodríguez, D. Marshall, Y. Liao, C. Herny, G. Arnold, C. Christou, S.K. Dadzie, O. Groussin, P. Hartogh, L. Jorda, E. Kührt, S. Mottola, O. Mousis, F. Preusker, F. Scholten, P. Theologou, J.S. Wu, K. Altwegg, R. Rodrigo, N. Thomas, A comparison of multiple Rosetta data sets and 3D model calculations of 67P/Churyumov-Gerasimenko coma around equinox (May 2015). Icarus **328**(March), 104–126 (2019). https://doi.org/10.1016/j.icarus.2019.02.008

R. Marschall, Y. Liao, N. Thomas, J.S. Wu, Limitations in the determination of surface emission distributions on comets through modelling of observational data – a case study based on Rosetta observations. Icarus **346**, 113742 (2020a). https://doi.org/10.1016/j.icarus.2020.113742. 2004.13199

R. Marschall, J. Markkanen, S.B. Gerig, O. Pinzón-Rodríguez, N. Thomas, J.S. Wu, The dust-to-gas ratio, size distribution, and dust fall-back fraction of comet 67p/Churyumov-Gerasimenko: inferences from linking the optical and dynamical properties of the inner comae. Front. Phys. **8**, 227 (2020b). https://doi.org/10.3389/fphy.2020.00227

D.W. Marshall, P. Hartogh, L. Rezac, P. von Allmen, N. Biver, D. Bockelée-Morvan, J. Crovisier, P. Encrenaz, S. Gulkis, M. Hofstadter, W.H. Ip, C. Jarchow, S. Lee, E. Lellouch, Spatially resolved evolution of the local $H_2O$ production rates of comet 67P/Churyumov-Gerasimenko from the MIRO instrument on Rosetta. Astron. Astrophys. **603**, A87 (2017). https://doi.org/10.1051/0004-6361/201730502

D. Marshall, O. Groussin, J.B. Vincent, Y. Brouet, D. Kappel, G. Arnold, M. Capria, G. Filacchione, P. Hartogh, M. Hofstadter, W.H. Ip, L. Jorda, E. Kuehrt, E. Lellouch, S. Mottola, L. Rezac, R. Rodrigo, S. Rodionov, P. Schloerb, N. Thomas, Thermal inertia and roughness of the nucleus of comet 67P/Churyumov-Gerasimenko from MIRO and VIRTIS observations. Astron. Astrophys. (2018)

S. Merouane, O. Stenzel, M. Hilchenbach, R. Schulz, N. Altobelli, H. Fischer, K. Hornung, J. Kissel, Y. Langevin, E. Mellado, J. Rynö, B. Zaprudin, Evolution of the physical properties of dust and cometary dust activity from 67P/Churyumov–Gerasimenko measured in situ by Rosetta/COSIMA. Mon. Not. R. Astron. Soc. **469**(2), S459–S474 (2017). https://doi.org/10.1093/mnras/stx2018

A. Migliorini, G. Piccioni, F. Capaccioni, G. Filacchione, D. Bockelée-Morvan, S. Erard, C. Leyrat, M.R. Combi, N. Fougere, J. Crovisier, F.W. Taylor, M.C. De Sanctis, M.T. Capria, D. Grassi, G. Rinaldi, G.P. Tozzi, U. Fink, Water and carbon dioxide distribution in the 67P/Churyumov-Gerasimenko coma from VIRTIS-M infrared observations. Astron. Astrophys. **589**, A45 (2016). https://doi.org/10.1051/0004-6361/201527661

S. Mottola, S. Lowry, C. Snodgrass, P.L. Lamy, I. Toth, A. Rożek, H. Sierks, M.F. A'Hearn, F. Angrilli, C. Barbieri, M.A. Barucci, J.L. Bertaux, G. Cremonese, V. Da Deppo, B. Davidsson, M. De Cecco, S. Debei, S. Fornasier, M. Fulle, O. Groussin, P. Gutiérrez, S.F. Hviid, W. Ip, L. Jorda, H.U. Keller, J. Knollenberg, D. Koschny, R. Kramm, E. Kührt, M. Küppers, L. Lara, M. Lazzarin, J.J. Lopez Moreno, F. Marzari, H. Michalik, G. Naletto, H. Rickman, R. Rodrigo, L. Sabau, N. Thomas, K.P. Wenzel, J. Agarwal, I. Bertini, F. Ferri, C. Güttler, S. Magrin, N. Oklay, C. Tubiana, J.B. Vincent, The rotation state of 67P/Churyumov-Gerasimenko from approach observations with the OSIRIS cameras on Rosetta. Astron. Astrophys. **569**, L2 (2014). https://doi.org/10.1051/0004-6361/201424590

K. Nanbu, Angular distributions of molecular flux from orifices of various thickness. Vacuum **35**(12), 573–576 (1985). https://doi.org/10.1016/0042-207X(85)90317-3

H. Okabe, M.J. Blunt, Prediction of permeability for porous media reconstructed using multiple-point statistics. Phys. Rev. E **70**, 066,135 (2004). https://doi.org/10.1103/PhysRevE.70.066135

M. Pajola, A. Lucchetti, M. Fulle, S. Mottola, M. Hamm, V. Da Deppo, L. Penasa, G. Kovacs, M. Massironi, X. Shi, C. Tubiana, C. Güttler, N. Oklay, J.B. Vincent, I. Toth, B. Davidsson, G. Naletto, H. Sierks, C. Barbieri, P.L. Lamy, R. Rodrigo, D. Koschny, H. Rickman, H.U. Keller, J. Agarwal, M.F. A'Hearn, M.A. Barucci, J.L. Bertaux, I. Bertini, G. Cremonese, S. Debei, M. De Cecco, J. Deller, M.R. El Maarry, S. Fornasier, E. Frattin, A. Gicquel, O. Groussin, P.J. Gutierrez, S. Höfner, M. Hofmann, S.F. Hviid, W.H. Ip, L. Jorda, J. Knollenberg, J.R. Kramm, E. Kührt, M. Küppers, L.M. Lara, M. Lazzarin, J.J.L. Moreno, F. Marzari, H. Michalik, F. Preusker, F. Scholten, N. Thomas, The pebbles/boulders size distributions on Sais: Rosetta's final landing site on comet 67P/Churyumov-Gerasimenko. Mon. Not. R. Astron. Soc. **469**, S636–S645 (2017). https://doi.org/10.1093/mnras/stx1620

A. Pommerol, N. Thomas, M.R. El-Maarry, M. Pajola, O. Groussin, A.T. Auger, N. Oklay, S. Fornasier, C. Feller, B. Davidsson, A. Gracia-Berná, B. Jost, R. Marschall, O. Poch, M.A. Barucci, J.L. Bertaux, F. La Forgia, H.U. Keller, E. Kührt, S.C. Lowry, S. Mottola, G. Naletto, H. Sierks, C. Barbieri, P.L. Lamy, R. Rodrigo, D. Koschny, H. Rickman, J. Agarwal, M.F. A'Hearn, I. Bertini, S. Boudreault, G. Cremonese, V. Da Deppo, M. De Cecco, S. Debei, C. Güttler, M. Fulle, P.J. Gutierrez, S.F. Hviid, W.H. Ip, L. Jorda, J. Knollenberg, G. Kovacs, J.R. Kramm, E. Küppers, L. Lara, M. Lazzarin, J.L. Lopez Moreno, F. Marzari, H. Michalik, F. Preusker, F. Scholten, C. Tubiana, J.B. Vincent, OSIRIS observations of meter-sized exposures of $H_2O$ ice at the surface of 67P/Churyumov-Gerasimenko and interpretation using laboratory experiments. Astron. Astrophys. **583**, A25 (2015). https://doi.org/10.1051/0004-6361/201525977

F. Preusker, F. Scholten, K.D. Matz, T. Roatsch, S.F. Hviid, S. Mottola, J. Knollenberg, E. Kührt, M. Pajola, N. Oklay, J.B. Vincent, B. Davidsson, M.F. A'Hearn, J. Agarwal, C. Barbieri, M.A. Barucci, J.L. Bertaux, I. Bertini, G. Cremonese, V. Da Deppo, S. Debei, M. De Cecco, S. Fornasier, M. Fulle, O. Groussin, P.J. Gutiérrez, C. Güttler, W.H. Ip, L. Jorda, H.U. Keller, D. Koschny, J.R. Kramm, M. Küppers, P. Lamy, L.M. Lara, M. Lazzarin, J.J. Lopez Moreno, F. Marzari, M. Massironi, G. Naletto, H. Rickman, R. Rodrigo, H. Sierks, N. Thomas, C. Tubiana, The global meter-level shape model of comet 67P/Churyumov-Gerasimenko. Astron. Astrophys. **607**, L1 (2017). https://doi.org/10.1051/0004-6361/201731798

F. Rabier, Recent experimentation on 4D-Var and first results from a Simplified Kalman Filter, in *ECMWF* (1997)

F. Rabier, H. Järvinen, E. Klinker, J.F. Mahfouf, A. Simmons, The ECMWF operational implementation of four-dimensional variational assimilation. I: Experimental results with simplified physics. Q. J. R. Meteorol. Soc. **126**(564), 1143–1170 (2000). https://doi.org/10.1002/qj.49712656415

H.J. Reitsema, W.A. Delamere, A.R. Williams, D.C. Boice, W.F. Huebner, F.L. Whipple, Dust distribution in the inner coma of comet Halley: comparison with models. Icarus **81**(1), 31–40 (1989). https://doi.org/10.1016/0019-1035(89)90123-1

L. Rezac, Y. Zhao, P. Hartogh, J. Ji, D. Marshall, X. Shi, Three-dimensional analysis of spatial resolution of Miro/Rosetta measurements at 67P/Churyumov-Gersimenko. Astron. Astrophys. **126**(564), 1143–1170 (2019). https://doi.org/10.1002/qj.49712656415

G. Rinaldi, U. Fink, L. Doose, G.P. Tozzi, F. Capaccioni, G. Filacchione, D. Bockelée-Morvan, C. Leyrat, G. Piccioni, S. Erard, Properties of the dust in the coma of 67P/Churyumov-Gerasimenko observed with VIRTIS-M. Mon. Not. R. Astron. Soc. **462**, S547–S561 (2016). https://doi.org/10.1093/mnras/stw3197

G. Rinaldi, V. Della Corte, M. Fulle, F. Capaccioni, A. Rotundi, S.L. Ivanovski, D. Bockelée-Morvan, G. Filacchione, E. D'Aversa, M.T. Capria, Cometary coma dust size distribution from in situ IR spectra. Mon. Not. R. Astron. Soc. **469**, S598–S605 (2017). https://doi.org/10.1093/mnras/stx1873

G. Rinaldi, D. Bockelée-Morvan, M. Ciarniello, G.P. Tozzi, F. Capaccioni, S.L. Ivanovski, G. Filacchione, U. Fink, L. Doose, F. Taylor, Summer outbursts in the coma of comet 67P/Churyumov-Gerasimenko as observed by Rosetta-VIRTIS. Mon. Not. R. Astron. Soc. **481**(1), 1235–1250 (2018). https://doi.org/10.1093/mnras/sty2266

C.D. Rodgers, *Inverse Methods for Atmospheric Sounding: Theory and Practice* (World Scientific, Singapore, 2000)

A.V. Rodionov, Simulation of the outflow of three-dimensional multiphase jets from the surface of a cometary nucleus. Sol. Syst. Res. **30**(3), 230 (1996)

A.V. Rodionov, J.F. Crifo, Time-dependent, three-dimensional fluid model of the outer coma, with application to the comet Hale-Bopp gas spirals. Adv. Space Res. **38**(9), 1923–1927 (2006). https://doi.org/10.1016/j.asr.2006.03.019

V.K. Rosenbush, O.V. Ivanova, N.N. Kiselev, L.O. Kolokolova, V.L. Afanasiev, Spatial variations of brightness, colour and polarization of dust in comet 67P/Churyumov-Gerasimenko. Mon. Not. R. Astron. Soc. **469**, S475–S491 (2017). https://doi.org/10.1093/mnras/stx2003

A. Rotundi, H. Sierks, V. Della Corte, M. Fulle, P.J. Gutierrez, L. Lara, C. Barbieri, P.L. Lamy, R. Rodrigo, D. Koschny, H. Rickman, H.U. Keller, J.J. López-Moreno, M. Accolla, J. Agarwal, M.F. A'Hearn, N. Altobelli, F. Angrilli, M.A. Barucci, J.L. Bertaux, I. Bertini, D. Bodewits, E. Bussoletti, L. Colangeli, M. Cosi, G. Cremonese, J.F. Crifo, V. Da Deppo, B. Davidsson, S. Debei, M. De Cecco, F. Esposito, M. Ferrari, S. Fornasier, F. Giovane, B. Gustafson, S.F. Green, O. Groussin, E. Grün, C. Güttler, M.L. Herranz, S.F. Hviid, W. Ip, S. Ivanovski, J.M. Jerónimo, L. Jorda, J. Knollenberg, R. Kramm, E. Kührt, M. Küppers, M. Lazzarin, M.R. Leese, A.C. López-Jiménez, F. Lucarelli, S.C. Lowry, F. Marzari, E.M. Epifani, J.A.M. McDonnell, V. Mennella, H. Michalik, A. Molina, R. Morales, F. Moreno, S. Mottola, G. Naletto, N. Oklay, J.L. Ortiz, E. Palomba, P. Palumbo, J.M. Perrin, J. Rodríguez, L. Sabau, C. Snodgrass, R. Sordini, N. Thomas, C. Tubiana, J.B. Vincent, P. Weissman, K.P. Wenzel, V. Zakharov, J.C. Zarnecki, Dust measurements in the coma of comet 67P/Churyumov-Gerasimenko inbound to the Sun. Science **347**, aaa3905 (2015). https://doi.org/10.1126/science.aaa3905

F.P. Schloerb, S. Keihm, P. von Allmen, M. Choukroun, E. Lellouch, C. Leyrat, G. Beaudin, N. Biver, D. Bockelée-Morvan, J. Crovisier, P. Encrenaz, R. Gaskell, S. Gulkis, P. Hartogh, M. Hofstadter, W.H. Ip, M. Janssen, C. Jarchow, L. Jorda, H.U. Keller, S. Lee, L. Rezac, H. Sierks, MIRO observations of subsurface temperatures of the nucleus of 67P/Churyumov-Gerasimenko. Astron. Astrophys. **583**, A29 (2015). https://doi.org/10.1051/0004-6361/201526152

M.I. Schmitt, C. Tubiana, C. Güttler, H. Sierks, J.B. Vincent, M.R. El-Maarry, D. Bodewits, S. Mottola, S. Fornasier, M. Hofmann, Long-term monitoring of comet 67P/Churyumov-Gerasimenko's jets with OSIRIS onboard Rosetta. Mon. Not. R. Astron. Soc. **469**, S380–S385 (2017). https://doi.org/10.1093/mnras/stx1780

R. Schulz, M. Hilchenbach, Y. Langevin, J. Kissel, J. Silen, C. Briois, C. Engrand, K. Hornung, D. Baklouti, A. Bardyn, H. Cottin, H. Fischer, N. Fray, M. Godard, H. Lehto, L. Le Roy, S. Merouane, F.R. Orthous-Daunay, J. Paquette, J. Rynö, S. Siljeström, O. Stenzel, L. Thirkell, K. Varmuza, B. Zaprudin, Comet 67P/Churyumov-Gerasimenko sheds dust coat accumulated over the past four years. Nature **518**, 216–218 (2015). https://doi.org/10.1038/nature14159

Z. Sekanina, Distribution and activity of discrete emission areas on the nucleus of periodic Comet Swift-Tuttle. Astron. J. **86**, 1741–1773 (1981). https://doi.org/10.1086/113059

Z. Sekanina, Anisotropic emission from comets: fans versus jets. I. Concept and modelling, in *ESA*, vol. SP-278 (1987), pp. 315–322

J.V. Sengers, Y.Y. Lin Wang, B. Kamgar-Parsi, J.R. Dorfman, Kinetic theory of drag on objects in nearly free molecular flow. Phys. A, Stat. Mech. Appl. **413**, 409–425 (2014). https://doi.org/10.1016/j.physa.2014.06.026. 1404.7826
X. Shi, X. Hu, S. Mottola, H. Sierks, H.U. Keller, M. Rose, C. Güttler, M. Fulle, S. Fornasier, J. Agarwal, M. Pajola, C. Tubiana, D. Bodewits, C. Barbieri, P.L. Lamy, R. Rodrigo, D. Koschny, M.A. Barucci, J.L. Bertaux, I. Bertini, S. Boudreault, G. Cremonese, V. Da Deppo, B. Davidsson, S. Debei, M. De Cecco, J. Deller, O. Groussin, P.J. Gutiérrez, S.F. Hviid, W.H. Ip, L. Jorda, J. Knollenberg, G. Kovacs, J.R. Kramm, E. Kührt, M. Küppers, L.M. Lara, M. Lazzarin, J.J. Lopez-Moreno, F. Marzari, G. Naletto, N. Oklay, I. Toth, J.B. Vincent, Coma morphology of comet 67P controlled by insolation over irregular nucleus. Nat. Astron. **2**, 562–567 (2018). https://doi.org/10.1038/s41550-018-0481-5
Y. Shou, M. Combi, G. Toth, V. Tenishev, N. Fougere, X. Jia, M. Rubin, Z. Huang, K. Hansen, T. Gombosi, A new 3D multi-fluid dust model: a study of the effects of activity and nucleus rotation on dust grain behavior at comet 67P/Churyumov-Gerasimenko. Astrophys. J. **850**, 72 (2017). https://doi.org/10.3847/1538-4357/aa91ca
Y. Skorov, H. Rickman, A kinetic model of gas flow in a porous cometary mantle. Planet. Space Sci. **43**(12), 1587–1594 (1995). https://doi.org/10.1016/0032-0633(95)00062-3
Y.V. Skorov, H. Rickman, Gas flow and dust acceleration in a cometary Knudsen layer. Planet. Space Sci. **47**, 935–949 (1999). https://doi.org/10.1016/S0032-0633(99)00008-2
Y.V. Skorov, R. van Lieshout, J. Blum, H.U. Keller, Activity of comets: gas transport in the near-surface porous layers of a cometary nucleus. Icarus **212**(2), 867–876 (2011). https://doi.org/10.1016/j.icarus.2011.01.018. 1101.2525
Y. Skorov, V. Reshetnyk, P. Lacerda, P. Hartogh, J. Blum, Acceleration of cometary dust near the nucleus: application to 67P/Churyumov-Gerasimenko. Mon. Not. R. Astron. Soc. **461**, 3410–3420 (2016a). https://doi.org/10.1093/mnras/stw1470
Y.V. Skorov, L. Rezac, P. Hartogh, A.T. Bazilevsky, H.U. Keller, A model of short-lived outbursts on the 67P from fractured terrains. Astron. Astrophys. **593**, A76 (2016b). https://doi.org/10.1051/0004-6361/201628365
Y.V. Skorov, L. Rezac, P. Hartogh, H.U. Keller, Is near-surface ice the driver of dust activity on 67P/Churyumov-Gerasimenko. Astron. Astrophys. **600**, A142 (2017). https://doi.org/10.1051/0004-6361/201630000
Y. Skorov, V. Reshetnyk, L. Rezac, Y. Zhao, R. Marschall, J. Blum, P. Hartogh, Dynamical properties and acceleration of hierarchical dust in the vicinity of comet 67P/Churyumov-Gerasimenko. Mon. Not. R. Astron. Soc. **477**, 4896–4907 (2018). https://doi.org/10.1093/mnras/sty1014
A.J. Steffl, M.F. A'Hearn, J.L. Bertaux, L.M. Feaga, P.D. Feldman, J. Noonan, J.W. Parker, E. Schindhelm, S.A. Stern, H.A. Weaver, Icy outbursts from comet 67P/Churyumov-Gerasimenko as observed by the Alice FUV spectrograph, in *AAS/Division for Planetary Sciences Meeting Abstracts #48*. AAS/Division for Planetary Sciences Meeting Abstracts (2016), p. 110.03
S.A. Stern, D.C. Slater, J. Scherrer, J. Stone, M. Versteeg, M.F. A'Hearn, J.L. Bertaux, P.D. Feldman, M.C. Festou, J.W. Parker, O.H.W. Siegmund, Alice: The Rosetta Ultraviolet Imaging Spectrograph. Space Sci. Rev. **128**(1–4), 507–527 (2007). https://doi.org/10.1007/s11214-006-9035-8. astro-ph/0603585
K. Szego, I. Toth, Z. Szatmary, B.A. Smith, A. Kondor, E. Merenyi, Dust photometry in the near nucleus region of Comet Halley. Adv. Space Res. **9**(3), 89–92 (1989). https://doi.org/10.1016/0273-1177(89)90246-9
V. Tenishev, M. Combi, B. Davidsson, A global kinetic model for cometary comae: the evolution of the coma of the Rosetta target comet Churyumov-Gerasimenko throughout the mission. Astrophys. J. **685**, 659–677 (2008). https://doi.org/10.1086/590376
V. Tenishev, M.R. Combi, M. Rubin, Numerical simulation of dust in a cometary coma: application to comet 67P/Churyumov-Gerasimenko. Astrophys. J. **732**, 104 (2011). https://doi.org/10.1088/0004-637X/732/2/104
N. Thomas, H.U. Keller, Interpretation of the inner coma observations of comet P/Halley by the Halley Multicolor Camera. Ann. Geophys. **8**, 147–165 (1990)
N. Thomas, D.C. Boice, W.F. Huebner, H.U. Keller, Intensity profiles of dust near extended sources on comet Halley. Nature **332**, 51–52 (1988). https://doi.org/10.1038/332051a0
N. Thomas, B. Davidsson, M.R. El-Maarry, S. Fornasier, L. Giacomini, A.G. Gracia-Berná, S.F. Hviid, W.H. Ip, L. Jorda, H.U. Keller, J. Knollenberg, E. Kührt, F. La Forgia, I.L. Lai, Y. Liao, R. Marschall, M. Massironi, S. Mottola, M. Pajola, O. Poch, A. Pommerol, F. Preusker, F. Scholten, C.C. Su, J.S. Wu, J.B. Vincent, H. Sierks, C. Barbieri, P.L. Lamy, R. Rodrigo, D. Koschny, H. Rickman, M.F. A'Hearn, M.A. Barucci, J.L. Bertaux, I. Bertini, G. Cremonese, V. Da Deppo, S. Debei, M. de Cecco, M. Fulle, O. Groussin, P.J. Gutierrez, J.R. Kramm, M. Küppers, L.M. Lara, M. Lazzarin, J.J. Lopez Moreno, F. Marzari, H. Michalik, G. Naletto, J. Agarwal, C. Güttler, N. Oklay, C. Tubiana, Redistribution of parti-

cles across the nucleus of comet 67P/Churyumov-Gerasimenko. Astron. Astrophys. **583**, A17 (2015a). https://doi.org/10.1051/0004-6361/201526049

N. Thomas, B. Davidsson, M.R. El-Maarry, S. Fornasier, L. Giacomini, A.G. Gracia-Berná, S.F. Hviid, W.H. Ip, L. Jorda, H.U. Keller, J. Knollenberg, E. Kührt, F. La Forgia, I.L. Lai, Y. Liao, R. Marschall, M. Massironi, S. Mottola, M. Pajola, O. Poch, A. Pommerol, F. Preusker, F. Scholten, C.C. Su, J.S. Wu, J.B. Vincent, H. Sierks, C. Barbieri, P.L. Lamy, R. Rodrigo, D. Koschny, H. Rickman, M.F. A'Hearn, M.A. Barucci, J.L. Bertaux, I. Bertini, G. Cremonese, V. Da Deppo, S. Debei, M. de Cecco, M. Fulle, O. Groussin, P.J. Gutierrez, J.R. Kramm, M. Küppers, L.M. Lara, M. Lazzarin, J.J. Lopez Moreno, F. Marzari, H. Michalik, G. Naletto, J. Agarwal, C. Güttler, N. Oklay, C. Tubiana, Redistribution of particles across the nucleus of comet 67P/Churyumov-Gerasimenko. Astron. Astrophys. **583**, A17 (2015b). https://doi.org/10.1051/0004-6361/201526049

N. Thomas, H. Sierks, C. Barbieri, P.L. Lamy, R. Rodrigo, H. Rickman, D. Koschny, H.U. Keller, J. Agarwal, M.F. A'Hearn, F. Angrilli, A.T. Auger, M.A. Barucci, J.L. Bertaux, I. Bertini, S. Besse, D. Bodewits, G. Cremonese, V. Da Deppo, B. Davidsson, M. De Cecco, S. Debei, M.R. El-Maarry, F. Ferri, S. Fornasier, M. Fulle, L. Giacomini, O. Groussin, P.J. Gutierrez, C. Güttler, S.F. Hviid, W.H. Ip, L. Jorda, J. Knollenberg, J.R. Kramm, E. Kührt, M. Küppers, F. La Forgia, L.M. Lara, M. Lazzarin, J.J.L. Moreno, S. Magrin, S. Marchi, F. Marzari, M. Massironi, H. Michalik, R. Moissl, S. Mottola, G. Naletto, N. Oklay, M. Pajola, A. Pommerol, F. Preusker, L. Sabau, F. Scholten, C. Snodgrass, C. Tubiana, J.B. Vincent, K.P. Wenzel, The morphological diversity of comet 67P/Churyumov-Gerasimenko. Science **347**, aaa0440 (2015c). https://doi.org/10.1126/science.aaa0440

F. Tosi, F. Capaccioni, M.T. Capria, S. Mottola, A. Zinzi, M. Ciarniello, G. Filacchione, M. Hofstadter, S. Fonti, M. Formisano, D. Kappel, E. Kührt, C. Leyrat, J.B. Vincent, G. Arnold, M.C. De Sanctis, A. Longobardo, E. Palomba, A. Raponi, B. Rousseau, B. Schmitt, M.A. Barucci, G. Bellucci, J. Benkhoff, D. Bockelée-Morvan, P. Cerroni, J.P. Combe, D. Despan, S. Erard, F. Mancarella, T.B. McCord, A. Migliorini, V. Orofino, G. Piccioni, The changing temperature of the nucleus of comet 67P induced by morphological and seasonal effects. Nat. Astron. **3**, 649–658 (2019). https://doi.org/10.1038/s41550-019-0740-0

C. Tubiana, C. Snodgrass, I. Bertini, S. Mottola, J.B. Vincent, L. Lara, S. Fornasier, J. Knollenberg, N. Thomas, M. Fulle, J. Agarwal, D. Bodewits, F. Ferri, C. Güttler, P.J. Gutierrez, F. La Forgia, S. Lowry, S. Magrin, N. Oklay, M. Pajola, R. Rodrigo, H. Sierks, M.F. A'Hearn, F. Angrilli, C. Barbieri, M.A. Barucci, J.L. Bertaux, G. Cremonese, V. Da Deppo, B. Davidsson, M. De Cecco, S. Debei, O. Groussin, S.F. Hviid, W. Ip, L. Jorda, H.U. Keller, D. Koschny, R. Kramm, E. Kührt, M. Küppers, M. Lazzarin, P.L. Lamy, J.J. Lopez Moreno, F. Marzari, H. Michalik, G. Naletto, H. Rickman, L. Sabau, K.P. Wenzel, 67P/Churyumov-Gerasimenko: activity between March and June 2014 as observed from Rosetta/OSIRIS. Astron. Astrophys. **573**, A62 (2015). https://doi.org/10.1051/0004-6361/201424735

C. Tubiana, G. Rinaldi, C. Güttler, C. Snodgrass, X. Shi, X. Hu, R. Marschall, M. Fulle, D. Bockelée-Morvan, G. Naletto, F. Capaccioni, H. Sierks, G. Arnold, M.A. Barucci, J.L. Bertaux, I. Bertini, D. Bodewits, M.T. Capria, M. Ciarniello, G. Cremonese, J. Crovisier, V. Da Deppo, S. Debei, M. De Cecco, J. Deller, M.C. De Sanctis, B. Davidsson, L. Doose, S. Erard, G. Filacchione, U. Fink, M. Formisano, S. Fornasier, P.J. Gutiérrez, W.H. Ip, S. Ivanovski, D. Kappel, H.U. Keller, L. Kolokolova, D. Koschny, H. Krueger, F. La Forgia, P.L. Lamy, L.M. Lara, M. Lazzarin, A.C. Levasseur-Regourd, Z.Y. Lin, A. Longobardo, J.J. López-Moreno, F. Marzari, A. Migliorini, S. Mottola, R. Rodrigo, F. Taylor, I. Toth, V. Zakharov, Diurnal variation of dust and gas production in comet 67P/Churyumov-Gerasimenko at the inbound equinox as seen by OSIRIS and VIRTIS-M on board Rosetta. Astron. Astrophys. **630**, A23 (2019). https://doi.org/10.1051/0004-6361/201834869. 1905.03022

J.B. Vincent, H. Böhnhardt, L.M. Lara, A numerical model of cometary dust coma structures. Application to comet 9P/Tempel 1. Astron. Astrophys. **512**, A60 (2010)

J.B. Vincent, L.M. Lara, G.P. Tozzi, Z.Y. Lin, H. Sierks, Spin and activity of comet 67P/Churyumov-Gerasimenko. Astron. Astrophys. **549**, A121 (2013). https://doi.org/10.1051/0004-6361/201219350

J.B. Vincent, M.F. A'Hearn, Z.Y. Lin, M.R. El-Maarry, M. Pajola, H. Sierks, C. Barbieri, P.L. Lamy, R. Rodrigo, D. Koschny, Summer fireworks on comet 67P. Mon. Not. R. Astron. Soc. **462**, S184–S194 (2016a). https://doi.org/10.1093/mnras/stw2409. 1609.07743

J.B. Vincent, M.F. A'Hearn, Z.Y. Lin, M.R. El-Maarry, M. Pajola, H. Sierks, C. Barbieri, P.L. Lamy, R. Rodrigo, D. Koschny, H. Rickman, H.U. Keller, J. Agarwal, M.A. Barucci, J.L. Bertaux, I. Bertini, S. Besse, D. Bodewits, G. Cremonese, V. Da Deppo, B. Davidsson, S. Debei, M. De Cecco, J. Deller, S. Fornasier, M. Fulle, A. Gicquel, O. Groussin, P.J. Gutiérrez, P. Gutiérrez-Marquez, C. Güttler, S. Höfner, M. Hofmann, S.F. Hviid, W.H. Ip, L. Jorda, J. Knollenberg, G. Kovacs, J.R. Kramm, E. Kührt, M. Küppers, L.M. Lara, M. Lazzarin, J.J. Lopez Moreno, F. Marzari, M. Massironi, S. Mottola, G. Naletto, N. Oklay, F. Preusker, F. Scholten, X. Shi, N. Thomas, I. Toth, C. Tubiana, Summer fireworks on comet 67P. Mon. Not. R. Astron. Soc. **462**, S184–S194 (2016b). https://doi.org/10.1093/mnras/stw2409. 1609.07743

J.B. Vincent, N. Oklay, M. Pajola, S. Höfner, H. Sierks, X. Hu, C. Barbieri, P.L. Lamy, R. Rodrigo, D. Koschny, H. Rickman, H.U. Keller, M.F. A'Hearn, M.A. Barucci, J.L. Bertaux, I. Bertini, S. Besse, D. Bodewits, G. Cremonese, V. Da Deppo, B. Davidsson, S. Debei, M. De Cecco, M.R. El-Maarry, S. Fornasier, M. Fulle, O. Groussin, P.J. Gutiérrez, P. Gutiérrez-Marquez, C. Güttler, M. Hofmann, S.F. Hviid, W.H. Ip, L. Jorda, J. Knollenberg, G. Kovacs, J.R. Kramm, E. Kührt, M. Küppers, L.M. Lara, M. Lazzarin, Z.Y. Lin, J.J. Lopez Moreno, S. Lowry, F. Marzari, M. Massironi, F. Moreno, S. Mottola, G. Naletto, F. Preusker, F. Scholten, X. Shi, N. Thomas, I. Toth, C. Tubiana, Are fractured cliffs the source of cometary dust jets? Insights from OSIRIS/Rosetta at 67P/Churyumov-Gerasimenko. Astron. Astrophys. **587**, A14 (2016c). https://doi.org/10.1051/0004-6361/201527159. 1512.03193

J.B. Vincent, T. Farnham, E. Kührt, Y. Skorov, R. Marschall, N. Oklay, M.R. El-Maarry, H.U. Keller, Local manifestations of cometary activity. Space Sci. Rev. **215**(4), 30 (2019). https://doi.org/10.1007/s11214-019-0596-8

M.K. Wallis, M.H.A. Hassan, Electrodynamics of submicron dust in the cometary coma. Astron. Astrophys. **121**(1), 10–14 (1983)

X. Xie, M.J. Mumma, Monte Carlo simulation of cometary atmospheres: application to comet P/Halley at the time of the Giotto spacecraft encounter. I. Isotropic model. Astrophys. J. **464**, 442 (1996). https://doi.org/10.1086/177335

Z. Yoldi, A. Pommerol, B. Jost, O. Poch, J. Gouman, N. Thomas, VIS-NIR reflectance of water ice/regolith analogue mixtures and implications for the detectability of ice mixed within planetary regoliths. Geophys. Res. Lett. **42**, 6205–6212 (2015). https://doi.org/10.1002/2015GL064780

T. Ytrehus, Kinetic theory description and experimental results for vapor motion in arbitrary strong evaporation. Tech. rep. (1975)

V.V. Zakharov, J.F. Crifo, G.A. Lukyanov, A.V. Rodionov, Modeling of the inner atmosphere of comets with aspherical nucleus of "apple" type. J. Math. Model. **15**, 48 (2003)

V. Zakharov, D. Bockelée-Morvan, N. Biver, J. Crovisier, A. Lecacheux, Radiative transfer simulation of water rotational excitation in comets Comparison of the Monte Carlo and escape probability methods. Astron. Astrophys. **473**, 303–310 (2007). https://doi.org/10.1051/0004-6361:20066715

V.V. Zakharov, A.V. Rodionov, G.A. Lukyanov, J.F. Crifo, Navier Stokes and direct Monte-Carlo simulations of the circumnuclear gas coma. III. Spherical, inhomogeneous sources. Icarus **194**, 327–346 (2008). https://doi.org/10.1016/j.icarus.2007.08.038

V.V. Zakharov, A.V. Rodionov, G.A. Lukianov, J.F. Crifo, Monte-Carlo and multifluid modelling of the circumnuclear dust coma II. Aspherical-homogeneous, and spherical-inhomogeneous nuclei. Icarus **201**, 358–380 (2009). https://doi.org/10.1016/j.icarus.2008.12.022

V.V. Zakharov, J.F. Crifo, A.V. Rodionov, M. Rubin, K. Altwegg, The near-nucleus gas coma of comet 67P/Churyumov-Gerasimenko prior to the descent of the surface lander PHILAE. Astron. Astrophys. **618**, A71 (2018a). https://doi.org/10.1051/0004-6361/201832883

V.V. Zakharov, S.L. Ivanovski, J.F. Crifo, V. Della Corte, A. Rotundi, M. Fulle, Asymptotics for spherical particle motion in a spherically expanding flow. Icarus **312**, 121–127 (2018b). https://doi.org/10.1016/j.icarus.2018.04.030

Space Sci Rev (2020) 216:44
https://doi.org/10.1007/s11214-020-00662-1

# Dust-to-Gas and Refractory-to-Ice Mass Ratios of Comet 67P/Churyumov-Gerasimenko from Rosetta Observations

**Mathieu Choukroun[1] · Kathrin Altwegg[2] · Ekkehard Kührt[3] · Nicolas Biver[4] · Dominique Bockelée-Morvan[4] · Joanna Drążkowska[5] · Alain Hérique[6] · Martin Hilchenbach[7] · Raphael Marschall[8] · Martin Pätzold[9] · Matthew G.G.T. Taylor[10] · Nicolas Thomas[2]**

Received: 17 May 2019 / Accepted: 19 March 2020 / Published online: 8 April 2020


**Abstract** This chapter reviews the estimates of the dust-to-gas and refractory-to-ice mass ratios derived from *Rosetta* measurements in the lost materials and the nucleus of 67P/Churyumov-Gerasimenko, respectively. First, the measurements by *Rosetta* instruments are described, as well as relevant characteristics of 67P. The complex picture of the activity of 67P, with its extreme North-South seasonal asymmetry, is presented. Individual estimates of the dust-to-gas and refractory-to-ice mass ratios are then presented and compared, showing wide ranges of plausible values. *Rosetta*'s wealth of information suggests that estimates of the dust-to-gas mass ratio made in cometary comae at a single point in time may not be fully representative of the refractory-to-ice mass ratio within the cometary nuclei being observed.

**Keywords** Comets · 67P/Churyumov-Gerasimenko · Rosetta · Dust-to-gas ratio · Refractory-to-Ice mass ratio

Comets: Post 67P / Churyumov-Gerasimenko Perspectives
Edited by Nicolas Thomas, Björn Davidsson, Laurent Jorda, Ekkehard Kührt, Raphael Marschall, Colin Snodgrass and Rafael Rodrigo

✉ M. Choukroun
mathieu.choukroun@jpl.nasa.gov

1 Jet Propulsion Laboratory, California Institute of Technology, Pasadena, CA, USA

2 University of Bern, Bern, Switzerland

3 DLR Institute of Planetary Research, Berlin, Germany

4 Observatoire de Paris-Meudon, Meudon, France

5 Ludwig Maximilian University of Munich, Munich, Germany

6 Université Grenoble Alpes, CNRS, CNES, IPAG, Grenoble, France

7 Max Planck Institute, Göttingen, Germany

8 Southwest Research Institute, Boulder, CO, USA

9 RIU-Planetary Research at University of Köln, Köln, Germany

10 ESA ESTEC, Noordwijk, Netherlands

# 1 Introduction

Comets are believed to be the least processed remnants of the early days of the outer Solar System because their low gravity prevented high internal pressure, and their birth place in the outer protosolar disk (with low mass densities and moderate Kepler velocities) only led to moderate modifications by collisions after their formation. Comets consist of a mixture of refractory materials in the classical sense (minerals), an organic "soup" that ranges from simple molecules more volatile than water ice, such as methane and methanol, to heavy and refractory molecules such as kerogen-like insoluble organic matter (IOM), and a number of frozen volatiles (ices) like $H_2O$, $CO_2$, CO, $H_2S$, $NH_3$, and multiple other minor species (Bockelée-Morvan et al. 2004; Mumma and Charnley 2011; Bockelée-Morvan and Biver 2017). Comet nuclei have a very low density/high porosity (Weissman et al. 2004; Weissman and Lowry 2008; Preusker et al. 2017), and the few samples available on Earth that likely originate from comets (chondritic-porous interplanetary dust particles, ultra-carbonaceous Antarctica micro-meteorites), as well as grains collected in the coma of 81P/Wild by the *Stardust* mission, all exhibit a very fine-grained texture (Joswiak et al. 2017). These compositional and physical characteristics point to a rather slow and gentle accretion under cold conditions $\ll$ 100 K, i.e. at large heliocentric distances within the protosolar disk.

The relative proportions of ices and refractory materials within comet nuclei (the refractory-to-ice mass ratio) is important to trace how and where these objects formed and evolved early on. So far it is unclear if ice and refractories are mixed on a microscopic scale, where ice forms an ice mantle around µm-sized dust grains as suggested for the condensation sequence in a molecular cloud or the interstellar medium, or if mixing takes place at a "macroscopic" scale, i.e. independent ice grains and refractory grains mixed together, most likely in the outer disk, or a combination of both. The refractory-to-ice ratio is one indirect clue to address this, in combination with the composition of outgassed materials and refractories. It may help discriminate between comet formation models (Weissman and Lowry 2008).

Comet formation processes are still a subject of intensive research and no fully self-consistent model is available yet. Dust delivered to the protosolar nebula from the interstellar medium was sub- µm to µm-sized. To form comet parent bodies (planetesimals), cometary dust had to accrete over many orders in magnitude in size. Dust coagulation models are based on collisional properties derived from laboratory experiments, which have been recently summarized by Blum (2018). Most of these experiments have been performed using silicate dust aggregates, which bounce at collision speeds of $\Delta v \sim 1$ cm/s and fragment at $\Delta v \sim 1$ m/s. This would suggest that no silicate aggregates larger than mm-sized were able to grow via this process (Zsom et al. 2010). Experiments including water ice suggested that it is considerably stickier, as the icy aggregates only fragment at $\Delta v \sim 10$ m/s (Aumatell and Wurm 2014; Gundlach and Blum 2014) consistent with earlier predictions of numerical models (Wada et al. 2011). However, more recent work suggests that the tensile strength of icy aggregates is lower than expected at low temperatures (Gundlach et al. 2018; Musiolik and Wurm 2019). Nevertheless, the potential impact of the refractory-to-ice ratio on the sticking properties of mixtures between ices and refractories has not been addressed by laboratory experiments to date.

Numerical experiments focused on global redistribution of dust in the protosolar disk showed that if the aggregate fragmentation threshold indeed depends on the refractory-to-ice ratio, a pile-up of solids inside of the water snow line could be triggered and planetesimals may preferentially form just outside of the snow line, where icy aggregates can grow and accumulate (Drążkowska and Alibert 2017; Schoonenberg and Ormel 2017). However, even if the icy dust sticks very efficiently, its direct growth to km-sized bodies seems unlikely, as the growing aggregates are removed from the outer parts of the disk by inward drift (Krijt et al. 2016; Homma and Nakamoto 2018). The most widely accepted scenario of planetesimal formation is a multifaceted process, in which the μm-sized dust first grows to mm- to cm-sized aggregates that subsequently undergo so-called streaming instability, forming overdense filaments that collapse to km-sized, gravitationally bound objects (Drążkowska and Dullemond 2014; Johansen et al. 2014; Lorek et al. 2018). The refractory-to-ice ratio of comets formed in this scenario would reflect that of dust aggregates participating in the streaming instability, expected from one model in the range 3-9 (Lorek et al. 2016).

The refractory-to-ice ratio has a most direct impact on the thermal properties and activity of comet nuclei. Outgassing of cometary materials is driven by sublimation of ices ($H_2O$ and $CO/CO_2$ mainly), mostly from subsurface layers, which in turn accelerates particles ("dust") as they leave the nucleus. These "dust" particles may be only composed of refractory materials, or may also contain ice. In the remainder of the manuscript, we refer to ice-free particles as "dry dust". The relative proportions of ice and dust in the near-surface of the nucleus directly impact its thermal properties, as the filling of pores by water ice, either primordial or due to recondensation of sublimated ice from below, would increase the thermal conductivity of this environment, easing propagation of heat from the Sun to greater depths and mobilization of ices (Prialnik et al. 2004, 2008).

For comets observed to date, the dust-to-gas ratio that can be measured in the coma was assumed to be directly representative of the refractory-to-ice ratio within the nucleus. The dust-to-gas ratio in the coma of 1P/Halley was first determined as 1.3-3 (McDonnell et al. 1991), although a re-evaluation of the dust size distribution (Levasseur-Regourd et al. 1999) suggests much larger dust-to-gas ratio values (Fulle et al. 2000). The dust-to-gas ratio of 9P/Tempel is greater than 1 (Küppers et al. 2005). However, these estimates were derived from observations acquired at a single time along their orbit. ESA's *Rosetta* mission (Glassmeier et al. 2007) escorted comet 67P/Churyumov-Gerasimenko (hereafter 67P) through 2014-2016, and was equipped with a comprehensive suite of instruments to survey the comet surface, its interior, its activity, and evolution over time. Thus, *Rosetta* provided the first opportunity to investigate dust-to-gas and refractory-to-ice mass ratios from up close, with multiple measurements, over an extended period of time. This review summarizes pertinent results obtained to date, their uncertainties, and potential ways to investigate this in the future.

## 2 67P as Seen by *Rosetta*

### 2.1 General Characteristics

67P/Churyumov-Gerasimenko, named after the two astronomers who discovered it in 1969, is a Jupiter-Family Comet (JFC). As such, its elliptical orbit around the Sun (current orbital period 6.44 years, perihelion distance 1.24 AU, aphelion distance 5.68 AU) is controlled by gravitational interactions with the largest planet of our Solar System. Dynamical models retracing the history of 67P's orbit have shown that a close encounter with Jupiter in 1959 injected it into its current orbit, only 9 revolutions before the arrival of *Rosetta* (Krolikowska 2003; Groussin et al. 2007; Maquet 2015). However, these dynamical history computations also showed that the evolution of 67P's orbit is highly chaotic. This stochastic source of uncertainty prevents from determining whether 67P was injected into the inner Solar System for the first time after the 1959 encounter with Jupiter, or if it could have further evolved as a JFC in the inner Solar System before that, possibly with an even closer perihelion distance than now. Nevertheless, all dynamical models agree that 67P's orbit must have been controlled by Jupiter for thousands of years, hence suggesting that it is a relatively evolved comet.

Despite 67P having likely passed through the inner Solar System multiple times in its geologically recent past, the volatiles emitted by its nucleus are consistent with their original ices being formed from outer protosolar disk and perhaps presolar materials, see hereafter and the review by Hoppe et al. (2018). Indeed, 67P exhibits the highest D/H ratio measured to date in cometary water (Altwegg et al. 2015), suggesting significant incorporation of outer protosolar disk materials. The abundances of highly volatile molecules like $N_2$ (Rubin et al. 2015) and $O_2$ (Bieler et al. 2015a, 2015b), and noble gases like Ar (Balsiger et al. 2015), are consistent with formation at extremely low temperatures found in the Kuiper Belt and beyond. The presence of sulfur allotropes and other sulfur-bearing molecules (Calmonte et al. 2016) is consistent with their formation via radiolytic processes in the presolar cloud and/or the outer protosolar disk, as shown in laboratory experiments (Mahjoub et al. 2017), prior to their incorporation in 67P. The isotopic composition of sulfur-bearing molecules (Calmonte et al. 2017) furthermore suggests a significant contribution of materials extraneous to the solar nebula. Thus, 67P appears to still retain at present-day the primordial presolar/protosolar materials from which it formed.

The 67P nucleus, revealed by the *Rosetta* mission's Optical, Spectroscopic, Infrared Remote Imaging System, OSIRIS (Keller et al. 2007), has a striking bilobate structure (Sierks et al. 2015) with a stable spin axis and strong seasonal characteristics. The high obliquity of 67P results in significant differences in insolation between the northern and southern "hemispheres" (Keller et al. 2015; Jorda et al. 2016; Keller et al. 2017). The southern regions experienced a relatively short summer around perihelion in August 2015 (with equinoxes on May 2015 and March 2016). Although the 67P spin orientation is stable, its rotation period changes over time owing to sublimation-induced torques (Kramer et al. 2018; Kramer and Läuter 2019). Hirabayashi et al. (2016) computed the stresses required to propagate a hundred-meters-long crack between the two lobes to the extent of splitting the nucleus, and found that the rotation period would need to reach values as low as $\sim$ 7 hrs. If the rotation period maintains its linear decrease with every orbit, it would take $\sim$ 15 orbits, almost 100 years, for 67P to reach the point where it may split into two (provided that no catastrophic events occur before that).

67P exhibits a low surface albedo $\sim$ 4% (Sierks et al. 2015), is dominated by organic materials (Capaccioni et al. 2015; Quirico et al. 2016; Bardyn et al. 2017; Fray et al. 2017) and has an extremely high porosity (Ciarletti et al. 2015; Kofman et al. 2015; Hérique et al. 2016; Pätzold et al. 2016, 2019). These compositional and structural characteristics result in a low thermal inertia (Capaccioni et al. 2015; Choukroun et al. 2015; Gulkis et al. 2015; Schloerb et al. 2015; Marshall et al. 2018), which promotes large temperature gradients over a few cm and extreme temperature swings following rapid changes in illumination. Little to no ice is directly detected on the surface, except: 1) in specific shadowed areas where ice can remain stable for extended periods of time (Filacchione et al. 2016a), 2) as surface frost associated to a diurnal cycle where ice is mobilized from the subsurface during the day and then is recondensed on the surface during the night (De Sanctis et al. 2015), and 3) when the activity level increased enough towards perihelion for nucleus erosion to expose fresher materials (Fornasier et al. 2016).

The *Philae* lander was deployed on November 12, 2014, and bounced off its initial landing site Agilkia on a hard subsurface layer beneath $\sim$10–20 cm of loose regolith (Biele et al. 2015), which was imaged at high resolution by the ROLIS camera (Mottola et al.

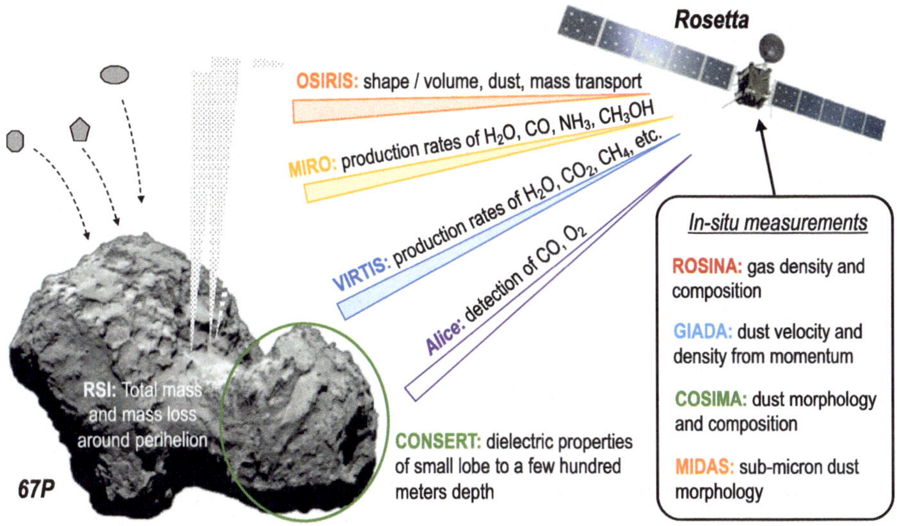


**Fig. 1** Schematic overview of measurements by *Rosetta* instruments that inform the parameters needed to derive the dust-to-gas and refractory-to-ice mass ratios of 67P

2015). At Abydos, *Philae*'s final resting place on 67P, the SESAME package (Seidensticker et al. 2007) made observations compatible with a meter-scale layer of compact materials in the subsurface via SESAME-PP (Lethuillier et al. 2016) and SESAME-CASSE (Knapmeyer et al. 2018). The locally high thermal inertia of $85 \pm 35\ \mathrm{J\,m^{-2}\,K^{-1}\,s^{-0.5}}$ ("high" by cometary standards only) measured around Abydos by the MUPUS-TM sensors (Spohn et al. 2015) suggests that this layer could be formed by sublimation and recondensation of water ice at depth within the pores, similarly to observations made in the KOmet-SImulationen (KOSI) laboratory experiments (Kochan et al. 1989; Grün et al. 1991).

## 2.2 Rosetta Measurements Relevant to Dust-to-Gas (Lost Materials) and Refractory-to-Ice (Nucleus) Mass Ratios

*Rosetta* was the first mission to Rendezvous with a Jupiter Family Comet and escort it from 3.6 AU pre-perihelion to 3.6 AU post-perihelion. It carried 11 instruments on its orbiter and deployed a fully instrumented lander on the surface of 67P. It was thus ideally equipped to study the surface, interior, and activity of 67P and their evolution through the active portion of this comet's orbit (Fig. 1).

The Radio Science Investigation, RSI (Pätzold et al. 2007), measured total mass and mass loss during the perihelion passage from radio tracking of the *Rosetta* spacecraft. OSIRIS (Keller et al. 2007) constrained the total volume of the nucleus, from which the nucleus density could be derived, and surveyed activity and mass transport processes. The Comet Nucleus Sounding Experiment by Radiowave Transmission, CONSERT (Kofman et al. 2007), used electromagnetic waves to sound the interior and derive dielectric properties of the nucleus. The Microwave Instrument on the Rosetta Orbiter, MIRO (Gulkis et al. 2007), the Visual InfraRed Thermal Imaging Spectrometer, VIRTIS (Coradini et al. 2007), and the *Alice* UV spectrograph (Stern et al. 2007) determined remotely the production rates of water and other volatiles, and their integration over the course of the mission. The Rosetta Orbiter Spectrometer for Ion and Neutral Analysis, ROSINA (Balsiger et al. 2007), monitored gas density and volatile abundances at the location of the spacecraft throughout the mission. The Grain Impact Analyzer and Dust Accumulator, GIADA (Colangeli et al. 2007), sensed dust grains and measured their velocity, cross-section, and mass. The COmetary Secondary Ion Mass Analyzer, COSIMA (Kissel et al. 2007), collected dust particles, imaged them and measured their composition using secondary ion mass spectrometry. The Micro-Imaging Dust Analysis System for the *Rosetta* Mission, MIDAS (Riedler et al. 2007), used atomic force microscopy to image the dust particles to sub-micron scale.

All these complementary measurements, illustrated in Fig. 1, provide various pieces of information needed to determine the dust-to-gas $\delta_{DG}$ mass ratio in the coma and the refractory-to-ice $\delta_{RI}$ mass ratio within the 67P nucleus. In the subsequent sections, we present the results and discuss sources of uncertainties related to the measurements themselves (calibration over long-term mission, potential sources of instrument bias) and the models that need to be used to derive these parameters from the data. Furthermore, *Rosetta* found that 67P's activity involves nucleus-scale mass transfer processes, see next section. These processes contribute to materials being emitted and redistributed around the nucleus. All these issues may affect the determination of the $\delta_{DG}$ and $\delta_{RI}$ mass ratios if they were not adequately considered.

## 2.3 Mass Transport Processes on 67P

The irregular shape and very high obliquity of 55° of 67P, combined with its highly elliptical orbit, are responsible for an asymmetric distribution of illumination from the Sun. This results in extreme seasonal variations between the Northern and Southern regions, with the Northern regions being subject to a long but mild summer around aphelion, while the Southern regions conversely undergo a brief but extremely intense summer around perihelion, e.g. Choukroun et al. (2015). Smooth plains observed mainly in the northern 67P hemisphere are evidence of dust redistribution on the nucleus, referred to as "airfall" (Thomas et al. 2015) and explained as a dominant mass transfer from South to North occurring mainly around perihelion (Keller et al. 2015, 2017).

Keller et al. (2015) investigated the complex shape of the nucleus and predicted the impact of insolation over the period of the mission, providing inferences of potential water production rates and surface erosion. Fougère et al. (2016a) used a Direct Simulation Monte Carlo (DSMC) model to invert the activity distribution from ROSINA data, shape model, and illumination, and found that $H_2O$ seemed to originate predominantly from the Northern regions, in particular Hapi, in the early phases of the mission (pre-perihelion).

As 67P approached perihelion, subsequent OSIRIS observations of the nucleus at visible wavelengths suggested a general increase in water ice content, likely associated with 67P shedding off its dust mantle and exposing fresher materials in particular in the Southern regions (Fornasier et al. 2016). This inference is consistent with previous comparisons between OSIRIS spectrophotometry data and VIRTIS-M IR spectra of exposed water ice (Filacchione et al. 2016a). These observations confirm strong erosion in the South during the perihelion passage.

The analysis of ROSINA/DFMS (Läuter et al. 2019) and MIRO (Biver et al. 2019) data throughout the *Rosetta* mission shows how the source locations of gases in the coma varied with seasons and insolation. The long-term monitoring of dust jets and their source location over the course of the mission (Shi et al. 2016; Vincent et al. 2016a; Schmitt et al. 2017; Shi et al. 2018) confirm that these features are driven by local illumination, and further supports the derivation of gas sources locations and their evolution. Violent outbursts have also been observed (Grün et al. 2016; Vincent et al. 2016b; Agarwal et al. 2017), however

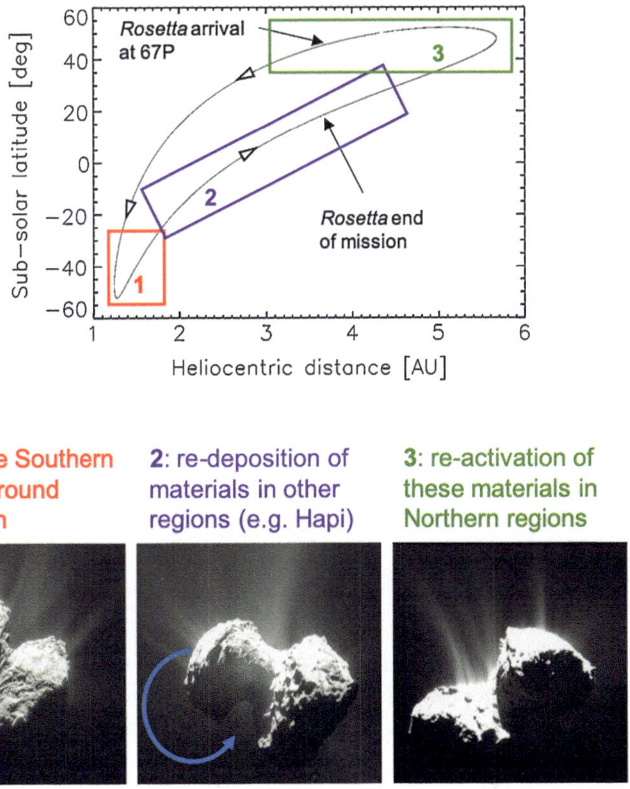


**Fig. 2** Schematic evolution in activity of 67P along its orbit. *Top:* evolution in sub-solar latitude of a sphere with the same orbital parameters and spin pole orientation as 67P to illustrate the extreme seasonal variability. Modified after (Choukroun et al. 2015). *Bottom:* Images of 67P's activity at locations along the orbit representative of the three periods mentioned (Credit: ESA/NAVCAM/OSIRIS). Note that the re-deposition of dm-size particles may occur within the perihelion window (1) and up to several months after perihelion (Bertini et al. 2018) within period (2). Period (3) consists of the Northern summer around aphelion and early during 67P's next approach to the Sun for particles deposited in the Hapi region. The period for reactivation of these large particles deposited elsewhere would depend on local illumination and their ice content

these features likely originate from other processes, such as thermal cracking, rupture of a pressurized gas reservoir, and/or large-scale collapse as observed by Pajola et al. (2017).

Hapi is also suggested as a location where sublimation from airfall particles (Thomas et al. 2015) emitted near perihelion would contribute significantly to the activity in the Northern regions around aphelion, during the northern summer (Keller et al. 2015). Large dm-size "chunks" of 67P materials could be emitted from the Southern regions near perihelion (Ott et al. 2017). These boulders could be transported and redeposited in Hapi and elsewhere, retain a significant portion of their ice content, and be later on activated once the northern regions exit polar night and are illuminated during the next approach towards the Sun (Fulle et al. 2017; Keller et al. 2017; Fulle et al. 2019). The enhanced backscattering ratio observed in dust particles in the inner coma of 67P during the post-perihelion phase by Bertini et al. (2018) is interpreted as consistent with this scenario.

Thus, 67P offers a picture of the complex processes playing a role in cometary activity, with a heterogeneity that is both spatial and temporal in nature. Figure 2 illustrates the evolution in subsolar latitude with heliocentric distance (in the approximation where 67P is a sphere, thus should not be taken literally), and evolution of activity and history of mass transport in this scenario. Potential sources of uncertainty and their impact on this scenario are discussed in Sect. 4.2.

# 3 Dust-to-Gas and Refractory-to-Ice Mass Ratios of Comet 67P

## 3.1 Derivation of the Dust-to-Gas and Refractory-to-Ice Mass Ratios

The dust-to-gas mass ratio in the lost materials of 67P, $\delta_{DG}$, can be derived from the total mass lost by the nucleus ($M_{Tot}$) over the course of the *Rosetta* mission, $(10.5 \pm 3.4) \times 10^9$ kg (Pätzold et al. 2019) and the integrated amounts of volatiles that left the nucleus ($M_{gas}$) over the same timespan:

$$\delta_{DG} = (M_{Tot} - M_{gas})/M_{gas} \tag{1}$$

Or, if considering measurements conducted at specific points in time $t$ (from instruments onboard *Rosetta* or ground- or space-based observations), then in the absence of any fallback materials $\delta_{DG}(t)$ would be related to the volatile and dust loss rates $Q_V$ and $Q_D$ at that time:

$$\delta_{DG}(t) = Q_D(t)/Q_V(t) \tag{2}$$

The latter approach has been most commonly used to derive the $\delta_{DG}$ (assumed constant) of comets to date from production rates at specific times, due to the observations available (Fulle et al. 2004; Sykes et al. 2004). However, the production rate approach is likely subject to large uncertainties at comets that exhibit strong seasonal variations in insolation, as $\delta_{DG}$ could change over time.

In addition, one must note that most of these previous measurements at other comets considered that the gas phase was only composed of water vapor. *Rosetta* enabled detailed measurements of the composition of several additional volatiles in the coma. For an appropriate comparison with previous measurements, the dust-to-water $\delta_{DG}^{H_2O}$ and dust-to-all-volatiles $\delta_{DG}^{V}$ ratios are treated separately hereafter.

In an ideal case where all dust particles in a given volume of the nucleus can be lifted by the sublimation of ices within that volume, and all these materials are lost, $\delta_{RI} = \delta_{DG} = (M_{Tot} - M_{gas})/M_{gas}$. However, deriving the refractory-to-ice mass ratio of the 67P nucleus, $\delta_{RI}$, is more difficult and prone to assumptions in the framework of a spatially and temporally varying activity. *Rosetta* showed that a substantial amount of materials emitted from the nucleus actually fall back onto the surface of 67P (Sect. 2.3).

The mass transfer processes observed by *Rosetta* at 67P strongly suggest a much more complex scenario, which is depicted in Fig. 3. We define the following sources of mass transfer (loss to space or fallback) around 67P: ${M_{gas}}^{PE}$ is the mass of gas primarily emitted directly from the nucleus and lost; ${M_{(d+i)}}^{PE}$ is the mass of dust (potentially still containing frozen volatiles) emitted directly then lost, which would after sublimation of those volatiles equate to a mass of refractory dust ${M_d}^{PE}$ and a mass of gas in extended sources ${M_{ext}}^{PE}$; ${M_{(d+i)}}^{FB}$ is the mass of dust (and possibly ices) that falls back on the nucleus; ${M_{ext}}^{FB}$ is the amount of gas emitted from large (~dm) particles that fall back onto the nucleus but sublimate a portion of their internal ices during their journey in the inner coma; ${M_{gas}}^{RE}$ is the mass of gas re-emitted subsequently from the sublimation of ices within the particles that have fallen back on the nucleus; ${M_d}^{RE}$ is the mass of refractory materials subsequently emitted upon reactivation.

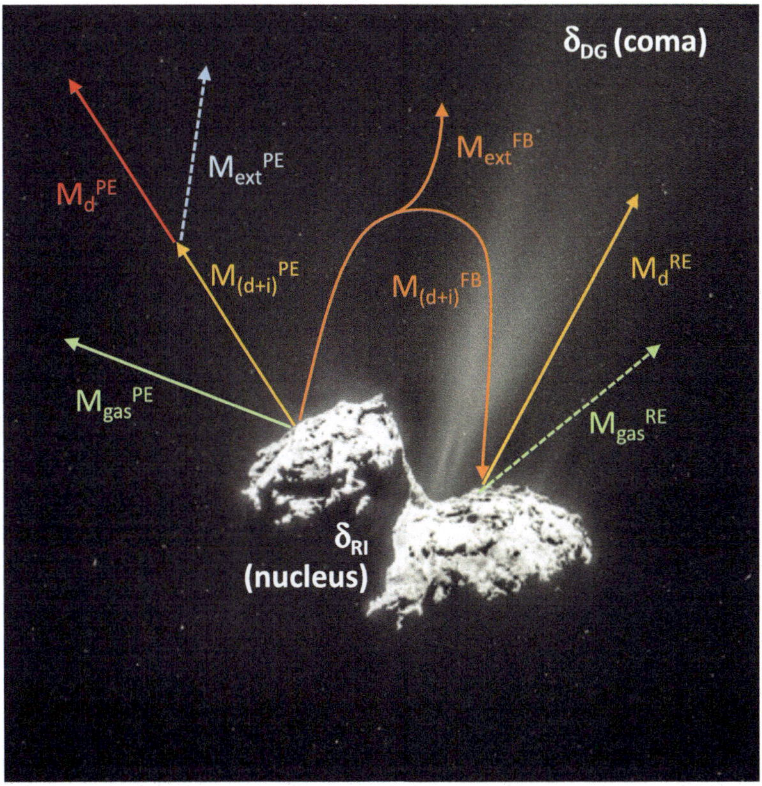


**Fig. 3** Sketch of the quantities involved in mass transfer processes around 67P along its orbit. All these quantities contribute to the dust-to-gas $\delta_{DG}$ and refractory-to-ice $\delta_{RI}$ mass ratios determined in the lost materials and the nucleus, respectively. Depending on times in the mission and activity, the *Rosetta* spacecraft may be located inward or outward of the location where ice embedded in decimeter-size particles (coma boulders) would be completely sublimated, i.e. where dust particles (perhaps with ice) $M_{(d+i)}$ turn into dry dust $M_d$ and possible extended sources $M_{ext}$, whether these particles are lost to space at primary emission (superscript "PE") or fall back on the nucleus and may be reactivated later (superscripts fallback "FB" and re-emitted "RE"). See text for details

None of these individual parameters are directly accessible from measurements made by *Rosetta*. What can be assessed, albeit with rather large uncertainties, is the sum of all material lost during the mission from the comet:

$$M_{Tot} = {M_{gas}}^{PE} + {M_d}^{PE} + {M_{ext}}^{PE} + {M_{ext}}^{FB} + {M_{gas}}^{RE} + {M_d}^{RE} \tag{3}$$

by considering the measured change in nucleus mass over the mission. In addition, the extent of fallback materials ${M_{(d+i)}}^{FB}$ can be estimated from the observed transport of dm-size particles around perihelion (Fulle et al. 2017; Ott et al. 2017; Fulle et al. 2019). The total mass loss and reasonable ranges of dust-to-gas and refractory-to-ice mass ratios suggest that fallback materials onto the nucleus may amount to multiple times the lost mass, see Sect. 3.2 (Pätzold et al. 2019). The total loss of volatiles ($M_{gas}$ from Eq. (1)) can also be determined by integrating the in-situ and remote sensing measurements made throughout the 2015 apparition of 67P, see Sect. 3.4.

Using production rates measured at discrete times in the mission to derive the refractory-to-ice mass ratios in the nucleus might be tempting, as suggested above for coma measurements, but this approach has the significant drawback that the respective mass contributions of fallback and re-emitted dust may take place at different times along the orbit of 67P. Thus, integrated mass losses in Eq. (3) cannot be simply replaced by production rates. However, as suggested in Fulle et al. (2019), one may do so near perihelion where most of the activity takes place, and then derive the amount of fallback materials to retrieve their refractory-to-ice mass fraction at that time.

## 3.2 Total Mass and Mass Loss Due to 2015 Perihelion Passage

67P's gravity field was measured from close distances in August-November 2014 by RSI, allowing to determine the total mass of the 67P nucleus: $(9.982 \pm 0.003) \times 10^{12}$ kg (Pätzold et al. 2016). Assuming a uniform nucleus, this mass and the 67P nucleus volume, first estimated by Preusker et al. (2015) and (Jorda et al. 2016), then refined by (Preusker et al. 2017) to $18.56 \pm 0.02\ \text{km}^3$, yield a bulk density of $537.8 \pm 0.6\ \text{kg m}^{-3}$ (Pätzold et al. 2019).

Additional RSI measurements have been conducted in July-September 2016 (after the August 2015 perihelion passage), which allow to constrain the total mass lost by the 67P nucleus over the course of the *Rosetta* mission: $(10.5 \pm 3.4) \times 10^9$ kg (Pätzold et al. 2019), i.e. 0.1% of the nucleus mass. The bulk density of the nucleus post-perihelion, $537.3 \pm 0.6\ \text{kg m}^{-3}$, remains within errors consistent with the value determined from pre-perihelion measurements.

Pätzold et al. (2019) discuss ranges of porosity and ice contents compatible with the bulk mass and density values from RSI data. For a range of compact dust material density from 2000 to 3500 kg/m3, the porosity varies between 65%–79% when the refractory-to-ice mass ratio $\delta_{RI}$ for the nucleus lies in the range 3–7 (Pätzold et al. 2019). Values as low as 1 cannot be excluded but are not favored in their analysis. The nucleus is interpreted as a highly porous dusty body with low ice content.

By comparing the total mass loss with water production rate from ROSINA and MIRO (Sect. 3.4), Pätzold et al. (2019) suggested a dust-to-gas mass ratio of 0.5 to 1. From this dust-to-gas mass ratio and the range of refractory-to-ice mass ratios, they derived a range of plausible mass of fallback materials of 1.8 to 4.8 times the mass lost by 67P around the 2015 perihelion passage.

## 3.3 Dielectric Properties of the 67P Nucleus

CONSERT probed 67P's small lobe ("head") in the vicinity of Abydos up to a propagation length (along the line-of-sight *Philae-Rosetta*) of about 1 km and a penetration depth of about 100 m. The measured average dielectric permittivity $\varepsilon = 1.27 \pm 0.05$ shows a highly porous nucleus (Kofman et al. 2015).

The composition of the nucleus interior deduced from this low permittivity (Hérique et al. 2016) suggests an organic-rich refractory part of the comet, consisting of at least 75% volume fraction of organics (66% mass fraction) and of less than 25% silicates, which is consistent with surface observations from other *Rosetta* instruments (Sect. 2.1). This leads to a nucleus with an ice volume fraction ranging from 6% to 11%, a refractory volume fraction from 16% to 21% and a porosity from 73% to 76%. It is however worth noting a discrepancy between the organic matter abundance from CONSERT data and that measured in dust grains collected in the coma by COSIMA, which is around 45% by mass (Bardyn et al. 2017; Fray et al. 2017), see Sect. 3.5.

**Table 1** Summary of water mass loss estimates derived from instrument measurements

| Reference | Total water mass loss (kg) | Peak production rate (molecules/s) | Method | Period |
|---|---|---|---|---|
| (Hansen et al. 2016) | $6.4 \times 10^9$ | $(3.5 \pm 0.5) \times 10^{28}$ | Interpolation of COPS data | 18-22 days after perihelion |
| (Läuter et al. 2019) | $(4.8 \pm 1.5) \times 10^9$ | $(2.1 \pm 0.1) \times 10^{28}$ | Analysis of DFMS data | 17-28 days after perihelion |
| (Marshall et al. 2017) | $(2.4 \pm 1.1) \times 10^9$ | $(1.42 \pm 0.51) \times 10^{28}$ | Analysis of MIRO data | 29/8/2015 |
| (Biver et al. 2019) | $(2.4 \pm 0.2) \times 10^9$ | $(0.8–1.1) \times 10^{28}$ | Analysis of MIRO data | 14-20 days after perihelion |
| (Shinnaka et al. 2017) | N/A | $(1.46 \pm 0.47) \times 10^{28}$ | Lyman $\alpha$ data | 7/9/2015 |
| (Fougère et al. 2016a) | N/A | $5.5 \times 10^{27}$ | DSMC calculations and VIRTIS-H data | 9/8/2015 |

CONSERT measurements suggest that the refractory-to-ice mass ratio of 67P, $\delta_{RI}$, is larger than 3. This is an average value within the portion of the 67P nucleus probed by CONSERT and cannot constrain local variation of the ratio (Hérique et al. 2019). It constitutes a lower limit, with all uncertainties included, and results from modelling. Large refractory-to-ice mass ratios are required to explain the low average permittivity measured by CONSERT: the constitutive materials must fit the 1.27 value and have a high enough density for small volume fractions to fit the 67P average density as derived from RSI. It is the case for some organics ($\varepsilon \approx 2$, $\rho \approx 2$), less for silicates ($\varepsilon \approx 5 \sim 7$, $\rho \approx 3.5$), and definitively not the case for ices ($\varepsilon \approx 3.1$, $\rho \approx 1$ for $H_2O$ ice; $\varepsilon \approx 2.7$, $\rho \approx 0.94$ for a mixture of 71% $H_2O$, 14% $CO_2$ and 15% CO in mass), as discussed in Fulle et al. (2019).

## 3.4 Mass Loss of Major Volatiles

### *3.4.1 $H_2O$*

Water was directly observed throughout the mission by the MIRO, VIRTIS and ROSINA instruments. ROSINA measured local water number densities at the spacecraft, while the remote sensing instruments provided line-of-sight integrated information of the water distribution. The interpretation of these measurements is complex and the determination of water sources and total production rate requires extensive modelling (Sect. 4.1.3). We summarize in Table 1 estimates of the peak water production rate and total water mass loss obtained by these instruments.

Hansen et al. (2016) analyzed ROSINA measurements from June 2014 to May 2016 using an empirical model of the water coma derived from the Direct Simulation Monte Carlo model (DMSC) of Fougere et al. (2016b), which was constrained by a restricted set of ROSINA data. Most recently, analyses of ROSINA data identified surface location of gas sources and derived production rates from August 2014 to September 2016, encompassing the entire comet close proximity operations phase of *Rosetta* (Läuter et al. 2019; Combi et al. 2020).

Marshall et al. (2017) used MIRO observations of $H_2^{16}O$ and $H_2^{18}O$ acquired in nadir viewing from August 2014 to April 2016. The $H_2^{16}O$ line and, to a lesser extent, the $H_2^{18}O$

line observed by MIRO, are optically thick and require radiative transfer calculations. Marshall et al. (2017) relied on the $H_2^{16}O/H_2^{18}O$ line ratio as the key variable to infer the water production rate. The analysis of MIRO observations by Biver et al. (2019) uses a different approach. The $\sim$ 100 inner coma maps obtained from July 2014 to June 2016 were analyzed using a radiative transfer model, describing the outgassing pattern by a Gaussian shape. Both the line intensity and spectral shape were combined to derive the latitude of the water emitting region and total water production rate. Water production rates derived from VIRTIS observations are still sparse and published values rely on DSMC calculations using activity maps constrained by ROSINA data (Bockelée-Morvan et al. 2015; Fougère et al. 2016a; Fougere et al. 2016b).

Remote observations from spacecraft far from the comet also constrained water mass loss rates from 67P at various points in time. *SOHO* (SOlar Heliospheric Observatory) SWAN (Solar Wind ANisotropies) Lyman $\alpha$ observations were presented in Bertaux (2015), and by Shinnaka et al. (2017) using *PROCYON*/LAICA (PRoximate Object Close flYby with Optical Navigation/Lyman Alpha Imaging Camera) (Funase et al. 2015).

Table 1 illustrates a disparity in water mass loss estimates, with results that appear incompatible when considering their error bars. These measurements are used to constrain the dust-to-water ratio in materials emitted by 67P in Sect. 3.4.3, after considering other volatiles, see next section. Further discussion of the approaches used to derive these values and their respective sources of uncertainties is provided in Sect. 4.1.3.

### 3.4.2 Other Volatiles

The analysis of ROSINA/DFMS data by Läuter et al. (2019) retrieved mission-integrated total mass losses for $CO_2$, CO, and $O_2$, amounting altogether to $(10.1 \pm 3.7) \times 10^8$ kg. These results are comparable to relative abundances to $H_2O$ determined previously at fixed points in time (Bieler et al. 2015a, 2015b; Gasc et al. 2017). The measured relative abundances to water of sulfur-bearing molecules, 5% (Calmonte et al. 2016), hydrocarbon molecules, 5% (Schuhmann et al. 2019), $CH_4$, $NH_3$ and $CH_3OH$ together $\sim$2%, amount to $(5.8 \pm 1.8) \times 10^8$ kg, where the uncertainty is primarily driven by that on water production. Adding the integrated masses of $CO_2$, CO, and $O_2$ (Läuter et al. 2019) yields a mass loss of volatiles other than water of $(15.9 \pm 5.5) \times 10^8$ kg. Hence, the total mass loss of volatiles from in-situ measurements is $(6.4 \pm 1.9) \times 10^9$ kg.

The recent analysis of MIRO data throughout the 67P rendezvous phase (Biver et al. 2019) included all compounds accessible at the fixed-tuned frequencies of the MIRO Chirp Transform Spectrometer: $H_2O$, $CH_3OH$, $NH_3$, and CO. In addition, VIRTIS-H data allowed to determine the relative abundances to water of $CO_2$, $CH_4$, and OCS around perihelion and their variations (Bockelée-Morvan et al. 2016). $CH_4$ and OCS have very low abundances relative to $H_2O$, thus would not contribute significantly to the total volatile losses. Conversely, the $CO_2$ abundance relative to $H_2O$ of $\sim$ 15%, with a large distinction between pre-perihelion and post-perihelion, makes it the second most abundant volatile at 67P (Bockelée-Morvan et al. 2016). By combining the MIRO and VIRTIS measurements, and including relative abundances of additional species to water as done above, Biver et al. (2019) report a total mass loss of volatiles from remote sensing and MIRO map data only of $(4.2 \pm 0.2) \times 10^9$ kg, and a possible range $(4.0–5.8) \times 10^9$ kg when including MIRO nadir observations. However, greater uncertainties are associated with retrievals from nadir observations, thus the high end of the range is deemed less likely. Biver et al. (2019) also derived a dust-to-gas ratio for 67P of $1.5 \pm 0.8$.

Analysis of *Alice* remote sensing UV data yielded much higher $O_2$ abundances relative to water (Keeney et al. 2017) than the ROSINA/DFMS analysis (Läuter et al. 2019), up to

60% $O_2/H_2O$. A more recent analysis of *Alice* stellar occultation data conversely yielded $O_2/H_2O$ abundances comparable to those measured by ROSINA (Keeney et al. 2019). The source of these differences is not understood yet, although high-energy electron impact not accounted for in the original analyses has been suggested.

### 3.4.3 Dust-to-Water and Dust-to-Gas Ratios $\delta_{DG}{}^{H_2O}$ and $\delta_{DG}{}^{V}$

We determine the dust-to-all-volatiles mass ratio $\delta_{DG}{}^{V}$ by combining the volatiles mass losses from in-situ measurements of $(6.4 \pm 1.9) \times 10^9$ kg, and from remote sensing of $(4.2 \pm 0.2) \times 10^9$ kg (MIRO maps) with a possible range $(4.0\text{–}5.8) \times 10^9$ kg (Biver et al. 2019), with the total nucleus mass loss of $(10.5 \pm 3.4) \times 10^9$ kg (Pätzold et al. 2019). From these values, we first derive the dry dust mass loss in the two cases, $M_d$(in-situ) of $(4.1 \pm 3.9) \times 10^9$ kg and $M_d$(remote sensing) of $(6.3 \pm 3.4) \times 10^9$ kg. Volatiles loss from remote sensing (MIRO maps) yields a $\delta_{DG}{}^{V}$(remote sensing) of 1.5 for average values and a range that spans 0.7–2.3 when accounting for 1-sigma uncertainties. Including nadir observations extends the low end of the range and yields 0.2–2.3. Volatiles loss from in-situ measurements yields an average $\delta_{DG}{}^{V}$ (in-situ) of 0.64 and a range 0.01–1.28.

We then determine the dust-to-water mass ratio $\delta_{DG}{}^{H_2O}$ by combining the water mass losses from in-situ measurements of $(4.8 \pm 1.5) \times 10^9$ kg (Läuter et al. 2019) and remote sensing of $(2.4 \pm 0.2) \times 10^9$ kg (Biver et al. 2019) with the dust mass loss as calculated above for in-situ and remote sensing measurements. Water mass loss from in-situ yields a $\delta_{DG}{}^{H_2O}$ (in-situ) of 0.85 for average values and a range that spans 0.01–1.71 accounting for 1-sigma uncertainties. Similarly, water mass loss from remote sensing yields a $\delta_{DG}{}^{H_2O}$ (remote sensing) of 2.6 for average values and a range that spans 1.2–4.0 when accounting for 1-sigma uncertainties.

## 3.5 Dust Particles: Properties and Fallback

### 3.5.1 Dust Particles Characteristics

Observations of dust particles by GIADA, COSIMA, and OSIRIS yield overlapping dust size distributions up to m-sizes (Fulle et al. 2016a, 2016b; Merouane et al. 2016; Ott et al. 2017). The reader is referred to Güttler et al. (2019) for a synthesis of dust particle types, substructures, characteristics, and nomenclature.

GIADA measured directly the momentum and the speed of individual dust particles with particle sizes ranging from about 50 μm up to about 1 mm. Combining these two measurements, the mass of each particle was derived. The cross-section of each particle was derived from calibrated optical measurements (Della Corte et al. 2016), and together with the mass measurement, the particle density was derived (Rotundi et al. 2015; Fulle et al. 2016a, 2016b; Fulle et al. 2017). In addition, the GIADA microbalance system measured the integrated mass (1 reading every 300s) of particles $<5$ μm (Della Corte et al. 2019).

GIADA observations have revealed two distinct dust populations: a 'fluffy' one with size range 0.2–2.5 mm and of density $<1\ \mathrm{kg\,m^{-3}}$, and more compact particles 80–800 μm of density $(1.9 \pm 1.1) \times 10^3\ \mathrm{kg\,m^{-3}}$ (Della Corte et al. 2015; Rotundi et al. 2015). Following acquisition of data throughout the *Rosetta* comet operations at 67P, the bulk density of dust particles collected by GIADA was subsequently updated to $785^{+520}{}_{-115}\ \mathrm{kg\,m^{-3}}$ by Fulle et al. (2017), from which a volumetric nucleus composition of $54 \pm 5\%$ hydrocarbons, $22 \pm 2\%$ silicates, $4 \pm 1\%$ sulfides and $20 \pm 8\%$ ices was deduced.

COSIMA measured directly the composition of dust particles via secondary ion mass spectrometry (SIMS) and found average relative proportions by weight of 45% for organic and 55% for inorganic matter, or 55% and 45% by volume, respectively, with about 30% variation between individual particles (Bardyn et al. 2017). Thus, the interpretations of GIADA data and the measurements made by COSIMA are compatible within their uncertainties.

Langevin et al. (2016) report a detailed typology of dust particles, enabled by COSIMA/COSISCOPE, identifying a variety of clusters or aggregates of particles in the range 10 – several 100 μm constituting 85% of their collected particles, while the remaining 15% are made up of >100 μm compact particles that did not fragment on impact with the collection targets. Some of these compact particles subsequently fragmented during SIMS analyses, owing to Lorentz forces caused by particle charging (Hilchenbach et al. 2017), while others were lifted and could not be analyzed.

A set of recent numerical simulations also suggested that part of the COSIMA collection is compatible with "fluffy" particle simulations, with either two populations with fractal dimensions <2 and about 2.6 and one impact velocity or one particle population with a fractal dimension of 2.6 and a spread of impact velocities on collection (Lasue et al. 2019).

The particles collected by COSIMA were assumed ice-free, as the instrument was kept at about 10 °C within the sample storage portion, and particles were analyzed days to weeks after their collection. The particles were illuminated in sequence with a red LED (640 nm) at phase angles between 72 and 84 degrees from two opposite directions. The particles showed a surprisingly diverse reflectance factor in the range 3–22% (Langevin et al. 2016, 2017). The measured reflectance, in particular the comparison between left and right illumination, was explained by scattering centers inside the agglomerate volume and an optical mean free path in the 20–25 μm range. The required porosity depends on the size of the scattering centers and is estimated to be in the 50–90% range (Langevin et al. 2017).

The aggregate nature of the particles from 67P extends below 10 μm size range according to measurements by MIDAS (Riedler et al. 2007; Bentley et al. 2016), and indicates a fractal-like structure in some particles (Mannel et al. 2016). COSIMA and MIDAS dust observations suggest similarities with IDPs (Schulz et al. 2015; Bentley et al. 2016).

### 3.5.2 $\delta_{DG}$ and $\delta_{RI}$ from Dust Particles

GIADA data and the derived density of dust particles have enabled several determinations of dust-to-gas and refractory-to-ice mass ratios of 67P while considering various time periods over the course of the *Rosetta* mission. Early estimates of the dust-to-water $\delta_{\mathrm{DG}}{}^{\mathrm{H_2O}}$ of 6 $\pm$ 2 and dust-to-all-volatiles $\delta_{\mathrm{DG}}{}^{\mathrm{V}}$ of 4 $\pm$ 2 were first derived from GIADA and MIRO measurements around 3.6 AU pre-perihelion in late 2014 (Rotundi et al. 2015). Updated dust and gas mass loss estimates from that time through perihelion suggested an average $\delta_{\mathrm{DG}}{}^{\mathrm{H_2O}}$ of 6, and a minimum value of 5 and maximum of 10 or greater (Fulle et al. 2016a, 2016b). Considering GIADA data obtained throughout the comet operations phase of *Rosetta*, Fulle et al. (2017) then derived an average refractory-to-ice mass ratio $\delta_{\mathrm{RI}}$ of 7.5 and within the range of 3 or greater.

COSIMA measurements also enabled estimating the density of dust particles (Hornung et al. 2016), from which the potential original ice content after collection and sublimation, assumed to be representative of the refractory-to-ice mass ratio within the nucleus, can be derived. The following analysis could not be made in that publication because the density of the 67P nucleus was only constrained later in the mission (Pätzold et al. 2016, 2019).

In the COSIMA dust particle collection, all particles can be broken up by impact or electrostatic forces into elements, and can be considered as agglomerates of smaller elements shaped by hierarchical growth. The particles show sub-structures down to the instrument optical resolution limit of 14 μm/pixel. This indicates that the sub-structures, possibly again agglomerates, stick together with higher forces than the larger-scale agglomerates. To infer mechanical properties from the impact fragmentation, Hornung et al. (2016) used this model of agglomerates and inferred the strength limits for the fragmented and non-fragmented particles. The strength limits are determined by the binding force between sub-agglomerate structures or elements, and are around 1000 Pa for elements in the 10–40 μm size range.

A macroscopic porosity of $0.5 \pm 0.1$ is derived from observations of sub-structures (or elements of non-fragmented agglomerates) in the 60–300 μm size range. These elements are then assumed to be porous with the smallest solid unit of $\sim 0.2$ μm diameter (Greenberg and Li 1999). This is formally described by their size-dependent filling factor $\varphi \propto r^{-0.4}$, as implied by a hierarchical cascade on scales down to the solid grain. The fragmentation model constrains strength boundaries rather than void spaces.

Evaluating impact dynamics, size distribution of the fragments, and underlying hierarchical cascade of sub-structures and porosities, the density of the ice-free dust particles $\rho_D$ collected by COSIMA is in the range 100–400 $kg\,m^{-3}$ (Hornung et al. 2016). This value is lower than the mean nucleus density, $\rho_{67P} = 537.8 \pm 0.6\ kg\,m^{-3}$ (Pätzold et al. 2019).

Modeling the filling of the COSIMA dry dust particles with ice and deriving the putative refractory-to-ice mass ratio of these particles, assumed to be representative of the 67P nucleus, is relatively straightforward in that context. Assuming an average density of the monomers within dry dust particles $\rho_M$, which are mixtures of silicates, sulfides, and organic materials as described above, the COSIMA-scale (microscopic) porosity of the dry particles is $\phi_P = 1 - \rho^P/\rho_M$, and the volume fraction (not to be confused with the filling factor) of compact dry dust monomers is $f_M = \rho_P/\rho_M$. Given the density of compact water ice $\rho_I$ and the density of the nucleus $\rho_{67P}$, the volumetric fraction of porous ice $f_I$ that needs to be added to the pores to match the density of the 67P nucleus is: $f_I = \frac{\rho_{67P} - \rho_P}{\rho_I}$.

The refractory-to-ice ratio of the particles can then be derived in the scenario of ice filling porosity within dust particles. Indeed, the relative volumetric fractions of solid ice $f_I$ and of dust monomers $f_M$ that match the bulk density of the 67P nucleus are known: $\rho_{67P} = f_M \rho_M + f_I \rho_I = \rho_P + f_I \rho_I$. So are their approximate densities, of $\rho_M \sim 2500\ kg\,m^{-3}$ and $\rho_I \sim 920\ kg\,m^{-3}$. In this scenario where ice fills the porosity within dust particles, themselves consisting of aggregates of compact monomers up to the size of the nucleus (i.e. not considering any macroporosity), the refractory-to-ice mass ratio $\delta_{RI}$ is:

$$\delta_{RI} = \frac{f_M}{f_I} \cdot \frac{\rho_M}{\rho_I} = \frac{\rho_P}{\rho_{67P} - \rho_P} \tag{4}$$

Adding ice to the ice-free porous dust particles, of density in the range 100–400 $kg\,m^{-3}$ (Hornung et al. 2016), results in a porosity of the icy particles in the range of 0.43 to 0.7, see Fig. 4. A direct transposition of these results to the 67P nucleus implies a refractory-to-ice mass ratio $\delta_{RI}$ in the range 0.2 to 3, in stark contrast with the estimates derived with dust densities in the mm size range larger than the mean nucleus density (Fulle et al. 2016a, 2016b; Fulle et al. 2017). The latter would then require further macroscopic porosity considerations to sizes much larger than the dust collected by COSIMA, in which case the ice content of the nucleus may not be reliably constrained from the ice-free dust particle density shown above.

The difference between the refractory-to-ice estimates from dust particles based on GIADA and COSIMA arises from the density of dust particles. GIADA data suggest dust particles denser than the nucleus itself (which then requires a substantial macroporosity),

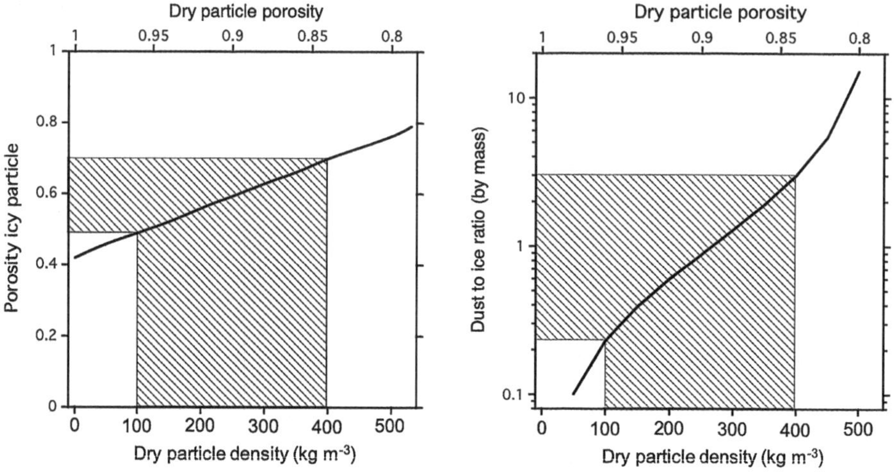


**Fig. 4** Porosity of icy particles as derived from the ice-free particle density and the mean nucleus density (left), and refractory-to-ice mass ratio as function of the dry dust particle density and the mean nucleus density (right). The potential parameter ranges based on the COSIMA dust collection are indicated by the shaded areas

while COSIMA data suggest dry dust particles less dense than the nucleus. In the latter case, ice is added until the density of dust particles prior to collection is equal to that of the 67P nucleus. Possible reasons for the discrepancy between GIADA and COSIMA dust density estimates are briefly mentioned in Levasseur-Regourd et al. (2018) and Güttler et al. (2019). Further discussion on potential collection biases and uncertainties associated to the density derivation principles can be found in Sect. 4.1.4.

### *3.5.3 Fallback Materials and Impact on Refractory-to-Ice Mass Ratio*

Early estimates of dust mass loss were around 15 $kg\,s^{-1}$ at 2.9 AU (Moreno et al. 2016), rising by a few orders of magnitude during perihelion, where dust mass loss rates of $(1.7 \pm 0.9) \times 10^3\ kg\,s^{-1}$ (Fulle et al. 2016a, 2016b) have been reported for boulders not all on radial trajectories. Subsequently, an estimate of up to $8.3 \times 10^3\ kg\,s^{-1}$ is found for boulders that are on radial trajectories (Ott et al. 2017). These high dust production rates are such that the total nucleus mass loss measured by RSI corresponds to 71.5 and 14.6 days of activity at the Fulle et al. (2016a, 2016b) and Ott et al. (2017) rates, respectively. The above estimates neglect any ice or water vapor emission, thus are upper bounds on the duration of the activity peak that would be commensurate with these data.

Given the commonly accepted overall activity scenario for 67P, see Fig. 2, and the OSIRIS coma dust data from Bertini et al. (2018), a significant fallback deposition that is still rich in volatiles seems needed to sustain the cometary activity of 67P (Fulle et al. 2019). One might expect the materials contained in this fallback to be partially depleted in the more volatile species, given the lower $CO_2/H_2O$ ratio emitted from the Northern regions as seen from early parts of the mission (Bockelée-Morvan et al. 2015; Hässig et al. 2015) through perihelion (Bockelée-Morvan et al. 2016).

However, the abundances of the super volatile CO and the volatile $CO_2$ relative to water derived from the South pre-perihelion in May 2015 (Rubin et al. 2019), probably representative for the overall nucleus composition, are not higher than in the northern regions when

**Table 2** Summary of estimates of $\delta_{DG}^{H_2O}$, $\delta_{DG}^{V}$, and $\delta_{RI}$ for comet 67P. Values in parentheses are lower bound estimates that cannot be excluded but are not preferred in the original analyses. A "+" indicates that a higher upper bound cannot be excluded. "$N/A$" = not applicable. "*inf.*" = results could be compatible with an infinitely high value, although a minimal fraction of ice is needed

| Reference | Average | Min | Max | Method | Period |
|---|---|---|---|---|---|
| ***Dust-to-Water mass ratio in the coma*** | | | | | |
| Rotundi et al. (2015) | 6 | 4 | 8 | GIADA dust + MIRO gas | 3.6 AU pre-perihelion |
| Fulle et al. (2016) | 6 | 5 | 10+ | GIADA dust densities | 3.6 AU through perihelion |
| Remote sensing | 2.6 | 1.2 | 4 | Biver et al. (2019), Section 3.4 | Mission-integrated |
| In-situ | 0.85 | 0.01 | 1.71 | Section 3.4 | Mission-integrated |
| ***Dust-to-Volatiles mass ratio in the coma*** | | | | | |
| Rotundi et al. (2015) | 4 | 2 | 6 | GIADA dust + MIRO gas | 3.6 AU pre-perihelion |
| Remote sensing | 1.5 | 0.7 (0.2) | 2.3 | Biver et al. (2019), Section 3.4 | Mission-integrated |
| In-situ | 0.64 | 0.01 | 1.28 | Section 3.4 | Mission-integrated |
| ***Refractory-to-Ice(s) mass ratio in the nucleus*** | | | | | |
| Hérique et al. (2016) | *N/A* | 3 (2.2) | 6+ | Nucleus dielectric permittivity | *Philae* lander (Nov. 2014) |
| Fulle et al. (2017) | 7.5 | 3.1 | *inf.* | GIADA dust densities | Mission-integrated |
| Fulle et al. (2019) | *N/A* | 3 | *N/A* | "Chunks" fallback estimates | Perihelion + fallback |
| Pätzold et al. (2019) | *N/A* | 3 (1) | 7 | Nucleus and gas mass loss | Mission-integrated |
| Alternate determination | *N/A* | 0.2 | 3 | COSIMA particles, Section 3.5 | Mission-integrated |

far from the Sun (Le Roy et al. 2015). This suggests that either ices more volatile than water have been retained in the fallback material, or the fallback layer in the north is not thick enough to dampen the release of CO and other volatiles from below. The high $CO/H_2O$ and $CO_2/H_2O$ in the South early in the mission could then be explained by low water emission due to frigid polar night temperatures, as suggested from MIRO observations (Biver et al. 2015; Choukroun et al. 2015).

Fallback materials are likely dominated in mass by dm-size particles, observed in the inner coma of 67P (Rotundi et al. 2015) and suggested to settle slowly during the post-perihelion portion of the orbit (Bertini et al. 2018). This large size would allow them to preserve ices within their interior, and enable their reactivation later on (Fulle et al. 2019). Fulle et al. (2019) suggest that the inferred original composition of these "chunks" may be representative of the interior of the 67P nucleus and indicates a refractory-to-ice mass ratio of 3 or more.

### 3.6 67P's Dust-to-Gas and Refractory-to-Ice Mass Ratio Estimates

Table 2 summarizes published values of the dust-to-water, dust-to-gas, and refractory-to-ice(s) mass ratios, and adds new determinations for these parameters based on recent literature data considered in this article (see previous sections). It describes the methods used to determine these parameters and the period at which they have been conducted. The wide ranges of values of individual determinations, as well as the very restricted regions where independent determinations overlap, are quite striking. Figure 5 is a graphical representation of the data shown in Table 2, in an attempt to tease out further information from these estimates.

The top portion of Fig. 5a presents dust-to-water mass ratios in the lost materials, while its bottom portion includes all volatiles. The total mass losses, integrated over the 2015 apparition of 67P, of $H_2O$, other volatiles, and nucleus as a whole (Sect. 3.4), yield the lowest dust-to-water and dust-to-volatiles mass ratios.

Figure 5b shows published values of the refractory-to-ice mass ratio within 67P, based on the analyses of GIADA and OSIRIS dust data (Fulle et al. 2017, 2019), of RSI data (Pätzold et al. 2019), CONSERT data (Hérique et al. 2016), and the analysis of dry dust

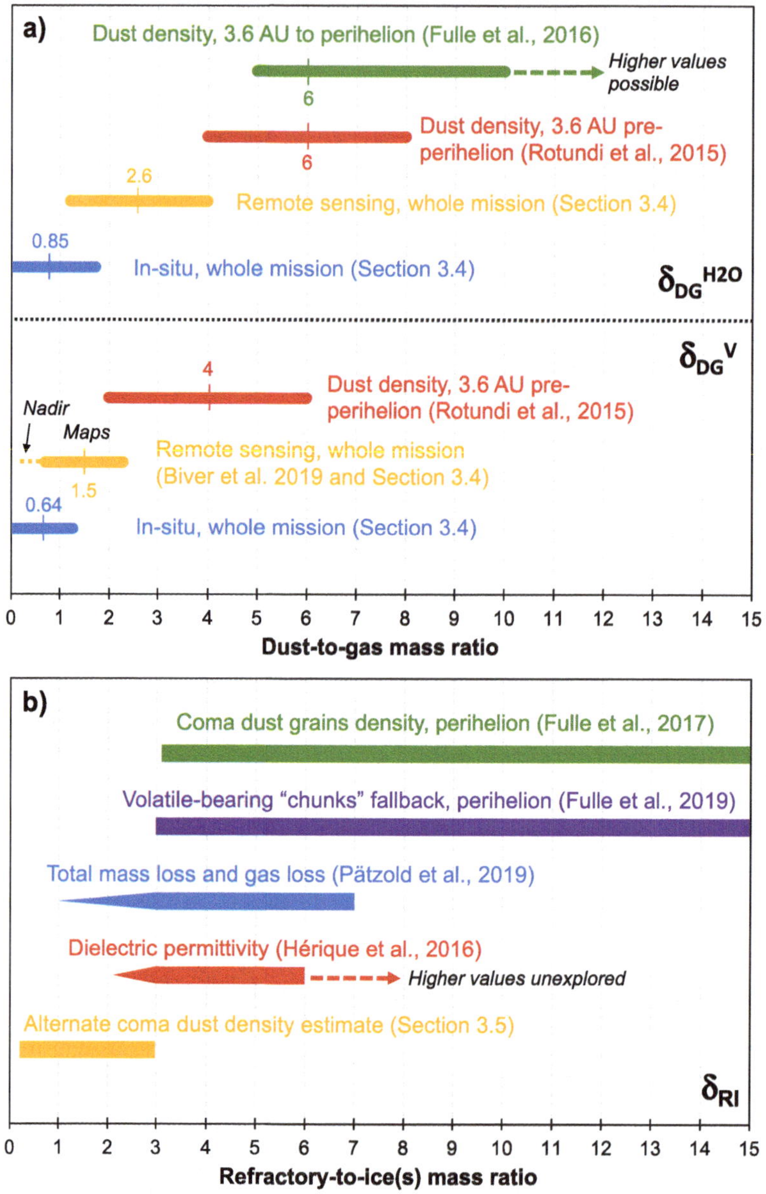


**Fig. 5** Summary of ranges of plausible values derived for the dust-to-gas $\delta_{DG}$ (panel **a**) and the refractory-to-ice $\delta_{RI}$ (panel **b**) mass ratios in 67P's lost materials and nucleus, respectively. In panel **a**) the mean value of dust-to-water or dust-to-all-volatiles mass ratios is indicated numerally, in addition to entire plausible ranges based on error bars of individual parameters. In panel **b**) the triangulated ends indicate possible lower values that are not favored in the interpretations. In both panels, the X-axis scale is limited at 15 for legibility. See text for details

density from COSIMA data based on the morphology of collected particles (Hornung et al. 2016), see Sect. 3.5.2. As in the case of the dust-to-gas ratio, the individual estimates differ substantially.

The dust-to-all-volatiles mass ratio $\delta_{DG}{}^{V}$ is the mass ratio that would actually be expected to correspond most directly to the refractory-to-ices mass ratio within the nucleus of 67P, except if ice is largely incorporated in coma boulders that might act as extended sources of water. Indeed, it includes all the volatile species that contribute to both dielectric properties as determined by CONSERT and total mass loss around the 2015 perihelion passage measured by RSI.

Furthermore, an additional complication arises from the timing of the derived values. The $4 \pm 2$ value of $\delta_{DG}{}^{V}$ (Rotundi et al. 2015) was obtained early in the mission, and differs from the mission-integrated $\delta_{DG}{}^{V}$ determinations. The dust-to-gas ratio may have changed over the course of the mission (Fulle et al. 2016a, 2016b). And/or the activity of 67P and volatile-bearing fallback materials could result in different values/ranges for $\delta_{DG}{}^{V}$ and $\delta_{RI}$ (Fulle et al. 2019).

Presently, there are two opposed interpretations of the refractory-to-ice mass ratio estimates shown in Table 2 and Fig. 5 within the community. One interpretation is that the uncertainties in the measurements themselves and in the steps needed to derive estimates of the refractory-to-ice mass ratio of *Rosetta* data analyzed to date are too large to draw a firm conclusion. Another interpretation is that a potential bias associated to the collection principle of COSIMA (see Sect. 4.1.4) justifies discarding those results and accepting a refractory-to-ice mass ratio of 3 or greater as consistent with all other *Rosetta* analyses to date. Such a minimum value would then be consistent with model predictions of 67P formation by gravitational collapse of pebbles driven by streaming instability in the protosolar disk (Lorek et al. 2016; Blum et al. 2017), albeit with the potential caveat that no predictions of refractory-to-ice mass ratios for alternate formation models have been published to date, to the best of our knowledge.

# 4 Discussion

## 4.1 Uncertainties on Instrument Measurements and Derived Parameters

Each and every parameter involved in the estimation of $\delta_{DG}$ and $\delta_{RI}$ is derived from instrument measurements with their own individual statistical and systematic errors that may not be completely known. Furthermore, a significant *a posteriori* modeling effort is often required to derive the parameters needed to constrain the mass ratios. Therefore, uncertainties and approximations in the models used to analyze the data should be considered, as well as potential biases associated to the nature and origin of materials being measured. This translates into the wide ranges of plausible dust-to-gas and refractory-to-ice mass ratios reported in Table 2 and Fig. 5.

### *4.1.1 Total Mass and Mass Loss*

The nucleus mass of 67P is amongst the best constrained values used in this work, because of the very low spacecraft velocity relative to 67P, high-precision X-band tracking, and extended stay at close distances that maximized spacecraft trajectory perturbations by the interactions of the gravity field of the comet with the spacecraft. The nucleus mass was determined both before and after the 2015 perihelion passage to a precision of 0.03% (Pätzold et al. 2016) and 0.01% (Pätzold et al. 2019), respectively. The mass loss due to the 2015 perihelion passage is $10.5 \times 10^9$ kg, or around 0.1% of the total nucleus mass (Pätzold et al. 2019), at an uncertainty of 32%. The total mass loss places constraints on the mass loss of volatiles determined by the other instruments.

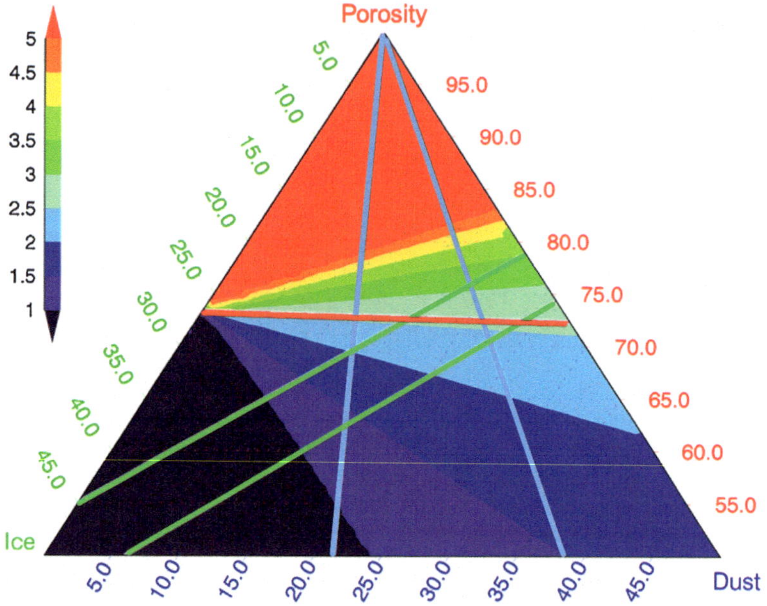


**Fig. 6** Ternary diagram of the dust permittivity as a function of the volume percentages of ice, dust (i.e. carbon with silicates), and porosity, with the constraints on the average density (green lines), on the refractory to ice ratio (blues lines) and from CONSERT measurements (red lines). Refractory-Ice-Porosity volume fractions that meet the three constraints are located within the triangle defined by the upper green line, the right blue line, and the red line. Modified after Hérique et al. (2016)

### *4.1.2 Dielectric Properties*

The composition estimation by CONSERT and, consequently, the $\delta_{RI}$ of the nucleus, are based on the modeling of the effective permittivity from models of ice and dust fractions (Hérique et al. 2016). This model uses the following physical quantities as inputs.

*Density*: The modelled cometary material (ice, refractory and porosity) has to be compliant with both the average permittivity and the average density which is assumed to be $\rho = 533 \pm 6\ \mathrm{kg\,m^{-3}}$, derived from mass measurements obtained by the *Rosetta* Radio Science Investigation (Pätzold et al. 2016) and the observed volume of 67P nucleus based on the 67P shape model (Preusker et al. 2015). In the retrieval process, the lower limit of $\delta_{RI}$ comes from the upper limit of the density (upper green line in Fig. 6).

*Refractory-to-Ice ratio*: The data retrieval approach developed in Hérique et al. (2016) takes a $\delta_{RI} = 4 \pm 2$ (blue lines in Fig. 6) as input (Rotundi et al. 2015). Model calculations providing a good fit to the CONSERT data fall in the range of $\delta_{RI}$ 3–6, which implies that for pure water ice, 3 is a lower limit. Values greater than 6 cannot be excluded as they have not been included in this analysis.

*Ice fraction*: The ice fraction considered consists of $H_2O$, $CO_2$ and CO ices, with a maximum of 14% mass fraction of $CO_2$ (8% molecular abundance) and 15% CO (13% mol.), with permittivity ranging from 2.7 to 3.3 for density values of 931 to 1171 $\mathrm{kg\,m^{-3}}$. Considering the ice fraction, the main uncertainties on the retrieved $\delta_{RI}$ is from the $CO_2$ and CO contents: a larger fraction of $CO_2$ and CO, up to 30% each, would decrease the permittivity of the ice fraction, and consequently may be consistent with $\delta$RI values as low as $\sim$2.2.

The dielectric constant of such an ice-dust-porosity mixture remains compatible with that derived from CONSERT measurements.

The portion of the 67P nucleus probed by CONSERT appears absent of heterogeneities at the few-meters scale or more (Hérique et al. 2019). The measurement approach is not sensitive to local variability in $\delta$RI. CONSERT measurements only allow for estimating the average $\delta$RI value within the probed volume.

It must be noted that the analyses presented in this article assume that the whole 67P nucleus is homogeneous and that its fraction probed by CONSERT at the southern border of the Hatmehit region is truly representative of the nucleus. The composition of the nucleus surface and its interior appear homogeneous at large scales, however no *Rosetta* data can fully validate this assumption.

### *4.1.3 Volatiles: Production Rates, Mass Loss, Extended Sources*

**Uncertainties on Production Rates and Mass Loss Derivation**

Deriving production rates and mass loss from measurements involves multiple modeling steps, typically with simplifying assumptions in order to keep the computation times manageable, and also calibration steps that will not be described here. Remote sensing instruments are sensitive to the molecular absorption or emission of photons at specific wavelengths (rotational modes of $H_2O$, $NH_3$, $CH_3OH$, and CO for MIRO, and vibrational modes excited by fluorescence for VIRTIS) along their line of sight, thus the determination of production rates relies on both coma models and excitation models. ROSINA measured in-situ, at the location of the spacecraft, the total gas density directly using its COma Pressure Sensor (COPS), and the nature and relative proportions of coma constituents ionized within the instrument for detection and counting.

Coma models are used to predict the entire outgassing pattern from the nucleus and determine the local density, temperature, and velocity of the gas within the portions of the coma probed along the line of sight for remote sensing instruments (Biver et al. 2019), or at the spacecraft for ROSINA (Bieler et al. 2015a, 2015b; Kramer et al. 2017; Läuter et al. 2019). These models typically use as basis either the physical interactions between gas molecules, such as the Direct Simulation Monte Carlo approach (Fougère et al. 2016a; Zakharov et al. 2018) or a simplified collosionless coma as used to invert source location from ROSINA/COPS (Kramer et al. 2017) and ROSINA/DFMS data (Läuter et al. 2019), or the geometric characteristics of the coma as observed (Biver et al. 2015, 2019). For all approaches, the 67P coma is found to be extremely asymmetric, with water activity originating primarily from the illuminated portions of the nucleus at any point in time, and additionally with latitudinal variability for some species (Hässig et al. 2015). The asymmetry of the coma is three dimensional, and although 1D simulations can mimic some of this complexity by adding plume contributions with very good performance (Biver et al. 2019), the actual structure of the coma may add unforeseen uncertainties.

The analysis of remote sensing spectral data requires excitation models, as well as radiative transfer models. Combined, they enable the retrieval of the column density along the line of sight of the molecule(s) of interest. When lines are optically thick, the retrievals are highly dependent on assumptions on the gas local density, temperature and velocity field. These parameters have been evaluated at best when interpreting MIRO observations (Biver et al. 2019). As far as possible, optically thin lines have been used to infer column densities and production rates from MIRO and VIRTIS data (Bockelée-Morvan et al. 2016; Biver et al. 2019).

In-situ volatile data acquired by ROSINA is also dependent on coma models and laboratory measurements. Coma models are needed to extrapolate the density of the molecule(s) of interest throughout the coma from measurements made locally (Fougère et al. 2016a) to derive production rates. Laboratory measurements are used to constrain the fragmentation pattern and ionization efficiency of these molecules (and/or their fragments) in ROSINA/DFMS and ROSINA/RTOF, and thus their individual contribution to specific peaks observed within the mass spectra obtained. Note that the analysis of ROSINA/COPS data does not require such laboratory efforts, as the COPS sensor solely measures gas density.

Lastly, all the individual sets of measurements need to be integrated over time to derive total volatile mass losses (Biver et al. 2019; Läuter et al. 2019), as was done in the initial empirical estimate by Hansen et al. (2016). This step adds further uncertainty to the results despite the high cadence of mapping measurements by MIRO and the regular acquisition of ROSINA data that have been planned throughout the comet operations phase (Taylor et al. 2017; Vallat et al. 2017).

**Extended Water Sources: Hidden Ice in Escaping Boulders?**

Extended sources may be associated to ice-bearing grains that start sublimating in the coma once they are beyond Rosetta's position (within the first few tens – hundreds of km around the nucleus). Such sources of gas could explain some OSIRIS dust observations (Agarwal et al. 2016; Gicquel et al. 2016). Extended sources of water are abundant at 103P/Hartley, a hyperactive comet (A'Hearn et al. 2011), and have been suggested on other comets such as 1P/Halley where distant sources of specific compounds (not water) were identified (Eberhardt 1999).

In the particular case of 67P, *Odin* (Frisk et al. 2003) conducted space-based observations of the 557 GHz rotational line of $H_2O$ at 67P, which is amongst the lines measured by MIRO, in November 2015. The line was not detected after 40 hrs of integration time and the 3-sigma noise yields an upper limit in production rate of $<3.3 \times 10^{27}$ molecules/s (Biver et al. 2019). From the MIRO measurements at the same time, the water production rate was $(2.2 \pm 0.1) \times 10^{27}$ molecules/s (Biver et al. 2019). Furthermore, the peak water production rate derived from *PROCYON* Lyman $\alpha$ measurements (Shinnaka et al. 2017) is also about 50% higher than determined by MIRO, see Table 1.

Even though extended sources are not required to explain the measurements, due to the measurement uncertainties it cannot be excluded that up to $\sim$ 50% of the water production (and presumably other volatiles) could originate from coma boulders that are escaping and do not have time to fully release their volatiles while in the vicinity of the comet explored by Rosetta. Since these volatiles would be undetected by *Rosetta*, this contribution of volatiles would be part of the dry dust mass loss calculated in Sect. 3.4.3. As a consequence, the dust-to-water $\delta_{DG}{}^{H_2O}$ and dust-to-all-volatiles $\delta_{DG}{}^{V}$ mass ratios determined in the inner coma by *Rosetta* (Table 2 and Fig. 5) could be somewhat overestimated.

### *4.1.4 Production Rates and Mass Loss of Dust*

Determining the production rate and mass loss of dust requires several elements: a good knowledge of the grain size distribution, composition and/or density, overall distribution around the nucleus to expand from local measurements to spatially-integrated production rates, and their evolution over time. The latter includes fallback materials and their uncertainties, which are discussed separately in Sect. 4.2.2.

GIADA and COSIMA used different approaches to constrain the density of dust particles. GIADA measured directly scattered light, velocity, and momentum of dust particles of a few μm to mm diameter. Particle mass and density was then derived from these measurements. COSIMA/COSISCOPE relied on optical observations to infer the properties of the particles. The density of particles collected by GIADA is on order of 800 kg m$^{-3}$ (Fulle et al. 2017), while the density of dust particles after fragmentation within COSIMA is significantly lower, in the range 100–400 kg m$^{-3}$ (Hornung et al. 2016).

COSIMA might have collected particles from one or two potential families, one with a fractal dimension of 2.5-3 and a spread in impact velocity, and the other $<2$ if all particles impacts with the same velocity (Lasue et al. 2019). A large fraction of the dust entering COSIMA is reflected, owing to a mass transfer efficiency estimated to 1 to 10% depending on size and velocity of incoming particles (Ellerbroek et al. 2019). This potential collection bias of COSIMA based on a collection or detection area suggests that a fractionation of the dust populations might be applicable. The GIADA measurement principle, which catches dust particles as they fly into the instrument, is not affected by this type of bias.

A coma model is then needed to extrapolate to the scale of the entire coma. Dust is lifted by gas drag, therefore all uncertainties on outgassing of volatiles propagate to the dust lifting, while additional sources of uncertainties exist due to the size, shape, density of the dust, potential ice content that is not measurable optically nor by any in-situ method conducted near room temperature, and processes through which it is lifted (see Vincent et al. 2019, this book).

For instance, this approach was used by Ott et al. (2017) to derive the total mass loss over 53 days around perihelion from parallax imaging of ~1–30 cm dust particles. The latter observations were restricted to 6 km from the *Rosetta* spacecraft in the few occurrences where such particles were imaged, implying that these dm-scale dust particles were on escape trajectories (radial velocity >1 m/s, 200 km distance from the nucleus). The extrapolation of the individual measurements over the 53-day period of observation yields a greater integrated amount of ejected (and lost) materials than the total 67P mass loss determined by RSI. Two lines of interpretation can explain this discrepancy. A fraction of these dm-scale coma particles, outside of the terminator plane where *Rosetta* was located, could fall back on the nucleus (Fulle et al. 2019). Or the sources of uncertainties associated to the derivation of volume, or subsequent extrapolation to particle mass then total mass, may be poorly constrained. These sources of uncertainties include distance/size from the parallax measurement (Drolshagen et al. 2017), density (a very high value of 1000 kg m$^{-3}$ was used), albedo, and statistical uncertainties. Note that the density does not affect the subsequent study by Fulle et al. (2019), and the assumed albedo of these particles is consistent with that measured by Bertini et al. (2018).

## 4.2 Uncertainties on Materials Observed, Fallback Extent, and Activity Scenario

### *4.2.1 Thermal Processing, Erosion, and Water Outgassing*

Often the dust-to-volatile ratio of a comet is measured in its coma, e.g. Fulle et al. (2019) and therein. But how representative is it of the refractory-to-ice ratio in the nucleus? In other words, what can we learn from coma measurements for the nucleus composition when accepting that:

1) The amount of hidden ice in the dm-scale escaping or falling back dust particles is widely unknown (Sect. 4.1.3). It is mostly inaccessible for remote sensing instruments and difficult to model because of the lack of detailed information on parameters needed.
2) The escaping materials may have been altered by thermal processing and may not represent the original bulk properties.

**Table 3** Thermal properties and corresponding thermal skin depths of the 67P surface

| Parameter | Units | Low inertia | High inertia |
|---|---|---|---|
| Thermal inertia[a] | $J\,m^{-2}\,K^{-1}\,s^{-0.5}$ | 10 | 170 |
| Density[b] | $kg\,m^{-3}$ | 538 | 538 |
| Specific heat[c] | $J\,kg\,K^{-1}$ | 540 | 540 |
| Thermal conductivity | $W\,m^{-1}\,K^{-1}$ | $3.4 \times 10^{-4}$ | 0.1 |
| Diurnal skin depth (12.4 hours period) | m | $4 \times 10^{-3}$ | $7 \times 10^{-2}$ |
| Seasonal skin depth (6.44 years period) | m | 0.28 | 4.7 |

[a]Groussin et al. (2019, this issue)

[b]Preusker et al. (2017)

[c]Corresponding to a dust/ice mass ratio of 2 at 120 K (Winter and Saari 1969; Herman and Weissman 1987) (Groussin et al. 2019, this issue). Higher dust/ice ratios would yield lower specific heat values and hence higher conductivity, resulting in larger skin depth values

**Thermal Properties**

Thermal and mechanical models can evaluate both the loss of volatiles in the escaping materials and the consequences of activity for the refractory-ice-ratio at the surface of the nucleus. For modeling such heat transfer processes, one has to know the thermal and mechanical properties of the materials involved. *Rosetta* instruments (MIRO, VIRTIS, OSIRIS, RSI, MUPUS, and SESAME) provided some physical characteristics that are related to the thermal inertia: density, strength, diurnal period, and thermal conductivity (for a review, see Marshall et al. (2018) and Groussin et al. (2019), this issue). The thermal inertia was derived from Rosetta measurements with considerable error bars. It controls the temperature profile at the surface and the thermal skin depth that characterizes the penetration depth of the diurnal and seasonal heat waves, respectively, e.g. Prialnik et al. (2004).

Table 3 offers simple calculations of the thermal conductivity of the nucleus material and the corresponding thermal skin depths for lower and upper limits of the most plausible range of thermal inertias (Groussin et al. 2019 this issue). The diurnal skin depth is extremely low, no more than a few cm, as shown from MIRO continuum observations (Schloerb et al. 2015), while the seasonal skin depth can be up to a few meters. Typically, the heat wave propagates to ~2-3 skin depths, as the skin depth only corresponds to a 1/e attenuation of the amplitude of temperature variations as compared to the surface.

**Nucleus Erosion**

These estimates of thermal processing length scale can then be compared to the length scale of nucleus erosion. Neglecting uncertainties on reported values for simplicity, the total mass loss of 67P around the 2015 perihelion passage is $10.5 \times 10^9$ kg (Pätzold et al. 2019), the nucleus volume is 18.56 $km^3$ and its surface area is 51.7 $km^2$ (Preusker et al. 2017). If we assume the erosion to be the same everywhere on the surface, the thickness of eroded materials is 37.7 cm. Non-uniform erosion due to the extreme seasonal asymmetry may further increase the thickness of this layer in the southern regions. Since a considerable amount of material that is lifted off later falls back to the surface, a near-surface eroded/redistributed layer is likely on order of ~ a meter thick, and perhaps more in the Southern regions (about 4 m suggested by Fulle et al. 2019).

Figure 7 compares the skin depths with the thickness of the erosion layer for several values of thermal inertia and percentage of fallback materials. The higher the inertia and the lower the fallback are, the more seasonal processing of the nucleus affects the near-surface

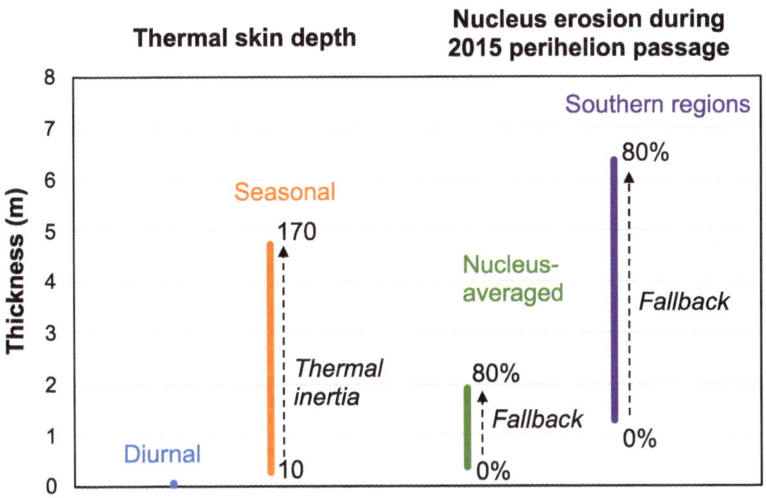


**Fig. 7** Comparison of the skin depth of thermal processing for diurnal and seasonal periods of 67P with the thickness of materials eroded and redistributed by the 2015 perihelion passage

materials that are later on measured in the coma. Seasonal processing is of the same order as nucleus erosion. Therefore, even the dm-size particles found in the coma have likely been already processed to some extent, perhaps even prior to their ejection. This eventuality has been incorporated in the recent analysis of coma boulders ejection timescale and re-deposition, and their implications for a possibly high ($>3$) refractory-to-ice mass ratio (Fulle et al. 2019).

**Refractory-to-Ice Mass Ratio, Dust Mantle Thickness, and Outgassing**

As illustrated in Table 2 and Fig. 5, and previously suggested (Fulle et al. 2019), the dust-to-gas ratio measured in the lost materials could differ substantially from the refractory-to-ice ratio within the 67P nucleus. If indeed the dust-to-gas ratio of the escaping materials is continuously lower than the refractory-to-ice in the nucleus, refractories would become progressively enriched at the surface of the nucleus either by losing volatiles or by re-deposition of fallback materials processed in the coma. This would lead to the formation of a dry lag deposit that could dampen activity, at least in some regions. Then, counter processes such as outbursts excavating deep-seated fresh materials, or extreme heterogeneities in distribution of the erosion and re-deposition of fallback materials, or mass wasting processes, would be necessary to sustain the activity over geologic timescales.

Detailed thermophysical modeling efforts further investigated the joint influence of the refractory-to-ice mass ratio and the thickness of a dry surface mantle on predicted erosion and water outgassing. A study assuming mm-size pebbles, comprised of a top dehydrated single layer over a dust-ice mixture with a $\delta_{RI}$ of 7.5, found that such a near-surface structure is compatible with multiple *Rosetta* and *Philae* measurements, and supports the interpretation that 67P formed through streaming instability (Blum et al. 2017). Model calculations with dust mantle thicknesses of 5 and 10 mm over a dust-ice mixture with $\delta_{RI}$ of 10 and 100 suggest that the water activity of 67P early on, well before perihelion, can be matched with very high $\delta_{RI}$ values (Hu et al. 2017a). In a follow-on study, the analysis of OSIRIS images for erosion rates in specific regions and dust cover re-deposition suggest that extremely high $\delta_{RI}$ values up to 100 cannot be excluded, and that much lower $\delta_{RI}$ as low as 4 may also be consistent with the results (Hu et al. 2017b). Computations using the complex *Rome* thermophysical model (Capria et al. 2017), which couples orbital evolution, starting composition, thermal properties, and mass balance for loss of volatiles and gas drag for loss of refractories, were run for two example locations that received intense insolation around the 2015 perihelion and $\delta_{RI}$ values of 4 and 10, including in the Anhur region where VIRTIS-M IR data suggested the temporary exposure of carbon dioxide ice (Filacchione et al. 2016b). These calculations suggest that a multi-decimeter erosion could occur around perihelion and preserve almost unaltered materials at relatively shallow depths $<1$ m after erosion (Capria et al. 2017).

Two aspects of the studies listed above are noteworthy. First, the refractory-to-ice mass ratio is an input parameter to the models, thus cannot be fully constrained by them unless inversion schemes were used to estimate sensitivity to, and uncertainties on, input parameters (which these studies did not investigate). Second, these works relied on published $\delta_{RI}$ values available at the time (Rotundi et al. 2015; Fulle et al. 2017), all of which were of 3 and above.

A recent study (Attree et al. 2019) constrains the activity of 67P by using both the ROSINA data of the water production and the non-gravitational forces derived from the spin change. They found that a complex pattern of the ice surface fraction that changes with time and surface region explains the measurements best. Further studies are needed to investigate the sensitivity of thermophysical models to their input parameters, to determine whether they could be used to reliably constrain $\delta_{RI}$, and to investigate if low $\delta_{RI}$ values commensurate with the mission-integrated $\delta_{DG}$ can be excluded.

### 4.2.2 Contribution of Re-Activation of Deposited Materials to Activity

The evidence for North-South asymmetry in morphology and dust cover seems naturally related to seasonal variability in activity, as discussed in Sect. 2.3. It is an unexpected feature of 67P, with high obliquity of its spin axis together with its elliptic orbit and the specific positions of the nodes between its orbital and equatorial planes. There are however several aspects that need confirmation to validate the proposed scenario, and its implication of a possibly extremely high $\delta_{RI}$ suggested by Fulle et al. (2019).

The main line of evidence for the scenario illustrated in Fig. 2 is a combination of factors: morphology consistent with fallback of material (originating from the Southern regions at perihelion) in the Northern regions, as well as predicted fallback in the North greater than mass loss of volatiles and dust associated to native activity, and lastly activity during the Northern summer originating presumably from Hapi where this fallback may dominate. However, this combination of factors could be at least in part circumstantial.

It must be noted that the coma models used to invert ROSINA DFMS data into activity maps (Fougère et al. 2016a) do not have unique solutions (in particular because each measurement is at one specific location within the coma at a time, and coverage can only be partial due to spacecraft trajectories). Zakharov et al. (2018) have shown that multiple other solutions to the activity distribution can yield the results obtained by ROSINA, some of these solutions having little to no $H_2O$ originating from the Hapi region itself. Furthermore, the surface activity maps recently derived from ROSINA/DFMS data (Läuter et al. 2019) show a widely varying distribution of activity, with Hapi and potential fallback materials only partially contributing to the activity in the North.

In addition, Vincent et al. (2016a) have suggested that jets observed in the northern regions during their summer are mostly consistent with activity occurring along cliffs and walls, where mass wasting would expose fresher materials and their ice content get sublimated fairly rapidly to form dust jets. A similar scenario has also been put forth to explain the large Ashwan cliff collapse (El-Maarry et al. 2017; Pajola et al. 2017). Nevertheless, some dust jets in the Northern regions clearly originate from the smooth areas within Hapi (Shi et al. 2018). Thus, the proposed re-activation of fallback material as a key contributor to the Northern activity is consistent with some observations, but may not be the only mechanism involved.

Lastly, the longevity of different ices within dm-scale particles that would be transported from the South to the North is advocated for (Fulle et al. 2019). Nevertheless, it remains to be fully evaluated, as function of the thermal and mechanical properties of these materials, the volatility of the ice species, how long their transport requires, how they rotate in the coma and are illuminated, how they lose their ices, and what conditions they experience following re-deposition.

## 4.3 Future Research, Instruments, and Missions

### *4.3.1 Future Research Using Rosetta Data*

Despite more than two years spent in orbit around 67P, the analysis of *Rosetta* data to date only allows to determine the dust-to-gas and refractory-to-ice mass ratios with very low precision, see Fig. 5. Further analysis of Rosetta data may help better constrain some of the parameters involved in these estimates. A few examples follow.

1) A quantification of the distribution of erosion and the possible extent of fallback materials from OSIRIS data could shed a new light on lateral transport processes on 67P. First erosion estimates have been conducted in selected areas from shadows cast by boulders and cliffs to estimate height, and from excavation of cavities (Fornasier et al. 2019; Hasselmann et al. 2019). In addition, high-resolution local Digital Terrain Models pre- and post-perihelion can help measure surface changes. The combination of Multi-resolution StereoPhotoClinometry by Deformation MSPCD (Capanna et al. 2015) with stereophotogrammetry shape modeling (Preusker et al. 2017), can achieve much higher resolution than currently available. These new approaches will likely improve the quantification of the mass transport processes in the near future.

2) Improving our knowledge of the history and thermal properties of the dm-scale particles before ejection/during flight/after re-deposition, as well as the relative contributions of fallback materials and fresher materials exposed by mass wasting to the 67P activity, may provide additional constraints on the original $\delta_{RI}$ of these materials.

3) An extensive set of thermophysical calculations that encompasses the entire surface of 67P, and uses inversion to constrain the dust mantle thickness and the refractory-to-ice ratio over extreme ranges of possible input values, would be a necessary step towards understanding how sensitive these models are to their input parameters. They could in particular test the hypothesis that $\delta_{DG}$ and $\delta_{RI}$ differ largely, whether at single points in time and/or over extended periods of time.

### *4.3.2 Potential Future Instruments and Missions*

Another factor that likely contributes to the low precision of the dust-to-gas and refractory-to-ice mass ratio estimates is the lack of direct observations of the deep interior and shallow subsurface of the nucleus. Onboard *Rosetta*, CONSERT probed a limited part of 67P's small lobe, while RSI did not conclusively find density differences between the two lobes. Direct observation of materials from the deep interior would be crucial to characterize the composition of pristine material and then to access directly $\delta_{RI}$. The direct observation of the shallow subsurface is also crucial to better understand comet activity processes including re-deposition and re-condensation.

Radar is the most mature technique to probe both the pristine interior and the shallow subsurface. Two different instruments or channels can achieve these objectives: 1) A low frequency radar to probe the deep interior, which could be a monostatic radar like MARSIS onboard *Mars Express* (Picardi et al. 2005) or bistatic radar with a Lander like CONSERT onboard *Rosetta* (Kofman et al. 1998). 2) A higher frequency radar to probe the first few tens of meters of the nucleus with decimeter resolution, in order to detect layers and undergrounds structures like HFR developed in the frame of AIDA mission (Michel et al. 2016; Hérique et al. 2018).

For example, two monostatic radars have been proposed to instrument the *Castalia* mission concept to Main Belt comet 133P/Elst-Pizzaro, proposed to ESA for the recent M5 cosmic vision round (Snodgrass et al. 2018).

Future comet missions carrying such radar instruments could better understand the vertical structure of the first ten to tens of meters of the subsurface with high vertical resolution. This would help better constrain the composition of the nucleus and its gradients over the thickness most affected by cometary activity (Fig. 7), which is necessary to fully address the $\delta_{RI}$ issue of 67P and other comets. In addition, better knowledge of the fine-scale structure of the most active layer of comet nuclei would likely enhance our understanding of cometary activity.

Lastly, the ideal way to determine the refractory-to-ice ratio within the nucleus of 67P and other comets would obviously be to analyze directly representative materials sampled at depth within the nucleus. *Rosetta* was intended to be a sample return mission at its inception (following the 1986 flyby of 1P/Halley by *Giotto*), but the concept was modified into bringing a laboratory to 67P because of the technological challenges associated with sampling, returning, and curating cryogenic samples.

The NRC Planetary Decadal survey report "Visions and Voyages" for the decade 2013-2022 advocated that a cryogenic comet sample return mission would have by far the highest scientific return for the Small Bodies community. That report also acknowledged the lack of maturity for the acquisition, return and curation of a cryogenic comet sample, then advocated for the present decade for a non-cryogenic comet surface sample return mission.

Cryogenic sampling and containment systems are being developed as part of other technology development efforts, for example as suggested for the Triple-F mission concept (Küppers et al. 2009). A cryogenic comet sample return mission equipped with instruments capable of assessing heterogeneities within the nucleus may be able to identify a most volatile-rich and porous, least processed region at depth. If the deep interior is homogeneous, a sample may be representative and inform the refractory-to-ice mass ratio. If the deep interios is heterogeneous at the scale of the sample, this may not enable a determination of the refractory-to-ice ratio of the nucleus to a precision better than *Rosetta*, however the sample would retain the textural information that holds the clues as to processes through which it formed in the outer protosolar disk. More than 30 years after the flyby of 1P/Halley by *Giotto* (Reinhard 1986), further efforts by space agencies around the world are needed to turn a cryogenic comet sample return mission concept into a reality.

## 5 Conclusions

*Rosetta* data obtained at 67P enabled multiple estimates of the dust-to-gas mass ratio in the coma and the refractory-to-ice mass ratio within the nucleus. These estimates are overall of low precision, because of the uncertainties associated to individual measurements, the numerous models that need to be used to derive global production rates and mass loss from measurements made at one location and one point in time (hence large extrapolations both spatially and over time), and complications arising from the redistribution of materials around the nucleus and the poor knowledge of their composition and ice content.

The mission-integrated dust-to-gas ratio $\delta_{DG}$, based on total mass loss of the nucleus and the total mass loss of volatiles, is probably the best constrained parameter to date. A $\delta_{DG}$ value of 2.3 or lower encompasses both remote sensing and in-situ determinations (considering the 1-sigma upper limit from the $1.5 \pm 0.8$ remote sensing determination, and although one cannot discard at this time the potential for systematic errors to affect this result).

However, it appears difficult to place firm lower or upper limits on the refractory-to-ice mass ratio. Opposing views are expressed to date on this topic within the community. Some argue that the data and analyses presently available have reached the level of maturity necessary to draw conclusions, while others caution that the numerous sources of uncertainties associated to the derivation of the refractory-to-ice mass ratio (retrievals, activity scenario, extent and composition of fallback materials, along with putative heterogeneities in composition within the nucleus and variability in comet activity through time) may not be understood and quantified well enough yet to support such conclusions.

It has been previously suggested that the dust-to-gas mass ratio measured in the coma may deviate significantly from the refractory-to-ice mass ratio in the nucleus, at least at a single point in time. One would expect that the integrated mass losses of dust and ice would help bound the refractory-to-ice mass ratio in the nucleus, but that is not the case at present.

Upcoming analyses of *Rosetta* data may shed a new light on the extent of fallback materials and erosion, and bring new quantified information on lateral transport processes. In turn, these new pieces of information may provide new constraints on the refractory-to-ice mass ratio within the 67P nucleus.

The most important lesson from *Rosetta* on the topic of dust-to-gas and refractory-to-ice mass ratios is probably that dust-to-gas mass ratio estimates from observations of cometary comae at limited points in time may not be representative of the refractory-to-ice mass ratio within the observed nuclei. This emphasizes the need for instrumentation tailored to investigating the interior and near-surface of comet nuclei and their contribution to cometary activity, and for future space missions to comets.

**Acknowledgements** *Rosetta* is an ESA mission with contributions from its Member States and NASA. *Rosetta's Philae* lander is provided by a consortium led by DLR, MPS, CNES and ASI. The authors are grateful to two anonymous reviewers for comments that helped improve on this manuscript, to Vincenzo Della Corte for valuable discussions of the GIADA collection and measurement principles, and to Marco Fulle for numerous discussions on determinations of the dust-to-gas and refractory-to-ice mass ratios, methods, and uncertainties. The *Rosetta* RSI experiment was funded by the Bundesministerium für Wirtschaft BMWi, Berlin, via the German Space Agency DLR, Bonn, under grants 50QM1401 and 50QM1802 for RIU Cologne, 50QM1302 for Max Planck Institute for Solar System Research, Göttingen, and 50QM1002 and 50QM1704 for UniBw München. The RSI team expresses deep appreciation for the critical support provided the ESTRACK and DSN ground station networks during the data acquisition periods. This project has received funding from the European Union's Horizon 2020 research and innovation program under grant agreement No 686709. This work was supported by the Swiss State Secretariat for Education, Research and Innovation (SERI) under contract number 16.0008-2. The opinions expressed and arguments employed herein do not necessarily reflect the official view of the Swiss Government. This work was supported in part by the H2020 MiARD project. Part of this work has been conducted at the Jet Propulsion Laboratory, California Institute of Technology, under contract to NASA. MC acknowledges support from the US Rosetta Project, which manages the NASA contributions to *Rosetta*. Government sponsorship acknowledged.

## References

J. Agarwal, M. A'Hearn, J.-B. Vincent, C. Güttler, S. Höfner, H. Sierks, C. Tubiana, C. Barbieri, P. Lamy, R. Rodrigo, Acceleration of individual, decimetre-sized aggregates in the lower coma of comet 67P/Churyumov–Gerasimenko. Mon. Not. R. Astron. Soc. **462**(Suppl 1), S78–S88 (2016)

J. Agarwal, V. Della Corte, P. Feldman, B. Geiger, S. Merouane, I. Bertini, D. Bodewits, S. Fornasier, E. Grün, P. Hasselmann, Evidence of sub-surface energy storage in comet 67P from the outburst of 2016 July 03. Mon. Not. R. Astron. Soc. **469**(Suppl 2), s606–s625 (2017)

M.F. A'Hearn, M.J. Belton, W.A. Delamere, L.M. Feaga, D. Hampton, J. Kissel, K.P. Klaasen, L.A. McFadden, K.J. Meech, H.J. Melosh, EPOXI at comet Hartley 2. Science **332**(6036), 1396–1400 (2011)

K. Altwegg, H. Balsiger, A. Bar-Nun, J.-J. Berthelier, A. Bieler, P. Bochsler, C. Briois, U. Calmonte, M. Combi, J. De Keyser, 67P/Churyumov-Gerasimenko, a Jupiter family comet with a high D/H ratio. Science **347**(6220), 1261952 (2015)

N. Attree, L. Jorda, O. Groussin, S. Mottola, N. Thomas, Y. Brouet, E. Kührt, M. Knapmeyer, F. Preusker, F. Scholten, J. Knollenberg, S. Hviid, P. Hartogh, R. Rodrigo, Constraining models of activity on comet 67P/Churyumov-Gerasimenko with Rosetta trajectory, rotation, and water production measurements (2019). ArXiv:e-prints

G. Aumatell, G. Wurm, Ice aggregate contacts at the nm-scale. Mon. Not. R. Astron. Soc. **437**(1), 690–702 (2014)

H. Balsiger, K. Altwegg, P. Bochsler, P. Eberhardt, J. Fischer, S. Graf, A. Jäckel, E. Kopp, U. Langer, M. Mildner, Rosina–Rosetta orbiter spectrometer for ion and neutral analysis. Space Sci. Rev. **128**(1–4), 745–801 (2007)

H. Balsiger, K. Altwegg, A. Bar-Nun, J.-J. Berthelier, A. Bieler, P. Bochsler, C. Briois, U. Calmonte, M. Combi, J. De Keyser, Detection of argon in the coma of comet 67P/Churyumov-Gerasimenko. Sci. Adv. **1**(8), e1500377 (2015)

A. Bardyn, D. Baklouti, H. Cottin, N. Fray, C. Briois, J. Paquette, O. Stenzel, C. Engrand, H. Fischer, K. Hornung, Carbon-rich dust in comet 67P/Churyumov-Gerasimenko measured by COSIMA/Rosetta. Mon. Not. R. Astron. Soc. **469**(Suppl 2), S712–S722 (2017)

M.S. Bentley, R. Schmied, T. Mannel, K. Torkar, H. Jeszenszky, J. Romstedt, A.-C. Levasseur-Regourd, I. Weber, E.K. Jessberger, P. Ehrenfreund, Aggregate dust particles at comet 67P/Churyumov–Gerasimenko. Nature **537**(7618), 73 (2016)

J.-L. Bertaux, Estimate of the erosion rate from $H_2O$ mass-loss measurements from SWAN/SOHO in previous perihelions of comet 67P/Churyumov-Gerasimenko and connection with observed rotation rate variations. Astron. Astrophys. **583**, A38 (2015)

I. Bertini, F. La Forgia, M. Fulle, C. Tubiana, C. Güttler, F. Moreno, J. Agarwal, O. Muñoz, S. Mottola, S. Ivanovsky, The backscattering ratio of comet 67P/Churyumov–Gerasimenko dust coma as seen by OSIRIS onboard Rosetta. Mon. Not. R. Astron. Soc. **482**(3), 2924–2933 (2018)

J. Biele, S. Ulamec, M. Maibaum, R. Roll, L. Witte, E. Jurado, P. Muñoz, W. Arnold, H.-U. Auster, C. Casas, The landing (s) of Philae and inferences about comet surface mechanical properties. Science **349**(6247), aaa9816 (2015)

A. Bieler, K. Altwegg, H. Balsiger, A. Bar-Nun, J.-J. Berthelier, P. Bochsler, C. Briois, U. Calmonte, M. Combi, J. De Keyser, Abundant molecular oxygen in the coma of comet 67P/Churyumov–Gerasimenko. Nature **526**(7575), 678 (2015a)

A. Bieler, K. Altwegg, H. Balsiger, J.-J. Berthelier, U. Calmonte, M. Combi, J. De Keyser, B. Fiethe, N. Fougere, S. Fuselier, Comparison of 3D kinetic and hydrodynamic models to ROSINA-COPS measurements of the neutral coma of 67P/Churyumov-Gerasimenko. Astron. Astrophys. **583**, A7 (2015b)

N. Biver, M. Hofstadter, S. Gulkis, D. Bockelee-Morvan, M. Choukroun, F. Lellouch, F.P. Schloerb, L. Rezac, W.H. Ip, C. Jarchow, P. Hartogh, S. Lee, P. von Allmen, J. Crovisier, C. Leyrat, P. Encrenaz, Distribution of water around the nucleus of comet 67P/Churyumov-Gerasimenko at 3.4 AU from the Sun as seen by the MIRO instrument on Rosetta. Astron. Astrophys. **583**, A3 (2015)

N. Biver, D. Bockelée-Morvan, M. Hofstadter, E. Lellouch, M. Choukroun, S. Gulkis, J. Crovisier, F. Schloerb, L. Rezac, P. Von Allmen, Long-term monitoring of the outgassing and composition of comet 67P/Churyumov-Gerasimenko with the Rosetta/MIRO instrument. Astron. Astrophys. **630**, A19 (2019)
J. Blum, Dust evolution in protoplanetary discs and the formation of planetesimals. Space Sci. Rev. **214**(2), 52 (2018)
J. Blum, B. Gundlach, M. Krause, M. Fulle, A. Johansen, J. Agarwal, I. von Borstel, X. Shi, X. Hu, M.S. Bentley, F. Capaccioni, L. Colangeli, V. Della Corte, N. Fougere, S.F. Green, S. Ivanovski, T. Mannel, S. Merouane, A. Migliorini, A. Rotundi, R. Schmied, C. Snodgrass, Evidence for the formation of comet 67P/Churyumov-Gerasimenko through gravitational collapse of a bound clump of pebbles. Mon. Not. R. Astron. Soc. **469**, S755–S773 (2017)
D. Bockelée-Morvan, N. Biver, The composition of cometary ices. Philos. Trans. R. Soc., Math. Phys. Eng. Sci. **375**(2097), 20160252 (2017)
D. Bockelée-Morvan, J. Crovisier, M. Mumma, H. Weaver, The composition of cometary volatiles, in *Comets II*, ed. by H.U. Keller, H.A. Weaver (2004), pp. 391–424
D. Bockelée-Morvan, V. Debout, S. Erard, C. Leyrat, F. Capaccioni, G. Filacchione, N. Fougere, P. Drossart, G. Arnold, M. Combi, First observations of $H_2O$ and $CO_2$ vapor in comet 67P/Churyumov-Gerasimenko made by VIRTIS onboard Rosetta. Astron. Astrophys. **583**, A6 (2015)
D. Bockelée-Morvan, J. Crovisier, S. Erard, F. Capaccioni, C. Leyrat, G. Filacchione, P. Drossart, T. Encrenaz, N. Biver, M.-C. de Sanctis, Evolution of $CO_2$, $CH_4$, and OCS abundances relative to $H_2O$ in the coma of comet 67P around perihelion from Rosetta/VIRTIS-H observations. Mon. Not. R. Astron. Soc. **462**(Suppl 1), S170–S183 (2016)
U. Calmonte, K. Altwegg, H. Balsiger, J.-J. Berthelier, A. Bieler, G. Cessateur, F. Dhooghe, E. Van Dishoeck, B. Fiethe, S. Fuselier, Sulphur-bearing species in the coma of comet 67P/Churyumov–Gerasimenko. Mon. Not. R. Astron. Soc. **462**(Suppl 1), S253–S273 (2016)
U. Calmonte, K. Altwegg, H. Balsiger, J.-J. Berthelier, A. Bieler, J. De Keyser, B. Fiethe, S. Fuselier, S. Gasc, T. Gombosi, Sulphur isotope mass-independent fractionation observed in comet 67P/Churyumov–Gerasimenko by Rosetta/ROSINA. Mon. Not. R. Astron. Soc. **469**(Suppl 2), S787–S803 (2017)
F.a. Capaccioni, A. Coradini, G. Filacchione, S. Erard, G. Arnold, P. Drossart, M. De Sanctis, D. Bockelee-Morvan, M. Capria, F. Tosi, The organic-rich surface of comet 67P/Churyumov-Gerasimenko as seen by VIRTIS/Rosetta. Science **347**(6220), aaa0628 (2015)
C. Capanna, L. Jorda, G. Gesquière, O. Groussin, P. Gutiérrez, S. Hviid, P. Lamy, S. Rodionov, D. Vibert, A new 3D shape reconstruction method for celestial bodies: multi-resolution stereophotoclinometry by deformation, in *EGU General Assembly Conference Abstracts* (2015)
M.T. Capria, F. Capaccioni, G. Filacchione, F. Tosi, M.C. De Sanctis, S. Mottola, M. Ciarniello, M. Formisano, A. Longobardo, A. Migliorini, E. Palomba, A. Raponi, E. Kührt, D. Bockelée-Morvan, S. Erard, C. Leyrat, A. Zinzi, How pristine is the interior of the comet 67P/Churyumov-Gerasimenko? Mon. Not. R. Astron. Soc. **469**, S685–S694 (2017)
M. Choukroun, S. Keihm, F.P. Schloerb, S. Gulkis, E. Lellouch, C. Leyrat, P. von Allmen, N. Biver, D. Bockelee-Morvan, J. Crovisier, P. Encrenaz, P. Hartogh, M. Hofstadter, W.-H. Ip, C. Jarchow, M. Janssen, S. Lee, L. Rezac, G. Beaudin, B. Gaskell, L. Jorda, H.U. Keller, H. Sierks, Dark side of comet 67P/Churyumov-Gerasimenko in Aug.-Oct. 2014 MIRO/Rosetta continuum observations of polar night in the southern regions. Astron. Astrophys. **583**, A28 (2015)
V. Ciarletti, A.C. Levasseur-Regourd, J. Lasue, C. Statz, D. Plettemeier, A. Hérique, Y. Rogez, W. Kofman, CONSERT suggests a change in local properties of 67P/Churyumov-Gerasimenko's nucleus at depth. Astron. Astrophys. **583**, A40 (2015)
L. Colangeli, J. Lopez-Moreno, P. Palumbo, J. Rodriguez, M. Cosi, V. Della Corte, F. Esposito, M. Fulle, M. Herranz, J. Jeronimo, The grain impact analyser and dust accumulator (GIADA) experiment for the Rosetta mission: design, performances and first results. Space Sci. Rev. **128**(1–4), 803–821 (2007)
M. Combi, Y. Shou, N. Fougere, V. Tenishev, K. Altwegg, M. Rubin, D. Bockelée-Morvan, F. Capaccioni, Y.-C. Cheng, U. Fink, The surface distributions of the production of the major volatile species, $H_2O$, $CO_2$, CO and $O_2$, from the nucleus of comet 67P/Churyumov-Gerasimenko throughout the Rosetta Mission as measured by the ROSINA double focusing mass spectrometer. Icarus **335**, 113421 (2020)
A. Coradini, F. Capaccioni, P. Drossart, G. Arnold, E. Ammannito, F. Angrilli, A. Barucci, G. Bellucci, J. Benkhoff, G. Bianchini, VIRTIS: an imaging spectrometer for the Rosetta mission. Space Sci. Rev. **128**(1–4), 529–559 (2007)
M. De Sanctis, F. Capaccioni, M. Ciarniello, G. Filacchione, M. Formisano, S. Mottola, A. Raponi, F. Tosi, D. Bockelée-Morvan, S. Erard, The diurnal cycle of water ice on comet 67P/Churyumov–Gerasimenko. Nature **525**(7570), 500 (2015)
V. Della Corte, A. Rotundi, M. Fulle, E. Gruen, P. Weissman, R. Sordini, M. Ferrari, S. Ivanovski, F. Lucarelli, M. Accolla, GIADA: shining a light on the monitoring of the comet dust production from the nucleus of 67P/Churyumov-Gerasimenko. Astron. Astrophys. **583**, A13 (2015)

V. Della Corte, R. Sordini, M. Accolla, M. Ferrari, S. Ivanovski, A. Rotundi, F.J.M. Rietmeijer, M. Fulle, E. Mazzotta-Epifani, P. Palumbo, L. Colangeli, J.J. Lopez-Moreno, J. Rodriguez, R. Morales, M. Cosi, GIADA - Grain Impact Analyzer and Dust Accumulator—onboard Rosetta spacecraft: extended calibrations. Acta Astronaut. **126**, 205–214 (2016)

V. Della Corte, A. Rotundi, V. Zakharov, S. Ivanovski, P. Palumbo, M. Fulle, A. Longobardo, Z. Dionnet, V. Liuzzi, GIADA microbalance measurements on board Rosetta: submicrometer- to micrometer-sized dust particle flux in the coma of comet 67P/Churyumov-Gerasimenko. Astron. Astrophys. **630**, A25 (2019)

J. Drążkowska, Y. Alibert, Planetesimal formation starts at the snow line. Astron. Astrophys. **608**, A92 (2017)

J. Drążkowska, C.P. Dullemond, Can dust coagulation trigger streaming instability? Astron. Astrophys. **572**, A78 (2014)

E. Drolshagen, T. Ott, D. Koschny, C. Güttler, C. Tubiana, J. Agarwal, H. Sierks, C. Barbieri, P.I. Lamy, R. Rodrigo, H. Rickman, M.F. A'Hearn, M.A. Barucci, J.-L. Bertaux, I. Bertini, G. Cremonese, V. da Deppo, B. Davidsson, S. Debei, M. de Cecco, J. Deller, C. Feller, S. Fornasier, M. Fulle, A. Gicquel, O. Groussin, P.J. Gutiérrez, M. Hofmann, S.F. Hviid, W.-H. Ip, L. Jorda, H.U. Keller, J. Knollenberg, J.R. Kramm, E. Kührt, M. Küppers, L.M. Lara, M. Lazzarin, J.J. Lopez Moreno, F. Marzari, G. Naletto, N. Oklay, X. Shi, N. Thomas, B. Poppe, Distance determination method of dust particles using Rosetta OSIRIS NAC and WAC data. Planet. Space Sci. **143**, 256–264 (2017)

P. Eberhardt, Comet Halley's gas composition and extended sources: results from the neutral mass spectrometer on Giotto, in *Composition and Origin of Cometary Materials*, (Springer, Berlin, 1999), pp. 45–52

M.R. El-Maarry, O. Groussin, N. Thomas, M. Pajola, A.-T. Auger, B. Davidsson, X. Hu, S.F. Hviid, J. Knollenberg, C. Güttler, Surface changes on comet 67P/Churyumov-Gerasimenko suggest a more active past. Science **355**(6332), 1392–1395 (2017)

L.E. Ellerbroek, B. Gundlach, A. Landeck, C. Dominik, J. Blum, S. Merouane, M. Hilchenbach, H. John, H.A. van Veen, The footprint of cometary dust analogues – II. Morphology as a tracer of tensile strength and application to dust collection by the Rosetta spacecraft. Mon. Not. R. Astron. Soc. **486**(3), 3755–3765 (2019)

G. Filacchione, M. De Sanctis, F. Capaccioni, A. Raponi, F. Tosi, M. Ciarniello, P. Cerroni, G. Piccioni, M. Capria, E. Palomba, Exposed water ice on the nucleus of comet 67P/Churyumov–Gerasimenko. Nature **529**(7586), 368 (2016a)

G. Filacchione, A. Raponi, F. Capaccioni, M. Ciarniello, F. Tosi, M. Capria, M. De Sanctis, A. Migliorini, G. Piccioni, P. Cerroni, Seasonal exposure of carbon dioxide ice on the nucleus of comet 67P/Churyumov-Gerasimenko. Science **354**(6319), 1563–1566 (2016b)

S. Fornasier, S. Mottola, H.U. Keller, M. Barucci, B. Davidsson, C. Feller, J. Deshapriya, H. Sierks, C. Barbieri, P. Lamy, Rosetta's comet 67P/Churyumov-Gerasimenko sheds its dusty mantle to reveal its icy nature. Science **354**, aag2671 (2016)

S. Fornasier, C. Feller, P.H. Hasselmann, M.A. Barucci, J. Sunshine, J.-B. Vincent, X. Shi, H. Sierks, G. Naletto, P.L. Lamy, R. Rodrigo, D. Koschny, B. Davidsson, J.-L. Bertaux, I. Bertini, D. Bodewits, G. Cremonese, V. Da Deppo, S. Debei, M. De Cecco, J. Deller, S. Ferrari, M. Fulle, P.J. Gutierrez, C. Güttler, W.-H. Ip, L. Jorda, H.U. Keller, M.L. Lara, M. Lazzarin, J.J. Lopez Moreno, A. Lucchetti, F. Marzari, S. Mottola, M. Pajola, I. Toth, C. Tubiana, Surface evolution of the Anhur region on comet 67P from high-resolution OSIRIS images (2019). arXiv:e-prints

N. Fougère, K. Altwegg, J.-J. Berthelier, A. Bieler, D. Bockelee-Morvan, U. Calmonte, F. Capaccioni, M.R. Combi, J. De Keyser, V. Debout, Three-dimensional direct simulation Monte-Carlo modeling of the coma of comet 67P/Churyumov-Gerasimenko observed by the VIRTIS and ROSINA instruments on board Rosetta. Astron. Astrophys. **588**, A134 (2016a)

N. Fougere, K. Altwegg, J.-J. Berthelier, A. Bieler, D. Bockelée-Morvan, U. Calmonte, F. Capaccioni, M.R. Combi, J. De Keyser, V. Debout, et al., Direct simulation Monte Carlo modelling of the major species in the coma of comet 67P/Churyumov-Gerasimenko. Mon. Not. R. Astron. Soc. **462**(1), S156–S169 (2016b)

N. Fray, A. Bardyn, H. Cottin, D. Baklouti, C. Briois, C. Engrand, H. Fischer, K. Hornung, R. Isnard, Y. Langevin, Nitrogen-to-carbon atomic ratio measured by COSIMA in the particles of comet 67P/Churyumov–Gerasimenko. Mon. Not. R. Astron. Soc. **469**(Suppl 2), S506–S516 (2017)

U. Frisk, M. Hagström, J. Ala-Laurinaho, S. Andersson, J.-C. Berges, J.-P. Chabaud, M. Dahlgren, A. Emrich, H.-G. Florén, G. Florin, The Odin satellite-I. Radiometer design and test. Astron. Astrophys. **402**(3), L27–L34 (2003)

M. Fulle, A.C. Levasseur-Regourd, N. McBride, E. Hadamcik, In situ dust measurements from within the coma of 1P/Halley: first-order approximation with a dust dynamical model. Astron. J. **119**, 1968–1977 (2000)

M. Fulle, C. Barbieri, G. Cremonese, H. Rauer, M. Weiler, G. Milani, R. Ligustri, The dust environment of comet 67P/Churyumov-Gerasimenko. Astron. Astrophys. **422**(1), 357–368 (2004)

M. Fulle, N. Altobelli, B. Buratti, M. Choukroun, M. Fulchignoni, E. Grün, M. Taylor, P. Weissman, Unexpected and significant findings in comet 67P/Churyumov–Gerasimenko: an interdisciplinary view. Mon. Not. R. Astron. Soc. **462**(Suppl 1), S2–S8 (2016a)

M. Fulle, F. Marzari, V. Della Corte, S. Fornasier, H. Sierks, A. Rotundi, C. Barbieri, P.L. Lamy, R. Rodrigo, D. Koschny, Evolution of the dust size distribution of comet 67P/Churyumov–Gerasimenko from 2.2 AU to perihelion. Astrophys. J. **821**(1), 19 (2016b)

M. Fulle, V. Della Corte, A. Rotundi, S. Green, M. Accolla, L. Colangeli, M. Ferrari, S. Ivanovski, R. Sordini, V. Zakharov, The dust-to-ices ratio in comets and Kuiper belt objects. Mon. Not. R. Astron. Soc. **469**(Suppl 2), S45–S49 (2017)

M. Fulle, J. Blum, S. Green, B. Gundlach, A. Herique, F. Moreno, S. Mottola, A. Rotundi, C. Snodgrass, The refractory-to-ice mass ratio in comets. Mon. Not. R. Astron. Soc. **482**(3), 3326–3340 (2019)

R. Funase, T. Inamori, S. Ikari, N. Ozaki, H. Koizumi, Initial operation results of a 50 kg-class deep space exploration micro-spacecraft PROCYON. In: *29th Annual AIAA/USU Conference on Small Satellites* (2015)

S. Gasc, K. Altwegg, H. Balsiger, J.-J. Berthelier, A. Bieler, U. Calmonte, B. Fiethe, S. Fuselier, A. Galli, T. Gombosi, Change of outgassing pattern of 67P/Churyumov–Gerasimenko during the March 2016 equinox as seen by ROSINA. Mon. Not. R. Astron. Soc. **469**(Suppl 2), S108–S117 (2017)

A. Gicquel, J.-B. Vincent, J. Agarwal, M. A'Hearn, I. Bertini, D. Bodewits, H. Sierks, Z.-Y. Lin, C. Barbieri, P. Lamy, Sublimation of icy aggregates in the coma of comet 67P/Churyumov–Gerasimenko detected with the OSIRIS cameras on board Rosetta. Mon. Not. R. Astron. Soc. **462**(Suppl 1), S57–S66 (2016)

K.-H. Glassmeier, H. Boehnhardt, D. Koschny, E. Kührt, I. Richter, The Rosetta mission: flying towards the origin of the solar system. Space Sci. Rev. **128**(1–4), 1–21 (2007)

J.M. Greenberg, A. Li, Morphological structure and chemical composition of cometary nuclei and dust. Space Sci. Rev. **90**(1–2), 149–161 (1999)

O. Groussin, G. Hahn, P.L. Lamy, R. Gonczi, G.B. Valsecchi, The long-term evolution and initial size of comets 46P/Wirtanen and 67P/Churyumov-Gerasimenko. Mon. Not. R. Astron. Soc. **376**, 1399–1406 (2007)

O. Groussin, N. Attree, Y. Brouet, V. Ciarletti, B. Davidsson, G. Filacchione, H.H. Fischer, B. Gundlach, M. Knapmeyer, J. Knollenberg, R. Kokotanekova, The thermal, mechanical, structural, and dielectric properties of cometary nuclei after Rosetta. Space Sci. Rev. **215**(4), 29 (2019)

E. Grün, H. Kochan, K. Seidensticker, Laboratory simulation, a tool for comet research. Geophys. Res. Lett. **18**(2), 245–248 (1991)

E. Grün, J. Agarwal, N. Altobelli, K. Altwegg, M.S. Bentley, N. Biver, V. Della Corte, N. Edberg, P. Feldman, M. Galand, The 2016 Feb 19 outburst of comet 67P/CG: an ESA Rosetta multi-instrument study. Mon. Not. R. Astron. Soc. **462**(Suppl 1), S220–S234 (2016)

S. Gulkis, M. Frerking, J. Crovisier, G. Beaudin, P. Hartogh, P. Encrenaz, T. Koch, C. Kahn, Y. Salinas, R. Nowicki, MIRO: microwave instrument for Rosetta orbiter. Space Sci. Rev. **128**(1–4), 561–597 (2007)

S. Gulkis, M. Allen, P. von Allmen, G. Beaudin, N. Biver, D. Bockelee-Morvan, M. Choukroun, J. Crovisier, B.J.R. Davidsson, P. Encrenaz, T. Encrenaz, M. Frerking, P. Hartogh, M. Hofstadter, W.-H. Ip, M. Janssen, C. Jarchow, S. Keihm, S. Lee, E. Lellouch, C. Leyrat, L. Rezac, F.P. Schloerb, T. Spilker, Subsurface properties and early activity of comet 67P/Churyumov-Gerasimenko. Science **347**(6220), aaa0709 (2015)

B. Gundlach, J. Blum, The stickiness of micrometer-sized water-ice particles. Astrophys. J. **798**(1), 34 (2014)

B. Gundlach, K.P. Schmidt, C. Kreuzig, D. Bischoff, F. Rezaei, S. Kothe, J. Blum, B. Grzesik, E. Stoll, The tensile strength of ice and dust aggregates and its dependence on particle properties. Mon. Not. R. Astron. Soc. **479**, 1273–1277 (2018)

C. Güttler, T. Mannel, A. Rotundi, S. Merouane, M. Fulle, D. Bockelée-Morvan, J. Lasue, A.C. Levasseur-Regourd, J. Blum, G. Naletto, Synthesis of the morphological description of cometary dust at comet 67P/Churyumov-Gerasimenko. Astron. Astrophys. **630**, A24 (2019)

K.C. Hansen, K. Altwegg, J.-J. Berthelier, A. Bieler, N. Biver, D. Bockelée-Morvan, U. Calmonte, F. Capaccioni, M. Combi, J. De Keyser, Evolution of water production of 67P/Churyumov–Gerasimenko: an empirical model and a multi-instrument study. Mon. Not. R. Astron. Soc. **462**(Suppl 1), S491–S506 (2016)

P. Hasselmann, A. Barucci, S. Fornasier, D. Bockelée-Morvan, J. Deshapriya, C. Feller, J. Sunshine, V. Hoang, H. Sierks, G. Naletto, P. Lamy, R. Rodrigo, D. Koschny, B. Davidsson, J.-L. Bertaux, I. Bertini, D. Bodewits, G. Cremonese, V. Da Deppo, S. Debei, M. Fulle, P. Gutierrez, C. Güttler, J. Deller, W.-H. Ip, H.U. Keller, L. Lara, M. De Cecco, M. Lazzarin, F. Marzari, X. Shi, C. Tubiana, Pronounced morphological changes in a southern active zone on comet 67P/Churyumov-Gerasimenko. Astron. Astrophys. **630**, A8 (2019). accepted

M. Hässig, K. Altwegg, H. Balsiger, A. Bar-Nun, J.-J. Berthelier, A. Bieler, P. Bochsler, C. Briois, U. Calmonte, M. Combi, Time variability and heterogeneity in the coma of 67P/Churyumov-Gerasimenko. Science **347**(6220), aaa0276 (2015)

A. Hérique, W. Kofman, P. Beck, L. Bonal, I. Buttarazzi, E. Heggy, J. Lasue, A.C. Levasseur-Regourd, E. Quirico, S. Zine, Cosmochemical implications of CONSERT permittivity characterization of 67P/CG. Mon. Not. R. Astron. Soc. **462**(Suppl 1), S516–S532 (2016)

A. Hérique, B. Agnus, E. Asphaug, A. Barucci, P. Beck, J. Bellerose, J. Biele, L. Bonal, P. Bousquet, L. Bruzzone, Direct observations of asteroid interior and regolith structure: science measurement requirements. Adv. Space Res. **62**(8), 2141–2162 (2018)

A. Hérique, W. Kofman, S. Zine, J. Blum, J. Vincent, V. Ciarletti, Homogeneity of 67P as seen by CONSERT: implication on composition and formation. Astron. Astrophys. **630**, A6 (2019)

G. Herman, P.R. Weissman, Numerical simulation of cometary nuclei: III. Internal temperatures of cometary nuclei. Icarus **69**(2), 314–328 (1987)

M. Hilchenbach, H. Fischer, Y. Langevin, S. Merouane, J. Paquette, J. Rynö, O. Stenzel, C. Briois, J. Kissel, A. Koch, Mechanical and electrostatic experiments with dust particles collected in the inner coma of comet 67P by COSIMA onboard Rosetta. Philos. Trans. R. Soc., Math. Phys. Eng. Sci. **375**(2097), 20160255 (2017)

M. Hirabayashi, D.J. Scheeres, S.R. Chesley, S. Marchi, J.W. McMahon, J. Steckloff, S. Mottola, S.P. Naidu, T. Bowling, Fission and reconfiguration of bilobate comets as revealed by 67P/Churyumov–Gerasimenko. Nature **534**(7607), 352 (2016)

K. Homma, T. Nakamoto, Collisional growth of icy dust aggregates in the disk formation stage: difficulties for planetesimal formation via direct collisional growth outside the snowline. Astrophys. J. **868**(2), 118 (2018)

P. Hoppe, M. Rubin, K. Altwegg, Presolar isotopic signatures in meteorites and comets: new insights from the Rosetta mission to comet 67P/Churyumov–Gerasimenko. Space Sci. Rev. **214**(6), 106 (2018)

K. Hornung, S. Merouane, M. Hilchenbach, Y. Langevin, E.M. Mellado, V. Della Corte, J. Kissel, C. Engrand, R. Schulz, J. Ryno, A first assessment of the strength of cometary particles collected in-situ by the COSIMA instrument onboard ROSETTA. Planet. Space Sci. **133**, 63–75 (2016)

X. Hu, X. Shi, H. Sierks, J. Blum, J. Oberst, M. Fulle, E. Kührt, C. Güttler, B. Gundlach, H.U. Keller, S. Mottola, M. Pajola, C. Barbieri, P.L. Lamy, R. Rodrigo, D. Koschny, H. Rickman, J. Agarwal, M.F. A'Hearn, M.A. Barucci, J.-L. Bertaux, I. Bertini, S. Boudreault, I. Büttner, G. Cremonese, V. Da Deppo, B. Davidsson, S. Debei, M. De Cecco, J. Deller, S. Fornasier, O. Groussin, P.J. Gutiérrez, P. Gutiérrez-Marqués, I. Hall, M. Hofmann, S.F. Hviid, W.-H. Ip, L. Jorda, J. Knollenberg, G. Kovacs, J.R. Kramm, M. Küppers, L.M. Lara, M. Lazzarin, J.J. López-Moreno, F. Marzari, G. Naletto, N. Oklay, M.L. Richards, J. Ripken, N. Thomas, C. Tubiana, J.-B. Vincent, Thermal modelling of water activity on comet 67P/Churyumov-Gerasimenko with global dust mantle and plural dust-to-ice ratio. Mon. Not. R. Astron. Soc. **469**, S295–S311 (2017a)

X. Hu, X. Shi, H. Sierks, M. Fulle, J. Blum, H.U. Keller, E. Kührt, B. Davidsson, C. Güttler, B. Gundlach, M. Pajola, D. Bodewits, J.-B. Vincent, N. Oklay, M. Massironi, S. Fornasier, C. Tubiana, O. Groussin, S. Boudreault, S. Höfner, S. Mottola, C. Barbieri, P.L. Lamy, R. Rodrigo, D. Koschny, H. Rickman, M. A'Hearn, J. Agarwal, M.A. Barucci, J.-L. Bertaux, I. Bertini, G. Cremonese, V. Da Deppo, S. Debei, M. De Cecco, J. Deller, M.R. El-Maarry, A. Gicquel, P. Gutierrez-Marques, P.J. Gutiérrez, M. Hofmann, S.F. Hviid, W.-H. Ip, L. Jorda, J. Knollenberg, G. Kovacs, J.-R. Kramm, M. Küppers, L.M. Lara, M. Lazzarin, J.J. Lopez-Moreno, F. Marzari, G. Naletto, N. Thomas, Seasonal erosion and restoration of the dust cover on comet 67P/Churyumov-Gerasimenko as observed by OSIRIS onboard Rosetta. Astron. Astrophys. **604**, A114 (2017b)

A. Johansen, J. Blum, H. Tanaka, C. Ormel, M. Bizzarro, H. Rickman, The multifaceted planetesimal formation process (2014). ArXiv preprint. arXiv:1402.1344

L. Jorda, R. Gaskell, C. Capanna, S. Hviid, P. Lamy, J. Ďurech, G. Faury, O. Groussin, P. Gutiérrez, C. Jackman, The global shape, density and rotation of comet 67P/Churyumov-Gerasimenko from preperihelion Rosetta/OSIRIS observations. Icarus **277**, 257–278 (2016)

D.J. Joswiak, D.E. Brownlee, A.N. Nguyen, S. Messenger, Refractory materials in comet samples. Meteorit. Planet. Sci. **52**, 1612–1648 (2017)

B.A. Keeney, S.A. Stern, M.F. A'hearn, J.-L. Bertaux, L.M. Feaga, P.D. Feldman, R.A. Medina, J.W. Parker, J.P. Pineau, E. Schindhelm, $H_2O$ and $O_2$ absorption in the coma of comet 67P/Churyumov–Gerasimenko measured by the Alice far-ultraviolet spectrograph on Rosetta. Mon. Not. R. Astron. Soc. **469**(Suppl 2), S158–S177 (2017)

B.A. Keeney, S.A. Stern, P.D. Feldman, M.F. A'Hearn, J.-L. Bertaux, L.M. Feaga, M.M. Knight, R.A. Medina, J. Noonan, J.W. Parker, J.P. Pineau, R.N. Schindhelm, A.J. Steffl, M. Versteeg, R.J. Vervack Jr., H.A. Weaver, Stellar occultation by comet 67P/Churyumov-Gerasimenko observed with Rosetta's Alice far-ultraviolet spectrograph. Astron. J. **157**, 5 (2019)

H.U. Keller, C. Barbieri, P. Lamy, H. Rickman, R. Rodrigo, K.-P. Wenzel, H. Sierks, M.F. A'Hearn, F. Angrilli, M. Angulo, OSIRIS–the scientific camera system onboard Rosetta. Space Sci. Rev. **128**(1–4), 433–506 (2007)

H.U. Keller, S. Mottola, B. Davidsson, S. Schröder, Y. Skorov, E. Kührt, O. Groussin, M. Pajola, S.F. Hviid, F. Preusker, Insolation, erosion, and morphology of comet 67P/Churyumov-Gerasimenko. Astron. Astrophys. **583**, A34 (2015)

H.U. Keller, S. Mottola, S.F. Hviid, J. Agarwal, E. Kührt, Y. Skorov, K. Otto, J.-B. Vincent, N. Oklay, S.E. Schröder, Seasonal mass transfer on the nucleus of comet 67P/Chuyumov–Gerasimenko. Mon. Not. R. Astron. Soc. **469**(Suppl 2), S357–S371 (2017)

J. Kissel, K. Altwegg, B. Clark, L. Colangeli, H. Cottin, S. Czempiel, J. Eibl, C. Engrand, H. Fehringer, B. Feuerbacher, COSIMA–high resolution time-of-flight secondary ion mass spectrometer for the analysis of cometary dust particles onboard Rosetta. Space Sci. Rev. **128**(1–4), 823–867 (2007)

M. Knapmeyer, H.-H. Fischer, J. Knollenberg, K. Seidensticker, K. Thiel, W. Arnold, C. Faber, D. Möhlmann, Structure and elastic parameters of the near surface of Abydos site on comet 67P/Churyumov–Gerasimenko, as obtained by SESAME/CASSE listening to the MUPUS insertion phase. Icarus **310**, 165–193 (2018)

H. Kochan, K. Roessler, L. Ratke, M. Heyl, H. Hellmann, G. Schwehm, Crustal strength of different model comet materials, in *Physics and Mechanics of Cometary Materials* (1989)

W. Kofman, Y. Barbin, J. Klinger, A.-C. Levasseur-Regourd, J.-P. Barriot, A. Herique, T. Hagfors, E. Nielsen, E. Grün, P. Edenhofer, Comet nucleus sounding experiment by radiowave transmission. Adv. Space Res. **21**(11), 1589–1598 (1998)

W. Kofman, A. Hérique, J.-P. Goutail, T. Hagfors, I. Williams, E. Nielsen, J.-P. Barriot, Y. Barbin, C. Elachi, P. Edenhofer, The comet nucleus sounding experiment by radiowave transmission (CONSERT): a short description of the instrument and of the commissioning stages. Space Sci. Rev. **128**(1–4), 413–432 (2007)

W. Kofman, A. Herique, Y. Barbin, J.-P. Barriot, V. Ciarletti, S. Clifford, P. Edenhofer, C. Elachi, C. Eyraud, J.-P. Goutail, Properties of the 67P/Churyumov-Gerasimenko interior revealed by CONSERT radar. Science **349**(6247), aab0639 (2015)

T. Kramer, M. Läuter, Outgassing induced acceleration of comet 67P/Churyumov-Gerasimenko (2019). arXiv:e-prints

T. Kramer, M. Läuter, M. Rubin, K. Altwegg, Seasonal changes of the volatile density in the coma and on the surface of comet 67P/Churyumov–Gerasimenko. Mon. Not. R. Astron. Soc. **469**(Suppl 2), S20–S28 (2017)

T. Kramer, M. Läuter, S. Hviid, L. Jorda, H.U. Keller, E. Kührt, Comet 67P/Churyumov-Gerasimenko rotation changes derived from sublimation induced torques (2018). arXiv:e-prints

S. Krijt, C.W. Ormel, C. Dominik, A.G. Tielens, A panoptic model for planetesimal formation and pebble delivery. Astron. Astrophys. **586**, A20 (2016)

M. Krolikowska, 67P/Churyumov-Gerasimenko – potential target for the Rosetta mission. Acta Astron. **53**, 195–209 (2003)

M. Küppers, I. Bertini, S. Fornasier, P.J. Gutierrez, S.F. Hviid, L. Jorda, H.U. Keller, J. Knollenberg, D. Koschny, R. Kramm, L.-M. Lara, H. Sierks, N. Thomas, C. Barbieri, P. Lamy, H. Rickman, R. Rodrigo, M.F. A'Hearn, F. Angrilli, M. Bailey, P. Barthol, M.A. Barucci, J.-L. Bertaux, J.A. Burns, G. Cremonese, W. Curdt, M. De Cecco, S. Debei, M. Fulle, F. Gliem, W.H. Ip, E. Huhrt, A. Llebaria, J.J. Lopez Moreno, F. Marzari, G. Naletto, L. Sabau, A. Sanz Andres, J.P. Sivan, G. Tondello, K.-P. Wenzel, A large dust/ice ratio in the nucleus of comet 9P/Tempel 1. Nature **437**, 987–990 (2005)

M. Küppers, H.U. Keller, E. Kührt, M.F. A'Hearn, K. Altwegg, R. Bertrand, H. Busemann, M.T. Capria, L. Colangeli, B. Davidsson, P. Ehrenfreund, J. Knollenberg, S. Mottola, A. Rathke, P. Weiss, M. Zolensky, E. Akim, A. Basilevsky, E. Galimov, M. Gerasimov, O. Korablev, I. Lomakin, M. Marov, M. Martynov, M. Nazarov, A. Zakharov, L. Zelenyi, A. Aronica, A.J. Ball, C. Barbieri, A. Bar-Nun, J. Benkhoff, J. Biele, N. Biver, J. Blum, D. Bockelée-Morvan, O. Botta, J.-H. Bredehöft, F. Capaccioni, S. Charnley, E. Cloutis, H. Cottin, G. Cremonese, J. Crovisier, S.A. Crowther, E.M. Epifani, F. Esposito, A.C. Ferrari, F. Ferri, M. Fulle, J. Gilmour, F. Goesmann, N. Gortsas, S.F. Green, O. Groussin, E. Grün, P.J. Gutiérrez, P. Hartogh, T. Henkel, M. Hilchenbach, T.-M. Ho, G. Horneck, S.F. Hviid, W.-H. Ip, A. Jäckel, E. Jessberger, R. Kallenbach, G. Kargl, N.I. Kömle, A. Korth, K. Kossacki, C. Krause, H. Krüger, Z.-Y. Li, J. Licandro, J.J. Lopez-Moreno, S.C. Lowry, I. Lyon, G. Magni, U. Mall, I. Mann, W. Markiewicz, Z. Martins, M. Maurette, U. Meierhenrich, V. Mennella, T.C. Ng, L.R. Nittler, P. Palumbo, M. Pätzold, D. Prialnik, M. Rengel, H. Rickman, J. Rodriguez, R. Roll, D. Rost, A. Rotundi, S. Sandford, M. Schönbächler, H. Sierks, R. Srama, R.M. Stroud, S. Szutowicz, C. Tornow, S. Ulamec, M. Wallis, W. Waniak, P. Weissman, R. Wieler, P. Wurz, K.L. Yung, J.C. Zarnecki, Triple F—a comet nucleus sample return mission. Exp. Astron. **23**, 809–847 (2009)

Y. Langevin, M. Hilchenbach, N. Ligier, S. Merouane, K. Hornung, C. Engrand, R. Schulz, J. Kissel, J. Rynö, P. Eng, Typology of dust particles collected by the COSIMA mass spectrometer in the inner coma of 67P/Churyumov Gerasimenko. Icarus **271**, 76–97 (2016)

Y. Langevin, M. Hilchenbach, M. Vincendon, S. Merouane, K. Hornung, N. Ligier, C. Engrand, R. Schulz, J. Kissel, J. Rynö, Optical properties of cometary particles collected by the COSIMA mass spectrometer on-board Rosetta during the rendezvous phase around comet 67P/Churyumov–Gerasimenko. Mon. Not. R. Astron. Soc. **469**(Suppl 2), S535–S549 (2017)

J. Lasue, I. Maroger, R. Botet, P. Garnier, S. Merouane, T. Mannel, A.-C. Levasseur-Regourd, M. Bentley, Flattened loose particles from numerical simulations compared to Rosetta collected particles (2019). ArXiv:e-prints

M. Läuter, T. Kramer, M. Rubin, K. Altwegg, Surface localization of gas sources on comet 67P/Churyumov-Gerasimenko based on DFMS/COPS data. Mon. Not. R. Astron. Soc. **483**, 852–861 (2019)

L. Le Roy, K. Altwegg, H. Balsiger, J.-J. Berthelier, A. Bieler, C. Briois, U. Calmonte, M.R. Combi, J. De Keyser, F. Dhooghe, Inventory of the volatiles on comet 67P/Churyumov-Gerasimenko from Rosetta/ROSINA. Astron. Astrophys. **583**, A1 (2015)

A. Lethuillier, A. Le Gall, M. Hamelin, W. Schmidt, K.J. Seidensticker, R. Grard, V. Ciarletti, S. Caujolle-Bert, H.-H. Fischer, R. Trautner, Electrical properties and porosity of the first meter of the nucleus of 67P/Churyumov-Gerasimenko-as constrained by the permittivity probe SESAME-PP/Philae/Rosetta. Astron. Astrophys. **591**, A32 (2016)

A.C. Levasseur-Regourd, N. McBride, E. Hadamcik, M. Fulle, Similarities between in situ measurements of local dust light scattering and dust flux impact data within the coma of 1P/Halley. Astron. Astrophys. **348**, 636–641 (1999)

A.-C. Levasseur-Regourd, J. Agarwal, H. Cottin, C. Engrand, G. Flynn, M. Fulle, T. Gombosi, Y. Langevin, J. Lasue, T. Mannel, Cometary dust. Space Sci. Rev. **214**(3), 64 (2018)

S. Lorek, B. Gundlach, P. Lacerda, J. Blum, Comet formation in collapsing pebble clouds-What cometary bulk density implies for the cloud mass and dust-to-ice ratio. Astron. Astrophys. **587**, A128 (2016)

S. Lorek, P. Lacerda, J. Blum, Local growth of dust-and ice-mixed aggregates as cometary building blocks in the solar nebula. Astron. Astrophys. **611**, A18 (2018)

A. Mahjoub, M.J. Poston, J. Blacksberg, J.M. Eiler, M.E. Brown, B.L. Ehlmann, R. Hodyss, K.P. Hand, R. Carlson, M. Choukroun, Production of sulfur allotropes in electron irradiated Jupiter Trojans ice analogs. Astrophys. J. **846**(2), 148 (2017)

T. Mannel, M.S. Bentley, R. Schmied, H. Jeszenszky, A.C. Levasseur-Regourd, J.R.K. Torkar, Fractal cometary dust–a window into the early Solar system. Mon. Not. R. Astron. Soc. **462**, stw2898 (2016)

L. Maquet, The recent dynamical history of comet 67P/Churyumov-Gerasimenko. Astron. Astrophys. **579**, A78 (2015)

D. Marshall, P. Hartogh, L. Rezac, P. von Allmen, N. Biver, D. Bockelée-Morvan, J. Crovisier, P. Encrenaz, S. Gulkis, M. Hofstadter, Spatially resolved evolution of the local $H_2O$ production rates of comet 67P/Churyumov-Gerasimenko from the MIRO instrument on Rosetta. Astron. Astrophys. **603**, A87 (2017)

D. Marshall, O. Groussin, J. Vincent, Y. Brouet, D. Kappel, G. Arnold, M. Capria, G. Filacchione, P. Hartogh, M. Hofstadter, et al., Thermal inertia and roughness of the nucleus of comet 67P/Churyumov-Gerasimenko from MIRO and VIRTIS observations. Astron. Astrophys. **616**, A122 (2018)

J. McDonnell, P. Lamy, G. Pankiewicz, *Physical Properties of Cometary Dust. International Astronomical Union Colloquium* (Cambridge University Press, Cambridge, 1991)

S. Merouane, B. Zaprudin, O. Stenzel, Y. Langevin, N. Altobelli, V. Della Corte, H. Fischer, M. Fulle, K. Hornung, J. Silén, Dust particle flux and size distribution in the coma of 67P/Churyumov-Gerasimenko measured in situ by the COSIMA instrument on board Rosetta. Astron. Astrophys. **596**, A87 (2016)

P. Michel, A. Cheng, M. Küppers, P. Pravec, J. Blum, M. Delbo, S. Green, P. Rosenblatt, K. Tsiganis, J.-B. Vincent, Science case for the asteroid impact mission (AIM): a component of the asteroid impact & deflection assessment (AIDA) mission. Adv. Space Res. **57**(12), 2529–2547 (2016)

F. Moreno, C. Snodgrass, O. Hainaut, C. Tubiana, H. Sierks, C. Barbieri, P.L. Lamy, R. Rodrigo, D. Koschny, H. Rickman, The dust environment of comet 67P/Churyumov-Gerasimenko from Rosetta OSIRIS and VLT observations in the 4.5 to 2.9 AU heliocentric distance range inbound. Astron. Astrophys. **587**, A155 (2016)

S. Mottola, G. Arnold, H.-G. Grothues, R. Jaumann, H. Michaelis, G. Neukum, J.-P. Bibring, S. Schröder, M. Hamm, K. Otto, The structure of the regolith on 67P/Churyumov-Gerasimenko from ROLIS descent imaging. Science **349**(6247), aab0232 (2015)

M.J. Mumma, S.B. Charnley, The chemical composition of comets—emerging taxonomies and natal heritage. Annu. Rev. Astron. Astrophys. **49**, 471–524 (2011)

G. Musiolik, G. Wurm, Contacts of water ice in protoplanetary disks—laboratory experiments. Astrophys. J. **873**, 1 (2019)

T. Ott, E. Drolshagen, D. Koschny, C. Güttler, C. Tubiana, E. Frattin, J. Agarwal, H. Sierks, I. Bertini, C. Barbieri, Dust mass distribution around comet 67P/Churyumov–Gerasimenko determined via parallax measurements using Rosetta's OSIRIS cameras. Mon. Not. R. Astron. Soc. **469**(Suppl 2), S276–S284 (2017)

M. Pajola, S. Höfner, J.-B. Vincent, N. Oklay, F. Scholten, F. Preusker, S. Mottola, G. Naletto, S. Fornasier, S. Lowry, The pristine interior of comet 67P revealed by the combined Aswan outburst and cliff collapse. Nat. Astron. **1**(5), 0092 (2017)

M. Pätzold, B. Häusler, K. Aksnes, J.D. Anderson, S.W. Asmar, J.-P. Barriot, M.K. Bird, H. Boehnhardt, W. Eidel, E. Grün, Rosetta radio science investigations (RSI). Space Sci. Rev. **128**(1–4), 599–627 (2007)

M. Pätzold, T. Andert, M. Hahn, S. Asmar, J.-P. Barriot, M. Bird, B. Häusler, K. Peter, S. Tellmann, E. Grün, A homogeneous nucleus for comet 67P/Churyumov–Gerasimenko from its gravity field. Nature **530**(7588), 63 (2016)

M. Pätzold, T.P. Andert, M. Hahn, J.-P. Barriot, S.W. Asmar, B. Häusler, M.K. Bird, S. Tellmann, J. Oschlisniok, K. Peter, The Nucleus of comet 67P/Churyumov–Gerasimenko–Part I: The global view–nucleus mass, mass-loss, porosity, and implications. Mon. Not. R. Astron. Soc. **483**(2), 2337–2346 (2019)

G. Picardi, J.J. Plaut, D. Biccari, O. Bombaci, D. Calabrese, M. Cartacci, A. Cicchetti, S.M. Clifford, P. Edenhofer, W.M. Farrell, Radar soundings of the subsurface of Mars. Science **310**(5756), 1925–1928 (2005)

F. Preusker, F. Scholten, K.-D. Matz, T. Roatsch, K. Willner, S. Hviid, J. Knollenberg, L. Jorda, P.J. Gutiérrez, E. Kührt, Shape model, reference system definition, and cartographic mapping standards for comet 67P/Churyumov-Gerasimenko–Stereo-photogrammetric analysis of Rosetta/OSIRIS image data. Astron. Astrophys. **583**, A33 (2015)

F. Preusker, F. Scholten, K.-D. Matz, T. Roatsch, S. Hviid, S. Mottola, J. Knollenberg, E. Kührt, M. Pajola, N. Oklay, The global meter-level shape model of comet 67P/Churyumov-Gerasimenko. Astron. Astrophys. **607**, L1 (2017)

D. Prialnik, J. Benkhoff, M. Podolak, Modeling the structure and activity of comet nuclei, in *Comets II* (2004), pp. 359–387

D. Prialnik, G. Sarid, E.D. Rosenberg, R. Merk, Thermal and chemical evolution of comet nuclei and Kuiper belt objects, in *Origin and Early Evolution of Comet Nuclei* (Springer, Berlin, 2008), pp. 147–164

E. Quirico, L. Moroz, B. Schmitt, G. Arnold, M. Faure, P. Beck, L. Bonal, M. Ciarniello, F. Capaccioni, G. Filacchione, Refractory and semi-volatile organics at the surface of comet 67P/Churyumov-Gerasimenko: insights from the VIRTIS/Rosetta imaging spectrometer. Icarus **272**, 32–47 (2016)

R. Reinhard, The Giotto encounter with comet Halley. Nature **321**(6067), 313–318 (1986)

W. Riedler, K. Torkar, H. Jeszenszky, J. Romstedt, H.S.C. Alleyne, H. Arends, W. Barth, J. Biezen, B. Butler, P. Ehrenfreund, MIDAS–the micro-imaging dust analysis system for the Rosetta mission. Space Sci. Rev. **128**(1–4), 869–904 (2007)

A. Rotundi, H. Sierks, V. Della Corte, M. Fulle, P.J. Gutierrez, L. Lara, C. Barbieri, P.L. Lamy, R. Rodrigo, D. Koschny, Dust measurements in the coma of comet 67P/Churyumov-Gerasimenko inbound to the Sun. Science **347**(6220), aaa3905 (2015)

M. Rubin, K. Altwegg, H. Balsiger, A. Bar-Nun, J.-J. Berthelier, A. Bieler, P. Bochsler, C. Briois, U. Calmonte, M. Combi, Molecular nitrogen in comet 67P/Churyumov-Gerasimenko indicates a low formation temperature. Science **348**, aaa6100 (2015)

M. Rubin, K. Altwegg, H. Balsiger, J.-J. Berthelier, M.R. Combi, J. De Keyser, M. Drozdovskaya, B. Fiethe, S.A. Fuselier, S. Gasc, T.I. Gombosi, N. Hänni, K.C. Hansen, U. Mall, H. Rème, I.R.H.G. Schroeder, M. Schuhmann, T. Sémon, J.H. Waite, S.F. Wampfler, P. Wurz, Elemental and molecular abundances in comet 67P/Churyumov-Gerasimenko. Mon. Not. R. Astron. Soc. **489**(1), 594–607 (2019)

F.P. Schloerb, S. Keihm, P. von Allmen, M. Choukroun, E. Lellouch, C. Leyrat, G. Beaudin, N. Biver, D. Bockelee-Morvan, J. Crovisier, P. Encrenaz, R. Gaske, S. Gulkis, P. Hartogh, M. Hofstadter, W.-H. Ip, M. Janssen, C. Jarchow, L. Jorda, H.U. Keller, S. Lee, L. Rezac, H. Sierks, MIRO observations of subsurface temperatures of the nucleus of 67P/Churyumov-Gerasimenko. Astron. Astrophys. **583**, A29 (2015)

M. Schmitt, C. Tubiana, C. Güttler, H. Sierks, J.-B. Vincent, M.R. El-Maarry, D. Bodewits, S. Mottola, S. Fornasier, M. Hofmann, Long-term monitoring of comet 67P/Churyumov–Gerasimenko's jets with OSIRIS onboard Rosetta. Mon. Not. R. Astron. Soc. **469**(Suppl 2), S380–S385 (2017)

D. Schoonenberg, C.W. Ormel, Planetesimal formation near the snowline: in or out? Astron. Astrophys. **602**, A21 (2017)

M. Schuhmann, K. Altwegg, H. Balsiger, et al., Aliphatic and aromatic hydrocarbons in comet 67P/Churyumov-Gerasimenko seen by ROSINA. Astron. Astrophys. **630**, A31 (2019)

R. Schulz, M. Hilchenbach, Y. Langevin, J. Kissel, J. Silen, C. Briois, C. Engrand, K. Hornung, D. Baklouti, A. Bardyn, Comet 67P/Churyumov-Gerasimenko sheds dust coat accumulated over the past four years. Nature **518**(7538), 216 (2015)

K. Seidensticker, D. Möhlmann, I. Apathy, W. Schmidt, K. Thiel, W. Arnold, H.-H. Fischer, M. Kretschmer, D. Madlener, A. Péter, Sesame–an experiment of the rosetta lander philae: objectives and general design. Space Sci. Rev. **128**(1–4), 301–337 (2007)

X. Shi, X. Hu, H. Sierks, C. Güttler, M. A'hearn, J. Blum, M.R. El-Maarry, E. Kührt, S. Mottola, M. Pajola, Sunset jets observed on comet 67P/Churyumov-Gerasimenko sustained by subsurface thermal lag. Astron. Astrophys. **586**, A7 (2016)

X. Shi, X. Hu, S. Mottola, H. Sierks, H. Keller, M. Rose, C. Güttler, M. Fulle, S. Fornasier, J. Agarwal, Coma morphology of comet 67P controlled by insolation over irregular nucleus. Nat. Astron. **2**, 562–567 (2018)

Y. Shinnaka, N. Fougere, H. Kawakita, S. Kameda, M.R. Combi, S. Ikezawa, A. Seki, M. Kuwabara, M. Sato, M. Taguchi, Imaging observations of the hydrogen coma of comet 67P/Churyumov–Gerasimenko in 2015 September by the Procyon/Laica. Astron. J. **153**(2), 76 (2017)

H. Sierks, C. Barbieri, P.L. Lamy, R. Rodrigo, D. Koschny, H. Rickman, H.U. Keller, J. Agarwal, M.F. A'hearn, F. Angrilli, On the nucleus structure and activity of comet 67P/Churyumov-Gerasimenko. Science **347**(6220), aaa1044 (2015)

C. Snodgrass, G. Jones, H. Boehnhardt, A. Gibbings, M. Homeister, N. Andre, P. Beck, M. Bentley, I. Bertini, N. Bowles, The Castalia mission to main belt comet 133P/Elst-Pizarro. Adv. Space Res. **62**(8), 1947–1976 (2018)

T. Spohn, J. Knollenberg, A.J. Ball, M. Banaszkiewicz, J. Benkhoff, M. Grott, J. Grygorczuk, C. Hüttig, A. Hagermann, G. Kargl, Thermal and mechanical properties of the near-surface layers of comet 67P/Churyumov-Gerasimenko. Science **349**(6247), aab0464 (2015)

S. Stern, D. Slater, J. Scherrer, J. Stone, M. Versteeg, M. A'hearn, J.-L. Bertaux, P. Feldman, M. Festou, J.W. Parker, Alice: the Rosetta ultraviolet imaging spectrograph. Space Sci. Rev. **128**(1–4), 507–527 (2007)

M.V. Sykes, E. Grün, W.T. Reach, P. Jenniskens, The interplanetary dust complex and comets, in *Comets II* (2004), pp. 677–693

M.G.G.T. Taylor, N. Altobelli, B.J. Buratti, M. Choukroun, The Rosetta mission orbiter science overview: the comet phase. Philos. Trans. R. Soc., Math. Phys. Eng. Sci. **375**(2097), 20160262 (2017)

N. Thomas, B. Davidsson, M.R. El-Maarry, S. Fornasier, L. Giacomini, A. Gracia-Berná, S. Hviid, W.-H. Ip, L. Jorda, H. Keller, Redistribution of particles across the nucleus of comet 67P/Churyumov-Gerasimenko. Astron. Astrophys. **583**, A17 (2015)

C. Vallat, N. Altobelli, B. Geiger, B. Grieger, M. Kueppers, C.M. Crego, R. Moissl, M.G. Taylor, C. Alexander, B. Buratti, M. Choukroun, The science planning process on the Rosetta mission. Acta Astronaut. **133**, 244–257 (2017)

J.-B. Vincent, N. Oklay, M. Pajola, S. Höfner, H. Sierks, X. Hu, C. Barbieri, P.L. Lamy, R. Rodrigo, D. Koschny, Are fractured cliffs the source of cometary dust jets? Insights from OSIRIS/Rosetta at 67P/Churyumov-Gerasimenko. Astron. Astrophys. **587**, A14 (2016a)

J.-B. Vincent, M.F. A'Hearn, Z.-Y. Lin, M.R. El-Maarry, M. Pajola, H. Sierks, C. Barbieri, P.L. Lamy, R. Rodrigo, D. Koschny, Summer fireworks on comet 67P. Mon. Not. R. Astron. Soc. **462**(Suppl 1), S184–S194 (2016b)

J. Vincent, T. Farnham, E. Kührt et al., Local manifestations of cometary activity. Space Sci. Rev. **215**, 30 (2019)

K. Wada, H. Tanaka, T. Suyama, H. Kimura, T. Yamamoto, The rebound condition of dust aggregates revealed by numerical simulation of their collisions. Astrophys. J. **737**(1), 36 (2011)

P.R. Weissman, S.C. Lowry, Structure and density of cometary nuclei. Meteorit. Planet. Sci. **43**(6), 1033–1047 (2008)

P. Weissman, E. Asphaug, S. Lowry, Structure and density of cometary nuclei, in *Comets II*, ed. by K. Festou, H.U. Keller, H.A. Weaver (University of Arizona Press, Tucson, 2004), pp. 337–357

D. Winter, J. Saari, A particulate thermophysical model of the lunar soil. Astrophys. J. **156**, 1135 (1969)

V. Zakharov, J.-F. Crifo, A. Rodionov, M. Rubin, K. Altwegg, The near-nucleus gas coma of comet 67P/Churyumov-Gerasimenko prior to the descent of the surface lander PHILAE. Astron. Astrophys. **618**, A71 (2018)

A. Zsom, C.W. Ormel, C. Güttler, J. Blum, C. Dullemond, The outcome of protoplanetary dust growth: pebbles, boulders, or planetesimals?-II. Introducing the bouncing barrier. Astron. Astrophys. **513**, A57 (2010)

Space Sci Rev (2020) 216:2
https://doi.org/10.1007/s11214-019-0627-5

# Nongravitational Effects of Cometary Activity

**Stefano Mottola[1] · Nicholas Attree[2] · Laurent Jorda[3] · Horst Uwe Keller[4,1] · Rosita Kokotanekova[5] · David Marshall[6] · Yuri Skorov[4]**

Received: 20 May 2019 / Accepted: 2 December 2019 / Published online: 6 January 2020


**Abstract** Nongravitational forces induced by sublimation of volatiles affect the rotational and orbital dynamics of comets. In addition to contributing to the improvement of ephemerides and of rotational models, nongravitational effects can help constrain the regional distribution and temporal evolution of cometary activity, which in turn provides input for the development of thermophysical and dust transport models. We review the progress that has been made in this field thanks to the Rosetta mission and we highlight the open questions.

**Keywords** Nongravitational acceleration · Orbital evolution · Cometary activity

## 1 Introduction

The very defining property of comets—the one that sets comets apart from other minor bodies—is the occurrence of gradual sublimation of their icy components under the influ-

Comets: Post 67P / Churyumov-Gerasimenko Perspectives
Edited by Nicolas Thomas, Björn Davidsson, Laurent Jorda, Ekkehard Kührt, Raphael Marschall, Colin Snodgrass and Rafael Rodrigo

✉ S. Mottola
stefano.mottola@dlr.de

1 Institute of Planetary Research, DLR, Rutherfordstrasse 2, 12489 Berlin, Germany

2 Earth and Planetary Observation Centre, Dept. of Biological and Environmental Sciences, University of Stirling, Stirling, UK

3 Laboratoire d'Astrophysique de Marseille, Aix Marseille Univ., CNRS, CNES, 13388 Marseille Cedex 13, France

4 Institute for Geophysics and Extraterrestrial Physics, TU Braunschweig, 38106 Braunschweig, Germany

5 European Southern Observatory (ESO), Karl-Schwarzschild Str. 2, 85748 Garching bei München, Germany

6 Max-Planck-Institut für Sonnensystemforschung, Justus-von-Liebig-Weg, 3, 37077 Göttingen, Germany

ence of solar irradiation. Sublimation results in the release of gas and debris (ranging in size from sub-μm to several meters, and often collectively called cometary dust) that form the comet's coma and tail. The complex manifestations that accompany sublimation, and that are generally referred to as cometary activity, have a profound influence on the comet evolution, contributing to sculpt and re-organize its shape, change its mass, modify its orbit and rotation state, and possibly lead it to its disintegration.

By studying the effects of activity it is possible to infer the physical as well as both rotational and orbital dynamical properties of a comet, to put independent constraints on the production rates of the dust and of different volatile species, and to map the distribution of the source regions on the surface.

In this paper we concentrate on the measurable effects of activity on the orbital and rotational dynamics of comets. These effects, which have been recognized and studied for two centuries, have now been measured with unprecedented accuracy for comet 67P/Churyumov-Gerasimenko (hereafter 67P) during the Rosetta mission.

### 1.1 Orbital Dynamics

The recoil forces from activity induce a nongravitational acceleration (NGA) that produces measurable deviations from a pure Keplerian orbit. Such an effect was first noticed for 2P/Encke (Enke 1823), although it was Bessel (1835) who first recognized the reaction by ejected material as a possible cause. The first interest in NGAs resulted therefore from the necessity of accurately describing the orbit of a comet for the purpose of recovering it during subsequent apparitions. Marsden (1969) laid out the problem—in what is now known as the standard model—by defining a nongravitational acceleration term that is summed to the Keplerian acceleration, and that is assumed to be proportional to an *activity function* $g(r)$. The latter is an empirical function seeking to describe the variation of the vaporization flux of water ice (which is the dominant volatile species) as a function of the heliocentric distance (Marsden et al. 1973). The NGA is decomposed into three orthogonal components defined in the orbit-plane reference system—the radial, transverse and normal component. The nongravitational acceleration $\boldsymbol{A}_{ng}$ can therefore be expressed as:

$$\boldsymbol{A}_{ng} = g(r)(A_1\boldsymbol{R} + A_2\boldsymbol{T} + A_3\boldsymbol{N}) \tag{1}$$

where $r$ is the heliocentric range and $\boldsymbol{R}$, $\boldsymbol{T}$, and $\boldsymbol{N}$ are the radial, transverse and normal unit vectors. For the purpose of orbit determination, the quantities $A_i$ are, in general, considered constant and derived as a fit to the astrometric data. This procedure generally produces results that are in very good agreement with the astrometric data, and certainly within the errors of ground-based measurements. Equation (1), however, is a crude simplification of the physics of cometary activity, especially in the assumption that the $A_i$ coefficients are constant over the comet's orbit. For this reason, the $A_i$ quantities are suitable as fitting parameters, but the corresponding nongravitational acceleration $\boldsymbol{A}_{ng}$ may considerably differ from the actual nongravitational acceleration at any given time (Yeomans et al. 2004).

Rickman (1986) showed that by integrating the Gauss' equation one could express the variation of the orbital period $\Delta P$ as a function of the NGA in the form:

$$\Delta P = 6\pi\frac{\sqrt{1-e^2}}{n^2}\left(\frac{e}{q(1+e)}\int_0^P A_1 g(r)\sin\nu\,dt + \int_0^P A_2\frac{g(r)}{r}\,dt\right) \tag{2}$$

where $P$ is the orbital period, $e$ is the eccentricity, $q$ is the perihelion distance, $\nu$ is the true anomaly, and $n$ is the mean motion. If the $A_i$ quantities are strictly constant over the orbit, then the integral of the radial acceleration component is zero, and the variation of the orbital period only depends on the transverse component $A_2$. Although it is known that many comets show a significant activity asymmetry with respect to perihelion—which implies that the $A_i$ quantities are *not* constant—the study of $A_2$ still provides some physical information, at least in a statistical sense. Rickman et al. (1991), for example, have correlated the $A_2$ parameter to the evolutionary age in a sample of short-period comets.

A more sophisticated approach seeks to determine the nongravitational force vector as a function of the heliocentric distance by estimating the momentum of the ejected gas by means of a realistic thermophysical model and by taking into account the nucleus kinematics. By using the change in orbital period as an observable, it is then possible to directly determine the nucleus mass and, if estimates of its size and shape are available, its bulk density. Rickman (1986, 1989) successfully applied this method to estimate the density of 1P/Halley, while Farnham and Cochran (2002) applied it to 19P/Borrelly. Davidsson and Gutierrez (2004, 2005) further refined the method by using the observed change in orbit, the rotational lightcurve, and the observed water production rate as constraints to derive estimates about the size, density and orientation in space of 19P and 67P. It is interesting to note that the densities estimated for 19P by Davidsson and Gutierrez (2004) are somewhat different from those determined by Farnham and Cochran (2002), which shows that those determinations significantly depend on the model assumptions. On the other hand, it is fair to state that most determinations of density obtained from orbital nongravitational models point towards cometary nuclei being underdense, a fact confirmed by direct measurements of the mass—and hence bulk density—of 67P by Rosetta.

## 1.2 Spin Period Changes

In addition to producing nongravitational accelerations that modify a comet's orbit, activity can also change the nucleus rotational state. The combination of asymmetries in the comet's shape and illumination pattern, and in the distribution of active regions, produces net torques that can modify the spin state of the nucleus, by changing its spin rate and possibly exciting forced precession (see Sect. 1.3). Although long postulated (Whipple 1950, 1982), the first evidence of a spin period change for a comet was only detected in 1994 (Mueller and Ferrin 1996) for 10P/Tempel 2.

As of writing, there are seven Jupiter-family comets (JFCs) for which a change in spin period has been detected, while upper limits on the period change have been placed for four further JFCs (Kokotanekova et al. 2018). Analyzing the known period changes, Mueller and Samarasinha (2018) have proposed a correlation between orbit, activity and rate of change of the spin period. Following this correlation, the largest JFC nuclei with radius $R \geq 3$ km exhibit the smallest period changes (typically $\Delta P < 10$ minutes), while significant rotation period changes of $\sim$2 hours per orbit (103P, Meech et al. 2011) and $>$ 20 hours (41P, Bodewits et al. 2018) have been detected for the smallest nuclei (Kokotanekova et al. 2018). Gutierrez et al. (2002) have modeled the spin changes of active cometary nuclei by integrating the Euler equations of motion under the influence of NGAs produced by random activity patterns controlled by a simplified thermal model. They used realistic random shapes for the nuclei, and found that, depending on the simulation parameters, the comet was subject to substantial spin-up or spin-down, while the onset of considerable non-principal-axis rotation was rare. They concluded that NGA-induced spin-up can therefore be considered an important mechanism for cometary splitting, as described by Davidsson (1999, 2001). Since both the orbital dynamics and the spin period changes are caused by the same nongravitational acceleration due to sublimation activity, these two properties are expected to be related. Rafikov (2018) used data for the seven comet nuclei for which changes in the spin rate and orbit have been determined and found a linear relationship between the change of the cometary rotation rate and the magnitude of its NGA.

## 1.3 Complex Rotation

Reaction torques from outgassing can have two effects on the rotational state of comets. The first one is a change in the direction and magnitude of the angular momentum vector. The second one is the possible onset of non-principal-axis (hereafter NPA) rotation (sometimes called "excited" or "complex" rotation in the literature). Assuming that a cometary nucleus is a rigid body, its dynamics are controlled by Euler's equations of motion, in which the ratios of the moments of inertia determine the different periods involved in the complex rotational motion of the nucleus. A thorough characterization of the rotational state of a nucleus can thus provide valuable information on its moments of inertia, which, in conjunction with the knowledge of its shape and mass, can be diagnostic of its internal mass distribution. Eventually, energy dissipation due to internal friction resulting from mechanical deformation of the nucleus, can lead to the relaxation of the NPA rotation down to the minimum energy level: a pure rotation around the axis of maximum moment of inertia. Monitoring the evolution of the excited rotation with time can therefore place constraints on the relaxation time of NPA rotation, which in turn can provide information on the mechanical properties of the nucleus (Samarasinha et al. 2011).

Two NPA modes can appear depending on the rotational energy level: *short-axis modes* (SAM) and *long-axis modes* (LAM, see Samarasinha and A'Hearn 1991). Timescales and equations are described in detail in Sect. 2 of the review on comet rotation by Samarasinha et al. (2004). Such complex NPA rotational states have been invoked in the past to reconcile lightcurves and imaging data of comet 1P/Halley (Belton 1991; Samarasinha and A'Hearn 1991), to explain the lightcurves of comets 2P/Encke (Belton et al. 2005) and possibly 29P/Schwassmann-Wachmann 1 (Meech et al. 1993), and to interpret the CN jets, spacecraft imaging and radar data of comet 103P/Hartley 2 (Samarasinha et al. 2011; Waniak et al. 2012; Belton et al. 2013). More recently, NPA rotation has been proposed to explain the lightcurves and morphology of comet 252P/2000 G1 LINEAR (Li et al. 2017) and of the interstellar object 1I/2017 U1 'Oumuamua (Belton et al. 2018).

## 1.4 Comet Splitting

Over the past two centuries, more than 40 comets have been observed to split into two or more components (Boehnhardt 2004) with a rate of $\sim 10^{-2}$ yr$^{-1}$ per nucleus (Chen and Jewitt 1994). One splitting mechanism is represented by tidal forces occurring when a nucleus experiences an encounter with a massive object. A notable case was comet Shoemaker-Levy 9, that split in a string of fragments in July 1992 as a consequence of a close encounter with Jupiter (Solem 1994).

Possible disintegration mechanisms for non-tidally split nuclei are the build-up of mechanical stresses due either to large temperature/pressure gradients on the nucleus or to the centrifugal pull as a consequence of fast rotation (Sekanina 1997; Boehnhardt 2004). Nucleus spin-up due to nongravitational acceleration therefore represents one viable mechanisms leading to comet splitting. In this scenario, if a comet spins up and reaches an instability limit, where the centrifugal force exceeds self-gravity and the material forces, the comet nucleus starts to shed mass and disintegrates. Davidsson (1999, 2001) developed analytical models to study the critical breakup spin rate as well as the splitting behavior of solid nuclei with spherical and ellipsoidal shapes. These models describe well the observed JFC population and have been used to study the tensile strength of JFCs (Kokotanekova et al. 2017; Toth and Lisse 2006; see also Groussin et al. 2019, in this issue).

## 2 Available Constraints for 67P

### 2.1 Orbital Evolution

The orbital parameters of 67P, including the Marsden $\boldsymbol{A}_{ng}$ terms introduced above, have been derived from ground-based observations and are available for each of its previous apparitions.

From August 2014 to September 2016, 67P was accompanied in its orbit by the Rosetta spacecraft, allowing high accuracy astrometry to be conducted. This was achieved through radio-tracking of the spacecraft combined with optical navigation, which allowed an accurate measurement of the location of the nucleus relative to Rosetta, by using the navigation camera and the OSIRIS instrument (Optical, Spectroscopic, and Infrared Remote Imaging System; Keller et al. 2007). The combined positioning allowed the ESA flight dynamics team to produce a model of the comet's trajectory, available as NASA SPICE kernels (Acton 1996), with a theoretical accuracy of $\sim$10 m in the Earth-comet range direction and better than $\sim$100 km in the perpendicular (across-track) direction (Godard et al. 2015, 2017). Unfortunately, as of the time of writing, the flight dynamics reconstruction is a purely gravitational solution, i.e. it does not include a nongravitational force model and deals with the resulting drift by "resetting" the modeled trajectory with the observed comet position at the beginning of each of the segments of the integration. Attree et al. (2019) showed that this resetting is convolved with the initial positional uncertainty in each segment and therefore manifests as a series of random "jumps" in the SPICE trajectory at intervals ranging from 3–4 days during perihelion, to $\sim$1 month during other phases of the mission.

As can be seen in Fig. 1, the discontinuities limit the accuracy of the reconstructed trajectory to between tens to over a hundred kilometers in three-dimensional heliocentric position, but are generally smaller than 1 km in the Earth-comet range direction.

The reconstructed trajectory still represents a gravitational solution plus the effects of the NGAs; theoretically, therefore, a gravitational integration could be subtracted from the SPICE positions and the residuals differentiated twice to obtain the accelerations. However, the discontinuities pose a problem for this differentiation. Two recent papers address this issue in two different ways. Attree et al. (2019) choose to model the NGAs with a full thermal/dynamics model (see Sect. 3, below) and fit the resulting trajectories to the relatively smooth Earth-comet range data. Kramer and Läuter (2019) instead, match the Marsden model to the SPICE positions before fitting the residuals with a series of exponential and polynomial functions up to fourth order. This provides a smooth, differentiable curve from which the NGAs can then be extracted (see Fig. 2). In order to determine the best initial conditions for the orbital integration, they iteratively varied the initial positions and velocities to minimize the residuals to the SPICE positions during a period—400 to 200 days before perihelion—during which the nongravitational acceleration was supposedly negligible. The determined NGAs, along with the SPICE positions themselves when considering the jump uncertainty, are the astrometry constraints to which the NGA models described below can be fitted.

Kramer and Läuter (2019) did not incorporate a thermophysical model of the cometary activity in their model, and therefore could not independently estimate gas production from

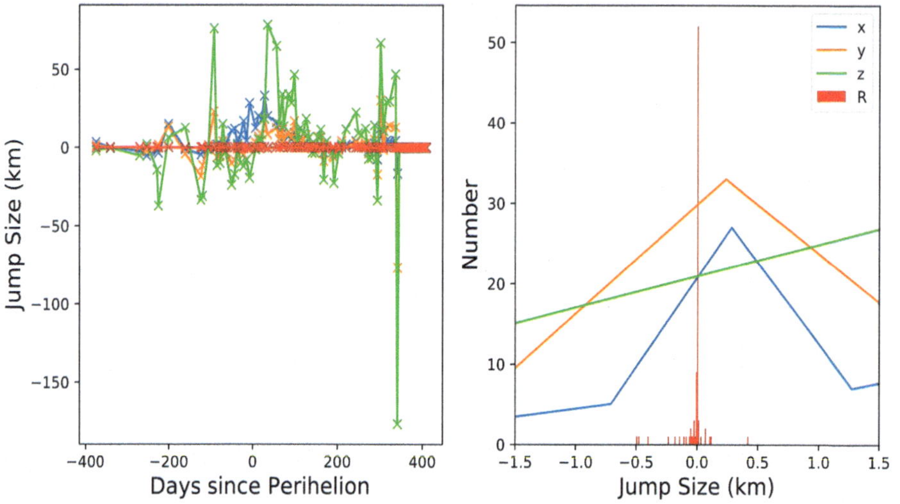


**Fig. 1** Discontinuities in the position of comet 67P from the SPICE kernels, in $(x, y, z)$ heliocentric J2000 coordinates, and Earth-comet range (R), from Attree et al. (2019). On the left, discontinuities are shown as a function of time and on the right as a histogram of jump sizes

**Fig. 2** The three components of the nongravitational acceleration in the terrestrial equatorial frame after Kramer and Läuter (2019). The shaded areas correspond to the error envelope obtained by perturbing the initial conditions of the orbital integration

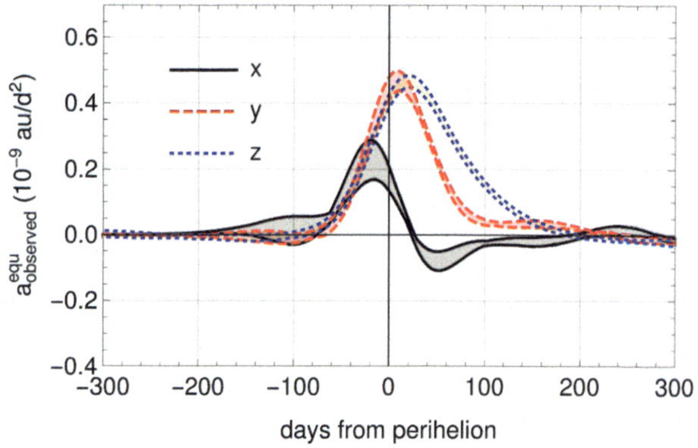


the NGAs. However, by studying the direction of the reaction force, they infer the presence of a diurnal lag in the direction of the NGAs with respect to the antisolar direction, up to 50° in the period between perihelion and 140 days post perihelion.

## 2.2 Spin Rate and Rotation State

Pre-Rosetta observations of 67P with the aim of determining the spin period were all performed between (and far from) the perihelion passages of Sep. 2002 and Feb. 2009. Lowry et al. (2012) used the best observational base available at that time to determine a rotation period $P = 12.76137 \pm 0.00006$ h. When Rosetta started the approach to the still unresolved 67P, in early 2014, it immediately became clear from OSIRIS disk-integrated lightcurves that the rotation period had changed to $P = 12.4043 \pm 0.0007$ h after the 2009 perihelion passage (Mottola et al. 2014). This was the first evidence that the rotation of 67P was strongly affected by nongravitational torques.

Prompted by this finding, Keller et al. (2015b) performed forward modeling of the nucleus kinematics subjected to sublimation torques, to predict the spin evolution around the 2015 perihelion passage. For this purpose they used a detailed (although preliminary) comet digital shape model (Preusker et al. 2015) and a two-layer thermophysical model (Keller et al. 2015a). The model correctly predicted that the comet would start to slow down its rotation shortly prior to the 2015 perihelion, and would rapidly spin up around perihelion. This study also highlighted the importance of a realistic thermal model and of a detailed shape model of the target for the description of the spin evolution, since both strongly affect the net torques.

The Rosetta mission also offered a unique opportunity to accurately monitor the rotational parameters of 67P with a high accuracy throughout the perihelion passage, and to search for possible NPA rotation. First hints of a complex rotation were indeed identified at the beginning of the mission during the stereophotogrammetric reconstruction of the nucleus shape (Preusker et al. 2015). Later, a periodogram analysis of the (RA, Dec) direction of the instantaneous angular velocity and nucleus Z-axis vectors provided by landmarks tracking data confirmed a slight NPA rotation (Jorda et al. 2016). Both data sets were compatible with a small amplitude SAM (0.15°), very close to the pure spin state of minimum energy. The period associated to this modulation was found to be close to 270 h in both cases. In an attempt to explain the observed NPA rotation, Gutierrez et al. (2016) tried to reproduce the data of Jorda et al. (2016) with the mathematical formalism of Samarasinha and Mueller (2015) for a torque-free nucleus. Although they could not derive unique ratios for the moments of inertia, they did find that the moments of inertia $I_X$, $I_Y$ and $I_Z$ are linked by the relationship: $I_Z \approx 0.96 I_X + 0.99 I_Y$.

## 2.3 Splitting

Hirabayashi et al. (2016) performed a Monte Carlo study of the orbital and spin evolution of 67P by integrating 1000 clones backwards in time for a period of 5000 years and simulating sublimation-induced torques. They found that the spin rate variation becomes completely random within the integration period, and that the period inverts its rate of variation several times in a random-walk way. They also found that when the spin exceeded the fission limit of $\sim$7 h, the two lobes composing the comet become temporally co-orbiting. Such a configuration, however, is short lived, after which the two bodies coalesce again into a re-configured body. The authors argue that 67P may have undergone several such reconfiguration phases since its formation, and that some of the large features visible today on the comet might be previous contact points between the two lobes, now exposed due to this rearrangement.

Further, by means of an elastic and plastic finite-element model, they interpret the presence of a long crack in the comet's Hapi region as the effect of tensile stresses during the phase in which the comet nearly reached—but did not exceed—the fission limit. The authors also argue that such evolution could be typical for most bilobate comets, whereas single bodies could either be primordial objects or surviving components of a pair.

## 2.4 Gas Production, Dust Production, Mass Loss

Because NGA is produced by the reaction force induced by the exchange of momentum between the comet and the ejected gas and dust, it is important to estimate the gas production, dust production and mass loss of 67P.

### 2.4.1 Water Production

The water production rate from 67P had been measured during pre-Rosetta apparitions in 1982/83 (Hanner et al. 1985; A'Hearn et al. 1995; Crovisier et al. 2002; Feldman et al. 2004; Schleicher 2006), 1995/6 (Schleicher 2006; Bertaux et al. 2014), 2002 (Bertaux et al. 2014) and 2008/9 (Ootsubo et al. 2012; Snodgrass et al. 2013; Bertaux et al. 2014). Some of these measurements were performed only at a given time and heliocentric distance, but some tracked the change in production rate around perihelion. Most of these estimates agree on a maximum production rate around perihelion of about $10^{28}\ s^{-1}$.

The Rosetta mission presented an opportunity to track the changes in water production in unprecedented detail for a long period of time and with a variety of different instruments. Some of the first measurements from MIRO (Microwave Instrument for the Rosetta Orbiter) estimated the water production from the nucleus between June and September 2014. Gulkis et al. (2015) reported the first estimate of $1 \times 10^{25}\ s^{-1}$ at 3.9 au in June and then a higher value of $4 \times 10^{25}\ s^{-1}$ in August at 3.5 au. Lee et al. (2015) looked in greater detail at the spectral lines measured in August 2014 and estimated water production rates varying by a factor 30, from $0.1 \times 10^{25}$ to $3 \times 10^{25}\ s^{-1}$. The authors attributed such a large variation in water production to a non-uniform distribution of sublimating water ice over the comet's surface.

Later, in September at 3.4 au, the production rate of the main water isotopologue ($H_2{}^{16}O$) was derived as $(4.9 \pm 2.5) \times 10^{25}\ s^{-1}$ (Biver et al. 2015). Further values for the water production were derived from observations by ROSINA (Rosetta Orbiter Spectrometer for Ion and Neutral Analysis). Between August and November 2014 (3.5 to 3.0 au), the ROSINA-COPS (Pressure Sensor) estimated the production as $8.7 \times 10^{25}$–$1.1 \times 10^{26}\ s^{-1}$, and then between November and January 2015 (3.0 to 2.5 au), the water outgassing rate was a factor of two higher (Bieler et al. 2015). From ROSINA-DFMS (Double Focussing Mass Spectrometer), Fougere et al. (2016) estimated the change in water production on approach to perihelion, finding that it increased from $< 10^{26}\ s^{-1}$ at 3.5 au, to $> 10^{27}\ s^{-1}$ at 1.5 au. Estimates on the water production rate in this time period were also made by Bockelee-Morvan (2015) using VIRTIS-H (Visible and Infrared Thermal Imaging Spectrometer). From November 2014 to January 2015 (2.9–2.5 au), the outgassing rate was between $6.9 \times 10^{25}$ and $8.7 \times 10^{25}\ s^{-1}$. Together, these measurements show that the water outgassing was approximately $10^{26}\ s^{-1}$ at the end of 2014 and beginning of 2015.

Fink et al. (2016) analyzed data cubes from VIRTIS-M measured in February 2015 (2.21 au) and April 2015 (1.76 au) and found water production rates of $2.5 \times 10^{26}$ and $4.65 \times 10^{26}\ s^{-1}$, respectively. They also examined the spatial distribution of outgassing in these observations, deriving that 83% of the water production comes from the illuminated side and 17% from the night side. In addition, the highly active neck region was found to contribute about 60% to the water production from an observation in February 2015. For comparison, in April 2015, Migliorini et al. (2016) found a higher estimate for the water outgassing of $10^{27}\ s^{-1}$ using measurements made by VIRTIS-M. Using the RPC-ICA (Rosetta Plasma Consortium Ion Composition Analyser), Simon Wedlund et al. (2016) estimated the change in water production rate to be $2 \times 10^{25}\ s^{-1}$ in September 2014 to $5 \times 10^{26}\ s^{-1}$ in April 2015.

The long-term evolution of the water production rate has also been assessed. Hansen et al. (2016) and Marshall et al. (2017) measured the water production rate from August 2014 to April 2016 with the ROSINA and MIRO instruments, respectively, and Läuter et al. (2019) assessed the water production up to September 2016. The maximum water production rates were found to be $(3.5 \pm 0.5) \times 10^{28}\ s^{-1}$ (Hansen et al. 2016), $(1.4 \pm 0.5) \times 10^{28}\ s^{-1}$ (Marshall et al. 2017) and $(2.0 \pm 0.1) \times 10^{28}\ s^{-1}$ (Läuter et al. 2019). The measurements from ROSINA produced slightly larger results than those from MIRO, and although a satisfactory explanation for this discrepancy is still not established, it might result from the basic difference

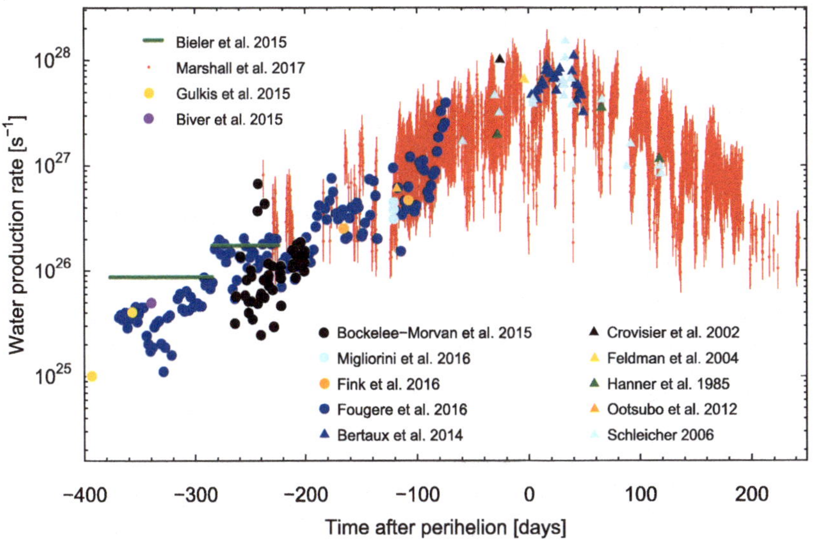


**Fig. 3** Water production rates for 67P calculated from Rosetta and ground-based measurements. Adapted from Marshall et al. (2017)

between the two instruments. ROSINA, being an *in situ* instrument, measures the gas density in the local vicinity of the spacecraft. MIRO—a remote-sensing instrument—measures the column density along the line of sight. The discrepancy may arise from different acquisition techniques and inversion models required to retrieve the gas densities.

The Rosetta measurements generally confirm the prior measurements for the water production, finding that the maximum production around perihelion is $> 10^{28}\ \mathrm{s}^{-1}$ and that the maximum production is shifted after perihelion. Additionally, the results indicate that there is an asymmetry in the outgassing pre- and post-perihelion (Hansen et al. 2016; Läuter et al. 2019). The production rate as a function of the heliocentric distance can be roughly approximated as a power law. The exponent of the power law has been measured by several authors: −7.06 (Simon Wedlund et al. 2016), −5.1 (Hansen et al. 2016), −3.8 (Marshall et al. 2017) and −7 (Läuter et al. 2019) pre-perihelion; and −7.15 (Hansen et al. 2016), −4.3 (Marshall et al. 2017) and −6.5 (Läuter et al. 2019) post-perihelion. As noted by (Marshall et al. 2019), these values must be treated with caution, though, as the power law exponent is in general not constant, but is also a function of heliocentric distance.

Läuter et al. (2019) attempted to determine the spatial distribution of the water production. Although the exact location of the active regions is not unique, generally the water production follows the seasonal illumination, originating from the Northern side far from perihelion and shifting to the Southern side at perihelion. The outgassing appears to follow the sub-solar point as it moves from North to South and then back to the Northern side on the outbound leg. This trend is also seen in the regional analysis by Marshall et al. (2017). The results of Fougere et al. (2016) and Hansen et al. (2016) point to a slightly more complicated picture: they find active regions in the Northern neck region and on the Southern lobe, and when these regions are well illuminated, they contribute significantly to the outgassing, but when they are not so well illuminated, water production is more uniformly distributed around the comet surface. The water production rates for 67P are summarized in Table 1 and Fig. 3.

**Table 1** Summary of production rates and mass loss from 67P during the 2015 apparition

| | 2.2 au pre-perihelion production | Perihelion production | Total mass loss (kg) |
|---|---|---|---|
| $H_2O$ | $2.5 \times 10^{26}$ $s^{-1}$ (6) | $(3.5 \pm 0.5) \times 10^{28}$ $s^{-1}$ (1)<br>$(1.42 \pm 0.51) \times 10^{28}$ $s^{-1}$ (2)<br>$(2.0 \pm 0.1) \times 10^{28}$ $s^{-1}$ (5) | $6.4 \times 10^9$ (1)<br>$(2.4 \pm 1.1) \times 10^9$ (2) |
| $CO_2$ | $1.18 \times 10^{25}$ $s^{-1}$ (6) | $10^{27}$ $s^{-1}$ (5) | |
| Dust | $(60 \pm 10)$ kg $s^{-1}$ (4) | $(1.7 \pm 0.9) \times 10^4$ kg $s^{-1}$ (4) | |
| All material | | | 4–8 $\times 10^{10}$ (1)<br>$(1.2 \pm 0.6) \times 10^{10}$ (2)<br>$(1.05 \pm 0.34) \times 10^{10}$ (3) |

(1) Hansen et al. (2016); (2) Marshall et al. (2017); (3) Pätzold et al. (2019); (4) Fulle et al. (2016); (5) Läuter et al. (2019); (6) Fink et al. (2016)

### *2.4.2 Other Species*

The production rates for other species have also been measured. For $CO_2$, Fink et al. (2016) estimated production rates in February 2015 and April 2015 of $1.18 \times 10^{25}$ and $1.13 \times 10^{25}$ $s^{-1}$, at 2.21 au and 1.76 au, respectively. In comparison to water, which shows a fairly predictable outgassing correlating to illumination, $CO_2$ in this work appeared to originate in isolated regions or spots. As a result of different outgassing behavior, it is difficult to interpret local measures in terms of $CO_2/H_2O$ ratios.

In April 2015, VIRTIS-M measurements implied an enhanced production rate of $CO_2$—$8 \times 10^{25}$ $s^{-1}$ at 1.9 au (Migliorini et al. 2016). In addition to water (described above) Läuter et al. (2019) also estimated the production of $CO_2$, CO and $O_2$, from DFMS/COPS data. These molecules peaked at perihelion with production rates of $10^{27}$ $s^{-1}$ ($CO_2$), $5 \times 10^{26}$ $s^{-1}$ (CO) and $3 \times 10^{26}$ $s^{-1}$ ($O_2$). All of these molecules are less abundant than water (1–10%), however their results indicate that on the outbound leg, beyond 3 au, it is possible that water in the coma becomes less abundant than $CO_2$ and even CO. The $O_2$ production is similar to $H_2O$ in that it shifts from North to South during perihelion passage but $CO_2$ and CO appear to be consistently originating from the Southern hemisphere, regardless of heliocentric distance.

Bockelee-Morvan (2015) estimated the change in abundance of $CO_2$, $CH_4$ and OCS for a few tens of days around perihelion from spectral observations made by VIRTIS-H. Relative to water, $CH_4$ and OCS have abundances of 0.23% and 0.12%, respectively, pre-perihelion, and 0.47% and 0.18% post-perihelion. $CO_2$ has an abundance of 14% pre-perihelion and then increases substantially post-perihelion to 32%. The maximum increase in $CO_2$ has a lag of 6 days, which compares to the behavior to water which also experienced peak production 20 days after perihelion. With ROSINA DFMS measurements made in May 2016, Rubin et al. (2018) measured the abundance of a number of minor species in the coma of 67P. Here, production rate ratios relative to water are reported (all in $s^{-1}$): $(8.9 \pm 2.4) \times 10^{-4}$ ($N_2$), $(4.9 \pm 1.9) \times 10^{-6}$ ($^{36}$A), $(5.8 \pm 2.2) \times 10^{-6}$ (Ar), $(4.9 \pm 2.2) \times 10^{-7}$ (Kr), $(2.4 \pm 1.1) \times 10^{-7}$ (Xe) and $< 5 \times 10^{-8}$ (Ne).

Around perihelion, the total gas production rate is about 30% larger than the one due to water production alone. Due to the large uncertainties in the derived water productions, it is generally justified to neglet the contribution of other species to the NGA. However, it must be considered that a possible different spatial distribution of the source regions of the different species could introduce small systematic deviations when modeling the orbital and spin evolution. Activity beyond 3 au is dominated by supervolatile species. However, the contribution of activity to the total NGA is negligible at these large heliocentric distances.

### 2.4.3 Dust Production

Several instruments onboard Rosetta have been used to estimate the amount of dust ejected from the surface. Rotundi et al. (2015) measured the dust production in August and September 2014 as the comet moved from 3.6 au to 3.4 au inbound using observations made by GIADA (Grain Impact Analyser and Dust Accumulator) and OSIRIS. They found a distribution for the ejected mass ranging from $10^{-10}$ to $10^{-2}$ kg which peaked at $10^{-2}$ kg with a differential power-law slope of $-2$ in the lower mass bins.

The mass loss in each mass bin varied from $1.5\ \mathrm{kg\,s^{-1}}$ to $9.9 \times 10^{-3}\ \mathrm{kg\,s^{-1}}$, giving a total mass loss of $(7 \pm 1)\ \mathrm{kg\,s^{-1}}$ for this time period. The combination of GIADA and OSIRIS was used again by Fulle et al. (2016) to derive the dust loss from 2.2 au to 1.2 au (perihelion). This evolved from $(60 \pm 10)\ \mathrm{kg\,s^{-1}}$ at 2.2 au, to $(70 \pm 30)\ \mathrm{kg\,s^{-1}}$ at 2.1 au and finally $(1.7 \pm 0.9) \times 10^4\ \mathrm{kg\,s^{-1}}$ at perihelion.

By excluding material larger than 1 kg, the mass loss at perihelion would be $(1.5 \pm 0.5) \times 10^3\ \mathrm{kg\,s^{-1}}$. This value can be compared to the maximum production of water, $(426 \pm 153)\ \mathrm{kg\,s^{-1}}$ measured by Marshall et al. (2017) at perihelion. The mass distribution reported by Fulle et al. (2016) steepens at perihelion compared to the distribution at larger heliocentric distances.

### 2.4.4 Mass Loss

Using water production rates from previous apparitions and some early Rosetta results, Bertaux et al. (2015) estimated the mass loss and erosion of material from 67P. The resulting total mass loss over one orbit was $(2.7 \pm 0.4) \times 10^9$ kg from observations performed by SWAN/SOHO, corresponding to an erosion depth of about 1 m.

Between 3.6 au inbound and 3.0 au outbound, Hansen et al. (2016) estimate the total water mass loss to be $6.4 \times 10^9$ kg. By making assumptions about the ratios of the minor gas species and the dust-gas ratio, a total mass loss of 4–8 $\times 10^{10}$ kg can be expected.

As stated before, lower values for the water production were estimated from MIRO observations. As a consequence, Marshall et al. (2017) also estimate a lower value for the mass loss, $(2.4 \pm 1.1) \times 10^9$ kg for water loss and $(1.2 \pm 0.6) \times 10^{10}$ kg for the total mass loss, although they do not account for the mass loss of minor species.

The radio science experiment RSI calculated the mass of 67P as $(9982 \pm 3) \times 10^9$ kg at the start of the Rosetta mission and $(9971.5 \pm 1.5) \times 10^9$ kg at the end of the mission, giving a total mass loss of $(10.5 \pm 3.4) \times 10^9$ kg during this apparition (Pätzold et al. 2019)[1].

Keller et al. (2017) estimate that a mass loss of this magnitude would erode to depths of 0.55 m for a uniformly outgassing, mass-equivalent sphere. This is much less than the theoretically calculated maximum capability of erosion from an ice/dust matrix covered by a thin inert layer, which was found to be 20 m (Keller et al. 2015a). This supports the notion that the active ice fraction (i.e. the fraction of the surface that can contribute to sublimation) must be much smaller than 1. It is worth restating that the total mass loss is only a fraction

[1] Note that in Pätzold et al. (2019) pre- and post-perihelion masses of $(9982 \pm 3) \times 10^{12}$ kg and $(9971.5 \pm 1.5) \times 10^{12}$ kg, respectively, are erroneously reported throughout their paper.

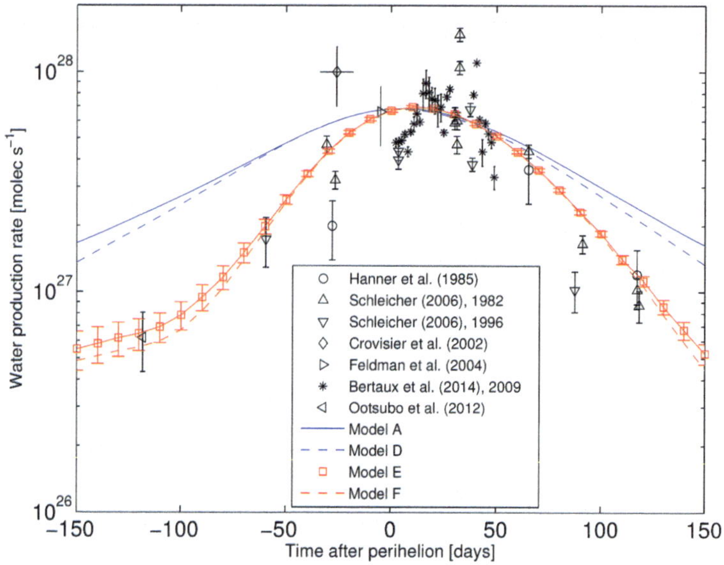


**Fig. 4** Observed and modeled water production rates for 67P after Keller et al. (2015a). Their models E and F provided the best fit to the data and were derived by adopting regionally variable activity levels for two different values of thermal inertia, respectively. Models A and D are models with constant properties over the surface. Model A is a simple stationary model with ice at the surface and D is a full thermodynamical model including thermal transport

of the total material that is ejected from the surface as some of the lifted material falls back onto the surface. Mass transport on 67P is described by Keller et al. (2017), with about 20% of the eroded material from the South estimated to fall back onto the Northern side.

## 3 Modeling Approaches

In addition to bearing interest for the study of orbital and spin evolution, and the occurrence of splitting events, NGAs provide powerful constraints for the determination of the thermophysical properties of the nucleus. The information content of the spin evolution and of the orbital evolution are quite complementary. The nongravitational resultant torque is the sum of the local reaction components, most of which mutually cancel out, with only the asymmetries in the shape, illumination, and activity distribution contributing to the spin evolution. A small change in the distribution of active regions (at constant average total production rate) can cause large changes (even a reversal) of the induced torque. Therefore, the spin evolution is most diagnostic for the spatial distribution of activity. Conversely, the orbital evolution is more diagnostic of the total production rate. Once the NGAs can be reliably retrieved from observations, they can therefore provide independent constraints for the thermophysics of cometary activity and its distribution on the nucleus.

One of the conclusions of the thermophysical modeling by Keller et al. (2015a) was that no *bona fide* thermal model that uses constant physical properties—over time and over the surface of the comet—can accurately reproduce the observed water production rate as a function of heliocentric distance, as measured by the Rosetta analyzers. Keller et al. (2015a) also showed that an *ad hoc* distribution of active ice fraction over the body could explain the observed results, even though it does not represent a unique solution (see Fig. 4).

However, the peculiarity of the 67P orbit and pole orientation—with its Southern solstice occurring close to perihelion, and its large obliquity—creates a strong coupling between sub-solar latitude and heliocentric distance. As a consequence, the observed water production as a function of heliocentric range could as well be explained by a regionally uniform comet, whose properties (e.g. dust layer thickness or active fraction) vary with time. Things are complicated by the fact that production rates were measured by Rosetta with different instruments that applied different measurement principles, measured on different spatial and temporal scales and required complex modeling in order for their data to be translated into a flux of molecules. As a result, as could be seen in Sect. 2.4, the measurements from different instruments can be difficult to compare to each other. Ideally, a model aiming to explain the physics of activity should be able to reproduce all of the available measurements of gas production, both on regional and on global scale, as well as its seasonal and secular variations. Using production data alone, however, the models are generally underconstrained, producing ambiguous results. Using in addition the nongravitational effects as constraints holds the potential of disentangling the contributions of a time-varying and regionally-varying activity, as highlighted in the following.

### 3.1 Regional Activity Modeling

Further to the *ad hoc* distributions of active ice fraction used by Keller et al. (2015a) and Kramer et al. (2019), several authors have attempted to link heterogeneous activity to regional surface variations on the nucleus. For example, in Marschall et al. (2016) and Marschall et al. (2017), the authors fit the pre-perihelion water production rates with a basic thermophysical model by varying local active fractions on the nucleus's cliffs versus its plains, Northern verses Southern hemispheres, and in six geographically grouped regions. Attree et al. (2019) fitted a similar model to the combined trajectory, rotation and production measurements by varying the active fraction in the Marschall et al. (2016, 2017) regions, as well as the 26 named geographical regions of Thomas et al. (2015). Their results, shown in Fig. 5, match the initial slope of production and NGA but break down around perihelion, suggesting temporal variations as discussed below.

The mapped active fraction is relatively consistent across several works (Marschall et al. 2016, 2017; Kramer et al. 2019; Attree et al. 2019) in showing the highest values in the Southern hemisphere, with intermediate values in the deposition regions of the Northern neck (Hapi), and the lowest values elsewhere in the North. The poor torque fit in the combined fit demonstrates that the spatial distribution of "torque efficiency", the propensity for outgassing in a particular area to influence the nucleus rotation, does not necessarily correlate with the geographically defined regions. Significant improvements to the torque fit could be made in Attree et al. (2019) by parametrizing the Southern hemisphere in terms of this torque efficiency, which led to a spotty activity distribution with enhanced peaks in Anhur, Bes, Khepry, and parts of Wosret.

### 3.2 Time-Dependent Activity Modeling

Attree et al. (2019) did not find adequate solutions to their combined fit to the 67P trajectory, rotation and production measurements with the above parametrization of a spatially varying active fraction. They therefore invoked a time-dependent active fraction, with the active fractions of two regions in the Southern hemisphere allowed to vary between two values. Only the Southern regions were varied in this way, as the discrepancies between the

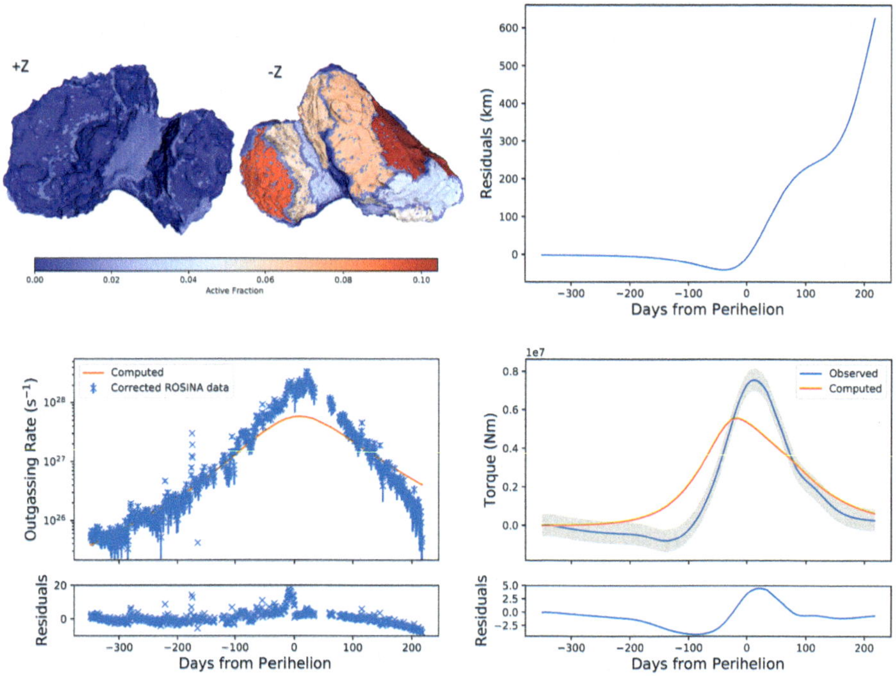


**Fig. 5** The best-fit solution of a thermophysical model from Attree et al. (2019) with varying active fractions in the 26 named comet regions (their solution B). Top left: Active fraction mapped onto the comet; top right: residuals in Earth-comet range relative to the SPICE solution; bottom left: total gas production relative to ROSINA; bottom right: torque relative to OSIRIS measurements

observations and time-independent models were greatest around perihelion, when the South dominates activity.

Attree et al. (2019) found good solutions, with significant improvement over time-independent models, by drastically increasing the Southern active fraction (from around 10% to 25–35%) in the ~hundred days around and just after perihelion (see Fig. 6). This allowed production to ramp-up to meet the high observed perihelion activity (right of Fig. 6) before falling off quickly again as seen by ROSINA.

The trajectory fit was likewise improved over a time-independent model, supporting the idea of a steeper slope of the activity at perihelion than at larger heliocentric distances, in agreement with the observed water production curve.

The interpretation for the cause of such a time-varying active fraction is that of changes and movement in surface dust cover. As has been suggested by several authors (e.g. Fornasier et al. 2016), strong activity lifts dust at perihelion with greater efficiency, exposing more volatile-enriched material and increasing the sublimation rate in a positive feedback. Dust would then be ejected or transported, mainly to the Northern hemisphere (Lai et al. 2016; Keller et al. 2017), before the decreasing activity post-perihelion allows it to once again be deposited in the South. Gas fluxes through a thicker dust mantle are then quenched, reducing the effective active fraction again.

Time-dependent activity is also implied by Kramer et al. (2019). In fitting the rotation rate, spin-pole direction and total gas production, they use a time-constant effective active fraction (fitting for both a spatially uniform fraction, and one varying in patches), but note

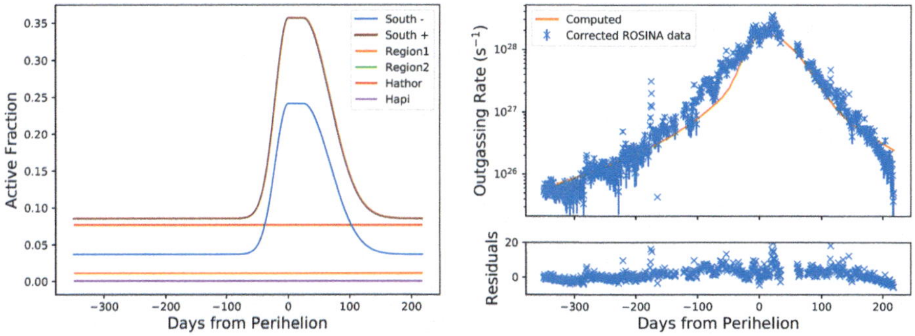


**Fig. 6** Best-fit time-dependent model of Attree et al. (2019) (their model C). Left: the active fractions of their six super regions with time. Right: the resulting fit and residuals to the total production rate, showing a marked improvement over the time-independent models of above. The active fraction for the Southern hemisphere is reported separately for regions that contribute to increase or decrease the spin rate, indicated with a plus and minus sign in the legend, respectively

that the gas production in particular can only be matched with an effective sublimation curve with an "enhanced response near perihelion", above the standard energy balance solution. Kramer et al. (2019) suggest the same thinning and thickening of a surface dust layer to explain this curve, citing the two thermal models of Keller et al. (2015a), who modeled the effects of different dust layer thicknesses on sublimation rates.

They raise the possibility of a mixed model that transitions between the two thermal regimes as the comet approaches the sun. Given the difficulty found in trying to fit time-independent models to the Rosetta data, this seems like a promising direction for future work.

It should be noted that, as discussed in Kramer and Läuter (2019), the nongravitational torques place the most stringent constraint on the spatial distribution of activity, whereas the accelerations and total production place more constraints on the temporal variability. Time-dependent activity appears necessary to reproduce not just the high peak activity seen at perihelion but also the quick falloff with heliocentric distance afterwards (i.e. simply having large active fractions in the South alone cannot match observations), while simultaneously improving the trajectory fit.

## 3.3 Thermophysics of Activity

For the purpose of understanding nongravitational forces, the variation of gas production of a comet as a function of heliocentric distance is of paramount importance. The close relationship between macroscopic characteristics and details of microphysical processes and/or properties of the cometary nucleus is nontrivial. While the overall picture of cometary activity, as outlined by Whipple (1950) is still valid today, the details are far from being understood.

Today it is clear that modern theoretical models aiming to describe cometary activity should explicitly take into account the fact that volatile sublimation does not occur at the surface of the nucleus, but takes place under a porous dusty crust. Such models were proposed in Keller et al. (2015a), Skorov et al. (2016), Hu et al. (2017a,b). The dust particles forming this crust have a complex chemical (Bardyn et al. 2017) and morphological structure (Langevin et al. 2016; Bentley et al. 2016). When vapor diffuses through a porous layer, a dramatic change in all its characteristics occurs. First, the effective rate of sublimation depends on the permeability of the layer, controlled by its thickness and pore size (Skorov et al. 2011; Gundlach et al. 2011). Secondly, the surface layer is an extremely non-isothermal region—during day-time, the temperature on the nucleus surface can exceed the temperature of sublimating ice by more than a hundred kelvins on centimeter scales. As a consequence, the sublimating gas is heated up when passing through the dust layer, which leads to the thermal velocity of the ejected molecules being noticeably higher than the kinetic velocity corresponding to the ice temperature. Finally, during diffusion, the angular distribution of the velocity of the escaping molecules may change (Skorov and Rickman 1995). The last two effects clearly affect the reactive force created by the sublimation products, and the resulting nongravitational perturbations. Elementary estimates show that ignoring these processes can lead to a relative error in the estimation of nongravitational forces of the order of 100%.

Nevertheless, this error is much smaller than the possible error caused by the fact that we know very little about the structure of the surface dust layer, and therefore its quenching effect. In all published theoretical models, without exception, it is assumed that the heat transport properties remain unchanged both as the heliocentric distance changes and for different regions of the nuclear surface.

This basic model assumption calls for an upper layer in a stationary state, in which dust particles on the upper boundary of the layer are constantly carried away (ablated) by the gas flow, and exactly replenished by the addition of new particles from the lower boundary of the layer.

Although it seems obvious that this assumption is artificial and non-physical, as of today, no models are available that are more adequate. It is not surprising, therefore, that none of the models mentioned can explain two very important effects we observe on 67P: the steep increase in gas production near perihelion and a delay in maximum gas production with respect to perihelion (see e.g. Hansen et al. 2016). Although in principle seasonal effects could contribute to the latter phenomenon, Keller et al. (2015a) have shown that the equatorial obliquity alone cannot account for the large delay of 20 days observed for 67P in the maximum of the activity.

On the other hand, it was shown above that these effects can be explained by the hypothesis of variability (either spatial or temporal) in activity (see Sect. 3). This concept was proposed for the analysis of observations at the beginning of the mission (Keller et al. 2015a) and was substantiated in Marshall et al. (2019).

Unfortunately, in the form in which it is proposed today, this hypothesis is essentially treated as model parametrization and does not contain a consistent microphysical model describing the observed changes. Nevertheless, it gives us important clues and constraints about what microphysical processes should be included in future models.

## 4 Conclusions

The long-term observations of 67P by Rosetta covering the onset, progression, maximum, and decay of activity—mainly measured by the varying gas (water) production rates, but also by the observed dust activity—provide a unique chain of results. The novel aspect of the mission is that these coma observations can be combined for the first time with a very detailed shape model of the nucleus. In addition, the magnitude and direction of the instantaneous spin vector and their subtle changes during the perihelion passage are well documented. Thermophysical modeling (Keller et al. 2015a) has shown that no spatially-uniform and time-constant activity model can explain the observed water production rates—especially the steep increase with decreasing heliocentric distance. Recent work (Fougere et al. 2016; Marschall et al. 2016, 2017; Marshall et al. 2019) has shown that this discrepancy can be reconciled by allowing for spatial and/or temporal variation of the physical properties of the comet's surface in the form of a variable active fraction. The resulting activity distribution patterns, however, are not unique.

Cometary activity is not only reflected in the properties of the resulting coma but also affects the NGA. The NGA induces torques that modify the spin state and an orbital acceleration of the whole nucleus that influences its orbit evolution. Modeling the effects of the NGA provides additional, complementary constraints for the parameters that describe the physics of activity and the spatial and temporal variation of the active fraction. First results have been very recently obtained in this direction by including the effects of NGA (Attree et al. 2019; Kramer et al. 2019; Kramer and Läuter 2019). However, the field is still in its infancy, with the active fraction being incorporated in the models as a pure fitting parameter, disconnected from the physics of activity. The next step will be to incorporate a more realistic model of the physics of activity, including dust removal and transport, which accounts for the observed variability. Further, the use of orbital data currently is limited by the accuracy of the orbit determination of 67P (Attree et al. 2019). It is therefore important that an effort be made in reconstructing the orbit of 67P to the full accuracy that the radio tracking data potentially allow, possibly by including an improved NGA model based on the activity measurements by Rosetta. In this way it will be possible to fully exploit the wealth of data returned by the mission.

**Acknowledgements** The authors are indebted to Nalin Samarasinha and to an anonymous referee, who, with their comments, helped improve this review paper. This project has received funding from the European Union's Horizon 2020 research and innovation programme under grant agreement no. 686709. This work was supported by the Swiss State Secretariat for Education, Research and Innovation (SERI) under contract number 16.0008-2. The opinions expressed and arguments employed herein do not necessarily reflect the official view of the Swiss Government. Y. S. thanks the Deutsche Forschungsgemeinschaft (DFG) for support under grant SK 264/2-1. N. A. was funded by the UK Space Agency, grant no. ST/R001375/2. R. K. acknowledges support through an ESO fellowship.



## References

C.H. Acton, Ancillary data services of NASA's navigation and ancillary information facility. Planet. Space Sci. **44**(1), 65 (1996)

M. A'Hearn et al., The ensemble properties of comets: results from narrowband photometry of 85 comets, 1976–1992. Icarus **118**(2), 223 (1995)

N. Attree et al., Constraining models of activity on comet 67P/Churyumov-Gerasimenko with Rosetta trajectory, rotation, and water production measurements. Astron. Astrophys. **630**, A18 (2019)

A. Bardyn et al., Carbon-rich dust in comet 67P/Churyumov-Gerasimenko measured by COSIMA/Rosetta. Mon. Not. R. Astron. Soc. **469**(S2), 712 (2017)

M.J.S. Belton, Observational and dynamical constraints on the rotation of Comet P/Halley. Icarus **93**, 183 (1991)

M.J.S. Belton et al., The excited spin state of Comet 2P/Encke. Icarus **175**, 181 (2005)

M.J.S. Belton et al., The complex spin state of 103P/Hartley 2: kinematics and orientation in space. Icarus **222**(2), 595 (2013)

M.J.S. Belton et al., The excited spin state of 1I/2017 U1 'Oumuamua. Astrophys. J. **856**(2), L21 (2018)

M.S. Bentley et al., Aggregate dust particles at comet 67P/Churyumov-Gerasimenko. Nature **537**(7618), 73 (2016)

J.-L. Bertaux et al., The water production rate of Rosetta target Comet 67P/Churyumov-Gerasimenko near perihelion in 1996, 2002 and 2009 from Lyman $\alpha$ observations with SWAN/SOHO. Planet. Space Sci. **91**, 14 (2014)

J.-L. Bertaux et al., Estimate of the erosion rate from $H_2O$ mass-loss measurements from SWAN/SOHO in previous perihelions of comet 67P/Churyumov-Gerasimenko and connection with observed rotation rate variations. Astron. Astrophys. **583**, A38 (2015)

F.W. Bessel, Astron. Nachr. **13**, 3 (1835)

A. Bieler et al., Comparison of 3D kinetic and hydrodynamic models to ROSINA-COPS measurements of the neutral coma of 67P/Churyumov-Gerasimenko. Astron. Astrophys. **583**, A7 (2015)

N. Biver et al., Distribution of water around the nucleus of comet 67P/Churyumov-Gerasimenko at 3.4 au from the Sun as seen by the MIRO instrument on Rosetta. Astron. Astrophys. **583**, A3 (2015)

D. Bockelee-Morvan, First observations of $H_2O$ and $CO_2$ vapor in comet 67P/Churyumov-Gerasimenko made by VIRTIS onboard Rosetta. Astron. Astrophys. **583**, A6 (2015)

D. Bodewits et al., A rapid decrease in the rotation rate of comet 41P/Tuttle-Giacobini-Kresak. Nature **553**, 186 (2018)

H. Boehnhardt, Split comets, in *Comets II* (University of Arizona Press, Tucson, 2004)

J. Chen, D. Jewitt, On the rate at which comets split. Icarus **108**(2), 265 (1994)

J. Crovisier et al., Observations at Nancay of the OH 18-cm lines in comets. The data base. Observations made from 1982 to 1999. Astron. Astrophys. **393**, 1053 (2002)

B.J.R. Davidsson, Tidal splitting and rotational breakup of solid spheres. Icarus **142**(2), 525 (1999)

B.J.R. Davidsson, Tidal splitting and rotational breakup of solid biaxial ellipsoids. Icarus **149**(2), 375 (2001)

B.J.R. Davidsson, P.J. Gutierrez, Estimating the nucleus density of Comet 19P/Borrelly. Icarus **168**, 392 (2004)

B.J.R. Davidsson, P.J. Gutierrez, Nucleus properties of Comet 67P/Churyumov-Gerasimenko estimated from non-gravitational force modeling. Icarus **176**, 453 (2005)

J.F. Enke, Fortgesetzte nachricht über den Pons'schen kometen, in *Berliner astronomische Jahrbuch für das jahr 1826* (1823), p. 135

T.L. Farnham, A.L. Cochran, A McDonald observatory study of Comet 19P/Borrelly: placing the deep space 1 observations into a broader context. Icarus **160**, 398 (2002)

P.D. Feldman et al., Observations of Comet 67P/Churyumov-Gerasimenko with the International Ultraviolet Explorer at Perihelion in 1982, in *The New Rosetta Targets* (Springer, Berlin, 2004)

U. Fink et al., Investigation into the disparate origin of $CO_2$ and $H_2O$ outgassing for Comet 67/P. Icarus **277**, 78 (2016)

S. Fornasier et al., Rosetta's comet 67P/Churyumov-Gerasimenko sheds its dusty mantle to reveal its icy nature. Science **354**(6319), 1566 (2016)

N. Fougere et al., Three-dimensional direct simulation Monte-Carlo modeling of the coma of comet 67P/Churyumov-Gerasimenko observed by the VIRTIS and ROSINA instruments on board Rosetta. Astron. Astrophys. **588**, A134 (2016)

M. Fulle et al., Evolution of the dust size distribution of Comet 67P/Churyumov-Gerasimenko from 2.2 au to Perihelion. Astrophys. J. **821**(1) (2016)

B. Godard et al., Orbit determination of Rosetta around Comet 67P/Churyumov-Gerasimenko, in *International Symposium on Space Flight Dynamics-25th ISSFD* (2015)

B. Godard et al., Multi-arc orbit determination to determine Rosetta trajectory and 67P physical parameters, in *International Symposium on Space Flight Dynamics-26th ISSFD* (2017)

O. Groussin et al., The thermal, mechanical, structural, and dielectric properties of cometary nuclei after Rosetta. Space Sci. Rev. **215**, 29 (2019)

S. Gulkis et al., Subsurface properties and early activity of comet 67P/Churyumov-Gerasimenko. Science **347**(6220), 0709 (2015)

B. Gundlach et al., Outgassing of icy bodies in the Solar System—I. The sublimation of hexagonal water ice through dust layers. Icarus **213**(2), 710 (2011)

P.J. Gutierrez et al., Evolution of the rotational state of irregular cometary nuclei. Earth Moon Planets **90**(1), 239 (2002)

P.J. Gutierrez et al., Possible interpretation of the precession of comet 67P/Churyumov-Gerasimenko. Astron. Astrophys. **590**, A46 (2016)

M.S. Hanner et al., The dust coma of periodic Comet Churyumov-Gerasimenko (1982 VIII). Icarus **64**(1), 11 (1985)

K.C. Hansen et al., Evolution of water production of 67P/Churyumov-Gerasimenko: an empirical model and a multi-instrument study. Mon. Not. R. Astron. Soc. **462**(7607), 491 (2016)

M. Hirabayashi et al., Fission and reconfiguration of bilobate comets as revealed by 67P/Churyumov-Gerasimenko. Nature **534**, 352 (2016)

X. Hu et al., Thermal modelling of water activity on comet 67P/Churyumov-Gerasimenko with global dust mantle and plural dust-to-ice ratio. Mon. Not. R. Astron. Soc. **469**(S2), 295 (2017a)

X. Hu et al., Seasonal erosion and restoration of the dust cover on comet 67P/Churyumov-Gerasimenko as observed by OSIRIS onboard Rosetta. Astron. Astrophys. **604**, A114 (2017b)

L. Jorda et al., The global shape, density and rotation of Comet 67P/Churyumov-Gerasimenko from preperihelion Rosetta/OSIRIS observations. Icarus **277**, 257 (2016)

H.U. Keller et al., OSIRIS the scientific camera system onboard Rosetta. Space Sci. Rev. **128**, 433 (2007)

H.U. Keller et al., Insolation, erosion, and morphology of comet 67P/Churyumov-Gerasimenko. Astron. Astrophys. **583**, A34 (2015a)

H.U. Keller et al., The changing rotation period of comet 67P/Churyumov-Gerasimenko controlled by its activity. Astron. Astrophys. **579**, L5 (2015b)

H.U. Keller et al., Seasonal mass transfer on the nucleus of comet 67P/Chuyumov-Gerasimenko. Mon. Not. R. Astron. Soc. **469**, 357 (2017)

R. Kokotanekova et al., Rotation of cometary nuclei: new light curves and an update of the ensemble properties of Jupiter-family comets. Mon. Not. R. Astron. Soc. **471**(3), 2974 (2017)

R. Kokotanekova et al., Implications of the small spin changes measured for large Jupiter-family comet nuclei. Mon. Not. R. Astron. Soc. **479**, 4665 (2018)

T. Kramer, M. Läuter, Outgassing induced acceleration of comet 67P/Churyumov-Gerasimenko. Astron. Astrophys. **630**, A4 (2019)

T. Kramer et al., Comet 67P/Churyumov-Gerasimenko rotation changes derived from sublimation-induced torques. Astron. Astrophys. **630**, A3 (2019)

I.-L. Lai et al., Gas outflow and dust transport of comet 67P/Churyumov-Gerasimenko. Mon. Not. R. Astron. Soc. **462**, S533 (2016)

Y. Langevin et al., Typology of dust particles collected by the COSIMA mass spectrometer in the inner coma of 67P/Churyumov-Gerasimenko. Icarus **271**, 76 (2016)

M. Läuter et al., Surface localization of gas sources on comet 67P/Churyumov-Gerasimenko based on DFMS/COPS data. Mon. Not. R. Astron. Soc. **483**, 852 (2019)

S. Lee et al., Spatial and diurnal variation of water outgassing on comet 67P/Churyumov-Gerasimenko observed from Rosetta/MIRO in August 2014. Astron. Astrophys. **583**, A5 (2015)

J.-Y. Li et al., The unusual apparition of comet 252P/2000 G1 (LINEAR) and comparison with comet P/2016 $BA_{14}$ (PanSTARRS). Astron. J. **154**(4), 136 (2017)

S. Lowry et al., The nucleus of Comet 67P/Churyumov-Gerasimenko. A new shape model and thermophysical analysis. Astron. Astrophys. **548**, A12 (2012)

R. Marschall et al., Modelling observations of the inner gas and dust coma of comet 67P/Churyumov-Gerasimenko using ROSINA/COPS and OSIRIS data: first results. Astron. Astrophys. **589**, A90 (2016)

R. Marschall et al., Cliffs versus plains: can ROSINA/COPS and OSIRIS data of comet 67P/Churyumov-Gerasimenko in autumn 2014 constrain inhomogeneous outgassing? Astron. Astrophys. **605**, A112 (2017)

B.G. Marsden, Comets and nongravitational forces. II. Astron. J. **74**(5), 720 (1969)

B.G. Marsden et al., Comets and nongravitational forces. V. Astron. J. **78**(2), 211 (1973)

D.W. Marshall et al., Spatially resolved evolution of the local $H_2O$ production rates of comet 67P/Churyumov-Gerasimenko from the MIRO instrument on Rosetta. Astron. Astrophys. **603**, A87 (2017)

D. Marshall et al., Interpretation of heliocentric water production rates of comets. Astron. Astrophys. **623**, A120 (2019)

K.J. Meech et al., Nucleus properties of P/Schwassmann-Wachmann 1. Astron. J. **106**(3), 1222 (1993)

K.J. Meech et al., Epoxy: Comet 103P/Hartley 2 observations from a worldwide campaign. Astrophys. J. **734**(1), L1 (2011)

A. Migliorini et al., Water and carbon dioxide distribution in the 67P/Churyumov-Gerasimenko coma from VIRTIS-M infrared observations. Astron. Astrophys. **589**, A45 (2016)

S. Mottola et al., The rotation state of 67P/Churyumov-Gerasimenko from approach observations with the OSIRIS cameras on Rosetta. Astron. Astrophys. **569**, L2 (2014)

B.E.A. Mueller, I. Ferrin, Change in the rotational period of Comet P/Tempel 2 between the 1988 and 1994 apparitions. Icarus **123**, 463 (1996)

B.E.A. Mueller, N.H. Samarasinha, Further investigation of changes in cometary rotation. Astron. J. **156**(3), 107 (2018)

T. Ootsubo et al., AKARI near-infrared spectroscopic survey for $CO_2$ in 18 comets. Astrophys. J. **752**(1), 15 (2012)

M. Pätzold et al., The nucleus of comet 67P/Churyumov-Gerasimenko—Part I: The global view—nucleus mass, mass-loss, porosity, and implications. Mon. Not. R. Astron. Soc. **483**, 2337 (2019)

F. Preusker et al., Shape model, reference system definition, and cartographic mapping standards for comet 67P/Churyumov-Gerasimenko. Stereo-photogrammetric analysis of Rosetta/OSIRIS image data. Astron. Astrophys. **583**, A33 (2015)
R.R. Rafikov, Non-Gravitational Forces and Spin Evolution of Comets (2018). arXiv:1809.05133
H. Rickman, in *Comet Nucleus Sample Return*, ed. by O. Melita. ESA, vol. SP-249, Noordwijk (1986)
H. Rickman, The nucleus of comet Halley: surface structure, mean density, gas and dust production. Adv. Space Res. **9**(3), 359 (1989)
H. Rickman et al., Nongravitational effects and the aging of periodic comets. Astron. J. **102**(4), 1446 (1991)
A. Rotundi et al., Dust measurements in the coma of comet 67P/Churyumov-Gerasimenko inbound to the Sun. Science **347**(6220), 03905 (2015)
M. Rubin et al., Krypton isotopes and noble gas abundances in the coma of comet 67P/Churyumov-Gerasimenko. Sci. Adv. **4**, 7 (2018)
N.H. Samarasinha, M.F. A'Hearn, Observational and dynamical constraints on the rotation of Comet P/Halley. Icarus **93**, 194 (1991)
N.H. Samarasinha, B.E.A. Mueller, Component periods of non-principal-axis rotation and their manifestations in the lightcurves of asteroids and bare cometary nuclei. Icarus **248**, 247 (2015)
N. Samarasinha et al., Rotation of cometary nuclei, in *Comets II* (University of Arizona Press, Tucson, 2004)
N.H. Samarasinha et al., Rotation of comet 103P/Hartley 2 from structures in the coma. Astrophys. J. **734**(1), L3 (2011)
D.G. Schleicher, Icarus **181**, 442 (2006)
Z. Sekanina, The problem of split comets revisited. Astron. Astrophys. **318**, L5 (1997)
C. Simon Wedlund et al., The atmosphere of comet 67P/Churyumov-Gerasimenko diagnosed by charge-exchanged solar wind alpha particles. Astron. Astrophys. **587**, A154 (2016)
Y. Skorov, H. Rickman, A kinetic model of gas flow in a porous cometary mantle. Planet. Space Sci. **42**(12), 1587 (1995)
Y. Skorov et al., Activity of comets: gas transport in the near-surface porous layers of a cometary nucleus. Icarus **212**(2), 867 (2011)
Y. Skorov et al., A model of short-lived outbursts on the 67P from fractured terrains. Astron. Astrophys. **593**, A76 (2016)
C. Snodgrass et al., Beginning of activity in 67P/Churyumov-Gerasimenko and predictions for 2014–2015. Astron. Astrophys. **557**, A33 (2013)
J.C. Solem, Density and size of Comet Shoemaker-Levy 9 deduced from a tidal breakup model. Nature **370**(6488), 349 (1994)
N. Thomas et al., The morphological diversity of comet 67P/Churyumov-Gerasimenko. Science **347**(6220), 03905 (2015)
I. Toth, C.M. Lisse, On the rotational breakup of cometary nuclei and centaurs. Icarus **181**(1), 162 (2006)
W. Waniak et al., Rotation-stimulated structures in the CN and C3 comae of comet 103P/Hartley 2 close to the EPOXI encounter. Astron. Astrophys. **543**, A32 (2012)
F.L. Whipple, A comet model. I. The acceleration of Comet Encke. Astrophys. J. **111**, 375 (1950)
F.L. Whipple, The rotation of comet nuclei, in *Comets* (University of Arizona Press, Tucson, 1982)
D.K. Yeomans et al., Cometary orbit determination and nongravitational forces, in *Comets II* (University of Arizona Press, Tucson, 2004)

Space Sci Rev (2020) 216:102
https://doi.org/10.1007/s11214-020-00718-2

# On the Origin and Evolution of the Material in 67P/Churyumov-Gerasimenko

**Martin Rubin[1] · Cécile Engrand[2] · Colin Snodgrass[3] · Paul Weissman[4] · Kathrin Altwegg[1] · Henner Busemann[5] · Alessandro Morbidelli[6] · Michael Mumma[7]**

Received: 9 September 2019 / Accepted: 3 July 2020 / Published online: 30 July 2020


**Abstract** Primitive objects like comets hold important information on the material that formed our solar system. Several comets have been visited by spacecraft and many more have been observed through Earth- and space-based telescopes. Still our understanding remains limited. Molecular abundances in comets have been shown to be similar to interstellar ices and thus indicate that common processes and conditions were involved in their formation. The samples returned by the Stardust mission to comet Wild 2 showed that the bulk refractory material was processed by high temperatures in the vicinity of the early sun. The recent Rosetta mission acquired a wealth of new data on the composition of comet 67P/Churyumov-Gerasimenko (hereafter 67P/C-G) and complemented earlier observations of other comets. The isotopic, elemental, and molecular abundances of the volatile, semi-volatile, and refractory phases brought many new insights into the origin and processing of the incorporated material. The emerging picture after Rosetta is that at least part of the volatile material was formed before the solar system and that cometary nuclei agglomerated over a wide range of heliocentric distances, different from where they are found today. Deviations from bulk solar system abundances indicate that the material was not fully homogenized at the location of comet formation, despite the radial mixing implied by the Stardust results. Post-formation evolution of the material might play an important role, which further

Comets: Post 67P / Churyumov-Gerasimenko Perspectives
Edited by Nicolas Thomas, Björn Davidsson, Laurent Jorda, Ekkehard Kührt, Raphael Marschall, Colin Snodgrass and Rafael Rodrigo

✉ M. Rubin

1 Physikalisches Institut, University of Bern, Sidlerstrasse 5, 3012 Bern, Switzerland

2 CNRS/IN2P3, IJCLab, Université Paris-Saclay, 91405 Orsay Cedex, France

3 Institute for Astronomy, University of Edinburgh, Royal Observatory, Edinburgh EH9 3HJ, UK

4 Planetary Science Institute, Tucson, AZ, 85719, USA

5 Institute of Geochemistry and Petrology, Department of Earth Sciences, ETH Zurich, Zurich, Switzerland

6 Observatoire de la Cote d'Azur, Nice, France

7 NASA Goddard Space Flight Center, 8800 Greenbelt Rd., Greenbelt, 20771, MD, USA

complicates the picture. This paper discusses these major findings of the Rosetta mission with respect to the origin of the material and puts them in the context of what we know from other comets and solar system objects.

**Keywords** Comets · Origin of solar system material · Rosetta mission · 67P/Churyumov-Gerasimenko · Icy bodies · Isotopic and molecular composition

## 1 Introduction

Comets are the least processed objects in the solar system. Studying them allows us to gather precious information about the early days of our solar system and possibly even beyond. The origin and processing of the material also holds important clues to our understanding of the formation of our solar system some 4.6 Ga ago. Early remote sensing observations of the anomalous acceleration of comets Encke, D'Arrest, and Wolf 1 led Whipple (1950, 1951) to the conclusion that comets are "dirty snowballs" emitting gas and dust which can either decelerate or accelerate the comet along its orbit around the sun. However, only in 1986 did the European Space Agency's (ESA) Giotto mission and the Soviet Vega 2 mission flying by comet 1P/Halley (Reinhard 1986) confirm that comets possess a solid nucleus composed of volatile and refractory materials. From these investigations Geiss (1987) established that comets have preserved the accreted and condensed materials better than other objects in the solar system.

One of the main goals of space missions to comets, and dedicated remote sensing campaigns, is to understand their environment and to link their composition to the early days of the solar system. Several spacecraft have since visited comets *in situ* or even brought back a sample: Giotto also flew-by comet 26P/Grigg-Skjellerup (Coates et al. 1997), Deep Space One passed by 19P/Borrelly (Soderblom et al. 2002), and 81P/Wild 2 was visited by the Stardust spacecraft before its extended mission, NExT, flew by 9P/Tempel 1 (Brownlee et al. 2006). Cometary dust from Wild 2 was brought back to Earth for in-depth analysis. Additional comets visited were 9P/Tempel 1 by Deep Impact (A'Hearn et al. 2005) and 103P/Hartley 2 by the (renamed) EPOXI spacecraft (A'Hearn et al. 2011), before ESA's Rosetta mission encountered comet 67P/C-G (Glassmeier et al. 2007). Furthermore numerous spectroscopic and photometric observation campaigns of cometary comae have been performed from Earth- and space-based telescopes (see e.g. Biver et al. 2002; Bockelée-Morvan et al. 2004; Mumma et al. 1996, 2003, 2005; Dello Russo et al. 2007, and Lis et al. 2019). These observations have greatly increased the sample of investigated comets and led to many discoveries and allowed for a detailed comparison between comets, other solar system objects, and all the way to the interstellar medium.

Lately the Rosetta mission to comet 67P/C-G came to its conclusion. One of the main tasks of the mission was to obtain relevant measurements in both the refractories and volatiles to address the origin of the material in 67P/C-G's nucleus. There are two main scenarios for the origin of the material incorporated into comets that are widely discussed in the community. First the inheritance from the interstellar medium (Greenberg 1982) where the chemistry occurs at very low temperatures ($<20$ K) driven by cosmic rays and UV radiation and/or on the surface of icy grains. The alternative scenario is the formation together with the solar system but in its cold, outer regions (Lunine and Stevenson 1985; Mousis et al. 2016a). The presence of crystalline silicates was first determined in comet 1P/Halley (Bregman et al. 1988; Campins and Ryan 1989). Analysis of IRTF spectra (Hanner et al. 1994) and later observations with the Infrared Space Observatory (ISO) (Wooden et al. 1999; Crovisier et al. 2000, and Wooden et al. 2004) suggested that a large fraction of the cometary minerals formed at high temperature in the inner solar system. The formation and incorporation of these minerals in comets thus required models for large scale mixing in the protoplanetary disk, as proposed by several authors (e.g. Bockelée–Morvan et al. 2002; Ciesla 2007). The refractory samples returned from comet 81P/Wild 2 by the Stardust mission confirmed that the comet contains at least 10% by mass of material that formed in the hot inner regions of the protosolar disk (Brownlee et al. 2006). However, the cometary volatile abundances observed in comet C/1995 O1 (Hale-Bopp) showed striking similarities to the volatile molecules from the cold interstellar medium (Bockelée-Morvan et al. 2000). Moreover, comparisons of HCN and $NH_3$ with $C_2H_6$ and $H_2O$ amongst 30 comets suggested that both HCN and $NH_3$ were better associated with $C_2H_6$ production rather than $H_2O$, and both showed enrichments in comets within 1 au of the Sun consistent with dissociation of the semi-volatile salt, ammonium cyanide (Altwegg et al. 2020; Hänni et al. 2019; Mumma et al. 2017, 2018, 2019). Evidently, multiple processes are at work, although their relative importance may differ from not only one comet to another but also among the various molecules. Furthermore, post-formation alteration of the ices in comets has to be taken into account, in particular the various heating processes and associated outgassing of species of high volatility. Critical information is thus found in a combination of the elemental, molecular, and isotopic abundances of volatile and refractory material in comets by comparison to the solar system bulk composition, other comets, and interstellar medium abundances.

In this paper we will discuss how the material found in comets and in particular 67P/C-G can be traced back to its origin (see also Levasseur-Regourd et al. 2018). Furthermore we refer to the paper by Weissman et al. (2020) on the formation and dynamical history of cometary nuclei.

## 2 Observation of Comets

The low level of evolution of comets makes them ideal targets to study the history of the material in our solar system. The details of the ices in a comet's nucleus are imprinted in the gas coma surrounding it and are thus accessible not only to *in situ* but also to remote sensing observations.

### 2.1 Abundances of Volatile Species in Comets and Classification

The composition of comets, and its variability among the different dynamical groups, is a crucial link to understanding the formation of our solar system. Remote observations measure composition via spectroscopy or narrowband photometry, with a long history in the ultraviolet and blue visible wavelength range, where emission bands from 'daughter' species such as OH, CN, and $C_2$ are seen. More recently, there are increasingly observations in the infrared and at sub-mm wavelengths, which are sensitive to rotational and vibrational excited modes in 'parent' molecules (Bockelée-Morvan and Biver 2017), and with space telescopes that can directly measure the most abundant trace species ($CO_2$), which cannot be observed through Earth's atmosphere, and other gases that require a significant Doppler shift for ground-based detections (e.g., $CH_4$, CO).

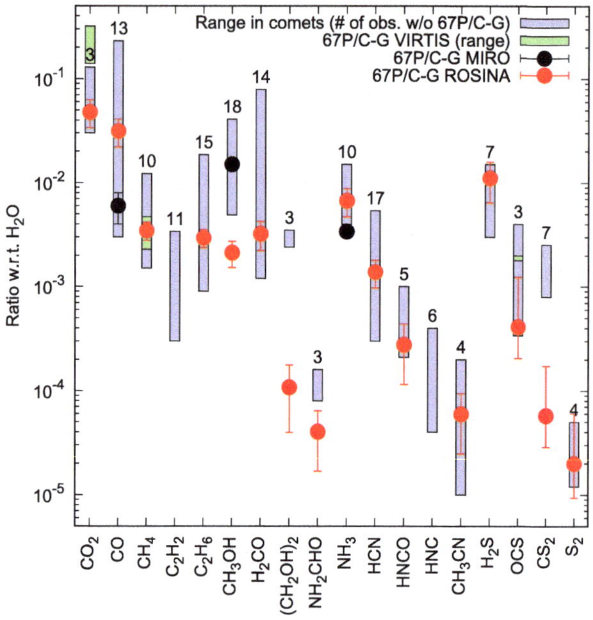


**Fig. 1** The blue boxes represent the observed range of volatile species w.r.t. $H_2O$ with at least the indicated number of detections in comets (aside 67P/C-G) within 2 au from the Sun (Biver et al. 2015; Bockelée-Morvan et al. 2004 and references therein). Separately listed are relative abundances of 67P/C-G from Table 2 measured by VIRTIS (range given by green boxes; Bockelée-Morvan et al. 2016), MIRO (black; Biver et al. 2019), and ROSINA (red, different isomers may be present, e.g. HCN cannot be distinguished from HNC with ROSINA; Rubin et al. 2019a)

Figure 1 shows in blue boxes the abundance range of commonly observed volatile species with respect to water in comets. Shown is a selected set of species detected in at least 3 comets aside from 67P/C-G. The most abundant species are $H_2O$, $CO_2$, and CO. Lower abundances are found for other CH- and CHO-bearing molecules and even lesser amounts for the CHS- and CHN-families of volatile species (Bockelée-Morvan et al. 2004; Dello Russo et al. 2016). These volatile species cover a wide range of sublimation temperatures and therefore relative abundances in the coma depend on the heliocentric distance. For comparability the observations in Fig. 1 are therefore limited to $<2$ au.

The first assessment of the average composition of a large sample of comets, based on narrowband photometry, was published by A'Hearn et al. (1995), who identified two broad taxonomic classes, 'typical' and 'carbon-chain depleted', the latter having lower production rates of $C_2$ and $C_3$ (relative to CN), and origins in the Kuiper belt. More recent work covering comets over a wider range of heliocentric distance has shown that these production rates vary significantly depending on where in its orbit the comet is observed (Opitom 2016), suggesting that this taxonomy must be used carefully. An updated review of the abundances of the major volatiles (CO, $CO_2$, $H_2O$) by A'Hearn et al. (2012), including the dedicated campaign for the $CO_2$ abundance in 18 comets by the Akari spacecraft (Ootsubo et al. 2012), found no evidence for significant bulk compositional variation between comets from the Kuiper belt or the Oort cloud.

Our knowledge of the minor constituent ices in comets primarily comes from relatively bright comets that can be observed at longer wavelengths (infra-red to radio), with many species first identified in observations of the great comets Hyakutake and Hale-Bopp (Bockelée-Morvan et al. 2004). Telescope technology has advanced sufficiently in the decades since these comets that there are now sufficient observations to attempt to group comets based on abundances of relatively rare parent molecules (e.g. hydrocarbons, methanol, and nitrogen-bearing compounds), with a general picture of greater abundance of more volatile ices in Oort cloud comets (Dello Russo et al. 2016; Mumma and Charnley 2011), thought to be less processed, but with an uncertain influence of the different evolution of short and long period comets (SPCs and LPCs). The utility of such taxonomies is still debated however, and no clear definitions have been broadly accepted; there is a very large diversity of relative abundance of different species among comets, with variation over orders of magnitude (Fig. 1; cf. Biver and Bockelée-Morvan 2015). An emerging view has some volatiles produced at least in part from semi-volatile ammonium salts (e.g., $NH_4^+CN^-$) (cf. Altwegg et al. 2020; Hänni et al. 2019; Mumma et al. 2019; Poch et al. 2020; Quirico et al. 2016).

High-resolution spectroscopy of brighter comets can also reveal compositional variation at an isotopic level (see Sect. 4); very recent results suggest differences in the isotopic composition between 'hyperactive' comets and those with more typical water production rates (Lis et al. 2019), in this case pointing to observed differences due to the current evolutionary state of the comet rather than its origin location. Finally, ground-based observation of large numbers of comets still reveals surprising cases with very unusual composition (e.g. the $N_2$-rich comet C/2016 R2; Biver et al. 2018; McKay et al. 2019; Opitom et al. 2019), or variation in relative abundance of different species with heliocentric distance (e.g., C/2009 P1; Feaga et al. 2013; Gicquel et al. 2015).

## 2.2 67P/C-G Abundances of Volatiles and Refractories

Ground-based observations indicated that 67P/C-G likely belongs to the carbon-chain depleted category of comets (Schulz et al. 2004), but *in situ* measurements were required to get a more detailed picture: despite a large campaign of observations supporting the Rosetta mission (Snodgrass et al. 2017), only the brightest emission features were detectable in remote data, and then only around the perihelion period while the Southern hemisphere was illuminated (Opitom et al. 2017; Snodgrass et al. 2016). Abundances of the major volatiles at comet 67P/C-G have been obtained by the Rosetta science instruments: ROSINA (Rosetta Orbiter Spectrometer for Ion and Neutral Analysis; Balsiger et al. 2007), VIRTIS (Visible and InfraRed Thermal Imaging Spectrometer; Coradini et al. (2007)), MIRO (Microwave Instrument for the Rosetta Orbiter; Gulkis et al. 2007), Alice (Stern et al. 2007) and the two mass spectrometers Ptolemy (Wright et al. 2007) and COSAC (Cometary Sampling and Composition experiment; Goesmann et al. 2007) on the Rosetta lander Philae. Most comet mission encounters lasted only for a short time, but ESA's Rosetta mission accompanied comet 67P/C-G for over two years along its orbit around the sun, accessing a much more detailed picture of the composition and activity of the comet.

A collection of volatile abundances measured with ROSINA are found in Table 1 (Bieler et al. 2015a; Le Roy et al. 2015). These measurements were obtained *in situ* and inbound from 3.1 au and are separated between the northern and southern hemispheres of the comet. Due to the axial tilt of the rotation axis of 52° with respect to its orbital plane (Sierks et al. 2015) the comet is subject to strong seasonal differences in the outgassing (Hässig et al. 2015). The northern hemisphere was in summer for the period of the measurements in October 2014 and thus dominated total outgassing. These early observations are compared to the relative abundances from the COSAC mass spectrometer on the Rosetta lander Philae (Goesmann et al. 2015). The measurements were obtained on 12 November 2014 at 3.0 au after the initial touchdown and rebound on the Agilkia landing site, some 25 minutes into the flight towards the final touchdown location in Abydos. The mass spectrometer Ptolemy was in operation during the same time; Wright et al. (2015) reported $-CH_2-$ and $-O-$ bearing compounds without providing relative abundances. The findings include the potential presence of polyoxymethylene (POM), a radiation-induced polymer, while other species, such as aromatic hydrocarbons including benzene and sulfur-bearing species, were either absent, or, in the case of nitrogen-bearing species, very low in abundance.

**Table 1** Subset of volatile species in the coma of comet 67P/Churyumov-Gerasimenko measured by in situ mass spectrometry ≥3 au by number (in % normalized to water). ROSINA measurements were obtained above the southern and northern hemispheres on 19 October 2014 and 20 October 2014, at approximately 10 km from the nucleus, respectively. References: Le Roy et al. (2015) and Bieler et al. (2015a). COSAC measurements from 12 November 2014, some 25 minutes after the initial touchdown at Agilkia on the way to Abydos, are likely a mixture of coma volatiles and excavated comet material from the initial landing (see the corresponding references for the full list of molecules). Note that for some molecules different isomers have been reported, e.g. HNC cannot be distinguished from HCN. Ptolemy: see text

| Molecule | ROSINA w.r.t. $H_2O$ [%] | | COSAC w.r.t. $H_2O$ [%] |
|---|---|---|---|
| | North | South | Philae/Agilkia |
| $H_2O$ | 100 | 100 | 100 |
| $CO_2$ | 2.5 | 80 | |
| CO | 2.7 | 20 | 1.2 |
| $O_2$ | 3.8 | 3.8 | |
| $CH_4$ | 0.13 | 0.56 | 0.5 |
| $C_2H_2$ | 0.045 | 0.55 | |
| $C_2H_6$ | 0.32 | 3.3 | |
| $CH_3OH$ | 0.31 | 0.55 | |
| $H_2CO$ | 0.33 | 0.53 | |
| HCOOH | 0.008 | 0.03 | |
| $(CH_2OH)_2$ | 0.0008 | 0.0025 | 0.2 |
| $HCOOCH_3$ | 0.004 | 0.023 | 0.4 |
| $CH_3CHO$ | 0.01 | 0.024 | 0.5 |
| $NH_2CHO$ | <0.0001 | <0.001 | 1.8 |
| $NH_3$ | 0.06 | 0.15 | |
| HCN | 0.09 | 0.62 | 0.9 |
| HNCO | 0.016 | 0.031 | 0.3 |
| $CH_3CN$ | 0.006 | 0.016 | 0.3 |
| $HC_3N$ | <0.00002 | <0.0005 | |
| $H_2S$ | 0.67 | 1.75 | |
| OCS | 0.017 | 0.098 | |
| SO | 0.004 | 0.0014 | |
| $SO_2$ | 0.011 | 0.041 | |
| $CS_2$ | 0.003 | 0.024 | |
| $S_2$ | 0.0004 | 0.0013 | |

A direct comparison of the relative abundances is difficult as the COSAC and Ptolemy mass spectra contain a mixture of coma volatiles and excavated surface material from the initial landing. Therefore, Altwegg et al. (2017b) compared the results of the two lander mass spectrometers with a dust impact event that occurred in ROSINA DFMS on September 5, 2016. As a result, the previously reported presence of methyl isocyanate ($CH_3NCO$), propanal ($C_2H_5CHO$), and glycol aldehyde ($CH_2OHCHO$) from COSAC was not supported. The signal of POM in Ptolemy, on the other hand, was attributed to toluene and hence an aromatic hydrocarbon, previously thought to be absent. Nevertheless, all three instruments together revealed a chemical complexity of the organics in, on, and around 67P/C-G that is much greater than expected.

Table 2, for comparison, lists relative abundances of a similar set of species but measured closer to perihelion or, when indicated, integrated over the whole Rosetta mission and hence dominated by the peak outgassing period around perihelion. *In situ* measurements by ROSINA were obtained at the end of May 2015, before perihelion passage (13 August 2015 at 1.24 au). This period, when Rosetta was passing rather closely over the then-active southern summer hemisphere and outbursts were still limited (Vincent et al. 2016), was identified by Calmonte et al. (2016) to be suitable to estimate bulk abundances. This approach assumes that the high erosion rate of the comet leads to relative abundances of the gases in

**Table 2** Measured coma abundances or abundance ranges by number (in % normalized to water) for a suite of volatile molecules from *in situ* ROSINA observations in May/June 2015 before perihelion (Rubin et al. (2019a) and references therein) and via remote sensing by MIRO, integrated over the 2 years Rosetta followed the comet (Biver et al. 2019), VIRTIS-H at perihelion (Bockelée-Morvan et al. 2016), and Alice from ~2 au inbound to 2.5 au outbound (Keeney et al. 2017). Species observed by mass spectrometry may (also) be present in the form of different isomers (cf. Fig. 1)

| Molecule | Abundances or abundance ranges relative to water [%] | | | |
|---|---|---|---|---|
| | ROSINA | VIRTIS-H | MIRO | Alice |
| $H_2O$ | 100 | 100 | 100 | 100 |
| $CO_2$ | $4.7 \pm 1.4$ | 14 – 32 | | |
| CO | $3.1 \pm 0.9$ | | $0.6 \pm 0.1$ | |
| $O_2$ | $3.1 \pm 1.1$ | | | 11 – 68 |
| $CH_4$ | $0.34 \pm 0.07$ | 0.23 – 0.47 | | |
| $C_2H_6$ | $0.29 \pm 0.06$ | | | |
| $CH_3OH$ | $0.21 \pm 0.06$ | | $1.5 \pm 0.1$ | |
| $H_2CO$ | $0.32 \pm 0.10$ | | | |
| HCOOH | $0.013 \pm 0.008$ | | | |
| $(CH_2OH)_2$ | $0.011 \pm 0.007$ | | | |
| $CH_3COOH$ | $0.0034 \pm 0.0020$ | | | |
| $CH_3CHO$ | $0.047 \pm 0.017$ | | | |
| $CH_3NO$ | $0.0040 \pm 0.0023$ | | | |
| $NH_3$ | $0.67 \pm 0.20$ | | $0.34 \pm 0.01$ | |
| $N_2$ | $0.089 \pm 0.024$ | | | |
| HCN | $0.14 \pm 0.04$ | | | |
| HNCO | $0.027 \pm 0.016$ | | | |
| $CH_3CN$ | $0.0059 \pm 0.0034$ | | | |
| $HC_3N$ | $0.00040 \pm 0.00023$ | | | |
| $H_2S$ | $1.10 \pm 0.46$ | | | |
| OCS | $0.041^{+0.082}_{-0.020}$ | 0.12 – 0.18 | | |
| SO | $0.071^{+0.142}_{-0.037}$ | | | |
| $SO_2$ | $0.127^{+0.254}_{-0.064}$ | | | |
| $CS_2$ | $0.0057^{+0.0114}_{-0.0028}$ | | | |
| $S_2$ | $0.0020^{+0.0040}_{-0.0010}$ | | | |
| S | $0.46 \pm 0.36$ | | | |
| $H_2CS$ | $0.0027^{+0.0058}_{-0.0024}$ | | | |
| $CH_3SH$ | $0.038^{+0.079}_{-0.028}$ | | | |
| $CH_3CH_2S$ $CH_3SCH_3$ | $0.00058^{+0.00123}_{-0.00049}$ | | | |

the coma that reflect the composition of ices inside the nucleus. Table 2 also lists a suite of relative abundances measured by the Rosetta remote sensing suite of instruments including VIRTIS-H (Bockelée-Morvan et al. 2016), MIRO (Biver et al. 2019), and Alice (Feldman et al. 2015; Keeney et al. 2017). The activity of water (Bieler et al. 2015b; Hansen et al. 2016; Kramer et al. 2017) and the major molecules (Biver et al. 2019; Bockelée-Morvan et al. 2016; Fougere et al. 2016; Läuter et al. 2019; Marshall et al. 2017) have been tracked by several instruments through large portions of the mission and their outgassing fluxes, both in absolute and relative numbers, exhibited a notable dependency on the heliocentric distance. The reported ratios were either derived at perihelion or represent relative abundances integrated over a large portion of the mission from pre- to post-perihelion. Table 2 lists some notable similarities as well as differences between *in situ* and remote sensing derived ratios, e.g. $NH_3$ and OCS agree within a factor of 2 between ROSINA and VIRTIS-H/MIRO, respectively. All these data are also represented in Fig. 1.

In a different approach, Combi et al. (2020) integrated ROSINA-derived gas production rates between inbound and outbound equinoxes and obtained $CO_2/H_2O = 7.4\%$ and $CO/H_2O = 2.7\%$, which are both somewhat closer to the corresponding MIRO and VIRTIS-H values. Combi et al. (2020) also showed that the derived gas activities are particularly sensitive to the used models which turn column densities from remote-sensing observations or *in situ* densities from ROSINA into production rates. This is particularly important for observations in the far-ultraviolet (FUV), where energetic electrons are an important source for the dissociation of molecules (e.g. $O_2$, CO, $CO_2$) and formation of excited H, C, and O atoms (Feldman et al. (2015), cf. Mumma et al. (1971, 1972). Such analysis requires a detailed simulation of neutral gas and plasma, driven by the corresponding measurements from Rosetta including the Rosetta Plasma Consortium (Carr et al. 2007). It is important to note that models have improved significantly and the resulting relative as well as absolute production rates are converging. Nevertheless, the efforts in understanding the differences among the reported values from the different instruments are still ongoing. This is also interesting in the view of 67P/C-G's next apparition in November 2021, which is much more favorable for a terrestrial observation campaign compared to summer 2015.

Figure 1 also reveals differences between 67P/C-G and other comets. This may either point at 67P/C-G exhibiting peculiar low abundances in species such as $CS_2$ or, as mentioned above, may derive from the difficulty to remotely detect species with low (relative) abundances in distant comets.

The instrument on Rosetta dedicated to the composition of the refractory phase was the Cometary Secondary Ion Mass Analyzer (COSIMA; Kissel et al. 2007). It was found that the dust is rich in high molecular weight organic matter (Fray et al. 2016). Elemental ratios show carbon-rich dust (Bardyn et al. 2017) with a C/Si ratio comparable to solar relative abundances and earlier measurements at 1P/Halley by the PUMA-1 instrument onboard Vega 1 (Jessberger et al. 1988). The authors concluded that by weight the dust is composed of nearly equal amounts of organic matter and anhydrous mineral phases, with up to 90% porosity (Langevin et al. 2017). The N/C atomic ratio of the organic matter in 67P/C-G matches that of insoluble organic matter (IOM) chemically extracted from primitive meteorites (Fray et al. 2017), but its higher H/C ratio compared to IOM suggests a more unsaturated state in 67P/C-G (Isnard et al. 2019).

The elemental composition of dust of 67P/C-G is broadly compatible with chondritic composition, but also exhibits notable differences, as Mg, Ca, and Fe are significantly depleted w.r.t. the chondritic composition (Bardyn et al. 2017). When compared to different types of meteorites, 67P/C-G dust bears more similarities to carbonaceous chondrite composition, than to other types of meteorites, but there is no perfect match (Stenzel et al. 2017).

ROSINA DFMS, as mentioned earlier, was dedicated to measure the neutral gas species in the coma. Early in the mission, DFMS measured several refractory elements in the coma (Wurz et al. 2015). It was established, that at 67P/C-G's low cometary activity beyond 3 au, the solar wind reached at least parts of the surface of the nucleus and subsequent sputtering released refractory elements into the coma. DFMS detected atomic Na, Si, K, and Ca, however, none of them were in molecular form. From these measurements their relative abundances have been derived. Figure 2 shows a comparison between COSIMA (Bardyn et al. 2017) and ROSINA (Wurz et al. 2015) derived ratios normalized to Si. With ROSINA, the small sputtered contribution of H, C, and O cannot be distinguished from the fragments of the abundant volatile organic and inorganic molecules formed by electron-impact inside the ion source of DFMS. Therefore, a meaningful comparison between these elements is not

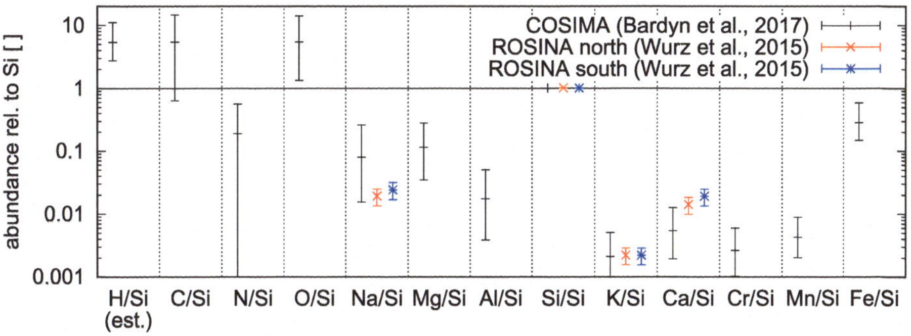


**Fig. 2** Abundances of the elements in the refractory phase relative to silicon measured by COSIMA (black; Bardyn et al. 2017) and ROSINA (Wurz et al. 2015) at comet 67P/C-G. The COSIMA measurement represents an average of the analyzed dust grains and the ROSINA observations are the individually averaged sputter-signals above the at the time more active northern summer hemisphere (red) and the lesser active southern winter hemisphere (blue). The H/Si ratio is an estimate from Bardyn et al. (2017)

possible. However, the relative abundances of the measured refractory elements overlap at the 1-$\sigma$ level of the corresponding COSIMA observations. ROSINA measurements show a high relative abundance of Si and therefore support the conclusion by Bardyn et al. (2017) that the mineral phase in 67P/C-G's dust is predominantly composed of anhydrous silicates.

## 3 Conditions at the Location of Formation of Comets

### 3.1 Comet Formation

The classic view has the solar system forming as a consequence of the gravitational collapse of an interstellar cloud, likely a giant molecular cloud. Such star-forming regions seen today often display multiple "hot cores" that are collapsing into individual stars and their planetary systems. The solar system's birth cloud likely featured many such cores, with one leading eventually to the Sun and planets of our evolved system. Most of the core mass wound up in the Sun but because the core was slowly rotating, a small fraction formed a protoplanetary disk and its surrounding gravitationally attached envelope of dust and gas (the ensemble being called the "Solar nebula"). The protoplanetary disk defined the equatorial plane of the nebula, with disk material and its gaseous environs rotating around the central star and flaring in height at large heliocentric distances. This disk was composed of gas plus submicron and larger particles of ice and dust (though the ice sublimated away close to the Sun or, upon arrival at the surface of the accretion disk, the infalling material may have crossed a shock and been evaporated (Chick and Cassen 1997)). This material slowly coalesced through a process known as hierarchical accretion, particles running into one another and sticking (Weidenschilling 1977, 1997). Thus, particles grew in size.

Numerical simulations show that micron-sized particles can grow through hierarchical accretion to centimeter size but as they approach meter-size, their inter-particle collisions increase in velocity and change from accretional to erosional (Blum 2010, 2018). Compaction of the growing particles during collisions also increases their strength and they begin to bounce off one another.

There are two hypotheses proposed to surmount this growth barrier. The first is that hierarchical accretion somehow continues to larger sizes creating kilometer-sized bodies.

The presence of water ice may increase the growth rate past the 1-meter barrier due to the stickiness of water ice (Ros and Johansen 2013). These are the forerunners of the comet nuclei we see today (Davidsson et al. 2016). The second hypothesis is that when particles grow to centimeter sizes they are brought together in streaming (hydrodynamic) instabilities in the solar nebula which then gravitationally collapse into macroscopic bodies on the order of 50 to 100 km in diameter (Youdin and Goodman 2005). Subsequent collisions create both smaller and larger bodies, all of which can be characterized as icy planetesimals (future comets), though the smaller ones are far more numerous.

Note that the distinguishing feature of comets is that they contain substantial amounts of volatiles, primarily water ice, aside from refractory material (Choukroun et al. 2020). Therefore, comets must have formed in colder regions of the solar nebula where water ice could be stable. In the current solar system, the "snowline" is found around 3-5 au (Min et al. 2011). But in the solar nebula it may have been closer to the Sun. In a disk solely illuminated by the star, the snowline would be inward of 1 au, because of the low optical depth (Chiang and Youdin 2010; Sasselov and Lecar 2000). However, the disk can be heated by viscous dissipation, so that the location of the snowline depends on the stellar accretion rate (Bitsch et al. 2015; Oka et al. 2011), with a snowline up to 4-5 au when the star accretes at $10^{-7}$ $M_{Sun}$/year (early disk). However, some of the more volatile species found in comets, e.g., $N_2$, CO, etc. suggest formation temperatures on the order of 30 K or less, and thus require formation and capture considerably farther out.

Regardless of which formation mechanism is correct, the comet nuclei likely continued to evolve collisionally before their dynamical dispersion to the Kuiper belt and Scattered disk, and/or to the Oort cloud (Morbidelli and Rickman 2015). Subtle physical processes, such as heating by nearby supernovae and cosmic rays will also modestly process the near-surface layers of the nuclei. These topics and related subjects are discussed in detail in the paper by Weissman et al. (2020) and later in Sect. 5.

## 3.2 On the Refractory (Dust) Phase of 67P/C-G and Other Comets

Cometary dust composition data are available for 67P/C-G (Bardyn et al. 2017; Fray et al. 2016, 2017; Isnard et al. 2019; Wurz et al. 2015) and a few other comets: 1P/Halley (Giotto and Vega), (e.g. Jessberger et al. 1988; Kissel et al. 1986a,b), 81P/Wild 2 (e.g. Brownlee 2014 and references therein), 9P/Temple 1 (Lisse et al. 2006), and C/1995 O1 (Hale-Bopp) (e.g. Crovisier et al. 1997). A detailed review can be found in Levasseur-Regourd et al. (2018).

The minerals found in these comets show variation in their nature from one comet to another, but mostly consist of crystalline minerals (refractory oxides and silicates, olivine, pyroxenes, metal, sulfides and accessory minerals) as well as amorphous silicates. The crystalline minerals identified in the comets were formed at high temperature close to the early Sun and subsequently transported to the external regions of the protoplanetary disk, in the comet forming region (e.g. Bockelée–Morvan et al. 2002; Ciesla 2007; Shu et al. 1997; Vinković 2009).

The most detailed mineralogical characterization was possible for 81P/Wild 2 samples brought back by the Stardust mission. These samples contain refractory minerals (Ca-Al-rich inclusions) as well as chondrule fragments and crystalline silicates similar to those identified in primitive meteorites (e.g. Brownlee et al. 2006; Nakamura et al. 2008; Zolensky et al. 2006). The olivines from the Stardust sample show a very large variation range of their major and minor element concentrations, relating them to olivines found in primitive (carbonaceous) chondrites, but also to olivines from ordinary chondrites (Frank et al.

2014). These very diverse olivine compositions and the lack of $^{26}$Al in refractory minerals in Wild 2 samples (e.g. Nakashima et al. 2015) could mean that comet 81P/Wild2 incorporated minerals that formed late (without $^{26}$Al) and that had been already processed (e.g. in an ordinary chondrite parent body). These minerals were therefore most likely accreted after the formation of the icy cometary body.

Hints for traces of aqueous alteration were found in comet Wild 2 minerals (e.g. Berger et al. 2011; Mikouchi et al. 2007; Hicks et al. 2017; Hanner and Zolensky 2010). Spitzer spectra of 9P/Tempel 1 acquired during the Deep Impact mission could contain a contribution from a few percent of hydrated silicates, although this is debated (Lisse et al. 2006).

Chondritic porous interplanetary dust particles (IDPs), that are probably of cometary origin, are mostly anhydrous, but they can contain trace amounts of hydrated minerals (e.g., Bradley 2003; Brownlee 2014). Ultracarbonaceous Antarctic Micrometeorites (UCAMMs) that are probably of cometary origin could also have experienced minimal aqueous alteration (Guérin et al. 2020; Yabuta et al. 2017). Whether these small amounts of aqueous alteration products found in comets were produced *in situ* or accreted from previous episodes is still debated (Suttle et al. 2020).

### 3.3 On the Ice Phase and Formation Temperature of 67P/C-G

Rubin et al. (2015a) reported $N_2/CO$ to be depleted in comet 67P/C-G by a factor of $25.4 \pm 8.9$ relative to protosolar N and C abundances. One interpretation is the slightly different trapping efficiency of $N_2$ with respect to CO in amorphous ice. Hence $N_2/CO$ is temperature-dependent and the measured ratio of $5.7 \cdot 10^{-3}$ at the comet, compared to the protosolar nebula value $0.145 \pm 0.048$ (Lodders et al. 2009) and assuming all N was in $N_2$ and C was in CO, led to a formation temperature below $\sim$30 K for 67P/C-G. Computations based on clathrate hydrates for $N_2/CO$ and Ar/CO reveal a somewhat higher formation temperature, though still below 50 K (Mousis et al. 2012, 2016a). However, this $N_2/CO$ ratio was obtained at a heliocentric distance beyond 3 au. Relative abundances of various volatiles including the major species $H_2O$, CO, $CO_2$, and $O_2$ changed over the course of comet 67P/C-G's eccentric orbit around the Sun (Biver et al. 2019; Bockelée-Morvan et al. 2016; Fougere et al. 2016; Keeney et al. 2017). In particular the relative abundance of CO with respect to water decreased due to the sharply increasing water production rate closer to the sun (Läuter et al. 2019). Furthermore, the relative abundance of other volatile molecules also increased towards perihelion, even for species of similar volatility such as $N_2$. A suitable period to derive bulk abundances has been identified to be May 2015, a few months before perihelion passage (Calmonte et al. 2016). If we combine the reported $N_2/H_2O = 8.9 \cdot 10^{-4}$ ratio from Rubin et al. (2018) and $CO/H_2O = 0.031$ ratio estimated from Rubin et al. (2019a) for the same period, a roughly 5 times larger $N_2/CO$ ratio and hence a lower depletion rate of $N_2$ with respect to CO is obtained. In the case of amorphous ice, such a ratio would move the formation temperature of 67P/C-G towards even lower values, possibly into the low 20 K range (Bar-Nun et al. 2007). However, such a scenario would also lead to an even higher Ar/CO ratio, which is not observed and is at odds with the absence of the noble gas neon in the coma of comet 67P/C-G (Bar-Nun et al. 2012; Rubin et al. 2018). What remains to be understood, however, is the impact of other ices aside from $H_2O$. For instance the addition of $CO_2$, the second most abundant species in the coma of comet 67P/C-G (Fougere et al. 2016; Läuter et al. 2019), has been shown to increase trapping efficiencies of $N_2$ and Ar (Greenberg et al. 2017). Late in the Rosetta mission, at heliocentric distances >3 au, when the noble gases Xe, Kr, and Ar were observed, $CO_2$ was the dominant volatile in the coma (Läuter et al. 2019). The jury is still out on whether highly volatile species are predominantly trapped in $CO_2$ as opposed to $H_2O$ in the case of amorphous ices.

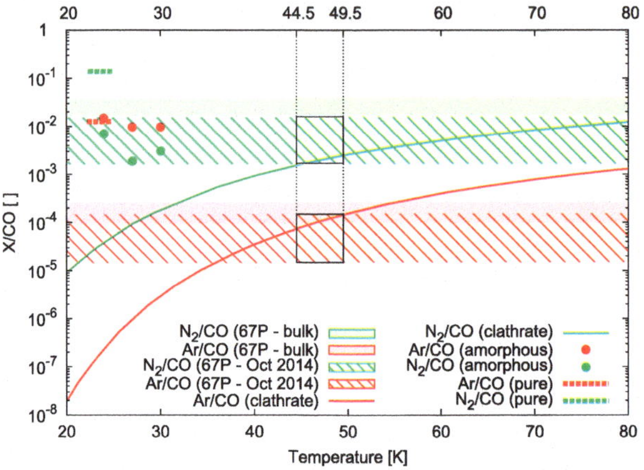


**Fig. 3** $N_2/CO$ (green) and Ar/CO (red) ratios measured at comet 67P/C-G and obtained from water ice experiments and statistical calculations of $H_2O$ guest clathrate adapted from Mousis et al. (2016a). The measured ranges of the ratios differ by approximately a factor 10 between October 2014 when Rosetta was beyond 3 au (hatched area, data from Balsiger et al. 2015 and Rubin et al. 2015a) and the comet's near-perihelion production rate ratios (solid area, numbers derived from Rubin et al. 2018 and Rubin et al. 2019a). Relative abundances of pure ice condensation are shown by the dashed lines at temperatures $<25$ K (Mousis et al. 2010) and trapping experiments in amorphous ices obtained at 24 K, 27 K, and 30 K (Bar-Nun et al. 2007) are represented by dots. Relative abundances from statistical thermodynamics modeling of clathrate hydrates along the equilibrium temperature curves are plotted in solid lines. The temperature range from 44.5 – 49.5 K simultaneously overlapping $N_2/CO$ and Ar/CO ratios modeled and the October 2014 Rosetta measurements are indicated by the black boxes

However, this discussion must be revisited in the context of the recent discovery of abundant ammonium salts in 67P, that suggest the 'missing' N in comets may be hidden in them (Altwegg et al. 2020; Poch et al. 2020). These semi-volatile salts can form in the interstellar medium at temperatures below 15 K and then remain stable until warmed to much higher temperatures (see later discussion). They can be carried into pre-cometary ices and survive until released and sublimated in the inner solar system, releasing $NH_3$, HCN, organic acids, and other products into the cometary coma (Mumma et al. 2019).

Scenarios of evaporation of amorphous ices (Chick and Cassen 1997; Mousis 2000) and recondensation in the protosolar nebula (Kouchi et al. 1994) have also been discussed in the literature. Figure 3 shows the measured $N_2/CO$ and Ar/CO ratios at comet 67P/C-G for heliocentric distances $>3$ au and derived production rate ratios near perihelion in comparison to ratios derived from relative trapping as pure ices, in amorphous ices, and in clathrate ices (adopted from Mousis et al. 2016a). While from the early measurements an overlapping formation temperature reproducing observed Ar/CO and $N_2/CO$ could be derived in the case of clathrate hydrates, this is no more possible for the measured bulk abundances. Further ice trapping experiments and statistical thermodynamics models (Lunine and Stevenson 1985) of $N_2$, noble gases, and CO trapping in comet-style ices, containing not only $H_2O$ but also $CO_2$, still have to be carried out. This mix of ices is especially important given that the noble gases were better correlated with $CO_2$ than with $H_2O$. Nevertheless, these results show that the ices of a comet cannot be represented by one single phase of water ice. This is consistent with the recent scenario proposed by Mousis et al. (2018b), where the ices are inherited from the presolar cloud but then undergo a phase transition from amorphous to crystalline/clathrate ices including retrapping of the more volatile species, with a few exceptions that will be discussed later in Sect. 4.2.

Most of these calculations were based on solar abundances (Lodders et al. 2009). Analysis including carbon monoxide and molecular nitrogen was performed on the assumption that C being predominantly incorporated in CO and N in $N_2$ (Fegley and Prinn 1989) and hence the relative amounts of these two molecules in the interstellar medium (ISM) and the protosolar nebula (PSN), respectively, were derived from the solar N/C ratio.

To complicate matters further, later evolutionary processes have to be considered for the comet's journey from the Scattered disk to the inner planetary system, i.e. the loss of preferentially highly volatile species by heating through collisional interactions among cometesimals or gradual warming of the nucleus during the intermediate Centaur stage. These topics will be discussed in Sect. 5.

# 4 Origin of Cometary Material

## 4.1 Isotopes

The isotopes found in cometary refractories and volatiles contain information on the origin and the processing of the material. In their work, Bockelée-Morvan et al. (2015) reviewed the suite of D/H, $^{14}N/^{15}N$, $^{16}O/^{18}O$, $^{12}C/^{13}C$, and $^{32}S/^{34}S$ ratios measured in cometary material and discussed their role as tracers for the physical chemical conditions responsible for the observed isotopic fractionation. In the following we will review some of these measurements with a focus on comet 67P/C-G and discuss their relevance with respect to their formation history. A complementary comparison of Rosetta-derived isotopic compositions with solar system materials and the ISM can be found in Hoppe et al. (2018) and Levasseur-Regourd et al. (2018).

### *4.1.1 The Variation in D/H*

Generally, due to the high relative mass difference between deuterium and hydrogen, the D/H ratio is most prone to fractionation effects. A key measurement is thus the D/H ratio in the cometary $H_2O$ molecule. It has been used to infer potential sources for the water on Earth and other objects in the solar system. It has been recognized that the degree of fractionation depends on the environmental conditions under which the ices formed, for example, on grain surfaces in molecular clouds. Grain surface reactions at low temperatures ($\sim$10 K) can enhance the deuterium fractionation considerably (Brown and Millar 1989a,b; Charnley et al. 1997). In particular, it has been predicted that the D/H ratio increases with heliocentric distance of the formation location of the ices (Aikawa and Herbst 2001). The cause for this effect is the difference in the zero-point energy between D- and H-bearing molecules. Thus in the cold environment D and H substitution reaction rates are not equal, favoring the substitution of H by D in a molecule (Ceccarelli et al. 2014).

The D/H ratio in water has been observed in more than a dozen comets, from both the Oort cloud and the Kuiper belt (for recent summaries see Paganini et al. 2017; Lis et al. 2019). The variation in the D/H ratio in water can vary by a factor of a few but most comets show deuterium enrichment with respect to the Earth, i.e. the Vienna Standard Mean Ocean Water (VSMOW) of D/H $= (1.558 \pm 0.001) \cdot 10^{-4}$, which is already much higher than the protosolar D/H ratio in $H_2$ of $2 \cdot 10^{-5}$ (Lodders et al. 2010). The early measurements of D/H in $H_2O$ were all performed on Oort cloud comets and showed enrichment compared to VSMOW. After the observation of a terrestrial D/H ratio in comet 103P/Hartley 2, however, it was hypothesized that Jupiter-family comets (JFCs) may have contributed sizeable amounts of water to the Earth (Hartogh et al. 2011). JFC 67P/C-G, on the other hand, yielded a much higher D/H ratio, more than 3.5 times the ratio on Earth (Altwegg et al. 2015). The two consequences are that the origin of terrestrial water is again unresolved and that dynamical families of comets cannot simply be distinguished by their D/H ratios. The D/H ratio in cometary water could instead be the result of a mixture of presolar, high D/H water with isotopically lighter water from the inner solar system. In summary, it is thus hypothesized that the measured D/H ratio rather reflects the formation location of the comet before being expelled through the migration of the giant planets (Gomes et al. 2005). A more recent study by Lis et al. (2019), after the observation of a D/H ratio in comet 46P/Wirtanen compatible with VSMOW, suggested a correlation of the D/H ratio with the size and activity of a comet. Hyperactive comets, such as 103P/Hartley 2 and 46P/Wirtanen may therefore contain more ices processed in the inner solar system, possibly locked-up in icy grains sublimating only after their release into the coma. A recent study by Schroeder et al. (2019) also showed that, within error, the water above both lobes shares the same D/H ratio. This hints at a formation of both lobes of the comet in the same region of the protoplanetary disk before the final collisional merger occurred.

Still, the question whether parts of the ice are even inherited from the presolar cloud is difficult to answer from the ratio of $HDO/H_2O$ alone. Furuya et al. (2016) showed that water can retain its high $HDO/H_2O$ ratio even if it is reprocessed in the protostellar disk. Nevertheless, the D/H ratios in comets are often lower than ratios observed near young stellar objects. Further complication arose from observations of much higher D/H ratios in doubly versus singly deuterated water vapor of $[D_2O/HDO]/[HDO/H_2O] \cong 7$ around low-mass protostar NGC1333 IRAS2A (Coutens et al. 2014). Furuya et al. (2016) therefore presented a model where the deuteration of water occurs in a two-step process: in stage 1, $H_2O$ forms on the surface of grains with a D/H ratio reflecting the surrounding environment; in stage 2, as the molecular cloud cools down, water production is reduced but deuteration becomes much more efficient due to the involvement of non-equilibrium chemistry at low temperatures. The model follows the earlier work by Dartois et al. (2003), based on observations of even higher D/H ratios in molecules freezing out at lower temperatures compared to water, e.g. in $NH_3$ (Lis et al. 2002; Roueff et al. 2000; Shah and Wootten 2001; Van der Tak et al. 2002), $CH_3OH$ (Parise et al. 2006), and $H_2S$ (Hatchell et al. 1999). In stage 2, where most of HDO and $D_2O$ is formed, the deposition of CO also leads to the formation of $H_2CO$ and $CH_3OH$ which are thus subject to higher D/H fractionation. Thus, Furuya et al. (2016) predicted D/H ratios from $CH_3OD/CH_3OH$ to be similar to $D_2O/HDO$ but larger than from $HDO/H_2O$, although the situation becomes more complicated once additional substitution and abstraction reactions are involved, e.g. in the case of $CH_2DOH/CH_3OH$.

A similar picture has been obtained for comet 67P/C-G when comparing the two different D/H ratios derived in water. $HDO/H_2O$ corresponds to $2 \cdot D/H$ as D in HDO can sit on either of the two positions of hydrogen in the water molecule and, likewise, $D_2O/HDO$ corresponds to $1/2 \cdot D/H$. From statistics one would therefore expect $1/4$ for $D_2O/HDO$ relative to $HDO/H_2O$ in an equilibrated system, where D/H matches in both ratios. Altwegg et al. (2017a), however, reported a $D_2O/HDO$ relative to $HDO/H_2O$ ratio of 17. Furthermore, the same authors found $HDS/H_2S = (1.2 \pm 0.3) \cdot 10^{-3}$, a ratio which overlaps that of $HDO/H_2O = (1.05 \pm 0.14) \cdot 10^{-3}$ within 1-$\sigma$ uncertainties, also in line with Furuya et al. (2016) where $H_2S$ is formed in the colder stage 2. Around the cold star-forming core IRAS 16293 an even higher ratio of $HDS/H_2S = 0.1$ has been derived (van Dishoeck et al. 1995). High D/H ratios have also been observed in other species such as HCN in comet Hale-Bopp with $D/H = (2.3 \pm 0.4) \cdot 10^{-3}$ (Meier et al. 1998a) and $D/H = (2.3 \pm 0.6) \cdot 10^{-3}$ (Crovisier et al. 2004). However, these ratios are still small compared to the ISM observations, which yield

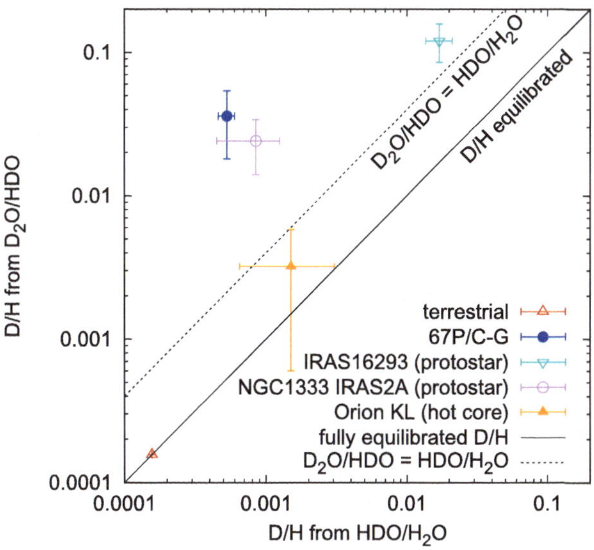


**Fig. 4** D/H ratio obtained from $HDO/H_2O$ versus D/H obtained from $D_2O/HDO$ in comet 67P/C-G (Altwegg et al. 2017a), the two protostars IRAS 16293 (Coutens et al. 2012; Le Roy et al. 2015) and NGC1333 IRAS 2A (Coutens et al. 2014; Rubin et al. 2019a), and the Orion KL hot core (Dello Russo et al. 2016; Neill et al. 2013) compared to the terrestrial value (Haynes 2013). The solid black line corresponds to equal D/H ratios in both $HDO/H_2O$ and $D_2O/HDO$. The dashed line represents where $HDO/H_2O$ is equal to $D_2O/HDO$ proposed by Furuya et al. (2016) to distinguish unprocessed from processed ices. Error bars are 1-$\sigma$

D/H in the range of $(0.4 - 7) \cdot 10^{-2}$ (Jørgensen et al. 2004; Roberts et al. 2002) and thus Bockelée-Morvan et al. (2015) argued for some degree of reprocessing in the solar nebula before incorporation into the comet.

Figure 4 shows the D/H ratios in singly and doubly deuterated water observed in several objects including the Earth, comet 67P/C-G, and in star forming regions. The non-equilibrated D/H ratio in 67P/C-G is similar to the observations around protostars IRAS 16293 and NGC1333 IRAS2A, while the warmer environment of the Orion KL hot core leads to a similar D/H ratio in both $D_2O/HDO$ and $HDO/H_2O$. In liquid water, hydrogen and deuterium atoms are rapidly exchanged, which leads to an equilibrated D/H ratio in the terrestrial oceans. Nevertheless, in the case of equilibration, a high D/H in $D_2O/HDO$ only marginally affects the original $HDO/H_2O$ ratio due to the low relative abundance of $D_2O$.

In their work, Furuya et al. (2016) concluded that the ratio of $[D_2O/HDO]/[HDO/H_2O]$ reveals much more about the formation of the ice, whether it is of presolar origin or has been reformed or reprocessed in the protosolar nebula: laboratory experiments show that H-D exchange in hydrogen-bonded molecules in mixed amorphous ices occurs efficiently on timescales of the order of $10^4$ years at temperatures $\gtrsim$70 K, well below the crystalline ice transition temperature (Faure et al. 2015; Lamberts et al. 2015; Mousis et al. 2016a; Ratajczak et al. 2009). On the other hand, given the two-stage setup of the model the highly deuterated water molecules HDO and $D_2O$ from the outer layers are not intimately mixed with the bulk $H_2O$ which reduces the efficiency of the H-D exchange and equilibration. Nevertheless, H-D isotope exchange would also affect other species and lead to the reprocessing of the DCN/HCN ratio.

In summary the model by Furuya et al. (2016) predicts $[D_2O/HDO]/[HDO/H_2O]$ ratios that are generally $\gg$1 around solar-type protostars while in stellar nebulae where ice is reprocessed the obtained ratios are $\lesssim$1 (cf. Altwegg et al. 2017a and references therein). The inheritance of two different D/H ratios in singly and doubly deuterated water therefore indicates that at least parts of the ices in comet 67P/C-G originated in the presolar cloud and survived the incorporation into the comet. Furthermore, different D/H ratios in singly versus doubly deuterated water emphasizes that isotopic equilibration in the ices of a comet are limited after its formation. To which extent D-H exchange can be inhibited in the alternative scenario of warmed-up ice agglomeration by Mousis et al. (2018b) remains to be investigated.

As Bockelée-Morvan et al. (2015) pointed out, the $HDO/H_2O$ ratio in comets is typically lower compared to the ISM. Possible explanations discussed are ion-molecule reactions in the outer solar nebula (Aikawa and Herbst 1999; Meier et al. 1998b) or the mixing of D-rich ices from the outer solar system with processed material transported to the formation location of the comet (Mousis 2000) as water vapor readily exchanges D and H atoms with the $H_2$, the main hydrogen reservoir (Geiss and Gloeckler 1998). It is not entirely clear, however, how a D/H ratio different in the two ratios $HDO/H_2O$ and $HDO/D_2O$ can be maintained. Another consideration is thus a consequence of the 2-stage model: HDO and $D_2O$ are associated to the layer of species more volatile than water. If co-desorption together with highly volatile molecules, such as CO, occurs, the D/H ratio exhibits a temperature-dependence with high D/H ratios below the water desorption temperature of $>140$ K (cf. Kouchi and Yamamoto 1995). This is in line with measurements in the ISM where cold gas exhibits much higher D/H ratios compared to warm gas (cf. Persson et al. 2014 and references therein). At the comet, unfortunately, observing such an effect, where outgassing occurs from a most likely thermally altered subsurface layer and locations of different temperatures, would be very difficult.

Also in the refractory material returned by the Stardust mission moderately elevated D/H ratios have been observed (McKeegan et al. 2006). The absence of hydrated minerals, however, led Bockelée-Morvan et al. (2015) to the conclusion that the measured D/H ratios in the Stardust samples are not representative of the D/H ratio in the ices of 81P/Wild 2. The D/H ratio in dust particles from 67P/C-G seems to be about ten times the terrestrial reference (Paquette et al. 2018a). However, this probably refers to the D/H ratio of the organic matter, as no hydrated minerals were identified on 67P/C-G's surface by the VIRTIS instrument (Quirico et al. 2016) and there was no evidence for the presence of hydrated silicates in the dust collected in the coma and analyzed by COSIMA (Bardyn et al. 2017).

#### *4.1.2 Nitrogen Isotopes*

The nitrogen isotope ratio shows remarkable variations across the solar system. Starting from the variation of the $^{14}N/^{15}N$ ratios observed in the interstellar medium to the more or less uniform value in the molecules observed in comets, to the Sun and the planets makes this an important object to study the provenance of the material in the solar system.

Nitrogen isotopes in the three molecules HCN, CN, and $NH_2$, have been measured in numerous comets (Bockelée-Morvan et al. 2008; Manfroid et al. 2009; Rousselot et al. 2014; Shinnaka et al. 2014) including C/1995 O1 (Hale-Bopp), 73P/Schwassmann-Wachmann 3, C/2012 F6 (Lemmon), and C/2012 S1 (ISON) among others. For 67P/C-G the nitrogen isotope ratios are still being analyzed. While nitrogen isotopic ratios in comets range from $^{14}N/^{15}N = 90-220$ (cf. Bockelée-Morvan et al. 2015 and Manfroid et al. 2009) the average ratio derived by Hily-Blant et al. (2017) of $^{14}N/^{15}N = 144 \pm 3$ is consistent through the three cometary molecules listed above.

However, this ratio is distinctly lower (i.e., $^{15}N$ enriched) than the solar wind and protosolar nebula values of $^{14}N/^{15}N = 441 \pm 6$ and $^{14}N/^{15}N = 459 \pm 4$, obtained and derived from the Genesis mission (Marty et al. 2011), which are consistent with the ratios obtained in Jupiter (Fletcher et al. 2014; Fouchet et al. 2000; Owen et al. 2001). Also the terrestrial ratio of $^{14}N/^{15}N = 273 \pm 1$ (Nier 1950) is significantly higher compared to cometary material. The degradation of ammonium salts can release all three species (HCN, CN, $NH_2$) in the coma, complicating the interpretation since isotopic ratios in stored ices and salts may differ.

The refractory samples returned by the Stardust mission to comet 81P/Wild 2 showed a large variation in individual grains, ranging from the protosolar value for a few grains which also showed enrichment in $^{13}C$ compared to terrestrial abundances all the way to a few $^{15}N$ hotspots, i.e. grains with $^{14}N/^{15}N$ ratios similar to cometary volatiles. The bulk composition was clustering somewhere between terrestrial and elevated $^{15}N$, i.e. $^{14}N/^{15}N = 180 - 270$ (McKeegan et al. 2006).

Unlike the volatile species in comets, the nitrogen isotope ratios in the interstellar medium around protostars differ significantly among the objects and the investigated molecules (Füri and Marty 2015). Evidence for multiple reservoirs of nitrogen have been found (Hily-Blant et al. 2017; Rodgers and Charnley 2008). Hily-Blant et al. (2013) presented a chemical network of gas-phase reactions where the two different reservoirs of molecular and atomic nitrogen have different isotope ratios (Furuya et al. 2018). From these two reservoirs the amines and nitriles form in separate pathways which thus leads to the differences of the $^{14}N/^{15}N$ ratios in NH, CN, $HCN^{(+)}$, HNC versus $N_2H^+$, $NH_x{}^{(+)}$ observed in the ISM (Füri and Marty 2015) and references therein), while in comets the $^{14}N/^{15}N$ ratios in HCN, CN, and $NH_2$ are quite uniform. Furuya et al. (2018) thus argue that the differences in N-bearing species between comets and interstellar ices indicates either a primordial variation or ice processing in the solar nebula (Furuya and Aikawa 2014; Lyons et al. 2009). Release from ammonium salts may further complicate this picture.

Two major fractionation mechanisms have been discussed in the literature. Chemical fractionation occurs when the rates for $^{15}N$-$^{14}N$ exchange reactions in nitrogen-bearing species deviate at low temperatures ($\sim$10 K), i.e. when the activation energy barriers quench the $^{14}N$ enrichment (Hily-Blant et al. 2013; Rodgers and Charnley 2008). Self-shielding, on the other hand, affects the dissociation of molecules in dense clouds as the attenuation of the incoming radiation is isotope-selective. Therefore, the $^{14}N/^{15}N$ ratio in molecules such as $N_2$ depends on the location within the cloud (Heays et al. 2014). The importance of selective dissociation of $N_2$ is supported in a recent work by Hily-Blant et al. (2019), who presented evidence for a gradient in the $HCN/HC^{15}N$ ratio as a function of radial distance in the protoplanetary disk of 8 Myr young T Tauri star TW Hya. Around 20 au, a typical region for the formation of comets, the resulting ratio is $HCN/HC^{15}N = 121 \pm 11$, similar to the one observed in the solar system comets, whereas farther out the ratio is close to the local interstellar medium.

### *4.1.3 Carbon and Oxygen Isotopes*

Hässig et al. (2017) presented a collection of measured oxygen and carbon isotopes in volatile species at various comets. Among the comets with known $^{16}O/^{18}O$ ratio in $H_2O$ and $CO_2$ are 1P/Halley (Balsiger et al. 1995; Eberhardt et al. 1995), 153P/Ikeya–Zhang, C/2002 T7 (LINEAR), C/2001 Q4 (NEAT), C/2004 Q2 (Machholz) (Biver et al. 2007; Hutsemékers et al. 2008), C/2012 F6 (Lemmon, Bockelée-Morvan et al. (2012)), and 67P/C-G (Schroeder et al. (2018), revised from Altwegg et al. (2015)). A non-exhaustive list of comets with measured $^{12}C/^{13}C$ ratios in $C_2$, CN, HCN $C_2H_x$, CO, and $CO_2$ contains 1P/Halley (Kleine et al. 1995), C/1995 O1 (Hale-Bopp, Jewitt et al. (1997)), and 67P/C-G (Hässig et al. 2017; Rubin et al. 2017).

Generally, the deviations of $^{16}O/^{18}O$ and $^{12}C/^{13}C$ measured at comets with respect to solar are smaller compared to D/H and $^{14}N/^{15}N$ (cf. Sects. 4.1.1 and 4.1.2), and often the associated accuracies do not assert a clear distinction from the solar reference ratios of $^{12}C/^{13}C = 98 \pm 2$ (Hashizume et al. 2004) and $^{16}O/^{18}O = 530$ (McKeegan et al. 2011). Thus carbon isotopes in most comets were shown to be consistent with the solar reference, with the notable exceptions of $C_2$ in comets West 1976 VI (Lambert and Danks 1983) and $CO_2$ in 67P/C-G (Hässig et al. 2017), and a similar picture holds for the oxygen isotopes. In 67P/C-G, Hässig et al. (2017) and Schroeder et al. (2018) derived a depletion of $^{16}O$ with respect to $^{18}O$ in $CO_2$ and $^{16}O$ with respect to both $^{17}O$ and $^{18}O$ in $H_2O$, respectively. Elevated $^{13}C$ and $^{18}O$ abundances can be the result of isotope-selective photodissociation due to self-shielding as discussed in Sect. 4.1.2. As Bockelée-Morvan et al. (2015) pointed out, this process was studied to understand the oxygen isotope ratios in primitive meteoritic matter (Lee et al. 2008), as the fractionation due to ion-molecule reactions proved to be negligible (Langer et al. 1984). In case of the volatiles, the measured effect in the oxygen isotopes of water is consistent with the predictions for primordial water from self-shielding models (Sakamoto et al. 2007; Yurimoto and Kuramoto 2004). This further strengthens the argument that at least some of the water in comet 67P/C-G is inherited from the presolar cloud.

The oxygen isotopic composition of cometary dust is usually compatible with that observed in carbonaceous chondrites, spanning a range from the solar value to the terrestrial value (e.g. Bockelée-Morvan et al. (2015), Levasseur-Regourd et al. (2018), Paquette et al. (2018b)). The carbon isotopic compositions of cometary dust show only small variations with regard to the terrestrial value (e.g. Levasseur-Regourd et al. (2018)).

### *4.1.4 Noble Gas Isotopes*

Noble gases have been detected in the coma of comet 67P/C-G. The measured isotope ratios of argon (Balsiger et al. 2015) and krypton (Rubin et al. 2018) have been shown to be consistent with solar isotope ratios. The xenon isotopes, on the other hand, show remarkable differences compared to the solar system bulk (Marty et al. 2017). In particular the heavy isotopes, $^{134}Xe$ and $^{136}Xe$, exhibit a strong depletion. As such the Xe-isotopes resemble the previously postulated primordial U-Xe (Pepin 2000), required to explain the xenon isotopic composition in the Earth's atmosphere. A $22 \pm 5\%$ contribution of cometary xenon was derived by Marty et al. (2017). The authors suggested that the xenon in 67P/C-G can be reproduced by a mixture of different nucleosynthetic processes. However, the same approach cannot easily be applied to krypton. Thus, similar to the case of meteoritic Kr and Xe isotopes (Gilmour 2010), Rubin et al. (2018) showed that a mix of two nucleosynthetic end-members, i.e. an exotic s-process krypton to a "normal" component resembling solar composition, can indeed reproduce the observed krypton and xenon isotopic ratios at comet 67P/C-G. However, this also implies an excess of an exotic s-process component at the location of comet formation and thus a non-homogenized protosolar disk.

### *4.1.5 Sulfur Isotopes*

Sulfur is an abundant species found in many comets (Altwegg 1995; Biver et al. 2016; Calmonte et al. 2016; Crovisier et al. 2004; Jewitt et al. 1997). Correspondingly, the isotopes of sulfur have also been observed in the volatiles of several comets, from *in situ* atomic S at comet 1P/Halley (Altwegg 1995) by the Giotto mission and in $H_2S$, CS, and OCS at comet 67P/C-G (Calmonte et al. 2017), as well as by remote sensing observations of $H_2S$ and CS in Hale-Bopp by Jewitt et al. (1997) and Crovisier et al. (2004), respectively, and CS in both comets C/2013 R1 (Lovejoy) and C/2012 F6 (Lemmon) (Biver et al. 2016). Furthermore, sulfur isotopes have also been measured in the refractory material of comet Wild 2 returned by the Stardust mission (Heck et al. 2012) and at 67P/C-G by the COSIMA dust mass spectrometer (Paquette et al. 2017). The reference standard is obtained from Vienna-Canyon

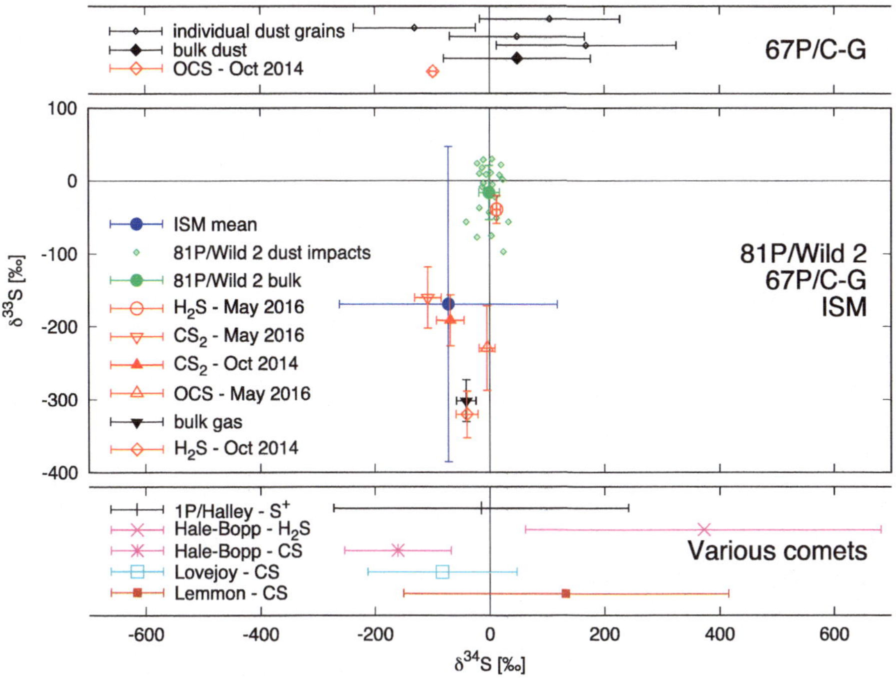


**Fig. 5** Sulfur isotope measurements in comets in $\delta$-notation. Bottom panel: $\delta^{34}S$ isotopic deviation at various comets with respect to V-CDT including atomic S in 1P/Halley: Altwegg (1995), $H_2S$ and CS in Hale-Bopp: Jewitt et al. (1997) and Crovisier et al. (2004), and CS in both Lovejoy & Lemmon: Biver et al. (2016). Middle panel: $\delta^{33}S$ and $\delta^{34}S$ isotopic deviations measured in $H_2S$, $CS_2$, and OCS during two different periods at 67P together with the computed bulk value (Calmonte et al. 2017) and bulk sulfur isotope measurements together with individual impact craters in the returned Stardust sample from comet 81P/Wild 2 (Heck et al. 2012) in comparison to the mean ISM value (Chin et al. 1996). Top panel: $\delta^{34}S$ in OCS (Calmonte et al. 2017) amended by the 4 measured dust grains and corresponding bulk value from the Rosetta/COSIMA instrument at 67P/C-G (Paquette et al. 2017)

Diablo Troilite (V-CDT) and amounts to $^{32}S/^{33}S = 126.948$ and $^{32}S/^{34}S = 22.6436$ (Ding et al. 2001). The relative sulfur abundances have never been measured in the Sun or the solar wind and thus the reference is based on meteoritic material.

Figure 5 shows the collection of cometary sulfur isotope measurements obtained thus far in $\delta$-notation (per mil deviation from V-CDT) and compared to the mean ISM value (Chin et al. 1996). The measurements in the volatiles of 67P/C-G show a depletion predominantly in $^{33}S$. Calmonte et al. (2017) showed that neither mass-dependent nor mass-independent fractionation from photodissociation is responsible for the depletion in $^{33}S$ and $^{34}S$. The authors also argued that the differences in the volatiles measured originate from different chemical pathways forming these molecules (cf. Sect. 4.2.2). In comparison, the depletion in $\delta^{34}S$ is within the 1-$\sigma$ error bars of the average ratio measured in the dust of 67P/C-G (Paquette et al. 2017).

Of the 24 analyzed impact craters in the dust impact residues on the aluminum foil returned by NASA's Stardust mission, Heck et al. (2012) reported only one crater to differ in both $\delta^{33}S$ and $\delta^{34}S$ by more than 2-$\sigma$ from the V-CDT value. Looking at individual ratios there are 5 impact craters depleted by $\geq$2-$\sigma$ in $\delta^{33}S$ compared to one depleted in $\delta^{34}S$. Generally, the scatter in $\delta^{33}S$ is larger compared to $\delta^{34}S$ and cannot be explained by the higher precision in $\delta^{34}S$ measurement alone. Nevertheless, the deviations from V-CDT in the 81P/Wild 2 data are small compared to the observations in 67P/C-G, pointing to either different processes and/or efficiencies in the isotope fractionation between volatiles and refractories. Another explanation is that the two comets formed at distinct locations in the early solar system and as a consequence from different source materials and at different temperatures (cf. also Sect. 4.1.1 on the D/H variation within the individual dynamical families of comets). In the case of 81P/Wild 2 Heck et al. (2012) concluded that most or all of the S-rich material formed in the solar system (see also Bullock et al. 2010; Tachibana and Huss 2005). Sulfur isotope ratios in 67P/C-G, on the other hand, are consistent with observations in the ISM (Chin et al. 1996) although the associated error bars are large. The degree of fractionation in 67P/C-G resembles presolar SiC grains from type II supernovae. Based on the Rauscher et al. (2002) 15 $M_{\odot}$ supernova model, Hoppe et al. (2018) showed that a late supernova contribution could indeed reproduce the observed S-isotope anomaly. This again would point to a non-homogenized distribution of the material. Thus, in the early solar system the situation is complicated as the sulfur incorporated in 67P/C-G appears to originate from various sources.

### *4.1.6 Silicon Isotopes in Refractories and Their Implications for Origins*

The ROSINA measurements showed a depletion in the heavy isotopes $^{29}Si$ and $^{30}Si$ with respect to the major isotope $^{28}Si$ (Rubin et al. 2017). The origin of the atomic silicon was most likely sputtering by solar wind protons, which early in the mission still reached the silicate-rich surface of the nucleus (Bardyn et al. 2017; Wurz et al. 2015). ROSINA detected no silicon bound in molecules. The reported deviations at 67P/C-G from solar (McSween and Huss 2010) of 92.230% for $^{28}Si$, 4.683% for $^{29}Si$, and 3.087% for $^{30}Si$ were $\delta^{29}Si = (-145 \pm 98)‰$ and $\delta^{30}Si = (-214 \pm 115)‰$ and contain 1-$\sigma$ errors (Rubin et al. 2017). Therefore, solar abundances could not be excluded (deviations between 1-$\sigma$ and 2-$\sigma$), however, several possible contributors to the depletion were put forward, including instrumental effects and fractionation in the sputtering process. On the other hand, together with the findings of other isotopes, a non-homogenized distribution of the material in the early solar system has to be considered too.

## 4.2 Molecular and Elemental Ratios in Refractories and Their Implications for Origins

Bockelée-Morvan et al. (2000) compared relative abundances of volatiles at comet Hale-Bopp with those derived in interstellar ices, hot molecular cores, and bipolar outflows around protostars. From the resemblance of the volatiles' abundance in these objects they concluded that similar chemical processes must be at work forming these compounds under comparable conditions, in particular grain-surface chemistry. In the vicinity of low-mass to massive protostars, i.e. from so-called hot corinos to hot cores, volatiles contained in these ice grain mantles evaporate and are then observed through their rotational lines using radio telescopes such as ALMA (Herbst and van Dishoeck 2009).

In the coma of 67P/C-G a multitude of molecules has been found by ROSINA (Le Roy et al. 2015) and in the near-surface environment of the comet by the lander mass spectrometers Ptolemy and COSAC (Goesmann et al. 2015; Wright et al. 2015). The combined analysis shows an inventory of volatiles, semi-volatiles, and refractories rich in organics (Altwegg et al. 2017b), in line with the findings in the cometary dust by the COSIMA mass spectrometer (Bardyn et al. 2017; Fray et al. 2016). Other key observations suggested the

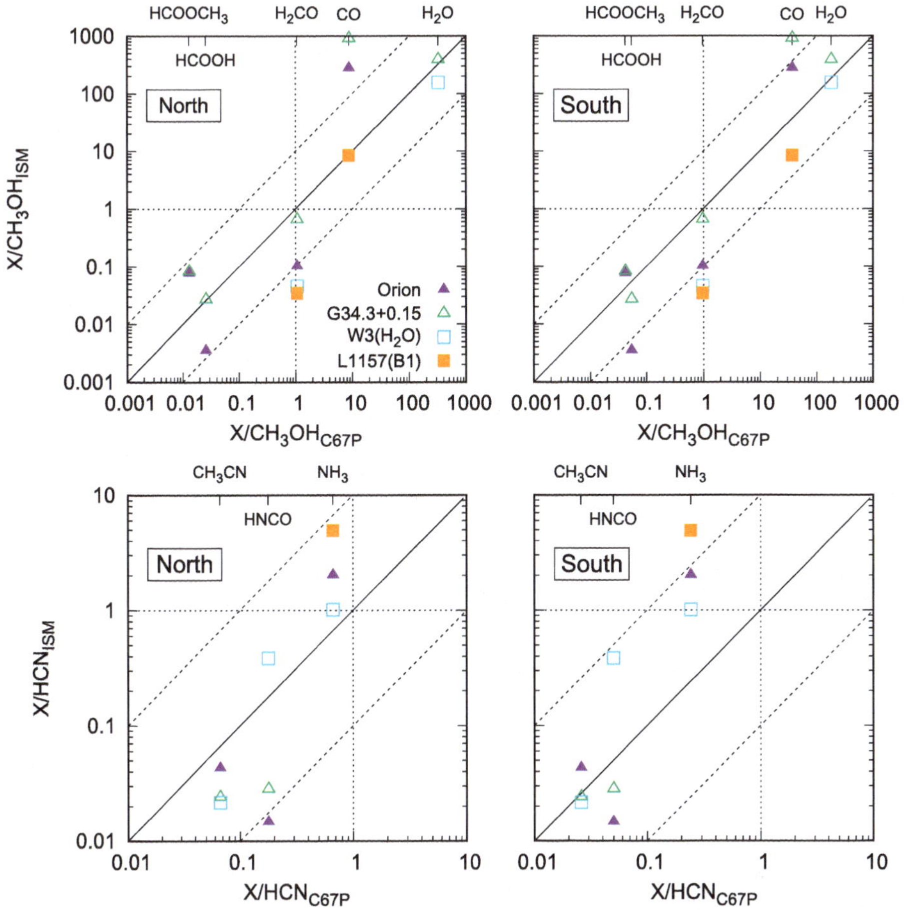


**Fig. 6** Relative abundances of oxygenated compounds with respect to $CH_3OH$ (top row) and N-bearing molecules with respect to HCN (bottom row) measured at comet 67P/C-G (x-axes, chemical formula on top) above the northern (left column) and southern (right column) hemispheres compared to relative abundances in the ISM (y-axes) after Bockelée-Morvan et al. (2000) and references therein (data from 67P/C-G: Le Roy et al. (2015), cf. Table 1). The solid black line denotes equal relative abundances while the two dashed lines represent deviations by a factor 10. Abundances equal to $CH_3OH$ and HCN are denoted by the vertical and horizontal dotted lines

presence of ammonium salts in 67P/C-G (Poch et al. 2020; Quirico et al. 2016) and other comets (Mumma et al. 2017, 2018, 2019). Indeed, the mass spectra obtained by ROSINA DFMS (Altwegg et al. 2020; Hänni et al. 2019) showed all possible sublimation products of five different ammonium salts. Significant amounts of nitrogen may be bound in ammonium salts and hence hidden from observations due to high sublimation temperatures, which lead to the well-known deficiency of nitrogen in comets (Geiss 1987).

Figure 6 shows a comparison of the relative abundances of oxygenated molecules (top row, normalized to methanol) and nitrogen-bearing molecules (bottom row, normalized to hydrogen cyanide) measured at comet 67P/C-G (x-axis) from Le Roy et al. (2015) and listed in Table 1 versus multiple objects in the ISM from Bockelée-Morvan et al. (2000) (y-axis). The measurements at 67P/C-G were performed above both the northern summer (left) and the southern winter hemispheres (right) at 3.1 au during the inbound part of the comet's orbit, which led to different relative abundances in the coma of the comet. On the other hand, the derived ISM ice composition varies among the different objects, possibly affected by different temperatures. However, close trends in the relative abundances can be observed: Schuhmann et al. (2019) showed that relative abundances of hydrocarbons in 67P/C-G are similar to those derived from a model of gas-grain chemistry in a dark interstellar cloud (Hasegawa et al. 1992), including abundant amounts of unsaturated hydrocarbons (Altwegg et al. 2017b). A detailed comparison of comet 67P/C-G with low-mass protostar IRAS 16293-2422 B was very recently performed by Drozdovskaya et al. (2019) and revealed a close match between the two objects for a suite of CHO-bearing species. Correlations were also obtained for N- and S-bearing molecules, albeit with larger scatter. Similar conclusions were obtained for other comets (Mumma and Charnley 2011), which supports the suggestion that at least parts of the ices originated in the interstellar medium and that chemical alteration during and after incorporation into the comet was inefficient.

### 4.2.1 The $O_2$ Story

One of the most surprising findings of the Rosetta mission was the detection of copious amounts of molecular oxygen, $O_2$, by ROSINA (Bieler et al. 2015a). The measurements showed a strong correlation with water despite $O_2$'s high reactivity and volatility compared to $H_2O$. Other species of similar volatility, such as $N_2$ and Ar, were found in much smaller amounts (cf. Sect. 3.3). Furthermore, the low abundances (or absence, cf. Taquet et al. (2018)) of $O_2$ in star forming regions, e.g. in Orion (Goldsmith et al. 2011) and in the dense core of $\rho$ Ophiuchus A (Larsson et al. 2007; Liseau et al. 2012) are in contrast to the $O_2/H_2O = 3.80 \pm 0.85\%$ measured at the comet. In the following, the presence of molecular oxygen in the coma of 67P/C-G has also been deduced from the OI 1356 Å/1304 Å ratio measured by the Rosetta Alice imaging spectrograph (Keeney et al. 2017). It was furthermore shown that a similar amount of $O_2$ (Rubin et al. 2015b) is consistent with the measurements of the Neutral Mass Spectrometer (NMS; Krankowsky et al. (1986)) on board of the Giotto mission to comet 1P/Halley (Reinhard 1986). Even though NMS could not resolve $O_2$ from species with close mass, i.e. methanol and atomic sulfur, the contribution of these species could be accounted for using an ion-neutral chemical network.

Several mechanisms for the formation of $O_2$ have been discussed in the literature. Bieler et al. (2015a) proposed two scenarios favoring a primordial origin of the $O_2$. The first is the formation of $O_2$ and by-products through radiolysis or photolysis of water ice, similar to the Galilean moons Europa, Callisto, and Ganymede (e.g. Johnson et al. (2004) and Teolis et al. (2006)) where energetic ions from Jupiter's magnetosphere irradiate the icy surfaces. However, typical radiolysis products $H_2O_2$ and $HO_2$ in 67P/C-G were only found in low relative abundances comparable to $\rho$ Ophiuchus A (Bergman et al. 2011; Parise et al. 2012) or were even absent in the case of $O_3$. The second possibility is the chemical formation of $O_2$ and subsequent trapping in water ice during a phase of rapid cooling. Possible formation scenarios have since been refined and shall be discussed below. Furthermore, alternative formation scenarios have been investigated in the literature, namely, the dismutation of $H_2O_2$ during the desorption of water ice from the nucleus (Dulieu et al. 2017) and Eley-Rideal reactions of energetic water-group ions and neutrals (Yao and Giapis 2017). In both cases $O_2$ is formed *in situ* and is not of primordial origin. The former scenario requires a high dismutation efficiency and the peroxide in the comet's nucleus to be primordial. The latter scenario, as pointed out by Heritier et al. (2018), can only account for up to $10^{-4}$ of the observed $O_2$ flux and furthermore the correlation of $O_2$ with $H_2O$ is lost. For a more detailed discussion of *in situ* processes we refer to the review by Luspay-Kuti et al. (2018). Given that $O_2$ has been detected unequivocally in 67P/C-G and tentatively in 1P/Halley, the question remains whether $O_2$ is a common constituent in comets and how it formed.

Mousis et al. (2016b) showed that large amounts of $O_2$ in the 1 – 10% range with respect to water can be formed through radiolysis of amorphous ice grains by galactic cosmic rays in low density environments such as the presolar cloud (cf. Zheng et al. 2006 and references therein). The authors also investigated the possibility of evaporation and re-trapping of volatile $O_2$ in crystalline ices and clathrates upon entering the PSN. A late formation of $O_2$ through irradiation in the midplane of the disk is difficult due to self-shielding of the disk material. Mousis et al. (2018a) therefore studied the impact of vertical transport of material from the midplane to the upper layers of the disk where the irradiation occurs. They conclude that even under favorable conditions the amount of $O_2$ produced is some two orders of magnitude too low with respect to the abundances measured at 67P/C-G.

Another possibility, the irradiation from the decay of endogenic short- and long-lived radionuclides located inside the refractories in the nucleus, has been studied by Bouquet et al. (2018) based on the model by Mousis et al. (2017a) and a dust to ice ratio inside the nucleus of 4 (Rotundi et al. 2015). While sizeable amounts of $O_2$ of up to 1% could be produced, other radiolysis products such as $H_2O_2$, are formed in amounts that are two orders of magnitude above the observed abundances.

Bieler et al. (2015a) also investigated the production of $O_2$ after the formation of the nucleus. The penetration depth of cosmic rays forming $O_2$ is on the order of a few meters which allows alteration of the uppermost layers of the comet. However, upon entry in the inner solar system after a close encounter with Jupiter in 1959 (Maquet 2015), the erosion of the comet's surface is estimated to be several meters per 6.5-year orbit (Keller et al. 2015). The altered layer would have long since gone. On the other hand, if $O_2$ is produced *in situ* through irradiation by solar wind, the $O_2/H_2O$ ratio must exhibit variations associated to the plasma interaction of the comet with the solar wind. Bieler et al. (2015a), however, reported a stable $O_2/H_2O$ ratio. As a result, a formation of $O_2$ in the low density environment of the presolar cloud remains their most plausible scenario (Bieler et al. 2015a; Mousis et al. 2016b).

Taquet et al. (2016) used a model of the chemical evolution around an evolving star coupled to an astrochemical model for the gas phase and depth-dependent ice-grain chemistry (cf. Furuya et al. 2015). The considered $O_2$ formation contains two main processes: first, neutral-neutral chemistry in the gas phase starting from abundant atomic O and molecular OH; and second, association reactions by atomic O recombination on or within the mantles of icy grains. Their model suggested that the formation of $O_2$ through radiolysis and photolysis is inefficient as $O_2$ is also converted to $H_2O$ through the recombination with H atoms.

Efficient formation of $O_2$ ($\geq$4% with respect to water ice) was achieved in high density regions with a lower gas-phase H/O ratios to decrease hydrogenation reactions (Taquet et al. 2017). Also, the measured amounts of $O_2$ in the comet favor intermediate temperatures of the interstellar ices of $\sim$20 K. At such temperatures H atoms are efficiently sublimated and the increased mobility of O atoms on the grain surface enhances O association reactions. Furthermore, an increased cosmic ray ionization rate removes H from the solid phase.

In their simulations $N_2$ and CO are formed later, on the outer layers of the icy grains. Thus the better correlation of $O_2/H_2O$ (Bieler et al. 2015a) as opposed to $N_2/H_2O$ (Rubin et al. 2018) and $CO/H_2O$ (Hässig et al. 2015) is a natural outcome of the model. On the other hand, the model predicted an $N_2/CO$ ratio of 50% in the ice, which is in contrast to the measurements at 67P/C-G (Rubin et al. 2015a). However, these outer layers of highly volatile species can be subject to evaporation upon heating during the transport from the dark cloud towards the protostellar disk. Subsequent recondensation is possible and the $N_2$/CO ratio depends again on ice phase and temperature as discussed in Sect. 3.3.

Taquet et al. (2016) also investigated the production of $O_2$ at the increased temperatures and UV fluxes expected during the protostellar collapse and protoplanetary disk formation. They find that $O_2$ is formed on the per cent level with respect to water only in the upper layers of the disk. However, there the $O_2$ seems rather associated to $CO_2$ than $H_2O$. Eistrup and Walsh (2018), on the other hand, found a sweet spot for the formation of $O_2$ via ice grain chemistry in the presolar nebula from 120 – 150 au. Nevertheless, given that these results depend strongly on the chemical parameters of the involved interactions, the authors still favor a primordial origin of $O_2$.

The third scenario investigated by Taquet et al. (2016) is the formation of $O_2$ through luminosity outbursts of the low-mass host star, which could trigger warm gas-phase formation following the sublimation of water ice at temperatures above $\sim$100 K. However, these outbursts are short and hence the amount of $O_2$ formed with respect to $H_2O$ remains $<0.1\%$. The authors conclude that $O_2$ must have a prestellar or molecular cloud origin.

In laboratory experiments, $O_2$ in water ice (Ioppolo et al. 2008; Miyauchi et al. 2008) is efficiently hydrogenated at low temperatures to form $H_2O$ and $H_2O_2$. This is in conflict with the measured amounts of $O_2$ and the low $H_2O_2/O_2$ (as well as $HO_2/O_2$) ratio of $10^{-3}$ in 67P/C-G from Bieler et al. (2015a). Mousis et al. (2016b), on the other hand, showed that $O_2$ can be efficiently incorporated and stabilized into the structure of crystalline ices.

Also, Laufer et al. (2018) showed that desorption fluxes of $O_2$ from amorphous water ice are much higher compared to other highly volatile species such as $N_2$ and Ar, even if the initial relative abundances in the gas phase before trapping were comparable. This could in part explain the difference in the abundance of these volatiles. Furthermore, their results also indicated that only 10 – 20% of the $O_2$ is released in the kinetic temperature range of 65 – 120 K derived for Orion (Goldsmith et al. 2011). At comet 67P/C-G the $O_2$ is correlated and hence co-desorbing with water (Bieler et al. 2015a) at a higher temperature compared to Orion. This further complicates matters and more work on the stability of $O_2$ in temperature ranges representative for the comet are required to obtain solid constraints for this problem.

### *4.2.2 Sulfur Chemistry*

Sulfur in dense cloud cores has been found to be depleted by roughly two orders of magnitude compared to the diffuse interstellar medium (Ruffle et al. 1999). Such a depletion is unique to sulfur (Drozdovskaya et al. 2018) and the missing sulfur has not been conclusively identified (e.g. Bilalbegović and Baranović 2014). Measurements at comet 67P/C-G revealed a wealth of sulfur-bearing volatiles (Calmonte et al. 2016). Sulfur has also been detected in the refractory phase (Paquette et al. 2017). Thus, the link from the diffuse interstellar medium, through the dense clouds, all the way to comets contains crucial links to the formation and processing of the material through the different evolutionary stages.

Formation of S-bearing volatiles have been attributed to ice grain chemistry: for instance, Ruffle et al. (1999) explained the formation of $H_2S$ by the adsorption of positively charged $S^+$ and $HS^+$ ions on negatively charged ice grains and subsequent hydrogen attachment. A suite of laboratory experiments have shown that $H_2S$ can then be transferred into other S-bearing molecules such as OCS and $CS_2$ in the presence of C-bearing species such as CO in the ice (see Calmonte et al. (2016) and references therein). In addition, the presence of other organosulfurs point to the importance of ice/dust-grain chemistry. Ice grain chemistry and radiolysis of $H_2S$ (Mousis et al. 2017b) and other sulfur-bearing species can form S-polymers up to the stable ring molecule $S_8$, out of which the possible fragments $S_2$, $S_3$, and

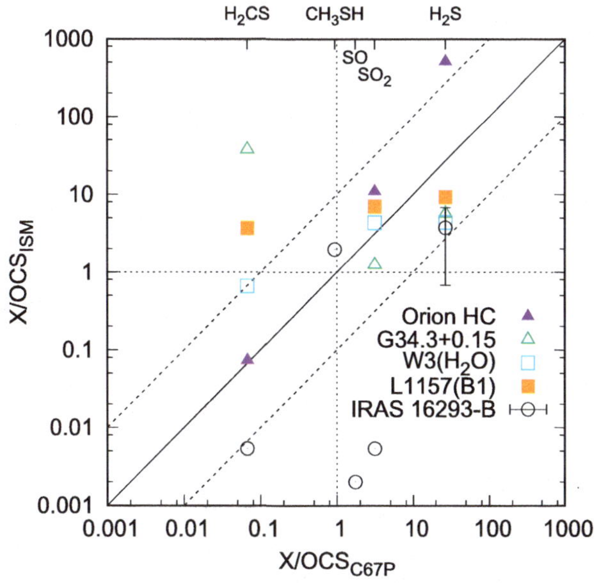


**Fig. 7** Similar to Fig. 6. Relative abundances of sulfur-bearing compounds normalized to OCS measured at comet 67P/C-G (x-axis) from the bulk abundances in Table 2 from Calmonte et al. (2016) compared to relative abundances in the ISM (y-axis) from Drozdovskaya et al. (2018) and Bockelée-Morvan et al. (2000) with references therein

$S_4$ have been identified at 67P/C-G. For $S_2$ two distinct populations have been suggested: first, $S_2$ in volatile form and second, $S_2$ as a dissociation product from larger S-bearing molecules up to solid $S_8$ associated to hot dust grains in the coma near-perihelion.

This S-residue formed by radiolysis and photolysis of interstellar ices is difficult to observe with radio astronomy. Its formation has been hypothesized to be the main cause for the observed depletion in dense cores, even though the model by Woods et al. (2015) suggested a contribution of up to 6% of the missing sulfur only. Nevertheless, the sulfur molecules $S_2$, $S_3$, and $S_4$ detected in the coma of comet 67P/C-G would support this theory. However, absolute abundances of these molecules in the refractory part could not be derived due to the strong temperature dependence of the outgassing. Furthermore, the mass range of the ROSINA DFMS sensor did not include the parent $S_8$.

Bulk abundances of the volatile S-bearing species, on the other hand, have been obtained (Calmonte et al. 2016). Figure 7 shows relative abundances of S-bearing volatiles in the comet versus various protostars (see Bockelée-Morvan et al. (2000) and Drozdovskaya et al. (2018)). Again, the spread among protostars can be considerable but with a few exceptions the trends seem similar to the observed cometary abundances. Furthermore, the recent study by Drozdovskaya et al. (2019) shows a good correlation of the sulfur-bearing species in comet 67P/C-G with low mass protostar IRAS 16293-2422B.

The presence of $S_2$ in the ices of the comet led Calmonte et al. (2016) to the conclusion that the ices of the comet predate the formation of the solar system. The reason for this is the very short lifetime of the sulfur dimer in the vapor phase when exposed to UV radiation. Mousis et al. (2017b) then also discussed a post solar system formation scenario where $S_2$ forms from $H_2S$ in ice grains that condensed in the protosolar nebula as long as they are sufficiently irradiated in the upper layers of the disk. Independent of the pre- or protosolar nebula origin, the radiolysis formation model predicts correlated outgassing of $S_2$ and $H_2S$. Furthermore, the radiation also forms voids in amorphous $H_2O$ ice in which $S_2$ can accumulate and remain stable even after crystallization. This suggests a correlation also between $S_2$ and $H_2O$ (Mousis et al. 2018b), while at 67P/C-G the correlation among these species was rather poor (Calmonte et al. 2016).

## 5 Post-Formation Evolution of 67P and Other Comets

A'Hearn et al. (2012) investigated the relative abundances of the major volatiles $CO/CO_2/H_2O$ in a multitude of comets. Any differences among the dynamical groups of comets, i.e. among Kuiper belt and Oort cloud comets, might link them to not only their formation location (Mumma and Charnley 2011) but also to their post-formation evolution. Several post-formation scenarios are discussed in the following.

### 5.1 Loss of Highly Volatile Molecules in 67P Through Evolutionary Processes?

The noble gases argon, krypton, and xenon have been identified in the coma of comet 67P/C-G (Balsiger et al. 2015; Marty et al. 2017; Rubin et al. 2018). 67P/C-G proved to be enriched in xenon over krypton and krypton over argon in comparison to the early Sun (Lodders et al. 2009), i.e. $^{84}Kr/^{36}Ar_{67P/C\text{-}G\text{ versus Sun}} = 144 \pm 58$ and $^{132}Xe/^{36}Ar_{67P/C\text{-}G\text{ versus Sun}} = 683 \pm 279$ and solar wind fluxes from the Genesis mission (Meshik et al. 2014), i.e. $^{84}Kr/^{36}Ar_{67P/C\text{-}G\text{ versus SW}} = 139 \pm 32$ and $^{132}Xe/^{36}Ar_{67P/C\text{-}G\text{ versus SW}} = 291 \pm 72$. For the noble gas neon only an upper limit could be derived. When comparing expected ratios of noble gases trapped in amorphous water ice (Bar-Nun et al. 1988; Dauphas 2003) it is not possible to match both ratios at the same time, $^{132}Xe/^{36}Ar_{67P/C\text{-}G}$ and $^{84}Kr/^{36}Ar_{67P/C\text{-}G}$. Similar issues, as discussed in Sect. 3.3, are raised for other types of ices including different crystalline ices formed in the protosolar nebula.

So, while comets belong to the most pristine objects in the solar system (Geiss 1987), post-formation alteration of the budget of highly volatile species cannot be excluded.

#### *5.1.1 The Centaur Stage*

One possibility is evolutionary processing occurring after the comet's formation, possibly through the preferential loss of highly volatile species during the comet's journey from the Scattered disk, through several Myr as a Centaur (Guilbert-Lepoutre et al. 2015), to the inner solar system (Maquet 2015). This is based on the idea that a dynamical cascade between Kuiper belt objects, Centaurs, and JFCs exists (Levison and Duncan 1997; Tiscareno and Malhotra 2003) which is different from the dynamical history of Oort cloud comets (Weissman et al. 2020).

According to Guilbert-Lepoutre et al. (2016), a comet such as 67P/C-G has its surface thermally altered down to a few hundred meters, where the temperature reaches up to 80 K. Depending on the orientation of the rotation axis even a transition of amorphous to crystalline ice may occur. In comparison, the erosion of 67P/C-G's surface on its current orbit is estimated to be a few meters per revolution (Keller et al. 2015). The comet reached the inner solar system after a close encounter with Jupiter in 1959 (Maquet 2015). If indeed the comet has not been that close to the Sun before 1959, the number of perihelion passages in the inner planetary system, which occur every $\sim$6.5 years, is limited and hence today's outgassing might still occur from a layer thermally altered during the Centaur stage. In comparison to Oort cloud comets, Schuhmann et al. (2019) find rather low abundances of the two simplest alkanes ($CH_4$ and $C_2H_6$) relative to water in 67P/C-G, and point out that other highly volatile species are also low including $N_2$ and CO. Furthermore, as shown in Fig. 8, the ratio of $CH_4$ to the lesser volatile $C_2H_6$ varies greatly among comets but is also low in 67P/C-G.

Nevertheless, more work is required in this respect. Nucleus models by Prialnik (1992) predict lower penetration depths of the heat wave of a few tens of meters due to the low

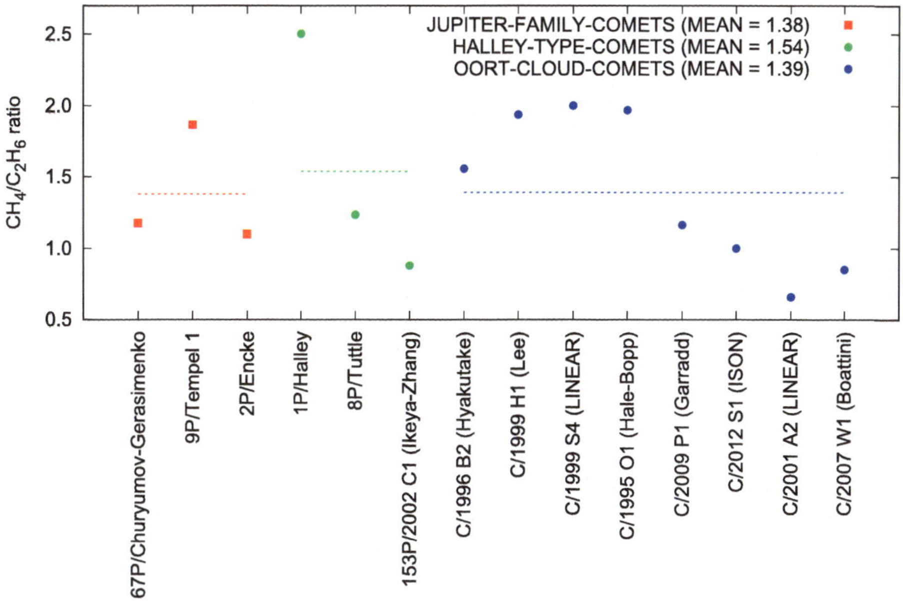


**Fig. 8** $CH_4/C_2H_6$ ratio in several comets including 67P/C-G. Strong variations within each family (JFCs, Halley-type comets (HTCs), and Oort cloud comets) are observed. References and heliocentric distance of the observations: 67P/C-G (1.5 au): Schuhmann et al. (2019), Tempel 1 (1.5 au), Encke (1.2 au), Lee (1.1 au), Ikeya-Zhang (0.8 au), and Hale-Bopp (0.9 au): Dello Russo et al. (2016) and references therein, S4 (LINEAR) (0.8 au), ISON (1.1 au), A2 (LINEAR) (1.2 au), and Boattini (0.9 au): Lippi et al. (2020), Halley (0.9 au): Eberhardt (1999); Garradd (1.6 au): Paganini et al. (2012), and Tuttle (1.2 au): Böhnhardt et al. (2008)

thermal conductivity of cometary material and hence the presence of amorphous ices within tens of meters from the surface. On top of that there is only limited information on the temperature-dependent mobility of highly volatile species (Lauck et al. 2015) in a highly porous icy and refractory material relevant to comet 67P/C-G.

### *5.1.2 Thermal Evolution Through Multiple Perihelion Passages*

Evolutionary processes and compositional differences among the different cometary families have been studied before, including the thermal evolution of the interior of comets and associated outgassing (e.g. Keller et al. (2015)). The numerical model by Prialnik et al. (2004) predicts the build-up of a dust layer affecting the outgassing behavior of the comet. Furthermore, also the volatiles recede to different depths based on their volatility (De Sanctis et al. 2006; Gortsas et al. 2011; Marboeuf and Schmitt 2014). However, Fulle et al. (2016) pointed out that numerical models exhibit problems reproducing observed outgassing profiles as a function of heliocentric distance.

A'Hearn et al. (2012) investigated the abundances of the volatile species $H_2O$, $CO_2$, and CO in comets. The results showed a wide range of abundance ratios in both JFCs and LPCs/HTCs. The authors concluded that this is unlikely the result of evolution through successive perihelion passages. Also, trends within comet families were not entirely defined, but it is clear that the volatile abundances have largely overlapping ranges in all the dynamical families. This suggests a largely overlapping region of origin, similar to the conclusion from the D/H measurements in cometary water (Sect. 4.1.1). With a two-order-of-magnitude spread in $CO/H_2O$, there is some indication that LPCs show higher values of $CO/H_2O$ and that JFCs exhibit somewhat lower values. It cannot therefore be concluded whether these compositional differences are the result of a different location of formation and/or subsequent thermal evolution, including as a Centaur in the case of JFCs (cf. Sect. 5.1.1).

### *5.1.3 Collisional Heating of Cometesimals*

Several of the comets visited to date show distinct bi-lobate shapes (Sunshine et al. 2016). These objects and their precursors can form through hierarchical accretion or collisionally induced disintegration of larger parent bodies. Jutzi and Asphaug (2015) investigated comets like 67P/C-G forming in sub-catastrophic collisional mergers leading to bi-lobate shapes and also the layering that has been observed in 67P/C-G (Massironi et al. 2015). There is still a debate as to how primordial these objects are (Davidsson et al. 2016; Fulle et al. 2016; Weissman 1986). Jutzi and Asphaug (2015) showed that the low density and high porosity can be preserved in such collisions, however, they point out that Kuiper belt objects larger than 5 km undergo multiple shape changing (Jutzi et al. 2017) and also catastrophic disruptions (Jutzi and Benz 2017) throughout the lifetime of the solar system. While the details of the origin of 67P/C-G and other comets are discussed in the accompanying paper by Weissman et al. (2020) we shall here focus on the presence of highly volatile species in these objects. This includes the gases of $CH_4$, CO, and others found in the comae of many comets (Bockelée-Morvan et al. 2004; Mumma and Charnley 2011). Hence, Schwartz et al. (2018) investigated the amount of impact heating caused by catastrophic collisions. Their results suggest that the reaccumulated comet is only marginally heated, i.e. $<1$ K, while material heated-up by several tens of K in sizeable amounts, i.e. up to 10%, can only be found in the ejecta. As a consequence, highly volatile species can be retained in cometary ices, even after a violent encounter. Hence the presence of highly volatile molecules and 67P/C-G's low density (Pätzold et al. 2016) alone is not proof for the nucleus itself to be primordial. Nevertheless, Jutzi et al. (2017) predict that a comet like 67P/C-G undergoes many shape changing collisions, critically depending on the differential size distribution of the objects in the disk. Davidsson et al. (2016) prefer a shallow size index, which reduces the number of collisional interactions of cometesimals and hence increases the survival rate in their primordial form. Jutzi and Benz (2017), on the other hand, prefer a steeper size distribution and the final object is hence the result of many collisions or could even form from the ejected material itself.

### *5.1.4 Impact of Radiogenic Heating on Comets*

Mousis et al. (2017a) investigated the consequences of the radioactive decay of $^{26}Al$ and $^{60}Fe$ on the budget of highly volatile molecules such as $N_2$ and CO in the comet. A thermal nucleus model has been employed to simulate heat transfer, phase transition and sublimation of ices, and the diffusion of volatiles through the porous material of the comet. The results depend on several critical parameters including the initial size of the object, the dust to ice ratio in the nucleus, and the initial abundance of radioactive isotopes, in particular the controversial amounts of $^{26}Al$ as Prialnik and Podolak (1995) pointed out. Nevertheless, their results suggested that either the formation of comet 67P/C-G was delayed by several million years with respect to the formation of Calcium-Aluminum-rich Inclusions in the PSN or the accretion time of the nucleus was on the order of a few million years to prevent formation of large devolatilized volumes inside the nucleus. Prialnik and Podolak (1995), on the other hand, reported that objects smaller than 20 km are mostly unaltered in the presence of the longer lived $^{40}K$, $^{232}Th$, $^{235}U$, and $^{238}U$ radioisotopes as long as the initial amounts of $^{26}Al$ are negligible. The addition of $^{26}Al$ reduces the critical radius of an object for crystallization to occur and also the time of its onset after formation is shorter. During crystallization highly volatile molecules, such as CO, are released before condensing again in the colder regions farther from the center of the comet. Heat advection by the flow of volatile species may even surpass the heat conduction of the solid phase in such a case. It is thus clear that an increase in the temperature resulting from any kind of radiogenic heating critically affects the composition of a comet.

## 6 Summary and Conclusions

The origin of the material in comets has been a main focus of research for a long time. The recent Rosetta mission provided numerous new insights to test contemporary theories and to verify our current understanding of the major processes that are involved.

- After the first measurement of the $HDO/H_2O$ ratio in a JFC (i.e. Hartley 2) it was hypothesized that there is an intrinsic difference between Oort cloud and Kuiper belt/Scattered disk objects (Hartogh et al. 2011). This rekindled the discussion on the Earth's water being of cometary origin. However, recent measurements of elevated D/H ratios in the water of additional JFCs, including 67P/C-G, countered this theory (Altwegg et al. 2015; Biver et al. 2016; Paganini et al. 2017). Thus, it seems unlikely that comets from the Kuiper belt and beyond were the major source for the water on Earth. Instead, the results suggest that the D/H ratio reflects the formation location of comets rather than where they are found today: comets formed over a wide range of heliocentric distances and were relocated afterwards. Both lobes of 67P/C-G share the same D/H ratio within error bars. This hints at a common location of formation before merging (Schroeder et al. 2019).
- Whether the ice itself is inherited from the ISM is more difficult to assess from the $HDO/H_2O$ ratio alone. Comparing the D/H ratio from both $HDO/H_2O$ and $D_2O/HDO$ seems to be the better measure (Furuya et al. 2016). The very different ratios in 67P/C-G indicate that at least parts of the ice must be inherited from the cold temperature chemistry occurring in the presolar cloud (Altwegg et al. 2017a) and that subsequent isotope-exchange reactions were not efficient.
- Apart from D/H, other isotopes in the volatile material of many comets show marked deviations from solar system bulk ratios. In particular the nitrogen isotopes are not yet understood. The three molecules observed thus far, CN, HCN, $NH_2$, show a consistent enrichment of $^{15}N$ in all comets for which we have measurements (Hily-Blant et al. 2017) whereas much larger variations are found in the ISM (Füri and Marty 2015 and references therein). Furthermore the isotopes of the volatile sulfur (Calmonte et al. 2017), oxygen (Hässig et al. 2017; Schroeder et al. 2018), xenon (Marty et al. 2017), and refractory silicon (Rubin et al. 2017) show or at least hint at deviations from bulk solar system abundances. Other volatiles, e.g. argon and krypton (Balsiger et al. 2015; Rubin et al. 2018) and refractory sulfur (Paquette et al. 2017) are within the error bars of solar system bulk abundances. The refractories in other comets, including the returned samples from comet Wild 2, contain thermally processed material from the inner solar system with only small amounts of presolar material. Thus, radial mixing occurs, but the protoplanetary disk remains only partially homogenized. Therefore, the isotopic and possibly also elemental and molecular abundances in at least the icy phase at the location of the comet's formation differed from bulk solar system.

- Comets may contain significant amounts of ammonium salts (Mumma et al. 2019) that are likely to account for the to-date observed deficiency of nitrogen in comets (Altwegg et al. 2020; Geiss 1987; Poch et al. 2020; Quirico et al. 2016).
- The relative abundances of the volatiles in comets exhibit similarities to the ISM (Bockelée-Morvan et al. 2000) and comet 67P/C-G is no exception (Drozdovskaya et al. 2019). Also the large amounts of unsaturated hydrocarbons and other organics (Altwegg et al. 2017b; Schuhmann et al. 2019) are consistent with the ISM expectations (Hasegawa et al. 1992). This further solidifies an ISM heritage of the volatile material in 67P/C-G. The complexity of the organic material found in comets is larger than previously known and requires time to form.
- Similar conclusions arise from the abundant amounts of $O_2$, which were well correlated to $H_2O$ in 67P/C-G (Bieler et al. 2015a). $O_2$ has a presolar or molecular cloud origin and formed either through cold temperature chemistry (Taquet et al. 2017) and/or radiolysis (Mousis et al. 2016b). Chemical formation requires slightly increased formation temperatures ($\sim$20 K) and gas densities ($n_H \gtrsim 10^5$ cm$^{-3}$), similar to the case of the dense core $\rho$ Ophiuchus A, but warmer compared to the lower mass environment expected for the early Sun. On the other hand, radiolysis and photolysis of water ice as a formation scenario for $O_2$ leads to the formation of byproducts such as $H_2O_2$, $HO_2$, and $O_3$ in amounts much larger compared to that observed at 67P/C-G.
- The presence of $S_2$, which has a very short lifetime when exposed to UV photons, requires preservation in the ices originating in the presolar cloud (Calmonte et al. 2016; Mousis et al. 2017b). However, the degree to which this also applies to other volatiles, apart from $H_2O$, $O_2$, and $S_2$ (Bar-Nun et al. 2012; Greenberg et al. 2017), or whether they were evaporated and then recondensed from the gas phase in the protosolar nebula (Mousis et al. 2018b) is still debated. It seems clear, however, that a single ice phase is not able to reproduce the observed relative abundances, and/or the material in the comet was subject to evolutionary processes, such as temperature-dependent fractionation, occurring on different time scales of the comet's journey through the solar system.
- Numerous molecular species are associated with life as we know it (Seager et al. 2016) but have an abiotic origin in 67P/C-G and other comets. Aside from $O_2$ this includes organohalogens (Fayolle et al. 2017), glycine, phosphorous (Altwegg et al. 2016) and various organo-sulfurs (Calmonte et al. 2016). Even if not for the water, comets could indeed have delivered sizeable amounts of material to the Earth. A contribution of $22 \pm 5\%$ to the terrestrial atmosphere was derived based on the xenon isotopic composition of 67P/C-G compared to the terrestrial atmosphere (Marty et al. 2017). As a consequence, also copious amounts of organic species would have found their way to the inner solar system and the Earth (Marty et al. 2016; Rubin et al. 2019b).

**Acknowledgements** Open access funding provided by University of Bern. We would like to acknowledge the hospitality and support of the International Space Science Institute (ISSI), Bern, Switzerland. This project has received funding from the European Union's Horizon 2020 research and innovation programme under grant agreement No 686709. This work was supported by the Swiss State Secretariat for Education, Research and Innovation (SERI) under contract number 16.0008-2. The opinions expressed and arguments employed herein do not necessarily reflect the official view of the Swiss Government. Rosetta is an ESA mission with contributions from its member states and NASA. ROSINA, MIRO, VIRTIS, Alice, COSIMA, COSAC, and Ptolemy data are the work of international teams which we herewith acknowledge.

MR was funded by the State of Bern and the Swiss National Science Foundation (200021_165869, 200020_182418). PRW thanks the U.S. Social Security Administration for financial support. MJM thanks the NASA Astrobiology Program for financial support under grant 13-13NAI7-0032. CE acknowledges the support of CNES, CNRS, ANR COMETOR, LabEx P2IO, and logistic support from IPEV-PNRA for the micrometeorite collections in Antarctica.

## References

M.F. A'Hearn, R.C. Millis, D.O. Schleicher, D.J. Osip, P.V. Birch, The ensemble properties of comets: results from narrowband photometry of 85 comets, 1976-1992. Icarus **118**(2), 223–270 (1995). https://doi.org/10.1006/icar.1995.1190

M.F. A'Hearn et al., Deep impact: excavating comet Tempel 1. Science **310**(5746), 258–264 (2005). https://doi.org/10.1126/science.1118923

M.F. A'Hearn et al., EPOXI at comet Hartley 2. Science **332**(6036), 1396–1400 (2011). https://doi.org/10.1126/science.1204054

M.F. A'Hearn et al., Cometary volatiles and the origin of comets. Astrophys. J. Lett. **758**(1), 29 (2012)

Y. Aikawa, E. Herbst, Deuterium fractionation in protoplanetary disks. Astrophys. J. Lett. **526**(1), 314 (1999)

Y. Aikawa, E. Herbst, Two-dimensional distributions and column densities of gaseous molecules in protoplanetary disks-II. Deuterated species and UV shielding by ambient clouds. Astron. Astrophys. **371**(3), 1107–1117 (2001)

K. Altwegg, *Sulfur in the Coma of Comet Halley: Results from in Situ Ion Measurements* (University of Bern, Bern, 1995), 42 pp

K. Altwegg et al., 67P/Churyumov-Gerasimenko, a Jupiter family comet with a high D/H ratio. Science **347**(6220), 1261952 (2015). https://doi.org/10.1126/science.1261952

K. Altwegg et al., Prebiotic chemicals-amino acid and phosphorus-in the coma of comet 67P/Churyumov-Gerasimenko. Sci. Adv. **2**(5), e1600285 (2016). https://doi.org/10.1126/sciadv.1600285

K. Altwegg et al., $D_2O$ and HDS in the coma of 67P/Churyumov–Gerasimenko. Philos. Trans. - Royal Soc. A, Math. Phys. Eng. Sci. **375**(2097), 1–10 (2017a). https://doi.org/10.1098/rsta.2016.0253

K. Altwegg et al., Organics in comet 67P – a first comparative analysis of mass spectra from ROSINA–DFMS, COSAC and Ptolemy. Mon. Not. R. Astron. Soc. **469**(Suppl 2), S130–S141 (2017b). https://doi.org/10.1093/mnras/stx1415

K. Altwegg, H. Balsiger, N. Hänni, M. Rubin, M. Schuhmann, I. Schroeder, T. Sémon, S. Wampfler, J.-J. Berthelier, C. Briois, Evidence of ammonium salts in comet 67P as explanation for the nitrogen depletion in cometary comae. Nat. Astron. **4**, 1–8 (2020)

H. Balsiger, K. Altwegg, J. Geiss, D/H and $^{18}O/^{16}O$ ratio in the hydronium ion and in neutral water from in situ ion measurements in comet Halley. J. Geophys. Res. Space Phys. **100**(A4), 5827–5834 (1995)

H. Balsiger et al., Rosina - Rosetta orbiter spectrometer for ion and neutral analysis. Space Sci. Rev. **128**(1–4), 745–801 (2007). https://doi.org/10.1007/s11214-006-8335-3

H. Balsiger et al., Detection of argon in the coma of comet 67P/Churyumov-Gerasimenko. Sci. Adv. **1**(8), e1500377 (2015). https://doi.org/10.1126/sciadv.1500377

A. Bar-Nun, I. Kleinfeld, E. Kochavi, Trapping of gas mixtures by amorphous water ice. Phys. Rev. B **38**(11), 7749–7754 (1988)

A. Bar-Nun, G. Notesco, T. Owen, Trapping of $N_2$, CO and Ar in amorphous ice—application to comets. Icarus **190**, 655–659 (2007)

A. Bar-Nun, D. Laufer, O. Rebolledo, S. Malyk, H. Reisler, C. Wittig, *Gas Trapping in Ice and Its Release upon Warming* (Springer, New York, 2012), pp. 487–499

A. Bardyn, D. Baklouti, H. Cottin, N. Fray, C. Briois, J. Paquette, O. Stenzel, C. Engrand, H. Fischer, K. Hornung, Carbon-rich dust in comet 67P/Churyumov-Gerasimenko measured by COSIMA/Rosetta. Mon. Not. R. Astron. Soc. **469**(Suppl 2), S712–S722 (2017)

E.L. Berger, T.J. Zega, L.P. Keller, D.S. Lauretta, Evidence for aqueous activity on comet 81P/Wild 2 from sulfide mineral assemblages in Stardust samples and CI chondrites. Geochim. Cosmochim. Acta **75**(12), 3501–3513 (2011)

P. Bergman, B. Parise, R. Liseau, B. Larsson, H. Olofsson, K. Menten, R. Güsten, Detection of interstellar hydrogen peroxide. Astron. Astrophys. **531**, L8 (2011)

A. Bieler et al., Abundant molecular oxygen in the coma of comet 67P/Churyumov-Gerasimenko. Nature **526**(7575), 678–681 (2015a). https://doi.org/10.1038/nature15707

A. Bieler et al., Comparison of 3D kinetic and hydrodynamic models to ROSINA-COPS measurements of the neutral coma of 67P/Churyumov-Gerasimenko. Astron. Astrophys. **583**, A7 (2015b). https://doi.org/10.1051/0004-6361/201526178

G. Bilalbegović, G. Baranović, Sulphur-bearing species in molecular clouds. Mon. Not. R. Astron. Soc. **446**(3), 3118–3129 (2014)

B. Bitsch, A. Johansen, M. Lambrechts, A. Morbidelli, The structure of protoplanetary discs around evolving young stars. Astron. Astrophys. **575**, A28 (2015). https://doi.org/10.1051/0004-6361/201424964

N. Biver, D. Bockelée-Morvan, Chemical diversity in the comet population. Proc. Int. Astron. Union **11**(A29A), 228–232 (2015)

N. Biver et al., The 1995–2002 long-term monitoring of comet C/1995 O1 (HALE–BOPP) at radio wavelength. Earth Moon Planets **90**(1/4), 5–14 (2002). https://doi.org/10.1023/a:1021599915018

N. Biver, D. Bockelée-Morvan, J. Crovisier, A. Lecacheux, U. Frisk, Å. Hjalmarson, M. Olberg, H.-G. Florén, A. Sandqvist, S. Kwok, Submillimetre observations of comets with Odin: 2001–2005. Planet. Space Sci. **55**(9), 1058–1068 (2007). https://doi.org/10.1016/j.pss.2006.11.010

N. Biver, D. Bockelée-Morvan, R. Moreno, J. Crovisier, P. Colom, D.C. Lis, A. Sandqvist, J. Boissier, D. Despois, S.N. Milam, Ethyl alcohol and sugar in comet C/2014 Q2 (Lovejoy). Sci. Adv. **1**(9), e1500863 (2015). https://doi.org/10.1126/sciadv.1500863

N. Biver et al., Isotopic ratios of H, C, N, O, and S in comets C/2012 F6 (Lemmon) and C/2014 Q2 (Lovejoy). Astron. Astrophys. **589**(A78), 1–11 (2016)

N. Biver, D. Bockelée-Morvan, G. Paubert, R. Moreno, J. Crovisier, J. Boissier, E. Bertrand, H. Boussier, F. Kugel, A. McKay, The extraordinary composition of the blue comet C/2016 R2 (PanSTARRS). Astron. Astrophys. **619**, A127 (2018)

N. Biver et al., Long-term monitoring of the outgassing and composition of comet 67P/Churyumov-Gerasimenko with the Rosetta/MIRO instrument. Astron. Astrophys. (2019). https://doi.org/10.1051/0004-6361/201834960

J. Blum, Dust growth in protoplanetary disks—a comprehensive experimental/theoretical approach. Res. Astron. Astrophys. **10**(12), 1199 (2010)

J. Blum, Dust evolution in protoplanetary discs and the formation of planetesimals. Space Sci. Rev. **214**(2), 52 (2018)

D. Bockelée–Morvan, D. Gautier, F. Hersant, J.-M. Huré, F. Robert, Turbulent radial mixing in the solar nebula as the source of crystalline silicates in comets. Astron. Astrophys. **384**(3), 1107–1118 (2002)

D. Bockelée-Morvan, N. Biver, The composition of cometary ices. Philos. Trans. R. Soc. A, Math. Phys. Eng. Sci. **375**(2097), 20160252 (2017)

D. Bockelée-Morvan, D. Lis, J. Wink, D. Despois, J. Crovisier, R. Bachiller, D. Benford, N. Biver, P. Colom, J. Davies, New molecules found in comet C/1995 O1 (Hale-Bopp). Investigating the link between cometary and interstellar material. Astron. Astrophys. **353**, 1101–1114 (2000)

D. Bockelée-Morvan, J. Crovisier, M.J. Mumma, H.A. Weaver, The composition of cometary volatiles, in *Comets II*, ed. by M.C. Festou, H.U. Keller, H.A. Weaver (University of Arizona Press, Tucson, 2004), pp. 391–423

D. Bockelée-Morvan, N. Biver, E. Jehin, A.L. Cochran, H. Wiesemeyer, J. Manfroid, D. Hutsemekers, C. Arpigny, J. Boissier, W. Cochran, Large excess of heavy nitrogen in both hydrogen cyanide and cyanogen from comet 17P/Holmes. Astrophys. J. Lett. **679**(1), L49 (2008)

D. Bockelée-Morvan, N. Biver, B. Swinyard, M. de Val-Borro, J. Crovisier, P. Hartogh, D.C. Lis, R. Moreno, S. Szutowicz, E. Lellouch, Herschel measurements of the D/H and $^{16}O/^{18}O$ ratios in water in the Oort-cloud comet C/2009 P1 (Garradd). Astron. Astrophys. **544**, L15 (2012)

D. Bockelée-Morvan et al., Cometary isotopic measurements. Space Sci. Rev. **197**(1–4), 47–83 (2015). https://doi.org/10.1007/s11214-015-0156-9

D. Bockelée-Morvan et al., Evolution of $CO_2$, $CH_4$, and OCS abundances relative to $H_2O$ in the coma of comet 67P around perihelion from Rosetta/VIRTIS-H observations. Mon. Not. R. Astron. Soc. **462**(Suppl 1), S170–S183 (2016). https://doi.org/10.1093/mnras/stw2428

H. Böhnhardt, M.J. Mumma, G.L. Villanueva, M.A. DiSanti, B.P. Bonev, M. Lippi, H.U. Käufl, The unusual volatile composition of the Halley-type comet 8P/Tuttle: addressing the existence of an inner Oort cloud. Astrophys. J. Lett. **683**(1), L71 (2008)

A. Bouquet, O. Mousis, B.D. Teolis, G. Nicolaou, O. Ozge, F. Pauzat, Y. Ellinger, T. Ronnet, H. Waite, Limits on the contribution of endogenic radiolysis to the presence of molecular oxygen in comet 67P/Churyumov–Gerasimenko. Astrophys. J. Lett. **864**(1), 9 (2018)

J.P. Bradley, *Interplanetary Dust Particles*. Treatise on Geochemistry, vol. 1 (2003), p. 711. https://doi.org/10.1016/B0-08-043751-6/01152-X

J. Bregman, H. Campins, F. Witteborn, D. Wooden, D. Rank, L. Allamandola, M. Cohen, A.G. Tielens, Airborne and groundbased spectrophotometry of comet P/Halley from 5–13 micrometers, in *Exploration of Halley's Comet* (Springer, Berlin, 1988), pp. 616–620

P.D. Brown, T.J. Millar, Grain-surface formation of multi-deuterated molecules. Mon. Not. R. Astron. Soc. **240**(1), 25P–29P (1989a). https://doi.org/10.1093/mnras/240.1.25P

P.D. Brown, T.J. Millar, Models of the gas–grain interaction – deuterium chemistry. Mon. Not. R. Astron. Soc. **237**(3), 661–671 (1989b). https://doi.org/10.1093/mnras/237.3.661

D. Brownlee, The Stardust mission: analyzing samples from the edge of the solar system. Annu. Rev. Earth Planet. Sci. **42**, 179–205 (2014)

D. Brownlee et al., Comet 81P/Wild 2 under a microscope. Science **314**(5806), 1711–1716 (2006). https://doi.org/10.1126/science.1135840

E.S. Bullock, K.D. McKeegan, M. Gounelle, M.M. Grady, S.S. Russell, Sulfur isotopic composition of Fe-Ni sulfide grains in CI and CM carbonaceous chondrites. Meteorit. Planet. Sci. **45**(5), 885–898 (2010)

U. Calmonte et al., Sulphur-bearing species in the coma of comet 67P/Churyumov–Gerasimenko. Mon. Not. R. Astron. Soc. **462**(Suppl 1), S253–S273 (2016). https://doi.org/10.1093/mnras/stw2601

U. Calmonte et al., Sulphur isotope mass-independent fractionation observed in comet 67P/Churyumov–Gerasimenko by Rosetta/ROSINA. Mon. Not. R. Astron. Soc. **469**(Suppl 2), S787–S803 (2017). https://doi.org/10.1093/mnras/stx2534

H. Campins, E.V. Ryan, The identification of crystalline olivine in cometary silicates. Astrophys. J. Lett. **341**, 1059–1066 (1989)

C. Carr et al., RPC: the Rosetta plasma consortium. Space Sci. Rev. **128**, 629–647 (2007)

C. Ceccarelli, P. Caselli, D. Bockelée-Morvan, O. Mousis, S. Pizzarello, F. Robert, D. Semenov, Deuterium Fractionation: the Ariadne's Thread from the Pre-collapse Phase to Meteorites and Comets today, arXiv preprint (2014). arXiv:1403.7143

S.B. Charnley, A.G.G.M. Tielens, S.D. Rodgers, Deuterated methanol in the Orion compact ridge. Astrophys. J. **482**(2), L203–L206 (1997). https://doi.org/10.1086/310697

E. Chiang, A.N. Youdin, Forming planetesimals in solar and extrasolar nebulae. Annu. Rev. Earth Planet. Sci. **38**(1), 493–522 (2010). https://doi.org/10.1146/annurev-earth-040809-152513

K.M. Chick, P. Cassen, Thermal processing of interstellar dust grains in the primitive solar environment. Astrophys. J. **477**(1), 398–409 (1997). https://doi.org/10.1086/303700

Y.-N. Chin, C. Henkel, J.B. Whiteoak, N. Langer, E.B. Churchwell, Interstellar sulfur isotopes and stellar oxygen burning. Astron. Astrophys. **305**, 960 (1996)

M. Choukroun et al., Dust-to-gas and refractory-to-ice mass ratios of comet 67P/Churyumov-Gerasimenko from Rosetta observations. Space Sci. Rev. **216**(3), 44 (2020). https://doi.org/10.1007/s11214-020-00662-1

F.J. Ciesla, Outward transport of high-temperature materials around the midplane of the solar nebula. Science **318**(5850), 613–615 (2007)

A.J. Coates, C. Mazelle, F.M. Neubauer, Bow shock analysis at comets Halley and Grigg-Skjellerup. J. Geophys. Res. Space Phys. **102**(A4), 7105–7113 (1997). https://doi.org/10.1029/96ja04002

M. Combi et al., The surface distributions of the production of the major volatile species, $H_2O$, $CO_2$, CO and $O_2$, from the nucleus of comet 67P/Churyumov-Gerasimenko throughout the Rosetta mission as measured by the ROSINA double focusing mass spectrometer. Icarus **335**, 113421 (2020). https://doi.org/10.1016/j.icarus.2019.113421

A. Coradini et al., Virtis: an imaging spectrometer for the Rosetta mission. Space Sci. Rev. **128**(1), 529–559 (2007). https://doi.org/10.1007/s11214-006-9127-5

A. Coutens, C. Vastel, E. Caux, C. Ceccarelli, S. Bottinelli, L. Wiesenfeld, A. Faure, Y. Scribano, C. Kahane, A study of deuterated water in the low-mass protostar IRAS 16293-2422. Astron. Astrophys. **539**, A132 (2012)

A. Coutens, J.K. Jørgensen, M.V. Persson, E.F. van Dishoeck, C. Vastel, V. Taquet, High $D_2O$/HDO ratio in the inner regions of the low-mass protostar Ngc 1333 IRAS2A. Astrophys. J. Lett. **792**(1), L5 (2014). https://doi.org/10.1088/2041-8205/792/1/L5

J. Crovisier, K. Leech, D. Bockelée-Morvan, T.Y. Brooke, M.S. Hanner, B. Altieri, H.U. Keller, E. Lellouch, The spectrum of comet Hale-Bopp (C/1995 O1) observed with the Infrared Space Observatory at 2.9 astronomical units from the Sun. Science **275**(5308), 1904–1907 (1997). https://doi.org/10.1126/science.275.5308.1904

J. Crovisier, T. Brooke, K. Leech, D. Bockelée-Morvan, E. Lellouch, M. Hanner, B. Altieri, H. Keller, T. Lim, S. Encrenaz, The thermal infrared spectra of comets Hale-Bopp and 103P/Hartley 2 observed with the Infrared Space Observatory, in *Thermal Emission Spectroscopy and Analysis of Dust, Disks, and Regoliths*, vol. 196 (2000), pp. 109–117

J. Crovisier, D. Bockelée-Morvan, P. Colom, N. Biver, D. Despois, D. Lis, The composition of ices in comet C/1995 O1 (Hale-Bopp) from radio spectroscopy-further results and upper limits on undetected species. Astron. Astrophys. **418**(3), 1141–1157 (2004)

E. Dartois, W.-F. Thi, T. Geballe, D. Deboffle, L. d'Hendecourt, E. van Dishoeck, Revisiting the solid HDO/$H_2O$ abundances. Astron. Astrophys. **399**(3), 1009–1020 (2003)

N. Dauphas, The dual origin of the terrestrial atmosphere. Icarus **165**(2), 326–339 (2003). https://doi.org/10.1016/s0019-1035(03)00198-2

B.J.R. Davidsson et al., The primordial nucleus of comet 67P/Churyumov-Gerasimenko. Astron. Astrophys. **592**(A63), 30–61 (2016)

M.C. De Sanctis, M.T. Capria, A. Coradini, 67P/Churyumov–Gerasimenko nucleus model: portrayal of the Rosetta target. Adv. Space Res. **38**(9), 1906–1910 (2006). https://doi.org/10.1016/j.asr.2006.02.052

N. Dello Russo, R. Vervack, H. Weaver, N. Biver, D. Bockelée-Morvan, J. Crovisier, C. Lisse, Compositional homogeneity in the fragmented comet 73P/Schwassmann–Wachmann 3. Nature **448**(7150), 172–175 (2007)

N. Dello Russo, H. Kawakita, R.J. Vervack, H.A. Weaver, Emerging trends and a comet taxonomy based on the volatile chemistry measured in thirty comets with high-resolution infrared spectroscopy between 1997 and 2013. Icarus **278**, 301–332 (2016). https://doi.org/10.1016/j.icarus.2016.05.039

T. Ding, S. Valkiers, H. Kipphardt, P. De Bievre, P. Taylor, R. Gonfiantini, R. Krouse, Calibrated sulfur isotope abundance ratios of three IAEA sulfur isotope reference materials and V-CDT with a reassessment of the atomic weight of sulfur. Geochim. Cosmochim. Acta **65**(15), 2433–2437 (2001)

M.N. Drozdovskaya et al., The ALMA-PILS survey: the sulphur connection between protostars and comets: IRAS 16293–2422 B and 67P/Churyumov–Gerasimenko. Mon. Not. R. Astron. Soc. Lett. **476**(4), 4949–4964 (2018). https://doi.org/10.1093/mnras/sty462

M.N. Drozdovskaya, E.F. van Dishoeck, M. Rubin, J.K. Jørgensen, K. Altwegg, Ingredients for solar-like systems: protostar IRAS 16293-2422 B versus comet 67P/Churyumov–Gerasimenko. Mon. Not. R. Astron. Soc. Lett. **490**(1), 50–79 (2019). https://doi.org/10.1093/mnras/stz2430

F. Dulieu, M. Minissale, D. Bockelée-Morvan, Production of $O_2$ through dismutation of $H_2O_2$ during water ice desorption: a key to understanding comet $O_2$ abundances. Astron. Astrophys. **597**, A56 (2017)

P. Eberhardt, Comet Halley's gas composition and extended sources: results from the neutral mass spectrometer on Giotto. Space Sci. Rev. **90**(1–2), 45–52 (1999). https://doi.org/10.1023/A:1005221309219

P. Eberhardt, M. Reber, D. Krankowsky, R.R. Hodges, The D/H and $^{18}O/^{16}O$ ratios in water from comet P/Halley. Astron. Astrophys. **302**, 301 (1995)

C. Eistrup, C. Walsh, Formation of cometary $O_2$ ice and related ice species on grain surfaces in the midplane of the pre-Solar nebula, arxiv preprint (2018). arXiv:1808.03329

A. Faure, M. Faure, P. Theulé, E. Quirico, B. Schmitt, Hydrogen isotope exchanges between water and methanol in interstellar ices. Astron. Astrophys. **584**, A98 (2015)

E.C. Fayolle et al., Protostellar and cometary detections of organohalogens. Nat. Astron. **1**(10), 703–708 (2017). https://doi.org/10.1038/s41550-017-0237-7

L.M. Feaga, M.F. A'Hearn, T.L. Farnham, D. Bodewits, J.M. Sunshine, A.M. Gersch, S. Protopapa, B. Yang, M. Drahus, D.G. Schleicher, Uncorrelated volatile behavior during the 2011 apparition of comet C/2009 P1 Garradd. Astron. J. **147**(1), 24 (2013)

B.J. Fegley, R.G. Prinn, Solar nebula chemistry - implications for volatiles in the solar system, in *The Formation and Evolution of Planetary Systems*, ed. by H. Weaver et al. (University of Arizona Press, Tucson, 1989), pp. 171–205

P.D. Feldman et al., Measurements of the near-nucleus coma of comet 67P/Churyumov-Gerasimenko with the Alice far-ultraviolet spectrograph on Rosetta. Astron. Astrophys. **583**(A8), 1–8 (2015). https://doi.org/10.1051/0004-6361/201525925

L.N. Fletcher, T.K. Greathouse, G.S. Orton, P.G.J. Irwin, O. Mousis, J.A. Sinclair, R.S. Giles, The origin of nitrogen on Jupiter and Saturn from the $^{15}N/^{14}N$ ratio. Icarus **238**, 170–190 (2014)

T. Fouchet, E. Lellouch, B. Bézard, T. Encrenaz, P. Drossart, H. Feuchtgruber, T. de Graauw, ISO-SWS observations of Jupiter: measurement of the ammonia tropospheric profile and of the $^{15}N/^{14}N$ isotopic ratio. Icarus **143**(2), 223–243 (2000)

N. Fougere et al., Direct simulation Monte Carlo modelling of the major species in the coma of comet 67P/Churyumov-Gerasimenko. Mon. Not. R. Astron. Soc. Lett. **462**(Suppl), S156–S169 (2016)

D.R. Frank, M.E. Zolensky, L. Le, Olivine in terminal particles of Stardust aerogel tracks and analogous grains in chondrite matrix. Geochim. Cosmochim. Acta **142**, 240–259 (2014). https://doi.org/10.1016/j.gca.2014.05.037

N. Fray, A. Bardyn, H. Cottin, K. Altwegg, D. Baklouti, C. Briois, L. Colangeli, C. Engrand, H. Fischer, A. Glasmachers, High-molecular-weight organic matter in the particles of comet 67P/Churyumov–Gerasimenko. Nature **538**(7623), 72 (2016)

N. Fray, A. Bardyn, H. Cottin, D. Baklouti, C. Briois, C. Engrand, H. Fischer, K. Hornung, R. Isnard, Y. Langevin, Nitrogen-to-carbon atomic ratio measured by COSIMA in the particles of comet 67P/Churyumov–Gerasimenko. Mon. Not. R. Astron. Soc. Lett. **469**(Suppl 2), S506–S516 (2017)

M. Fulle, N. Altobelli, B. Buratti, M. Choukroun, M. Fulchignoni, E. Grün, M.G.G.T. Taylor, P. Weissman, Unexpected and significant findings in comet 67P/Churyumov–Gerasimenko: an interdisciplinary view. Mon. Not. R. Astron. Soc. Lett. **462**(Suppl 1), S2–S8 (2016). https://doi.org/10.1093/mnras/stw1663

E. Füri, B. Marty, Nitrogen isotope variations in the Solar System. Nat. Geosci. **8**(7), 515 (2015)

K. Furuya, Y. Aikawa, Reprocessing of ices in turbulent protoplanetary disks: carbon and nitrogen chemistry. Astrophys. J. Lett. **790**, 2 (2014)

K. Furuya, Y. Aikawa, U. Hincelin, G. Hassel, E. Bergin, A. Vasyunin, E. Herbst, Water deuteration and ortho-to-para nuclear spin ratio of $H_2$ in molecular clouds formed via the accumulation of H I gas. Astron. Astrophys. **584**, A124 (2015)

K. Furuya, E.F. van Dishoeck, Y. Aikawa, Reconstructing the history of water ice formation from HDO/$H_2O$ and $D_2O$/HDO ratios in protostellar cores. Astron. Astrophys. **586**, A127 (2016)

K. Furuya, Y. Watanabe, T. Sakai, Y. Aikawa, S. Yamamoto, Depletion of 15N in the center of L1544: early transition from atomic to molecular nitrogen? Astron. Astrophys. **615**, L16 (2018)

J. Geiss, Composition measurements and the history of cometary matter. Astron. Astrophys. **187**, 859–866 (1987)

J. Geiss, G. Gloeckler, Abundances of deuterium and helium-3 in the protosolar cloud, in *Primordial Nuclei and Their Galactic Evolution* (Springer, Berlin, 1998), pp. 239–250

A. Gicquel, S. Milam, I. Coulson, G. Villanueva, M. Cordiner, S. Charnley, M. DiSanti, M. Mumma, S. Szutowicz, The evolution of volatile production in comet C/2009 P1 (Garradd) during its 2011–2012 apparition. Astrophys. J. Lett. **807**(1), 19 (2015)

J.D. Gilmour, "Planetary" noble gas components and the nucleosynthetic history of solar system material. Geochim. Cosmochim. Acta **74**(1), 380–393 (2010). https://doi.org/10.1016/j.gca.2009.09.015

K.H. Glassmeier, H. Boehnhardt, D. Koschny, E. Kührt, I. Richter, The ROSETTA mission: flying towards the origin of the solar system. Space Sci. Rev. **128**(1–4), 1–21 (2007). https://doi.org/10.1007/s11214-006-9140-8

F. Goesmann, H. Rosenbauer, R. Roll, C. Szopa, F. Raulin, R. Sternberg, G. Israel, U. Meierhenrich, W. Thiemann, G. Munoz-Caro, Cosac, the cometary sampling and composition experiment on Philae. Space Sci. Rev. **128**(1), 257–280 (2007). https://doi.org/10.1007/s11214-006-9000-6

F. Goesmann et al., Organic compounds on comet 67P/Churyumov-Gerasimenko revealed by COSAC mass spectrometry. Science **349**(6247), aab0689 (2015). https://doi.org/10.1126/science.aab0689

P.F. Goldsmith, R. Liseau, T.A. Bell, J.H. Black, J.-H. Chen, D. Hollenbach, M.J. Kaufman, D. Li, D.C. Lis, G. Melnick, Herschel measurements of molecular oxygen in Orion. Astrophys. J. Lett. **737**(2), 96 (2011)

R. Gomes, H.F. Levison, K. Tsiganis, A. Morbidelli, Origin of the cataclysmic Late Heavy Bombardment period of the terrestrial planets. Nature **435**(7041), 466–469 (2005). https://doi.org/10.1038/nature03676

N. Gortsas, E. Kührt, U. Motschmann, H.U. Keller, Thermal model of water and CO activity of comet C/1995 O1 (Hale-Bopp). Icarus **212**, 858–866 (2011). https://doi.org/10.1016/j.icarus.2011.01.017

J.M. Greenberg, What are comets made of – a model based on interstellar dust, in *Comets* (University of Arizona Press, Tucson, 1982), pp. 131–163

A.N. Greenberg, D. Laufer, A. Bar-Nun, The effect of $CO_2$ on gases trapping in cometary ices. Mon. Not. R. Astron. Soc. Lett. (2017). https://doi.org/10.1093/mnras/stx2017

B. Guérin et al., STEM and STXM-XANES analysis of FIB sections of ultracarbonaceous Antarctic micrometeorites (UCAMMs), abstract #2117, in *51st Lunar and Planet and Sci.*, The Woodlands, Texas, (2020)

A. Guilbert-Lepoutre, S. Besse, O. Mousis, M. Ali-Dib, S. Höfner, D. Koschny, P. Hager, On the evolution of comets. Space Sci. Rev. **197**(1–4), 271–296 (2015)

A. Guilbert-Lepoutre, E.D. Rosenberg, D. Prialnik, S. Besse, Modelling the evolution of a comet subsurface: implications for 67P/Churyumov–Gerasimenko. Mon. Not. R. Astron. Soc. Lett. **462**(Suppl 1), S146–S155 (2016). https://doi.org/10.1093/mnras/stw2371

S. Gulkis et al., MIRO: microwave instrument for Rosetta orbiter. Space Sci. Rev. **128**(1), 561–597 (2007). https://doi.org/10.1007/s11214-006-9032-y

M.S. Hanner, M.E. Zolensky, The mineralogy of cometary dust, in *Astromineralogy* (Springer, Berlin, 2010), pp. 203–232. https://doi.org/10.1007/978-3-642-13259-9_4

M.S. Hanner, D.K. Lynch, R.W. Russell, The 8-13 micron spectra of comets and the composition of silicate grains. Astrophys. J. Lett. **425**, 274–285 (1994)

N. Hänni, S. Gasc, A. Etter, M. Schuhmann, I. Schroeder, S.F. Wampfler, S. Schuerch, M. Rubin, K. Altwegg, Ammonium salts as a source of small molecules observed with high-resolution electron-impact ionization mass spectrometry. J. Phys. Chem. A **123**, 5805–5814 (2019)

K.C. Hansen et al., Evolution of water production of 67P/Churyumov–Gerasimenko: an empirical model and a multi-instrument study. Mon. Not. R. Astron. Soc. Lett. **462**(Suppl 1), S491–S506 (2016). https://doi.org/10.1093/mnras/stw2413

P. Hartogh et al., Ocean-like water in the Jupiter-family comet 103P/Hartley 2. Nature **478**(7368), 218–220 (2011). https://doi.org/10.1038/nature10519

T.I. Hasegawa, E. Herbst, C.M. Leung, Models of gas-grain chemistry in dense interstellar clouds with complex organic molecules. Astrophys. J. Suppl. Ser. **82**, 167–195 (1992)

K. Hashizume, M. Chaussidon, B. Marty, K. Terada, Protosolar carbon isotopic composition: implications for the origin of meteoritic organics. Astrophys. J. Lett. **600**(1), 480 (2004)

M. Hässig et al., Time variability and heterogeneity in the coma of 67P/Churyumov-Gerasimenko. Science **347**(6220), aaa0276 (2015). https://doi.org/10.1126/science.aaa0276

M. Hässig et al., Isotopic composition of $CO_2$ in the coma of 67P/Churyumov-Gerasimenko measured with ROSINA/DFMS. Astron. Astrophys. **605**(A50), 51–58 (2017). https://doi.org/10.1051/0004-6361/201630140

J. Hatchell, H. Roberts, T. Millar, Limits on HDS/$H_2S$ abundance ratios in hot molecular cores. Astron. Astrophys. **346**, 227–232 (1999)

W.M. Haynes, *CRC Handbook of Chemistry and Physics: A Ready-Reference Book of Chemical and Physical Data*, 94th edn. (CRC Press, Boca Raton, 2013)

A.N. Heays, R. Visser, R. Gredel, W. Ubachs, B.R. Lewis, S.T. Gibson, E.F. van Dishoeck, Isotope selective photodissociation of $N_2$ by the interstellar radiation field and cosmic rays. Astron. Astrophys. **562**, A61 (2014)

P.R. Heck, P. Hoppe, J. Huth, Sulfur four isotope NanoSIMS analysis of comet-81P/Wild 2 dust in impact craters on aluminum foil C2037N from NASA's Stardust mission. Meteorit. Planet. Sci. **47**(4), 649–659 (2012)

E. Herbst, E.F. van Dishoeck, Complex organic interstellar molecules. Annu. Rev. Astron. Astrophys. **47**, 427–480 (2009)

K.L. Heritier et al., On the origin of molecular oxygen in cometary comae. Nat. Commun. **9**(1), 2580 (2018). https://doi.org/10.1038/s41467-018-04972-5

L.J. Hicks, J.L. MacArthur, J.C. Bridges, M.C. Price, J.E. Wickham-Eade, M.J. Burchell, G.M. Hansford, A.L. Butterworth, S.J. Gurman, S.H. Baker, Magnetite in comet Wild 2: evidence for parent body aqueous alteration. Meteorit. Planet. Sci. **52**(10), 2075–2096 (2017). https://doi.org/10.1111/maps.12909

P. Hily-Blant, L. Bonal, A. Faure, E. Quirico, The $^{15}$N-enrichment in dark clouds and Solar System objects. Icarus **223**(1), 582–590 (2013)

P. Hily-Blant, V. Magalhaes, J. Kastner, A. Faure, T. Forveille, C. Qi, Direct evidence of multiple reservoirs of volatile nitrogen in a protosolar nebula analogue. Astron. Astrophys. **603**, L6 (2017)

P. Hily-Blant, V. Magalhaes de Souza, J. Kastner, T. Forveille, Multiple nitrogen reservoirs in a protoplanetary disk at the epoch of comet and giant planet formation. Astron. Astrophys. **632**, L12 (2019)

P. Hoppe, M. Rubin, K. Altwegg, Presolar isotopic signatures in meteorites and comets: new insights from the Rosetta mission to comet 67P/Churyumov-Gerasimenko. Space Sci. Rev. **214**(6), 106 (2018). https://doi.org/10.1007/s11214-018-0540-3

D. Hutsemékers, J. Manfroid, E. Jehin, J.M. Zucconi, C. Arpigny, The $^{16}$OH/$^{18}$OH and OD/OH isotope ratios in comet C/2002 T7 (LINEAR). Astron. Astrophys. **490**(3), L31–L34 (2008)

S. Ioppolo, H. Cuppen, C. Romanzin, E. van Dishoeck, H. Linnartz, Laboratory evidence for efficient water formation in interstellar ices. Astrophys. J. Lett. **686**(2), 1474 (2008)

R. Isnard et al., H/C elemental ratios of the refractory organic matter in cometary particles of 67P/Churyumov-Gerasimenko. Astron. Astrophys. **630**, A27 (2019)

E.K. Jessberger, A. Christoforidis, J. Kissel, Aspects of the major element composition of Halley's dust. Nature **332**(6166), 691 (1988)

D.C. Jewitt, H.E. Matthews, T. Owen, R. Meier, Measurements of $^{12}$C/$^{13}$C, $^{14}$N/$^{15}$N, and $^{32}$S/$^{34}$S ratios in comet Hale-Bopp (C/1995 O1). Science **278**(5335), 90 (1997)

R. Johnson, R. Carlson, J. Cooper, C. Paranicas, M. Moore, M. Wong, Radiation effects on the surfaces of the Galilean satellites, in *Jupiter: The Planet, Satellites and Magnetosphere* (2004), pp. 485–512

J.K. Jørgensen, F.L. Schöier, E.F. van Dishoeck, Molecular inventories and chemical evolution of low-mass protostellar envelopes. Astron. Astrophys. **416**(2), 603–622 (2004). https://doi.org/10.1051/0004-6361:20034440

M. Jutzi, E. Asphaug, The shape and structure of cometary nuclei as a result of low-velocity accretion. Science **348**(6241), 1355–1358 (2015). https://doi.org/10.1126/science.aaa4747

M. Jutzi, W. Benz, Formation of bi-lobed shapes by sub-catastrophic collisions. Astron. Astrophys. **597**, A62 (2017)

M. Jutzi, W. Benz, A. Toliou, A. Morbidelli, R. Brasser, How primordial is the structure of comet 67P? Astron. Astrophys. **597**, A61 (2017)

B.A. Keeney et al., $H_2O$ and $O_2$ absorption in the coma of comet 67P/Churyumov–Gerasimenko measured by the Alice far-ultraviolet spectrograph on Rosetta. Mon. Not. R. Astron. Soc. Lett. **469**(Suppl 2), S158–S177 (2017). https://doi.org/10.1093/mnras/stx1426

H.U. Keller et al., Insolation, erosion, and morphology of comet 67P/Churyumov-Gerasimenko. Astron. Astrophys. **583**(16), A34 (2015). https://doi.org/10.1051/0004-6361/201525964

J. Kissel et al., Composition of comet Halley dust particles from Giotto observations. Nature **321**(6067), 336–337 (1986a). https://doi.org/10.1038/321336a0

J. Kissel, R. Sagdeev, J. Bertaux, V. Angarov, J. Audouze, J. Blamont, K. Büchler, E. Evlanov, H. Fechtig, M. Fomenkova, Composition of comet Halley dust particles from Vega observations. Nature **321**(6067), 280 (1986b)

J. Kissel et al., Cosima – high resolution time-of-flight secondary ion mass spectrometer for the analysis of cometary dust particles onboard Rosetta. Space Sci. Rev. **128**(1), 823–867 (2007). https://doi.org/10.1007/s11214-006-9083-0

M. Kleine, S. Wyckoff, P.A. Wehinger, B.A. Peterson, The carbon-isotope abundance ratio in comet Halley. Astrophys. J. **439**(2), 1021–1033 (1995). https://doi.org/10.1086/175239

A. Kouchi, T. Yamamoto, Cosmoglaciology: evolution of ice in interstellar space and the early solar system. Prog. Cryst. Growth Charact. Mater. **30**(2–3), 83–107 (1995). https://doi.org/10.1016/0960-8974(95)00005-r

A. Kouchi, T. Yamamoto, T. Kozasa, T. Kuroda, J.M. Greenberg, Conditions for condensation and preservation of amorphous ice and crystallinity of astrophysical ices. Astron. Astrophys. **290**(3), 1009–1018 (1994)

T. Kramer, M. Läuter, M. Rubin, K. Altwegg, Seasonal changes of the volatile density in the coma and on the surface of comet 67P/Churyumov–Gerasimenko. Mon. Not. R. Astron. Soc. Lett. **469**(Suppl 2), S20–S28 (2017). https://doi.org/10.1093/mnras/stx866

D. Krankowsky et al., In situ gas and ion measurements at comet Halley. Nature **321**, 326–329 (1986)

D. Lambert, A. Danks, High-resolution spectra of C2 Swan bands from comet West 1976 VI. Astrophys. J. Lett. **268**, 428–446 (1983)

T. Lamberts, S. Ioppolo, H. Cuppen, G. Fedoseev, H. Linnartz, Thermal H/D exchange in polar ice–deuteron scrambling in space. Mon. Not. R. Astron. Soc. Lett. **448**(4), 3820–3828 (2015)

W. Langer, T. Graedel, M. Frerking, P. Armentrout, Carbon and oxygen isotope fractionation in dense interstellar clouds. Astrophys. J. Lett. **277**, 581–590 (1984)

Y. Langevin, M. Hilchenbach, M. Vincendon, S. Merouane, K. Hornung, N. Ligier, C. Engrand, R. Schulz, J. Kissel, J. Rynö, Optical properties of cometary particles collected by the COSIMA mass spectrometer on-board Rosetta during the rendezvous phase around comet 67P/Churyumov–Gerasimenko. Mon. Not. R. Astron. Soc. Lett. **469**(Suppl 2), S535–S549 (2017)

B. Larsson, R. Liseau, L. Pagani, P. Bergman, P. Bernath, N. Biver, J. Black, R. Booth, V. Buat, J. Crovisier, Molecular oxygen in the $\rho$ Ophiuchi cloud. Astron. Astrophys. **466**(3), 999–1003 (2007)

T. Lauck, L. Karssemeijer, K. Shulenberger, M. Rajappan, K.I. Öberg, H.M. Cuppen, CO diffusion into amorphous $H_2O$ ices. Astrophys. J. Lett. **801**(2), 118 (2015)

D. Laufer, A. Bar-Nun, A. Ninio Greenberg, Trapping mechanism of $O_2$ in water ice as first measured by Rosetta spacecraft. Mon. Not. R. Astron. Soc. Lett. **469**(Suppl 2), S818–S823 (2018)

M. Läuter, T. Kramer, M. Rubin, K. Altwegg, Surface localization of gas sources on comet 67P/Churyumov–Gerasimenko based on DFMS/COPS data. Mon. Not. R. Astron. Soc. Lett. **483**(1), 852–861 (2019). https://doi.org/10.1093/mnras/sty3103

L. Le Roy et al., Inventory of the volatiles on comet 67P/Churyumov-Gerasimenko from Rosetta/ROSINA. Astron. Astrophys. **583**, A1 (2015). https://doi.org/10.1051/0004-6361/201526450

J.-E. Lee, E.A. Bergin, J.R. Lyons, Oxygen isotope anomalies of the Sun and the original environment of the solar system. Meteorit. Planet. Sci. **43**(8), 1351–1362 (2008)

A.-C. Levasseur-Regourd, J. Agarwal, H. Cottin, C. Engrand, G. Flynn, M. Fulle, T. Gombosi, Y. Langevin, J. Lasue, T. Mannel, Cometary dust. Space Sci. Rev. **214**(3), 64 (2018)

H.F. Levison, M.J. Duncan, From the Kuiper belt to Jupiter-family comets: the spatial distribution of ecliptic comets. Icarus **127**, 13–32 (1997)

M. Lippi, G.L. Villanueva, M.J. Mumma, M.N. Camarca, S. Faggi, L. Paganini, New insights into the chemical composition of five Oort cloud comets after re-analysis of their infrared spectra. Astron. J. **159**(4), 16 (2020). https://doi.org/10.3847/1538-3881/ab7206

D. Lis, E. Roueff, M. Gerin, T. Phillips, L. Coudert, F. van der Tak, P. Schilke, Detection of triply deuterated ammonia in the Barnard 1 cloud. Astrophys. J. Lett. **571**(1), L55 (2002)

D.C. Lis, D. Bockelée-Morvan, R. Güsten, N. Biver, J. Stutzki, Y. Delorme, C. Durán, H. Wiesemeyer, Y. Okada, Terrestrial deuterium-to-hydrogen ratio in water in hyperactive comets. Astron. Astrophys. **625**, L5 (2019)

R. Liseau, P. Goldsmith, B. Larsson, L. Pagani, P. Bergman, J. Le Bourlot, T. Bell, A. Benz, E. Bergin, P. Bjerkeli, Multi-line detection of $O_2$ toward $\rho$ Ophiuchi A. Astron. Astrophys. **541**, A73 (2012)
C. Lisse, J. VanCleve, A. Adams, M. A'hearn, Y. Fernández, T. Farnham, L. Armus, C. Grillmair, J. Ingalls, M. Belton, Spitzer spectral observations of the Deep Impact ejecta. Science **313**(5787), 635–640 (2006)
K. Lodders, H. Palme, H.P. Gail, Abundances of the elements in the Solar System, in *The Solar System*, ed. by J.E. Trümper (Springer, Berlin, 2009), pp. 560–598
K. Lodders, H. Palme, H.P. Gail, Solar system abundances of the elements, in *Principles and Perspectives in Cosmochemistry*, ed. by A. Goswami, B.E. Reddy (Springer, Heidelberg, 2010), pp. 379–417. https://doi.org/10.1007/978-3-642-10352-0
J.I. Lunine, D.J. Stevenson, Thermodynamics of clathrate hydrate at low and high-pressures with application to the outer Solar-System. Astrophys. J. Suppl. Ser. **58**(3), 493–531 (1985). https://doi.org/10.1086/191050
A. Luspay-Kuti et al., Origin of molecular oxygen in comets: current knowledge and perspectives. Space Sci. Rev. **214(8), 115**, 111–124 (2018). https://doi.org/10.1007/s11214-018-0541-2
J.R. Lyons, E.A. Bergin, F.J. Ciesla, A.M. Davis, S.J. Desch, K. Hashizume, J.-E. Lee, Timescales for the evolution of oxygen isotope compositions in the solar nebula. Geochim. Cosmochim. Acta **73**, 4998–5017 (2009)
J. Manfroid, E. Jehin, D. Hutsemékers, A. Cochran, J.-M. Zucconi, C. Arpigny, R. Schulz, J. Stüwe, I. Ilyin, The CN isotopic ratios in comets. Astron. Astrophys. **503**(2), 613–624 (2009)
L. Maquet, The recent dynamical history of comet 67P/Churyumov-Gerasimenko. Astron. Astrophys. **579**(75), A78 (2015). https://doi.org/10.1051/0004-6361/201425461
U. Marboeuf, B. Schmitt, How to link the relative abundances of gas species in coma of comets to their initial chemical composition? Icarus **242**(0), 225–248 (2014). https://doi.org/10.1016/j.icarus.2014.07.001
D.W. Marshall et al., Spatially resolved evolution of the local $H_2O$ production rates of comet 67P/Churyumov-Gerasimenko from the MIRO instrument on Rosetta. Astron. Astrophys. **603**(A87), 1–11 (2017)
B. Marty, M. Chaussidon, R.C. Wiens, A.J. Jurewicz, D.S. Burnett, A $^{15}N$-poor isotopic composition for the solar system as shown by Genesis solar wind samples. Science **332**(6037), 1533–1536 (2011). https://doi.org/10.1126/science.1204656
B. Marty, G. Avice, Y.J. Sano, K. Altwegg, H. Balsiger, M. Hässig, A. Morbidelli, O. Mousis, M. Rubin, Origins of volatile elements (H, C, N, noble gases) on Earth and Mars in light of recent results from the ROSETTA cometary mission. Earth Planet. Sci. Lett. **441**, 91–102 (2016). https://doi.org/10.1016/j.epsl.2016.02.031
B. Marty et al., Xenon isotopes in 67P/Churyumov-Gerasimenko show that comets contributed to Earth's atmosphere. Science **356**(6342), 1069 (2017)
M. Massironi et al., Two independent and primitive envelopes of the bilobate nucleus of comet 67P. Nature **526**(7573), 402–405 (2015). https://doi.org/10.1038/nature15511
A.J. McKay et al., The peculiar volatile composition of CO-dominated comet C/2016 R2 (PanSTARRS). Astron. J. **158**(3), 128 (2019). https://doi.org/10.3847/1538-3881/ab32e4
K.D. McKeegan et al., Isotopic compositions of cometary matter returned by Stardust. Science **314**(5806), 1724 (2006)
K. McKeegan, A. Kallio, V. Heber, G. Jarzebinski, P. Mao, C. Coath, T. Kunihiro, R. Wiens, J. Nordholt, R. Moses, The oxygen isotopic composition of the Sun inferred from captured solar wind. Science **332**(6037), 1528–1532 (2011)
H.Y. McSween, G.R. Huss, *Cosmochemistry* (Cambridge University Press, Cambridge, 2010), p. 549
R. Meier, T.C. Owen, D.C. Jewitt, H.E. Matthews, M. Senay, N. Biver, D. Bockelée-Morvan, J. Crovisier, D. Gautier, Deuterium in comet C/1995 O1 (Hale-Bopp): detection of DCN. Science **279**(5357), 1707–1710 (1998a)
R. Meier, T.C. Owen, H.E. Matthews, D.C. Jewitt, D. Bockelée-Morvan, N. Biver, J. Crovisier, D. Gautier, A determination of the $HDO/H_2O$ ratio in comet C/1995 O1 (Hale-Bopp). Science **279**(5352), 842–844 (1998b)
A. Meshik, C. Hohenberg, O. Pravdivtseva, D. Burnett, Heavy noble gases in solar wind delivered by Genesis mission. Geochim. Cosmochim. Acta **127**, 326–347 (2014). https://doi.org/10.1016/j.gca.2013.11.030
T. Mikouchi, O. Tachikawa, K. Hagiya, K. Ohsumi, Y. Suzuki, K. Uesugi, A. Takeuchi, M. Zolensky, Mineralogy and crystallography of comet 81P/Wild 2 particles, abstract #1946, in *38th Lunar and Planetary Science Conference* (2007)
M. Min, C.P. Dullemond, M. Kama, C. Dominik, The thermal structure and the location of the snow line in the protosolar nebula: axisymmetric models with full 3-D radiative transfer. Icarus **212**(1), 416–426 (2011). https://doi.org/10.1016/j.icarus.2010.12.002

N. Miyauchi, H. Hidaka, T. Chigai, A. Nagaoka, N. Watanabe, A. Kouchi, Formation of hydrogen peroxide and water from the reaction of cold hydrogen atoms with solid oxygen at 10 K. Chem. Phys. Lett. **456**(1–3), 27–30 (2008)

A. Morbidelli, H. Rickman, Comets as collisional fragments of a primordial planetesimal disk. Astron. Astrophys. **583**, A43 (2015)

O. Mousis, Constraints on the formation of comets from D/H ratios measured in $H_2O$ and HCN. Icarus **148**(2), 513–525 (2000). https://doi.org/10.1006/icar.2000.6499

O. Mousis, J.I. Lunine, S. Picaud, D. Cordier, Volatile inventories in clathrate hydrates formed in the primordial nebula. Faraday Discuss. **147**(0), 509–525 (2010). https://doi.org/10.1039/C003658G

O. Mousis, A. Guilbert-Lepoutre, J.I. Lunine, A.L. Cochran, J.H. Waite, J.M. Petit, P. Rousselot, The dual origin of the nitrogen deficiency in comets: selective volatile trapping in the nebula and postaccretion radiogenic heating. Astrophys. J. **757**(2), 146 (2012). https://doi.org/10.1088/0004-637x/757/2/146

O. Mousis et al., A protosolar nebula origin for the ices agglomerated by comet 67P/Churyumov-Gerasimenko. Astrophys. J. Lett. **819**(2), L33 (2016a). https://doi.org/10.3847/2041-8205/819/2/L33

O. Mousis et al., Origin of molecular oxygen in comet 67p/Churyumov-Gerasimenko. Astrophys. J. Lett. **823**(2), L41 (2016b). https://doi.org/10.3847/2041-8205/823/2/L41

O. Mousis et al., Impact of radiogenic heating on the formation conditions of comet 67P/Churyumov–Gerasimenko. Astrophys. J. Lett. **839**(1), L4 (2017a)

O. Mousis, O. Ozgurel, J. Lunine, A. Luspay-Kuti, T. Ronnet, F. Pauzat, A. Markovits, Y. Ellinger, Stability of sulphur dimers (S2) in cometary ices. Astrophys. J. Lett. **835**(2), 134 (2017b)

O. Mousis, T. Ronnet, J. Lunine, R. Maggiolo, P. Wurz, G. Danger, A. Bouquet, Synthesis of molecular oxygen via irradiation of ice grains in the protosolar nebula. Astrophys. J. Lett. **858**(1), 66 (2018a)

O. Mousis et al., Noble gas abundance ratios indicate the agglomeration of 67P/Churyumov–Gerasimenko from warmed-up ice. Astrophys. J. Lett. **865**(1), L11 (2018b)

M.J. Mumma, S.B. Charnley, The chemical composition of comets—emerging taxonomies and natal heritage. Annu. Rev. Astron. Astrophys. **49**(1), 471–524 (2011). https://doi.org/10.1146/annurev-astro-081309-130811

M.J. Mumma, E.J. Stone, E.C. Zipf, Excitation of the CO fourth positive band system by electron impact on carbon monoxide and carbon dioxide. J. Chem. Phys. **54**(6), 2627–2634 (1971). https://doi.org/10.1063/1.1675223

M.J. Mumma, E.J. Stone, W.L. Borst, E.C. Zipf, Dissociative Excitation of Vacuum Ultraviolet emission features by electron impact on molecular gases. III. $CO_2$. J. Chem. Phys. **57**(1), 68–75 (1972). https://doi.org/10.1063/1.1678019

M.J. Mumma, M.A. DiSanti, N. Dello Russo, M. Fomenkova, K. Magee-Sauer, C.D. Kaminski, D.X. Xie, Detection of abundant ethane and methane, along with carbon monoxide and water, in comet C/1996 B2 Hyakutake: evidence for interstellar origin. Science **272**(5266), 1310–1314 (1996). https://doi.org/10.1126/science.272.5266.1310

M.J. Mumma, M.A. DiSanti, N. Dello Russo, K. Magee-Sauer, E. Gibb, R. Novak, Remote infrared observations of parent volatiles in comets: a window on the early solar system. Adv. Space Res. **31**(12), 2563–2575 (2003). https://doi.org/10.1016/s0273-1177(03)00578-7

M.J. Mumma et al., Parent volatiles in comet 9P/Tempel 1: before and after impact. Science **310**(5746), 270–274 (2005). https://doi.org/10.1126/science.1119337

M.J. Mumma, S.B. Charnley, M. Cordiner, L. Paganini, G.L. Villanueva, The relationship of HCN, $C_2H_6$, & $H_2O$ in comets: a key clue to origins?, in *AAS/Division for Planetary Sciences Meeting Abstracts #49* (2017)

M.J. Mumma, S.B. Charnley, M. Cordiner, L. Paganini, G.L. Villanueva, S. Faggi, L. Paganini, M. Lippi, M.A. DiSanti, The relationship of HCN, $NH_3$, $C_2H_6$, $H_2O$, and ammoniated salts in comets: a key clue to origins?, in *American Astronomical Society, 50th DPS Meeting*, Knoxville, Tennessee (2018)

M. Mumma, S. Charnley, M. Cordiner, G. Villanueva, S. Faggi, M. Lippi, L. Paganini, The relationship of HCN, $NH_3$, $C_2H_6$, $H_2O$, and ammoniated salts in comets: a key clue to origins, in *EPSC-DPS Joint Meeting* (2019)

T. Nakamura et al., Chondrulelike objects in short-period comet 81P/Wild 2. Science **321**(5896), 1664–1667 (2008). https://doi.org/10.1126/science.1160995

D. Nakashima, T. Ushikubo, N.T. Kita, M.K. Weisberg, M.E. Zolensky, D.S. Ebel, Late formation of a comet Wild 2 crystalline silicate particle, pyxie, inferred from Al–Mg chronology of plagioclase. Earth Planet. Sci. Lett. **410**, 54–61 (2015)

J.L. Neill, S. Wang, E.A. Bergin, N.R. Crockett, C. Favre, R. Plume, G.J. Melnick, The abundance of $H_2O$ and HDO in ORION KL from Herschel/HIFI. Astrophys. J. Lett. **770**(2), 142 (2013)

A.O. Nier, A redetermination of the relative abundances of the isotopes of carbon, nitrogen, oxygen, argon, and potassium. Phys. Rev. **77**(6), 789 (1950)

A. Oka, T. Nakamoto, S. Ida, Evolution of snow line in optically thick protoplanetary disks: effects of water ice opacity and dust grain size. Astrophys. J. Lett. **738**(2), 141 (2011). https://doi.org/10.1088/0004-637x/738/2/141

T. Ootsubo et al., Akari near-infrared spectroscopic survey for $CO_2$ in 18 comets. Astrophys. J. **752**(1), 15 (2012). https://doi.org/10.1088/0004-637x/752/1/15

C. Opitom, *Monitoring of the Chemical Composition of Comets in the Framework of the TRAPPIST Survey* (Université de Liège, Liège, 2016)

C. Opitom, C. Snodgrass, A. Fitzsimmons, E. Jehin, J. Manfroid, G. Tozzi, S. Faggi, M. Gillon, Ground-based monitoring of comet 67P/Churyumov–Gerasimenko gas activity throughout the Rosetta mission. Mon. Not. R. Astron. Soc. **469**(Suppl 2), S222–S229 (2017)

C. Opitom, D. Hutsemékers, E. Jehin, P. Rousselot, F. Pozuelos, J. Manfroid, Y. Moulane, M. Gillon, Z. Benkhaldoun, High resolution optical spectroscopy of the $N_2$-rich comet C/2016 R2 (PanSTARRS). Astron. Astrophys. **624**, A64 (2019)

T. Owen, P.R. Mahaffy, H.B. Niemann, S. Atreya, M. Wong, Protosolar nitrogen. Astrophys. J. **553**(1), L77–L79 (2001). https://doi.org/10.1086/320501

L. Paganini, M. Mumma, G. Villanueva, M. DiSanti, B. Bonev, M. Lippi, H. Boehnhardt, The chemical composition of CO-rich comet C/2009 P1 (Garradd) at Rh = 2.4 and 2.0 AU before perihelion. Astrophys. J. Lett. **748**(1), L13 (2012)

L. Paganini, M.J. Mumma, E.L. Gibb, G.L. Villanueva, Ground-based detection of deuterated water in comet C/2014 Q2 (Lovejoy) at IR wavelengths. Astrophys. J. Lett. **836**(2), L25 (2017). https://doi.org/10.3847/2041-8213/aa5cb3

J.A. Paquette, K. Hornung, O.J. Stenzel, J. Rynö, J. Silen, J. Kissel, M. Hilchenbach (the COSIMA team), The $^{34}S/^{32}S$ isotopic ratio measured in the dust of comet 67P/Churyumov–Gerasimenko by Rosetta/COSIMA. Mon. Not. R. Astron. Soc. **469**(Suppl 2), S230–S237 (2017). https://doi.org/10.1093/mnras/stx1623

J.A. Paquette, N. Fray, H. Cottin, C. Engrand, A. Bardyn, O.J. Stenzel, S. Merouane, M. Hilchenbach, C.M.O'D. Alexander, Y. Langevin, The D/H ratio in cometary dust measured by Rosetta/COSIMA, abstract #1982, in *49th Lunar and Planetary Science Conference* (2018a)

J.A. Paquette, C. Engrand, M. Hilchenbach, N. Fray, O.J. Stenzel, J. Silen, J. Rynö, J. Kissel, J. Silen (the COSIMA team), The oxygen isotopic composition ($^{18}O/^{16}O$) in the dust of comet 67P/Churyumov-Gerasimenko measured by COSIMA on-board Rosetta. Mon. Not. R. Astron. Soc. **477**(3), 3836–3844 (2018b). https://doi.org/10.1093/mnras/sty560

B. Parise, C. Ceccarelli, A. Tielens, A. Castets, E. Caux, B. Lefloch, S. Maret, Testing grain surface chemistry: a survey of deuterated formaldehyde and methanol in low-mass class 0 protostars. Astron. Astrophys. **453**(3), 949–958 (2006)

B. Parise, P. Bergman, F. Du, Detection of the hydroperoxyl radical $HO_2$ toward $\rho$ Ophiuchi A - additional constraints on the water chemical network. Astron. Astrophys. **541**, L11 (2012)

M. Pätzold et al., A homogeneous nucleus for comet 67P/Churyumov–Gerasimenko from its gravity field. Nature **530**(7588), 63–65 (2016). https://doi.org/10.1038/nature16535

R.O. Pepin, On the isotopic composition of primordial xenon in terrestrial planet atmospheres, in *From Dust to Terrestrial Planets: Proceedings of an ISSI Workshop*, ed. by W. Benz, R. Kallenbach, G.W. Lugmair 15–19 February 1999, Bern, Switzerland (Springer, Dordrecht, 2000), pp. 371–395. https://doi.org/10.1007/978-94-011-4146-8_24.

M.V. Persson, J.K. Jørgensen, E. van Dishoeck, D. Harsono, The deuterium fractionation of water on solar-system scales in deeply-embedded low-mass protostars. Astron. Astrophys. **563**, A74 (2014)

O. Poch et al., Ammonium salts are a reservoir of nitrogen on a cometary nucleus and possibly on some asteroids. Science **367**(6483), eaaw7462 (2020). https://doi.org/10.1126/science.aaw7462

D. Prialnik, Crystallization, sublimation, and gas release in the interior of a porous comet nucleus. Astrophys. J. Lett. **388**, 196–202 (1992)

D. Prialnik, M. Podolak, Radioactive heating of porous comet nuclei. Icarus **117**(2), 420–430 (1995)

D. Prialnik, J. Benkhoff, M. Podolak, Modeling the structure and activity of comet nuclei, in *Comets II* (2004), pp. 359–387

E. Quirico et al., Refractory and semi-volatile organics at the surface of comet 67P/Churyumov-Gerasimenko: Insights from the VIRTIS/Rosetta imaging spectrometer. Icarus **272**, 32–47 (2016). https://doi.org/10.1016/j.icarus.2016.02.028

A. Ratajczak, E. Quirico, A. Faure, B. Schmitt, C. Ceccarelli, Hydrogen/deuterium exchange in interstellar ice analogs. Astron. Astrophys. **496**(2), L21–L24 (2009)

T. Rauscher, A. Heger, R. Hoffman, S. Woosley, Nucleosynthesis in massive stars with improved nuclear and stellar physics. Astrophys. J. Lett. **576**(1), 323 (2002)

R. Reinhard, The Giotto encounter with comet Halley. Nature **321**(6067), 313–318 (1986). https://doi.org/10.1038/321313a0

H. Roberts, G. Fuller, T. Millar, J. Hatchell, J. Buckle, Molecular D/H ratios in the dense gas surrounding low-mass protostars. Planet. Space Sci. **50**(12–13), 1173–1178 (2002)

S. Rodgers, S. Charnley, Nitrogen isotopic fractionation of interstellar nitriles. Astrophys. J. Lett. **689**(2), 1448 (2008)

K. Ros, A. Johansen, Ice condensation as a planet formation mechanism. Astron. Astrophys. **552**, A137 (2013)

A. Rotundi et al., Dust measurements in the coma of comet 67P/Churyumov-Gerasimenko inbound to the Sun. Science **347**(6220), aaa3905 (2015). https://doi.org/10.1126/science.aaa3905

E. Roueff, S. Tiné, L. Coudert, G. Pineau des Forêts, E. Falgarone, M. Gerin, Detection of doubly deuterated ammonia in L134N. Astron. Astrophys. **354**, L63–L66 (2000)

P. Rousselot et al., Toward a unique nitrogen isotopic ratio in cometary ices. Astrophys. J. Lett. **780**(2), L17 (2014). https://doi.org/10.1088/2041-8205/780/2/L17

M. Rubin et al., Molecular nitrogen in comet 67P/Churyumov-Gerasimenko indicates a low formation temperature. Science **348**(6231), 232–235 (2015a). https://doi.org/10.1126/science.aaa6100

M. Rubin, K. Altwegg, E.F. van Dishoeck, G. Schwehm, Molecular oxygen in Oort cloud comet 1P/Halley. Astrophys. J. Lett. **815**(1), L11 (2015b). https://doi.org/10.1088/2041-8205/815/1/L11

M. Rubin et al., Evidence for depletion of heavy silicon isotopes at comet 67P/Churyumov-Gerasimenko. Astron. Astrophys. **601**(A123), 121–129 (2017). https://doi.org/10.1051/0004-6361/201730584

M. Rubin et al., Krypton isotopes and noble gas abundances in the coma of comet 67P/Churyumov-Gerasimenko. Sci. Adv. **4**(7), eaar6297 (2018). https://doi.org/10.1126/sciadv.aar6297

M. Rubin et al., Elemental and molecular abundances in comet 67P/Churyumov-Gerasimenko. Mon. Not. R. Astron. Soc. **489**(1), 594–607 (2019a). https://doi.org/10.1093/mnras/stz2086

M. Rubin, D.V. Bekaert, M.W. Broadley, M.N. Drozdovskaya, S.F. Wampfler, Volatile species in comet 67P/Churyumov-Gerasimenko: investigating the link from the ISM to the terrestrial planets. ACS Earth Space Chem. **3**(9), 1792–1811 (2019b). https://doi.org/10.1021/acsearthspacechem.9b00096

D. Ruffle, T. Hartquist, P. Caselli, D. Williams, The sulphur depletion problem. Mon. Not. R. Astron. Soc. **306**(3), 691–695 (1999)

N. Sakamoto, Y. Seto, S. Itoh, K. Kuramoto, K. Fujino, K. Nagashima, A.N. Krot, H. Yurimoto, Remnants of the early solar system water enriched in heavy oxygen isotopes. Science **317**(5835), 231–233 (2007)

D.D. Sasselov, M. Lecar, On the snow line in dusty protoplanetary disks. Astrophys. J. **528**(2), 995–998 (2000). https://doi.org/10.1086/308209

I.R.H.G. Schroeder et al., The $^{16}O/^{18}O$ ratio in water in the coma of comet 67P/Churyumov-Gerasimenko measured with the Rosetta/ROSINA double-focusing mass spectrometer. Astron. Astrophys. (2018). https://doi.org/10.1051/0004-6361/201833806

I.R.H.G. Schroeder, K. Altwegg, H. Balsiger, J.-J. Berthelier, M.R. Combi, J. De Keyser, B. Fiethe, S.A. Fuselier, T.I. Gombosi, K.C. Hansen, A comparison between the two lobes of comet 67P/Churyumov–Gerasimenko based on D/H ratios in $H_2O$ measured with the Rosetta/ROSINA DFMS. Mon. Not. R. Astron. Soc. **489**(4), 4734–4740 (2019)

M. Schuhmann et al., Aliphatic and aromatic hydrocarbons in comet 67P/Churyumov-Gerasimenko seen by ROSINA. Astron. Astrophys. (2019). https://doi.org/10.1051/0004-6361/201834666

R. Schulz, J. Stüwe, H. Boehnhardt, Rosetta target comet 67P/Churyumov-Gerasimenko-postperihelion gas and dust production rates. Astron. Astrophys. **422**(1), L19–L21 (2004)

S.R. Schwartz, P. Michel, M. Jutzi, S. Marchi, Y. Zhang, D.C. Richardson, Catastrophic disruptions as the origin of bilobate comets. Nat. Astron. (2018). https://doi.org/10.1038/s41550-018-0395-2

S. Seager, W. Bains, J.J. Petkowski, Toward a list of molecules as potential biosignature gases for the search for life on exoplanets and applications to terrestrial biochemistry. Astrobiology **16**(6), 465–485 (2016)

R.Y. Shah, A. Wootten, Deuterated ammonia in galactic protostellar cores. Astrophys. J. Lett. **554**(2), 933 (2001)

Y. Shinnaka, H. Kawakita, H. Kobayashi, M. Nagashima, D.C. Boice, $^{14}NH_2/^{15}NH_2$ ratio in comet C/2012 S1 (ISON) Observed during its Outburst in 2013 November. Astrophys. J. Lett. **782**(2), L16, (2014).

F.H. Shu, H. Shang, A.E. Glassgold, T. Lee, X-rays and fluctuating X-winds from protostars. Science **277**(5331), 1475–1479 (1997)

H. Sierks et al., On the nucleus structure and activity of comet 67P/Churyumov-Gerasimenko. Science **347**(6220), aaa1044 (2015). https://doi.org/10.1126/science.aaa1044

C. Snodgrass et al., The perihelion activity of comet 67P/Churyumov–Gerasimenko as seen by robotic telescopes. Mon. Not. R. Astron. Soc. **462**(Suppl 1), S138–S145 (2016). https://doi.org/10.1093/mnras/stw2300

C. Snodgrass, M.F. A'Hearn, F. Aceituno, V. Afanasiev, S. Bagnulo, J. Bauer, G. Bergond, S. Besse, N. Biver, D. Bodewits, The 67P/Churyumov–Gerasimenko observation campaign in support of the Rosetta mission. Philos. Trans. R. Soc. A, Math. Phys. Eng. Sci. **375**(2097), 20160249 (2017)

L.A. Soderblom et al., Observations of comet 19P/Borrelly by the miniature integrated camera and spectrometer aboard deep space 1. Science **296**(5570), 1087 (2002)

O.J. Stenzel, M. Hilchenbach, S. Merouane, J. Paquette, K. Varmuza, C. Engrand, F. Brandstätter, C. Koeberl, L. Ferrière, P. Filzmoser, Similarities in element content between comet 67P/Churyumov–Gerasimenko coma dust and selected meteorite samples. Mon. Not. R. Astron. Soc. **469**(Suppl 2), S492–S505 (2017)

S.A. Stern et al., Alice: the Rosetta ultraviolet imaging spectrograph. Space Sci. Rev. **128**(1–4), 507–527 (2007). https://doi.org/10.1007/s11214-006-9035-8

J.M. Sunshine, N. Thomas, M.R. El-Maarry, T.L. Farnham, Evidence for geologic processes on comets. J. Geophys. Res., Planets **121**(11), 2194–2210 (2016)

M.D. Suttle, L. Folco, M.J. Genge, S.S. Russell, Flying too close to the Sun – the viability of perihelion-induced aqueous alteration on periodic comets. Icarus **351**, 113956, (2020). https://doi.org/10.1016/j.icarus.2020.113956

S. Tachibana, G.R. Huss, Sulfur isotope composition of putative primary troilite in chondrules from Bishunpur and Semarkona. Geochim. Cosmochim. Acta **69**(12), 3075–3097 (2005)

V. Taquet, K. Furuya, C. Walsh, E.F. van Dishoeck, A primordial origin for molecular oxygen in comets: a chemical kinetics study of the formation and survival of $O_2$ ice from clouds to discs. Mon. Not. R. Astron. Soc. **462**(Suppl), S99–S115 (2016)

V. Taquet, K. Furuya, C. Walsh, E.F. van Dishoeck, On the origin of $O_2$ and other volatile species in comets, in *IAU Symposium No. 332*, (2017). https://doi.org/10.1017/S1743921317007414

V. Taquet, E.F. van Dishoeck, M. Swayne, D. Harsono, J.K. Jørgensen, L. Maud, N.F. Ligterink, H.S. Müller, C. Codella, K. Altwegg, A. Bieler, A. Coutens, M.N. Drozdovskaya, M.V. Furuya, K. Persson, M.L.R. van't Hoff, C. Walsh, S.F. Wampfler, Linking interstellar and cometary $O_2$: a deep search for $^{16}O^{18}O$ in the solar-type protostar IRAS 16293–2422. Astron. Astrophys. **618**, A11 (2018). https://doi.org/10.1051/0004-6361/201833175

B. Teolis, M. Loeffler, U. Raut, M. Fama, R. Baragiola, Ozone synthesis on the icy satellites. Astrophys. J. Lett. **644**(2), L141 (2006)

M.S. Tiscareno, R. Malhotra, The dynamics of known Centaurs. Astron. J. **126**(6), 3122 (2003)

F. Van der Tak, P. Schilke, H. Müller, D. Lis, T. Phillips, M. Gerin, E. Roueff, Triply deuterated ammonia in NGC 1333. Astron. Astrophys. **388**(3), L53–L56 (2002)

E.F. van Dishoeck, G.A. Blake, D.J. Jansen, T. Groesbeck, Molecular abundances and low mass star formation II. Organic and deuterated species towards IRAS 16293-2422. Astrophys. J. **447**, 760–782 (1995)

J.-B. Vincent, M.F. A'Hearn, Z.-Y. Lin, M.R. El-Maarry, M. Pajola, H. Sierks, C. Barbieri, P.L. Lamy, R. Rodrigo, D. Koschny, Summer fireworks on comet 67P. Mon. Not. R. Astron. Soc. **462**(Suppl), S184–S194 (2016)

D. Vinković, Radiation-pressure mixing of large dust grains in protoplanetary disks. Nature **459**(7244), 227 (2009)

S. Weidenschilling, Aerodynamics of solid bodies in the solar nebula. Mon. Not. R. Astron. Soc. **180**(2), 57–70 (1977)

S. Weidenschilling, The origin of comets in the solar nebula: a unified model. Icarus **127**(2), 290–306 (1997)

P.R. Weissman, Are cometary nuclei primordial rubble piles? Nature **320**(6059), 242–244 (1986)

P. Weissman, A. Morbidelli, B. Davidsson, J. Blum, Origin and evolution of cometary nuclei. Space Sci. Rev. **216**(1), 6 (2020). https://doi.org/10.1007/s11214-019-0625-7

F.L. Whipple, A comet model. I. The acceleration of comet Encke. Astrophys. J. Lett. **111**, 375–394 (1950)

F.L. Whipple, A comet model. II. Physical relations for comets and meteors. Astrophys. J. Lett. **113**, 464–474 (1951)

D.H. Wooden, D.E. Harker, C.E. Woodward, H.M. Butner, C. Koike, F.C. Witteborn, C.W. McMurtry, Silicate mineralogy of the dust in the inner coma of comet C/1995 01 (Hale-Bopp) pre-and postperihelion. Astrophys. J. Lett. **517**(2), 1034 (1999)

D.H. Wooden, C.E. Woodward, D.E. Harker, Discovery of crystalline silicates in comet C/2001 Q4 (NEAT). Astrophys. J. Lett. **612**(1), L77 (2004)

P.M. Woods, A. Occhiogrosso, S. Viti, Z. Kanuchova, M.E. Palumbo, S.D. Price, A new study of an old sink of sulphur in hot molecular cores: the sulphur residue. Mon. Not. R. Astron. Soc. **450**(2), 1256–1267 (2015). https://doi.org/10.1093/mnras/stv652

I.P. Wright et al., Ptolemy – an instrument to measure stable isotopic ratios of key volatiles on a cometary nucleus. Space Sci. Rev. **128**(1), 363–381 (2007). https://doi.org/10.1007/s11214-006-9001-5

I.P. Wright, S. Sheridan, S.J. Barber, G.H. Morgan, D.J. Andrews, A.D. Morse, CHO-bearing organic compounds at the surface of 67P/Churyumov-Gerasimenko revealed by Ptolemy. Science **349**(6247), aab0673 (2015). https://doi.org/10.1126/science.aab0673

P. Wurz et al., Solar wind sputtering of dust on the surface of 67P/Churyumov-Gerasimenko. Astron. Astrophys. **583**(A22), 21–29 (2015). https://doi.org/10.1051/0004-6361/201525980

H. Yabuta et al., Formation of an ultracarbonaceous Antarctic micrometeorite through minimal aqueous alteration in a small porous icy body. Geochim. Cosmochim. Acta **214**, 172–190 (2017). https://doi.org/10.1016/j.gca.2017.06.047
Y. Yao, K.P. Giapis, Dynamic molecular oxygen production in cometary comae. Nat. Commun. **8**(15298), 15291–15298 (2017). https://doi.org/10.1038/ncomms15298
A.N. Youdin, J. Goodman, Streaming instabilities in protoplanetary disks. Astrophys. J. Lett. **620**(1), 459 (2005)
H. Yurimoto, K. Kuramoto, Molecular cloud origin for the oxygen isotope heterogeneity in the solar system. Science **305**(5691), 1763–1766 (2004)
W. Zheng, D. Jewitt, R.I. Kaiser, Formation of hydrogen, oxygen, and hydrogen peroxide in electron-irradiated crystalline water ice. Astrophys. J. Lett. **639**(1), 534 (2006)
M.E. Zolensky et al., Mineralogy and petrology of comet 81P/Wild 2 nucleus samples. Science **314**(5806), 1735–1739 (2006). https://doi.org/10.1126/science.1135842

Space Sci Rev (2020) 216:6
https://doi.org/10.1007/s11214-019-0625-7

# Origin and Evolution of Cometary Nuclei

**Paul Weissman[1] · Alessandro Morbidelli[2] · Björn Davidsson[3] · Jürgen Blum[4]**

Received: 8 January 2019 / Accepted: 27 November 2019 / Published online: 14 January 2020


**Abstract** One of the key goals of the Rosetta mission was to understand how, where and when comets formed in our solar system. There are two major hypotheses for the origin of comets, both pre-Rosetta: (1) hierarchical accretion of dust and ice grains in the Solar Nebula and (2) the growth of pebbles, which are then brought together by streaming instabilities in the Solar Nebula to form larger bodies. Rosetta provided a wealth of new information on comet nuclei and confirmed many past ideas on comets, e.g., high volatile content, lack of aqueous alteration of grains, and the low bulk density of the nucleus. Rosetta also provided new data on the nature of cometary activity, the active geology on the nucleus surface and the interior structure and bulk density of the nucleus. Supporters of the above-mentioned origin hypotheses each find confirmation of their ideas in the Rosetta results. But the question of which hypothesis is preferred, or if there are other, better hypotheses that could be invoked, could not be answered. Theoretical studies suggest that comet nuclei were collisionally processed in the Primordial Disk though it is not clear that the nuclei we see today display the effects of that process. Both theoretical and observational studies suggest that the major end-states for cometary nuclei are dynamical ejection, random disruption and disintegration, and/or evolution of nuclei to inactive, asteroidal-appearing objects. Rosetta has provided us with many new insights that will help to guide future cometary missions, observations, experiments and theoretical investigations that will lead to answers to the fundamental questions with regard to cometary origin.

**Keywords** Comet, origin · Hierarchical accretion · Agglomeration · Pebbles · Streaming instabilities · Evolution · End-states · Primordial disk · Collisional evolution

---

Comets: Post 67P / Churyumov-Gerasimenko Perspectives
Edited by Nicolas Thomas, Björn Davidsson, Laurent Jorda, Ekkehard Kührt, Raphael Marschall, Colin Snodgrass and Rafael Rodrigo

---

✉ P. Weissman

[1] Planetary Science Institute, Tucson, USA

[2] Nice Observatory, Nice, France

[3] Jet Propulsion Laboratory, California Institute of Technology, Pasadena, USA

[4] Institute for Geophysics and Extraterrestrial Physics, Technische Universität, Braunschweig, Germany

## 1 Introduction

Comets are among the most primitive bodies in the solar system. They likely formed at large solar distances where volatile ices could condense and remain stable, and they retain a cosmo-chemical record of the composition of, and the processes in, the Solar Nebula when the planets were forming. Comets have been stored in cold, distant orbits, either in the Kuiper belt region (Kuiper 1951) at ∼35–100 AU, where the mean temperature is ∼40 K, or in the Oort cloud at ∼3–150 × $10^3$ AU (Oort 1950) where the mean temperature is ∼10 K. Note that the Kuiper belt region contains two distinct dynamical populations. The first is the Classical Kuiper belt at ∼35–100 AU, with objects in low eccentricity, low inclination orbits beyond Neptune (Duncan et al. 1988). The second is the Scattered disk at ∼30–∼$10^3$ AU (Duncan and Levison 1997; Levison and Duncan 1997) consisting of objects with perihelia close to Neptune's orbit but with larger orbital eccentricities, inclinations and aphelia than the Classical Kuiper belt. The Scattered disk is the source of most of the Jupiter-family comets (JFCs).

The primitive and cold nature of cometary nuclei is confirmed by the highly volatile ices that are released when the nuclei come close to the Sun. These include: CO ice with a sublimation point, $T = 25$ K; $CH_3OH$ ice, $T = 99$ K; and $CO_2$ ice, $T = 72$ K (corresponding to a saturation density of $10^{13}$ $cm^{-3}$; Yamamoto 1985).

Unlike earlier flyby missions, ESA's Rosetta mission (Glassmeier et al. 2007) was designed to be the first truly intensive space mission to a comet, following it from ∼3.6 AU pre-perihelion through perihelion at 1.24 AU and out to ∼4 AU post-perihelion—over a period of more than two years. The target comet, 67P/Churyumov-Gerasimenko (hereafter, 67P), is a typical Jupiter-family comet, though it proved far more interesting than anyone had expected. The mission yielded a wealth of data that is only now beginning to be fully digested and analyzed.

One of the key questions that Rosetta wanted to answer is how, where and when did comets form? The known presence of substantial volatile ices in cometary nuclei showed that the formation zone(s) had to be, at a minimum, moderately far from the Sun, beyond the so-called "snowline" where ices would be stable. In the present solar system that distance is taken to be 3–5 AU but in the dusty Solar Nebula it may have been closer to (or farther from) the Sun (Sasselov and Lecar 2000; Min et al. 2011).

For a long time the generally accepted hypothesis for comet (and asteroid and planet) formation was hierarchical accretion (Weidenschilling 1977, 1997). Grains of micron-sized dust and ice were expected to collide in the Solar Nebula and stick, leading to larger and larger particles. This led Donn and Hughes (1986) to propose the *fractal aggregate* model in which hierarchical accretion continued to occur at all size scales, leading to an irregular nucleus perhaps kilometers in diameter. A similar model was proposed by Weissman (1986), based more on phenomenological observations of comets, such as outbursts and splitting events, the *primordial rubble pile*. In Weissman's model the final stages of accretion led to kilometer-sized nuclei but there was no energy sources to mold this primitive material into a single coherent body. However, studies of hierarchical accretion had problems in growing particles larger than ∼1 meter (Weidenschilling 1977, 1997). Particles might begin to bounce off one another, or to be collisionally eroded and disrupted. How could proto-nuclei grow beyond this size range?

In 2005 Youdin and Goodman proposed a different scenario. They argued that hierarchical accretion would lead to dust aggregates on the order of 1 cm in size; these particles have come to be known as "pebbles." Streaming (hydrodynamic) instabilities in the Solar Nebula would bring the pebbles together until they gravitationally coalesced into larger bodies on the order of 100 km in diameter. However, this result was also problematic because such large bodies would be warmed internally by short-lived radio-nuclides such as $^{26}$Al, leading to temperatures that would have driven off many of the more volatile ices seen in comets (Prialnik et al. 1987; Merk and Prialnik 2003). Also, internal pressures in 100-km nuclei would lead to compaction and possible destruction of the micron-sized agglomerates collected by Rosetta (Merouane et al. 2016) and seen in Interplanetary Dust Particles (IDPs, Brownlee 1985), which are believed to have a cometary origin.

In this paper we will review the arguments for and against the two leading formation mechanisms and discuss what new information the Rosetta results have brought to this issue. In Sect. 2 we review hierarchical accretion while in Sect. 3 we cover accretion of pebbles in streaming instabilities. Section 4 presents a discussion of collisional processes in the Solar Nebula and the early solar system. Sections 5 and 6 are brief discussions of the evolution of cometary nuclei and the likely end-states for cometary nuclei, respectively. Finally, in Sect. 7 we discuss all these issues and suggest ways that they can be addressed by future space missions to comets, as well as by ground-based observations and laboratory and theoretical studies.

## 2 Comet Nucleus Formation Through Hierarchical Agglomeration

In this section we describe the dynamical aspects of hierarchical agglomeration (Sect. 2.1), the internal structure expected for bodies formed by hierarchical agglomeration (Sect. 2.2), and how these expectations compare with observations of 67P and other comets (Sect. 2.3).

### 2.1 Dynamics of Hierarchical Agglomeration

Hierarchical agglomeration is the gradual merging of smaller bodies (cometesimals) through gentle hit-and-stick collisions where gravity has little to no influence on the formation of bodies smaller than ∼10 km (Weidenschilling 1997). At larger sizes gravity-assisted accretion becomes increasingly important. In order for collisions to occur, bodies of different sizes need non-zero relative velocities with respect to each other. In the Solar Nebula this is caused by gas drag which promotes mergers between bodies that differ greatly in size (e.g., Adachi et al. 1976; Weidenschilling 1977). In the Primordial Disk, the collection of planetesimals left behind when the gas in the Solar Nebula is dissipated after ∼3 Myr (Zuckerman et al. 1995; Haisch et al. 2001; Sicilia-Aguilar et al. 2006), this is caused by viscous stirring (excitation of low-mass body orbits by more massive bodies), which allows for mergers between bodies of any size ratio. This means that hierarchical growth changes character as the Solar Nebula evolves into the Primordial Disk.

Detailed accounts for the dynamics of hierarchical agglomeration are found in, e.g., Weidenschillling (1977, 1997), Kenyon and Luu (1998), and Windmark et al. (2012a, 2012b). Davidsson et al. (2016) proposed a two-step hybrid scenario where streaming instabilities first create pebble swarms that collapse gravitationally to form $D > 50$ km bodies (see Sect. 3). Since this population would later grow into Pluto-sized objects they were referred to as "TNOs". When the number density of remaining pebbles falls below ∼15% of the initial value the pebble swarms can no longer reach the Roche density needed for gravitational collapse. This growth phase lasts $\sim 10^3$ yr (Johansen et al. 2007). At that point a second growth mode sets in where small comet nuclei are built slowly through hierarchical agglomeration. The slow growth means that heat produced by $^{26}$Al is dissipated from these small bodies without modifying the composition or the physical structure of the forming

**Table 1** Growth time scales, target/projectile diameters, and relative velocities in hierarchical growth

| Time $t$ (Myr) | $D$ (m) | $d$ (m) | $v_{rel}$ (m $s^{-1}$) |
|---|---|---|---|
| 0.1 | 0.1 | $4 \times 10^{-6}$ | 10 |
| 0.3–0.6 | 1 | 0.01–0.1 | 30 |
| 0.35–0.7 | 7 | 2–3 | 2 |
| 0.4–0.8 | 70 | 20 | 0.8 |
| 1.2–2.4 | 500 | 200 | 0.2 |
| 1.7–3.6 | 6,000 | 1,000 | 2 |

cometesimals. Because the bodies quickly grow larger than the ~1 cm diurnal thermal skin depth, into 10–100 m bodies, most of the ice is kept at substantially lower temperature than experienced by pebbles, facilitating the preservation of volatiles. Growth to the final size would have to proceed at a much slower rate to avoid radiogenic heating.

Davidsson et al. (2016) proposed the time scale of growth shown in Table 1, where $D$ is the diameter of bodies being formed, $d$ is the diameter of the smaller bodies being primarily consumed, and $v_{rel}$ is the relative velocity during collisions, taken from Weidenschilling (1997). Here $t = 0$ is the formation time of CAIs (calcium-aluminum inclusions), taken to coincide with the transition from a Class 0 to Class I protostar; the protosun would reach its final mass and the Solar Nebula (last remains of the accretion disk) would be born at $t \sim 10^5$ yr (Andre and Montmerle 1994; Montmerle et al. 2006). The ranges in $t$ reflect differences in growth rates at 15 AU (faster) and 30 AU (slower).

An important feature of Table 1 is the peak in the relative velocity, $v_{rel} = 30$ m $s^{-1}$, reached when growing bodies are a few meters in diameter. This is because meter-sized bodies, which have very slow radial drift velocity at 15–30 AU, primarily consume cm-dm sized bodies that have the highest radial drift velocities at these heliocentric distances (Weidenschilling 1997). It is therefore expected that the highest level of compaction (porosity ~60%) and therefore the strongest and most resilient structures in these nuclei have diameters of a few meters. Also, note that $d \ll D$ which means that these are single-lobe objects. The expected bulk density is in the range 200–400 kg $m^{-3}$, while the meter-sized compact building blocks have densities of 300–700 kg $m^{-3}$ (Davidsson et al. 2016).

At the end of the Solar Nebula lifetime the vast majority of hierarchically grown objects are small with $D = 0.1$–1 km. However, growth continues in the Primordial Disk under new conditions. Davidsson et al. (2016) estimate that the average $v_{rel}$ increases from 2 m $s^{-1}$ to 40 m $s^{-1}$ in the first 25 Myr because of viscous stirring, and it is not uncommon that similarly sized nuclei collide. This is therefore an era of adding a more compact outer shell to each body, pushing the bulk densities towards ~500 kg $m^{-3}$, and merging objects into bilobate nuclei. After 25 Myr have passed the largest hierarchically grown nuclei have $D \sim 50$ km and because of mergers and accretion onto the population of large previously formed TNOs the number density of comet nuclei is small enough to avoid substantial collisional processing even when $v_{rel}$ becomes destructive. However, if the disk mass was higher than assumed by Davidsson et al. (2016) the initiation of collisional grinding may be unavoidable.

## 2.2 The Internal Structure of Hierarchically Grown Bodies

What does a body grown through hierarchical agglomeration look like? The low $v_{rel}$ in Table 1 led some researchers to suggest that they would be highly porous structures with large internal voids, such as the *fractal aggregate model* of cometary nuclei proposed by Donn and Hughes (1986). The *primordial rubble pile model* of Weissman (1986) emphasized local melting and weak bonds at contact surfaces between the building blocks and was illustrated as a non-fractal yet porous and irregular object. See Weissman (1990) for graphical representations (also reproduced in Donn (1991)).

The images of Comet 19P/Borrelly obtained with Deep Space 1 seemed to confirm the primordial rubble-pile concept: "the shape of and structures in the nucleus are proof of its formation by coalescence of multiple bodies," (Soderblom et al. 2002). The discovery of steep walls on 81P/Wild 2 imaged by Stardust, which requires some nucleus cohesion, led Weaver (2004) to question a rubble-pile structure that he (incorrectly) thought of as completely cohesionless. The discovery of pervasive layering on Comet 9P/Tempel 1 in Deep Impact images (Thomas et al. 2007) inspired new concepts and ideas.

Lasue et al. (2009, 2011) performed numerical simulations where $5 \times 10^4$ hard spheres (monodisperse or with a size distribution) were allowed to collide, aggregate through sticking, form even stronger bonds over time through sintering, and break apart during collisions or compaction that leads to restructuring. In their view, layering occurs since the cohesive strength distribution becomes organized over time, forming a strong core and gradually weaker shells at larger distance from the center. However, Lasue et al.'s results largely reflect the input assumptions of their model, and the results of their simulations may not be useful. Belton et al. (2007) proposed an alternative view, where layers represent flattened cometesimals that became strongly deformed at impact—this is the *layered pile model* or the "talps" model (the word "splat" spelled backwards).

Davidsson et al. (2016) speculated that the rather high $v_{rel}$ during final accretion in their scenario would break the projectiles and release their most cohesive meter-sized building blocks, which would stand some chance of remaining intact. These small constituents could spread out locally and form a layer of finite size, analogous to the talps model. However, if bouncing takes place (illustrated by the jumping Philae lander) one can imagine that a global "particle atmosphere" forms during a few hours after the impact, before settling to form a global layer.

If so, the interior of comets could be dominated by heterogeneity on the meter-scale. The final merger of the lobes would not result in such disintegration because the impact energy would be dissipated in the deformation zone at the contact area that is small compared to the size of the lobes. de Niem et al. (2018) find that equal-sized planetesimals at kilometer size-scales can hit and stick at velocities as high as $30\ \mathrm{m\,s^{-1}}$. This would not be the case when a small projectile impacts a much larger body. Note that meter-sized heterogeneities in the interior of 67P could not have been detected by the CONSERT radar experiment on Rosetta/Philae because the radar wavelength was 3 meters.

The biggest challenge lying ahead is to model the bodies formed by hierarchical agglomeration realistically in order to produce qualitative predictions that can be compared with observations. The Smoothed Particle Hydrodynamics code applied by Jutzi and Asphaug (2015) and the Backward Euler hydrocode used by de Niem et al. (2018), that can consider low-velocity collisions between highly porous bodies with realistic material parameters, are steps in the right direction. However, such codes need to consider *multiple consecutive* collisions between objects resolved down to the meter-scale at relevant velocities in order to produce model nuclei that better represent hierarchically grown objects.

### 2.3 Is Hierarchical Agglomeration Consistent with Observed Properties of Comet Nuclei?

Hierarchical agglomeration predicts continuous comet nucleus growth during $\sim$25 Myr (Davidsson et al. 2016). Such long time scales could be consistent with the substantial fraction of crystalline silicates in comets (Kelley and Wooden 2009) originating from the inner

Solar System and gradually mixed out to the comet-forming region during the Solar Nebula lifetime. The presence of chondrule fragments in Stardust samples from 81P/Wild 2 (Nakamura et al. 2008), of which some have an age of at least 3 Myr (Ogliore et al. 2012), is consistent with slow growth. Furthermore, the lack of phyllosilicates in 81P/Wild 2 samples (e.g., Brownlee et al. 2012), the lack of the phyllosilicate 0.7 μm absorption in 67P (except possibly in a handful of small boulders (Oklay et al. 2016), and the absence of amino acids other than glycine in 67P (Altwegg et al. 2016) suggest that these comets did not contain liquid water. If $^{26}Al$, with a half life of $7 \times 10^5$ years (Norris et al. 1983), was present at large heliocentric distances this means that these comets formed sufficiently late not to become aqueously altered.

The bilobate shapes of 8P/Tuttle (Harmon et al. 2010), 19P/Borrelly (Oberst et al. 2004), 103P/Hartley 2 (A'Hearn et al. 2011), 67P/Churyumov-Gerasimenko (Sierks et al. 2015), and possibly 1P/Halley (Merenyi et al. 1990; Keller et al. 2004) show that low-velocity mergers between similarly-sized cometesimals are common. Although such mergers would not be common in the drag-dominated Solar Nebula (Weidenschilling 1997), they would be common in the gently viscously stirred Primordial Disk (Davidsson et al. 2016). Such nucleus shapes are therefore consistent with hierarchical agglomeration.

The scenario proposed by Davidsson et al. (2016) suggests a higher core porosity (low-velocity growth in the Solar Nebula) and a lower shell porosity (growth in the Primordial Disk at higher velocities) for each lobe. This is consistent with CONSERT observations indicating a 15% decrease of porosity over the top 150 meters in the small lobe (Ciarletti et al. 2015). The low bulk density of 67P, $\rho = 533 \pm 6\ \mathrm{kg\,m^{-3}}$, (Jorda et al. 2016; Pätzold et al. 2016) is consistent with the low comet bulk densities obtained primarily from nongravitational force modeling (e.g., Rickman 1986; Rickman et al. 1987; Weissman et al. 2004; Davidsson and Gutiérrez 2004, 2005, 2006; Davidsson et al. 2007; Richardson et al. 2007; Sosa and Fernández 2009), and the high level of porosity this implies is consistent with low-velocity mergers.

Cometesimals with $D = 2$ and 3 km (similar to the small and large lobes of 67P) are expected to be built by objects with diameters in the 0.3–0.6 km and 0.5–1 km ranges, respectively (Weidenschilling 1997). Structures in the Bastet region on the small lobe shaped as spherical caps with 320–480 m diameter called Positive Relief Features are possible examples of such cometesimals (Davidsson et al. 2016). According to CONSERT observations reported by Kofman et al. (2015): "Two or three well-defined propagation paths could indeed be potentially due to the presence of a large structure inside the nucleus." In addition, the 500-m size "accumulation basins" in Imhotep on the large lobe (particularly F, see Fig. 2 of Auger et al. 2015) with flat floors and raised rims are not high-velocity impact craters. They could be the scars of low-velocity impacts of objects roughly the size of the features. Simulations #18 and #19 by de Niem et al. (2018), where highly porous 1 km objects impact 3 km targets at low velocity (5–13 $\mathrm{m\,s^{-1}}$) appear to show similar morphologies. Or they could be sublimation basins where the walls are active and slowly grow outward as a non-volatile residue is left to form the smooth flat floors of these features.

The widespread layering on 67P (Massironi et al. 2015; Penasa et al. 2017), revealed by intricate systems of terraces carved by sublimation, suggest that the lobes were formed separately and that some sort of accretion process that acted over time was active during their formation. The discovery that the reflectance of the layers decreases systematically with increasing distance from the lobe center (Ferrari et al. 2018) could be the result of gradual accretion of material that changed compositional and/or textural properties over time. These properties are at least conceptually consistent with hierarchical agglomeration (Lasue et al. 2009, 2011; Belton et al. 2007; Davidsson et al. 2016), though we emphasize again that this process needs to be modeled in detail to be fully understood.

Finally, the consolidated terrains on the nucleus of 67P, including walls of pits that extend $\sim$100 m into the nucleus, are rich in small spheroidal structures called "goose-bumps" (Sierks et al. 2015). Having a typical diameter of $2.5 \pm 1$ m, they were interpreted by Davidsson et al. (2016) as the most cohesive structures formed during hierarchical agglomeration. However, this interpretation is controversial and it has been suggested (El-Maarry et al. 2015; Auger et al. 2018) that they are thermal contraction crack polygons further evolved by sublimation. The two views may coexist in case fractures follow pre-existing weak planes in the nucleus being the result of a primordial, meter-scale nucleus heterogeneity. A crucially important question is if these goose-bumps exist at depth or exclusively as a surface phenomenon. CONSERT measurements of signal broadening because of volume scattering are consistent with structures with diameters up to 9 meters as long as the dielectric constant contrast between boulders and matrix is at most $\Delta\varepsilon \leq 0.25$ (Ciarletti et al. 2017a). This is consistent with a boulder porosity of 60%, as predicted by Davidsson et al. (2016), as long as the surrounding material has a porosity of 95% (Ciarletti et al. 2017a).

Taken together, the properties of hierarchically grown comet nuclei (as predicted by direct numerical modeling of the process with codes accounting realistically for material behavior in the limits of high porosity and low collision velocity), as well as measured key properties of comet nuclei (such as the internal structure) are yet not known well enough to allow for a direct comparison. At this point it is not possible to prove or refute the hypothesis that comet nuclei grew through hierarchical agglomeration—however, the things we think we know about this process at least appear consistent with the things we think we know about comet nuclei.

## 3 Formation and Accretion of "Pebbles"

In this section we describe a formation model of planetesimals in which the protoplanetary dust grains first form mm- to cm-sized aggregates ("pebbles"), followed by pebble concentration and finally the gravitational collapse of the pebble cloud. Rosetta has delivered a wealth of new information about comet 67P, and we will utilize this information to test whether the gravitational-collapse model can describe its properties.

### 3.1 The Formation of Dust Aggregates

The physical processes that lead to the growth of aggregates from the initial (sub-) μm-sized dust and ice grains have been extensively described in the literature (Weidenschilling 1977; Dominik and Tielens 1997; Blum et al. 2000; Blum and Wurm 2008; Wada et al. 2008, 2009; Güttler et al. 2010; Schräpler et al. 2018). When dispersed in the gas of the protoplanetary disk, dust grains experience a variety of relative motions to the gas, e.g. Brownian motion, radial, azimuthal and vertical drift motions, or decoupling from turbulent eddies. Thus, the dust particles collide frequently.

If the binding energies of the dust particles experiencing a collision are strong enough, the kinetic collision energy can be dissipated and the particles stick together (Dominik and Tielens 1997; Wada et al. 2008, 2009; Güttler et al. 2010), thus forming loose aggregates. It has been shown numerically as well as experimentally that the initial stage of dust-aggregate growth is dominated by the formation of fractal aggregates (see Blum 2006 for details).

Due to the increasing collision energy with increasing dust-aggregate mass, collisional compaction occurs when the rolling threshold is overcome and results in the formation of non-fractal, but porous aggregates. For non-fractal dust aggregates collision speeds typically increase with increasing dust-aggregate mass, due to a size-dependent decoupling from the gas motion (Weidenschilling 1977). Numerical studies suggest that growth stops when the sticking threshold of the dust aggregates is reached, whereupon the particles are compacted in bouncing collisions (Weidling et al. 2009; Zsom et al. 2010; Lorek et al. 2018).

The growth timescale, $\tau_g$, until the bouncing barrier is hit, is on the order of a few hundred to a few thousand orbital periods (Zsom et al. 2010; Lorek et al. 2018). The final aggregate size depends on the nebula model (Zsom et al. 2010) as well as on the monomer-grain properties (size, material), the heliocentric distance, the strength of turbulence and details of the collision physics, and typically falls into the 1 mm to 1 cm range (Lorek et al. 2018). These dust aggregates have been termed "pebbles".

Growth to larger sizes might be possible at the iceline or beyond as suggested by Ros and Johansen (2013). However, laboratory experiments suggest that any further growth well beyond pebble sizes is impeded by a variety of processes, among which the most prominent are bouncing, fragmentation and erosion. A recent study by Schräpler et al. (2018) has revealed that erosion alone limits the maximum achievable pebble size to ~10 cm, even under the most favorable growth conditions.

Growth to pebble sizes has been observed in protoplanetary disks, with a tendency toward decreasing maximum pebble size with increasing distance from the central star. However, this result may be observationally incomplete for meter-sized bodies or larger, i.e., larger objects could not be detected by the observational techniques employed (see, e.g., van Boekel et al. 2004; D'Alessio et al. 2006; Natta et al. 2007; Birnstiel et al. 2010; Ricci et al. 2010; Pérez et al. 2012; Trotta et al. 2013; Testi et al. 2014; Pérez et al. 2015; Tazzari et al. 2016; Liu et al. 2017).

To describe the motion of dust particles (or pebbles) in the Solar Nebula, the Stokes number is the relevant property to consider. It describes, in dimensionless units, the efficiency of the frictional coupling of the dust particles to the gas motion. If the Stokes number is much smaller than unity, the dust almost perfectly follows the gas motion, whereas for very large Stokes numbers, the dust is decoupled from the gas and moves on Keplerian orbits. Formally, the Stokes number is given by: $St = \tau_f \Omega$, where $\tau_f$ and $\Omega$ are the gas-grain friction time and the orbital frequency, respectively. While the former depends on the dust-aggregate size and porosity as well as on the nebula-gas density, the latter is only a function of the heliocentric distance.

The Stokes number determines the drift motion of particles with respect to the gas. In particular, the radial drift motion limits the lifetime of the pebbles before they spiral into their central star. Typical drift timescales for $St \lesssim 1$ are $\tau_d = a/v_d \approx a/(2St v_{\max})$, where $a$, $v_d$ and $v_{\max}$ are the heliocentric distance, the drift velocity, and the maximum deviation of the orbital gas velocity from the Keplerian speed (typically 50 m s$^{-1}$, see Weidenschilling 1977; Weissman 1990), respectively. The ratio of growth timescale to drift timescale becomes $\tau_g/\tau_d \approx 2 \ldots 20\, St$ for $a = 1$ AU and $\tau_g/\tau_d \approx 20 \ldots 200\, St$ for $a = 100$ AU. Thus, drift becomes important when the dust aggregates reach pebble sizes; see Lorek et al. 2018 for a detailed discussion on timescales. It is interesting to note that the pebble sizes at the transition from sticking to bouncing, as discussed above, are equivalent to Stokes numbers in the range: $St = 0.001 \ldots 1$.

### 3.2 Concentration of Dust Aggregates and the Gravitational Collapse of a Pebble Cloud

Once dust aggregates have reached pebble sizes, they partly decouple from the nebula gas, which can lead to their spatial concentration by, e.g., sedimentation to the disk midplane, inside or between turbulent eddies or in pressure bumps (see Johansen et al. 2014 for a detailed review of concentration mechanisms). If any one of these concentration mechanisms leads to a local dust-to-gas mass ratio greater than unity, the so-called streaming instability can further enhance the pebble concentration considerably on a few orbital timescales (Youdin and Goodman 2005).

The streaming instability works the following way: with increasing dust concentration, the pebble cloud is slowed down less and less by gas drag due to the decreasing surface-to-mass ratio of the collectively acting pebble ensemble. Once the dust concentration locally exceeds that of the gas, the pebble cloud moves with Keplerian velocity and not at the sub-Keplerian speed valid for isolated pebbles, which also greatly reduces its radial drift. In turn, the nebula gas inside and around the pebble cloud is dragged along with the pebbles and forced to move at Keplerian velocity. Thus, single pebbles or smaller pebble assemblies on the same orbit as the pebble cloud are overtaken and incorporated into the pebble cloud, increasing its mass. In addition, pebbles or pebble assemblies drifting radially inwards are also captured, adding their mass to the pebble cloud.

By this process, the pebble concentration is strongly enhanced until it reaches the gravitational-collapse limit. Johansen et al. (2007) showed that gravitationally bound planetesimals can form by this process. However, besides the required pre-concentration of pebbles, the streaming instability only reaches sufficiently high concentrations for a gravitational collapse if the metallicity is high enough and the Stokes numbers fall into the range $St \approx 10^{-3} \ldots 5$ (Yang et al. 2017). The minimum required metallicity is $Z \approx 0.015$ for $St \approx 0.1$ and $Z > 0.015$ for smaller or larger Stokes numbers. Thus, the "optimal" Stokes number occurs for $Z \approx 0.015$, which refers to pebble sizes on the order of $\sim$10 cm at 1 AU and $\sim$1 mm at 100 AU. Also, note that the solar metallicity is $Z \approx 0.0134$ (Asplund et al. 2009) and, thus, slightly smaller than the required minimum metallicity. To achieve a higher metallicity and thus to trigger the streaming instability, partial dissipation of the nebula gas might be required (see Davidsson et al. 2016 for a discussion).

The outcomes of a gravitational collapse triggered by the streaming instability have been extensively discussed in recent years. Wahlberg Jansson and Johansen (2014, 2017) and Wahlberg Jansson et al. (2017) describe the fate of the dust aggregates during the collapse. Following the fragmentation model of Bukhari Syed et al. (2017), Wahlberg Jansson et al. (2017) find that pebbles only survive the collapse intact for final planetesimal radii smaller than about 50 km. Thus, only smaller planetesimals consist of primordial dust aggregates. In a follow-up work, Wahlberg Jansson and Johansen (2017) found some size sorting during the collapse, which leads to an internal pebble-size stratification of the planetesimals.

With numerical resolution of the hydrodynamic models increasing over time, it has become evident that the planetesimals follow a differential power law mass-frequency distribution $\mathrm{d}N/\mathrm{d}M_p \propto M_p^{-x}$, with $N$ and $M_p$ being the number of planetesimals with mass $M_p$ and a maximum planetesimal mass, $M_{p.max}$, corresponding to radii of $\sim 10^3$ km. The exponent is approximately $x = 1.6$ (Simon et al. 2016; Schäfer et al. 2017; Abod et al. 2018), which corresponds to a cumulative power-law size distribution exponent of $-2.8$. An even better fit to the data can be achieved for an exponentially truncated mass power law with an exponent, $x = 1.3$ (Abod et al. 2018), which corresponds to a cumulative size distribution exponent, $= -1.9$. As the current spatial resolution of the numerical models does not allow the formation of planetesimals with diameters $\lesssim$60 km, it is not known whether planetesimals of sizes comparable to cometary nuclei can form by the gentle gravitational collapse of pebble clouds.

The physical properties of planetesimals formed by the streaming instability with a final gravitational collapse have been analyzed by Skorov and Blum (2012), Blum et al. (2014), Blum et al. (2017) and Blum (2018). For planetesimal radii below $\sim$50 km, for which the pebbles should survive the collapse intact, the central lithostatic pressure is also not strong enough to structurally destroy the pebbles. Thus, we expect relatively homogeneous bodies (apart from the above-mentioned size-sorting effect) with two porosity size scales, i.e., on the length scale of the monomer grains (∼μm) and on the length scale of the pebbles (∼cm).

From experimental work, the packing fraction within the pebbles should be ∼0.4 (Weidling et al. 2009), following a period of bouncing collisions before their accretion into planetesimals. On the other hand, pebbles pack inside the planetesimal with a packing fraction of ∼0.6 (see Blum et al. 2017 for details) so that an overall porosity of $p \approx 1 - (0.4 \times 0.6) = 0.76$ should result. For larger planetesimals, for which the pebbles get destroyed during the collapse or by lithostatic compression, the inter-pebble void space should be absent, resulting in a porosity: $p \approx 1 - 0.4 = 0.60$. In terms of internal cohesion, the difference between small and large planetesimals is even more pronounced. While small planetesimals should have tensile strengths on the order of ∼1 Pa (Skorov and Blum 2012; Blum et al. 2014), large planetesimals should possess tensile strengths on the order of $\sim 10^3$ Pa (Blum 2018; Gundlach et al. 2018).

### 3.3 Comets as Test Cases for Planetesimal Formation Models

With extensive observational data, several cometary flybys and the vast quantity of Rosetta observations of comet 67P/Churyumov-Gerasimenko, we can now start to perform empirical tests of the above formation scenario. The most model-independent Rosetta measurements of 67P are those of the mass, $\sim 10^{13}$ kg, the volume, $\sim 2 \times 10^{10}\ m^3$, and the bulk density, $533\ kg\,m^{-3}$, (Pätzold et al. 2016; Preusker et al. 2015). For the determination of the packing fraction or porosity, one needs to know the composition of the cometary matter (Levasseur-Regourd et al. 2018), which is controversially debated, in particular the dust-to-ice mass ratio (Fulle et al. 2019). Due to its large sampled volume, the best material constraint at this time comes from the Rosetta CONSERT instrument, which arrived at a dust-to-ice mass ratio of 3–5 and a porosity, $p > 0.65$, most likely around $p \approx 0.75$ (Herique et al. 2016), in good agreement with a pebble-collapse model and a small planetesimal size (see above).

Blum et al. (2017) performed an analysis of whether comet 67P consists of pebbles, based on five different sets of Rosetta instrument data: (1) They modeled the thermophysical behavior of the near-surface layers with a pebble model, determined the subsurface temperature stratification and the resulting thermal radiation fluxes at the two Rosetta MIRO wavelengths, 0.5 and 1.6 mm. For reasonable optical properties of the cometary matter in the mm wavelength regime, Blum et al. (2017) could reproduce the Rosetta MIRO fluxes for pebble radii between 1 and 7 mm. (2) Similarly, the surface temperatures measured by Philae MUPUS-TM could be modeled with pebble radii between 0.2 and 55 mm. (3) From an analysis of the size-frequency distributions measured by various Rosetta instruments and from the ground, Blum et al. (2017) derived that most of the ejected dust mass was in particles with radii between 1 and $10^3$ mm. (4) With the published tensile strength data for 67P (Groussin et al. 2015; Thomas et al. 2015; Hirabayashi et al. 2016), Blum et al. (2017) arrived at pebble radii of up to 6 mm using the model of Skorov and Blum (2012) and measurements by Blum et al. (2014) and Brisset et al. (2016). However, newer results from Attree et al. (2018) suggest tensile strengths of <1 Pa on the pebble size scale, which can be achieved with pebble radii larger than ∼1 mm. Low strength values close to the surface of a comet nucleus are supported by impact-crater modeling by Housen and Holsapple (2007) who predicted the observed crater size (150 m diameter) of the Deep Impact mission on comet Tempel 1 for a tensile-strengths of ∼1 Pa. (5) Finally, images taken with the Philae/CIVA camera at the final Philae landing site with a spatial resolution better than 1 mm, showed a rugged terrain in which Poulet et al. (2016) identified 695 pebbles. The cumulative size-frequency distribution of these pebbles, as seen in Fig. 6 of Poulet et al. (2016), shows a clear mass peak at pebble radii between 3.5 and 6 mm. Interestingly, this corresponds to Stokes numbers, $St \approx 0.1$ at 30 AU with a minimum-mass solar-nebula model, which is the minimum required metallicity for the streaming instability to occur (see above).

A comet nucleus made of pebbles can also explain the sublimation-driven dust activity, particularly if the pebbles have sizes on the order of $\sim$1 cm. Only then is the sublimation pressure below the desiccated dust-aggregate layer strong enough to lift the pebbles off the surface (Skorov and Blum 2012; Blum et al. 2014, 2015; Gundlach et al. 2015; Bischoff et al. 2019). Also, the steeply increasing sublimation rate of water ice with decreasing heliocentric distance can be explained by the pebble model of comet 67P (Blum et al. 2017; Hu et al. 2019).

Moreover, the inactivity of asteroids in comet-like orbits (ACOs) can be explained by the pebble model, as Gundlach and Blum (2016) have shown. Although the orbital parameters of ACOs are almost indistinguishable from JFCs (Kim et al. 2014), ACOs do not generally show dust activity. Gundlach and Blum (2016) explain this by the larger sizes of the ACOs compared to the JFCs. While JFCs and ACOs possess a very similar size-frequency distribution for radii between 1 and 2 km, there is an excess of ACOs at larger sizes. Due to the larger size and the correspondingly larger lithostatic pressure inside the ACOs and a memory effect described by Blum et al. (2014) and Gundlach and Blum (2016), the sub-surface tensile strength of ACOs is higher than that of JFCs. Bodies become inactive whenever the tensile strength can no longer be overcome by the internal gas pressure of the sublimating volatiles. As a result, the depth at which inactivity happens is shallower for larger bodies (i.e., for ACOs) than for the smaller JFCs. Thus, assuming a constant depth of erosion per orbit (as suggested by the similar orbital parameters), JFCs can turn into ACOs after a finite number of orbits, which is smaller for large bodies than for small ones.

Also observation of very large dust particles with masses up to 100 kg in the coma of comet 67P (Ott et al. 2017) can be explained by the pebble model. Blum et al. (2017) speculated that most of the dust mass loss in the southern hemisphere of 67P is emitted in particles between 1 and $10^3$ mm in size, and that these particles are either single pebbles or clusters of pebbles. In a recent study, Bischoff et al. (2019) showed that the lateral size of an emitted dust patch slightly exceeds its depth so that one can expect that the chunks are clusters of pebbles with similar dimensions in all directions. Gundlach et al. (personal communication) showed that chunks with sizes exceeding 10 cm in depth can be emitted during the southern hemisphere summer owing to the sublimation of $CO_2$ ice.

The reported measurement of extremely fluffy dust particles by Rosetta GIADA (Fulle et al. 2015) with porosities $p \gtrsim 0.99$ and the discovery of a fractal dust aggregate by Rosetta MIDAS (Mannel et al. 2016) were explained by Fulle and Blum (2017). In their model, these high-porosity or fractal particles are primordial aggregates from the early stages of dust growth, which coexisted with the pebbles and were stored safely away inside the inter-pebble pores as the nucleus of comet 67P formed by the gentle gravitational collapse of a pebble cloud.

To summarize, there is strong evidence for an abundance of mm- to cm-sized particles in 67P, which might be identified as the pebbles grown in the Solar Nebula. The observed bulk density and tensile strength of 67P appear consistent with a gentle collapse of a pebble cloud. However, further modeling is needed to demonstrate that kilometer-sized bodies do form by this mechanism, that the properties of comets are unique to this mode of formation, and that they are not ruined by high-velocity collisional evolution, should that have taken place. Also the existence of extremely fluffy grains is not in line with the hierarchical growth model. However, the dominance of pebbles in cometary observations is far from being the final proof for the formation of planetesimals by the streaming instability. For instance, there exists the possibility that follow-up collisional processes (see Sect. 4) might have turned the planetesimals into rubble-pile bodies with typical size scales of mm to cm. This needs to be investigated in the future before the uniqueness of one or the other formation model for cometary nuclei can be confirmed.

## 4 Collisional Evolution of Comets

The comets that we see today as ecliptic or nearly-isotropic comets (see Levison and Duncan 1997 for definitions) come from the Scattered disk and the Oort cloud, respectively. But, before joining these low-density reservoirs, the comets remained for some time in their original orbits, i.e., the massive trans-Neptunian planetesimal disk, earlier referred to as the Primordial Disk. According to the most accurate models of solar system dynamical evolution (Nesvorný and Morbidelli 2012; Nesvorný 2015a, 2015b; Nesvorný and Vokrouhlický 2016; Nesvorný et al. 2017) this planetesimal disk was located between an inner edge at ~15–20 AU and an outer edge at ~30–40 AU and comprised 20 to 30 Earth masses of material. The dispersal of this disk during a phase of instability of the giant planets generated the hot population of the Kuiper belt (Nesvorný 2015a), the Scattered disk and the Oort cloud (Brasser and Morbidelli 2013; Nesvorný et al. 2017), the Trojan populations of Jupiter and Neptune (Nesvorný et al. 2013; Nesvorný and Vokrouhlický 2009) and the clouds of irregular satellites of the giant planets (Nesvorný et al. 2007). The size distribution of the disk at the time it was dispersed, deduced by combining a series of observational constraints from these daughter populations, was presented in Nesvorný and Vokrouhlický (2016) and is depicted in Fig. 1.

This size distribution is similar to that of the current Kuiper belt deduced in Robbins et al. (2017) in the 200 m–10 km range from the crater size-frequency distributions on Pluto and Charon, if one assumes that a crater on these bodies is roughly 10× the size of the projectile that caused it, as estimated from crater impact models (Melosh 1989). In particular the turnover of the cumulative impactor size distribution from one steeper than $q = -2$ for $D > 1.5$ km to one shallower than $q = -2$ for $D < 1.5$ km seems to be a recurrent feature on most imaged terrains.

A question arises because the cumulative slope of $-3$ found for KBOs with $1 < D < 10$ km in Fig. 1 (left) is steeper than any slope derived from direct observations of JFCs, whose cumulative slopes range from $-1.9$ to $-2.7$ (Snodgrass et al. 2011; Belton 2014; Tancredi et al. 2006; Meech et al. 2004). This issue was recognized by Nesvorný and Vokrouhlický (2016). They suggested that it may be due to the surface mass loss of JFCs after they leave the Kuiper belt. Still, it is an open question that needs to be resolved.

At diameters, $D \leq 100$ km, the distribution in Nesvorný and Vokrouhlický (2016) is similar to that measured in the asteroid belt, which is the result of a collisional evolution that has reached equilibrium (Bottke et al. 2005). Unless coincidental, this suggests that the trans-Neptunian planetesimal disk had time to evolve to collisional equilibrium before being dispersed. If this is the case, comet-size objects (typically a few km in diameter) should have experienced many collisions. Presumably, most of those we see today should be fragments of originally larger objects.

Reasoning by the absurd, Morbidelli and Rickman (2015) assumed that the size-frequency distribution in the planetesimal disk is primordial (i.e., was determined by the

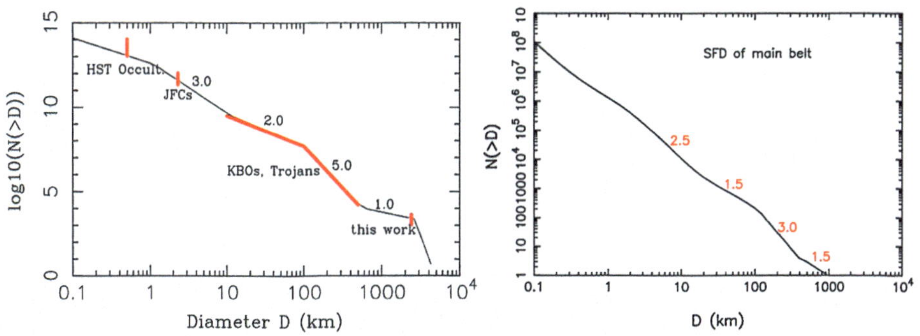


**Fig. 1** Left panel: The cumulative size distribution in the trans-Neptunian disk at the time it was dispersed, according to Nesvorný and Vokrouhlický (2016). The numbers indicate the local slopes of the distribution. Note that the numbers for the local slope values should be negative numbers, the number of objects decreasing with increasing size. The main constraints used for this reconstruction are also indicated. Right panel: The cumulative size distribution in the main asteroid belt, from Bottke et al. (2005). Numbers again indicate local slopes and should be taken as negative numbers

accretion process, not by collisional evolution) and computed the probability that a cometesimal with $D = 4$ km is collisionally disrupted. If the resulting probability is high, clearly the assumption is invalid. The size-frequency distribution assumed by Morbidelli and Rickman was very similar to that found in Nesvorný and Vokrouhlický (2016), comprising $2 \times 10^{11}$ bodies with $D > 2.3$ km, with a slope of $-2$ in the cumulative size distribution at less than 2.3 km. Morbidelli and Rickman found that, if the planetesimal disk remained massive for $\sim$50 Myr or more before being dispersed by the giant planets instability, the probability that a $D = 4$ km body is disrupted by a catastrophic collision exceeds 1. That is, all the original planetesimals at this size should have been disrupted and therefore those existing today would be fragments of larger bodies. Only in the case where the disk was dispersed immediately after planetesimal formation would comet-size planetesimals have had a chance (nevertheless smaller than 50%; Jutzi et al. 2017) to avoid catastrophic disruptions.

Focusing on this last possibility, Jutzi et al. (2017) computed the probability that a comet nucleus like 67P preserved its shape. Preserving the bilobate shape is less probable than avoiding all catastrophic collisions, because the collisional energy sufficient at smashing one of the lobes is about two orders of magnitude less than the catastrophic disruption energy for the comet nucleus as a whole (Jutzi et al., Fig. 7). Assuming a size distribution like that in Morbidelli and Rickman (2015), Jutzi et al. concluded that during disk dispersal a 67P-like object would have suffered 10–50 reshaping collisions, concluding that the morphology of the comet that we observe today cannot be primordial.

These results raise a number of interesting questions: First, if the bilobate shape of 67P is not primordial, how was it acquired? Notice also that bilobate shapes are common for JFC nuclei, so one cannot invoke any "singular" argument to explain the shape of 67P. Jutzi and Benz (2017) showed that the shape could have been acquired in a sub-catastrophic collision of a somewhat larger progenitor. A collection of bilobate shapes obtained in Jutzi and Benz sub-catastrophic collisions is shown in Fig. 2.

Schwartz et al. (2018) showed that the bilobate structure can be obtained also in catastrophic collisions, although the bilobate shapes they could generate are less pronounced than in Jutzi and Benz (2017). This is not necessarily a problem because the lower is the collisional energy the higher is the probability for it to happen (because a smaller projectile is required) so the last relevant collision that 67P suffered was most likely a sub-catastrophic one.

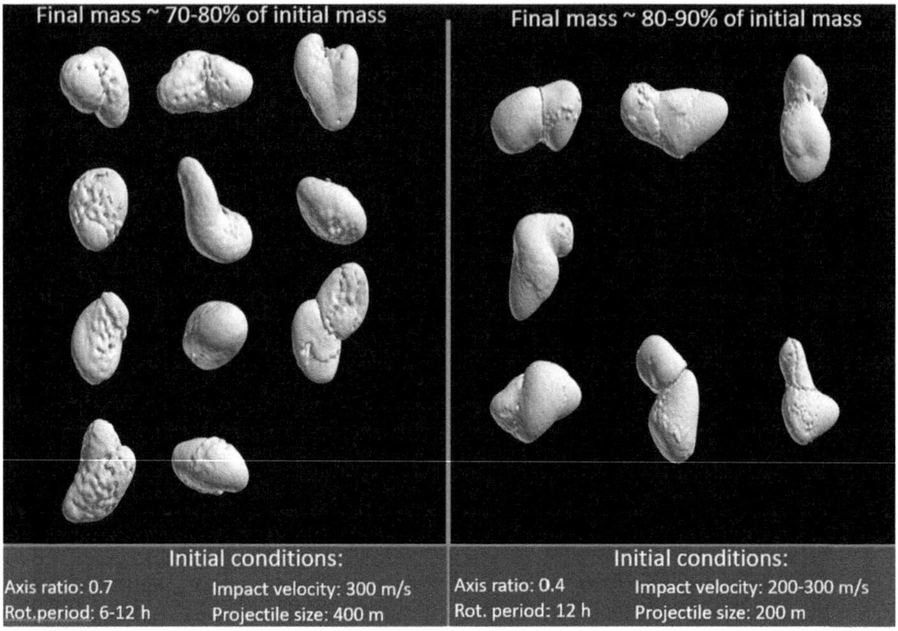


**Fig. 2** Examples of bilobate shapes obtained in the SPH sub-catastrophic collision experiments of Jutzi and Benz (2017)

A second important question is whether catastrophic fragmentation would lead to compaction and heating. If this were the case, the high porosity and low-temperature chemistry observed for 67P and other comets would become strong constraints against comets being fragments of significantly larger objects. Schwartz et al. (2018) showed that only a tiny fraction of a comet would have been heated by more than few degrees by the catastrophic collision that generated it. A larger fraction of the parent body might have been substantially heated, but the heated material is typically not incorporated in macroscopic fragments; see Fig. 3.

An analogous result was found for the porosity. It should be noted, however, that the collisions in Schwartz et al. (2018) are just above the catastrophic limit and the parent bodies considered are quite small: about 7 km in diameter. The streaming instability simulations show that the most typical planetesimal produced in the accretion process should have had $D = 50$–$100$ km. Catastrophic collisions of parent bodies of these sizes have not yet been simulated, so we don't know whether the same results concerning heating and compaction apply. Simulations of this kind can provide constraints on how large the progenitors of the collisional cascade leading to comet-sized objects could be. The specific energy in catastrophic collisions for bodies with $D = 16$, 32, and 64 km are 3, 7.2, and 17 $\mathrm{kJ\,kg^{-1}}$, respectively. If distributed uniformly these energies would correspond to global temperature enhancements of 12, 29 and 69 K. At 69 K, $CO_2$ would sublimate throughout the warm interior and recondense at the rapidly cooling surface, releasing the CO trapped in the cages of the original $CO_2$ ice (Gasc et al. 2017). CO would therefore escape into space, leading to CO-depleted fragments. However, 67P/CG and most comets are rich in CO. Therefore this argument suggests that they cannot be fragments of progenitors with $D = 20$ km or larger. Thus, it is essential to quantify how uniform the temperature increase would be in the pro-

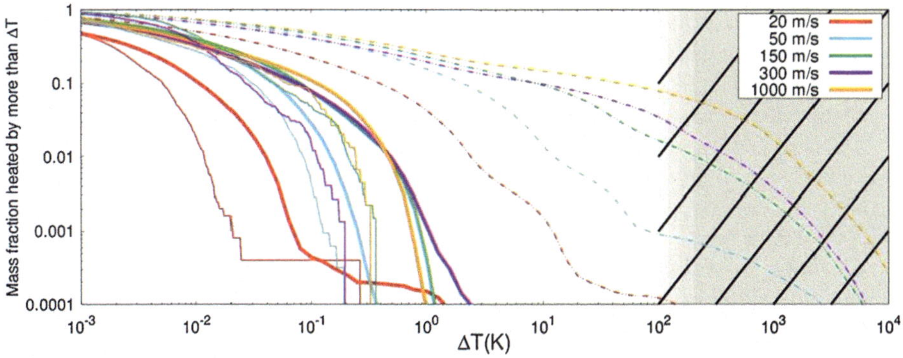


**Fig. 3** Solid curves show the fraction of the mass of the resulting comet that has been heated above a given temperature in a catastrophic collision of a larger parent body. Different colors relate to different impact speeds, the projectile being smaller at higher speed. The dashed curves show the same fraction for the parent body material. At high speed (1 km/s), 20% of the parent body can be heated by more than 10 degrees, but none of the over-heated material is incorporated in the largest remnant. From Schwartz et al. (2018)

genitor and the peak temperatures in the main fragments in simulations like that of Schwartz et al. (2018) applied to larger-scale collisions.

A third relevant question is whether the original pebbles, potentially detected in 67P (see Sect. 3), would be preserved in a catastrophic collision, or if re-accumulated collisional fragments have sizes less than a few meters as constrained by CONSERT. The SPH simulations of Schwartz et al. do not have the resolution to address these questions and there is still a long way to go before this can be done.

## 5 Comet Nucleus Evolution

As noted in the Introduction, comet nuclei are among the most primitive bodies in the solar system. Because of their small size and storage in two cold, distant reservoirs they have likely experienced the least physical processing of any known solar system bodies. Yet it has long been recognized that nuclei do evolve and change with time. The Rosetta mission provided absolute proof that 67P was a body undergoing "active" geological processes, in particular mass-wasting and fracturing, that have modified the nucleus surface and possibly the interior (Thomas et al. 2015).

"Primitive" is a relative term and it is important to define what we mean by that. Precursor elements and molecules that were incorporated in comets were processed in the interstellar medium and before that in the cores of massive stars, possibly multiple times. Volatile ices are deposited onto interstellar grains where UV radiation and cosmic rays provide chemical evolution. Further processing occurs as the natal interstellar cloud collapses and material is brought together in denser and warmer environments.

For the purposes of this paper we will define "primitive" as the chemical state of cometary materials as they began to agglomerate/accrete into macroscopic bodies. As the surfaces of these bodies are modified by the physical processes described below they lose their primitive nature, but the nucleus interior likely remains primitive. Given the earlier processes mentioned above, it would be wrong to ever refer to these bodies as "pristine" as some researchers mistakenly do.

As noted in Sect. 4, the initial processing of comet nuclei was accomplished by collisional evolution in the planetesimal disk and subsequently as the giant planets cleared their zones of remaining planetesimals. The current Nice Model (named for the Observatoire de Nice where the initial work was done; Tsiganis et al. 2005; Gomes et al. 2005) for the solar system's early evolution suggests that this could have occurred very violently if Jupiter and Saturn passed through a mean-motion resonance. According to the Nice Model a relatively compact giant planets region was pushed outward due to planetary migration and huge swarms of icy planetesimals were dispersed to orbits in the Kuiper belt region or to the distant Oort cloud. Simulations of these processes by Stern and Weissman (2001) and Charnoz and Morbidelli (2003, 2007) suggested that under some starting conditions, comet nuclei might have ground themselves to dust during this dispersal, though this result is very dependent on what was assumed for the typical size of a comet nucleus as well as the number of comets in the Disk.

### 5.1 Processing in the Kuiper Belt and Oort Cloud

Although we think of the Kuiper belt and Oort cloud as very benign environments, there are processes that would have modified the comet nuclei there. These include: (1) internal heating by short-lived radio-nuclides, (2) irradiation by galactic cosmic rays and solar protons, (3) heating by stars passing through the Oort cloud and by nearby supernovae, (4) accretion of interstellar molecules and dust grains, (5) erosion by hyper-velocity interstellar dust grain impacts.

(1) Even as they are first accreting in the Primordial Disk, comet nuclei are warmed by short-lived radio-nuclides, in particular $^{26}$Al. Prialnik et al. (1987) showed that the central temperature of cometary nuclei with radii of 1, 5, and 10 km would be heated to 15, 60 and 90 K respectively in $10^6$ to $10^7$ years. This was for nuclei composed solely of amorphous water ice, contaminated with several of the expected short-lived radioactive isotopes in their solar system abundances. None of these temperature values are high enough to convert the amorphous water ice into crystalline water ice. These central temperatures would be even lower if we consider the nuclei to be an even (1:1) mixture of amorphous water ice and meteoritic non-volatiles since the non-volatiles would absorb some of the heat from the radio-nuclides. Some of the most volatile ices would sublimate at the temperatures above and then would diffuse through the porous nucleus until they reached colder layers where they could recondense. Note that these modeling results are sensitive to the size of the nuclei, their porosity, and the initial $^{26}$Al content (Prialnik and Podolak 1995, 1999). The Prialnik and Podolak models use an $^{26}$Al initial mass fraction of $5 \times 10^{-8}$ of the non-volatile mass, which is somewhat lower than the value found from studies of primitive meteorites.

Giotto measurements (McDonnell et al. 1989) suggested that the dust-to-ice mass ratio in the 1P/Halley coma was $\sim$2:1. Rosetta measurements (Rotundi et al. 2015; Herique et al. 2016) were even higher, with the mass of non-volatile dust in 67P being 3–5 times the mass of the sublimating ices, even when such ices as CO and $CO_2$ were included. This creates a new problem because the mass fraction of $^{26}$Al and other short-lived radio-nuclides would provide a larger internal heat source. Future modeling of the internal temperatures of comet nuclei will need to take this into account.

(2) Irradiation of volatile ices and ice-dust mixtures by high-energy particles has been studied in the laboratory (Johnson 1991; Cooper et al. 1998; Strazzulla 1999). The interaction breaks chemical bonds producing volatile free radicals, some of which recombine to form a dense, dark polymer which is far less volatile than the original material. Polymerization occurs particularly if hydrocarbons are present, which we know to be the case for

cometary nuclei. At the nucleus surface, the high energy particle bombardment results in a net erosion by sputtering (observed at 67P by the ROSINA instrument on Rosetta; Wurz et al. 2015) and probable escape of the more volatile species, leaving behind a low-volatility residue.

The depth to which the particles penetrate the nucleus surface depends strongly on the energy of the particles and the bulk density of the surface layers. Low-energy solar protons of a few to a few hundred keV penetrate only the top millimeter; cosmic ray protons with typical energies of 2 MeV penetrate 1–2 meters. This raises the interesting possibility that comets from the Kuiper belt and Oort cloud may have already developed thick non-volatile crusts before they are perturbed to smaller and warmer perihelion distances.

(3) Random passing stars occasionally penetrate the Oort cloud, heating the comets along their path and ejecting them to interstellar space. For solar mass stars, comets out to about 500 AU from the star's path will be dynamically ejected. This is much greater than the distances at which significant heating occurs, so all the heated comets will be lost. But for very massive and luminous OB giants, the heating radius is larger than the ejection radius. Stern (1987) estimated that all comets in the Oort cloud have been heated to 27 K and that 20% to 40% had at least one episode of heating to 50 K. These warming events occur over periods of several thousand years and thus should be sufficient to warm most of each nucleus to these temperatures. As a result the more volatile ices will have been mobilized within the nucleus and some fraction can be expected to be lost from near-surface layers.

Such heating events in the Kuiper belt will be far less frequent because of its smaller cross-section relative to random passing stars. Whereas the Oort cloud has a radius of $\sim 10^5$ AU or more, the (known) Kuiper belt extends only about 50–100 AU. The Scattered disk, the source of the JFCs, extends $\sim 10^3$ AU so it too is relatively unlikely to have experienced significant external heating.

(4) and (5) The competing processes of erosion by hypervelocity interstellar dust impacts and accretion of interstellar gas and dust grains only affect a very thin layer at the nucleus surface. Stern (1986) estimated that micro-cratering by grain impacts will remove a layer only 5–80 mm in thickness over a comet's lifetime in the Oort cloud. O'Dell (1971) estimated that accretion of interstellar gas and dust would add a layer only 10–100 microns in thickness to a nucleus surface. Although dust gains can be captured in impacts with very low density media such as the aerogel used on the Stardust mission (Tsou et al. 1984) the much higher bulk density of 67P would result in grain impacts being erosional rather than accretionary.

## 5.2 Solar Heating

The major processing of comet nuclei occurs when they are perturbed back into the planetary region and are subject to the heat of the Sun. Weissman (1979) and Fernández (1981) showed that the typical long-period comet (LPC) from the Oort cloud makes only 5–11 returns before it is dynamically ejected or randomly disrupted (though some comets may survive for tens or even hundreds of returns). Thus the physical evolution of these LPCs is probably not very significant during their relatively short dynamical and disruption lifetimes in the planetary region.

For Jupiter-family comets like 67P, also known as "ecliptic comets," their history can be much more complex. Comets in the Kuiper belt and the Scattered disk can make close approaches to Neptune and be perturbed to orbits with smaller perihelia. Neptune can gravitationally scatter these comets to Uranus which will then pass them on to Saturn, and eventually Jupiter.

Levison and Duncan (1997) studied this dynamical process using Monte Carlo codes. They found that a typical Kuiper belt or Scattered disk comet took $4.5 \times 10^7$ years to reach one of two possible end-states: dynamical ejection from the solar system or striking a planet. Thirty percent of these comets became Jupiter-crossing with perihelia $\leq 2.5$ AU, where Levison and Duncan defined them to be "visible comets." By comparing the predicted orbital element distributions from their simulations to that of known JFCs, they estimated that the typical JFC spent $1.2 \times 10^4$ years as a visible comet. If we take a typical JFC orbital period of 6 years, this suggests the lifetime of ecliptic comets in the active region ($q < 2.5$ AU) is $\sim 2 \times 10^3$ returns.

These planet-crossing orbits are not very stable and can change on time scales of a few to tens of orbits. As the most massive planet, Jupiter dominates the motion of the JFCs. Close approaches to Jupiter will random walk perihelion distances closer to and farther from the Sun, as well as changing the orbital periods and inclinations.

For example, 67P itself was discovered in 1969. Integration of its orbit backward in time indicated that the comet made a close approach to Jupiter, $d = 0.052$ AU, in February 1959. That encounter reduced its perihelion distance from 2.74 AU to 1.28 AU (Kronk 1984; Krolikówska 2003). Around the year 2200 67P will make another close approach to Jupiter that will likely raise the perihelion distance back to $\sim 2.6$ AU. Because of nongravitational forces caused by jetting of volatiles from the nucleus surface, no JFC can have its past or future orbit reliably predicted past one or two of these close Jupiter approaches.

The activity of 67P during the Rosetta mission is documented in detail in many of the other papers in this volume and will only be summarized herein. One of the main results from Rosetta is the variety of geological processes that were observed on the nucleus. These included airfall, thermal fracturing, surface dust transport, mass wasting, insolation weathering, orbiting debris and mass loss through the ejection of large chunks of material (Thomas et al. 2015). Much of the activity of the nucleus appears to come from vertical cliffs that were observed to crumble over time (Vincent et al. 2016a), exposing the volatile rich interior. This result was also suggested to explain the mesa-like structures observed on the nucleus of comet 19P/Borrelly during the Deep Space 1 flyby (Soderblom et al. 2002), and possibly on comet 9P/Tempel 1 (Veverka et al. 2013).

Another major result from Rosetta is that seasonal effects on 67P were particularly strong with important consequences (Keller et al. 2015). Because of the eccentricity of 67P's orbit, $e = 0.64$, and the obliquity and orientation of its rotation pole, most of its orbital period is spent with the Sun over the northern hemisphere. However, close to perihelion the Sun moves quickly to southern latitudes that have not seen any sunlight for the previous $\sim 6$ years. This leads to a sharp rise in activity that lasts only $\sim$6–9 months before the Sun once again moves to northern latitudes.

This extreme seasonal behavior was proposed by Weissman (1987) who was trying to explain the post-perihelion brightening of Halley's Comet. The southern "nuclear summer" that 67P experienced in mid- to late-2015 led to significant differences between the two hemispheres. The southern hemisphere outgassing close to perihelion produced huge quantities of dust, much of which fell back onto the relatively inactive northern hemisphere, while preventing significant dust deposition on the southern hemisphere. It is noted that, like 1P/Halley, 67P reaches maximum activity and brightness post-perihelion. When the Sun is over the northern hemisphere, 67P is farther from the Sun and thus the activity is not high enough to remove the dust coatings obtained during southern hemisphere summer. This situation has likely repeated itself every perihelion passage since 67P was perturbed to its current orbit circa 1960.

A very important Rosetta result for our purposes comes from the Radio Science Investigation, which measured the mass of the 67P nucleus before and after perihelion (Pätzold et al. 2016, 2017). It was found that the comet lost $\sim 0.1\% \pm 0.03\%$ of its mass during the perihelion passage. First-order thermal models suggest that the average decrease in radius of a JFC on each orbit is a constant, i.e., the thickness of the removed layer is the same from orbit to orbit. This in turn suggests that 67P has a remaining physical lifetime of $\sim 10^3$ returns, if the current orbit and activity level were to remain unchanged. As noted above, the orbit will vary considerably. Interestingly, this is, to first order, similar to the $2 \times 10^3$ returns found by Levison and Duncan (1997) for the JFC population as a whole.

If we assume that the mass loss from the nucleus is uniform over the measured surface area of 51.7 $\mathrm{km}^2$ (Preusker et al. 2017) and we assume the nucleus bulk density = 533 $\mathrm{kg\,m^{-3}}$ (Pätzold et al. 2016, 2017) then the mean thickness of the removed layer is $\sim 3.3$ m ($\pm 30\%$). Note that more material is lofted off the nucleus surface but some fraction of it falls back, in particular onto the northern hemisphere, which is not illuminated around perihelion.

Another interesting Rosetta result is that the rotation period of the nucleus changed from one orbit to the next. Lowry et al. (2012) determined a rotation period of 12.76 hours for 67P from ground-based observations during the 2006 aphelion passage. However, at spacecraft rendezvous in 2014, the Rosetta OSIRIS (Imaging) Team found a period of only 12.41 hours (Sierks et al. 2015), a difference of approximately 21 minutes. Following 67P's perihelion passage and return to larger solar distances in 2015–2016, the rotation period decreased further to 12.07 hours, another change of $\sim 20$ minutes (Gaskell et al. 2016).

This behavior was seen previously for comet 9P/Tempel 1 (Chesley et al. 2013). It is caused by outgassing torques from the irregularly shaped nucleus, as well as the differences in seasonal outgassing rates. It is essentially the same nongravitational forces that change the comet's orbital parameters, in particular the orbital period.

Comet 67P also exhibited outbursts close to the time of perihelion on an almost daily basis (Vincent et al. 2016b). These outbursts tended to be located close to the sub-solar latitude at the time. The cause of the outbursts is still under investigation. One possibility is the sudden escape of volatile gases from a sealed sub-surface cavity. Another is the transition of amorphous water ice to crystalline water ice at depth, which is an exothermic reaction that occurs at temperatures between 110 and 150 K. A third possibility is that dust and gas are suddenly released when a vertical cliff collapses and exposes fresh cometary volatiles.

## 5.3 Sub-Surface Heating

Sunlight falling on the surface of a cometary nucleus is divided between three uses. Because of the low albedos of nucleus surfaces, $\sim 3$–7% (Lamy et al. 2004) only a small fraction is reflected back into space. The rest of the energy flows into the nucleus surface where it is divided between heating the cometary materials and the sublimation of cometary ices. As stated above, a result of this heating is that amorphous water ice will convert to crystalline water ice, which is an irreversible process.

Sublimation rates are highly dependent on the local temperature; they are a direct function of the vapor pressure of gas molecules over their respective ices, which vary exponentially with temperature (e.g., Lide 2004). The conversion rate of amorphous ice to crystalline ice is also highly temperature dependent.

The "thermal skin depth" is a measure of how far surface heat will penetrate below the surface (Rickman 1991). It is defined as:

$$d = (KP/\pi\rho C)^{1/2} \tag{5.1}$$

where $K$ is the thermal conductivity, $P$ is the period (rotation period for diurnal skin depth, orbital period for orbital skin depth), $\rho$ is the density, and $C$ is the specific heat. The thermal skin depth is the distance over which a temperature perturbation at the surface will decrease by a factor of 1/e. For conductivities typical of solid water ice, $d = 20$ cm for a rotation period of 24 hours or 9.2 meters for an orbital period of 6 years. However, for the low conductivities actually measured for nucleus surfaces, the diurnal skin depth is $\sim$1 cm while the orbital skin depth is about 1 meter.

As the thermal wave penetrates beneath the nucleus surface it will sublimate and mobilize ices. For ices more volatile than water ice, this occurs at substantially lower temperatures and greater depths than for water ice. The result is that volatile ices will be mobilized at depths where water ice is still relatively stable. Because of the low bulk density and high porosity of the nuclei the evolved gases can diffuse both upward toward the surface and inward to greater (cooler) depths where they will recondense (Espinasse et al. 1991). This will make for a complex composition in the near-surface layers of the nucleus.

Rosetta showed that the actual effects of sunlight on the nucleus surface were far more complex than can be described herein. As noted above, we leave those discussions of active geology to the other papers in this volume that specifically address them.

## 6 Nucleus End-States

There are a variety of possible end-states for cometary nuclei, several which have actually been observed. The end-states we identify are:

1. Random disintegration/disruption.
2. Tidal disruption.
3. Loss/burial of all volatiles leading to an asteroidal-appearing object.
4. Collision with the Sun, a planet or a satellite.

### 6.1 Random Disintegration/Disruption

Comet nuclei are often seen to spontaneously disrupt (Weissman 1980; Boehnhardt 2004). For example, comet C/1999 S4 LINEAR was seen to break into at least 16 pieces as it approached perihelion in July 2000 at a heliocentric distance of $\sim$0.76 AU (Weaver et al. 2001; Fig. 4). All trace of the comet disappeared within about three weeks.

The reasons for random disruption are not known but a leading candidate is rotational spin-up due to asymmetric outgassing torques from the irregular nucleus. Weissman and Lowry (2008) showed that no comet nucleus with a measured rotation period spins faster than about 5.5 hours. If comet nuclei are strengthless rubble piles held together only by self-gravity, this sets an upper limit on the nucleus bulk density of $\sim$600 kg m$^{-3}$. The Radio Science Investigation and the OSIRIS imaging team on Rosetta determined a bulk density for the 67P nucleus of $533 \pm 6$ kg m$^{-3}$ (Preusker et al. 2015; Pätzold et al. 2016).

Random disruption does not appear to be correlated with any known orbital parameters such as perihelion distance, orbital inclination, pre-/post-perihelion time, etc., suggesting that it is some internal property(s) of the nucleus that leads to this loss state. Another suggested mechanism for random disruption is thermal stresses due to the strong thermal gradients as cometary surface materials rotate in and out of sunlight. Rosetta images of the 67P nucleus show considerable evidence of thermal stressing and cracking (Auger et al. 2018). This included cracking on the walls of the cylindrical pits that extend 100 m or more below the 67P nucleus surface (Vincent et al. 2015).

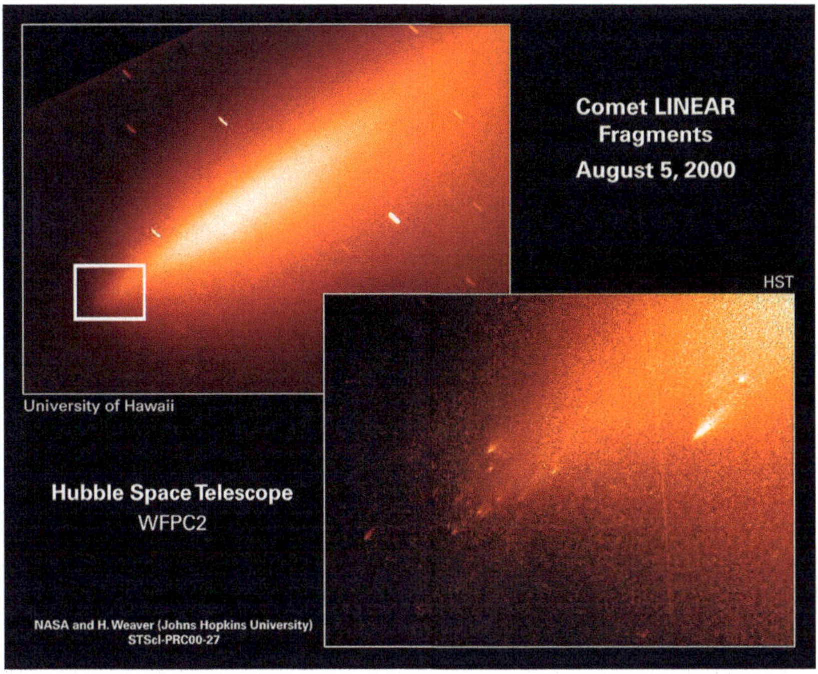


**Fig. 4** Images of the disintegrating comet C/1999 S4 from a ground-based telescope (upper left) and from the Hubble Space Telescope (lower right). The approximate field of view of the HST image is shown by the small white box on the upper left image. Many individual fragments of the comet are seen in the HST image

Lastly, the development of sealed pockets of volatile gases below the nucleus surface may have some effect on splitting. Rosetta observed many outbursts around perihelion and the outbursts clustered along the sub-solar latitudes (Lin et al. 2017). A problem with this scenario is how do sealed pockets form in such a porous medium as the cometary nucleus?

Random disruption may not always lead to complete destruction of the nucleus. Comet 73P/Schwassmann-Wachmann 3 was observed to break into four separate nuclei at its perihelion passage in 1995 (Boehnhardt et al. 1996). For identification purposes these fragments were labeled A, B, C and D. All four fragments returned in 2006. Observations with HST showed that the fragments were individually shedding many smaller pieces on the order of 50 m or less in diameter (Weaver et al. 2006, see Fig. 5).

Another interesting case is periodic comet 3D/Biela which was first discovered in 1772 and observed on several returns (Kronk 1984). During the 1846 return it was found to be a double, comet with two independent nuclei that slowly moved apart. The double-comet was observed again in 1852. Observing conditions were unfavorable in 1859 but were good in 1865. However neither comet was recovered. The comets were not found at their predicted return in 1872 but an unexpected and intense meteor shower, $\sim$3,000 meteors per hour, was observed that year consistent with particles in the comets' orbits. Additional intense meteor showers were observed at the comets' predicted returns in 1885 (15,000 meteors per hour) and 1892 (6,000 meteors per hour), and a smaller shower in 1899 (150 meteors per hour). Neither the comets nor the meteor showers were found on subsequent returns.

Yet another possible example of a surviving split comet pair may be periodic comets 42P/Neujmin 3 and 53P/Van Biesbroeck. Integration of the comets' orbits backward in time show them to be remarkably similar prior to a close Jupiter encounter in 1850 (Carusi et al.

**Fig. 5** HST images of the disintegrating B-fragment of comet Schwassmann-Wachmann 3 in 2006. The numerous small fragments breaking off of the B-fragment are readily seen and can even be traced from day to day. An upper limit of 50 m each was determined for the size of the fragments. (Weaver et al. 2006)

1985). They may have once been a single comet that broke apart and then experienced slightly different orbital evolutions. Matonti et al. (2019) noted evidence of fractures near the "neck" of 67P that may indicate enhanced erosion and the possibility that 67P will break into two comets in the future.

Weissman (1980) found that dynamically new long-period comets split ∼10% of the time during their discovery apparitions, with older, returning LPCs splitting ∼4% per perihelion passage, and JFCs splitting ∼1% per perihelion passage. A review of the split comets listed in Boehnhardt (2004) suggests that these numbers are ∼4.5% and ∼3.5% per perihelion passage for the dynamically new and returning LPCs, respectively.

Weissman (1979) modeled the physical and dynamical evolution of LPCs from the Oort cloud and showed that random disruption accounted for the loss of ∼ 27% of LPCs entering the planetary system, the second most common loss mechanism after dynamical ejection.

## 6.2 Tidal Disruption

Cometary nuclei can be gravitationally disrupted if they pass within the Roche limits of large bodies such as the Sun or the giant planets. Since such close approaches are typically rare, this is not a commonly observed phenomenon. However, the Kreutz group of sungrazing comets was recognized by Kreutz (1888) and eight naked-eye Kreutz comets were discovered between 1843 and 1970 (Marsden 1967, 1989). Kreutz group members typically have perihelia within one solar radius of the solar photosphere, well within the solar Roche radius. They are likely the result of tidal splitting events on previous perihelion distances. Their high orbital inclinations, ∼139–144°, do not allow them to pass close to any of the planets.

Comet Ikeya-Seki (C/1965 S1) was a bright Kreutz group comet discovered in 1965. It passed perihelion at a distance of 0.007786 AU, or 1.67 solar radii (Kronk 1984). Just before perihelion it appeared to split into at least three pieces. One fragment faded rapidly but the second fragment persisted until both it and the main nucleus faded from view.

In 1979 the orbiting SOLWIND coronagraph discovered six Kreutz group comets (Michels et al. 1982; Sheeley et al. 1982) that were too faint to be observed from the ground. None of those comets were seen to survive perihelion passage. This was followed by the Solar Maximum and SOHO (Solar and Heliospheric Observatory) missions, launched in 1980 and 1995, respectively. SOHO has discovered over 3,000 comets, mostly in Kreutz-type orbits (Biesecker et al. 2002; Knight et al. 2010). Most of the small comets discovered by

SOHO did not survive perihelion passage. Other comets discovered in sun-grazing orbits were the Marsden and Kracht groups but these are far less numerous than the Kreutz group.

Although the Kreutz group progenitor nucleus was almost certainly disrupted by tidal forces, one must also realize that the sun-grazing comets suffer intense heating during their very close approaches to the Sun. This heating is great enough to vaporize iron and mineral grains. Weissman (1983) showed that a typical comet would lose at least ~15 m in radius during a close passage with the Sun. Since many of the sun-grazing comets appear to be smaller than that (Biesecker et al. 2002; Knight et al. 2010), their total destruction can be achieved through thermal effects alone, not even considering gravitational tides.

Extreme nucleus heating could be ruled out in the case of two comets that tidally disrupted due to close passages to Jupiter. Comet 16P/Brooks 2 (1889 N1) was discovered in 1889 and multiple nuclei were observed near the main comet. An integration of its orbit backward in time showed that it had passed within 0.001 AU of Jupiter in 1886, or ~2.1 Jupiter radii, well inside the orbit of Io (Kronk 1984). This strongly suggested that the nucleus was tidally disrupted during its close passage with Jupiter.

Comet Shoemaker-Levy 9 was discovered in 1993 and was found to be at least 9 separate nuclei (later increased to 21) in a straight line (Shoemaker et al. 1993). Orbit determinations showed that the multiple nuclei were in orbit around Jupiter and had most likely been captured as a single parent nucleus around 1929 ± 9 years (Chodas and Yeomans 1996). Moreover, the orbit reconstruction showed that the comet had passed within Jupiter's Roche limit at 1.6 Jupiter radii in 1992, and that the multiple nuclei would impact Jupiter on their next perijove passage. Asphaug and Benz (1994, 1996) showed that the multiple nuclei could be explained if the nucleus was a weakly-bonded rubble pile that disrupted at perijove and then gravitationally reassembled into multiple nuclei. Interestingly, the final number of reassembled nuclei depended on the bulk density of the progenitor nucleus and 21 fragments suggested an original bulk density of ~600 kg $m^{-3}$ (other densities were possible depending on the nucleus rotation state). This was again pre-Rosetta evidence of the low bulk density of cometary nuclei. Subsequent reconstruction of the impact events on Jupiter suggested that the progenitor nucleus was ~1.5 km in diameter (Asphaug and Benz 1996).

## 6.3 Loss/Burial of All Volatiles

The pre-Rosetta view of comets was that they were conglomerates of ice, silicate dust, and organic dust grains brought together in the Solar Nebula into small bodies several kilometers in diameter (Whipple 1950). These objects slowly lost their ices and dust until they dwindled down to nothing, or they left deposits of large non-volatile grains on their surfaces that eventually choked off all activity, not allowing insolation to penetrate to depths where ices might still be present. Thermal models showed that such deposits needed to be only tens of meters thick to turn active comets into asteroidal-appearing objects (Brin and Mendis 1979; Fanale and Salvail 1984).

In the case of active comets with known nucleus radii, one could define the "active fraction," the percentage of the nucleus surface that needed to be active in order to provide the observed water production rate for that comet (A'Hearn et al. 1995; Gutiérrez et al. 2003). Active comets like 1P/Halley had an active fraction of ~30% of the sunlit hemisphere, whereas typical JFCs had active fractions on the order of only ~1%. Spacecraft flyby images of comet nuclei showed that this dichotomy of apparently active and inactive regions was seen on several nuclei. This was taken as proof that comets could eventually evolve to completely inactive objects.

Studies of the possible evolution of comets into asteroids (Weissman et al. 2002) showed that there were many objects in the near-Earth asteroid (NEA) population that might be dormant cometary nuclei, and that these objects usually had primitive-type taxonomic identifications, i.e., C, D, F, or low albedos. Weissman et al. (2002) estimated that 6% of the NEA population could be dormant comets. Dynamical simulations of the possible sources of the NEO population (Bottke et al. 2002) showed that NEOs with a high probability of a JFC origin tended to be primitive taxonomic types. Similarly, Fernández et al. (2001) compared measured albedos for 14 comet nuclei and 10 NEOs with Tisserand parameters,[1] $T < 3$, and 34 NEOs with $T > 3$. They showed that all of the comets and 9 out of 10 of the NEOs with $T < 3$ had albedos $\leq 0.07$, while most of the NEOs with $T > 3$ had albedos $>0.15$.

DeMeo and Binzel (2006) obtained spectra of 32 NEOs with $T < 3$ and found that 17 of 32 objects had primitive taxonomic types. They estimated that ~16% of the NEO population had taxonomic types and Tisserand parameters consistent with a cometary origin.

It appears likely that cometary nuclei can slowly evolve to an asteroidal state. But the fraction of comets that do so, versus the fraction that disintegrates completely cannot yet be determined. Nonetheless, this is likely to be one of the major end states for cometary nuclei.

## 6.4 Collision with a Planet or the Sun

The probability of a comet striking a planet in a circular orbit is given by Öpik (1951) by

$$p = s^2 U / \left(\pi \sin i |U_x|\right) \tag{6.1}$$

where $s$ is the capture radius of the target planet (including gravitational focusing), $U$ is the encounter velocity of the comet with the planet (also known as *V-infinity*), $i$ is the inclination of the comet's orbit relative to the planet's orbit and $U_x$ is the component of the encounter velocity in the radial direction.

Kessler (1981) provided similar equations for estimating the impact probability for planets in elliptical orbits. Weissman (1982) used these equations to estimate a mean impact probability of $2.2 \times 10^{-9}$ for Earth-crossing LPCs (per perihelion passage). Similar calculations find that Jupiter, the largest planet, also has the largest impact probability for LPCs, $p = 9.5 \times 10^{-8}$, approximately four times that for the other seven planets combined.

Impact probabilities for JFCs are more difficult to assess, in particular for the giant planets, because we do not know the orbital distributions of planet-crossing objects in that region of the planetary system. Based on calculations for the known JFCs and the terrestrial planets, typical JFC impact probabilities are about three times greater than for LPCs (Weissman 1982; Shoemaker et al. 1995; Levison et al. 2000).

Given these very low impact probabilities, collisions with planets is an unlikely end-state for cometary nuclei. Given the small size of the targets, collisions with satellites, asteroids and comets will be even less probable than for the planets.

Cometary orbits that impact the Sun are rare and are dominated by the Kreutz group comets (see Sect. 6.2). Technically, most of the Kreutz group members have perihelia greater than the Sun's photospheric radius and are destroyed by tidal disruption and intense heating. There do exist dynamical paths for sending comets onto sun-impacting trajectories (Jones et al. 2018) but few have been observed to do so. Solar impact is an unlikely end-state for comets.

[1] The Tisserand parameter, $T$, is a pseudo-constant of the motion in the restricted 3-body problem (Sun-Jupiter-comet) that is used to identify returning JFCs even though their orbits may have been changed by a close approach to Jupiter. Typically, comets have $T$ values $< 3$, while asteroids have $T > 3$.

## 7 Discussion

ESA's Rosetta/Philae mission to comet 67P/Churyumov-Gerasimenko has yielded a bonanza of new information on comet nuclei, comet comae, and their interaction with the solar wind. Rosetta/Philae has provided detailed information on the composition of dust and ice in the coma, the bulk density, porosity, rotational properties and mass of the nucleus, the morphology of the nucleus surface and the active processes acting on it, and much more. But has Rosetta provided us new insights into how comet nuclei formed? Yes, we know far more about a typical comet nucleus, information that constrains models of nucleus formation. But, no, we do not yet know enough to discriminate between the currently competing hypotheses.

Pre-Rosetta models for cometary nuclei (Weissman 1986; Donn and Hughes 1986) predicted that the nuclei would be irregular objects with substantial internal voids. The first prediction has been confirmed by several nucleus flyby missions and by Rosetta. But the CONSERT radar instrument on Rosetta/Philae detected no large voids, at least none larger than $\sim$9 meters in size (Ciarletti et al. 2017a). That was a very unexpected result. Note that CONSERT only obtained two relatively shallow radar scans through the "head" (small lobe) of comet 67P. The spatial resolution limits of CONSERT cannot tell us about internal structure at scales $\lesssim$ 9 meters. So if there is macro-porosity in the nucleus, the only evidence we have is that it does not appear to be present on size scales $>$ 9 meters.

Yet the high porosity numbers determined for the bulk nucleus, 70% to 85% (Pätzold et al. 2016), suggests that there must be substantial micro-porosity. We see that micro-porosity in the COSIMA images of dust aggregates (Merouane et al. 2016), the MIDAS high resolution examination of dust grains (Bentley et al. 2016; Mannel et al. 2016; Mannel et al. 2019), and the possible existence of extremely porous dust aggregates in the 67P coma (Fulle et al. 2015). Can these examples of micro-porosity account completely for the bulk porosity of the 67P nucleus or must there also be some macro-porosity on size scales of 1 mm to 9 meters? The answer is, we don't know.

An important fact to consider is that both the hierarchical accretion and pebble growth hypotheses describe hierarchical accretion as the first step in producing particles up to 1 mm to 10 cm in size. For our purposes they are the same up to that size range. The question is what happens afterwards that leads to comet nuclei several kilometers in radius. In Sects. 2 and 3 we described these two competing hypotheses. Advocates for each hypothesis point to Rosetta/Philae results that they regard as supporting their ideas. But there are also critics of each process and we begin here by describing those criticisms.

### 7.1 Issues with the Hierarchical Growth of Planetesimals

Although the hierarchical growth of planetesimals described in Sect. 2 can explain a variety of observables among comets and Kuiper belt objects, it faces at least four serious issues:

**Fragmentation** Empirical evidence from laboratory experiments primarily performed for aggregates consisting of μm-sized silica monomers shows that fragmentation, i.e., a net mass loss and not mass gain (see Blum and Wurm 2008 and Güttler et al. 2010) is the dominant outcome in the majority of the collisions in the Davidsson et al. (2016) model. Bukhari Syed et al. (2017) derived a dust-aggregate collision model for macroscopic aggregates and collision velocities in the $\mathrm{m\,s^{-1}}$ regime. If we compare the impact energy per unit target mass, $E_{\mathrm{imp}}/m_{\mathrm{t}}$, from the Davidsson et al. (2016) model (see Table 1) with the dust-aggregate strength $Q^*$ extrapolated from the Bukhari Syed et al. (2017) model (see their Eq. (10)), we see that $E_{\mathrm{imp}}/m_{\mathrm{t}} \gg Q^*$ for all bodies with $D \gtrsim 1$ m. Thus, if no consolidation (e.g., by sintering or melting) occurs, which would considerably increase $Q^*$ above the values given by Bukhari et al. (2017), collisions will result in fragmentation. As Blum (2018) showed, the parameter space for mass gain of the target body is very limited.

**Erosion** The situation is even worse if one considers another physical process that only recently has been identified as a growth limiter. Schräpler et al. (2018) experimentally investigated the effect of impacts of very small dust particles and aggregates into large target aggregates. They showed that erosion, i.e., mass loss from the target, is the main result for impact speeds above a few $\mathrm{m\,s^{-1}}$. Based on their empirical laboratory data, Schräpler et al. (2018) showed in a numerical simulation that even under the assumptions that all other (non-eroding) collisions would result in perfect sticking, the maximum achievable aggregate size is on the order of 10 cm. This is because any erosional impact releases more small particles than it consumes so that an avalanche effect sets in. In an earlier work, Gundlach and Blum (2015) showed that erosion can also occur for µm-sized water-ice particles at low temperatures. However, see the discussion on "Ices" in Sect. 7.2 concerning the possibility that the stickiness of ice in general, and amorphous water ice in particular, may remove the fragmentation and erosion barriers.

**Timescale** The two most important timescale limitations with respect to mass growth in the pre-planetesimal phase are the radial drift, caused by friction with the sub-Keplerian gas disk, and the dissipation of the nebula gas. While the former is already important for aggregates in the mm- to cm-size range (see Lorek et al. 2018), the latter limits the formation of planetesimals by hierarchical growth. With realistic growth physics, Garaud et al. (2013) showed that bodies with sizes of $\sim$100 m can form at 1 AU, but farther out, at $\sim$30 AU in the disk, no growth beyond $\sim$10 m size was possible within $\sim 10^6$ years. Similarly, Windmark et al. (2012a, 2012b) calculated that $\sim$100 m-sized bodies can form within $\sim 10^6$ years at 3 AU, but nothing bigger than that. Note however that the Primordial Disk likely survived much longer than the $10^6$ years used in these simulations so there was time to grow larger bodies, though accretion of pebbles is eventually limited by pebble drift into the Sun.

**Strength** If bodies formed by hierarchical growth, impact pressures are always high enough to compact the material to porosities of $\sim$60% (Weidling et al. 2009) and turn the growing planetesimals into homogeneous bodies. It was shown experimentally that the tensile strength (inner cohesion) of such bodies is on the order of 1–10 kPa (Gundlach et al. 2018). Such high strength values are a severe problem for the activity of comets.

## 7.2 Issues with the Growth of Planetesimals from Pebbles

Here we identify some serious issues with the growth of planetesimals from pebbles.

**Numerical Simulations** of the gravitational collapse of pebble swarms show that this process typically produces two bodies of similar size that orbit about their common center of mass at distances that are very large compared to the body diameters (Nesvorný et al. 2010). The high fraction of ultrawide binaries (diameter $D = 100$ km class bodies separated by $10^3$–$10^5$ km) among the dynamically cold Kuiper belt objects (Stephens and Noll 2006) suggests that this mechanism indeed was active in the early Solar System. The collapse time of a $D = 100$ km body is $\sim$25 yr but the collapse time of a comet-sized $D = 1$ km body is several times $10^3$ yr (Wahlberg Jansson and Johansen 2014). The time needed to form comet-sized objects through pebble swarm collapse is so long that these swarms may disperse or merge into much larger swarms, before they have the time to form small objects.

**Computational Limits** Magnetohydrodynamical computer models of Solar Nebula gas and plasma interacting with pebbles do not yet have the resolution to demonstrate that small pebble swarms actually form and that they survive long enough against dissipation or growth to much larger size before collapsing to $D = 1$ km objects. The smallest swarms forming in the highest-resolution simulations published thus far can create $D = 60$ km objects (Johansen et al. 2015). Such bodies would likely not preserve the low bulk densities and high porosities we see for cometary nuclei, as shown in Sect. 3.2. Therefore, it is not yet known whether pebble swarm collapse is a viable mechanism for forming bodies the size of known comet nuclei.

**Heating** Because the pebble swarm collapse mechanism allows bodies to reach their final size on short time scales, heating from decaying short-lived radio-nuclides like $^{26}$Al may be strong. The high bulk density of some small, dynamically-cold Kuiper belt objects, like (66652) Borasisi with primary diameter $D_1 = 126^{+25}_{-51}$ km and $\rho = 2100^{+2600}_{-1200}\ \mathrm{kg\,m^{-3}}$ (Vilenius et al. 2014) suggest that loss of porosity due to radiogenic heating and ice melting probably occurred. Comet nuclei could not have experienced such heating without losing their observed rich inventory of supervolatiles (e.g., Prialnik et al. 1987; Merk and Prialnik 2003, 2006; Podolak and Prialnik 2006). This suggests that comets may have grown slowly and reached their final sizes late.

Johansen and Mac Low (2009) show that pebble swarm formation and collapse may require high disk metallicities, only achieved at the end of the ~3 Myr Solar Nebula lifetime (Zuckerman et al. 1995; Haisch et al. 2001; Sicilia-Aguilar et al. 2006). This would solve the $^{26}$Al problem but introduces new difficulties because the pebbles have rapid radial drift that cannot proceed for millions of years without emptying the outer planets region. Furthermore, direct exposure to protosolar radiation during excursions to high disk scale heights, likely to happen at some point during millions of years of evolution, leads to supervolatile loss in hours. Icy materials need to be protected inside a body much larger than the diurnal and orbital thermal skin depths of pebbles.

**Ices** Ros and Johansen (2013) showed that the role of ices could be very important in increasing the growth rate and the eventual sizes of pebbles. In their model, water ice vapor condenses onto particles beyond the snowline (see Introduction). Decimeter refractory particles can grow to meter sizes or larger in only a few times $10^3$ years. This model can thus help hierarchically grown particles jump the 1 meter barrier and continue to grow to larger sizes. That would make streaming instabilities inefficient in creating pebble swarms because of the difficulty of concentrating such larger bodies. Bridges et al. (1996) showed that particles covered with an ice frost could have increased stickiness as compared with refractory particles. Wettlaufer (2010) proposed that amorphous ice particles, warmed by collisions, could also increase stickiness.

### 7.3 Issues with Collisional Evolution

As described in Sect. 4, the best current models for the collisional environment of comets in the Primordial Disk and during their dispersal to the Kuiper belt, Scattered disk and Oort cloud (Morbidelli and Rickman 2015; Stern and Weissman 2001; Charnoz and Morbidelli 2003, 2007) suggest that comets suffered serious collisional evolution during that time. However, those estimates are based on some poorly known values, such as the total number of comet nuclei in the Primordial Disk, their orbital distributions and their size distributions.

The best estimates for the slope of the cumulative size distribution for the JFCs give values near −2 (Snodgrass et al. 2011; Meech et al. 2004) and are in good agreement with slope estimates for the size distributions found in theoretical predictions of the collisional evolution of planetesimals in the strength regime by O'Brien and Greenberg (2005). However, the agreement is not good for the predictions in Fig. 1 (left), which find a cumulative slope of −3 for the JFCs. Note that the JFC size distribution is only well known for radii between ∼1.2 and ∼10 km. Both observational and theoretical investigations need to continue in this area to further improve our understanding of its effects on comet nuclei.

### 7.4 Future Missions and Investigations

Rosetta has clearly told us more about comets than any previous space mission. It has demonstrated the importance of long-duration rendezvous missions with the capability to study cometary nuclei and cometary phenomena over extended periods of time. Below we discuss what future missions may be able to accomplish.

The next logical mission to a comet is Comet Surface Sample Return (CSSR). Such a mission, CAESAR (Comet Astrobiology Exploration Sample Return) was under consideration by NASA for its New Frontiers Program but it was not selected for development.. The goal of CSSR is to return a comet surface sample of 80–800 grams to Earth for study in terrestrial laboratories. The current capabilities of those laboratories far surpass that of any instruments that could be flown on a comet rendezvous mission. Those capabilities would have an additional 15–20 years to continue to improve while the mission is formulated, assembled, launched and flown. Returned cometary samples have the potential to tell us far more about the composition of cometary nuclei, the physical process of accretion and the formation of the low-volatility nucleus surface crusts. The samples might even contain "pebbles" if they exist.

The next most discussed comet mission after CSSR is Comet Cryogenic Sample Return (CCSR), like the Triple-F mission proposed to ESA several years ago (Küppers et al. 2009). This mission would dig below the nucleus surface to obtain samples that contain cometary ices. The collection depth would need to be at least several orbital thermal skin depths, $\gtrsim 3$ meters. The sample would need to be maintained at its cryogenic temperature, <80 K (or colder) for the duration of the return flight to Earth, including atmospheric entry and capsule retrieval. Both the cryogenic sample collection below the nucleus surface and maintaining the sample at its collected temperature are major technological challenges that require substantial investments by national space agencies in the coming decade. Also, because of its complexity the CCSR mission would need to be a flagship mission with a multi-billion dollar/euro budget. A CCSR mission would tell us far more about the complete composition of cometary nuclei and the accretion process.

Both CSSR and CCSR will greatly improve our knowledge of comet nucleus composition and possibly its chemical evolution over time. It is unlikely that either mission could address nucleus stratigraphy. CSSR concepts usually collect samples with semi-violent methods, i.e., nitrogen bursts into the surface or a mechanical brush-wheel sampler. The collection method for CCSR is not yet determined. However, the deceleration forces of re-entry of the sample return capsule into the Earth's atmosphere would likely destroy any existing structures within the low-strength comet materials. This problem could be partially overcome if samples could be separately obtained from different depths below the nucleus surface and stored isolated from one another. Smaller structures may survive re-entry and so the returned samples may provide important clues to the accretion process at microscopic scales.

CCSR might also be valuable in answering one of the major questions concerning the origin of icy planetesimals: whether water ice existed in amorphous or crystalline form in the Solar Nebula, or whether the water-ice grains were formed in the precursor molecular cloud or by condensation in the Solar Nebula. Another question of great interest for planetesimal-formation studies is the composition of the organic component and its stickiness in grain-grain collisions. If the organic particles were stickier than the silicate or water-ice particles, a direct collisional growth path from dust to planetesimals seems feasible with no need for pebbles and gravitational collapse. However, a stickier material than silicates and water ice would possibly increase internal cohesion within the comet nucleus so that dust activity might be impossible. The same issue may also be addressed regarding the stickiness of amorphous water-ice grains.

Like CSSR, CCSR may need to be so focused on sample collection and preservation that it will be unable to carry other scientific instruments that could address comet origin and evolution. So what can we do to address the issues discussed herein? It may require a separate rendezvous mission, like Rosetta, but carrying specific instruments to address origin and evolution. One choice would be a powerful radar instrument such as the Shallow Radar (SHARAD, Seu et al. 2007) on the Mars Reconnaissance Orbiter spacecraft. SHARAD could probe the interior structure of a cometary nucleus at all locations and not just the limited ray paths possible for CONSERT. SHARAD has a vertical resolution of 15 meters and a penetration depth of $\sim$1 km in the Martian soil; the latter number should be significantly greater in the ice-hydrocarbon-silicate mix of a comet nucleus. Thus, SHARAD may be able to map the interior structure of the entire nucleus for a typical 2–4 km diameter body.

In its current configuration SHARAD does not have the spatial resolution necessary to detect pebbles. Thus, another instrument worth considering is the WISDOM (Water Ice and Subsurface Deposit On Mars) radar on the ExoMars Rover, planned for launch in 2020. WISDOM (Ciarletti et al. 2017b) operates at several wavelengths, the shortest of which is 0.1 m (30 times finer than CONSERT) and can penetrate 3 m into the rocky soil of Mars. As noted above, it should be able to penetrate significantly deeper into the ice- and carbon-rich interior of a comet nucleus. Note that WISDOM requires a landed package or preferably a surface rover to accomplish its investigation. Also, the spatial resolution of WISDOM is not sufficient to resolve individual pebbles if the pebbles are less than tens of cm in diameter. Thus even higher resolution radars may be needed.

A landed instrument package could also address many of the Philae objectives that could not be addressed because of Philaes's final landed orientation. With the knowledge gained from study of Philae's bouncing, a lander package could be designed that would more likely remain at its targeted landing site. Again, the capabilities of the landed package would be greatly enhanced if it could be a rover rather than a fixed lander. Updated versions of all of the instruments on Philae would be highly desirable. This would include thermal probes to measure the temperature versus depth, penetrator instruments that could measure the strength of the surface materials, a drill, and compositional instruments to measure the composition of the near-surface materials. Finally, a high resolution camera system might be able to resolve individual pebbles on the surface.

Current thinking for CSSR is to return to 67P because of Rosetta's extensive mapping and the nucleus' known qualities. However, we propose that any future rendezvous mission as described above should visit a different comet nucleus to address questions of nucleus diversity as well as to provide a second target choice for the future CCSR mission. Having seen one cometary nucleus in great detail, 67P, it is too tempting to apply the Rosetta and Philae results in general to all comet nuclei. The several flyby missions to comets so far have shown them to be a very diverse population with different degrees of activity and surface evolution, and highly different topographies. A priority scientific objective is to visit as many nuclei as possible and to observe each with the intensity of Rosetta/Philae. As of this date there have already been six asteroid orbiters: NEAR, Hayabusa, Dawn at Vesta, Dawn at Ceres, Haybusa 2, and Osiris-REX. We need many more orbiter missions to comets.

In addition to new comet missions, future theoretical and laboratory investigations as well as computer simulations and telescopic programs will continue to improve our understanding of comet nuclei and how they formed. It will be valuable to see if pebble accretion scenarios can be extended down to actual comet nucleus sizes, i.e., radii $< 10$ km, while maintaining reasonable formation times. It will be interesting to understand better the role of amorphous water (and other) ices in possibly achieving more rapid accretion times as well as improving the ability of hierarchical accretion to cross the $\sim 1$ meter barrier that has stymied researchers for decades. Increasingly deep small body surveys of the solar system will improve our understanding of the size distributions of comets in the two major reservoirs as well as their orbital and spatial distributions. We are not foolish enough to try and predict all or even some of the future discoveries about comets and their origin but we do look eagerly toward them.

**Acknowledgements** We thank the two anonymous reviewers for their helpful and constructive comments and suggestions. PRW thanks the U.S. Social Security Administration for financial support while writing this paper. BD's part of this paper was performed at the Jet Propulsion Laboratory under contract with NASA. JB thanks the Deutsche Forschungsgemeinschaft (DFG) and the Deutsches Zentrum für Luft- und Raumfahrt (DLR) for continuous support.

**Publisher's Note** Springer Nature remains neutral with regard to jurisdictional claims in published maps and institutional affiliations.

## References

C.P. Abod, J.B. Simon, R. Li, P.J. Armitage, A.N. Youdin, K.A. Kretke, The mass and size distribution of planetesimals formed by the streaming instability. II. The effect of the radial gas pressure gradient. Astrophys. J. **883**(2), 192 (2018). http://adsabs.harvard.edu/abs/2018arXiv181010018A

I. Adachi, C. Hayashi, K. Nakazawa, The gas drag effect on the elliptic motion of a solid body in the primordial solar nebula. Prog. Theor. Phys. **56**(6), 1756–1771 (1976)

M.F. A'Hearn, R.C. Millis, D.G. Schleicher, D.J. Osip, P.V. Birch, The ensemble properties of comets: results from narrowband photometry of 85 comets. Icarus **118**, 223–270 (1995)

M.F. A'Hearn, M.J.S. Belton, W.A. Delamere, L.M. Feaga, D. Hampton, J. Kissel, K.P. Klaasen, L.A. McFadden, K.J. Meech, H.J. Melosh, P.H. Schultz, J.M. Sunshine, P.C. Thomas, J. Veverka, D.D. Wellnitz, D.K. Yeomans, S. Besse, D. Bodewits, T.J. Bowling, B.T. Carcich, S.M. Collins, T.L. Farnham, O. Groussin, B. Hermalyn, M.S. Kelley, M.S. Kelley, J.-Y. Li, D.J. Lindler, C.M. Lisse, S.A. McLaughlin, F. Merlin, S. Protopapa, J.E. Richardson, J.L. Williams, EPOXI at comet Hartley 2. Science **332**, 1396–1400 (2011)

K. Altwegg, H. Balsiger, A. Bar-Nun, J.-J. Berthelier, A. Bieler, P. Bochsler, C. Briois, U. Calmonte, M.R. Combi, H. Cottin, J. De Keyser, F. Dhooghe, B. Fiethe, S.A. Fuselier, S. Gasc, T.I. Gombosi, K.C. Hansen, M. Haessig, A. Jäckel, E. Kopp, A. Korth, L. Le Roy, U. Mall, B. Marty, O. Mousis, T. Owen, H. Reme, M. Rubin, T. Semon, C.-Y. Tzou, J.H. Waite, P. Wurz, Prebiotic chemicals—amino acids and phosphorus—in the coma of comet 67P/Churyumov-Gerasimenko. Sci. Adv. **2**, e1600285 (2016)

P. Andre, T. Montmerle, From T Tauri stars to protostars: circumstellar material and young stellar objects in the Ophiuchi cloud. Astron. J. **420**, 837–862 (1994)

E. Asphaug, W. Benz, Density of comet Shoemaker-Levy 9 deduced by modeling of the parent rubble pile. Nature **370**, 120–124 (1994)

E. Asphaug, W. Benz, Size, density and structure of comet Shoemaker-Levy 9 inferred from the physics of tidal breakup. Icarus **121**, 225–248 (1996)

M. Asplund, N. Grevesse, A.J. Sauval, P. Scott, The chemical composition of the Sun. Annu. Rev. Astron. Astrophys. **47**, 481–522 (2009)

N. Attree, O. Groussin, L. Jorda, D. Nébouy, N. Thomas, Y. Brouet, E. Kührt, F. Preusker, F. Scholten, J. Knollenberg, P. Hartogh, H. Sierks, C. Barbieri, P. Lamy, R. Rodrigo, D. Koschny, H. Rickman, H.U. Keller, M.F. A'Hearn, A.-T. Auger, M.A. Barucci, J.-L. Bertaux, I. Bertini, D. Bodewits, S. Boudreault, G. Cremonese, V. Da Deppo, B. Davidsson, S. Debei, M. De Cecco, J. Deller, M.R. El-Maarry, S. Fornasier, M. Fulle, P.J. Gutiérrez, C. Güttler, S. Hviid, W.-H. Ip, G. Kovacs, J.R. Kramm, M. Küppers, L.M. Lara, M. Lazzarin, J.J. Lopez Moreno, S. Lowry, S. Marchi, F. Marzari, S. Mottola, G. Naletto, N. Oklay, M. Pajola, I. Toth, C. Tubiana, J.-B. Vincent, X. Shi, Tensile strength of 67P/Churyumov-Gerasimenko nucleus material from overhangs. Astron. Astrophys. **611**, A33 (2018)

A.-T. Auger, O. Groussin, L. Jorda, S. Bouley, R. Gaskell, P.L. Lamy, C. Capanna, N. Thomas, A. Pommerol, H. Sierks, C. Barbieri, R. Rodrigo, D. Koschny, H. Rickman, H.U. Keller, J. Agarwal, M.F. A'Hearn, M.A. Barucci, J.-L. Bertaux, I. Bertini, G. Cremonese, V. Da Deppo, B. Davidsson, S. Debei, M. De Cecco, M.R. El-Maarry, S. Fornasier, M. Fulle, P.J. Gutierrez, C. Güttler, S. Hviid, W.-H. Ip, J. Knollenberg, J.-R. Kramm, E. Kührt, M. Küppers, F. La Forgia, L.M. Lara, M. Lazzarin, J.J. Lopez Moreno, S. Marchi, F. Marzari, M. Massironi, H. Michalik, G. Naletto, N. Oklay, M. Pajola, L. Sabau, C. Tubiana, J.-B. Vincent, K.-P. Wenzel, Geomorphology of the Imhotep region on comet 67P/Churyumov-Gerasimenko from OSIRIS observations. Astron. Astrophys. **583**, A35 (2015)

A.-T. Auger, O. Groussin, L. Jorda, M.R. El-Maarry, S. Bouley, A. Sejourne, R. Gaskell, C. Capanna, B. Davidsson, S. Marchi, S. Höfner, P.L. Lamy, H. Sierks, C. Barbieri, R. Rodrigo, D. Koschny, H. Rickman, H.U. Keller, J. Agarwal, M.F. A'Hearn, M.A. Barucci, J.-L. Bertaux, I. Bertini, G. Cremonese, V. Da Deppo, S. Debei, M. De Cecco, S. Fornasier, M. Fulle, P.J. Gutierrez, C. Güttler, S. Hviid, W.-H. Ip, J. Knollenberg, J.-R. Kramm, E. Kührt, M. Küppers, L.M. Lara, M. Lazzarin, J.J. Lopez Moreno, F. Marzari, M. Massironi, H. Michalik, G. Naletto, N. Oklay, A. Pommerol, L. Sabau, N. Thomas, C. Tubiana, J.-B. Vincent, K.-P. Wenzel, Meter-scale thermal contraction crack polygons on the nucleus of comet 67P/Churyumov-Gerasimenko. Icarus **301**, 173–188 (2018)

M.J.S. Belton, The size-distribution of scattered disk TNOs from that of JFCs between 0.2 and 15 km effective radius. Icarus **231**, 168–182 (2014)

M.J.S. Belton, P. Thomas, J. Veverka, P. Schultz, M.F. A'Hearn, L. Feaga, T. Farnham, O. Groussin, J.-Y. Li, C. Lisse, L. McFadden, J. Sunshine, K.J. Meech, W.A. Delamere, J. Kissel, The internal structure of Jupiter family cometary nuclei from Deep Impact observations: the "talps" or "layered pile" model. Icarus **187**, 332–344 (2007)

M.S. Bentley, R. Schmied, T. Mannel, . co-authors, Aggregate dust particles at comet 67P/Churyumov-Gerasimenko. Nature **537**, 73–75 (2016)

D.A. Biesecker, P. Lamy, O.C. St. Cyr, A. Llebaria, R.A. Howard, Sungrazing comets discovered with the SOHO/LASCO coronagraphs 1996–1998. Icarus **157**, 323–348 (2002)

T. Birnstiel, L. Ricci, F. Trotta, C.P. Dullemond, A. Natta, L. Testi, C. Dominik, T. Henning, C.W. Ormel, A. Zsom, Testing the theory of grain growth and fragmentation by millimeter observations of protoplanetary disks. Astron. Astrophys. **516**, L14 (2010)

D. Bischoff, B. Gundlach, M. Neuhaus, J. Blum, Experiments on cometary activity: ejection of dust aggregates from an evaporating water–ice surface. Mon. Not. R. Astron. Soc. **483**, 1202–1210 (2019)

J. Blum, Dust agglomeration. Adv. Phys. **55**, 881–947 (2006)

J. Blum, Dust evolution in protoplanetary discs and the formation of planetesimals. What have we learned from laboratory experiments? Space Sci. Rev. **214**, 52 (2018)

J. Blum, G. Wurm, The growth mechanisms of macroscopic bodies in protoplanetary disks. Annu. Rev. Astron. Astrophys. **46**, 21–56 (2008)

J. Blum, G. Wurm, S. Kempf, T. Poppe, H. Klahr, T. Kozasa, M. Rott, T. Henning, J. Dorschner, R. Schräpler, H.U. Keller, W.J. Markiewicz, I. Mann, B.A. Gustafson, F. Giovane, D. Neuhaus, H. Fechtig, E. Grün, B. Feuerbacher, H. Kochan, L. Ratke, A. El Goresy, G. Morfill, S.J. Weidenschilling, G. Schwehm, K. Metzler, W.-H. Ip, Growth and form of planetary seedlings: results from a microgravity aggregation experiment. Phys. Rev. Lett. **85**, 2426–2429 (2000)

J. Blum, B. Gundlach, S. Mühle, J.M. Trigo-Rodriguez, Comets formed in solar nebula instabilities!—An experimental and modeling attempt to relate the activity of comets to their formation process. Icarus **235**, 156–169 (2014)

J. Blum, B. Gundlach, S. Mühle, J.M. Trigo-Rodriguez, Corrigendum to "Comets formed in solar-nebula instabilities!—An experimental and modeling attempt to relate the activity of comets to their formation process" [Icarus 235 (2014) 156–169]. Icarus **248**, 135–136 (2015)

J. Blum, B. Gundlach, M. Krause, M. Fulle, A. Johansen, J. Agarwal, I. von Borstel, X. Shi, X. Hu, M.S. Bentley, F. Capaccioni, L. Colangeli, V. Della Corte, N. Fougere, S.F. Green, S. Ivanovski, T. Mannel, S. Merouane, A. Migliorini, A. Rotundi, R. Schmied, C. Snodgrass, Evidence for the formation of comet 67P/Churyumov- Gerasimenko through gravitational collapse of a bound clump of pebbles. Mon. Not. R. Astron. Soc. **469**, S755–S773 (2017)

H. Boehnhardt, Split comets, in *Comets II*, ed. by M.C. Festou, U. Keller, H.A. Weaver (Univ. Arizona Press, Tucson, 2004), pp. 301–316

H. Boehnhardt, H.U. Kaeufl, P. Goudgrooij, J. Storm, J. Manfroid, K. Reinsch, The break-up of periodic comet Schwassmann-Wachmann 3: image documents from La Silla telescopes. Messenger **84**, 26–29 (1996)

W.F. Bottke, A. Morbidelli, R. Jedicke, J.-M. Petit, H.F. Levison, P. Michel, T.S. Metcalfe, Debiased orbital and absolute magnitude distribution of the near-Earth objects. Icarus **156**, 399–433 (2002)

W.F. Bottke, D.D. Durda, D. Nesvorný, R. Jedicke, A. Morbidelli, D. Vokrouhlický, H. Levison, The fossilized size distribution of the main asteroid belt. Icarus **175**, 111–140 (2005)

R. Brasser, A. Morbidelli, Oort cloud and Scattered Disc formation during a late dynamical instability in the Solar System. Icarus **225**, 40–49 (2013)

F.G. Bridges, K.D. Supulver, D.N.C. Lin, R. Knight, M. Zafra, Energy loss and sticking mechanisms in particle aggregation in planetesimal formation. Icarus **123**, 422–435 (1996)

G.D. Brin, D.A. Mendis, Dust release and mantle development in comets. Astron. J. **229**, 1095–1108 (1979)

J. Brisset, D. Heißelmann, S. Kothe, R. Weidling, J. Blum, Submillimetre-sized dust aggregate collision and growth properties. Experimental study of a multi-particle system on a suborbital rocket. Astron. Astrophys. **593**, A3 (2016)

D.E. Brownlee, Cosmic dust: collection and research. Annu. Rev. Earth Planet. Sci. **13**, 147–173 (1985)

D. Brownlee, D. Joswiak, G. Matrajt, Overview of the rocky component of Wild 2 comet samples: insight into the early solar system, relationship with meteoritic materials and the differences between comets and asteroids. Meteorit. Planet. Sci. **47**(4), 453–470 (2012)

M. Bukhari Syed, J. Blum, K. Wahlberg Jansson, A. Johansen, The role of pebble fragmentation in planetesimal formation. I. Experimental study. Astrophys. J. **834**, 145 (2017)

A. Carusi, L. Kresak, E. Perozzi, G.B. Valsecchi, First results of the integration of motion of short-period comets over 800 years, in *Dynamics of Comets: Their Origin and Evolution*, ed. by A. Carusi, G.B. Valsechhi. (D. Reidel, Dordrecht, 1985), pp. 319–340

S. Charnoz, A. Morbidelli, Coupling of dynamical and collisional evolution of small bodies: an application to the early ejection of planetesimals from the Jupiter-Saturn reion. Icarus **166**, 141–156 (2003)

S. Charnoz, A. Morbidelli, Coupling dynamical and collisional evolution of small bodies. II. Forming the Kuiper belt, the Scattered Disk and the Oort Cloud. Icarus **188**, 468–480 (2007)

S.R. Chesley, M.J.S. Belton, B. Carcich, P.C. Thomas, J. Pittichova, K.P. Klaasen, J.-Y. Li, T.L. Farnham, S.D. Gillam, A.W. Harris, J. Veverka, An updated rotation model for Comet 9P/Tempel 1. Icarus **222**, 516–525 (2013)

P.W. Chodas, D.K. Yeomans, The orbital motion and impact circumstances of comet Shoemaker-Levy 9, in *The Collision of Comet Shoemaker-Levy 9 and Jupiter*, ed. by K.S. Noll, H.A. Weaver, P.D. Feldman. Space Tel. Sci. Inst. (1996), pp. 1–30

V. Ciarletti, A.C. Levasseur-Regourd, J. Lasue, C. Statz, D. Plettemeier, A. Herique, Y. Rogez, W. Kofman, CONSERT suggests a change in local properties of 67P/Churyumov-Gerasimenko's nucleus at depth. Astron. Astrophys. **583**, A40 (2015)

V. Ciarletti et al., The WISDOM radar: unveiling the subsurface beneath the ExoMars rover and identifying the best locations for drilling. Astrobiology **17**, 565–584 (2017b)

V. Ciarletti, A. Herique, J. Lasue, A.-C. Levasseur-Regourd, D. Plettemeier, F. Lemmonier, C. Guiffaut, P. Pasquero, W. Kofman, CONSERT constrains the internal structure of 67P at a few meters size scale. Mon. Not. R. Astron. Soc. **469**, S805–S817 (2017a)

J.F. Cooper, E.R. Christian, R.R. Johnson, Heliospheric cosmic ray irradiation of Kuiper belt comets. Adv. Space Res. **21**, 1611–1614 (1998)

P. D'Alessio, N. Calvet, L. Hartmann, R. Franco-Hernández, H. Servín, Effects of dust growth and settling in T Tauri disks. Astrophys. J. **638**, 314–335 (2006)

B.J.R. Davidsson, P.J. Gutiérrez, Estimating the nucleus density of Comet 19P/Borrelly. Icarus **168**(2), 392–408 (2004)

B.J.R. Davidsson, P.J. Gutiérrez, Nucleus properties of Comet 67P/Churyumov-Gerasimenko estimated from non-gravitational force modeling. Icarus **176**(2), 453–477 (2005)

B.J.R. Davidsson, P.J. Gutiérrez, Non-gravitational force modeling of comet 81P/Wild 2. I. A nucleus bulk density estimate. Icarus **180**, 224–242 (2006)

B.J.R. Davidsson, P.J. Gutierrez, H. Rickman, Nucleus properties of Comet 9P/Tempel 1 estimated from non-gravitational force modeling. Icarus **187**, 306–320 (2007)

B.J.R. Davidsson, H. Sierks, C. Güttler, F. Marzari, M. Pajola, H. Rickman, M.F. A'Hearn, A.-T. Auger, M.R. El-Maarry, S. Fornasier, P.J. Gutiérrez, H.U. Keller, M. Massironi, C. Snodgrass, J.-B. Vincent, C. Barbieri, P.L. Lamy, R. Rodrigo, D. Koschny, M.A. Barucci, J.-L. Bertaux, I. Bertini, G. Cremonese, V. Da Deppo, S. Debei, M. De Cecco, C. Feller, M. Fulle, O. Groussin, S.F. Hviid, S. Höfner, W.-H. Ip, L. Jorda, J. Knollenberg, G. Kovacs, J.-R. Kramm, E. Kührt, M. Küppers, F. La Forgia, L.M. Lara,

M. Lazzarin, J.J. Lopez Moreno, R. Moissl-Fraund, S. Mottola, G. Naletto, N. Oklay, N. Thomas, C. Tubiana, The primordial nucleus of comet 67P/Churyumov-Gerasimenko. Astron. Astrophys. **592**, A63 (2016)

D. de Niem, E. Kührt, S. Hviid, B. Davidsson, Low velocity collisions of porous planetesimals in the early solar system. Icarus **301**, 196–218 (2018)

F. DeMeo, R.P. Binzel, Comets in the near-Earth object population. Bull. Am. Astron. Soc. **38**, 581 (2006)

C. Dominik, A.G.G.M. Tielens, The physics of dust coagulation and the structure of dust aggregates in space. Astrophys. J. **480**, 647–673 (1997)

B. Donn, The accumulation and structure of comets, in *Comets in the Post-Halley Era*, vol. 1, ed. by R.L. Newburn, M. Neugebauer, J. Rahe (Kluwer Academic Publishers, Dordrecht, 1991), pp. 335–355

B. Donn, D. Hughes, A fractal model of a cometary nucleus formed by random accretion, in *Proc. 20th ESLAB Symposium on the Exploration of Halley's Comet* (1986), pp. 523–524. ESA SP-250

M. Duncan, H.F. Levison, A scattered comet disk and the origin of Jupiter family comets. Science **276**, 1670–1672 (1997)

M. Duncan, T. Quinn, S. Tremaine, The origin of short-period comets. Astrophys. J. Lett. **328**, L69–L73 (1988)

M.R. El-Maarry, N. Thomas, A. Garcia Berna, R. Marschall, A.-T. Auger, O. Groussin, M. Massironi, S. Marchi, F. Preusker, F. Scholten, L. Jorda, E. Kührt, M. Hofmann, S. Hoefner, J. Deller (the OSIRIS team), Fractures on comet 67P/Churyumov-Gerasimenko observed by the Rosetta/OSIRIS camera. Geophys. Res. Lett. **42**(13), 5170–5178 (2015)

S. Espinasse, J. Klinger, C. Ritz, B. Schmitt, Modeling of the thermal behavior and of the chemical differentiation of cometary nuclei. Icarus **92**, 350–365 (1991)

F.P. Fanale, J.R. Salvail, An idealized short-period comet model: surface insolation, H2O flux, and mantle development. Icarus **60**, 476–511 (1984)

J.A. Fernández, New and evolved comets in the solar system. Astron. Astrophys. **96**, 26–35 (1981)

Y.R. Fernández, D.C. Jewitt, S.S. Shephard, Low albedos among extinct comet candidates. Astrophys. J. Lett. **553**, L197–L200 (2001)

S. Ferrari, L. Penasa, F. La Forgia, M. Massironi, G. Naletto, M. Lazzarin, S. Fornasier, P.H. Hasselmann, A. Lucchetti, M. Pajola, F. Ferri, P. Cambianica, N. Oklay, C. Tubiana, H. Sierks, P.L. Lamy, R. Rodrigo, D. Koschny, B. Davidsson, M.A. Barucci, J.-L. Bertaux, I. Bertini, D. Bodewits, G. Cremonese, V. Da Deppo, S. Debei, M. De Cecco, J. Deller, M. Franceschi, E. Frattin, M. Fulle, O. Groussin, P.J. Gutiérrez, C. Güttler, S.F. Hviid, W.-H. Ip, L. Jorda, H.U. Keller, J. Knollenberg, E. Kührt, M. Küppers, L.M. Lara, J.J. López-Moreno, F. Marzari, X. Shi, E. Simioni, N. Thomas, J.-B. Vincent, The big lobe of 67P/Churyumov-Gerasimenko comet: morphological and spectrophotometric evidences of layering as from OSIRIS data. Mon. Not. R. Astron. Soc. **479**(2), 1555–1568 (2018)

M. Fulle, J. Blum, Fractal dust constrains the collisional history of comets. Mon. Not. R. Astron. Soc. **469**, S39–S44 (2017)

M. Fulle, V. Della Corte, A. Rotundi, P. Weissman, A. Juhasz, K. Szego, R. Sordini, M. Ferrari, S. Ivanovski, F. Lucarelli, M. Accolla, S. Merouane, V. Zakharov, E. Mazzotta Epifani, J.J. L_opez-Moreno, J. Rodríguez, L. Colangeli, P. Palumbo, E. Grün, M. Hilchenbach, E. Bussoletti, F. Esposito, S.F. Green, P.L. Lamy, J.A.M. McDonnell, V. Mennella, A. Molina, R. Morales, F. Moreno, J.L. Ortiz, E. Palomba, R. Rodrigo, J.C. Zarnecki, M. Cosi, F. Giovane, B. Gustafson, M.L. Herranz, J.M. Jerónimo, M.R. Leese, A.C. López-Jiménez, N. Altobelli, Density and charge of pristine fluffy particles from comet 67P/Churyumov-Gerasimenko. Astrophys. J. Lett. **802**, L12 (2015)

M. Fulle, J. Blum, S.F. Green, B. Gundlach, A. Henrique, F. Moreno, S. Mottola, A. Rotundi, C. Snodgrass, The refractory-to-ice mass ratio in comets. Mon. Not. R. Astron. Soc. **482**, 3326–3340 (2019)

P. Garaud, F. Meru, M. Galvagni, C. Olczak, From dust to planetesimals: an improved model for collisional growth in protoplanetary disks. Astrophys. J. **764**, 146 (2013)

S. Gasc, et al., Change of outgassing pattern of 67P/Churyumov-Gerasimenko during the March 2016 equinox as seen by ROSINA. Mon. Not. R. Astron. Soc. **469**, S108–S117 (2017)

R.W. Gaskell, L. Jorda, H. Sierks, P. Gutierrez, S. Faurschou Hviid, H.U. Keller, S. Mottola, C. Campanna (the OSIRIS team), Changes in Comet 67P/Churyumov-Gerasimenko during the ROSETTA Era—Shape, Topography and Rotation. AAS/DPS Meet. **48**, 11603 (2016)

K.-H. Glassmeier, H. Boehnhardt, D. Koschny, e. Kuhrt, I. Richter, The Rosetta mission: flying towards the origin of the solar system. Space Sci. Rev. **128**, 1–21 (2007)

R. Gomes, H.F. Levison, K. Tsiganis, A. Morbidelli, Origin of the cataclysmic Late Heavy Bombardment of the terrestrial planets. Nature **435**, 466–469 (2005)

O. Groussin, L. Jorda, A.-T. Auger, E. Kührt, R. Gaskell, C. Capanna, F. Scholten, F. Preusker, P. Lamy, S. Hviid, J. Knollenberg, U. Keller, C. Huettig, H. Sierks, C. Barbieri, R. Rodrigo, D. Koschny, H. Rickman, M.F. A'Hearn, J. Agarwal, M.A. Barucci, J.-L. Bertaux, I. Bertini, S. Boudreault, G. Cremonese, V. Da Deppo, B. Davidsson, S. Debei, M. De Cecco, M.R. El-Maarry, S. Fornasier, M. Fulle,

P.J. Gutiérrez, C. Güttler, W.-H. Ip, J.-R. Kramm, M. Küppers, M. Lazzarin, L.M. Lara, J.J. Lopez Moreno, S. Marchi, F. Marzari, M. Massironi, H. Michalik, G. Naletto, N. Oklay, A. Pommerol, M. Pajola, N. Thomas, I. Toth, C. Tubiana, J.-B. Vincent, Gravitational slopes, geomorphology, and material strengths of the nucleus of comet 67P/Churyumov-Gerasimenko from OSIRIS observations. Astron. Astrophys. **583**, A32 (2015)

B. Gundlach, J. Blum, The stickiness of micrometer-sized water-ice particles. Astrophys. J. **798**, 34 (2015)

B. Gundlach, J. Blum, Why are Jupiter-family comets active and asteroids in cometary-like orbits inactive? How hydrostatic compression leads to inactivity. Astron. Astrophys. **589**, A111 (2016)

B. Gundlach, J. Blum, H.U. Keller, Y.V. Skorov, What drives the dust activity of comet 67P/Churyumov-Gerasimenko? Astron. Astrophys. **583**, A12 (2015)

B. Gundlach, K.P. Schmidt, C. Kreuzig, D. Bischoff, F. Rezaei, S. Kothe, J. Blum, B. Grzesik, E. Stoll, The tensile strength of ice and dust aggregates and its dependence on particle properties. Mon. Not. R. Astron. Soc. **479**, 1273–1277 (2018)

P.J. Gutiérrez, R. Rodrigo, J.L. Ortiz, B.J.R. Davidsson, An investigation of errors in estimates of the cometary nuclei active area fractions. Astron. Astrophys. **401**, 755–761 (2003)

C. Güttler, J. Blum, A. Zsom, C.W. Ormel, C.P. Dullemond, The outcome of protoplanetary dust growth: pebbles, boulders, or planetesimals? I. Mapping the zoo of laboratory collision experiments. Astron. Astrophys. **513**, A56 (2010)

K.E. Haisch, E.A. Lada, C.J. Lada, Disk frequencies and lifetimes in young clusters. Astrophys. J. **553**, L153–L156 (2001)

J.K. Harmon, M.C. Nolan, J.D. Giorgini, E.S. Howell, Radar observations of 8P/Tuttle: a contact-binary comet. Icarus **207**, 499–502 (2010)

A. Herique, W. Kofman, P. Beck, L. Bonal, I. Buttarazzi, E. Heggy, J. Lasue, A.C. Levasseur- Regourd, E. Quirico, S. Zine, Cosmochemical implications of CONSERT permittivity characterization of 67P/CG. Mon. Not. R. Astron. Soc. **462**, S516–S532 (2016)

M. Hirabayashi, D. Scheeres, S.R. Chesley, S. Marchi, J.W. McMahon, J. Steckloff, S. Mottola, S.P. Naidu, T. Bowling, Fission and reconfiguration of bilobate comets as revealed by 67P/Churyumov-Gerasimenko. Nature **534**, 352–355 (2016)

K.M. Housen, K. Holsapple, A crater and its ejecta: an interpretation of Deep Impact. Icarus **187**, 345–356 (2007)

X. Hu, B. Gundlach, I. von Borstel, J. Blum, X. Shi, Effect of radiative heat transfer in porous comet nuclei: case study of 67P/Churyumov-Gerasimenko. Astron. Astrophys. **630**, A5 (2019)

A. Johansen, A.N.Y.M.-M. Mac Low, Particle clumping and planetesimal formation depend strongly on metallicity. Astrophys. J. **704**, L75–L79 (2009)

A. Johansen, J.S. Oishi, M.-M. Mac Low, H. Klahr, T. Henning, A. Youdin, Rapid planetesimal formation in turbulent circumstellar disks. Nature **448**, 1022–1025 (2007)

A. Johansen, J. Blum, H. Tanaka, C. Ormel, M. Bizzarro, H. Rickman, The multifaceted planetesimal formation process, in *Protostars and Planets VI* (2014), pp. 547–570

A. Johansen, M.-M. Mac Low, P. Lacerda, M. Bizzarro, Growth of asteroids, planetary embryos, and Kuiper belt objects by chondrule accretion. Sci. Adv. **1**, e1500109 (2015)

R.E. Johnson, Irradiation effects in a comet's outer layers. J. Geophys. Res. **96**, 17553–17557 (1991)

G. Jones, M.M. Knight, K. Battams, . co-athors, The science of sungrazers, sunskirters, and other near-Sun comets. Space Sci. Rev. **214**(1), 20 (2018)

L. Jorda, R. Gaskell, C. Capanna, S. Hviid, P. Lamy, J. Durech, G. Faury, O. Groussin, P. Gutierrez, C. Jackman, S.J. Keihm, H.U. Keller, J. Knollenberg, E. Kührt, S. Marchi, S. Mottola, E. Palmer, F.P. Schloerb, H. Sierks, J.-B. Vincent, M.F. A'Hearn, C. Barbieri, R. Rodrigo, D. Koschny, H. Rickman, M.A. Barucci, J.L. Bertaux, I. Bertini, G. Cremonese, V.D. Deppo, B. Davidsson, S. Debei, M. De Cecco, S. Fornasier, M.F.A.C. Güttler, W.-H. Ip, J.R. Kramm, M.K. an, L.M. Lara, M. Lazzarin, J.J. Lopez Moreno, F. Marzari, G. Naletto, N. Oklay, N. Thomas, C. Tubiana, K.-P. Wenzel, The global shape, density and rotation of comet 67P/Churyumov-Gerasimenko from preperihelion Rosetta/OSIRIS observations. Icarus **277**, 257–278 (2016)

M. Jutzi, E. Asphaug, The shape and structure of cometary nuclei as a result of low-velocity accretion. Science **348**(6241), 1355–1358 (2015)

M. Jutzi, W. Benz, Formation of bi-lobed shapes by sub-catastrophic collisions. A late origin of comet 67P's structure. Astron. Astrophys. **597**, A62 (2017)

M. Jutzi, W. Benz, A. Toliou, A. Morbidelli, R. Brasser, How primordial is the structure of comet 67P? Combined collisional and dynamical models suggest a late formation. Astron. Astrophys. **597**, A61 (2017)

H.U. Keller, D. Britt, B.J. Buratti, N. Thomas, In situ observations of cometary nuclei, in *Comets II*, ed. by M.C. Festou, U. Keller, H.A. Weaver (Univ. Arizona Press, Tucson, 2004), pp. 211–222

H.U. Keller, S. Mottola, B. Davidsson, . co-authors, Insolation, erosion and morphology of comet 67P/Churyumov-Gerasimenko. Astron. Astrophys. **583**, A34 (2015)

M.S. Kelley, D.H. Wooden, The composition of dust in Jupiter-family comets inferred from infrared spectroscopy. Planet. Space Sci. **57**, 1133–1145 (2009)

S.J. Kenyon, J.X. Luu, Accretion in the early Kuiper belt. I. Coagulation and velocity evolution. Astron. J. **115**, 2136–2160 (1998)

D.J. Kessler, Derivation of the collision probability between orbiting objects: the lifetimes of Jupiter's outer moons. Icarus **48**, 39–48 (1981)

Y. Kim, M. Ishiguro, F. Usui, Physical properties of asteroids in comet-like orbits in infrared asteroid survey catalogs. Astrophys. J. **789**, 151 (2014)

M. Knight, M.F. A'Hearn, D.A. Biesecker, G. Faury, D.P. Hamilton, P. Lamy, A. Llebaria, Photometric study of the Kreutz comets observed by SOHO from 1996–2005. Astron. J. **139**, 926–949 (2010)

W. Kofman, A. Herique, Y. Barbin, J.-P. Barriot, V. Ciarletti, S. Clifford, P. Edenhofer, C. Elachi, C. Eyraud, J.-P. Goutail, E. Heggy, L. Jorda, J. Lasue, A.-C. Levasseur-Regourd, E. Nielsen, P. Pasquero, F. Preusker, P. Puget, D. Plettemeier, Y. Rogez, H. Sierks, C. Statz, H. Svedhem, I. Williams, S. Zine, J. Van Zyl, Properties of the 67P/Churyumov-Gerasimenko interior revealed by CONSERT radar. Science **249**, aab0639-1 (2015)

H. Kreutz, Investigations about the cometary system 1843 I, 1880 I and 1882 II Part I. Publ. Kiel Observatory. Printed by C. Schaidt, C.F. Mohr Nachfl. (1888)

M. Krolikówska, 67P/Churyumov-Gerasimenko—potential target for the Rosetta mission. Acta Astron. **54**, 195–209 (2003)

G.W. Kronk, *Comets, A Descriptive Catalog* (Enslow, Hillside, 1984). 331 pp

G.P. Kuiper, On the origin of the solar system, in *Astrophysics*, ed. by J.A. Hynek (McGraw Hill, New York, 1951), pp. 357–424

M. Küppers, H.U. Keller, E. Kührt, M.F. A'Hearn, K. Altwegg, R. Bertrand, H. Busemann, M.T. Capria, L. Colangeli, B. Davidsson, P. Ehrenfreund, J. Knollenberg, S. Mottola, A. Rathke, P. Weiss, M. Zolensky, E. Akim, A. Basilevsky, E. Galimov, M. Gerasimov, O. Korablev, I. Lomakin, M. Marov, M. Martynov, M. Nazarov, A. Zakharov, L. Zelenyi, A. Aronica, A.J. Ball, C. Barbieri, A. Bar-Nun, J. Benkhoff, J. Biele, N. Biver, J. Blum, D. Bockelée-Morvan, O. Botta, J.-H. Bredehöft, F. Capaccioni, S. Charnley, E. Cloutis, H. Cottin, G. Cremonese, J. Crovisier, S.A. Crowther, E.M. Epifani, F. Esposito, A.C. Ferrari, F. Ferri, M. Fulle, J. Gilmour, F. Goesmann, N. Gortsas, S.F. Green, O. Groussin, E. Grün, P.J. Gutiérrez, P. Hartogh, T. Henkel, M. Hilchenbach, T.-M. Ho, G. Horneck, S.F. Hviid, W.-H. Ip, A. Jäckel, E. Jessberger, R. Kallenbach, G. Kargl, N.I. Kömle, A. Korth, K. Kossacki, C. Krause, H. Krüger, Z.-Y. Li, J. Licandro, J.J. Lopez-Moreno, S.C. Lowry, I. Lyon, G. Magni, U. Mall, I. Mann, W. Markiewicz, Z. Martins, M. Maurette, U. Meierhenrich, V. Mennella, T.C. Ng, L.R. Nittler, P. Palumbo, M. Pätzold, D. Prialnik, M. Rengel, H. Rickman, J. Rodriguez, R. Roll, D. Rost, A. Rotundi, S. Sandford, M. Schönbächler, H. Sierks, R. Srama, R.M. Stroud, S. Szutowicz, C. Tornow, S. Ulamec, M. Wallis, W. Waniak, P. Weissman, R. Wieler, P. Wurz, K.L. Yung, J.C. Zarnecki, Triple F—a comet nucleus sample return mission. Exp. Astron. **23**, 809–847 (2009)

P.L. Lamy, I. Toth, Y.R. Fernández, H.A. Weaver, The sizes, shapes albedos, and colors of cometary nuclei, in *Comets II*, ed. by M.C. Festoru, H.U. Keller, H.A. Weaver (Univ. Arizona Press, Tucson, 2004), pp. 223–264

J. Lasue, R. Botet, A.C. Levasseur-Regourd, E. Hadamcik, Cometary nuclei internal structure from early aggregation simulations. Icarus **203**, 599–609 (2009)

J. Lasue, R. Botet, A.C. Levasseur-Regourd, E. Hadamcik, W. Kofman, Appearance of layered structures in numerical simulations of polydisperse bodies accretion: application to comet nuclei. Icarus **213**, 369–381 (2011)

A.-C. Levasseur-Regourd et al., Cometary dust. Space Sci. Rev. **214**, 64 (2018)

H.F. Levison, M.J. Duncan, From the Kuiper belt to Jupiter-family comets: the spatial distribution of ecliptic comets. Icarus **127**, 13–32 (1997)

H.F. Levison, M.J. Duncan, K. Zahnle, M. Holman, L. Dones, Planetary impact rates from ecliptic comets. Icarus **143**, 415–420 (2000)

D.R. Lide (ed.), *CRC Handbook of Chemistry and Physics*, 85th edn. (CRC Press, Boca Raton, 2004), pp. 6–8. ISBN 978-0-8493-0485-9

Z.Y. Lin, et al., Investigating the physical properties of outbursts on comet 67P/Churyumov-Gerasimenko. Mon. Not. R. Astron. Soc. **469**, S731–S740 (2017)

Y. Liu, T. Henning, C. Carrasco-González, C.J. Chandler, H. Linz, T. Birnstiel, R. van Boekel, L.M. Pérez, M. Flock, L. Testi, L.F. Rodríguez, R. Galván-Madrid, The properties of the inner disk around HL Tau: multi-wavelength modeling of the dust emission. Astron. Astrophys. **607**, A74 (2017)

S. Lorek, P. Lacerda, J. Blum, Local growth of dust- and ice-mixed aggregates as cometary building blocks in the solar nebula. Astron. Astrophys. **611**, A18 (2018)

S. Lowry, S.R. Duddy, B. Rozitis, S.F. Green, A. Fitzsimmons, C. Snodgrass, H. Hsieh, O. Hainaut, The nucleus of Comet 67P/Churyumov-Gerasimenko. A new shape model and thermophysical analysis. Astron. Astrophys. **548**, A12, 15pp (2012)

T. Mannel, M.S. Bentley, R. Schmied, H. Jeszenszky, A.C. Levasseur-Regourd, J. Romstedt, K. Torkar, Fractal cometary dust—a window into the early Solar system. Mon. Not. R. Astron. Soc. **462**, S304–S311 (2016)

T. Mannel, M.S. Bentley, P.D. Boakes, H. Jeszenszky, P. Ehrenfreund, C. Engrand, C. Koeberl, A.C. Levasseur-Regourd, J. Romstedt, R. Schmied, K. Torkar, I. Weber, Dust of comet 67P/Churyumov-Gerasimenko collected by Rosetta/MIDAS: classification and extension to the nanometre scale. Mon. Not. R. Astron. Soc. **630**, A26 (2019)

B.G. Marsden, The sungrazing comet group. Astron. J. **72**, 1170–1183 (1967)

B.G. Marsden, The sungrazing comet group. II. Astron. J. **98**, 2306–2321 (1989)

M. Massironi, E. Simoni, F. Marzari, et al., Two independent and primitive envelopes of the bilobate nucleus of comet 67P. Nature **526**, 402–405 (2015)

C. Matonti, N. Attree, O. Groussin, et al., Bilobate comet morphology and internal structure controlled by shear deformation. Nat. Geosci. **12**, 157–162 (2019)

J.A.M. McDonnell, P.L. Lamy, G.S.A. Pankiewucz, S.F. Green, C.H. Perry, The comet nucleus: Ice and dust morphological balancesin a production surface of comet P/Halley. LPSC **20**, 658 (1989)

K.J. Meech, O.R. Hainaut, B.G. Marsden, Comet nucleus size distributions from HST and Keck telescopes. Icarus **170**, 463–491 (2004)

H.J. Melosh, *Impact Cratering: A Geologic Process.* Oxford Monographs on Geology and Geophysics, vol. 11 (Oxford University Press, New York, 1989). 253 pp

E. Merenyi, L. Földy, K. Szegö, I. Toth, A. Kondor, The landscape of comet Halley. Icarus **86**, 9–20 (1990)

R. Merk, D. Prialnik, Early thermal and structural evolution of small bodies in the trans-Neptunian zone. Earth Moon Planets **92**, 359–374 (2003)

R. Merk, D. Prialnik, Combined modeling of thermal evolution and accretion of trans-neptunian objects—occurrence of high temperatures and liquid water. Icarus **183**, 283–295 (2006)

S. Merouane, B. Zaprudin, O. Stenzel, . co-auhtos, Dust particle flux and size distribution in the coma of 67P/Churyumov-Gerasimenko measured in situ by the COSIMA instrument on board Rosetta. Astron. Astrophys. **596**, A87 (2016)

D.J. Michels, N.R. Sheeley, R.A. Howard, M.J. Koomen, Observations of a comet on collision course with the Sun. Science **215**, 1097–1102 (1982)

M. Min, C.P. Dulllemond, M. Kama, C. Dominik, The thermal structure and the location of the snow line in the protosolar nebula: axisymmetric models with full 3-D radiative transfer. Icarus **212**, 416–426 (2011)

T. Montmerle, J.-C. Augereau, M. Chaussidon, M. Gounelle, B. Marty, A. Morbidelli, 3. Solar system formation and early evolution: the first 100 million years. Earth Moon Planets **98**, 39–95 (2006)

A. Morbidelli, H. Rickman, Comets as collisional fragments of a primordial planetesimal disk. Astron. Astrophys. **583**, A43 (2015)

T. Nakamura, T. Noguchi, A. Tsuchiyama, T. Ushikubo, N.T. Kita, J.W. Valley, M.E. Zolensky, Y. Kakazu, K. Sakamoto, E. Mashio, K. Uesugi, T. Nakano, Chondrule-like objects in short-period comet 81P/Wild 2. Science **321**, 664–667 (2008)

A. Natta, L. Testi, N. Calvet, T. Henning, R. Waters, D. Wilner, Dust in protoplanetary disks: properties and evolution, in *Protostars and Planets V* (2007), pp. 767–781

D. Nesvorný, Jumping Neptune can explain the Kuiper belt kernel. Astron. J. **150**, 68 (2015a)

D. Nesvorný, Evidence for slow migration of Neptune from the inclination distribution of Kuiper belt objects. Astron. J. **150**, 73 (2015b)

D. Nesvorný, A. Morbidelli, Statistical study of the early solar system's instability with four, five, and six giant planets. Astron. J. **144**, 117 (2012)

D. Nesvorný, D. Vokrouhlický, Chaotic capture of Neptune Trojans. Astron. J. **137**, 5003–5011 (2009)

D. Nesvorný, D. Vokrouhlický, Neptune's orbital migration was grainy, not smooth. Astrophys. J. **825**, 94 (2016)

D. Nesvorný, D. Vokrouhlický, A. Morbidelli, Capture of irregular satellites during planetary encounters. Astron. J. **133**, 1962–1976 (2007)

D. Nesvorný, A.N. Youdin, D.C. Richardson, Formation of Kuiper belt binaries by gravitational collapse. Astron. J. **140**, 785–793 (2010)

D. Nesvorný, D. Vokrouhlický, A. Morbidelli, Capture of Trojans by jumping Jupiter. Astrophys. J. **768**, 45 (2013)

D. Nesvorný, D. Vokrouhlický, L. Dones, H.F. Levison, N. Kaib, A. Morbidelli, Origin and evolution of short-period comets. Astrophys. J. **845**, 27 (2017)

T.L. Norris, A.J. Gancarz, D.J. Rokop, K.W. Thomas, Half like of Al-26. J. Geophys. Res. **88**, B331–B333 (1983)

J. Oberst, B. Giese, E. Howington-Kraus, R. Kirk, L. Soderblom, B. Buratti, M. Hicks, R. Nelson, D. Britt, The nucleus of comet Borrelly: a study of morphology and surface brightness. Icarus **167**, 70–79 (2004)

D.P. O'Brien, R. Greenberg, The collisional and dynamical evolution of the main-belt and NEA size distributions. Icarus **178**, 179–212 (2005)

C.R. O'Dell, A new model for cometary nuclei. Icarus **19**, 137–146 (1971)

R.C. Ogliore, G.R. Huss, N. Nagashima, A.L. Butterworth, Z. Gainsforth, J. Stodolna, A.J. Westphal, D. Joswiak, T. Tyliszczak, Incorporation of a late-forming chondrule into comet Wild 2. Astrophys. J. Lett. **745**, L19 (2012)

N. Oklay, J.-B. Vincent, S. Fornasier, M. Pajola, S. Besse, B.J.R. Davidsson, L.M. Lara, S. Mottola, G. Naletto, H. Sierks, A.M. Barucci, F. Scholten, F. Preusker, A. Pommerol, N. Masoumzadeh, M. Lazzarin, C. Barbieri, P.L. Lamy, R. Rodrigo, D. Koschny, H. Rickman, M.F. A'Hearn, J.-L. Bertaux, I. Bertini, G. Cremonese, V. Da Deppo, S. Debei, M. De Cecco, M. Fulle, O. Groussin, P.J. Gutierrez, C. Güttler, I. Hall, M. Hofmann, S.F. Hviid, W.-H. Ip, L. Jorda, H.U. Keller, J. Knollenberg, G. Kovacs, J.-R. Kramm, E. Kührt, M. Küppers, Z.-Y. Lin, F. Marzari, F. Moreno, X. Shi, N. Thomas, I. Toth, C. Tubiana, Variegation on comet 67P/Churyumov-Gerasimenko in the regions showing activity. Astron. Astrophys. **586**, A80 (2016)

J.H. Oort, The structure of the cloud of comets surrounding the solar system and a hypothesis concerning its origin. Bull. Astron. Inst. Neth. **11**, 91–110 (1950)

E.J. Öpik, Collision probabilities with the planets and the distribution of interplanetary matter. Proc. R. Ir. Acad. A **54**, 165–199 (1951)

T. Ott, E. Drolshagen, D. Koschny, C. Güttler, C. Tubiana, E. Frattin, J. Agarwal, H. Sierks, I. Bertini, C. Barbieri, P.I. Lamy, R. Rodrigo, H. Rickman, M.F. A'Hearn, M.A. Barucci, J.-L. Bertaux, S. Boudreault, G. Cremonese, V. Da Deppo, B. Davidsson, S. Debei, M. De Cecco, J. Deller, C. Feller, S. Fornasier, M. Fulle, B. Geiger, A. Gicquel, O. Groussin, P.J. Gutiérrez, M. Hofmann, S.F. Hviid, W.-H. Ip, L. Jorda, H.U. Keller, J. Knollenberg, G. Kovacs, J.R. Kramm, E. Kührt, M. Küppers, L.M. Lara, M. Lazzarin, Z.-Y. Lin, J.J. López-Moreno, F. Marzari, S. Mottola, G. Naletto, N. Oklay, M. Pajola, X. Shi, N. Thomas, J.-B. Vincent, B. Poppe, Dust mass distribution around comet 67P/Churyumov-Gerasimenko determined via parallax measurements using Rosetta's OSIRIS cameras. Mon. Not. R. Astron. Soc. **469**, S276–S284 (2017)

M. Pätzold, T. Andert, M. Hahn, S.W. Asmar, J.-P. Barriot, M.K. Bird, B. Häusler, K. Peter, S. Tellmann, E. Grün, P.R. Weissman, H. Sierks, L. Jorda, R. Gaskell, F. Preusker, F. Scholten, A homogeneous nucleus for comet 67P/Churyumov-Gerasimenko from its gravity field. Nature **530**, 63–65 (2016)

M. Pätzold, T. Andert, J.-P. Barriot, M. Hahn, M. Bird, B. Häusler, S.A. Tellemann, The mass loss of comet 67P/Churyumov-Gerasimenko. AAS/DPS Meet. **49**, 509.01 (2017)

L. Penasa, M. Massironi, G. Naletto, E. Simioni, S. Ferrari, M. Pajola, A. Lucchetti, F. Preusker, F. Scholten, L. Jorda, R. Gaskell, F. Ferri, F. Marzari, B. Davidsson, S. Mottola, H. Sierks, C. Barbieri, P.L. Lamy, R. Rodrigo, D. Koschny, H. Rickman, H.U. Keller, J. Agarwal, M.F. A'Hearn, M.A. Barucci, J.L. Bertaux, I. Bertini, G. Cremonese, V. Da Deppo, S. Debei, M. De Cecco, J. Deller, C. Feller, S. Fornasier, E. Frattin, M. Fulle, O. Groussin, P.J. Gutierrez, C. Guttler, M. Hofmann, S.F. Hviid, W.H. Ip, J. Knollenberg, J.R. Kramm, E. Kuhrt, M. Kuppers, F. La Forgia, L.M. Lara, M. Lazzarin, J.-C. Lee, J.J. Lopez Moreno, N. Oklay, X. Shi, N. Thomas, C. Tubiana, J.B. Vincent, A threedimensional modelling of the layered structure of comet 67P/Churyumov-Gerasimenko. Mon. Not. R. Astron. Soc. **469**, S741–S754 (2017)

L.M. Pérez, J.M. Carpenter, C.J. Chandler, A. Isella, S.M. Andrews, L. Ricci, N. Calvet, S.A. Corder, A.T. Deller, C.P. Dullemond, J.S. Greaves, R.J. Harris, T. Henning, W. Kwon, J. Lazio, H. Linz, L.G. Mundy, A.I. Sargent, S. Storm, L. Testi, D.J. Wilner, Constraints on the radial variation of grain growth in the AS 209 circumstellar disk. Astrophys. J. Lett. **760**, L17 (2012)

L.M. Pérez, C.J. Chandler, A. Isella, J.M. Carpenter, S.M. Andrews, N. Calvet, S.A. Corder, A.T. Deller, C.P. Dullemond, J.S. Greaves, R.J. Harris, T. Henning, W. Kwon, J. Lazio, H. Linz, L.G. Mundy, L. Ricci, A.I. Sargent, S. Storm, M. Tazzari, L. Testi, D.J. Wilner, Grain growth in the circumstellar disks of the young stars CY Tau and DoAr 25. Astrophys. J. **813**, 41 (2015)

M. Podolak, D. Prialnik, The conditions for liquid water in cometary nuclei, in *Comets and the Origin and Evolution of Life*, ed. by P.J. Thomas, C. Chyba, C. McKay (Springer, Berlin, 2006)

F. Poulet, A. Lucchetti, J.-P. Bibring, J. Carter, B. Gondet, L. Jorda, Y. Langevin, C. Pilorget, C. Capanna, G. Cremonese, Origin of the local structures at the Philae landing site and possible implications on the formation and evolution of 67P/Churyumov-Gerasimenko. Mon. Not. R. Astron. Soc. **462**, S23–S32 (2016)

F. Preusker, F. Scholten, K.-D. Matz, T. Roatsch, K. Willner, S.F. Hviid, J. Knollenberg, L. Jorda, P.J. Gutiérrez, E. Kührt, S. Mottola, M.F. A'Hearn, N. Thomas, H. Sierks, C. Barbieri, P. Lamy, R. Rodrigo, D. Koschny, H. Rickman, H.U. Keller, J. Agarwal, M.A. Barucci, J.-L. Bertaux, I. Bertini, G. Cremonese, V. Da Deppo, B. Davidsson, S. Debei, M. De Cecco, S. Fornasier, M. Fulle, O. Groussin,

C. Güttler, W.-H. Ip, J.R. Kramm, M. Küppers, L.M. Lara, M. Lazzarin, J.J. Lopez Moreno, F. Marzari, H. Michalik, G. Naletto, N. Oklay, C. Tubiana, J.-B. Vincent, Shape model, reference system definition, and cartographic mapping standards for comet 67P/Churyumov-Gerasimenko—Stereo-photogrammetric analysis of Rosetta/OSIRIS image data. Astron. Astrophys. **583**, A33 (2015)
F. Preusker, F. Scholten, K.-D. Matz, et al., The global meter-level shape model of comet 67P/Churyumov-Gerasimenko. Astron. Astrophys. **607**, L1 (2017)
D. Prialnik, M. Podolak, Radioactive heating of porous comet nuclei. Icarus **117**, 420–430 (1995)
D. Prialnik, M. Podolak, Changes in the structure of comet nuclei due to radioactive heating. Space Sci. Rev. **90**, 169–178 (1999)
D. Prialnik, A. Bar-Nun, M. Podolak, Radiogenic heating of comets by $^{26}$Al and implications for their time of formation. Astrophys. J. **319**, 993–1002 (1987)
L. Ricci, L. Testi, A. Natta, K.J. Brooks, Dust grain growth in $\rho$-Ophiuchi protoplanetary disks. Astron. Astrophys. **521**, A66 (2010)
J.E. Richardson, H.J. Melosh, C.M. Lisse, B. Carcich, A ballistics analysis of the Deep Impact ejecta plume: determining comet Tempel 1's gravity, mass, and density. Icarus **190**(2), 357–390 (2007)
H. Rickman, Masses and densities of Comets Halley and Kopff, in *The Comet Nucleus Sample Return Mission*, ed. by O. Melita (ESA Publications Division, ESTEC, Noordwijk, 1986), pp. 195–205
H. Rickman, The thermal history and structure of cometary nuclei, in *Comets in the Post-Halley Era*, vol. 2, ed. by R.L. Newburn, M. Neugebauer, J. Rahe (Kluwer Academic Publishers, Dordrecht, 1991), pp. 733–760
H. Rickman, L. Kamel, M.C. Festou, C. Froeschle, Estimates of masses, volumes and densities of short-period comet nuclei, in *Symposium on the Diversity and Similarity of Comets*, ed. by E.J. Rolfe, B. Battrick (ESA Publications Division, ESTEC, Noordwijk, 1987), pp. 471–481
S.J. Robbins et al., Craters of the Pluto-Charon system. Icarus **287**, 187–206 (2017)
K. Ros, A. Johansen, Ice condensation as a planet formation mechanism. Astrophys. J. **552**, A113 (2013)
A. Rotundi, H. Sierks, V. Della Corte, . co-authors, Dust measurements in the coma of comet 67P/Churyumov-Gerasimenko inbound to the Sun. Science **347**, aaa3905-06 (2015)
D.D. Sasselov, M. Lecar, On the snow line in dusty protoplanetary disks. Astrophys. J. **528**, 995–998 (2000)
U. Schäfer, C.-C. Yang, A. Johansen, Initil mass function of planetesimals formed by the streaming instability. Astron. Astrophys. **597**, A69 (2017)
R. Schräpler, J. Blum, S. Krijt, J.-H. Raabe, The physics of protoplanetary dust agglomerates. X. High-velocity collisions between small and large dust agglomerates as a growth barrier. Astrophys. J. **853**, 74 (2018)
S.R. Schwartz, P. Michel, M. Jutzi, S. Marchi, Y. Zhang, D.C. Richardson, Catastrophic disruptions as the origin of bilobate comets. Nat. Astron. **2**, 379–382 (2018)
R. Seu, R.J. Phillips, D. Biccari, R. Orosei, A. Masadea, G. Pcardi, A. Safaenilli, B.A. Campbell, J.J. Plaut, L. Marinangeli, S.E. Smrekar, D.C. Nunes, SHARAD sounding radar on the Mars Reconnaissance Orbiter. J. Geophys. Res., Planets **112**(E5), Issue (2007)
N.R. Sheeley Jr., R.A. Howard, M.J. Koomen, D.J. Michels, Coronoagraphic observations of two new sungrazing comets. Nature **300**, 239–242 (1982)
C.S. Shoemaker, E.M. Shoemaker, D. Levy (1993). IAU Circular 5725
E.M. Shoemaker, P.R. Weissman, C.S. Shoemaker, The flux of periodic comets near Earth, in *Hazards Due to Comets and Asteroids*, ed. by T. Gehrels (University of Arizona Press, Tucson, 1995), pp. 313–335
A. Sicilia-Aguilar, L. Hartmann, N. Calvet, S.T. Megeath, J. Muzerolle, L. Allen, P. D'Alessio, B. Merlin, J. Stauffer, E. Young, C. Lada, Disk evolution in CEP OB2: results from the Spitzer Space Telescope. Astrophys. J. **638**, 897–919 (2006)
H. Sierks, et al., On the nucleus structure and activity of comet 67P/Churyumov-Gerasimenko. Science **347**, aa1043 (2015)
J.B. Simon, P.J. Armitage, R. Li, A.N. Youdin, The mass and size distribution of planetesimals formed by the streaming instability. I. The role of self-gravity. Astrophys. J. **822**, 55 (2016)
Y. Skorov, J. Blum, Dust release and tensile strength of the non-volatile layer of cometary nuclei. Icarus **221**, 1–11 (2012)
C. Snodgrass, A. Fitzsimmons, S.C. Lowry, P. Weissman, The size distribution of Jupiter family comets. Mon. Not. R. Astron. Soc. **414**, 458–469 (2011)
L.A. Soderblom, T.L. Becker, G. Bennett, D.C. Boice, D.T. Britt, R.H. Brown, B.J. Buratti, C. Isbell, B. Giese, T. Hare, M.D. Hicks, E. Howington-Kraus, R.L. Kirk, M. Lee, R.M. Nelson, J. Oberst, T.C. Owen, M.D. Rayman, B.R. Sandel, S.A. Stern, N. Thomas, R.V. Yelle, Observations of Comet 19P/Borrelly by the Miniature Integrated Camera and Spectrometer aboard Deep Space 1. Science **296**, 1087–1091 (2002)
A. Sosa, J.A. Fernández, Cometary masses derived from non-gravitational forces. Mon. Not. R. Astron. Soc. **393**, 192–214 (2009)

D.C. Stephens, K.S. Noll, Detection of six trans-neptunian binaries with NICMOS: a high fraction of binaries in the cold classical disk. Astron. J. **131**, 1142–1148 (2006)

S.A. Stern, The effects of mechanical interaction between the interstellar medium and comets. Icarus **68**, 276–283 (1986)

S.A. Stern, Two important mechanisms contributing to cometary evolution in the Oort cloud. LPSC **18**, 951 (1987). (Abstract)

S.A. Stern, P.R. Weissman, Rapid collisional evolution of comets during the formation of the Oort cloud. Nature **409**, 589–591 (2001)

G. Strazzulla, Ion irradiation and the origin of cometary materials. Space Sci. Rev. **90**, 269–274 (1999)

G. Tancredi, J.A. Fernández, H. Rickman, J. Licandro, Nuclear magnitudes and the size distribution of Jupiter family comets. Icarus **187**, 527–549 (2006)

M. Tazzari, L. Testi, B. Ercolano, A. Natta, A. Isella, C.J. Chandler, L.M. Pérez, S. Andrews, D.J. Wilner, L. Ricci, T. Henning, H. Linz, W. Kwon, S.A. Corder, C.P. Dullemond, J.M. Carpenter, A.I. Sargent, L. Mundy, S. Storm, N. Calvet, J.A. Greaves, J. Lazio, A.T. Deller, Multiwavelength analysis for interferometric (sub-)mm observations of protoplanetary disks. Radial constraints on the dust properties and the disk structure. Astron. Astrophys. **588**, A53 (2016)

L. Testi, T. Birnstiel, L. Ricci, S. Andrews, J. Blum, J. Carpenter, C. Dominik, A. Isella, A. Natta, J.P. Williams, D.J. Wilner, Dust evolution in protoplanetary disks, in *Protostars and Planets VI* (2014), pp. 339–361

P.C. Thomas, J. Veverka, M.J.S. Belton, A. Hidy, M.F. A'Hearn, T.L. Farnham, O. Groussin, J.-Y. Li, L.A. McFadden, J. Sunshine, D. Wellnitz, C. Lisse, P. Schultz, K.J. Meech, W.A. Delamere, The shape, topography, and geology of Tempel 1 from Deep Impact observations. Icarus **187**, 4–15 (2007)

N. Thomas, H. Sierks, C. Barbieri, . co-authors, The morphological diversity of comet 67P/Churyumov-Gerasimenko. Science **347**, aaa0440 (2015)

F. Trotta, L. Testi, A. Natta, A. Isella, L. Ricci, Constraints on the radial distribution of the dust properties in the CQ Tauri protoplanetary disk. Astron. Astrophys. **558**, A64 (2013)

K. Tsiganis, R. Gomes, A. Morbidelli, H.F. Levison, Origin of the orbital architecture of the giant planets of the solar system. Nature **435**, 459–461 (2005)

P.D. Tsou et al., Experiments on intact capture of hypervelocity particles. LPSC **15**, 866–867 (1984). (Abstract)

R. van Boekel, M. Min, C. Leinert, L.B.F.M. Waters, A. Richichi, O. Chesneau, C. Dominik, W. Jaffe, A. Dutrey, U. Graser, T. Henning, J. de Jong, R. Köhler, A. de Koter, B. Lopez, F. Malbet, S. Morel, F. Paresce, G. Perrin, T. Preibisch, F. Przygodda, M. Schöller, M. Wittkowski, The building blocks of planets within the 'terrestrial' region of protoplanetary disks. Nature **432**, 479–482 (2004)

J. Veverka, K. Klaasen, M. A'Hearn, . co-authors, Return to comet Tempel 1: overview of Stardust-NExT results. Icarus **222**, 424–435 (2013)

E. Vilenius, C. Kiss, T. Müller, M. Mommert, P. Santos-Sanz, A. Pal, J. Stansberry, M. Mueller, N. Peixinho, E. Lellouch, S. Fornasier, A. Delsanti, A. Thirouin, J.L. Ortiz, R. Duffard, D. Perna, F. Henry, "TNOs are cool": a survey of the trans-Neptunian region. X. Analysis of classical Kuiper belt objects from Herschel and Spitzer observations. Astron. Astrophys. **564**, A35 (2014)

J.-B. Vincent, D. Bodewits, S. Besse, . co-authors, Large heterogeneities in comet 67P as revealed by active pits from sinkhole collapse. Nature **523**, 63–68 (2015)

J.-B. Vincent, M.F. A'Hearn, Z.-Y. Lin, et al., Summer fireworks on comet 67P. Mon. Not. R. Astron. Soc. **462**, 184–194 (2016b)

J.-B. Vincent, N. Oklay, M. Pajola, . co-authors, Are fractured cliffs the source of cometary dust jets? Insights from OSIRIS/Rosetta at 67P/Churyumov-Gerasimenko. Astron. Astrophys. **587**, A14 (2016a)

K. Wada, H. Tanaka, T. Suyama, H. Kimura, T. Yamamoto, Numerical simulation of dust aggregate collisions. II. Compression and disruption of three-dimensional aggregates in head-on collisions. Astrophys. J. **677**, 1296–1308 (2008)

K. Wada, H. Tanaka, T. Suyama, H. Kimura, T. Yamamoto, Collisional growth conditions for dust aggregates. Astrophys. J. **702**, 1490–1501 (2009)

K. Wahlberg Jansson, A. Johansen, Formation of pebble-pile planetesimals. Astron. Astrophys. **570**, A47 (2014)

K. Wahlberg Jansson, A. Johansen, Radially resolved simulations of collapsing pebble clouds in protoplanetary discs. Mon. Not. R. Astron. Soc. **469**, S149–S157 (2017)

K. Wahlberg Jansson, A. Johansen, M. Bukhari Syed, J. Blum, The role of pebble fragmentation in planetesimal formation. II. Numerical simulations. Astrophys. J. **835**, 109 (2017)

H.A. Weaver, Not a rubble pile? Science **304**, 1760–1762 (2004)

H.A. Weaver, Z. Sekanina, I. Toth, . co-authors, HST and VLT investigations of the fragments of Comet C/1999 S4 (LINEAR). Science **292**, 1329–1334 (2001)

H.A. Weaver, C.M. Lisse, M.J. Mutchler, P. Lamy, I. Toth, W.T. Reach, Hubble Space Telescope investigation of the disintegration of 73P/Schwassmann-Wachmann 3. Bull. Am. Astron. Soc. **38**, 490 (2006)

S.J. Weidenschilling, Aerodynamics of solid bodies in the solar nebula. Mon. Not. R. Astron. Soc. **180**, 57–70 (1977)

S.J. Weidenschilling, The origin of comets in the solar nebula: a unifed model. Icarus **127**, 290–306 (1997)

R. Weidling, C. Güttler, J. Blum, F. Brauer, The physics of protoplanetesimal dust agglomerates. III. Compaction in multiple collisions. Astrophys. J. **696**, 2036–2043 (2009)

P.R. Weissman, Physical and dynamical evolution of long-period comets, in *Dynamics of the Solar System*, ed. by R.L. Duncombe (D. Reidel, Dordrecht, 1979), pp. 277–282

P.R. Weissman, Physical loss of long-period comets. Astron. Astrophys. **85**, 191–196 (1980)

P. Weissman, Cometary impacts on the terrestrial planets, in *Conference on Planetary Volatiles*. LPI Contribution, vol. 488 (1982), p. 109

P.R. Weissman, Cometary impacts with the Sun: physical and dynamical considerations. Icarus **55**, 448–454 (1983)

P.R. Weissman, Are cometary nuclei primordial rubble piles? Nature **320**, 242–244 (1986)

P.R. Weissman, Post-perihelion brightening of Halley's Comet: spring time for Halley. Astron. Astrophys. **187**(873), 878 (1987)

P.R. Weissman, in *Comet Halley, Investigations, Results, Interpretations*, vol. 2, ed. by J. Mason (1990), pp. 241–257

P.R. Weissman, S.C. Lowry, Structure and density of cometary nuclei. Meteorit. Planet. Sci. **43**, 1033–1047 (2008)

P.R. Weissman, W.F. Bottke, H.F. Levison, Evolution of comets into asteroids, in *Asteroids III*, ed. by W. Bottke, A. Cellino, P. Paolicchi, R. Binzel (Univ. Arizona Press, Tucson, 2002), pp. 669–686

P.R. Weissman, E. Asphaug, S.C. Lowry, Structure and density of cometary nuclei, in *Comets II*, ed. by M.C. Festou, U. Keller, H.A. Weaver (Univ. Arizona Press, Tucson, 2004), pp. 337–357

J.S. Wettlaufer, Accretion in protoplanetary disks by collisional fusion. Astrophys. J. **719**, 540–549 (2010)

F.L. Whipple, A comet model, I, the acceleration of comet Encke. Astrophys. J. **110**, 375–394 (1950)

F. Windmark, T. Birnstiel, C. Güttler, J. Blum, C.P. Dullemond, T. Henning, Planetesimal formation by sweep-up: how the bouncing barrier can be beneficial to growth. Astron. Astrophys. **540**, A73 (2012a)

F. Windmark, T. Birnstiel, C.W. Ormel, C.P. Dullemond, Breaking through: the effects of a velocity distribution on barriers to dust growth. Astron. Astrophys. **544**, L16 (2012b)

P. Wurz, M. Rubin, K. Altwegg, . co-authors, Solar wind sputtering from the surface of comet Churyumov-Gerasimenko. Astron. Astrophys. **583**, A22 (2015)

T. Yamamoto, Formation history and environment of cometary nuclei, in *Ices of the Solar System*, ed. by J. Klinger, D. Benest, A. Dollfus, R. Smoluchowski. NATO ASI Series C: Mathematical and Physical Sciences, vol. 156 (1985), pp. 205–219

C.-C. Yang, A. Johansen, D. Carrera, Concentrating small particles in protoplanetary disks through the streaming instability. Astron. Astrophys. **606**, A80 (2017)

A.N. Youdin, J. Goodman, Streaming instabilities in protoplanetary disks. Astrophys. J. **620**, 459–469 (2005)

A. Zsom, C.W. Ormel, C. Güttler, J. Blum, C.P. Dullemond, The outcome of protoplanetary dust growth: pebbles, boulders, or planetesimals? II. Introducing the bouncing barrier. Astron. Astrophys. **513**, A57 (2010)

B. Zuckerman, T. Foreille, J.H. Kastner, Inhibition of giant-planet formation by rapid gas depletion around young stars. Nature **373**, 494–496 (1995)

Space Sci Rev (2019) 215:47
https://doi.org/10.1007/s11214-019-0611-0

# Towards New Comet Missions

**N. Thomas[1] · S. Ulamec[2] · E. Kührt[3] · V. Ciarletti[4] · B. Gundlach[5] · Z. Yoldi[1] · G. Schwehm[6] · C. Snodgrass[7,8] · S.F. Green[7]**

Received: 10 April 2019 / Accepted: 25 September 2019 / Published online: 7 November 2019


**Abstract** The Rosetta observations have greatly advanced our knowledge of the cometary nucleus and its immediate environment. However, constraints on the mission (both planned and unplanned), the only partially successful Philae lander, and other instrumental issues have inevitably resulted in open questions. Surprising results from the many successful Rosetta observations have also opened new questions, unimagined when Rosetta was first planned. We discuss these and introduce several mission concepts that might address these issues. It is apparent that a sample return mission as originally conceived in the 1980s during the genesis of Rosetta would provide many answers but it is arguable whether it is technically feasible even with today's technology and knowledge. Less ambitious mission concepts are described to address the suggested main outstanding scientific goals.

Comets: Post 67P / Churyumov-Gerasimenko Perspectives
Edited by Nicolas Thomas, Björn Davidsson, Laurent Jorda, Ekkehard Kührt, Raphael Marschall, Colin Snodgrass and Rafael Rodrigo

✉ N. Thomas
nicolas.thomas@space.unibe.ch

1 Physikalisches Institut, University of Bern, Sidlerstr. 5, 3012 Bern, Switzerland

2 Deutsches Zentrum für Luft- und Raumfahrt (DLR), RB-MUSC, Linder Höhe 1, 51147 Cologne, Germany

3 Deutsches Zentrum für Luft- und Raumfahrt (DLR), Institut für Planetenforschung, Asteroiden und Kometen, Rutherfordstraße 2, 12489 Berlin, Germany

4 LATMOS/IPSL, UVSQ (Université Paris-Saclay), UPMC (Sorbonne Univ.), CNRS, 11, boulevard D'Alembert, 78280 Guyancourt, France

5 Institut für Geophysik und extraterrestrische Physik, Technische Universität Braunschweig, Mendelssohnstr. 3, 38106 Braunschweig, Germany

6 ESA-ESTEC, Keplerlaan 1, 2200 AG Noordwijk, The Netherlands

7 Planetary and Space Sciences, School of Physical Sciences, The Open University, Milton Keynes, MK7 6AA, UK

8 Institute for Astronomy, Royal Observatory, University of Edinburgh, Edinburgh EH9 3HJ, UK

## 1 Introduction

The original Rosetta mission was a comet-nucleus sample return and significant important studies were carried out towards this goal. A mission definition study was performed by Matra Espace and a final report issued in 1987. Two "nominal" missions were proposed one of which was to comet 67P/Churyumov-Gerasimenko. The requirements on the mission were [sic]

- Rendez-vous to an active and fresh comet with a large perihelion distance
- Characterise the surface of the nucleus into active and inactive regions, perform high resolution mapping of the sampling site and provide in-situ characterisation thereof,
- Acquire three classes of samples: (1) one which preserves stratigraphy to a depth of at least one metre and preferably three metres, (2) one containing the most volatile components, (3) surface samples intended to provide a larger volume of non-volatile component
- Store the samples until return to Earth at the temperature ambient at the sampling site, but in any case below 160 K.

In hindsight, these goals seem remarkably ambitious and it is arguable whether we are able to achieve these objectives with the knowledge that we have now post-Rosetta.

The requirements on the sample acquisition and storage were equally severe.

- The core sample shall form a continuous sample from the nucleus surface down to a desirable depth of 3 m, but not less than 1 m. It should be sub-divided and stored in several segments of approximately 10 cm in diameter in a way that preserve coarse stratigraphy [sic]. Nucleus material should not be significantly altered by the sampling process. Total mass of the core sample should be about 10 kg.
- The volatile sample should preferably be obtained from a location where the most volatile components can be expected, e.g. from the bottom of the core sample or below, down to 5 m depth. They should be stored in a totally sealed container holding 10 to 100 g.
- The bulk surface sample might be collected from the surface of the nucleus. Total mass is 1 to 5 kg.

Following the reconfiguration of Rosetta into a rendezvous mission with a landing element in 1993 (with a somewhat similar profile to that proposed for NASA's Comet Rendezvous and Asteroid Flyby mission, CRAF), the ambitions were reined back. However, it is often assumed that a sample return, recovering the science descoped from the original Rosetta, would form the ultimate goal for the next phase of cometary investigation and indeed the losing mission in the recent (2018) NASA selection for its New Frontiers programme, CAESAR (Comet Astrobiology Exploration Sample Return), had the goal of returning a sample (100–800 g) from 67P/Churyumov-Gerasimenko to Earth in 2038.

Other, perhaps less ambitious, missions to comets have been proposed in recent years. The most notable was CHopper (Comet Hopper) that became one of three missions studied in detail in the final selection round of NASA's Discovery programme in 2011, eventually losing out to the InSight mission to Mars. (There is relatively little published/refereed information about CHopper. The web site https://nssdc.gsfc.nasa.gov/planetary/text/discovery_pr_20110505.txt indicates its selection or see Hand 2012.)

The current planning for the European Space Agency science programme indicates that large scale missions to planetary targets (beyond those already selected such as JUICE) are unlikely to be feasible programmatically or financially until the late 2030s at the earliest. However, smaller, lower cost, missions looking at very specific aspects of cometary science may be possible if the technology proves feasible.

**Table 1** Rosetta's comet related goals according to the Science Management Plan and a statement of achievement

| | Goal | Achievement |
|---|---|---|
| 1 | Global characterisation of the nucleus, determination of dynamic properties, surface morphology and composition | Mostly, yes. The surface composition remains poorly defined |
| 2 | Chemical, mineralogical and isotopic compositions of volatiles and refractories in a cometary nucleus | The chemical and isotopic composition of volatiles has been well established. The non-volatile material considerably less so |
| 3 | Physical properties and interrelation of volatiles and refractories in a cometary nucleus | The density ratios of some gas species to bulk species were observed to be near-invariant indicating a fundamental relationship (although detailed analysis is beginning to suggest that this is too simplistic). However, this seems to be the only progress towards this goal |
| 4 | Study the development of cometary activity and the processes in the surface layer of the nucleus and in the inner coma (dust-gas interaction) | The time series of cometary activity has been reasonably well defined (with some exceptions) and the dust-gas interaction in the coma is probably close to being understood but the processes in the surface layer were not established |
| 5 | Origin of comets, relationship between cometary and interstellar material. Implications for the origin of the solar system | Compositionally, Rosetta achieved its goals for the volatiles only. However, the discussion of the shape of the nucleus and its origin/evolution has provoked significant debate |

Given the range of possibilities, it is worthwhile looking at what Rosetta actually achieved, the questions that it left open, and what mission concepts could reasonably fill those gaps in our knowledge. It is not the purpose of this paper to advocate one mission concept but rather to identify possible future mission scenarios that directly address outstanding scientific issues.

We begin by looking at the major questions that have been answered by Rosetta and illustrating their importance for future missions. We then describe what Rosetta did not achieve—even in scientific topics where we expected significant steps forward. The main section follows where we look at key observations for the future. The final product from this discussion is in Sect. 5 where mission concepts are compared against the scientific return we would expect these mission concepts to achieve. It should be noted that while this plasma interaction is an interesting physical problem and its historical importance (e.g. the deduction of the existence of the solar wind through studies of cometary plasma tails by Biermann 1951) cannot be underestimated, the balance of this work is tipped towards the source of this material because it is the wish to access the "unprocessed" nucleus that forms the driving goal in cometary physics today.

## 2 Review of Big Questions We Have Answered

The aims of the Rosetta mission that was finally implemented were described in the Rosetta Science Management Plan (RO-EST-PL-001). The comet-related goals are listed in Table 1.

While these objectives are very broad, it is apparent that Rosetta has answered some of these goals well but others not and that the deficits are important when trying to place comets in the more general framework of solar system formation and evolution. In the sub-sections here, we look at a few of the major successes with respect to these goals.

## 2.1 Density and Bulk Properties

One of the major difficulties in designing the Rosetta mission, and the Philae lander in particular, was that the bulk properties of cometary nuclei were essentially unknown. Following the rapid improvement in technology in the 1990s, the combination of visible and thermal infrared measurements from Earth-orbit of cometary nuclei when remote from the Sun has subsequently provided good measures of the sizes and rotation periods which are key parameters in any rendezvous or landed mission. It might be argued that the three dimensional shape of 67P determined prior to rendezvous (Lamy et al. 2007; Lowry et al. 2012) was not really close to the remarkable bi-lobate structure actually observed. However, the importance of non-gravitational torques on solutions for the rotation period were not well recognised in advance and caution will certainly be exercised in future studies. The change of spin period through the perihelion passage was measured with high accuracy and the analysis of these data is proving to be an interesting means of analysing the total loss rate and its distribution over the nucleus (Mottola et al. 2019). The precession seems to be rather small and unlike that inferred for comet 1P/Halley, for example (Preusker et al. 2017; Gutierrez et al. 2016). Given the torques required to spin up the comet, this is perhaps a little surprising and might require some additional study. Associated work on the orbital dynamics and the relationship to the activity distribution is in progress (Attree et al. 2019) but may result in a non-unique solution. However, there is good reason to believe that the dynamical properties have been well characterised by Rosetta.

While the case for comets having low densities was very strong and values in the range of 200–600 $kg\,m^{-3}$ were widely expected for 67P, the confirmation of a density in this range was extremely useful, for both modelling of cometesimal growth and planning of future exploration. The gravitational potential resulting from the bi-lobate structure and the density is obviously complex close to the nucleus, with variations in surface gravity of $>60\%$.

The low surface reflectance found at 1P/Halley, 9P/Tempel 1 and 19P/Borrelly was confirmed and also shown to be valid down to fairly small scales. Exposures of higher reflectance material were restricted to rather small blocks or chunks (Pommerol et al. 2015b) and occasional small scale transient features (e.g. Fornasier et al. 2017).

## 2.2 Surface Morphology

The enormous diversity of morphology on the surface of 67P was unexpected and one of the major contributions of Rosetta to cometary science (Thomas et al. 2015b, 2018) indicating that no single process dominates the evolution of the surface.

At 67P, the evidence of airfall (also known as "dust hail" or simply "returning particles") was overwhelming (Möhlmann 1994; Thomas et al. 2015a). It is quite apparent from the data that large areas of the surface in the northern hemisphere were covered by an almost conformal coating of dust that originated from elsewhere on the comet. The depth of this coating is unknown and almost certainly non-uniform. Much of the material in the coating has probably originated from the southern hemisphere (Keller et al. 2017).

The surface texture also indicates significant variability even in areas that should have seen the same temporal variation in insolation. This suggests that there are fundamental differences in the structure of the surface layer over relatively small length scales.

## 2.3 The Importance of Slow Moving, Large Particles

The presence of airfall deposits is related to a further important result. The presence of "neck-lines" in ground-based observations of dust at 67P were thought to be caused by particles, slow moving with respect to the nucleus and remaining in the vicinity of the nucleus for one or more orbits around the Sun (Fulle et al. 2004). The true significance of these observations was revealed by Rosetta. Large particles, moving at velocities close to the escape velocity, were indeed detected (Thomas et al. 2015a, 2015b; Lin et al. 2015; Agarwal et al. 2016; Ott et al. 2017) but the vast number of individual slow particles seen had not been predicted. While their presence cannot be disputed, the means by which these particles are lifted from the surface remains a subject of significant debate.

## 2.4 Importance and Distribution of Major Volatiles

The infrared spectrometer on Deep Impact at both 9P/Tempel 1 and 103P/Hartley 2 showed that $CO_2$ was strongly emitting from areas that were comparatively weak in $H_2O$ outgassing and vice versa (e.g. Feaga et al. 2007). Observations from VIRTIS-M at 67P showed remarkably similar results considering the different objects involved (Fink et al. 2016; Migliorini et al. 2016). The ROSINA DFMS system provided further evidence that the local $CO_2$ outgassing rates are not in simple relationships to the $H_2O$ outgassing rate (e.g. Gasc et al. 2017). One should not expect a linear relationship between $H_2O$ and $CO_2$ outgassing simply from the thermodynamics. However, the complexity is evident from the difficulty in fitting physically and chemically more complex models to the data. Nonetheless, it has been established that both $H_2O$ and $CO_2$ need to be accounted for in cometary models and that dominance of one ice over the other is probably both position and season dependent.

## 2.5 The Volatile Composition

The ROSINA data set of chemical species has not yet been fully exploited. However, most species previously identified in cometary comae through in situ and ground-based remote sensing have been detected (Le Roy et al. 2015). The ROSINA observations should be considered as providing the baseline for composition studies for several decades to come. Isotopic abundances in the gas phase have been measured to high accuracy (e.g. Altwegg et al. 2015; Hässig et al. 2017; Marty et al. 2017; Calmonte et al. 2017). The presence of larger than expected amounts of $O_2$ in the coma (Bieler et al. 2015) provides further clues to the surface formation and evolution and challenges associated models.

The ROSINA data indicates that, typically, minor species "follow" one of the bulk molecules in that the density ratios are roughly invariant with respect to either $H_2O$, $CO_2$, or $O_2$ although detailed investigation suggests more subtle differences are present (M. Rubin, pers. comm.). While this goes towards goal 3 in Table 1, the physical interrelationships have not been addressed at this stage although there are good grounds to assume that this can be derived from the existing data.

## 2.6 Thermal Inertia

Modelling of the surface temperatures of cometary nuclei were pointing towards low thermal inertia although there was potential for misinterpretation in observations acquired during fast flybys (Groussin et al. 2013; Davidsson et al. 2013). The 67P data clearly show that low values of thermal inertia are present and that it is probably varying across the object (Attree et al. 2018). To first order this variability ought to be unsurprising given the observed morphology. Airfall should lead to relatively fluffy, porous, surfaces that should be thermally insulating whereas the more rock-like appearance of other surface elements would be consistent with a higher thermal conductivity. However, the low bulk density of the nucleus suggests that even the consolidated material is highly porous (>70% perhaps) and thus, even at shallow depths below the surface (e.g. >20 cm), diurnal temperature variations should be very modest.

## 2.7 Tensile and Compressive Strength

All indications are that the larger scale tensile strength of nucleus material is very low and probably below 20 Pa in most cases (Attree et al. 2018). However, it is not clear how this value increases as the scale length is reduced. Local compressive strengths are potentially much higher as indicated by the shallowness of the depressions in the surface made by the Philae lander on first impact (Biele et al. 2015). The MUPUS results indicate a much higher compressive strength material just below the surface (Spohn et al. 2015). There is no universal relationship between the compressive and tensile strengths of materials but the difference between them as measured at 67P is remarkable if confirmed. The low bulk tensile strength of comets had been inferred from the break-up of comet Shoemaker-Levy 9 (Asphaug and Benz 1996) but the evidence of the stresses on the nucleus arising from the rotation and spin-up, seen in the form of cracks in the neck region of the nucleus, has clearly illustrated a mechanism for cometary splitting and other phenomena associated with comet break-up (El-Maarry et al. 2017).

## 2.8 Seasonal Processes

The concept of cometary obliquity playing a role in coma and surface evolution was widely discussed after the Giotto encounter with comet 1P/Halley when it was observed that the major dust jet was not directed towards the Sun but roughly aligned with the pole of initially derived rotation axis and around 60 degrees from the comet-Sun vector (e.g. Keller 1990). The Rosetta observations have been instrumental in further demonstrating the importance of seasonal processes. In addition to the importance of south-north dust transport, Fornasier et al. (2015) demonstrated surface colour variations dependent upon the annual cycle. The changes in the relative abundance of the major species when the Sun crossed the equator (Gasc et al. 2017) also points to more complex relationships than heretofore discussed.

## 2.9 Some Knowledge of Depth Structure

While the internal structure of the comet beneath the visible surface remains largely unknown (see below), some measurements have increased our knowledge. Brouet et al. (2016) showed that CONSERT and SESAME-PP data indicate that the porosity increases with the increasing depth in the small lobe of the nucleus (see also Ciarletti et al. 2017).

There is strong evidence of a layering in the uppermost parts of the nucleus from optical remote sensing. A controversial aspect of this observation is whether this layering is a large-scale property of the nucleus and related to its formation (Massironi et al. 2015) or whether it is a shallow phenomenon produced through cometary evolutionary processes such as dust sedimentation and thermal loading (Sunshine et al. 2016).

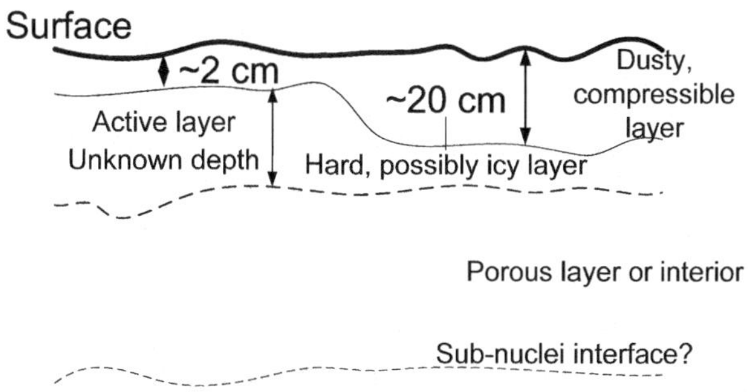


**Fig. 1** Schematic diagram attempting to reconcile the information we currently have of the interiors of cometary nuclei

Combined with the MUPUS measurements and the absence of evidence of large scale internal voids, obtained with the radio science experiment (Pätzold et al. 2016, 2019), we obtain a picture of the shallow interior as seen in Fig. 1. There are many uncertainties in this diagram but the key thing we have learnt is that more porous material with limited thermal processing is close to the surface. Hence, the original Rosetta requirement of 3 metre penetration into the interior to access "fresh" material is unlikely to be necessary. However, the possible presence of a less porous and possibly icy sub-surface layer with considerable compressive strength implies that the fundamental problem of anchoring a lander system remains challenging.

### 2.10 The Relative Contributions of Quasi-Continuous and Outburst-Like Activity

Outburst phenomena have been studied in detail by several authors (e.g. Agarwal et al. 2017) in several ways (e.g. Shi et al. 2017). However, it needs to be emphasised that the most of the emission from the nucleus is quasi-continuous. The dust and gas emission is roughly reproduced on a diurnal cycle with smaller outbursts (with notable exceptions) superposed on this distribution. The similarity of the gas density measurements in the coma to models of a purely insolation-driven coma (Bieler et al. 2015) served initially as a reasonable approximation showing that outburst phenomena are not dominant. As details were more rigorously investigated, it became clear that this model was too simplistic and that non-uniqueness in solutions to the distribution of production at the surface itself was a major problem for the interpretation of the observations. Nonetheless, the basic reproducibility, the consistency with analytical descriptions (Gerig et al. 2018; Zakharov et al. 2018), and the agreement when linking space-based and ground-based measurements of the dust emission (Gerig et al. 2018) gives considerable confidence that the physics of the outflow beyond 5 km from the nucleus is broadly understood. Closer to the surface, the situation is far less clear.

## 3 What Did We Not Get That We Should Have

Rosetta was a successful mission that collected about 220 GB of scientific data. However, as indicated in Table 1, not all physical and chemical properties that would be essential for the deep understanding of the comet could be derived. There are several reasons for this situation.

(a) Although there were 21 experiments onboard of Rosetta and its lander Philae (Glassmeier et al. 2007) they did not cover all desirable measurements. For example, a thermal IR detector or spectrometer, as has been flown on other planetary missions (e.g. Christensen et al. 2004) was not part of the Rosetta payload. (An instrument was proposed but not flown because of budget constraints.) Therefore, low surface temperatures could not be measured and well-resolved temperature and thermal inertia maps could not be derived.
(b) Caution exercised because of the dust environment of the comet affecting the star trackers and the lack of suitable compromises between the different requirements of the instruments, limited the outcome of some experiments. Near perihelion distances of several 100 km from the nucleus and near terminator orbits led to high phase angles and poor resolution that was not optimal for most of the experiments onboard.
(c) Not all scientific instruments worked as planned. VIRTIS-M failed in April 2015 (Ciarniello et al. 2016), before perihelion. Possibly more significant, however, was the malfunction of the Philae lander. Unfortunately, when Philae first touched the comet surface the anchoring harpoons did not fire and a cold gas system, intended to press the lander to the surface, did not work. As a consequence, Philae lifted off again and only came to rest after a "hop" of about 2 hours at a location almost 1 km from the originally targeted site, "Agilkia", at a place subsequently named Abydos (Biele et al. 2015). Philae could be operated, telemetry was received, but the lander was not anchored and in an undefined attitude relative to the local surface (Ulamec et al. 2016).

When looking in more detail the following topics have been compromised because of limits on the available instrumentation or their performance.

### 3.1 Small-Scale Surface Temperature and Thermal Properties

Several Rosetta experiments were designed to measure surface and near surface temperatures (VIRTIS, MIRO on the spacecraft, MUPUS and SESAME on the lander, see Table 2) to derive thermal properties of the ground.

Generally, temperature measurements in space are challenging. With remote sensing one detects the IR flux (or equivalently the brightness temperature) and has to retrieve kinetic temperatures by model assumptions taking the emissivity and roughness of the surface into account. This can result in considerable errors. Additionally, measurements in the near infrared (VIRTIS-M) are contaminated by reflected solar radiation (Keihm et al. 2012). Furthermore, its ground resolution at a distance of 100 km was several tens of metres which resulted in a nonlinear average of very different temperatures in the field of view with large and small scale topographic features and compositional heterogeneities. The same is true for MIRO measurements and, additionally here, both electrical and thermal properties control the signal in such a way as to make the interpretation even more challenging. MUPUS results are limited since the penetration of PEN into the surface failed (Spohn et al. 2015) and because the complex environment of the accidental landing site is not fully understood (Groussin et al. 2019).

The kinetic temperature is a basic physical quantity that gives information on thermal properties of a surface. But Rosetta did not provide precise high resolution maps of kinetic temperature at the surface over a diurnal cycle. The absence of night time measurements, as a consequence of the orbit and the lack of a thermal infrared instrument that could measure both low temperatures and that was not affected by sunlight reflected from the surface, contribute to this issue. Consequently, thermal conductivity or thermal inertia were derived with large error bars and at low spatial resolution only (Groussin et al. 2019 and references therein).

**Table 2** Rosetta instruments to measure temperatures on the comet

| | VIRTIS-M | MIRO | MUPUS-TM | MUPUS-PEN | MUPUS-ANC-T | SESAME |
|---|---|---|---|---|---|---|
| Measurement principle | IR spectral flux from surface | Microwave flux | IR flux from surface | 16 sensors for contact temperature up to 32 cm depth | Contact temperature sensors in Philae anchors up to 1.5 m depth | Contact temperature sensors in Philae's feet |
| Wavelength range | 0.95–5.0 μm | 0.5 mm and 1.6 mm | 5–25 μm | – | – | – |
| Resolution | 0.86 arcmin | 7.5 arcmin and 23.8 arcmin | 40 deg | – | – | – |
| Challenges/ Issues to derive kinetic temperatures | Instrument failure in 03/2015, Model dependent kinetic temperature, contamination by reflected light | Model dependent kinetic temperature, dependence on (unknown) electrical soil parameters | Model dependent kinetic temperature environmental conditions only partly known | Failed to penetrate | Failed due to malfunction of the anchors | Limitation of usable temperature range due to HW issues |
| Range | Global | Global | Local | Local | Local | Local |
| References | Coradini et al. (2007) | Gulkis et al. (2007) | Spohn et al. (2007) | | | Groussin et al. (2019) |

## 3.2 Deep Interior

For the first time, a mission was able to collect first-hand information about the internal structure of a comet nucleus. Unfortunately, Philae's unfavourable landing site and attitude strongly limited the quantity and quality of the data collected by the two instruments—the CONSERT radar, see Kofman et al. (1998, 2007) and the permittivity probe SESAME-PP, see Seidensticker et al. (2007)—designed to characterise the nucleus electrical properties at depth (Groussin et al. 2019). Of the five electrodes of SESAME-PP (three in the lander's feet, one on MUPUS PEN, and one on APXS), only the ones in the feet could be used because APXS did not reach the comet's surface and there were operational constraints at Abydos.

Only very few exploitable measurements were acquired at Abydos (Lethuillier et al. 2016) during the First Science Sequence (FSS) with only three electrodes and a questionable contact of one of the lander feet with the nucleus surface. In addition, no absolute reference measurements were performed correctly during the SDL (Separation Descent Landing) phase because of electromagnetic interference. Despite all these shortcomings, a lower bound of the real permittivity value for the first metre below the lander was estimated that provided a constraint on the porosity of the sounded area (Lethuillier et al. 2016). A comprehensive series of measurements performed for different electrode configurations would have allowed characterisation and analysis of spatial variations of the electrical properties around the lander down to 1 or 2 metres, possibly including the imaginary part of the permittivity (linked to losses in the matter).

The CONSERT radar was designed to perform tomography of the whole nucleus thanks to a series of orbiter's trajectories optimised with respect to the lander's position. However, this was not achieved during the FSS (First Science Sequence). Furthermore, even though circular polarisation was used to minimise the mismatch between the orbiter's and lander's antennae, the radiation pattern of Philae's antennae was significantly modified by the non-optimal attitude of Philae and the fact that one antenna was probably in contact with the nucleus. Consequently, only part of the nucleus's small lobe has been sounded and characterised. On the detected signals, only the propagation delays and the shape of the received pulses have been analysed and provided an average value for the real part of the permittivity (Kofman et al. 2015) and some hints about the spatial variations (Brouet et al. 2016; Ciarletti et al. 2015) and the limited heterogeneity inside the sounded area (Ciarletti et al. 2017). The interpretation of the received amplitudes would have brought further constraints on the nucleus composition that have already been obtained (Herique et al. 2016) but would require a better knowledge of the Philae's antenna radiation pattern in the actual local environment.

### 3.3 Activity

While outburst phenomena have been catalogued (Fornasier et al. 2018), there remains little understanding of how the (dominant) quasi-continuous activity proceeds. It is this topic above all that has been the subject of most frustration within the Rosetta community.

The "symptoms" of cometary activity could be observed by Rosetta over many months but the microscopic observations to be achieved by Philae and designed to shed light on the mechanisms of activity did not take place. Close-up images by CIVA and ROLIS show the surface structure of Abydos and (although from a greater distance) Agilkia in great detail. However, no direct sign of ongoing activity was found (Bibring et al. 2015; Mottola et al. 2015; Schröder et al. 2016). In any case, all images taken by the lander cameras were acquired from the time between November 12 and 14, 2014 at a heliocentric distance of 3 AU when activity was not expected to have been significant.

The absence of observations constraining the processes leading to activity is arguably the biggest single issue left unaddressed by Rosetta. The uncertainty has, if anything, been increased by the observations of particles, considered to be larger than liftable by gas drag alone, above the surface and evidence that large (20 metre diameter) boulders have been moved across the surface (Agarwal et al. 2017; El-Maarry et al. 2017). Concrete progress in this area is totally absent and it appears that the only way to address this properly is to study the microphysics of the active region/layer.

### 3.4 Innermost Gas and Dust Comae Properties and Surface Relationships

The proximity of the Rosetta spacecraft to the nucleus was expected to provide detailed information on the gas and dust dynamics in the inner coma. Unlike previous fly-by missions, the rendezvous nature of the mission was expected to ensure a steady stream of data at different positions in the coma leading to strong constraints on the gas and dust flow. Indeed, significant steps forward were achieved. However, it has also become apparent that interpretation of the acquired data is not entirely straightforward.

Three instruments were expected to make the most significant contributions to the definition of the neutral gas flow field of the major species. The COPS sensor on ROSINA determined the gas density at the spacecraft and there remains some hope that velocities might also be derived in the near future (Tzou 2017). The VIRTIS instrument, and particularly

VIRTIS-M, obtained 2D maps of the $H_2O$ and $CO_2$ distributions using IR imaging spectroscopy until the cryo-cooler failure in March 2015. The MIRO microwave spectrometry obtained line-of-sight measurements of the $H_2O$ density, temperature, and velocity (Gulkis et al. 2007, 2015).

In all cases, there are difficulties in inversion of the data to provide unambiguous descriptions of the flow. COPS provides accurate measurements but only at the spacecraft and, for much of the mission, the spacecraft was in near-terminator orbits. In the terminator geometry, the contribution to the gas flow field from lateral expansion of gas from the near sub-solar regions is significant with strong gradients. Furthermore, it has been shown (Liao 2017) that lateral expansion will smooth out source inhomogeneities on scales up to the mean free path and hence the COPS data cannot be used to infer surface source distributions accurately. MIRO can make up for this deficiency, in principle (Marschall et al. 2019), but the inversion of the measured lines is far more complicated than in a bound atmosphere application and is compromised by the relatively broad beam width. One important aspect that might yet be resolved is the temperature of the gas at source which may be higher than the free sublimation temperature as a result of interaction between the gas and the surface layer(s). The VIRTIS measurements were limited in number and do not give flow velocities.

It is necessary to use these data sets together to constrain models (e.g. Marschall et al. 2019; Tenishev et al. 2016) of the surface activity distribution. However, from the work so far, it is not obvious that a satisfactory unique solution can be obtained even for optimum cases.

The dust distribution within the inner coma relies almost exclusively on the 2D imaging of OSIRIS with models using the size distribution derived from GIADA and COSIMA measurements (with OSIRIS looking at the largest particles). While the dust outflow has been shown to be equivalent to a free-radial outflow at distances greater than 12 km from the nucleus on average (Gerig et al. 2018), there are significant deviations in some cases that remain to be investigated. Furthermore, the dominant factors influencing the deviations from 1/r closer than 12 km to the surface are still unclear although acceleration is obviously of major importance (Zakharov et al. 2018). The surface source relationship between the gas and dust remains to be clarified although Tenishev et al. (2016) has suggested that the dust source distribution is different from the gas source distribution on larger scales which is clearly a significant problem for activity modelling.

## 3.5 Surface Composition

Rosetta included several instruments to analyse cometary matter. The ROSINA and COSIMA instruments obtained numerous excellent mass spectra for both gas in the coma and individual dust grains captured in the immediate vicinity of the nucleus (Altwegg et al. 2017; Hilchenbach et al. 2016). The imaging IR spectrometer, VIRTIS, mapped the entire surface of 67P in the wavelength range 0.25 to 5 μm (Filacchione et al. 2019) and found dehydrated, refractory, and organic-rich material on the surface—something that had been expected on the basis of previous observations going back to the 1980s (e.g. Chyba and Sagan 1987; Chyba et al. 1989). However, the exact composition or geometry of the molecules (notably any chiral inhomogeneity) could not be distinguished by remote IR spectroscopy and remains undetermined. Furthermore, VIRTIS results characterise the surface. But this is not necessarily representative of the bulk composition of the nucleus.

One of the main reasons leading to the proposal and development of a lander for Rosetta, was the aim of analysing (quasi-)pristine material from the comet surface and sub-surface (Ulamec et al. 1997; Wittmann et al. 1999; Biele and Ulamec 2008). The payload of Philae included four instruments, dedicated to this aim: APXS (an alpha-particle fluorescence spectrometer, Klingelhöfer et al. 2007), two evolved gas analysers (EGA's), COSAC and Ptolemy, each including a mass spectrometer and a gas-chromatograph (Goesmann et al. 2007, 2014; Wright et al. 2007), and $SD^2$, a drill and sampling device, to feed the ovens of the EGA's (Ercoli-Finzi et al. 2007). The camera system CIVA included a visible and IR microscope (CIVA-M), intended to image the sampled material in dedicated ovens with windows (Bibring et al. 2007).

Immediately after the first touchdown at Agilkia, the Philae system started a pre-programmed FSS, including measurements with the mass spectrometers of both evolved gas analysers, COSAC and Ptolemy, in "sniffing mode". As it appears, some surface material was introduced into the venting systems of the instruments, where excellent mass spectra could be obtained with both instruments (Goesmann et al. 2015; Wright et al. 2015). However, no gas chromatography was performed in this phase.

During an adapted FSS, an attempt was made to sample surface material, deliver it into a COSAC oven, and perform a measurement of the composition of the volatile fraction matter close to the surface with the gas-chromatograph. Although the system worked well, no sample could be obtained and the oven stayed empty as the $SD^2$ drill penetrated into a cavity and never touched ground (Di Lizia et al. 2016).

Ptolemy was operated after about 50 hours at the comet surface, using an oven designed to capture cometary gas onto a cold molecular sieve reagent (Comet Atmosphere Sample Experiment, CASE). The oven was heated to 200 °C, but unfortunately, the sample gas pressure was too low to flow through the GC columns (Morse et al. 2016).

As the ovens stayed empty, no attempt was made to operate CIVA-M, so no close-up IR imaging, which could have given information on the mineralogy of individual surface grains, was performed. As with $SD^2$, the APXS instrument did not touch the surface of the comet and no useful alpha/x-ray spectra have been obtained.

Shortly after the FSS, Philae went into an unplanned hibernation. Despite "waking-up" in June 2015, all attempts to command the instruments and start long term science investigations, turned out to be unsuccessful (Ulamec et al. 2016, 2017). Hence, the expected results regarding analyses of the cometary surface composition failed to materialise.

### 3.6 Some of the Properties of Cometary Dust

Cometary dust particles have been analysed in the laboratory following the Stardust mission to comet Wild 2. The atomic composition has been established in situ on mission to 1P/Halley and by the COSIMA experiment on Rosetta. Hence the mineralogical composition of cometary material has been fairly well established and the mineral to organic ratio (55:45) has been determined (Bardyn et al. 2017). However, the compositional studies have not provided details on the molecular composition of the refractory component and it may yet prove difficult to deduce complex molecules in the gas phase if the fracture patterns in the mass spectrometers prove challenging to determine.

Perhaps more fundamentally, the Rosetta observations have opened issues that might have been thought to be closed. The Giotto dust detection systems provided a complete set of observations from around $10^{-20}$ kg, corresponding to radii of around 0.02 microns, upwards (e.g. McDonnell et al. 1991). The Rosetta instrumentation provided no direct measurements of particles densities for particle sizes less than about 14 microns but also indicated the presence of non-escaping particles in the centimetre size range. This has led to enormous debate on the size distribution of both escaping and non-escaping particles within the nucleus and the relative significance of different particle regimes. Furthermore, the physical properties (e.g. porosity and structure) remain issues of discussion. The in situ measurements do suggest that large particles dominate the mass loss (Merouane et al. 2016) but there are considerable uncertainties. And while in situ measurements are admittedly challenging, the complete absence of more straightforward high phase angle (low scattering angle) remote sensing data requested in proposals in 1994 is purely down to the implemented spacecraft operational profile. This is not a moot point because the dust distribution function is a necessary component of refractory to volatile ratio calculations.

### 3.7 Summary

With respect to the goals seen in Table 1, we can conclude that several have been broadly met while others were compromised by various issues with the spacecraft and its instrumentation. The nucleus of 67P has been characterised, much has been learnt of morphology and the dynamical properties have been established. The surface composition, however, is not well understood and one can only really say that organics are present—something that was clearly established 30 years ago at 1P/Halley. It is improbable that the composition can be accurately determined by remote sensing leaving a repeated attempt at in situ analysis or sample return as the two operational techniques available to make further progress. The lack of solid information on the mineralogical and chemical composition of refractories is also present as part of goal 2 (Table 1) and similar statements apply. Here, however, the excellent knowledge of the gas composition gives us confidence that the basic volatile composition is known although the variability and the changing composition with illumination conditions remains to be understood in detail. To a large extent, this is a consequence of the major deficiency in Rosetta data—namely the absence of meaningful constraints on the physical properties of the surface layer and the interrelationship of volatiles and refractories at depth. This forms an important input to solar system formation models and is also directly related to how cometary activity (and subsequent evolution) develops. Our knowledge of the processes in the surface layer (goal 4) have not really been advanced by the mission and we continue to suppose that effects such as sub-surface sublimation, recondensation in the interior and amorphous-crystalline transitions could have significance without having major additional input to the discussion. Hence, the physical structure and composition of the surface layer to a depth of 1 metre and the inhomogeneity of this layer across the surface remain key questions after the mission.

## 4 Key Observations for the Future

### 4.1 The Depth Profile (the First 2 Metres)

While the Rosetta observations have provided hints on the structure of surface layers and specifically the porosity, the difficulties with Philae essentially compromised the goal of looking at the interrelation of volatiles and refractories and their variation with depth at intermediate scales. The primary goals would be

- to measure the density of cometary material with depth
- to establish the refractory to volatile ratio with depth
- to determine the volatile composition with depth

These three quantities would provide us with several pieces of information that are blocking study of cometary activity, namely

- the depth and density of the uppermost inert layer (if it exists)
- the changing volatile mixing ratio with depth and detection of a sub-surface recondensation layer
- the determination of the positions of species-specific sublimation fronts (should they exist)
- the depth at which internal properties are no longer influenced by the irradiation of the surface

It should be noted that models have been developed that purport to derive these quantities but none of these have actually been measured to confirm and/or constrain the models. It is also notable that coring down to at least 1 metre and preferably 3 metres was the initial target of the Rosetta mission when it was still a sample return mission in the 1980s. It is perhaps telling that we have not been able to refine this significantly other than to confirm what we already knew.

## 4.2 Microphysics of the Surface Layers

It is evident from the work on Rosetta data that our understanding of the microphysics of the surface and sub-surface layers remains rudimentary. This has left two major uncertainties in our understanding of comets.

First, we do not know how the volatile component is associated with the refractory component of cometary material. This is vital to understanding how comets formed and may ultimately provide key information on the solar system formation process. One can envisage ices surrounding or encapsulating the dust component, ices within dust matrices, ices as a component isolated from the refractories and so on. There is no strong evidence supporting any of these models.

Second, the thermophysical behaviour of the surface and sub-surface layers remain inadequately constrained. We have values for the thermal inertia but the thermal conductivity and heat capacity are unknown, the importance of gas transport and re-condensation are unknown, and the refractory to volatile ratio inside the nucleus remains a subject of fierce debate. Without these, models of cometary evolution are, at best, poorly constrained.

Hence, one of the primary objectives for future missions must be improved knowledge of thermophysical parameters since data interpretation and our understanding of cometary activity heavily depends on the output of thermophysical computer simulations of the upper cometary surface layers ($\leq 1$ m). These parameters strongly depend on microphysical details of the cometary surface, such as e.g., the material composition, the morphology of the material (structure, arrangement of the particles, coordination number, void space) and the mixing of the different components (silicates, organics and ices). Besides the thermal modelling aspect, also the macroscopic behaviour of the cometary surface is influenced by the microphysical properties of the material, such as the tensile strength (Gundlach et al. 2018), compressive strength (Lorek et al. 2016; Schräpler et al. 2015), and the thermal conductivity (Gundlach and Blum 2012). However, the degree to which we can provide data on these topics needs to be traded against the complexity of method required to get that data.

Possible scenarios aiming to investigate the microphysical properties of cometary surfaces are illustrated in Fig. 2. The concepts are given in increasing order of complexity. Level A aims at measuring the material composition and improving upon Rosetta (Bardyn et al. 2017) and could dramatically reduce the number of free parameters of thermophysical models. In level B, the morphology of the upper surface layers (0.25 m) could be investigated by carefully acquiring samples without destruction. Mechanical alteration can be avoided by

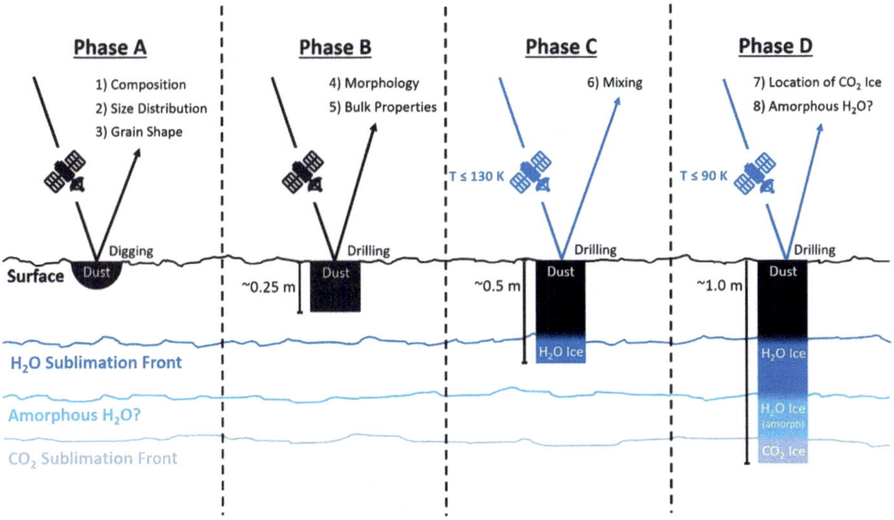


**Fig. 2** Possible concepts designed to study the microphysical properties of cometary surfaces. Level A and B could be conducted without cooling of the sample and would therefore be simplest to implement. In contrast, the aim to study the microphysical properties of the volatile components (levels C and D) requires cooling of the samples during sampling, transport and measurements

careful and slow drilling. Levels A and B could be performed without the need for cryogenic sampling, whereas sampling the cometary surface (0.5 m) at low temperatures ($\leq 130$ K) to avoid thermal alteration is a further step in complexity. Because of the relatively high temperature, this scenario would only enable the investigation of the non-volatile components and water ice. The final sample acquisition level, level D requires very low temperatures ($\leq 90$ K) to provide material from a depth of $>1$ m. This scenario provides the possibility to study the location of $CO_2$ ice beneath the surface and the existence of amorphous water ice.

The microphysical properties of the cometary surface to be studied can be summarised as follows:

(1) *Composition*[1] of the material (level A)
(2) *Size distribution* of the grains (level A)
(3) *Shape* of the grains (level A)
(4) *Morphology* of the material, i.e., porosity, coordination number and void space (level B)
(5) *Bulk properties* of the building blocks, e.g., specific surface energy, Poisson's ratio, Young's modulus (level B)
(6) *Mixing* of the different components (non-volatile materials and water ice) (level C)
(7) Location of *carbon-dioxide ice* (level D)
(8) Existence of *amorphous water ice* (level D)

These properties and the difficulty of determining them illustrate the challenges that face surface landed missions.

[1]Knowledge of the material composition can be used to infer the bulk properties of the material, such as the specific surface energy, the Poisson's ratio and the Young's modulus by laboratory experiments conducted with analogue materials.

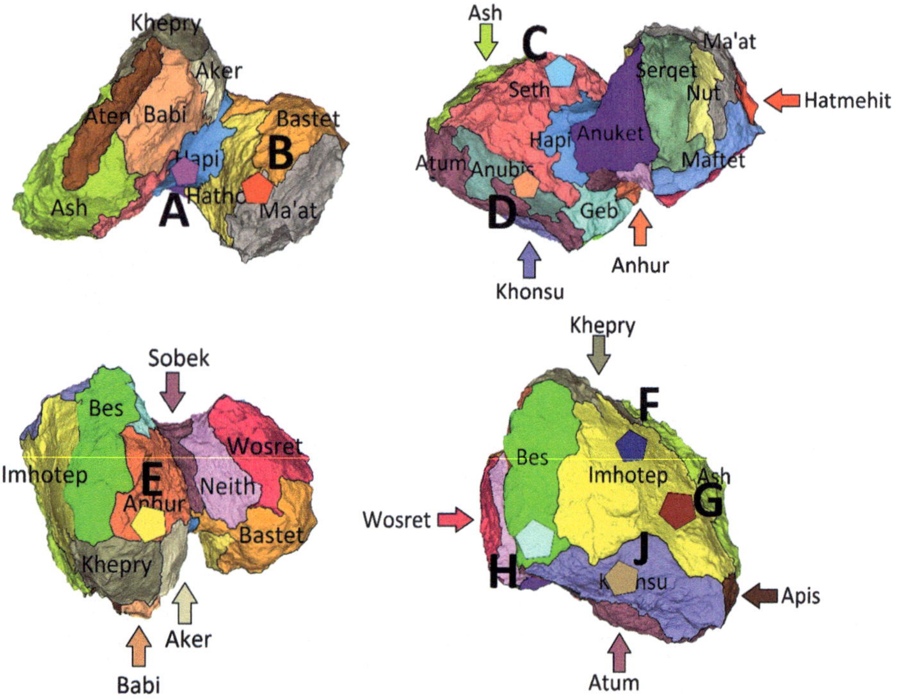


**Fig. 3** The region definition on 67P with general areas of major local interest marked as polygons

## 4.3 Local Inhomogeneity: Comparisons at Several Local Sites

It is apparent from the surface morphology of 67P and the variations in activity (both in magnitude and composition) with position that inhomogeneity in longitude and latitude on the nucleus are present. The variability evident at 9P/Tempel 1 was perhaps less extreme but nonetheless present. There remains considerable doubt whether this inhomogeneity is primordial or evolutionary. Hence comparisons of areas of different morphology and/or morphological context are highly desirable.

For 67P, the diversity is so large that selecting a limited number of regions of interest becomes a challenge in itself. In Fig. 3, 10 sites have been marked that could justify mobility. A comparison of the surface layers of the smooth, relatively inactive, Imhotep region (site F) with the smooth, highly active, Hapi region (A) would clearly be of interest while comparing the dust-covered plains areas of the north (site B) with the dust-poor region of the southern hemisphere (sites G and J) would also have considerable scientific merit. Site B is near a region now known to have become strongly active for multiple rotations of the nucleus. Site C is the Aswan area where a cleaving and collapse of a cliff was seen (Pajola et al. 2017). Site D is in Anubis and is an area that showed surface changes similar to those seen in Imhotep and would therefore provide a comparison with site F. The Anhur region (site E) shows multiple morphologies and activity that can be compared with the adjacent Bes region (site H). H is where ices/bright material was exposed (Fornasier et al. 2017) during the approach to perihelion while G is an area that shows circular structures similar to those observed on Tempel 1 and is very different from the adjacent smooth site F.

All of these sites can provide a good case for in situ investigation because one must expect that there are physico-chemical reasons to explain their morphological diversity. The conclusion here is that mobility or the use of multiple-landed assets would have a strong scientific rationale.

## 4.4 The Deep Interior

The results obtained by the CONSERT radar (even if operated in very unfavourable conditions) demonstrate the unique capability of radars to provide information about the internal structure of a comet nucleus. CONSERT was operating at low frequency (90 MHz) to maximise the penetration inside the nucleus at good spatial resolution while still keeping the antennas size reasonable. This frequency range allowed analysis of potential structures larger than 10 metres in size. The use of higher frequencies would give access to smaller structures and thus provide better constraints on the nucleus building blocks sizes. Since the data collected by Rosetta indicate a contrast in physical properties between the shallow subsurface at Abydos and Agilkia, and the deeper internal structure, it would be quite interesting to get a global description of the shallow subsurface (over a few metres/tens of metres) which could be provided by a monostatic radar operating in reflection from orbit with no need of a lander (see, for example, Herique et al. 2018).

## 4.5 The Nature of Complex Organics and Maintaining Integrity

The isotopic, elemental, chemical, and chiral composition of non-volatile material, including complex organic material was not established. The elemental composition of ejected dust was found with COSIMA but the composition of complex organics, for example, remains unknown. Clearly, any discussion of comets bringing complex organics to the inner solar system and their participation in the evolution of life requires knowledge of the composition of these organics and it is fully apparent that the optimum way to address composition is by using the power of Earth-based analytical laboratories. Remote-sensing approaches are totally inadequate whereas in situ analysis remains technically challenging and limited by spacecraft resources. It is this aspect that forms the strongest argument for sample return.

Analytically, we are in a far better position now than 20 years ago with recent decades having seen both progressive increases in performance of instrumentation, huge increases in computer power and ease of use, and breakthroughs such as the introduction of nano-SIMS technology, the spread of con-focal microscopy and sub-micron resolution X-ray microscopes as well as the greater availability of synchrotron-based techniques.

Laboratory analysis (especially of the mineralogy, chemistry and isotopic composition) of solar-system and extra-solar material has been enormously important in furthering our understanding of the origin and evolution of the solar system, as well as shedding light on the nucleosynthetic processes in several types of star. Using samples returned from the Moon, as well as meteorites likely to have come from a variety of parent bodies (including Mars and perhaps the asteroid Vesta), we have been able to determine the age of the Earth and the solar system, the timescales for the dynamic evolution of the solar system, and the timescale over which liquid water existed on early planetesimals. The power of laboratory instrumentation, and the advantage of having samples on Earth, is shown by the way in which lunar samples returned almost 50 years ago are still leading to fresh insights. If we had instead relied solely on in situ analysis of the Moon with 1960's technology, our understanding would be a fraction of what it is today.

The two major issues with returning to Earth are connected to evolution of the sample during the sampling process and the return trip and the effects on the sample during reentry. The latter in particular is a significant problem for attempts to acquire undisturbed cores. The forces due to deceleration on re-entry will undoubtedly compromise the sample's integrity structurally, leading to mixing between products from different depths. It seems apposite to conclude that, for the foreseeable future at least, an ambition to return to Earth, a cored cometary sample is highly challenging and possibly untenable. Perhaps in the future some aspects of the problem would be overcome if there was a working "Deep Space Gateway" or "Lunar Orbital Platform" of some description that would allow the sampling to be carried out in an off-world environment (where conditions of low pressures, low temperatures and low gravity would be available for exploitation). But for now, such a possibility is certainly not feasible and the original Rosetta definition team, in considering a return to the Space Station, concluded that such a scenario was potentially even more complex than return to Earth. The CAESAR approach, of sample grabbing and limiting evolution by cooling and removal of the gaseous component, was a valid compromise.

## 4.6 Non-local Thermal Equilibrium in the Immediate Proximity of the Nucleus and Coma Structure

Rosetta has demonstrated that constraining the spatial distribution of gas and dust sources uniquely is vastly more challenging than originally thought. And yet this remains a key question in the efforts to relate surface structure (thermal, mechanical, chemical) to activity and evolution. The observational requirements for future missions needed to break the degeneracy and relate coma gas and dust densities to surface properties are challenging.

The single point density measurements by COPS have shown the importance of this type of measurement. Local gas density measurements at multiple positions within the inner coma would have provided far more rigorous constraints on the gas flow field. In particular, continuous measurements near the sub-solar point (over periods greater than the rotation period at a cadence of $<1$/min) would have been particularly valuable. When coupled with a local velocity and temperature measurement (and we note here the additional importance of determining the temperature anisotropy as a constraint on the energy distributions)

Remote sensing observations of the gas distribution were obtained by Rosetta but their interpretation remains complex. In a future mission concept, the inversion of density, temperature, and velocity from remote sensing should be traded against local direct multipoint measurements to establish the optimum strategy. The critical aspect is whether the gas flow field can be inverted to produce global maps of source strength at a resolution that is sufficient to relate the production rate to surface morphologies and properties. Given the variation in surface morphology seen on the surface of 67P, the required resolution is potentially $<100$ m and comparable to the mean free path for modest activity cases.

The composition of evolving gases and their interaction is not widely discussed in the literature. Finklenburg (2013) looked at $CO_2$-$H_2O$ interactions in highly simplified geometries and showed the importance of understanding effects in terms of the Knudsen penetration number and the momentum transfer between $CO_2$ and $H_2O$. Where simultaneous remote sensing observations of multiple species in proximity to the nucleus have been obtained, the different spatial distributions of the different parent molecules are clearly evident (Feaga et al. 2007; Filacchione et al. 2019). The relative dominance of $CO_2$ at some points in the coma of 67P in local measurements was also apparent (Hässig et al. 2015). As a result, obtaining density, temperature, and velocity of the major species ($H_2O$, $CO_2$ and CO) will be important in most cases.

It is usually assumed that the dust has no influence on the gas flow field when the dust optical depth is low (as was usually the case at 67P—outburst phenomena produced exceptions). Reports of sublimation above the surface have been made (Gicquel et al. 2016) and evidence of significant ice particles in an inner coma have been reported for 103P/Hartley 2 (Hermalyn et al. 2013) and hence this may not be an accurate assumption—at least in some cases. The lifetimes of pure water ice particles in the innermost coma are quite long (Lien 1990) and this may explain the evidence of sublimating grains seen in ground-based observations of C/1996 B2 (Harris et al. 1997). Consequently the volatile content of the dust within a few kilometres of the source is a key parameter for future measurements.

There are several properties of the dust that affect the dust flow field pattern in addition to the source. It is well known that particle size and shape are important. The dust size distribution remains a critical parameter in all studies of the dust flow field and its brightness at all wavelengths. The absence of direct (unambiguous) measurements of small ($<14$ micron) particles is a significant source of concern in establishing accurate models for 67P and it can be argued that the size distributions of larger particles are biased as a consequence of the spacecraft being almost always in a terminator orbit configuration. It is almost mandatory to resolve this problem in future missions.

Individual scattering properties of particles are also important for extrapolating local measurements to 3D distributions using remote sensing instrument. One could envisage some form of measurement that would constrain this property and provide information on the refractive index. Ivanovski et al. (2014) have shown that the rotation of particles can influence the direction of motion but it is hard to envisage experiments to establish mean rotation rates and their influence on the bulk flow.

### 4.7 Cross-Comet Calibration

While we have so far considered the questions that need to be answered to better understand individual comets, the wider context of using comets to understand planet formation requires comparison between comets. We need to understand what the average comet is like, which properties are the results of the unique history of individual objects, and what can be attributed to conditions in their formation regions.

The comets that we see today come from three reservoirs: The Oort cloud is the most distant, and the source of long period and dynamically new comets, while short period Jupiter family comets (JFCs) come from the scattered disc of the Kuiper Belt. The modern distinction is based on the comet's orbit (specifically its Tisserand parameter with respect to Jupiter), and splits comets into "ecliptic" comets from the trans-Neptunian region and "nearly isotropic comets" (NICs) from the Oort cloud, the latter including both long period and high inclination short period comets of the Halley type (Levison 1996). The third reservoir is the main asteroid belt, which has recently been recognised to contain a significant population of icy bodies, including the so-called "Main Belt Comets" (MBCs), whose relatively circular orbits stay within the belt (Jewitt et al. 2015; Snodgrass et al. 2017b). The MBCs have stable orbits over the age of the Solar System, and likely formed in the same region they are found now, or at least were emplaced in the belt as the giant planets were forming. We note here that China is currently working on a mission to 2016 HO3 and 133P/Elst-Pizarro that may provide further data on this somewhat bizarre class of objects (Zhang et al. 2019).

Kuiper Belt objects also likely formed close to their current distance from the Sun, although the scattered disc (as the name implies) has a mixed origin, while the Oort cloud is populated by bodies scattered out of the original Solar System proto-planetary disc. It was thought that Oort cloud objects (and therefore long period comets) had their origin in the giant planet region, and therefore formed closer to the Sun than JFCs from the Kuiper Belt (e.g. Dones et al. 2004), but the latest dynamical models suggest a greater mixing and less clear distinctions between the original source region for different comet classes (e.g. Nesvorny et al. 2017).

The isotopic ratios of various elements, and especially the Deuterium to Hydrogen ratio (D/H), can be used as a tracer of the original formation location of a comet, as it is expected that these ratios were set by the temperature when their ices condensed out of the gas in the protoplanetary disc, and therefore their location within this disc (Robert 2006). Measurement of the isotopic ratios in a number of common elements (e.g. H, C, O, N, S) for a given comet can more uniquely constrain models (Bockelée-Morvan et al. 2015). To date we only have in situ measurements of D/H in 1P and 67P (Altwegg et al. 2015), and a few other ratios from Rosetta. Instead most isotopic measurements, and indeed nearly all compositional information on cometary comae, come from remote observations with telescopes on Earth and in space. D/H measurements via high-resolution spectroscopy have been limited to the brightest comets, which are nearly all NICs, as they tend to be more active than returning short period comets. Remote D/H measurements were possible for the JFC 103P during a close approach to Earth at perihelion, using the ESA Herschel space telescope (Hartogh et al. 2011). The D/H measured in this JFC and that found by Rosetta at 67P are very different, implying a wide range within the same comet type, or possibly systematic differences between remote and in situ observation. Measurements at a larger number of comets, and at different types of comet, would be highly desirable. Lis et al. (2019) have proposed an explanation based on their observation that hyperactive comets, such as 46P/Wirtanen, require an additional source of water vapour in their comae, explained by the presence of subliming icy grains expelled from the nucleus and that these particular objects have D/H ratios in water consistent with the terrestrial value. They propose that the isotopic properties of water outgassed from the nucleus and that of icy grains may be different because of fractionation effects during the sublimation process. There clearly remain issues of interpretation here that need to be resolved.

So far we have only visited short period comets: one NIC (1P) and five JFCs (not including the spacecraft that interacted only with the distant ion tails of comets; see e.g. Neugebauer et al. 2007). It would be very interesting to visit a long period comet, especially a dynamically new one entering the inner Solar System for the first time since it was scattered to the Oort cloud, to see a more 'pristine' surface, but these comets pass through the accessible region of the Solar System on much shorter timescales than missions can be planned and launched. One concept that may overcome this difficulty is to have a spacecraft launched to a suitable 'parking' orbit, from where it can then be sent to intercept a newly discovered comet (e.g. Hewagama et al. 2018), which becomes more feasible with the expected early discoveries of inbound objects by next generation sky surveys like the Large Synoptic Survey Telescope (LSST). An implementation of this concept, Comet Interceptor, has recently been proposed to ESA as a 'fast' mission (a relatively low budget class of mission, launched as a secondary payload), and was selected for a 2028 launch in June 2019.[2] The other unexplored population, the MBCs, are far more accessible, with relatively circular orbits and low activity levels. Concepts to visit one of these objects have been proposed to various space agencies (e.g. Meech and Castillo-Rogez 2015; Snodgrass et al. 2018; Jones et al. 2018), but have not yet been selected for flight.

[2]Comet Interceptor is not yet described in the literature, but more details can be found at http://www.cometinterceptor.space.

There is also significant diversity within the same population of comets. Vincent et al. (2017) suggest different evolutionary ages for the JFC nuclei that have been imaged by spacecraft, but, as pointed out by Kokotanekova et al. (2018), the comets visited by spacecraft so far are all most likely 'young', as younger comets are more active and therefore brighter and easier to discover. Very evolved comets are expected to lose their volatiles and/or build up an insulating mantle, and it would also be interesting to visit a much lower activity comet, or even one of the population of 'asteroids in cometary orbits' (ACOs), which are thought to be extinct comets at the end of their evolution. As survey telescopes are becoming more sensitive, more of these faint targets are being discovered.

Modern astronomical surveys are a powerful tool in understanding comets, as they allow us to study them in very large numbers. While it is clear that spacecraft encounters provide much more detailed information, we can only ever hope to visit a tiny fraction of the total population, while telescopes can provide a broad overview. Survey telescopes themselves provide useful information on orbits of very large numbers (LSST expects to discover $> 10^4$ comets; LSST Science Collaboration 2009), allowing their dynamics to be studied, and some information on sizes and activity levels to be derived. Surveys also discover unusual objects that tell us about the variety within the population. Targeted surveys of large numbers of comets tell us about the variation in composition, through either narrowband photometry or spectroscopy, and have allowed some taxonomic descriptions to be created (e.g. carbon-depleted vs. normal abundance—A'Hearn et al. 1995). Surveys of inactive nuclei have produced size distributions (e.g. Fernandez et al. 2013) and constrain strength and density (see Kokotanekova et al. 2017, Groussin et al. 2019, this issue). The occasional appearance of a particularly bright comet allows further detail to be extracted from telescopic observations, as a variety of techniques can be employed over a wide range of wavelengths, including higher resolution spectroscopy to study gas species or investigation of the thermal emission from dust. Bright comets can also be studied for a larger part of their orbit, allowing the evolution with changing solar heating to be investigated, for example looking at the changing relative abundance of coma species (the famous 'Christmas tree' plot for Hale-Bopp; Biver et al. 1997). While much of our knowledge comes from a few particularly bright comets, which tend to be dynamically new NICs, advances in telescope technology mean that similar studies are now possible for more typical objects, especially with the anticipated launch of JWST (Kelley et al. 2016). Other well-studied comets include those visited by spacecraft, as the opportunity to make simultaneous telescopic and in situ measurements meant that large observing campaigns could be justified (Meech et al. 2011; Snodgrass et al. 2017a). Although there is still some work to do to join the details returned by spacecraft to the very large-scale view from telescopes, these observations provide the link between the focussed spacecraft results and the general properties of the broader population of comets.

## 4.8 Analogue Materials for Testing

It is interesting that the measurements during the KOSI experiments (Kochan et al. 1998) are still occasionally referred to. This is in part a consequence of the rather limited amount of high quality analogue testing performed in the 15–20 years that followed KOSI. This is however changing. Groups in Braunschweig (e.g. Gundlach et al. 2011), Graz and Bern (e.g. Pommerol et al. 2015a) have been particularly active in the past 10 years in looking at cometary analogues. The behaviour of analogues is sometimes rather surprising. For example, there are significant differences in the opto-mechanical behaviour of dust encapsulated in water ice as opposed to dust mixed with water ice. Analogue testing can be used to look at the release of material from a surface, how volatiles diffuse through a surface layer, how the surface texture evolves, and how different chemical species behave with respect to water ice and its outgassing. The difference between $CO_2$ and $H_2O$ as a driving volatile can also be investigated.

Constructing laboratory models in a 1 g environment applicable for low g cometary cases remains challenging and may ultimately be limited in applicability. However, these laboratory models do show that the physics can be far more complicated that the simplified assumptions used in even the most sophisticated numerical studies (e.g. Marboeuf et al. 2012).

A further use of laboratory models is in evaluating sample integrity. Even if analysis is conducted in situ, the sampling process itself may affect the sample before it reaches the analytical instrument. The maintenance of sample integrity before measurement is therefore potential of importance in constructing an accurate result. Hence, analogue testing is required and should be a component of any future comet mission.

## 5 Cross-Table of Missions v. Goals

We refer here to the summary in Table 3.

### 5.1 The Coma Swarm

One of the major issues with the Rosetta mission profile as flown was the persistent use of terminator orbits combined with the large cometocentric distances flown during the perihelion passage. This has resulted in limited information in several specific areas associated with the gas and dust dynamics of the outflow and subsequent controversy.

While single point measurements of the gas density and composition by the ROSINA instrument provided a constraint on the gas flow field and composition, the non-uniqueness of the results when extending them to 3 dimensions around the nucleus has been shown to be extreme. Incorporation of other data (from the MIRO experiment for example) has been shown to be difficult because of the non-LTE nature of the flow (Marschall et al. 2019). The observations of VIRTIS-M were of insufficient resolution to provide meaningful constraints.

One approach to addressing these issues is to use a swarm of small satellites to surround the nucleus and make multi-point simultaneous measurements of the coma. The aim would be to fully characterise the 3D distribution of gas and dust including density, temperature, and composition. Each small satellite could also carry low resolution camera systems (visible and/or infrared) to support the 3D analyses. Such a mission would also allow higher accuracy measurements of production rates.

Swarms could also be useful in resolving the dust production rate issues encountered by Rosetta. It would remain to be seen how much payload could be incorporated onto each element of any one member of a swarm but the need for further investigations of dust particle properties in situ remains paramount after Rosetta. It should also be noted that sample return may lead to further information but transfer to and re-entry into the Earth's atmosphere will have influence on potentially delicate particles including thermal evolution.

### 5.2 The Surface Network

Inhomogeneity of the nucleus in morphology and (probably) chemistry is evident. Furthermore, there are seasonal variations that require considerably more investigation than Rosetta could afford. Hence, the scientific need to investigate and compare several areas on a comet is still there. One approach is use a surface "network". Here, multiple static landers are placed on the surface and conduct several experiments in a coordinated manner. The advantage of the network is that sounding can be conducted from one network element to another and seismic studies can also be carried out to determine the internal properties of the object. A CONSERT style system would still have advantages, however, in providing a moving element on an orbiter. However, the experiment package for each landed element might also include instruments dedicated to comparisons of the physical properties between the landing sites. Instrumentation could include microscopic imaging, miniaturised mass spectrometers and/or Raman spectrometers and penetrating thermal sensors. The technical challenges would include the anchoring to the surface and the need to study the seasonal variations by ensuring network element lifetimes are adequate to perform observations for extended periods in darkness at high latitudes.

### 5.3 The Surface "Rover"

In the early phase of the conceptual design of the Rosetta Lander (RoLand), the possibility of mobility by hopping was investigated (RoLand Proposal 1995; Ulamec et al. 1997). The idea was given up though, as any movement from a safe landing site was considered an unnecessary risk and mobility is difficult to combine with firm anchoring. However, any high precision landing at a specific site (e.g. within proposed active areas) is highly attractive. Consequently, mobile comet landers have been proposed. One particular concept studied in detail was the Comet Hopper, CHopper, proposed for the opportunity #12 within the NASA Discovery Program, selected as one of the three finalists but eventually not selected for implementation in 2012.

The CHopper mission would have attempted to measure cometary activity of comet 46P/Wirtanen at several (up to six) locations and various heliocentric distances. CHopper would have been a flexible spacecraft, powered by two Advanced Stirling Radioisotope Generators, ASRG's, rendezvousing comet 46P/Wirtanen and investigating several surface areas at various heliocentric distances using a propulsion system based "hopping" (Clark et al. 2008). Another Discovery proposal to investigate cometary activity was CHagall, based on ideas as proposed for CHopper. As neither CHopper nor CHagall have been selected, and Philae hardly contributed to this particular aspect of cometary science, these measurements are still to be done in the course of future missions. It is worth noting that comet sample return missions will not necessarily allow new results regarding the mechanisms of cometary activity. For this, long term high-resolution in-situ observations with cameras, mass spectrometers/pressure sensors and possibly drills are almost certainly the best approach.

Generally, mobility strategies as considered or designed for other low gravity bodies like asteroids or the Martian moon Phobos, are relevant also for comets. An overview of small bodies hoppers, including the Soviet Phobos Hopper from 1988 (PROP-F) is given e.g. by Ulamec et al. (2011); a description of MASCOT, a small mobile surface package which has been delivered by the Japanese Hayabusa 2 spacecraft to asteroid (162173) Ryugu is given by Ho et al. (2017).

### 5.4 The Multi-Object Fly-by

The diversity of comets has been demonstrated by imaging of nuclei during fly-bys. Comparison of Hartley 2, Wild 2, and 67P is sufficient to show this clearly. However, chemical differences are also evident. This suggests that there remains substantial justification for multi-object studies.

The juxtaposition of broad surveys versus detailed investigations of individual bright comets is analogous to the arguments for missions to fly-by many comets versus a Rosetta-like rendezvous mission. While many of the questions discussed herein require another rendezvous mission, there is also a strong desire to increase the breadth of our understanding of the comet population by visiting a larger number. The ill-fated CONTOUR mission would have visited three or four comets (Cochran et al. 2002), but failed shortly after launch. Multiple asteroid fly-by tours within the main belt have been proposed (Rivkin et al. 2015; Bowles et al. 2018), which could potentially visit MBCs, while the extended missions of the Giotto (1P/Halley, 26P/Grigg-Skjellerup), Stardust (82P/Wild 2, 9P/Tempel 2), and Deep Impact (9P/Tempel 2, 103P/Hartley 2) spacecraft show that it is also quite feasible to encounter multiple comets in near-Earth space although none of these concepts have been selected at the time of writing. Fly-by missions can address nucleus science through remote sensing, even if they return only a snap-shot at a single time, and have the potential to return compositional information on the more abundant coma species if they pass close enough with suitable mass spectrometer instrumentation. Fly-by missions necessarily trade the scientific value of a closer approach against the safety of the spacecraft making a high-speed pass through a dusty environment, but with careful orbit planning and a payload designed to return results from a safe distance, many comets could potentially be studied by the same spacecraft. The common instrumentation for each fly-by would be extremely beneficial in making detailed comparisons. Sensitive instruments are increasingly being designed to be small enough for use in micro-satellites (particularly 'CubeSats'), raising the possibility of deployable probes designed to get much closer during a fly-by, taking a higher risk but probing the inner coma. The Comet Interceptor concept (see above) makes use of such deployable probes to enable close approach investigation during a potentially very-high speed fly-by of a long period comet, while keeping the 'mothership' at a safer distance.

## 5.5 Advanced Impact Exhumation

The Deep Impact experiment demonstrated a method to gain access to the interior of a comet without the complexity of drilling. Impactor technologies are being further advanced through the Asteroid Impact and Deflection Assessment (AIDA) mission studies. As part of AIDA (Cheng et al. 2015), two independent spacecraft would be sent to the asteroid, Didymos. An asteroid impactor—the NASA Double Asteroid Redirection Test (DART) spacecraft—would be sent to the target and a follow-up asteroid rendezvous spacecraft, Hera (Michel et al. 2018), would observe the consequence of the impact.

The Deep Impact mission was not able to assess the impact crater accurately and while much was learnt about the impact itself and the surface of the comet target (9P/Tempel 1), there was relatively little learnt of the exposed interior. However, it is apparent that this is a means of accessing primitive material in a cost effective manner. It is also apparent that the exposure of the interior results in a relative slow change in the new surface layer properties (as shown by Pajola et al. 2017 following the Aswan cleaving event). Hence, this approach is of significant scientific interest. It would need to be established, however, how much the exposed material is modified by the impact itself and whether the properties we are most interested in (e.g. the volatile-refractory inter-relationship) are influenced by the impact process.

## 5.6 Sample Return

Sample return from a comet has been a "Holy Grail" for more than 30 years. Early concepts were discussed in Eberhardt et al. (1986). This is now being studied in detail with a modern approach (although coincidentally using the same acronym) through the CAESAR mission that was recently pre-selected by NASA as part of its New Frontiers programme but which lost out in the final down-selection in July 2019. The concept proposed therein was focussed very strongly on returning the sample with only additional observations that are necessary to complete the task.

The sampling approach differs significantly from the original (1980s) Rosetta concept and illustrates that "sample return" can mean many different things. In the case of CAESAR, the approach was strongly focussed on the chemistry. The sampling system is currently intended to get a sample from the surface, if necessary, by using a rotational grinding-like system. This would reduce the integrity of any sample for studies of the physical nature of the surface layer and its structure with depth. The volatile component is then allowed to outgas with the gaseous products being captured and analysed. The refractory materials are maintained at low temperature and returned to Earth.

This clearly answers one of the main outstanding issues from previous comet missions in that it determines the surface composition (under the obvious assumption that post-sampling reactions are of no importance) while the volatile composition (about which we have rather accurate knowledge already) is also assessed. However, it is not obvious that such an approach can place further constraints on the structural and physical nature of the surface layer. Hence, there will remain considerable scope for future sample return missions even if CAESAR is eventually completed successfully. On the other hand, as noted above, the maintenance of sample integrity in a physical sense through the return to Earth is likely to be extremely challenging.

For completeness, we should also note that Albee et al. (1994) proposed a coma sample return mission called Soccer as a joint NASA-ISAS mission which indirectly led to the successful STARDUST coma sample return mission within NASA's Discovery programme.

## 5.7 Other Targets of Relevance

While focussing on missions to active comets, there are other objects that are of relevance for cometary research that deserve some level of attention. The existence of "main-belt comets" has provoked proposals for future missions because they are easier to get to and it is clear that understanding their outgassing would contribute to cometary science although the exact relationship may not be completely obvious.

Investigation of the surface layers of "dead" comets (3200 Phaethon and 2015 TB145 have been discussed as possible examples with Phaethon now being the target of JAXA's Destiny+) would provide an important description of how activity eventually ceases. Our observations at 67P would suggest that, in some of these objects, transport and airfall of material has finally choked all activity but that relatively pristine material is close to the surface. This hypothesis could be tested. The absence of activity would make the mission far simpler if there is concern about outgassing and large dust particles in the vicinity of the object. Hence, getting at pristine material in these objects may be more straightforward than for active comets.

Observations of Centaurs, which could be objects in transition from the Kuiper Belt to the inner Solar System, would also be of major interest in view of their limited thermal evolution. 2060 Chiron is a remarkable object that could easily yield significant scientific knowledge of primitive materials. It is, of course, also known to outgas. The major drawback is that Centaurs are challenging objects to study in situ, requiring an energy source beyond solar power and significant delta-V to reach.

Trojan asteroids may also be relevant and are now the subject of NASA's Discovery mission, Lucy. Their relationship to cometary objects is, however, unclear.

**Table 3** Comparison table

| Mission concept | Description | Goals | Comments |
|---|---|---|---|
| The coma swarm | Multiple small satellites or cube-sats orbiting or manoeuvering around the nucleus making local measurements of the gas and dust coma | – Detailed evaluation of the gas and dust dynamics in 3D<br>– Dust size distributions and their variation within the coma<br>– Simultaneous multi-directional remote-sensing of the nucleus to monitor activity | |
| The surface network | Multiple small landed packages placed on the surface of the nucleus | – Determination of the interior structure through low frequency tomography<br>– Seismic sensing of the interior<br>– Local investigation of the surface layer structure at sub-centimetre scales | It is assumed that this concept would not allow a drilling system |
| The surface rover | Single mobile station traversing the cometary nucleus | – Determination of the diversity of cometary material within one object<br>– Local investigation of the surface layer structure at sub-metre scales | It is assumed that this concept would carry a drilling system |
| The multi-object fly-by | Single spacecraft making multiple fast fly-bys of cometary and/or comet-like objects | – Characterisation of the diversity of comet-like objects<br>– Imaging and detailed measurements of composition | |
| Advanced impact exhumation | Dual spacecraft system with an impactor and a monitoring spacecraft. Improved version of Deep Impact to view in detail the exhumed material through a rendezvous | – Evaluation of the internal properties of the target<br>– Determination of the layer structure (if present)<br>– Determination of the internal composition | AIDA: DART/Hera concept might be considered |
| Sample return | Spacecraft rendezvous and acquisition of cometary material followed by return to Earth | – Detailed laboratory analysis of cometary material | Studies through Phase A in the CAESAR New Frontiers mission but subsequently rejected. Good chemical and isotopic studies but perhaps limited physical knowledge of the material |

Finally, New Horizons's observations of 2014 MU69 (Stern et al. 2019) have illustrated the importance of visiting pristine objects outside the orbit of Neptune although travel times and power sources remain challenging.

As noted in the introduction, we have effectively ignored the physics of a comet's interaction with the solar wind herein. While the study of this interaction is interesting and, for example, multi-point observations of the interaction would undoubtedly be of benefit in constraining details, it seems unlikely to us that the next mission(s) to comets will be driven by this sub-topic.

## 6 Conclusion

The in situ investigation of comets has taken a giant leap forward with Rosetta. On the other hand, the mission did not constrain well the activity mechanism and our knowledge of the physical properties of the surface layer remain somewhat limited. We also have major questions about the localisation of activity and the gas and dust comae within 2–3 kilometres of the surface.

It is arguable whether the original comet nucleus sample return objectives laid out in the early studies for what became Rosetta could be achieved today—even with improved technology and knowledge of the target properties. Hence, the more limited focus of NASA's CAESAR, was undoubtedly a wise approach. However, in its current form, there will still remain unaddressed issues such as the relationship between volatile and refractory material in the nucleus interior and the variability between different objects.

Various mission profiles can be considered to answer the outstanding questions. It is not the purpose of this paper to propose one over the rest (something that would be best addressed by requesting full proposals) but it is evident that several mission concepts could be competed against each other leading to optimised scientific return.

**Acknowledgements** The team from the University of Bern is supported through the Swiss National Science Foundation and through the NCCR PlanetS. The project has also received funding from the European Union's Horizon 2020 research and innovation programme under grant agreement No 686709. This work was supported by the Swiss State Secretariat for Education, Research and Innovation (SERI) under contract number 16.0008-2. The opinions expressed and arguments employed herein do not necessarily reflect the official views of the Swiss Government.



## References

J. Agarwal, M.F. A'Hearn, J.-B. Vincent, C. Guettler, S. Hoefner, H. Sierks, C. Tubiana, C. Barbieri, P.L. Lamy, R. Rodrigo, D. Koschny, H. Rickman, M.A. Barucci, J.-L. Bertaux, I. Bertini, S. Boudreault, G. Cremonese, V. Da Deppo, B. Davidsson, S. Debei, M. De Cecco, J. Deller, S. Fornasier, M. Fulle, A. Gicquel, O. Groussin, P.J. Gutierrez, M. Hofmann, S.F. Hviid, W.-H. Ip, L. Jorda, H.U. Keller, J. Knollenberg, J.-R. Kramm, E. Kuehrt, M. Kueppers, L.M. Lara, M. Lazzarin, J.J. Lopez Moreno, F. Marzari, G. Naletto, N. Oklay, X. Shi, N. Thomas, Acceleration of individual, decimetre-sized aggregates in the lower coma of comet 67P/Churyumov-Gerasimenko. Mon. Not. R. Astron. Soc. **462**, S78–S88 (2016)

J. Agarwal et al., Evidence of sub-surface energy storage in comet 67P from the outburst of 2016 July 03. Mon. Not. R. Astron. Soc. **469**, s606–s625 (2017)

M.F. A'Hearn, R.C. Millis, D.O. Schleicher, D.J. Osip, P.V. Birch, The ensemble properties of comets: Results from narrowband photometry of 85 comets, 1976–1992. Icarus **118**, 223–270 (1995)

A.L. Albee, K.T. Uesugi, P. Tsou, SOCCER: Comet coma sample return mission, in *Lunar and Planetary Inst., Workshop on Particle Capture, Recovery and Velocity/Trajectory Measurement Technologies* (1994), pp. 7–11

K. Altwegg, H. Balsiger, A. Bar-Nun, J.J. Berthelier, A. Bieler, P. Bochsler, C. Briois, U. Calmonte, M. Combi, J. De Keyser, P. Eberhardt, B. Fiethe, S. Fuselier, S. Gasc, T.I. Gombosi, K.C. Hansen, M. Hässig, A. Jäckel, E. Kopp, A. Korth, L. LeRoy, U. Mall, B. Marty, O. Mousis, E. Neefs, T. Owen, H. Reme, M. Rubin, T. Semon, C.-Y. Tzou, H. Waite, P. Wurz, 67P/Churyumov-Gerasimenko, a Jupiter family comet with a high D/H ratio. Science **347**, 1261952 (2015)

K. Altwegg, H. Balsiger, J. Berthelier et al., Organics in comet 67P—A first comparative analysis of mass spectra from ROSINA–DFMS, COSAC and Ptolemy. Mon. Not. R. Astron. Soc. **469**, S130–S141 (2017)

E. Asphaug, W. Benz, Size, density, and structure of comet Shoemaker-Levy 9 inferred from the physics of tidal breakup. Icarus **121**, 225–248 (1996)

N. Attree et al., Tensile strength of 67P/Churyumov-Gerasimenko nucleus material from overhangs. Astron. Astrophys. **611**, A33 (2018)

N. Attree, L. Jorda, O. Groussin, S. Mottola, N. Thomas, Y. Brouet, E. Kührt, M. Knapmeyer, F. Preusker, F. Scholten, J. Knollenberg, S. Hviid, P. Hartogh, R. Rodrigo, Constraining models of activity on comet 67P/Churyumov-Gerasimenko with Rosetta trajectory, rotation, and water production measurements. Astron. Astrophys. **630**, A18 (2019)

O. Bardyn et al., Carbon-rich dust in comet 67P/Churyumov-Gerasimenko measured by COSIMA/Rosetta. Mon. Not. R. Astron. Soc. **469**, S712 (2017)

J.-P. Bibring, P. Lamy, Y. Langevin, A. Soufflot, M. Berthé, J. Borg, F. Poulet, S. Mottola, Civa. Space Sci. Rev. **128**, 397–412 (2007)

J.-P. Bibring, Y. Langevin, J. Carter et al., 67P/Churyumov–Gerasimenko surface properties as derived from CIVA panoramic images. Science **349**, aab0671 (2015)

J. Biele, S. Ulamec, Capabilities of Philae, the Rosetta Lander. Space Sci. Rev. **138**, 275–289 (2008)

J. Biele, S. Ulamec, M. Maibaum, R. Roll, L. Witte, E. Jurado, P. Munoz, W. Arnold, H.-U. Auster, C. Casas, C. Faber, C. Fantinati, F. Finke, H.-H. Fischer, K. Geurts, C. Güttler, P. Heinisch, A. Herique, S. Hviid, G. Kargl, M. Knapmeyer, J. Knollenberg, W. Kofman, N. Kömle, E. Kührt, V. Lommatsch, S. Mottola, R. Pardo de Santayana, E. Remetean, F. Scholten, K.J. Seidensticker, H. Sierks, T. Spohn, The landing(s) of Philae and inferences about comet surface mechanical properties. Science **349**, aaa9816 (2015)

A. Bieler, K. Altwegg, H. Balsiger, A. Bar-Nun, J.-J. Berthelier, P. Bochsler, C. Briois, U. Calmonte, M. Combi, J. de Keyser, E.F. van Dishoeck, B. Fiethe, S.A. Fuselier, S. Gasc, T.I. Gombosi, K.C. Hansen, M. Hässig, A. Jäckel, E. Kopp, A. Korth, L. Le Roy, U. Mall, R. Maggiolo, B. Marty, O. Mousis, T. Owen, H. Reme, M. Rubin, T. Semon, C.-Y. Tzou, J.H. Waite, C. Walsh, P. Wurz, Abundant molecular oxygen in the coma of comet 67P/Churyumov-Gerasimenko. Nature **526**, 678–681 (2015)

L. Biermann, Kometenschweife und solare Korpuskularstrahlung. Z. Astrophys. **29**, 274 (1951)

N. Biver, D. Bockelée-Morvan, P. Colom, J. Crovisier, B. Germain, E. Lellouch, J.K. Davies, W.R.F. Dent, R. Moreno, G. Paubert, J. Wink, D. Despois, D.C. Lis, D. Mehringer, D. Benford, M. Gardner, T.G. Phillips, M. Gunnarsson, H. Rickman, A. Winnberg, P. Bergman, L.E.B. Johansson, H. Rauer, Long-term evolution of the outgassing of comet Hale-Bopp from radio observations. Earth Moon Planets **78**, 5–11 (1997)

D. Bockelée-Morvan, U. Calmonte, S. Charnley, J. Duprat, C. Engrand, A. Gicquel, M. Hässig, E. Jehin, H. Kawakita, B. Marty, S. Milam, A. Morse, P. Rousselot, S. Sheridan, E. Wirström, Cometary isotopic measurements. Space Sci. Rev. **197**, 47–83 (2015)

N.E. Bowles, C. Snodgrass, A. Gibbings, J.P. Sanchez, J.A. Arnold, P. Eccleston, T. Andert, A. Probst, G. Naletto, A.C. Vandaele, J. de Leon, A. Nathues, I.R. Thomas, N. Thomas, L. Jorda, V. Da Deppo, H. Haack, S.F. Green, B. Carry, K.L. Donaldson Hanna, J. Leif Jorgensen, A. Kereszturi, F.E. DeMeo, M.R. Patel, J.K. Davies, F. Clarke, K. Kinch, A. Guilbert-Lepoutre, J. Agarwal, A.S. Rivkin, P. Pravec, S. Fornasier, M. Granvik, R.H. Jones, N. Murdoch, K.H. Joy, E. Pascale, M. Tecza, J.M. Barnes, J. Licandro, B.T. Greenhagen, S.B. Calcutt, C.M. Marriner, T. Warren, I. Tosh, CASTAway: An asteroid main belt tour and survey. Adv. Space Res. **62**, 1998–2025 (2018)

Y. Brouet, A.C. Levasseur-Regourd, P. Sabouroux, L. Neves, P. Encrenaz, O. Poch, A. Pommerol, N. Thomas, W. Kofman, A. Le Gall, V. Ciarletti, A. Herique, A. Lethuillier, N. Carrasco, C. Szopa, A porosity gradient in 67P/C-G nucleus suggested from CONSERT and SESAME-PP results: An interpretation based on new laboratory permittivity measurements of porous icy analogues. Mon. Not. R. Astron. Soc. **462**, S89–S98 (2016)

U. Calmonte, K. Altwegg, H. Balsiger, J.-J. Berthelier, A. Bieler, J. De Keyser, B. Fiethe, S.A. Fuselier, S. Gasc, T.I. Gombosi, L. Le Roy, M. Rubin, T. Sémon, C.-Y. Tzou, S.F. Wampfler, Sulphur isotope mass-independent fractionation observed in comet 67P/Churyumov-Gerasimenko by Rosetta/ROSINA. Mon. Not. R. Astron. Soc. **469**, S787–S803 (2017)

A.F. Cheng, J. Atchison, B. Kantsiper, A.S. Rivkin, A. Stickle, C. Reed, A. Galvez, I. Carnelli, P. Michel, S. Ulamec, Asteroid impact & deflection assessment mission. Acta Astronaut. **115**, 262–269 (2015)

P.R. Christensen, B.M. Jakosky, H.H. Kieffer, M.C. Malin, H.Y. McSween Jr., K. Nealson, G.L. Mehall, S.H. Silverman, S. Ferry, M. Caplinger, M. Ravine, The Thermal Emission Imaging System (THEMIS) for the Mars 2001 Odyssey Mission. Space Sci. Rev. **110**, 85–130 (2004)

C. Chyba, C. Sagan, Infrared emission by organic grains in the coma of comet Halley. Nature **330**, 350–353 (1987)

C.F. Chyba, C. Sagan, M.J. Mumma, The heliocentric evolution of cometary infrared spectra—Results from an organic grain model. Icarus **79**, 362–381 (1989)

V. Ciarletti, A.C. Levasseur-Regourd, J. Lasue, C. Statz, D. Plettemeier, A. Herique, Y. Rogez, W. Kofman, CONSERT suggests a change in local properties of 67P/Churyumov-Gerasimenko's nucleus at depth. Astron. Astrophys. **583**, A40 (2015)

V. Ciarletti, A. Herique, J. Lasue, A.-C. Levasseur-Regourd, D. Plettemeier, F. Lemmonier, C. Guiffaut, P. Pasquero, W. Kofman, CONSERT constrains the internal structure of 67P at a few metres size scale. Mon. Not. R. Astron. Soc. **469**, S805–S817 (2017)

M. Ciarniello, A. Raponi, F. Capaccioni, G. Filacchione, F. Tosi, M.C. De Sanctis, D. Kappel, B. Rousseau, G. Arnold, M.T. Capria, M.A. Barucci, E. Quirico, A. Longobardo, E. Kuehrt, S. Mottola, S. Erard, D. Bockelee-Morvan, C. Leyrat, A. Migliorini, A. Zinzi, E. Palomba, B. Schmitt, G. Piccioni, P. Cerroni, W.-H. Ip, G. Rinaldi, M. Salatti, The global surface composition of 67P/Churyumov-Gerasimenko nucleus by Rosetta/VIRTIS. II Diurnal and seasonal variability. Mon. Not. R. Astron. Soc. **462**, S443–S458 (2016)

B.C. Clark, J.M. Sunshine, M.F. A'Hearn, A.L. Cochran, T.L. Farnham, W.M. Harris, T.J. McCoy, J. Veverka, Comet Hopper: A mission concept for exploring the heterogeneity of comets, in *Asteroids, Comets, Meteors*, vol. 1405 (2008), p. 8131

A. Cochran, J. Veverka, J. Bell, M. Belton, J. Benkhoff, A. Benkhoff, B. Clark, P. Feldman, J. Kissel, P. Mahaffy, M. Malin, S. Murchie, H. Neimann, T. Owen, M. Robinson, G. Schwehm, S. Squyres, P. Thomas, F. Whipple, D. Yeomans, The comet nucleus tour (contour); a NASA discovery mission. Earth Moon Planets **89**, 289–300 (2002)

A. Coradini, F. Capaccioni, P. Drossart, G. Arnold, E. Ammannito, F. Angrilli, A. Barucci, G. Bellucci, J. Benkhoff, G. Bianchini, J.P. Bibring, M. Blecka, D. Bockelee-Morvan, M.T. Capria, R. Carlson, U. Carsenty, P. Cerroni, L. Colangeli, M. Combes, M. Combi, J. Crovisier, M.C. De Sanctis, E.T. Encrenaz, S. Erard, C. Federico, G. Filacchione, U. Fink, S. Fonti, V. Formisano, W.H. Ip, R. Jaumann, E. Kuehrt, Y. Langevin, G. Magni, T. McCord, V. Mennella, S. Mottola, G. Neukum, P. Palumbo, G. Piccioni, H. Rauer, B. Saggin, B. Schmitt, D. Tiphene, G. Tozzi, Virtis: An imaging spectrometer for the Rosetta mission. Space Sci. Rev. **128**, 529–559 (2007)

B.J.R. Davidsson, P.J. Gutiérrez, O. Groussin, M.F. A'Hearn, T. Farnham, L.M. Feaga, M.S. Kelley, K.P. Klaasen, F. Merlin, S. Protopapa, H. Rickman, J.M. Sunshine, P.C. Thomas, Thermal inertia and surface roughness of Comet 9P/Tempel 1. Icarus **224**, 154–171 (2013)

P. Di Lizia, F. Bernelli-Zazzera, A. Ercoli-Finzi, S. Mottola, C. Fantinati, E. Remetean, B. Dolives, Planning and implementation of the on-comet operations of the instrument SD2 onboard the lander Philae of Rosetta mission. Acta Astronaut. **125**, 183–195 (2016)

L. Dones, P.R. Weissman, H.F. Levison, M.J. Duncan, in *Comets II*, ed. by M.C. Festou, H.U. Keller, H.A. Weaver (University of Arizona Press, Tucson, 2004), p. 153

P. Eberhardt, J.A.M. McDonnell, R. Reinhard, G.H. Schwehm, E. Bussoletti, M. Hechler, A. Johnstone, A.-C. Levasseur-Regourd, H.U. Keller, G.E.N. Scoon, CAESAR mission: A comet atmosphere encounter and sample return, in *ESA, Proceedings of the 20th ESLAB Symposium on the Exploration of Halley's Comet, vol. 3: Posters* (1986), pp. 243–248

M.R. El-Maarry et al., Surface changes on comet 67P/Churyumov-Gerasimenko suggest a more active past. Science **355**, 1392–1395 (2017)

A. Ercoli-Finzi, F.B. Zazzera, C. Dainese, F. Malnati, P.G. Magnani, E. Re, P. Bologna, S. Espinasse, A. Olivieri, $SD^2$—How to sample a comet. Space Sci. Rev. **128**, 281 (2007)

L.M. Feaga, M.F. A'Hearn, J.M. Sunshine, O. Groussin, T.L. Farnham, Asymmetries in the distribution of $H_2O$ and $CO_2$ in the inner coma of Comet 9P/Tempel 1 as observed by Deep Impact. Icarus **191**, 134–145 (2007)

Y.R. Fernandez, M.S. Kelley, P.L. Lamy, I. Toth, O. Groussin, C.M. Lisse, M.F. A'Hearn, J.M. Bauer, H. Campins, A. Fitzsimmons, J. Licandro, S.C. Lowry, K.J. Meech, J. Pittichova, W.T. Reach, C. Snodgrass, H.A. Weaver, Thermal properties, sizes, and size distribution of Jupiter-family cometary nuclei. Icarus **226**, 1138–1170 (2013)

G. Filacchione, O. Groussin, C. Herny, D. Kappel, S. Mottola, N. Oklay, A. Pommerol, I. Wright, Z. Yoldi, M. Ciarniello, L. Moroz, A. Raponi, Comet 67P/CG nucleus composition and comparison to other comets. Space Sci. Rev. **215**, 19 (2019)

U. Fink, L. Doose, G. Rinaldi, A. Bieler, F. Capaccioni, D. Bockelee-Morvan, G. Filacchione, S. Erard, C. Leyrat, M. Blecka, M.T. Capria, M. Combi, J. Crovisier, M.C. De Sanctis, N. Fougere, F. Taylor, A.

Migliorini, G. Piccioni, Investigation into the disparate origin of $CO_2$ and $H_2O$ outgassing for Comet 67/P. Icarus **277**, 78–97 (2016)
S. Finklenburg Ph.D. thesis, University of Bern (2013)
S. Fornasier, P.H. Hasselmann, M.A. Barucci, C. Feller, S. Besse, C. Leyrat, L. Lara, P.J. Gutierrez, N. Oklay, C. Tubiana, F. Scholten, H. Sierks, C. Barbieri, P.L. Lamy, R. Rodrigo, D. Koschny, H. Rickman, H.U. Keller, J. Agarwal, M.F. A'Hearn, J.-L. Bertaux, I. Bertini, G. Cremonese, V. Da Deppo, B. Davidsson, S. Debei, M. De Cecco, M. Fulle, O. Groussin, C. Güttler, S.F. Hviid, W. Ip, L. Jorda, J. Knollenberg, G. Kovacs, R. Kramm, E. Kührt, M. Küppers, F. La Forgia, M. Lazzarin, J.J. Lopez Moreno, F. Marzari, K.-D. Matz, H. Michalik, F. Moreno, S. Mottola, G. Naletto, M. Pajola, A. Pommerol, F. Preusker, X. Shi, C. Snodgrass, N. Thomas, J.-B. Vincent, Spectro-photometric properties of the 67P/Churyumov-Gerasimenko's nucleus from the OSIRIS instrument onboard the ROSETTA spacecraft. Astron. Astrophys. **583**, A30 (2015). https://doi.org/10.1051/0004-6361/201525901
S. Fornasier et al., The highly active Anhur-Bes regions in the 67P/Churyumov-Gerasimenko comet: Results from OSIRIS/ROSETTA observations. Mon. Not. R. Astron. Soc. **469**, S93–S107 (2017)
S. Fornasier, V.H. Hoang, P.H. Hasselmann, C. Feller, M.A. Barucci, J.D.P. Deshapriya, H. Sierks, G. Naletto, P.L. Lamy, R. Rodrigo, D. Koschny, B. Davidsson, J. Agarwal, C. Barbieri, J.-L. Bertaux, I. Bertini, D. Bodewits, G. Cremonese, V. Da Deppo, S. Debei, M. De Cecco, J. Deller, S. Ferrari, M. Fulle, P.J. Gutierrez, C. Guttler, W.-H. Ip, H.U. Keller, M. Küppers, F. La Forgia, M.L. Lara, M. Lazzarin, Z. Lin, J.J. Lopez Moreno, F. Marzari, S. Mottola, M. Pajola, X. Shi, I. Toth, C. Tubiana, Linking surface morphology, composition, and activity on the nucleus of 67P/Churyumov-Gerasimenko (2018). arXiv:1809.03997
M. Fulle, C. Barbieri, G. Cremonese, H. Rauer, M. Weiler, G. Milani, R. Ligustri, The dust environment of comet 67P/Churyumov-Gerasimenko. Astron. Astrophys. **422**, 357–368 (2004)
S. Gasc, K. Altwegg, H. Balsiger, J.-J. Berthelier, A. Bieler, U. Calmonte, B. Fiethe, S. Fuselier, A. Galli, T. Gombosi, M. Hoang, J. De Keyser, A. Korth, L. Le Roy, U. Mall, H. Rème, M. Rubin, T. Sèmon, C.-Y. Tzou, J.H. Waite, P. Wurz, Change of outgassing pattern of 67P/Churyumov-Gerasimenko during the March 2016 equinox as seen by ROSINA. Mon. Not. R. Astron. Soc. **469**, S108–S117 (2017)
S.-B. Gerig, R. Marschall, N. Thomas, I. Bertini, D. Bodewits, B. Davidsson, M. Fulle, W.-H. Ip, H.U. Keller, M. Küppers, F. Preusker, F. Scholten, C.C. Su, I. Toth, C. Tubiana, J.-S. Wu, H. Sierks, C. Barbieri, P.L. Lamy, R. Rodrigo, D. Koschny, H. Rickman, J. Agarwal, M.A. Barucci, J.-L. Bertaux, G. Cremonese, V. Da Deppo, S. Debei, M. De Cecco, J. Deller, S. Fornasier, O. Groussin, P.J. Gutierrez, C. Güttler, S.F. Hviid, L. Jorda, J. Knollenberg, J.-R. Kramm, E. Kührt, L.M. Lara, M. Lazzarin, J.J. Lopez Moreno, F. Marzari, S. Mottola, G. Naletto, N. Oklay, J.-B. Vincent, On deviations from free-radial outflow in the inner coma of comet 67P/Churyumov-Gerasimenko. Icarus **311**, 1–22 (2018)
A. Gicquel et al., Sublimation of icy aggregates in the coma of comet 67P/Churyumov-Gerasimenko detected with the OSIRIS cameras on board Rosetta. Mon. Not. R. Astron. Soc. **462**, S57–S66 (2016)
K.-H. Glassmeier, H. Boehnhardt, D. Koschny, E. Kührt, I. Richter, The Rosetta mission: Flying towards the origin of the solar system. Space Sci. Rev. **128**, 1–21 (2007)
F. Goesmann, H. Rosenbauer, R. Roll, C. Szopa, F. Raulin, R. Sternberg, G. Israel, U. Meierhenrich, W. Thiemann, G. Munoz-Caro, Cosac, the cometary sampling and composition experiment on Philae. Space Sci. Rev. **128**, 257 (2007)
F. Goesmann, F. Raulin, J.H. Bredehöft, M. Cabane, P. Ehrenfreund, A.J. MacDermott, S. McKenna-Lawlor, U.J. Meierhenrich, G.M. Munoz Caro, C. Szopa, R. Sternberg, R. Roll, W.H.-P. Thiemann, S. Ulamec, COSAC prepares for sampling and in situ analysis of cometary matter from comet 67P/Churyumov-Gerasimenko. Planet. Space Sci. **103**, 318–330 (2014)
F. Goesmann, H. Rosenbauer, J.H. Bredehöft, M. Cabane, P. Ehrenfreund, T. Gautier, C. Giri, H. Krüger, L. Le Roy, A.J. MacDermott, S. McKenna-Lawlor, U.J. Meierhenrich, G.M.M. Caro, F. Raulin, R. Roll, A. Steele, H. Steininger, R. Sternberg, C. Szopa, W. Thiemann, S. Ulamec, Organic compounds on comet 67P/Churyumov-Gerasimenko revealed by COSAC mass spectrometry. Science **349**, aab0689 (2015)
O. Groussin, J.M. Sunshine, L.M. Feaga, L. Jorda, P.C. Thomas, J.-Y. Li, M.F. A'Hearn, M.J.S. Belton, S. Besse, B. Carcich, T.L. Farnham, D. Hampton, K. Klaasen, C. Lisse, F. Merlin, S. Protopapa, The temperature, thermal inertia, roughness and color of the nuclei of Comets 103P/Hartley 2 and 9P/Tempel 1. Icarus **222**, 580–594 (2013)
O. Groussin, N. Attree, Y. Brouet, V. Ciarletti, B. Davidsson, G. Filacchione, H.-H. Fischer, B. Gundlach, M. Knapmeyer, J. Knollenberg, R. Kokotanekova, E. Kührt, C. Leyrat, D. Marshall, I. Pelivan, Y. Skorov, C. Snodgrass, T. Spohn, F. Tosi, The thermal, mechanical, structural, and dielectric properties of cometary nuclei after Rosetta. Space Sci. Rev. **215**, 29 (2019)
S. Gulkis, M. Frerking, J. Crovisier, G. Beaudin, P. Hartogh, P. Encrenaz, T. Koch, C. Kahn, Y. Salinas, R. Nowicki, R. Irigoyen, M. Janssen, P. Stek, M. Hofstadter, M. Allen, C. Backus, L. Kamp, C. Jarchow, E. Steinmetz, A. Deschamps, J. Krieg, M. Gheudin, D. Bockelee-Morvan, N. Biver, T. Encrenaz, D. Despois, W. Ip, E. Lellouch, I. Mann, D. Muhleman, H. Rauer, P. Schloerb, T. Spilker, MIRO: Microwave instrument for Rosetta orbiter. Space Sci. Rev. **128**, 561–597 (2007)

S. Gulkis, M. Allen, P. von Allmen, G. Beaudin, N. Biver, D. Bockelee-Morvan, M. Choukroun, J. Crovisier, B.J.R. Davidsson, P. Encrenaz, T. Encrenaz, M. Frerking, P. Hartogh, M. Hofstadter, W.-H. Ip, M. Janssen, C. Jarchow, S. Keihm, S. Lee, E. Lellouch, C. Leyrat, L. Rezac, F.P. Schloerb, T. Spilker, Subsurface properties and early activity of comet 67P/Churyumov-Gerasimenko. Science **347**, aaa0709 (2015)

B. Gundlach, J. Blum, Outgassing of icy bodies in the Solar System—II: Heat transport in dry, porous surface dust layers. Icarus **219**, 618 (2012)

B. Gundlach, S. Kilias, E. Beitz, J. Blum, Micrometer-sized ice particles for planetary-science experiments—I. Preparation, critical rolling friction force, and specific surface energy. Icarus **214**, 717–723 (2011)

B. Gundlach, K.P. Schmidt, C. Kreuzig, D. Bischoff, F. Rezaei, S. Kothe, J. Blum, B. Grzesik, E. Stoll, The tensile strength of ice and dust aggregates and its dependence on particle properties. Mon. Not. R. Astron. Soc. **479**, 1273–1277 (2018)

P.J. Gutierrez et al., Possible interpretation of the precession of comet 67P/Churyumov-Gerasimenko. Astron. Astrophys. **590**, A46 (2016)

E. Hand, NASA set to choose low-cost Solar System mission. Nature **487**, 151 (2012)

W.M. Harris, M.R. Combi, R.K. Honeycutt, B.E.A. Mueller, F. Scherb, Evidence for interacting gas flows and an extended volatile source distribution in the coma of Comet C/1996 B2 (Hyakutake). Science **277**, 676 (1997)

P. Hartogh, D.C. Lis, D. Bockelee-Morvan, M. de Val-Borro, N. Biver, M. Küppers, M. Emprechtinger, E.A. Bergin, J. Crovisier, M. Rengel, R. Moreno, S. Szutowicz, G.A. Blake, Ocean-like water in the Jupiter-family comet 103P/Hartley 2. Nature **478**, 218–220 (2011)

M. Hässig, K. Altwegg, H. Balsiger, A. Bar-Nun, J.J. Berthelier, A. Bieler, P. Bochsler, C. Briois, U. Calmonte, M. Combi, J. De Keyser, P. Eberhardt, B. Fiethe, S.A. Fuselier, M. Galand, S. Gasc, T.I. Gombosi, K.C. Hansen, A. Jäckel, H.U. Keller, E. Kopp, A. Korth, E. Kührt, L. Le Roy, U. Mall, B. Marty, O. Mousis, E. Neefs, T. Owen, H. Rème, M. Rubin, T. Sémon, C. Tornow, C.-Y. Tzou, J.H. Waite, P. Wurz, Time variability and heterogeneity in the coma of 67P/Churyumov-Gerasimenko. Science **347**, aaa0276 (2015)

M. Hässig, K. Altwegg, H. Balsiger, J.J. Berthelier, A. Bieler, U. Calmonte, F. Dhooghe, B. Fiethe, S.A. Fuselier, S. Gasc, T.I. Gombosi, L. Le Roy, A. Luspay-Kuti, K. Mandt, M. Rubin, C.-Y. Tzou, S.F. Wampfler, P. Wurz, Isotopic composition of $CO_2$ in the coma of 67P/Churyumov-Gerasimenko measured with ROSINA/DFMS. Astron. Astrophys. **605**, A50 (2017)

A. Herique, W. Kofman, P. Beck, L. Bonal, I. Buttarazzi, E. Heggy, J. Lasue, A.C. Levasseur-Regourd, E. Quirico, S. Zine, Cosmochemical implications of CONSERT permittivity characterization of 67P/CG. Mon. Not. R. Astron. Soc. **462**, S516–S532 (2016)

A. Herique et al., Direct observations of asteroid interior and regolith structure: science measurement requirements. Adv. Space Res. **62**, 2141–2162 (2018)

B. Hermalyn, T.L. Farnham, S.M. Collins, M.S. Kelley, M.F. A'Hearn, D. Bodewits, B. Carcich, D.J. Lindler, C. Lisse, K. Meech, P.H. Schultz, P.C. Thomas, The detection, localization, and dynamics of large icy particles surrounding Comet 103P/Hartley 2. Icarus **222**, 625–633 (2013)

T. Hewagama, S. Aslam, P. Clark, M. Daly, L. Feaga, D. Folta, N. Gorius, T. Hurford, T.A. Livengood, B. Malphrus, M. Mumma, C. Nixon, J. Sunshine, G. Villanueva, A. Zucherman, Primitive Object Volatile Explorer (PrOVE)—Waypoints and opportunistic deep space missions to comets, in *Lunar and Planetary Science Conference, 49* (2018), p. 2800

M. Hilchenbach, J. Kissel, Y. Langevin, C. Briois, H. von Hoerner, A. Koch, R. Schulz, J. Silén, K. Altwegg, L. Colangeli, H. Cottin, C. Engrand, H. Fischer, A. Glasmachers, E. Grün, G. Haerendel, H. Henkel, H. Höfner, K. Hornung, E.K. Jessberger, H. Lehto, K. Lehto, F. Raulin, L. Le Roy, J. Rynö, W. Steiger, T. Stephan, L. Thirkell, R. Thomas, K. Torkar, K. Varmuza, K.-P. Wanczek, N. Altobelli, D. Baklouti, A. Bardyn, N. Fray, H. Krüger, N. Ligier, Z. Lin, P. Martin, S. Merouane, F.R. Orthous-Daunay, J. Paquette, C. Revillet, S. Siljeström, O. Stenzel, B. Zaprudin, Comet 67P/Churyumov-Gerasimenko: Close-up on dust particle fragments. Astrophys. J. **816**, L32 (2016)

T.-M. Ho, V. Baturkin, C. Grimm, J.T. Grundmann, C. Hobbie, E. Ksenik, C. Lange, K. Sasaki, M. Schlotterer, M. Talapina, N. Termtanasombat, E. Wejmo, L. Witte, M. Wrasmann, G. Wübbels, J. Rössler, C. Ziach, R. Findlay, J. Biele, C. Krause, S. Ulamec, M. Lange, O. Mierheim, R. Lichtenheldt, M. Maier, J. Reill, H.-J. Sedlmayr, P. Bousquet, A. Bellion, O. Bompis, C. Cenac-Morthe, M. Deleuze, S. Fredon, E. Jurado, E. Canalias, R. Jaumann, J.-P. Bibring, K.H. Glassmeier, D. Hercik, M. Grott, L. Celotti, F. Cordero, J. Hendrikse, T. Okada, MASCOT—The mobile asteroid surface scout onboard the Hayabusa2 mission. Space Sci. Rev. **208**, 339–374 (2017)

S. Ivanovski, V. Zakharov, V. Della Corte, F. Lucarelli, J.-F. Crifo, A. Rotundi, M. Fulle, Aspherical dust dynamics code for GIADA experiment in the coma of 67P/Churyumov-Gerasimenko, in *Proceedings EGU General Assembly Conference Abstracts, 16*, 15664 (2014)

D. Jewitt, H.H. Hsieh, J. Agarwal, in *Asteroids III*, ed. by P. Michel, F. Demeo, W. Bottke (University of Arizona Press, Tucson, 2015), p. 221

G.H. Jones, J. Agarwal, N. Bowles, M. Burchell, A.J. Coates, A. Fitzsimmons, A. Graps, H.H. Hsieh, C.M. Lisse, S.C. Lowry, A. Masters, C. Snodgrass, C. Tubiana, The proposed Caroline ESA M3 mission to a Main Belt Comet. Adv. Space Res. **62**, 1921–1946 (2018)

S. Keihm, F. Tosi, L. Kamp, F. Capaccioni, S. Gulkis, D. Grassi, M. Hofstadter, G. Filacchione, S. Lee, S. Giuppi, M. Janssen, M. Capria, Interpretation of combined infrared, submillimeter, and millimeter thermal flux data obtained during the Rosetta fly-by of Asteroid (21) Lutetia. Icarus **221**, 395–404 (2012)

H.U. Keller, The nucleus, in *Physics and Chemistry of Comets*, ed. by W.F. Huebner (Springer, Berlin, 1990)

H.U. Keller, S. Mottola, S.F. Hviid, J. Agarwal, E. Kührt, Y. Skorov, K. Otto, J.-B. Vincent, N. Oklay, S.E. Schröder, B. Davidsson, M. Pajola, X. Shi, D. Bodewits, I. Toth, F. Preusker, F. Scholten, H. Sierks, C. Barbieri, P. Lamy, R. Rodrigo, D. Koschny, H. Rickman, M.F. A'Hearn, M.A. Barucci, J.-L. Bertaux, I. Bertini, G. Cremonese, V. Da Deppo, S. Debei, M. De Cecco, J. Deller, S. Fornasier, M. Fulle, O. Groussin, P.J. Gutierrez, C. Güttler, M. Hofmann, W.-H. Ip, L. Jorda, J. Knollenberg, J.R. Kramm, M. Küppers, L.-M. Lara, M. Lazzarin, J.J. Lopez-Moreno, F. Marzari, G. Naletto, C. Tubiana, N. Thomas, Seasonal mass transfer on the nucleus of comet 67P/Chuyumov-Gerasimenko. Mon. Not. R. Astron. Soc. **469**, S357–S371 (2017)

M.S.P. Kelley, C.E. Woodward, D. Bodewits, T.L. Farnham, M.S. Gudipati, D.E. Harker, D.C. Hines, M.M. Knight, L. Kolokolova, A. Li, I. de Pater, S. Protopapa, R.W. Russell, M.L. Sitko, D.H. Wooden, Cometary science with the James Webb Space Telescope. Publ. Astron. Soc. Pac. **128**, 018009 (2016)

G. Klingelhöfer, J. Brückner, C. D'uston et al., The Rosetta alpha particle X-ray spectrometer (APXS). Space Sci. Rev. **128**, 383–396 (2007)

H.W. Kochan, W.F. Huebner, D.W.G. Sears, Simulation experiments with cometary analogous material. Earth Moon Planets **80**, 369–411 (1998)

W. Kofman, Y. Barbin, J. Klinger, A.-C. Levasseur-Regourd, J.-P. Barriot, A. Herique, T. Hagfors, E. Nielsen, E. Grün, P. Edenhofer, H. Kochan, G. Picardi, R. Seu, J. van Zyl, C. Elachi, J. Melosh, J. Veverka, P. Weissman, L.H. Svedhem, S.E. Hamran, I.P. Williams, Comet nucleus sounding experiment by radiowave transmission. Adv. Space Res. **21**, 1589–1598 (1998)

W. Kofman, A. Herique, J.-P. Goutail, T. Hagfors, I.P. Williams, E. Nielsen, J.-P. Barriot, Y. Barbin, C. Elachi, P. Edenhofer, A.-C. Levasseur-Regourd, D. Plettemeier, G. Picardi, R. Seu, V. Svedhem, The comet nucleus sounding experiment by radiowave transmission (CONSERT): A short description of the instrument and of the commissioning stages. Space Sci. Rev. **128**, 413–432 (2007)

W. Kofman, A. Herique, Y. Barbin, J.-P. Barriot, V. Ciarletti, S. Clifford, P. Edenhofer, C. Elachi, C. Eyraud, J.-P. Goutail, E. Heggy, L. Jorda, J. Lasue, A.-C. Levasseur-Regourd, E. Nielsen, P. Pasquero, F. Preusker, P. Puget, D. Plettemeier, Y. Rogez, H. Sierks, C. Statz, H. Svedhem, I. Williams, S. Zine, J. Van Zyl, Properties of the 67P/Churyumov-Gerasimenko interior revealed by CONSERT radar. Science **349**, aab0639 (2015)

R. Kokotanekova, C. Snodgrass, P. Lacerda, S.F. Green, S.C. Lowry, Y.R. Fernandez, C. Tubiana, A. Fitzsimmons, H.H. Hsieh, Rotation of cometary nuclei: New light curves and an update of the ensemble properties of Jupiter-family comets. Mon. Not. R. Astron. Soc. **471**, 2974–3007 (2017)

R. Kokotanekova, C. Snodgrass, P. Lacerda, S.F. Green, P. Nikolov, T. Bonev, Implications of the small spin changes measured for large Jupiter-family comet nuclei. Mon. Not. R. Astron. Soc. **479**, 4665–4680 (2018)

P.L. Lamy, I. Toth, B.J.R. Davidsson, O. Groussin, P. Gutierrez, L. Jorda, M. Kaasalainen, S.C. Lowry, A portrait of the nucleus of Comet 67P/Churyumov-Gerasimenko. Space Sci. Rev. **128**, 23–66 (2007)

L. Le Roy, K. Altwegg, H. Balsiger, J.-J. Berthelier, A. Bieler, C. Briois, U. Calmonte, M.R. Combi, J. De Keyser, F. Dhooghe, B. Fiethe, S.A. Fuselier, S. Gasc, T.I. Gombosi, M. Hässig, A. Jäckel, M. Rubin, C.-Y. Tzou, Inventory of the volatiles on comet 67P/Churyumov-Gerasimenko from Rosetta/ROSINA. Astron. Astrophys. **583**, A1 (2015)

A. Lethuillier, A. Le Gall, M. Hamelin, W. Schmidt, K.J. Seidensticker, R. Grard, V. Ciarletti, S. Caujolle-Bert, H.-H. Fischer, R. Trautner, Electrical properties and porosity of the first meter of the nucleus of 67P/Churyumov-Gerasimenko. As constrained by the Permittivity Probe SESAME-PP/Philae/Rosetta. Astron. Astrophys. **591**, A32 (2016)

H.F. Levison, in *Completing the Inventory of the Solar System*, ed. by T. Rettig, J.M. Hahn. Astronomical Society of the Pacific Conference Series, vol. 107, San Francisco (1996), p. 173

Y. Liao, Global explorations of inner neutral gas coma of Comet 67P/Churyumov-Gerasimenko with DSMC approach. Ph.D. thesis, University of Bern (2017)

D.J. Lien, Dust in comets. I. Thermal properties of homogeneous and heterogeneous grains. Astrophys. J. **355**, 680 (1990)

Z.-Y. Lin, W.-H. Ip, I.-L. Lai, J.-C. Lee, J.-B. Vincent, L.M. Lara, D. Bodewits, H. Sierks, C. Barbieri, P.L. Lamy, R. Rodrigo, D. Koschny, H. Rickman, H.U. Keller, J. Agarwal, M.F. A'Hearn, M.A. Barucci, J.-L. Bertaux, I. Bertini, G. Cremonese, V. Da Deppo, B. Davidsson, S. Debei, M. De Cecco, S. Fornasier, M. Fulle, O. Groussin, P.J. Gutièrrez, C.G. üttler, S.F. Hviid, L. Jorda, J. Knollenberg, G. Kovacs, J.-R. Kramm, E. Kührt, M. Küppers, F. La Forgia, M. Lazzarin, J.J. Lopez-Moreno, S. Lowry, F. Marzari, H. Michalik, S. Mottola, G. Naletto, N. Oklay, M. Pajola, A. Rozek, N. Thomas, Y. Liao, C. Tubiana, Morphology and dynamics of the jets of comet 67P/Churyumov-Gerasimenko: early-phase development. Astron. Astrophys. **583**, A11 (2015)

D.C. Lis, D. Bockelée-Morvan, R. Güsten, N. Biver, J. Stutzki, Y. Delorme, C. Duràn, H. Wiesemeyer, Y. Okada, Terrestrial deuterium-to-hydrogen ratio in water in hyperactive comets. Astron. Astrophys. **625**, L5 (2019)

S. Lorek, B. Gundlach, P. Lacerda, J. Blum, Comet formation in collapsing pebble clouds. What cometary bulk density implies for the cloud mass and dust-to-ice ratio. Astron. Astrophys. **587**, A128 (2016)

S. Lowry, S.R. Duddy, B. Rozitis, S.F. Green, A. Fitzsimmons, C. Snodgrass, H.H. Hsieh, O. Hainaut, The nucleus of Comet 67P/Churyumov-Gerasimenko. A new shape model and thermophysical analysis. Astron. Astrophys. **548**, L12 (2012)

LSST Science Collaboration, LSST Science Book (2009). arXiv:0912.0201

U. Marboeuf, B. Schmitt, J.-M. Petit, O. Mousis, N. Fray, A cometary nucleus model taking into account all phase changes of water ice: Amorphous, crystalline, and clathrate. Astron. Astrophys. **542**, A82 (2012)

R.M. Marschall, L. Rezac, D. Kappel, C.C. Su, S.-B. Gerig, M. Rubin, O. Pinzon-Rodriguez, D. Marshall, Y. Liao, C. Herny, G. Arnold, C. Christou, S.K. Dadzie, O. Groussin, P. Hartogh, L. Jorda, E. Kührt, S. Mottola, O. Mousis, F. Preusker, F. Scholten, P. Theologou, J.-S. Wu, K. Altwegg, R. Rodrigo, N. Thomas, A comparison of multiple Rosetta data sets and 3D model calculations of 67P/Churyumov-Gerasimenko coma around equinox. Icarus **328**, 104 (2019)

B. Marty, K. Altwegg, H. Balsiger, A. Bar-Nun, D.V. Bekaert, J.-J. Berthelier, A. Bieler, C. Briois, U. Calmonte, M. Combi, J. De Keyser, B. Fiethe, S.A. Fuselier, S. Gasc, T.I. Gombosi, K.C. Hansen, M. Hässig, A. Jäckel, E. Kopp, A. Korth, L. Le Roy, U. Mall, O. Mousis, T. Owen, H. Rème, M. Rubin, T. Sèmon, C.-Y. Tzou, J.H. Waite, P. Wurz, Xenon isotopes in 67P/Churyumov-Gerasimenko show that comets contributed to Earth's atmosphere. Science **356**, 1069–1072 (2017)

M. Massironi et al., Two independent and primitive envelopes of the bilobate nucleus of comet 67P. Nature **526**, 402–405 (2015)

J.A.M. McDonnell, P.L. Lamy, G.S. Pankiewicz, Physical properties of cometary dust, in *IAU Colloq. 116: Comets in the Post-Halley Era*, vol. 167 (1991), pp. 1043–1073

K.J. Meech, J.C. Castillo-Rogez, in *IAU General Assembly*, vol. 22 (2015), 2257859

K.J. Meech et al., EPOXI: Comet 103P/Hartley 2 observations from a worldwide campaign. Astrophys. J. **734**, L1 (2011)

S. Merouane, B. Zaprudin, O. Stenzel, Y. Langevin, N. Altobelli, V. Della Corte, H. Fischer, M. Fulle, K. Hornung, J. Silén, N. Ligier, A. Rotundi, J. Ryno, R. Schulz, M. Hilchenbach, J. Kissel (Cosima Team), Dust particle flux and size distribution in the coma of 67P/Churyumov-Gerasimenko measured in situ by the COSIMA instrument on board Rosetta. Astron. Astrophys. **596**, A87 (2016)

P. Michel, M. Kueppers, H. Sierks, I. Carnelli, A.F. Cheng, K. Mellab, M. Granvik, A. Kestilä, T. Kohout, K. Muinonen, A. Näsilä, A. Penttila, T. Tikka, P. Tortora, V. Ciarletti, A. Hérique, N. Murdoch, E. Asphaug, A. Rivkin, O. Barnouin, A.C. Bagatin, P. Pravec, D.C. Richardson, S.R. Schwartz, K. Tsiganis, S. Ulamec, O. Karatekin, European component of the AIDA mission to a binary asteroid: characterization and interpretation of the impact of the DART mission. Adv. Space Res. **62**, 2261 (2018)

A. Migliorini, G. Piccioni, F. Capaccioni, G. Filacchione, D. Bockelée-Morvan, S. Erard, C. Leyrat, M.R. Combi, N. Fougere, J. Crovisier, F.W. Taylor, M.C. De Sanctis, M.T. Capria, D. Grassi, G. Rinaldi, G.P. Tozzi, U. Fink, Water and carbon dioxide distribution in the 67P/Churyumov-Gerasimenko coma from VIRTIS-M infrared observations. Astron. Astrophys. **589**, A45 (2016)

D. Möhlmann, Surface regolith and environment of comets. Planet. Space Sci. **42**, 933–937 (1994)

A. Morse, D. Andrews, G. Morgan et al., Ptolemy operations at the surface of a comet, from planning to reality. Acta Astron. **125**, 196–211 (2016)

S. Mottola, G. Arnold, H.-G. Grothues et al., The structure of the regolith on 67P/Churyumov-Gerasimenko from ROLIS descent imaging. Science **349**, aab0232 (2015)

Mottola (2019, in preparation)

D. Nesvorny, D. Vokrouhlicky, L. Dones, H.F. Levison, N. Kaib, A. Morbidelli, Origin and evolution of short-period comets. Astrophys. J. **845**, 27 (2017)

M. Neugebauer, G. Gloeckler, J.T. Gosling, A. Rees, R. Skoug, B.E. Goldstein, T.P. Armstrong, M.R. Combi, T. Mäkinen, D.J. McComas, R. von Steiger, T.H. Zurbuchen, E.J. Smith, J. Geiss, L.J. Lanzerotti, Encounter of the Ulysses spacecraft with the ion tail of Comet McNaught. Astrophys. J. **667**, 1262 (2007)

T. Ott et al., Dust mass distribution around comet 67P/Churyumov-Gerasimenko determined via parallax measurements using Rosetta's OSIRIS cameras. Mon. Not. R. Astron. Soc. **469**, S276–S284 (2017)

M. Pajola et al., The pristine interior of comet 67P revealed by the combined Aswan outburst and cliff collapse. Nat. Astron. **1**, 0092 (2017)

M. Pätzold et al., in *AAS/Division for Planetary Sciences Meeting Abstracts, 48*, 116.27 (2016)

M. Pätzold, T.P. Andert, M. Hahn, J.-P. Barriot, S.W. Asmar, B. Häusler, M.K. Bird, S. Tellmann, J. Oschlisniok, K. Peter, The nucleus of comet 67P/Churyumov-Gerasimenko—Part I: The global view—Nucleus mass, mass-loss, porosity, and implications. Mon. Not. R. Astron. Soc. **483**, 2337–2346 (2019)

A. Pommerol, B. Jost, O. Poch, M.R. El-Maarry, B. Vuitel, N. Thomas, The SCITEAS experiment: Optical characterizations of sublimating icy planetary analogues. Planet. Space Sci. **109**, 106–122 (2015a)

A. Pommerol, N. Thomas, M.R. El-Maarry, M. Pajola, O. Groussin, A.-T. Auger, N. Oklay, S. Fornasier, C. Feller, B. Davidsson, A. Gracia, B. Jost, R. Marschall, O. Poch, M.A. Barucci, J.-L. Bertaux, F. La Forgia, H.U. Keller, E. Kuehrt, S.C. Lowry, S. Mottola, G. Naletto, H. Sierks, C. Barbieri, P.L. Lamy, R. Rodrigo, H. Rickman, D. Koschny, J. Agarwal, M.F. A'Hearn, I. Bertini, G. Cremonese, V. Da Deppo, M. De Cecco, S. Debei, C. Guettler, M. Fulle, P.J. Gutierrez, S.F. Hviid, W.-H. Ip, L. Jorda, J. Knollenberg, G. Kovacs, J.-R. Kramm, M. Küppers, L. Lara, M. Lazzarin, J.L. Lopez Moreno, F. Marzari, H. Michalik, F. Preusker, F. Scholten, C. Tubiana, J.-B. Vincent, OSIRIS observations of metre-size exposures of $H_2O$ ice at the surface of 67P/Churyumov-Gerasimenko and interpretation using laboratory experiments. Astron. Astrophys. **583**, A25 (2015b). https://doi.org/10.1051/0004-6361/201525977

F. Preusker, F. Scholten, K.-D. Matz, T. Roatsch, S.F. Hviid, S. Mottola, J. Knollenberg, E. Kührt, M. Pajola, N. Oklay, J.-B. Vincent, B. Davidsson, M.F. A'Hearn, J. Agarwal, C. Barbieri, M.A. Barucci, J.-L. Bertaux, I. Bertini, G. Cremonese, V. Da Deppo, S. Debei, M. De Cecco, S. Fornasier, M. Fulle, O. Groussin, P.J. Gutièrrez, C. Güttler, W.-H. Ip, L. Jorda, H.U. Keller, D. Koschny, J.R. Kramm, M. Küppers, P. Lamy, L.M. Lara, M. Lazzarin, J.J. Lopez Moreno, F. Marzari, M. Massironi, G. Naletto, H. Rickman, R. Rodrigo, H. Sierks, N. Thomas, C. Tubiana, The global meter-level shape model of comet 67P/Churyumov-Gerasimenko. Astron. Astrophys. **607**, L1 (2017)

A. Rivkin, B.A. Cohen, O.S. Barnouin, N.L. Chabot, C.M. Ernst, R.L. Klima, J. Helbert, Z. Sternovsky, The Main-belt Asteroid and NEO Tour with Imaging and Spectroscopy (MANTIS), AGU Fall Meeting Abstracts, P11B-2087 (2015)

F. Robert, in *Meteorites and the Early Solar System II*, ed. by D.S. Lauretta, H.Y. McSween (2006), p. 341

RoLand, Cometary Lander for the Rosetta Mission. Proposal to ESA (1995)

R. Schräpler, J. Blum, I. von Borstel, C. Güttler, The stratification of regolith on celestial objects. Icarus **257**, 33 (2015)

S.E. Schröder, S. Mottola, G. Arnold et al., Close-up images of the final Philae landing site on comet 67P/Churyumov-Gerasimenko acquired by the ROLIS camera. Icarus **285**, 263–274 (2016)

K.J. Seidensticker, D. Möhlmann, I. Apathy, W. Schmidt, K. Thiel, W. Arnold, H.-H. Fischer, M. Kretschmer, D. Madlener, A. Peter, R. Trautner, S. Schieke, Sesame—An experiment of the Rosetta Lander Philae: Objectives and general design. Space Sci. Rev. **128**, 301–337 (2007)

X. Shi, X. Hu, S. Mottola, M. Rose, H. Sierks, C. Güttler, H.U. Keller, Observing and modeling the near-nucleus coma structure around terminators on 67P/Churyumov-Gerasimenko, in *European Planetary Science Congress, 11*, EPSC2017-487 (2017)

C. Snodgrass et al., The 67P/Churyumov-Gerasimenko observation campaign in support of the Rosetta mission. Philos. Trans. R. Soc. Lond. Ser. A **375**, 20160249 (2017a)

C. Snodgrass, J. Agarwal, M. Combi, A. Fitzsimmons, A. Guilbert-Lepoutre, H.H. Hsieh, M.-T. Hui, E. Jehin, M.S.P. Kelley, M.M. Knight, C. Opitom, R. Orosei, M. de Val-Borro, B. Yang, The main belt comets and ice in the solar system. Astron. Astrophys. Rev. **25**, 5 (2017b)

C. Snodgrass, G.H. Jones, H. Boehnhardt, A. Gibbings, M. Homeister, N. Andre, P. Beck, M.S. Bentley, I. Bertini, N. Bowles, M.T. Capria, C. Carr, M. Ceriotti, A.J. Coates, V. Della Corte, K.L. Donaldson Hanna, A. Fitzsimmons, P.J. Gutiérrez, O.R. Hainaut, A. Herique, M. Hilchenbach, H.H. Hsieh, E. Jehin, O. Karatekin, W. Kofman, L.M. Lara, K. Laudan, J. Licandro, S.C. Lowry, F. Marzari, A. Masters, K.J. Meech, F. Moreno, A. Morse, R. Orosei, A. Pack, D. Plettemeier, D. Prialnik, A. Rotundi, M. Rubin, J.P. Sanchez, S. Sheridan, M. Trieloff, A. Winterboer, The Castalia mission to Main Belt Comet 133P/Elst-Pizarro. Adv. Space Res. **62**, 1947–1976 (2018)

T. Spohn, K. Seiferlin, A. Hagermann, J. Knollenberg, A.J. Ball, M. Banaszkiewicz, J. Benkhoff, S. Gadomski, W. Gregorczyk, J. Grygorczuk, M. Hlond, G. Kargl, E. Kührt, N. Kömle, J. Krasowski, W. Marczewski, J.C. Zarnecki, Mupus—a thermal and mechanical properties probe for the Rosetta lander Philae. Space Sci. Rev. **128**, 339 (2007)

T. Spohn, J. Knollenberg, A.J. Ball, M. Banaszkiewicz, J. Benkhoff, M. Grott, J. Grygorczuk, C. Hüttig, A. Hagermann, G. Kargl, E. Kaufmann, N. Kömle, E. Kührt, K.J. Kossacki, W. Marczewski, I. Pelivan, R. Schrödter, K. Seiferlin, Thermal and mechanical properties of the near-surface layers of comet 67P/Churyumov-Gerasimenko. Science **349**, aab0464 (2015)

S.A. Stern et al., Initial results from the New Horizons exploration of 2014 $MU_{69}$, a small Kuiper Belt object. Science **364**, aaw9771 (2019)

J. Sunshine, N. Thomas, M.R. El Maarry, T. Farnham, J. Geophys. Res., Planets **121**, 2194–2210 (2016)

V. Tenishev, N. Fougere, D. Borovikov, M.R. Combi, A. Bieler, K.C. Hansen, T.I. Gombosi, A. Migliorini, F. Capaccioni, G. Rinaldi, G. Filacchione, L. Kolokolova, U. Fink, Analysis of the dust jet imaged by Rosetta VIRTIS-M in the coma of comet 67P/Churyumov-Gerasimenko on 2015 April 12. Mon. Not. R. Astron. Soc. **462**, S370 (2016)

N. Thomas, B. Davidsson, M.R. El-Maarry, S. Fornasier, L. Giacomini, A.G. Gracia Berna, S.F. Hviid, W.-H. Ip, L. Jorda, H.U. Keller, J. Knollenberg, E. Kührt, F. La Forgia, I.L. Lai, Y. Liao, R. Marschall, M. Massironi, S. Mottola, M. Pajola, O. Poch, A. Pommerol, F. Preusker, F. Scholten, C.C. Su, J.S. Wu, J.-B. Vincent, H. Sierks, C. Barbieri, P.L. Lamy, R. Rodrigo, H. Rickman, D. Koschny, M.F. A'Hearn, M.A. Barucci, J-L. Bertaux, I. Bertini, G. Cremonese, V. Da Deppo, S. Debei, M. Fulle, O. Groussin, P.J. Gutierrez, J.-R. Kramm, M. Küppers, L.M. Lara, M. Lazzarin, J.J. Lopez Moreno, F. Marzari, H. Michalik, G. Naletto, C. Güttler, Redistribution of particles across the nucleus of comet 67P/Churyumov-Gerasimenko. Astron. Astrophys. **583**, A17 (2015a). https://doi.org/10.1051/0004-6361/201526049

N. Thomas, H. Sierks, C. Barbieri, P.L. Lamy, R. Rodrigo, H. Rickman, D. Koschny, H.U. Keller, J. Agarwal, M.F. A'Hearn, F. Angrilli, A.-T. Auger, M.A. Barucci, J.-L. Bertaux, I. Bertini, S. Besse, D. Bodewits, G. Cremonese, V. Da Deppo, B. Davidsson, M. De Cecco, S. Debei, M.-R. El-Maarry, F. Ferri, S. Fornasier, M. Fulle, L. Giacomini, O. Groussin, P.J. Gutierrez, C. Güttler, S.F. Hviid, W.-H. Ip, L. Jorda, J. Knollenberg, J.-R. Kramm, E. Kührt, M. Küppers, F. La Forgia, L.M. Lara, M. Lazzarin, J.J. Lopez Moreno, S. Magrin, S. Marchi, F. Marzari, M. Massironi, H. Michalik, R. Moissl, S. Mottola, G. Naletto, N. Oklay, M. Pajola, A. Pommerol, F. Preusker, L. Sabau, F. Scholten, C. Snodgrass, C. Tubiana, J.-B. Vincent, K.-P. Wenzel, The morphological diversity of Comet 67P/Churyumov-Gerasimenko. Science **347**, aaa0440 (2015b)

N. Thomas, M.R. El Maarry, P. Theologou, F. Preusker, F. Scholten, L. Jorda, S.F. Hviid, R. Marschall, E. Kührt, G. Naletto, H. Sierks, P.L. Lamy, R. Rodrigo, D. Koschny, B. Davidsson, M.A. Barucci, J.L. Bertaux, I. Bertini, D. Bodewits, G. Cremonese, V. Da Deppo, S. Debei, M. De Cecco, S. Fornasier, M. Fulle, O. Groussin, P.J. Gutierrez, C. Güttler, W.H. Ip, H.U. Keller, J. Knollenberg, L.M. Lara, M. Lazzarin, J.J. Lopez-Moreno, F. Marzari, C. Tubiana, J.B. Vincent, Regional unit definition for the nucleus of comet 67P/Churyumov-Gerasimenko on the SHAP7 model. Planet. Space Sci. **164**, 19–36 (2018)

C.-Y. Tzou Ph.D. thesis, Univ. Bern (2017)

S. Ulamec, J. Block, M. Fenzi et al., RoLand, a long term Lander for the Rosetta Mission. Space Technol. **17**(1), 59–64 (1997)

S. Ulamec, V. Kucherenko, J. Biele, A. Bogatchev, A. Makurin, S. Matrossov, Hopper concepts for small body landers. Adv. Space Res. **47**, 428–439 (2011)

S. Ulamec, C. Fantinati, M. Maibaum, K. Geurts, J. Biele, S. Jansen, O. Küchemann, B. Cozzoni, F. Finke, V. Lommatsch, A. Moussi-Soffys, C. Delmas, L. O'Rourke, Rosetta Lander—Landing and operations on comet 67P/Churyumov-Gerasimenko. Acta Astronaut. **125**, 80–91 (2016)

S. Ulamec, L. O'Rourke, J. Biele, B. Grieger, R. Andrés, S. Lodiot, P. Munoz, A. Charpentier, S. Mottola, J. Knollenberg, M. Knapmeyer, E. Kührt, F. Scholten, K. Geurts, M. Maibaum, C. Fantinati, O. Küchemann, V. Lommatsch, C. Delmas, E. Jurado, R. Garmier, T. Martin, Rosetta Lander—Philae: Operations on comet 67P/Churyumov-Gerasimenko, analysis of wake-up activities and final state. Acta Astronaut. **137**, 38–43 (2017)

J.-B. Vincent et al., Constraints on cometary surface evolution derived from a statistical analysis of 67P's topography. Mon. Not. R. Astron. Soc. **469**, S329–S338 (2017)

K. Wittmann, B. Feuerbacher, S. Ulamec, H. Rosenbauer, J.P. Bibring, D. Moura et al., Rosetta Lander in situ characterization of a comet nucleus. Acta Astronaut. **45**, 389 (1999)

I.P. Wright, S. Barber, G. Morgan et al., Ptolemy—An instrument to measure stable isotopic ratios of key volatiles on a cometary nucleus. Space Sci. Rev. **128**, 363–381 (2007)

I.P. Wright, S. Sheridan, S.J. Barber, G.H. Morgan, D.J. Andrews, A.D. Morse, CHO-bearing organic compounds at the surface of 67P/Churyumov-Gerasimenko revealed by Ptolemy. Science **349**, aab0673 (2015)

V.V. Zakharov, S.L. Ivanovski, J.-F. Crifo, V. Della Corte, A. Rotundi, M. Fulle, Asymptotics for spherical particle motion in a spherically expanding flow. Icarus **312**, 121–127 (2018)

H. Zhang, B.C. Li, J.J. Yin, W.G. Wang, J.C. Huang, C.-H. Hsia, Scientific objectives and design concept of the spectrometers for proposed flyby mission to Main Belt Comet 133p/Elst-Pizarro, in *50th Lunar and Planetary Science Conference*, 2462 (2019)

Printed by Printforce, the Netherlands